ROBERT G. WETZEL

Professor of Botany
Kellogg Biological Station
Michigan State University

Limnology

SAUNDERS COLLEGE PUBLISHING
Philadelphia

Saunders College Publishing
West Washington Square
Philadelphia, PA   19105

Library of Congress Cataloging in Publication Data

Wetzel, Robert G

Limnology.

Bibliography: p.

Includes index.

1. Limnology.   I. Title.

QH96.W47 1975   551.4'82   74–31839

ISBN 0–7216–9240–0

Limnology   ISBN   0-7216-9240-0

2 3 4   026   18 17 16 15 14 13 12 11 10

To Theodore Jones, who took the time to inspire a fledgling at an early, critical time.

preface

The ensuing small volume evolved from many years of frustration with the inadequacy of available summary texts in limnology used for instruction at the undergraduate level. Existing texts in the English language are quite traditional in approach and strongly orient the student to the descriptive characteristics of the flora and fauna of aquatic systems. A basic understanding of the biota of fresh waters is necessary, and one can only be saddened by those students of the subject who have not witnessed firsthand or been able to appreciate the beauty and enormous diversity and complexity of freshwater organisms. However, there is much more to be learned. I have attempted here to integrate the major organisms of fresh waters in their functional relationships to each other and to the dynamics of their environment.

Interest factions wax and wane in any field of science. Limnology is far from exempt from these changes. Today the approaches to the ecology of freshwater organisms vary greatly, but limnology has evolved through a long history of descriptive analyses. In North America, for many years emphasis had focused on the fauna, and until very recently limnology was taught almost entirely by zoology faculty. Many instructors have taught and continue to teach the necessary balanced perspective. However, within the field there exists a distinct bias towards the animals as operational dictators of the metabolism of fresh waters. As a result of an exaggerated quantitative emphasis, sight has often been lost of their qualitative importance to the metabolism of the system. This area of intensive interest is just one of several that are undergoing a reorientation in perspective in a rapidly developing field of study.

Descriptive analyses in limnology are now receiving much needed physiological and biochemical augmentation. Gradually, a shift from largely correlative analyses to experimentally verified causal mechanisms of metabolic controls of natural populations is emerging. It is only through such combined approaches that true operational insight will emerge.

In the search for order in complex systems and for ultimate generalization, two distinct problems exist. First, there is a tendency to amass physical and biotic data on aquatic systems based simply on convenient, available methods, often with little or no rationale for their accrual or with very little appreciation of their interrelationships. While staged progress must be based on sound facts, the relative importance of these dynamic characteristics must be understood before generalization, and ultimately abstraction, can be meaningful. Second, the tendency to reverse this sequence, attempting generalization without a solid and reasoned foundation, can lead to an impediment to progress. One observes a significant fraction of students attempting to express operation of freshwater systems in mathematical terms without appreciable experienced contact with the components of the systems. The enormous predictive potential of the theoretical modeling of freshwater components can be damaged if it is based on lack of experience with the subject and use of superficial samplings of data appearing in the literature.

Limnology has become a rigorous and sophisticated field of science. The expanding use and misuse of finite freshwater resources by man demands not only concern but concerted efforts to expand our knowledge of the function of fresh waters. Some detail is required, especially in such a broad interdisciplinary subject. Both the serious biology major and the student who is casually interested in fresh waters should realize that true insights into a subject can come only from an in-depth probe. Therefore my intent is to demonstrate thoroughly the fascinating integrity and operational integration of the biota and their metabolism within fresh waters.

In attempting to summarize the salient features of limnology, one is confronted with the great range of interacting subjects that affect the metabolism of organisms and their productivity. No one person can be versed in all of these areas and be acquainted with the wealth of existing information. In the comparative approach I have taken, rigorous selection of both subject areas and examples was necessary. Should the result prove useful for instruction, and revisions and updating prove desirable in the future, I urge my colleagues and interested workers to call my attention to gaps, to better examples than those given, or to express any constructive criticism that might improve the book's usefulness in presenting the rudiments of this important field to others.

A preface permits an author to express his appreciation to those who have influenced his thoughts and assisted in the arduous task of drawing the pieces together. One's knowledge and biases are formulated over years of interaction with instructors, colleagues, and friends. Some of these more conspicuous influences are to be found in the writings of others, as, for example, all of us have been influenced to some degree by the perceptiveness of G. E. Hutchinson. Fruitful exchange has taken place with Charles R. Goldman, back in the days when more time was available from his schedule, with Wilhelm Rodhe, W. Ohle, T. T. Macan, G. W. Saunders, Jr., D. G. Frey, and G. E. Likens, and more recently with my immediate working colleagues, especially M. J. Klug, G. H. Lauff, D. J. Hall, K. W. Cummins, E. E. Werner, P. A. Werner, A. Otsuki, S. F. Mitchell, D. McGregor, and P. A. Lane. Constant exchange of ideas and problems with my former and present graduate students has been and continues to be most stimulating. I would especially like to mention P. H. Rich, B. A. Manny, M. C. Miller, H. L. Allen, R. A. Hough, W. S. White, A. K. Ward, K. R. McKinley, and G. L. Godshalk. C. H. Mortimer was especially helpful with the penultimate draft of the chapter on water movements. C. W. Burns, E. E. Werner, and D. J. Hall offered many comments on portions of the section on zooplankton. Many other colleagues read and commented on various portions of all chapters in all stages of preparation. Gordon L. Godshalk read the entire draft, offering numerous suggestions, and assisted with the indexing, verification of the bibliography, and proofing. Amelia K.

Ward also assisted extensively with the proofing and indexing tasks in the final stages. Our librarian M. Shaw was most helpful in filling my unending requests for literature, often from most obscure sources. M. G. Hewitt and J. Holt prepared the final figures from my often rough original drawings. A. J. Johnson and M. Hughes were most patient in typing the deluge from my pen. My wife Carol was always tolerant of the innumerable hours that the work purloined from family responsibilities. To these persons, as well as others too numerous to mention, I express my profound appreciation. Portions of the unpublished results cited in the text were supported by subventions from the U.S. Atomic Energy Commission and the National Science Foundation. Finally, the assistance and patience of R. Lampert, Evelyn Weiman, and other staff at W. B. Saunders Company contributed significantly to the smooth execution of the many redactory tasks.

The delightful scene of a shoreline along a boreal forest depicted in the frontispiece emanates from the pen of Stjepko Golubić.

Robert G. Wetzel

contents

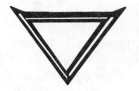

prologue

A fundamental feature of the earth is an abundance of water, which covers 71 per cent of its surface to an average depth of 3800 meters. Thus we find an immense quantity of water constituting the hydrosphere, over 99 per cent of which is deposited in the ocean depressions (Table 1–1). The relatively small amounts of water that occur in freshwater lakes and rivers belie their basic importance in the maintenance of terrestrial life.

Our Freshwater Resources

Human growth and utilization of fresh water on a sustained exponential basis are preeminent components of any analysis of inland water resources.

TABLE 1–1   Water in the Biosphere[a]

	Volume (thousands of km³)	Per Cent of Total	Renewal Time
Oceans	1370000	97.61	37000 years[b]
Polar ice, glaciers	29000	2.08	16000 years
Groundwater (actively exchanged)[c]	4000	0.29	300 years
Freshwater lakes	125	0.009	1–100 years[d]
Saline lakes	104	0.008	10–1000 years[d]
Soil and subsoil moisture	67	0.005	280 days
Rivers	1.2	0.00009	12–20 days[e]
Atmospheric water vapor	14	0.0009	9 days

[a]From Vallentyne, J. R., after Kalinin and Bykov. *In* The Environmental Future. London, Macmillan Publishers Ltd. Reprinted by permission of Macmillan London and Basingstoke.

[b]Based on net evaporation from the oceans.

[c]Kalinin and Bykov (1969) estimated that the total groundwater to a depth of 5 km in the earth's crust amounts to $60 \times 10^6 km^3$. This is much greater than the estimate by the U.S. Geological Survey of 8.3×10^6 km³ to a depth of 4 km. Only the volume of the upper, actively exchanged groundwater is included here.

[d]Renewal times for lakes vary directly with volume and mean depth, and inversely with rate of discharge. The absolute range for saline lakes is from days to thousands of years.

[e]Twelve days for rivers with relatively small catchment areas of less than 100,000 km²; 20 days for major rivers that drain directly to the sea.

1

Man must be recognized for what he is: an animal whose population growth is in an exponential phase. In spite of its absurdity, a belief prevails that the earth's supply of finite water resources can be increased constantly to meet these exponential demands. Fresh waters are a finite resource that can be increased only slightly, e.g., via desalinization, and at the present time at a tremendous energy cost. Society as a whole, and many freshwater ecologists, have tended to ignore man, and his use and misuse of freshwaters, as an influential factor in the maintenance of lake ecosystems. The issue immediately concerns resource utilization, governed by the spiraling relationships of an expanding situation in which supply is constantly made to respond to demand. The unfortunate effect of these conditions of essentially uncontrolled growth is that consumption increases in response to rising supply. Every increase in supply is met by a corresponding increase in consumption, because in contemporary society voluntary control over consumption is ineffective.

Demophoric Growth

As discussed in a particularly penetrating analysis by J. R. Vallentyne, the impending environmental crisis is not only the result of population growth. It is also a result of technological growth, both directly in the sense of increased per caput production and consumption, and indirectly in that technology has furthered the growth of population and urbanization. This concept of growth, encompassing the combined effects of population in a biological sense, and of production-consumption in a technological sense, has been termed *demophora*. The importance of the concept of demophoric growth lies in its emphasis on both production and consumption. It describes the cycle in which biospheric degradation occurs as a result of production utilization of the environment and consumption of technological products, both leading to pollution of water resources. In this way, attention is properly focused on all of the aspects of technological growth, i.e., the technological metabolism of man.

The demands that demophoric growth have imposed upon freshwater resources are monumental. Although about 105,000 km^3 of precipitation, the ultimate source of the freshwater supply, fall on the land surface per year, only about one-third of it (ca 37,500 km^3 $year^{-1}$) reaches the oceans as river discharge (Vallentyne, 1972). About two-thirds of the annual water supply is returned to the atmosphere by evaporation and plant transpiration. If the potential water supply of 37,500 km^3 $year^{-1}$ were divided evenly among the 3.7 billion humans now existing on earth (1971; cf. Clark, 1967), each person would have potentially 10,000 m^3 $year^{-1}$ or 27,000 l day^{-1}. These values would be halved at the projected population level of 7.0 billion humans in the year 2000. Even though these quantities seem large in comparison to the human physiological requirement of 2 l $caput^{-1}$ day^{-1} they are insufficient in view of modern technological demands. Domestic consumption averages 250 l $caput^{-1}$ day^{-1}, the average industrial consumption is 1500 l $caput^{-1}$ day^{-1} in developed countries, and agriculture uses up to several thousand l $caput^{-1}$ day^{-1} in countries with hot, dry climates (Vallentyne, 1972).

Human Impact on Freshwater Ecosystems

The potential freshwater supply cited earlier is in reality much less because of many factors. First, rainfall is not evenly distributed over land surfaces, nor has man distributed himself proportionally in relation to water availability. Hence, this disparity results in a great expense of energy for distribution systems. Second, total consumption has increased exponentially with demophoric growth. In the expansion of distribution systems to areas of low precipitation, such as irrigation of semiarid regions, the use of water for these purposes is disproportionately high because of very high losses from the system by evapotranspiration. Third, the most potentially serious factor stemming from demophoric growth is the severe degradation of the quality of water by pollution. The effect is a severe reduction of the water supply available for other purposes.

We can still look to fresh waters for purposes other than water supply, e.g., for recreation, transportation systems, and the like. However, it is clear that the demands of exponential demophoric growth have been given total precedence over the use of fresh waters for any other purposes. The most fundamental laws of resource utilization may be recognized by most agencies and industries, but they are not being implemented to a significant degree. The recent fuel 'crisis' (1973–74), one of the consequences of demophoric growth, demonstrated the prevailing human behavioral pattern of virtually ignoring all diagnoses or prognostications of impending crisis. When the crisis finally occurs, the public response is not to seek corrective measures which strike at the root of the problem, such as stabilization of population and technological growth, but rather to expand the distribution system to more remote and rapidly decreasing sources of energy. Only when disaster is imminent, when risk of survival of a large population group is glaringly obvious to the majority, does a unified global response occur to save man from himself.

While unfortunately, the above remarks are pessimistic, they are an accurate assessment of existing patterns of utilization of our water resources. It is clear that demophoric growth will continue to impose increasing demands upon freshwater supplies until either inefficient utilization creates a disaster situation threatening the survival of a major segment of the human race, or until the expenditures of energy needed to obtain water exceed tolerable operational levels. Looking back at the history of repetitious responses to impending environmental disasters, we can maintain optimism about the future only until such time as our understanding of the operation of the biosphere, and our knowledge of freshwater ecosystems in particular, will be adequate to allow us to recognize the point of irreversibility. Reflecting on the progress that has been made in limnology since its inception nearly a century ago, it becomes apparent that our time is disconcertingly limited; we need more time to learn to understand freshwater ecosystems sufficiently to judge their resiliency and capacity for change in response to exponential demophoric loading. There is a need to extend the existing knowledge of fresh waters to a greater percentage of the population being educated at the college level. Further, there are significant voids in contemporary limnological studies that require attention.

The Study of Limnology

It is of major importance, therefore, that we fully understand the structure and function of fresh waters. Man is a very real component of these ecosystems, and his effects on them will increase markedly until stabilization of demophoric growth is imposed. Emotionalism and alarmist reactions to the momentum of exploitation of the finite biosphere by the technological system accomplish little, and, as has been repeatedly demonstrated, are often antagonistic to improvement. Strict conservation and isolation of resource parcels, in the belief that such areas are exempt from technological alterations of the atmosphere and water supply, are naive and likewise contribute little to solution of the overall problem.

Sufficient understanding of functional metabolic responses of aquatic ecosystems is necessary in order to confront and offset the effects of these alterations, and in order to achieve maximum, meaningful management of freshwater resources. All waters, of course, cannot be managed directly. Rather, an integration of man's demophora with the metabolism of fresh waters is required to minimize or improve changes. A well-documented effect of human impact upon aquatic ecosystems is eutrophication, a multifaceted term generally associated with increased productivity, structural simplification of biotic components, and a reduction in the ability of the metabolism of the organisms to adapt growth responses to imposed changes (reduced stability). In this condition, excessive inputs commonly seem to exceed the capacity of the ecosystem to be balanced, but in reality the systems are out of equilibrium only with respect to the freshwater chemical and biotic characteristics desired by man for specific purposes. In order to have any hope of effectively integrating man as a component of lake ecosystems, and of monitoring his utilization of these resources, it is mandatory that we comprehend in some detail the functional properties of fresh waters. Only then can we evaluate, with reasonable certainty, the influence that man's activities will have on the metabolic characteristics of these systems.

Limnology is, in broad terms, the study of the functional relationships and productivity of freshwater biotic communities as they are affected by the dynamics of physical, chemical, and biotic environmental parameters. The systematic analyses of the reasonably distinct topics that follow focus primarily on standing (lentic) waters. The rationale for this approach comprises several facets. The lake ecosystem is a system intimately coupled with the land surrounding it in its drainage area and its running (lotic) waters that transport, and metabolize enroute, components of the land to the lake. The limnology of running waters has been given a masterful, eloquent, and succinct treatment in a recent introductory text by H. B. N. Hynes (1970). No attempt will be made to encapsulate his review here, except to emphasize tangentially the significant roles that inflowing waters have in lake dynamics.

In many sectors of the earth, a majority of the lakes are of glacial origins. Because most limnological research has been concentrated in northern temperate regions, a strong bias has been created in current instruction by the disproportionate knowledge of natural temperate lakes. Understanding of tropical and warm water lakes of other regions is emerging now, and is being expanded. Furthermore, the number of manmade reservoirs has increased to the point where they form major lake systems over large areas of the world. Although reservoirs possess many characteristics fundamentally different from lakes which have longer retention times, a firm grasp of the dynamics of lakes permits a relatively easy transition to the understanding of the much more variable and individual characteristics of reservoirs. Underlying all of these systems are basic similarities in the dynamics of metabolism. In this treatment of lake ecosystems, we will attempt to introduce these fundamental, functional similarities without becoming mired in the plethora of individual detail.

Selecting material to include in such a study is difficult not only because of individual biases, but because there is a prevalent lack of understanding of the subject. Some background detail is needed to appreciate even the rudiments of the field. It is hoped that, in the end, the exemplary choices presented here are balanced and provide a basic overview of contemporary comparative limnology, and a basic minimal understanding of freshwater ecology at the undergraduate level. A serious major in limnology will realize that he or she needs much greater depth of understanding to truly comprehend the subject. Many of the reference works cited, such as G. E. Hutchinson's classical, perceptive treatise (1957, 1967, and forthcoming volumes), are only initial introductory summaries of specialized subjects. It is hoped that the forefronts of contemporary limnology, as well as the gaps in need of more intensive investigation, will be evident from the ensuing discussions.

Hegel (1807) stated that *"Das Wahre ist das Ganze"* (The truth is the whole), yet holism alone is inadequate for a comprehensive understanding of limnology. Integration of our knowledge about the individual operational components and environmental factors regulating composite productivity of freshwater ecosystems is fundamental. Previous treatments have been deficient in this respect and almost totally directed towards open water pelagic communities, to the neglect of the major components of lake systems, i.e., microbial decomposition of dissolved and particulate detritus; the intensive metabolism of the littoral zone; metabolism in the sediments; influxes of nutrients and organic matter from outside the lake; the rapidity of nutrient cycling; and regulatory feedback mechanisms of fauna on production and decomposition processes. Integrated study of the dynamics of all components of lake ecosystems requires alterations in former perspectives on freshwater ecology.

chapter 2

water as a substance

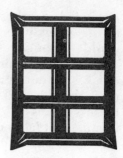

THE CHARACTERISTICS OF WATER

Water is the essence of life as found on earth, and totally dominates the chemical composition of all organisms. The ubiquity of water in biota as the fulcrum of biochemical metabolism rests on its unique physical and chemical properties.

Specifically, the characteristics of water are effective in the regulation of lake metabolism in that modifications of the aquatic milieu occur in response to climatic and geographic variations. In particular, the unique thermal-density properties, high specific heat, and liquid-solid characteristics of water form a stratified environment that controls the chemical and biotic dynamics of lakes to a marked degree. Water provides a tempered milieu in which extreme fluctuations in water availability and temperatures are ameliorated relative to aerial life. Coupled with a relatively high degree of viscosity, these properties have enabled the development of a large number of biotic adaptations that have served to improve sustained productivity.

Molecular Structure and Properties

The unique properties of water center upon its atomic structure and bonding, and the association of water molecules in solid, liquid, and gaseous phases. In a state of equilibrium, the nuclei of a water molecule form an isosceles triangle, with a slightly obtuse bond angle of 104.5° at the oxygen nucleus (Eisenberg and Kauzmann, 1969). The bond length from the center of the oxygen atom to that of each hydrogen atom is 0.96×10^{-8} cm. The nuclei of the molecules are in a continual state of vibration. As the molecule resonates, an electrical dipole moment occurs such that a complex wave function results in a valence electron density pattern in which electron density is highest near the atoms and along the bonds (Fig. 2–1). The electronic charge of the molecule is not restricted to the planar configuration depicted, but is distributed in multiple directions.

Although a majority (ca 56 per cent) of the water molecules have their valances balanced, excitation and ionization do occur. The most common

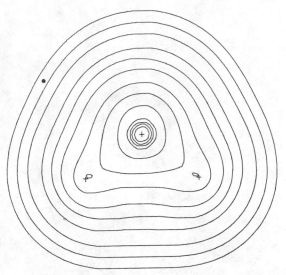

Figure 2–1 Contour map of the electronic charge distribution of the water molecule in the plane of the nuclei (positions indicated by crosses). Contours increase in value of atomic units (1 au = 6.7e$^-$ per Å^3) from the outermost contour in steps of $2 \times 10^n, 4 \times 10^n, 8 \times 10^n$. The smallest contour value is 0.002 with n increasing in steps of unity to yield a maximum value of 20. (Unpublished figure courtesy of Dr. R. F. W. Bader, Department of Chemistry, McMaster University.)

ionized state is removal of one of the hydrogen atoms which is charged positively, and is essentially functioning as a proton. Higher ionization potentials, that is, the energy required to remove electrons from the molecule, are needed to reach further ionization states of water, and occur less frequently. The weak Coulombic characteristics of the bonding of hydrogen atoms to the weakly electronegative oxygen atom result in both ionized and covalent states, and maintain the integrity of water simultaneously. Water is nearly the only known compound that possesses these characteristics.

The density properties of water that are so germane to limnology and life in fresh waters center upon the aggregation and bonding characteristics of water molecules. It is instructive to look first at the structure of ice, whose physical properties are understood better than those of liquid water. Every oxygen atom is at the center of a tetrahedron formed by four oxygen atoms, each about 2.76×10^{-8} cm distant (Eisenberg and Kauzmann, 1969; Horne, 1972). Every water molecule is hydrogen-bonded to its four nearest neighbors; its O-H bonds are directed towards lone pairs of electrons on two of these adjacent molecules, forming two O-H—O hydrogen bonds. In turn, each of its lone pairs is directed towards an O-H bond on one of the other adjacent molecules, forming two O—H-O hydrogen bonds. This arrangement leads to an open lattice in which intermolecular cohesion is great (Fig. 2–2). This structure results in both parallel and perpendicular voids between the molecules, an open tetrahedral structure that permits ice to float upon liquid water.

Molecules in liquid water near 0°C experience about 10^{11} or 10^{12} reorientation and translational movements per second, whereas ice molecules near 0°C experience only about 10^5 to 10^6 movements per second. Increasing the temperature of water increases the rate of reorientations and molecular dis-

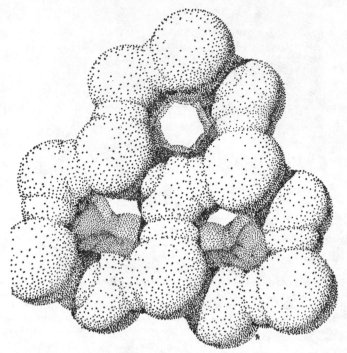

Figure 2–2 A diagrammatic representation of an ice crystal showing the van der Waals radii of the atoms and open voids between the aggregated molecules. (From Pimental, G. C., and McClellan, A. L.: The Hydrogen Bond. San Francisco, W. H. Freeman and Co., copyright © 1960.)

placements and results in decreasing viscosity, decreased molecular relaxation times, and increased rates of self-diffusion. By virtue of their dipolar nature, water molecules interact to form quasi-stable polymers.

Although we have viewed water molecules as diffusionally averaged structures both in vibrational time and in space between molecules, variations in temperature change the intermolecular distances. Thermal increases result in increased agitation of molecules, which distorts or breaks down the hydrogen-bonded networks. As the ice melts increased bond dislocation and rupture occur, which disrupt and fill in the open spaces of the ice lattice structure. The result is an increase in density, and this effect predominates and reaches a maximum at 4°C. As water is heated above 4°C, intermolecular vibrations increase in amplitude and increase interatomic distances. The result is expansion of the liquid and decreased density. It is a characteristic of water that minimum volume and hence maximum density of water occurs at 3.94°C, a point at which competition between the negative configurational contribution of hydrogen bonding and its positive vibrational contribution is maximized.

Isotopic Content

Known isotopes of hydrogen are ^1H, ^2H or D (deuterium), and ^3H (tritium). Tritium is radioactive with a half-life of 12.5 years, sufficiently short that after its decay to ^{3}He, most of which is lost from the atmosphere into space, concentrations of tritium in uncontaminated natural water are very low (ca 1 ^{3}H atom per 10^{18} ^{1}H atoms) and do not accumulate. Six isotopes of oxygen are known: ^{14}O, ^{15}O, ^{16}O, ^{17}O, ^{18}O, and ^{19}O. The isotopes ^{14}O, ^{15}O, and ^{19}O are radioactive but are short-lived, and do not occur significantly in natural water. The precise isotopic content of natural water depends on the origin of the sample but, within the limits of variation, the abundances of $H_2{}^{18}O$, $H_2{}^{17}O$, and $HD^{16}O$ are 0.20 mole per cent, 0.04 per cent, and 0.03 per cent, respectively (Eisenberg and Kauzmann, 1969). Other isotopic combinations are exceedingly rare in natural water (Hutchinson, 1957).

Changes in the ratios of these isotopes in compounds and remains of organisms formed within a lake over geological time permit an estimation of paleotemperatures. The technique is based on the temperature dependence of fractionation of the oxygen isotopes in the carbon dioxide-water-carbonate system (Stuiver, 1968). During slow precipitation of the carbonate, the temperature of the solution is reflected in the ratio of ^{18}O to ^{16}O in the carbonates of sediments or mollusks. The ratio of oxygen isotopes in the precipitated carbonates depends not only on the temperature, but also on the ratio of ^{18}O to ^{16}O isotopes in the water in which carbonate is formed. The measured difference in the ratio of ^{18}O to ^{16}O isotopes between a carbonate fossil and a contemporaneous sample is the result of changes in both temperature and composition of water isotopes. Before any paleotemperature determinations can be made, an estimate of the change in composition of oxygen isotopes of the water is needed; this is more difficult to do in lakes where variations over time are greater than in the oceans. However, estimates can be done with reasonable precision.

Specific Heat

The specific heat (that amount of heat in calories that is required to raise the temperature 1°C of a unit weight of a substance) of liquid water is very high (1.0), exceeded only by a few substances such as liquid ammonia (1.23), liquid hydrogen (3.4), and lithium at high temperatures. Other substances of the biosphere, such as many rocks, have much lower specific heats (ca 0.2). The high specific heat of water, as well as a high latent heat of evaporation, is a function of the relatively large amounts of heat energy required to disrupt the hydrogen bonding of liquid water.

These heat-requiring and heat-retaining properties of water provide a much more stable environment than is found in terrestrial situations. Fluctuations in water temperature occur very gradually, and seasonal and diurnal extremes are small in comparison to those of aerial habitats. The high specific heat of water bodies has profound effects on the climatic conditions of adjacent air and land masses. This thermal inertia of the hydrosphere (Hutchinson, 1957) occurs on either a large or a small scale, in relation to the volume of water body. Examples are many, including the warm Gulf Stream currents of the Atlantic Ocean which improve the climate of Western Europe, and the prevailing air movements across the Great Lakes which moderate the climate of Michigan and other states adjacent to and east of the lakes. Mild winters with higher precipitation rates are found in these areas than in the continental interiors, since the water masses cool and yield heat to the air more slowly. Similarly, areas adjacent to large water masses experience moist, cool, summer periods. Fall fogs are common over and adjacent to lakes, since water vapor from evaporation of the warm lake water condenses and stagnates in the cold overlying air. This phenomenon is described in common parlance as "the lake is steaming."

The specific heat of ice is half (0.5) that of water. As a result, ice forms relatively quickly with the cooling of liquid water at 0°C, that is, it requires the loss of only small amounts of heat (80 cal g^{-1}) for fusion of molecules of 0°C water to ice, as compared to the amount of energy needed (540 cal g^{-1}) to disrupt hydrogen bonding in evaporation. Conversely, ice melts relatively quickly and requires less energy to return to liquid water than is needed to increase the temperature of liquid water.

Density Relationships

Without question, the regulation of the entire physical and chemical dynamics of lakes and the resultant metabolism is governed to a very great extent by differences in density. Throughout the remaining chapters of this book, we will refer repeatedly to the fundamental importance of this unique property of water.

The density or specific gravity of pure ice at 0°C is 0.9168, about 8.5 per cent lighter than liquid water at 0°C (0.99987). The density of water increases to a maximum of 1.0000 at 3.94°C, beyond which molecular expansion and decreasing density occur at a progressively increasing rate (Fig. 2–3). The

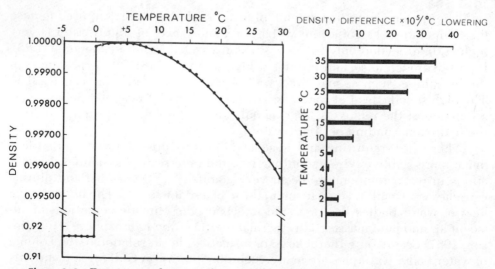

Figure 2–3 Density as a function of temperature for distilled water at 1 atm. The density difference per °C lowering is shown in the righthand portion of the figure at various temperatures. (Modified from Vallentyne, 1957.)

TABLE 2–1   Changes in the Density of Water with Salt Content[a]

Salinity (‰)	Density (at 4°C)
0	1.00000
1	1.00085
2	1.00169
3	1.00251
10	1.00818
35 (mean, sea water)	1.02822

[a] After Ruttner, 1963.

density differences are small but highly significant. It is important to examine the difference in density between water at a given temperature and water at a temperature 1°C lower, referred to as the *density difference per degree lowering* (Vallentyne, 1957). The magnitudes of this density difference for water of different temperatures are shown on the right side of Figure 2–3. The density difference per degree lowering increases markedly as the temperature goes above or below 4°C. Physical work is required to mix fluids of differing density as, for example, when mixing cream into milk, and the amount of energy input required is proportional to the difference in density. The amount of work required to mix layered water masses between 29 and 30°C is 40 times, and between 24° and 25°C 30 times that required for the same masses between 4° and 5°C.

Density also increases with increasing concentrations of dissolved salts in an approximately linear fashion (Table 2–1). The salinity of a majority of inland waters is within a range of 0.01 and 1.0 g l^{-1}, usually between 0.1 and 0.5 g l^{-1}. Inorganic salinity of highly saline lakes can commonly exceed 60 g l^{-1} (Rawson and Moore, 1944; International Association of Limnology, 1959; Wetzel, 1964). However, in a majority of lakes, the salinity is very low and varies less than 0.1 g l^{-1} spatially and seasonally. Consequently, salinity-induced variations in density are small, but they cannot be ignored since under certain conditions of lake stratification inorganic salts can accumulate temporarily or permanently. We shall return to this subject later on (Chapter 6).

Salinity also decreases the temperature of maximum density of water at a rate of ca 0.2°C per g l^{-1} increase. The temperature of maximum density of sea water (mean 35 g l^{-1}) is −3.52°C, which is below the freezing point (−1.91°C). In most lakes, however, the change in point of maximum density is very small.

Hydrostatic pressure can be high enough to compress water sufficiently to lower the temperature of maximum density. Pressure increases one atmosphere per 10 m depth. The temperature of maximum density decreases about 0.1°C per 100 m of depth (Strøm, 1945). In very deep lakes temperatures below 4°C are found to be partially, but not totally, related to the relatively high pressures encountered at great depth (Eklund, 1963, 1965; Johnson, 1964, 1966).

Viscosity–Density Relationships

The density of water is 775 times greater than air at standard temperature and pressure (0°C, 760 mm Hg). The greater density of water exerts a marked buoyancy on organisms, against gravitational pull, and consequently reduces the amount of energy that an organism must expend in order to support or maintain its position. Although a marked reduction in supporting tissue is seen in many freshwater animals, especially among the lower invertebrates, these adaptations are particularly conspicuous among aquatic vascular plants. A good example are the truly submersed angiosperms, almost totally limited to fresh waters, which have immigrated and adapted to aquatic conditions relatively recently. Many modifications have been observed in these plants, particularly with respect to the reduction of their vascular tissue. Submersed organs exhibit a reduction in the extent of lignification of the xylem or water-conducting and supporting elements, and most of the vascular strands are condensed into a weakly developed central cylinder. The supportive tissue of these hydrophytes develops strongly only in those plants or parts that are adapted to aerial existence as surface-floating or emergent forms.

The viscosity of water is influenced to a minor extent by the salinity of the water, but to a considerable extent by temperature. Viscosity decreases as temperature increases; water viscosity doubles as the temperature is lowered from 25°C to 0°C. The viscosity of water offers roughly 100 times the frictional resistance to a moving organism or particle as does air, depending upon surface exposure area, speed, and the temperature and chemical composition of the fluid. As will be discussed further on, organisms with locomotion must expend considerable energy to overcome changes in viscosity. The sinking rates and distribution of passive organisms, such as planktonic algae or sedimenting particles, are influenced by density-related changes in viscosity.

Surface Tension

The quasi-polymeric bonding properties of liquid water molecules discussed above are disturbed at the interface with air. At the interface plane the molecular attractions are unbalanced and exert an inward adhesion to the liquid phase. The result is an interface surface or film under some tension. The surface tension at the air-water interface of pure water is higher than that for any other liquid except mercury. Surface tension decreases with increasing temperature, and increases slightly with dissolved salts (Table 2–2).

The surface tension of water is reduced markedly by the addition of organic compounds, a phenomenon reflected in lakes or portions of fresh waters where dissolved organic matter concentrates (Table 2–2). For example, the surface tensions of relatively pristine waters containing few algae differed little from those of distilled water. Bog lakes, which are heavily stained with dissolved organic compounds, exhibit depressions in surface tension of 6 to 7 dynes cm^{-1}. Where growth of floating algae or higher aquatic plants is par-

TABLE 2–2   Surface Tension of Waters with Changes in Temperature and Its
Depression in Natural Waters under Various Conditions[a]

°C	Water Dynes cm^{-1}	Condition	Depression of Surface Tension, Range in Dynes cm^{-1}
Pure Water			
0	75.6	Oligotrophic lakes	0–2
5	74.9	Eutrophic lakes	0–20
10	74.4	Bog lakes	0–20
15	73.5	Lake water with foam	2–9
20	72.7	Near floating-leaved angiosperms	5–20
25	72.0	Near submersed angiosperms	1–2
30	71.2	During a blue-green algal bloom	0–20
35	70.4	Open sea	<1
40	69.6	Plymouth Sound, near muddy beach	6–20
Sea Water			
ca 5	75.0	Harbor, heavy boat traffic	15–>20

[a]From data of Adam, 1937, and Hardman, 1941.

ticularly abundant, the surface tension is depressed by about 20 dynes cm^{-1}.
It will be shown further on that the concentrations of dissolved organic matter
in the more productive waters are high. Additionally, natural populations of
algae and submersed angiosperms secrete large quantities of organic compounds during active photosynthesis, as well as during senescence and lysis.
The effects of artificial introduction of organic pollutants into fresh waters on
surface tension are readily apparent (cf. Jarvis, 1967; Jarvis, et al., 1967).

   The air-water interface forms a special habitat for organisms adapted to
living in surface film. This community is collectively referred to as neuston,
and will be discussed in some detail further on (Chapter 14). The surface tension is sufficient to serve as a supporting surface for many organisms of
considerable size, for example for gyrinid and other beetles, and can destroy
others, such as cladoceran microcrustacea, that happen to become caught in
the interface through wave action, and cannot reenter their normal submersed
habitat.

chapter 3

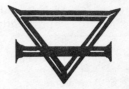

lakes — their distribution, origins, forms

The amount of fresh water on earth is very small in comparison to the water of the oceans, but has much more rapid renewal times (cf. Table 1–1). Distribution on a volumetric basis is concentrated in the large, deep basins of several great lakes and about 20 per cent in Lake Baikal, USSR. The number of individual depressions of smaller lakes and reservoirs is extremely large, however, and a majority are concentrated in temperate and subarctic regions of the Northern Hemisphere.

Catastrophic events of glacial, volcanic, and tectonic activity resulting in the formation of lakes have aggregated many fresh waters into lake districts. The morphometry and geological substrates of the lake basins are of major importance in determining sediment-water interactions, the resultant productivity, and the significance of littoral productivity to that of the entire lake. Increased percentage exposure of sediment area to water volume generally is concordant with increased productivity and an increase in the proportion of total productivity by the attached littoral communities.

DISTRIBUTION OF FRESH WATERS

Inland waters cover less than 2 per cent of the earth's surface, approximately 2.5×10^6 km². About 20 lakes are extremely deep (in excess of 400 m), and a significant portion of the world's fresh water lies within these basins. For example, approximately 20 per cent of the world's fresh water is contained in Lake Baikal, Siberian USSR ($A = 31,500$ km²; $z_m = 1,620$ m; $\bar{z} = 740$ m), the deepest lake of enormous volume (23,000 km³). Like Lake Baikal, nearly all other extremely deep lakes are tectonic and volcanic in origin or have formed from fjords that have subsequently become fresh. Although a few lakes of glacial origin, such as the Laurentian Great Lakes and Great Slave Lake of Canada, have basins of great depth, few exceed 300 m deep. Most of the very deep lakes are found in mountainous regions along the western

14

portions of North and South America, of Europe, and in the mountainous areas of central Africa and Asia. Lakes Baikal and Tanganyika in Africa are the only lakes known to have maximum depths in excess of 1000 m and mean depths over 500 m.

On an areal basis, only a few lakes are of very large surface area; most of these are compared in Figure 3–1. If we exclude the Black Sea, which is an inland water really connected to the ocean, the Caspian Sea, a large (436,400 km²) salt lake, is the largest inland basin separated from the ocean. The Laurentian Great Lakes of North America—lakes Superior, Huron, Michigan, Ontario, and Erie—constitute the greatest continuous mass of fresh water on earth, with a collective area of 245,240 km² and a volume of 24,620 km³. Lake Superior has the greatest area (83,300 km²) of any purely freshwater lake.

While these very large lakes constitute a major freshwater resource, most lakes are much smaller. Most lakes are of catastrophic origin, formed by glacial, volcanic, or tectonic processes. As a result, many lakes are localized into lake districts in which large numbers of lake basins are concentrated. Glacial activity during the most recent period of major ice advance and retreat was instrumental in creating literally millions of small depressions that subsequently were filled and modified. Thus large numbers of lakes are found in the Northern Hemisphere, where large land masses of North America and Europe-Asia interfaced with the glacial movements. Concentrations of small, shallow basins are found throughout the arctic, subarctic, and northern temperate zones.

A recent event of lesser but increasing significance is the creation of large numbers of reservoirs and ponds by man. A majority of both natural and man-made lakes are very small and relatively shallow, usually <20 m in depth.

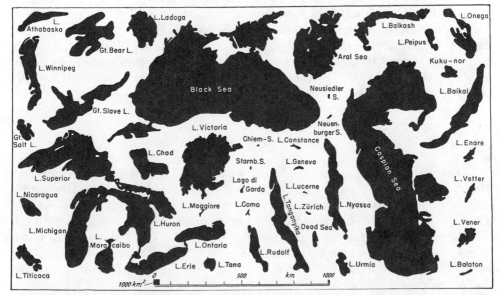

Figure 3–1 An approximate comparison of the surface areas of many of the larger inland waters of the world, all drawn to the same scale. (After Ruttner, F.: Fundamentals of Limnology. Toronto, University of Toronto Press, 1963.)

Because of these morphometric characteristics, an increasing proportion of the lake volume is exposed to and interacts with the chemical and metabolic processes of soil and sediments. A greater proportion of shallow lakes possess properties necessary for the development of sessile littoral flora, which generally results in markedly increased productivity.

GEOMORPHOLOGY OF LAKE BASINS

The origins of lake basins and their morphometry are of much more than casual interest. The geomorphology of lakes is intimately reflected in physical, chemical, and biological events within the basins and plays a major role in the control of a lake's metabolism, within the climatological constraints of its location. The geomorphology of a lake controls the nature of its drainage, inputs of nutrients to the lake, and the volume of influx in relation to flushing-renewal time. Thermal and stratification patterns are, as discussed in the following chapters, markedly influenced by basin morphometry and volume of inflow. These patterns in turn govern the distribution of dissolved gases, nutrients, and organisms; the entire metabolism of freshwater systems is influenced to varying degrees by the geomorphology of the basin and how it has been modified throughout its subsequent history.

The shape of a lake basin often is reflected in its productivity. Steep-sided U- or V-shaped basins, often formed by tectonic forces, are usually deep and unproductive. In such lakes a proportionally smaller volume of water is contiguous with sediments. Shallow depressions with greater percentage contact of water with the sediments generally exhibit intermediate to high productivity.

The following brief résumé on the origin of lakes is based on Hutchinson's (1957) superb detailed summary of the subject which was drawn from an array of global sources. Hutchinson differentiates 76 types of lakes on the basis of geomorphological inception. Despite this detailed classification, numerous variant though minor exceptions have been cited since it appeared (e.g., Horie, 1962, among others). Discussion here will be limited to nine distinct groups of lakes, each formed by different processes.

Tectonic Basins

Tectonic basins are depressions formed by movements of deeper portions of the earth's crust, and are differentiated from lake basins resulting from volcanic activity. The major type of tectonic basin forms as the result of faulting in which depressions occur between the masses of a single fault displacement or in downfaulted troughs (Fig. 3–2). The latter type of basin is referred to as a *graben*, and is the mode of origin of a large number of the most spectacular relict lakes of the world. Foremost among these is Lake Baikal of eastern Siberia, the deepest lake in the world ($z_m = 1620$ m from most recent measurements; Kozhov, 1963), which has a continuous lacustrine history from at least the early Tertiary period. Lake Baikal, and many other

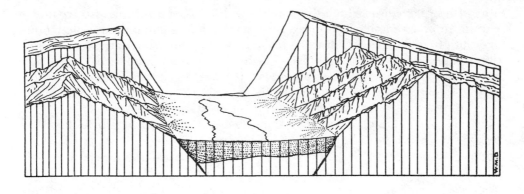

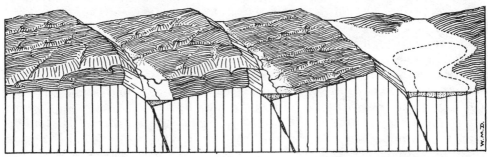

Figure 3–2 Tectonic lake basins. *Upper:* In the background, a depressed fault block between two upheaved fault blocks; in foreground, the same after a considerable period of erosion and deposition. *Lower:* Diagram of the great fault blocks of the northern Sierra Nevada Mountains with the plain of Honey Lake to the east. (From Davis, W. M.: Calif. J. Mines Geol., 29:175, 1933.)

relict lakes, most of which lie in tectonic grabens, are of particular interest because they contain a large number of relict endemic species. For example, of 1200 animal species and at least half that number of plant species known from Lake Baikal, over 80 per cent of those occurring in the open water are endemic to this lake. Lake Tanganyika of equitorial Africa is a similar deep (z_m = 1435 m) graben lake formed in rift valley displacements along crustal fractures, and contains a large number of relict endemic species of plants and animals. Pyramid Lake of Nevada and Lake Tahoe of California are familiar examples of American graben lakes.

Tectonic movements causing moderate uplifting of the marine sea bed have isolated several very large lake basins. The relict marine basins of Eastern Europe, which include the Caspian Sea and the Sea of Aral, were separated by the formation of uplifted mountain ranges in the Miocene period. Upwarping of the earth's crust in lesser degrees also has resulted in the formation of many large lake systems. Lake Okeechobee in Florida resulted from minor depressions (z_m = ca 4 m) in the sea floor as it uplifted in the Pliocene epoch to form the Floridian peninsula, and it forms one of the largest lakes (ca 1880 km²) within the United States, exceeded only by Lake Michigan. Similarly Lake Victoria in central Africa resulted from upwarping of the margins of the plateau in which the lake basin lies. This uptilting also created a natural impoundment of the major river system flowing into the valley. Drainage outlets of the Great Salt Lake basin of Utah were eliminated by uplifted deformations, forming a closed lake basin. Upwarping of the earth's crust also contributed, along with the primary glacial scouring activity, to the formation of the Great Lakes of America. As the massive glacial sheet receded, the release of pressure resulted in a significant crustal rebound.

Lake basins are occasionally formed in areas of localized subsidence that result from earthquake activity. Many of these depressions are dry or temporarily contain water, depending on the porosity of the basin material, while others become permanent lakes, usually open basins with outlet drainage.

Lakes Formed by Volcanic Activity

The catastrophic events associated with volcanic activity generate lake basins in several different ways. As volcanic materials are ejected upward and create a void, or as released magma cools and is distorted in various ways, depressions and cavities are created. If these voids are undrained, they may contain a lake (Fig. 3–3). Because of the granitic nature of these lake basins and their drainage areas, which are often very restricted, many lakes associated with volcanic activity contain low concentrations of nutrients and are relatively unproductive.

Small crater lakes are occasionally found occupying unmodified cinder cones of quiescent volcanic peaks. However, crater lakes in depressions formed by the violent ejection of magma, or by collapse of overlying materials where underlying magma has been ejected to create a cavity, are more common in all areas of recent volcanic activity. Craters of explosive origin, termed *maars*, are generally very small depressions with diameters less than

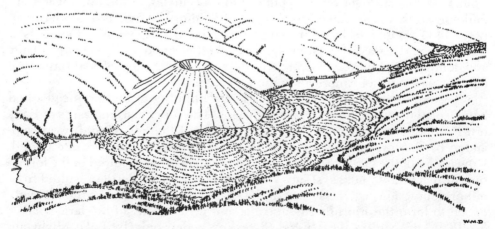

Figure 3–3   Volcanic lakes. A caldera lake within the volcanic cone and several lakes within the valleys dammed by lava flows. (From Davis, W. M.: Calif. J. Mines Geol., 29:175, 1933.)

2 km, and result from lava coming into contact with ground water or from degassing of magma. Maars are usually nearly circular in shape and can be extremely deep (>100 m) in relation to their small surface area. The basins formed by the subsidence of the roof of a partially emptied magmatic chamber are termed *calderas*, and can be somewhat larger than maars (minimum diameter about 5 km). Among the most spectacular of lakes formed by the collapse of the center of a volcanic cone is Crater Lake, Oregon, with an area of 64 km² and a depth of 608 m (seventh deepest lake in the world). Caldera lakes can be modified in various ways by secondary peaks partially filling the original caldera depressions, or by the occurrence of faulting over an emptied magma chamber.

Some volcanic lakes originated by a combination of large-scale volcanic and tectonic processes. In these situations caldera collapse occurred on such a large scale that extensive portions of the surrounding land subsided, in addition to the central portion of the volcano. Such subsidence usually takes place along prexisting fault fractures. Some of the largest lakes associated with volcanic activity were formed in this manner. Many examples of these lakes exist in equatorial Asia and New Zealand (cf. Bayly and Williams, 1973).

Extruded lava flows from volcanic activity can form lakes in several ways. As lava streams flow, cool, and solidify, surface lava frequently collapses into voids created by the continued flow of the underlying molten lava. A lake basin may be formed in this manner when the unsupported overstory crust collapses, and may be filled when the depression extends below groundwater level. Lava streams also commonly flow into a preexisting river valley and form a dam, behind which a lake can collect (Fig. 3–3). If the dam is of sufficient magnitude, the entire hydrography of the region can be changed by the resultant reversal of the river system. In some cases the river flow is forced underground in order to pass beyond the lava obstruction.

Lakes Formed by Landslides

Sudden movements of large quantities of unconsolidated material in the form of landslides into the floors of stream valleys can cause dams and create lakes, often of very large size (Fig. 3–4). Such landslide dams may result from rockfalls, mudflows, iceslides, and even flows of large amounts of peat, but they are usually found in glaciated mountains. The landslides usually are brought about by abnormal meteorological events, such as excessive rains acting on unstable slopes. More spectacular slides are occasionally initiated by earthquake activity. Lakes formed behind landslide dams are often transitory, existing only for a few weeks to several months. This is because unless the slide is very massive, the unconsolidated dam is susceptible to rapid erosion by the effluent of the newly formed lake. Many disastrous floods have resulted from the rapid erosion of such damming material by the effluent which, once started, can quickly empty the accumulated basin. As with damming by lava flows, if the landslide dam is sufficiently large, the lake can become permanent and reverse the direction of drainage flow from the river valley.

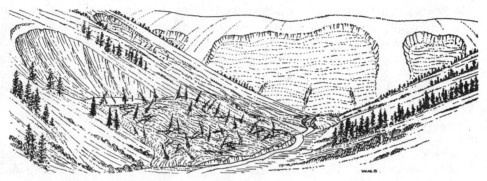

Figure 3–4 Lakes formed by a large landslide into a steep-sided stream-eroded canyon (*upper*) and in a hollow behind a recent slide with tilted trees (*lower*). Four mountain landslides are shown in the background. (From Davis, W. M.: Calif. J. Mines Geol., 29:175, 1933.)

Lakes Formed by Glacial Activity

By far the most important agents in the formation of lakes are the gradual but nonetheless catastrophic corrosion and deposition effects of glacial ice movements. Land surfaces that are now glaciated include Greenland and several smaller arctic islands, Antarctica, and numerous small centers of high mountains throughout the world. These contemporary glacial activities are small, however, in comparison to the massive Pleistocene glaciation that advanced and receded in four major episodes of activity in the Northern Hemisphere. With the retreat of the last stages of glaciation, an immense number of small lakes were created which numerically far exceed those lakes formed by other processes. The action of glaciers in mountainous regions of high relief usually produces lake basins that are quite different from those resulting from the movements of large ice sheets on regions of more mature and gentle relief.

A number of lakes, often temporary in nature, occur on the surfaces, within, or beneath existing glacial ice masses in areas of transitory thaw. For example, numerous meltwater subice lakes several kilometers in diameter are known to form below 3000 to 4000 m of ice of the Antarctic ice sheet

(Oswald and Robin, 1973). In high mountain regions the fronts and lateral arms of glaciers, as well as terminal morainal deposits supported by the ice, often will function as effective dams in river valleys that originate almost totally from meltwater.

Glacial ice-scour lakes refer to the vast number of small lakes formed by ice moving over relatively flat mature rock surfaces that are jointed and contain fractures. The ice-scour lakes are particularly common in mountainous regions where glacial movements have removed loosened rock material along fractures. Upon deglaciation, the rock basins thus formed fill with meltwater. Such glacial scour lakes may be found on the upland peneplains of Scandinavia, in the United Kingdom, and in the great Canadian shield region.

A frequently occurring type of ice-scour lake forms in the upper portion of glaciated valleys of mountainous areas, where the valleys are shaped into structures resembling amphitheaters by freezing and thawing ice action (Fig. 3–5). The amphitheater-like formation is referred to as a *cirque*, and lakes within such depressions as *cirque lakes*. Water is held in the cirque depression either by a rock lip above the general level of the depression, or by morainal deposits. Cirque lakes, which are generally small and relatively shallow (<50 m), often are found in tandem arrangement within a glaciated

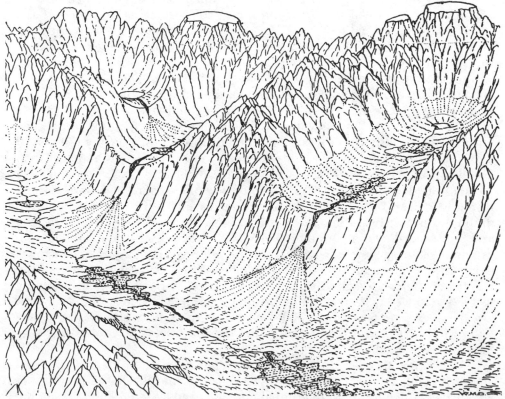

Figure 3–5   Several small cirque lakes within a mountain group from which several converging, cirque-headed branch troughs all join the same trunk trough. (From Davis, W. M.: Calif. J. Mines Geol., 29:175, 1933.)

trough of a mountain valley, with the higher lake 'hanging' above the lower succeeding lake in a stairway-like fashion. True cirque lakes, which are common to all major mountainous regions, are formed at approximately the snow line in glaciated valleys. When the glaciers extend well below the snow line of constant freezing and thawing, the corrosive action of the ice can form rock basins within the glacial valley. When such valley rock basins form a chain of small lakes in a glaciated valley resembling a string of beads, they are referred to as *paternoster lakes*. Where there are mountains around the sea, as in many areas of Norway and western Canada, fjord lakes can be formed in glacially deepened valleys in narrow, deep troughs.

The English Lake District, a small group of lakes in a roughly circular area of northwestern England, is worthy of special notice because of the long history of limnological investigations that have centered in this region (recently reviewed in detail by Macan, 1970). The geological formations of the Lake District consist of an elevated dome of Ordovician and Silurian slates and volcanic intrusions. Rock formations overlying the early uplifted region were later eroded into nine major valleys. In the process of widening and deepening the troughs, glacial corrosive activity scoured piedmont basins in each of these valleys. The result was the formation of a series of major valleys radiating from the dome, most of which contain a major lake depression and several smaller basins constituted from deposits of morainal till.

Where continental ice sheets encountered weak areas in primary basal rock formations, glacial scouring of lake basins occurred on a massive scale in nonmountainous piedmont areas. Great Slave Lake and Great Bear Lake in the central Canadian subarctic region are examples of scouring of preexisting valleys to very great depths by the ice sheet movements. The most impressive example, however, of large rock basins produced by glacial continental ice erosion are the Great Lakes of the St. Lawrence drainage, which collectively form the largest continuous volume of liquid fresh water in the world. The entire Great Lakes region was covered by continental ice during the glacial maxima. Three major events led to the formation of the contemporary Great Lakes during the retreat of the Wisconsin ice sheet (Fig. 3–6), approximately 15,000 years B. P. [1] (Hough, 1958). In the early phases of retreat, lakes formed in the Michigan and Huron-Erie basins against the retreating ice lobes and drained southwestward into the Mississippi valley. As the ice sheet retreated and readvanced over a period of nearly 8000 years, discharge channels from previous scouring were uncovered and altered, resulting in drainage changes to an easterly direction. Another factor that contributed to changes in drainage was the crustal rebound uplifting that took place as the weight of the overlying ice was removed. The major flow patterns and configurations of the modern Great Lakes were fixed about 5000 years B. P., except for a lowering of lake levels and diversion of all discharge through the Erie-Ontario basins to the St. Lawrence drainage.

As continental glacial ice sheets retreated in the late Pleistocene, vast amounts of rock debris, moved and incorporated into the ice during former advances, were deposited in terminal moraines and lateral to the lobes of the

---

[1]Before Present

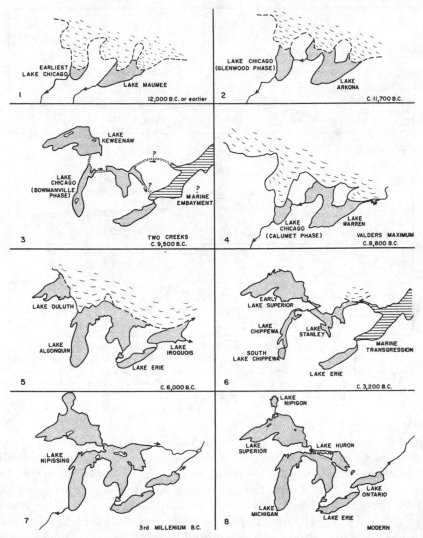

Figure 3-6 The history of the Laurentian Great Lakes. *1*, Cary substage; *2*, Late Cary; *3*, Cary-Valders interstadial, low water in eastern basins, marine transgression in Ontario Basin; *4*, Valders maximum; *5*, Post-Mankato retreat; *6*, Postglacial thermal maximum; *7*, Lake Nipissing with triple drainage; *8*, Modern lakes. (From Hutchinson, G. E.: A Treatise on Limnology. Vol. 1, New York, John Wiley & Sons, Inc., 1957, after various sources; see Hough, 1958, for details.)

retreating glacier. These deposits dammed up valleys and depressions in a highly irregular way and formed lake basins. Some depressions were below later groundwater levels; others were filled by meltwater and drainage from the surrounding topography. Morainal damming of preglacial valleys, usually at one end but also less frequently at both ends, created many of the numerous lakes of the glaciated northern United States. Lake Mendota and many other Wisconsin lakes were formed in this way, and the Finger Lakes of New York underwent complex glacial modification in which morainic damming occurred at both ends of their deep, narrow valley basins (von Engeln, 1961).

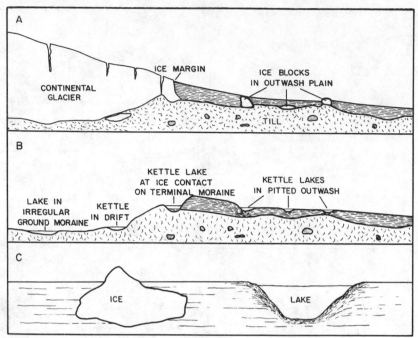

Figure 3–7   The formation of various types of kettle lakes. *A*, An outwash plain of retreating continental ice containing ice blocks; *B*, Lakes formed in the outwash plain and morainal till; *C*, Irregular slopes and shelves of a kettle lake formed by deposition of overburden till on an irregular block of ice. (From Hutchinson, G. E.: A Treatise on Limnology. Vol. 1, New York, John Wiley & Sons, Inc., 1957, after Zumberge, 1952.)

A very common process of formation of lakes in regions glaciated by continental ice sheets was associated with deposits of meltwater outwash left at the border of the retreating ice mass, and blocks of ice buried in this debris. The major features of basins formed in this manner, which accounts for the origin of thousands of small lakes of northern North America, are shown in Figure 3–7. Glacial drift material deposited terminally and beneath the ice washed out into plains and often contained large segments of ice broken from the decaying glacier. Based on paleolimnological evidence, we know that these blocks of ice, deposited in glacial drift or in the outwash plain, often took several hundred years to melt completely, and resulted in the formation of the *kettle lakes.* As the ice block melted, the morphometry of the resulting depressions was modified variously by the extent of overburden of rock debris (Fig. 3–7C). The ensuing lake basins are highly irregular in shape, size, and slopes, corresponding to the irregularities of the original ice blocks. Kettle lakes often exhibit variable underwater relief, with multiple depressions separated by irregular ridges and mounds. These lakes rarely exceed 50 m in depth, their shallow basin being related to the limiting depth of crevasse formation in the fracturing of the terminal glacier portions.

Most arctic lake basins result from glacial activity, as discussed above, or are *cryogenic lakes* formed from the effects of permafrost (Hobbie, 1973). The most abundant type of water body found in the Arctic is the shallow pond formed inside of an ice-wedge polygon which grows in the permafrost from water seepage through cracks in the surface of the ground. Eventually, polygonal networks of ridges are formed that often contain ponds from 10 to 50 m in diameter. Millions of these ponds exist in coastal northern Alaska, Canada, and flat regions of Siberia. The shallow ponds may coalesce into large ponds, or large amounts of ice deeper in the permafrost may melt, especially if the plant cover is disturbed or destroyed, and form a shallow *thermokarst lake.*

A very large number of Arctic thaw lakes are elliptical, with their long axis oriented in a northeast-southwest direction across the prevailing winds. It is apparent, at least among the smaller elliptical lakes of this region of nearly constant wind, that the prevailing system of currents would tend to erode and thaw permafrost at the ends of the long axis of the ellipse lying across the wind (Livingstone, 1954, 1963, Livingstone, et al., 1958).

Solution Lakes

Lake depressions can be created in any area by deposits of soluble rock that are slowly dissolved by percolating water. While many rock formations are readily soluble (salts such as sodium chloride [NaCl], calcium sulfate [$CaSO_4$], and ferric and aluminum hydroxides), most solution lakes are formed in depressions resulting from the solution of limestone (calcium carbonate, $CaCO_3$) by slightly acidic water containing carbon dioxide (CO_2). Solution lakes are very common in limestone regions of the World, notably the karst regions of the Adriatic, especially in Yugoslavia, the Balkan Peninsula, the Alps of Central Europe, and in Indiana, Kentucky, Tennessee, and particularly Florida in North America.

Solution basins are usually very circular and conically shaped sinks, termed *dolines,* which are developed from the solution and gradual erosion of the soluble rock stratum. Percolating surface or ground water dissolves limestone most readily at the joints and points of fault fracture from which it drains. Adjacent dolines may eventually fuse to form compound depressions that are more irregular in conformity. Alternately, solution of limestone frequently occurs in large subterranean caves. Continued solution of the superstructure by ground water results in its weakening, to the point where the roof collapses, forming a reasonably regular conical doline.

The level of water in the solution basin is often highly variable. Usually, the depressions are of sufficient depth to extend well into the groundwater table and permanently contain water. Other basins that just reach the water table can undergo fluctuations in water level in response to seasonal and long-term variations in groundwater levels.

Lake Basins Formed by River Activity

The running waters of rivers possess considerable corrosive power that may create lake basins in the flow from elevated land to large lakes or the sea.

In the upper reaches of rivers where gradients are steep, excavation by water can produce rock basins that may persist as lakes after the course of the river has been diverted. *Plunge-pool lakes,* excavated at the foot of waterfalls, provide a rare but spectacular example of such destructive fluviatile action. Several large lakes of the Grand Coulee system of the State of Washington were surely plunge pools shaped by the former interglacial course of the Columbia River system.

A combination of destructive erosional and obstructive depositional processes occurs as rivers flow through more gentle gradients of mature lower reaches to form many lakes in the river flood plains. Many lakes were formed along major rivers when sediments of the main stream were deposited as levees across the mouths of tributary streams. In this way, the obstruction of the tributary flow continued until the side valley was flooded and a lateral lake was formed. Lateral lakes are frequently found in tributary valleys along major river systems in all continents, especially in the upper portion of the drainage. The reverse situation of fluviatile dams holding lakes in the main river channel occasionally occurs as a result of deposition by a lateral tributary, either temporarily or perennially, of more sediment than the main stream can remove.

Where rivers enter the relatively quiescent waters of a lake or the sea, sedimentation results in the formation of deltas, often of very large size. Occasionally, a delta is sufficiently large when entering a long, narrow lake on one side to form a barrier that divides the original lake in half. Much more common, however, are lakes that are formed in the deltas of all major rivers of the world. As the river velocity becomes reduced and sediments are deposited, the water tends to flow around the sediments in a U-shaped pattern

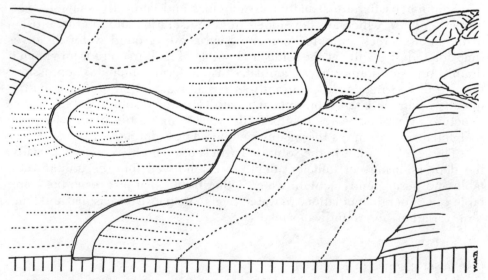

Figure 3–8   Diagram of lakes, including an isolated oxbow lake, and sloughs resulting from river activity. (From Davis, W. M.: Calif. J. Mines Geol., 29:175, 1933.)

where the open end extends seaward. As alluvial deposition extends further seaward, the inward depressions eventually are isolated as shallow lakes, often of large size. Subject to the influence of tides and other water movements of the sea, these *deltaic lakes* often receive salt water and frequently are brackish.

As rivers meander within the irregularities of the topography, greater turbulence and erosion occur on the outside, concave side of the river bend, while deposition occurs on the inside, convex side of slower current and less turbulence. With time, continued erosion and concavity takes place until the U-shaped meander of the river closes in upon itself (Fig. 3–8). The main course of the river cuts a channel through the initial portion of the meander, and levee deposits eventually isolate the loop, referred to as an *oxbow lake*, from the river channel. The outer erosional side of oxbow lakes is usually deeper than the inner concave side.

Wind-Formed Lake Basins

Wind action operates in arid regions to create lake basins by deflation or erosion of broken rock, or by redistribution of sand, resulting in the formation of *dune lakes*. Such lake depressions may be solely or partially the result of wind action, and the water they contain is often temporary and dependent upon fluctuations in climate.

Bayly and Williams (1973) differentiate several types of dune lakes based on those found in Australia and New Zealand, which occur in many other parts of the world as well. Dune barrage lakes form behind sand dunes that are moving inland to block a river valley draining toward a coast. These lakes are typically triangular in shape, with the deepest part close to the sand dune, and are found inland in desert regions as well as in coastal regions. If a large amount of deflation occurs so that little or no sand is left on the floor between the trailing arms of parabolic dunes and an impervious rock floor is exposed, a lake may develop. Organic additions from vegetation assist in creating impervious, organically bonded sand-rock in dune depressions. Deflation depressions that are permeable to water movement can form lakes when they extend below an extensive water table. Numerous small, shallow lakes of this type occur along the eastern shore of Lake Michigan, among wind-blown depressions of the sand dunes. All of these lakes are extremely transitory.

Deflation basins also are common where material is moved and eroded from horizontal strata of rock or clay. The wind-eroded material may be deposited in crescent-shaped mounds downwind or removed completely from the area, permitting the formation of large pans or nearly level areas, often called *playas*. Deflation basins commonly fill with water during rainy seasons or wet periods, and become increasingly saline with evaporation and dry during opposing seasons or dry periods. These ephemeral lakes are common in large portions of Australia, South Africa, endorheic regions of Asia and South America, and the plains and arid regions of the United States.

Basins Formed by Shoreline Activity

When the coast line of a large body of water, such as the sea or a large lake, possesses some irregularity or indentation the potential exists for the formation of a bar across the depression to form a coastal lake. When disturbed by wave action, a longshore current flowing along the shoreline and carrying sediment will, on encountering a bay, deposit the material in the form of a bar or spit across the mouth of the indentation (Fig. 3–9). Often the spit eventually can separate the bay from the sea or large lake to form a coastal lake.

Marine coastal lakes are commonly the result of bar formation across the mouths of old estuaries that have been inundated by rising water levels or slight subsidence. Often river discharge and tidal currents are insufficient to prevent complete separation of the lake from the sea. The result is an alternation between fresh and brackish water in the lake, in relation to the ratio

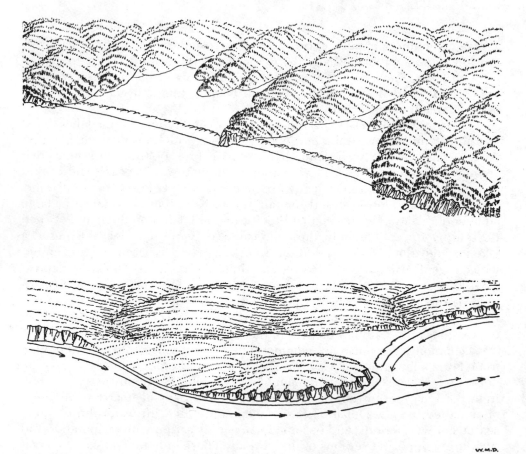

Figure 3–9   *Upper:* Coastal lakes formed by enclosure of lagoons by wave and current-built bars on a shore line embayed by slight subsidence. *Lower:* A land-bound island with a bay behind it, closed by a sand-reef beach formed by reverse eddy currents. (After Davis, W. M.: Calif. J. Mines Geol., 29:175, 1933.)

of freshwater inputs and salt water intrusions. Other coastal lakes are completely separated from the sea.

Numerous coastal lakes are found inland, adjacent to large lakes. Formation of these water bodies occurs in analogous fashion, by the deposition of bars across bays and river valleys. Spits are known to have formed on two sides of a lake, as a result of current patterns in relation to specific morphometry. They may then join in the middle, dividing the lake into two.

Lakes of Organic Origin

The full magnitude of lakes created by the damming action of plant growth and associated detritus is incompletely known. It is clear that plant growth can be sufficiently profuse to dam the outlet of shallow depressions, and create small lakes of alternate drainage patterns. The effectiveness of this method of water retention in flat regions, such as the arctic tundra, is unknown.

Two mammals, the beaver and man, are particularly effective in constructing dams across river valleys to impound water into lakes. The American beaver created numerous large, long-lived lakes, many of which became permanent by means of sediment deposited against the dams. However, it is rapidly being relegated to a lesser importance as a result of man's recent exploitation of the behaviorial activity involved in the making of lakes. Man has created artificial lakes by damming streams for at least 4000 years. Only in the last two centuries, however, has this activity become highly significant for the purposes of flood control, and the provision of power and water supplies for urban concentrations.

Reservoirs are being constructed on an unprecedented scale in response to the exponential demophoric demands of man. Plans for the construction of enormous reservoirs, approaching the size of the Laurentian Great Lakes, are underway in Canada and the Soviet Union. Such massive alterations of large drainage systems will result in major modifications in topography and regional climate that are not yet fully recognized or even partially understood.

Small, shallow reservoirs, in the form of farm ponds and moderate-sized inundations, have created literally millions of small lakes. Particularly characteristic of the morphometry of these lakes is that they are generally shallow and possess large areas where macrophytic vegetation can grow; such plant life radically alters the productivity of the lake system. Moreover, the small reservoirs generally receive high nutrient inputs in relation to their volume, which further increases their productive capacity. Most reservoirs are relatively short-lived because of the high rates of sediment load delivered by the influents.

MORPHOLOGY OF LAKE BASINS

The morphology of a lake basin has important effects on nearly all of the major physical, chemical, and biological parameters of lakes, as will be repeatedly shown in the ensuing chapters. The forms of lake basins are highly varied, and reflect their modes of origin, subsequent modifying events of water movements within the basin, and the degree of loading from the drainage basin.

The morphology of a lake is described by a detailed bathymetric map which is required for evaluation of all major morphometric parameters. The preparation of such a map requires a survey of the shore line by standard methods, often in combination with aerial photography. From the map of the shore line a detailed bathymetric map of depth contours must be constructed by a combination of accurate soundings along intersecting transects. Methods commonly employed for both lake and stream mapping are discussed by Welch (1948). Accurate sonar transects, which permit acquisition of much greater detail than was previously possible by manual sounding methods, now are used frequently.

Morphometric Parameters

The many morphometric parameters that can be determined from a detailed bathymetric map are discussed at length by Hutchinson (1957). The most commonly used parameters are defined here.

Maximum Length (1).   The distance on the lake surface between the most distant points on the lake shore. This length is the maximum effective length or fetch for wind to interact on the lake without land interruption.

Maximum Width or Breadth (b).   The maximum distance on the lake surface at a right angle to the line of maximum length between the shores. The mean width $(\bar{b})$ is equal to the area divided by the maximum length; $\bar{b} = A/1$.

Area (A).   The area of the surface and each contour at depth z is best determined by planimetry (cf. Welch, 1948) or, less precisely, by a grid enumeration analysis (Olson, 1960).

Volume (V).   The volume of the basin is the integral of the areas of each stratum at successive depths from the surface to the point of maximum depth. The volume is closely approximated by plotting the areas of contours, as closely spaced as possible, against depth and the area of this curve integrated by planimetry. Alternatively, the volume can be estimated by summation of a series of truncated cones of the strata:

$$V = \frac{h}{3} (A_1 + A_2 + \sqrt{A_1 A_2})$$

where h is the vertical depth of the stratum, A_1 the area of the upper surface, and A_2 the area of the lower surface of the stratum whose volume is to be determined.

Maximum Depth (z_m).   The greatest depth of the lake.

Mean Depth $(\bar{z})$.   The volume divided by its surface area; $\bar{z} = V/A_0$.

Relative Depth (z_r).   The maximum depth as a percentage of the mean diameter.

$$z_r = \frac{50\ z_m\ \sqrt{\pi}}{\sqrt{A_o}}.$$

Most lakes have a z_r of less than 2 per cent, whereas deep lakes with a small surface area exhibit greater stability and usually have z_r >4 per cent.

Shore Line (L).   The intersection of the land with permanent water is nearly constant in most natural lakes. The shore line, however, can fluctuate widely in ephemeral lakes and especially in reservoirs in response to variations in precipitation and discharge. The length of the shore line can be determined directly or from maps with a map measurer (chartometer; rotometer; cf. Welch, 1948).

Shoreline Development (D_L).   The ratio of the length of the shore line (L) to the length of the circumference of a circle of area equal to that of the lake

$$D_L = \frac{L}{2\sqrt{\pi\ A_o}}.$$

Very circular lakes such as crater lakes and some kettle lakes approach the minimum shoreline development value of unity. The conformation of most lakes, however, deviates strongly from the circular. Many are subcircular and elliptical in form, with D_L values of about 2. A more elongated morphometry increases the value of D_L markedly, as for example is found in the dendritic outlines of lakes occupying flooded river valleys. Shoreline development is of considerable interest because it reflects the potential for greater development of littoral communities in proportion to the volume of the lake.

Hypsographic and Volume Curves

The hypsographic curve, or depth-area curve, is a graphic representation of the relationship between the surface area of a lake and its depth. This relationship may be expressed in terms of per cent of the lake area which is above water of a given depth, or in absolute units such as m^2, hectares, or km^2 (Fig. 3–10). The depth-volume curve is closely related to the hypsographic curve, and represents the relationship of lake volume to depth. Similarly, the units of expression can be either in per cent of total lake volume above a specific depth, or in volume units against depth.

The importance of these area and volume curves in limnological investigations stems from the relationship between lake morphology and biological productivity. The hypsographic curve represents the relative proportion of the bottom area of the lake, which is included between the strata under consideration. However, it is only an approximation of the area

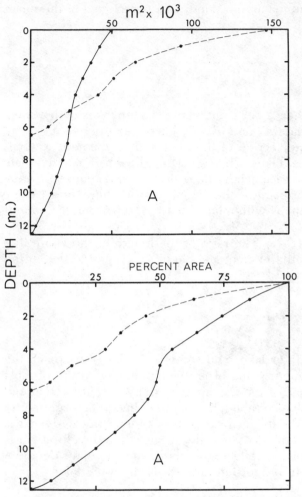

Figure 3-10 Hypsographic (depth-area) curves (*A*) and depth-volume curves (*B*) of oligotrophic Lawrence Lake (_____) and eutrophic Wintergreen Lake (– – – –), southwestern Michigan. (From Wetzel, unpublished data.)

Illustration continued on opposite page.

of the exposed lake bottom, since areal measurements are related to the plane of the lake surface, whereas the actual area of the sediments is greater. In lakes with otherwise comparable conditions, biological productivity is generally greater in those with greater superposition of zones of photosynthetic production and of decomposition (Thienemann, 1927; Strøm, 1933; Rawson, 1955, 1956). The extent of shallow water in a lake is a determining factor in the interrelationship of these zones (cf. Chapter 6), as well as in determining the area available for growth of rooted aquatic plants and associated littoral communities.

The ratio of mean to maximum depth ($\bar{z}:z_m$) is an expression similar to the ratio of the volume of the lake to that of a cone of basal area A and height z_m $\left[A\bar{z} \Big/ \left(\frac{1}{3} z_m A \right) = 3\ \bar{z}/z_m \right]$. The ratio $\bar{z}:z_m$ thus gives a comparative value of the form of the basin in terms of volume development. For most lakes the value of this ratio $\bar{z}:z_m$ is >0.33, the value that would be given by a conical depression.

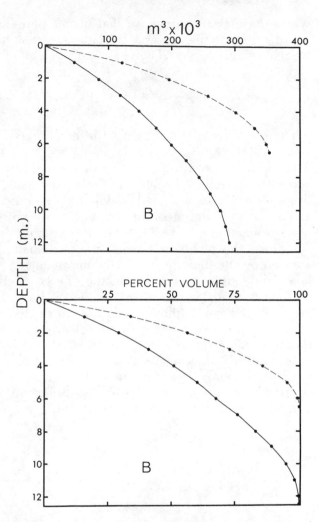

Figure 3–10 *Continued*

The ratio exceeds 0.5 in many caldera, graben, and fjord lakes, whereas most lakes in easily eroded rock have ratios between 0.33 and 0.5. Very low values of $\bar{z}:z_m$ occur only in lakes with deep holes, such as solution or kettle lakes.

In his examination of the morphometry of a large number of lakes, Neumann (1959) has shown that the average shape of lake basins approximates an elliptic sinusoid (Fig. 3–11). The elliptic sinusoid is a geometric body

Figure 3–11 Vertical cross-sections through three forms of lake basins: *a,* half an ellipsoid of revolution; *b,* elliptic sinu-soid; and *c,* right elliptic cone. (Modified from Neumann, 1959.)

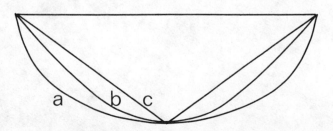

whose base is an ellipse, so that planes perpendicular to the base ellipse passing through the center of the latter intersect the surface of the body along troughs of sine curves. The volume of such an elliptic sinusoid is:

$$V = 4 \left(1 - \frac{2}{\pi}\right) abz_m = 1.456 \ abz_m,$$

where a and b are the half-axes of the lake surface ellipse. Since the area of the lake surface ellipse concerned is πab, $\bar{z} = V/A = 0.464z_m$, the ratio $\bar{z}:z_m = 0.464$. This value is very close to the average value (0.467) of the ratio for over 100 lakes that have been evaluated (cf. also Anderson, 1961). Therefore the elliptic sinusoid serves as a good model for an average lake in which irregularities and submerged depression individuality are not severe. The mean depth of lakes is about one-half of (0.46) their maximum depth.

Since the pioneering work of Thienemann, much attention has been focused on the importance of lake morphometry, especially mean depth, to lake productivity in relation to the effects of climatic and edaphic factors (Hutchinson, 1938; Rawson, 1952, 1955, 1956; Edmondson, 1961; Patalas, 1961; Hayes and Anthony, 1964). Mean depth, regarded as the best single index of morphometric conditions, clearly shows a general inverse correlation to productivity at all trophic levels among large lakes. This relationship deteriorates among small lakes and indicates, as will become apparent in later discussions, that regulation of the dynamics of metabolism and productivity in aquatic ecosystems is multifaceted. Morphometry is only one, although important, interacting parameter.

water economy

The balance of water in lakes is expressed by the basic hydrological relationship in which change in water storage is governed by inputs from all sources less the rates of water losses. Water income from precipitation, surface influents, and groundwater sources is balanced by outflow from surface effluents, seepage to ground water, and evapotranspiration. Each of these incomes and losses varies seasonally and geographically, and is governed by the characteristics of the lake basin, its drainage basin, and the climate.

The distribution of water over continental land masses is governed by the global hydrological cycle, in which excessive oceanic evaporation is counterbalanced by greater precipitation over land. The hydrological cycle, which can be altered by extensive man-induced changes of surface water systems, determines the distribution of lakes regionally in relation to the distribution of suitable catchment lake basins.

THE HYDROLOGICAL CYCLE

Consideration of the hydrological cycle of the earth and biosphere is basically an evaluation of the cyclical budgetary processes of water movement. This includes both its movement from the atmosphere, its inflow, and temporary storage on land, and its outflow to the primary reservoir, the oceans. The cycle consists of three principal phases: precipitation, evaporation, and surface and groundwater runoff. Each phase involves transport, temporary storage, and a change in the state of the water.

Evaporation and Precipitation

Evaporation of water into the atmosphere occurs from land, the oceans, and other water surfaces. Major sites include evaporation from precipitation, from precipitation intercepted by vegetation, from the oceans, lakes, and streams, from soils, and from transpiration of plants. The evaporation rates of each of these and other sources of water are governed by an array of dynamic environmental parameters (cf. Meinzer, 1942; Grey, 1970; and others). The atmospheric vapor, although it is smallest in relation to the amount of total

35

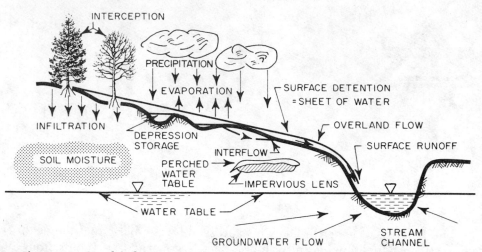

Figure 4–1 Simplified representation of the major pathways of the runoff phase of the hydrological cycle. (From Grey, D. M., ed.: Handbook on the Principles of Hydrology. Ottawa, National Research Council of Canada, 1970. Reprinted by permission of Water Information Center, Inc., Manhasset Isle, Port Washington, New York.)

global water, is stored for the least amount of time (average renewal time of 8.9 days) before returning to the earth as rain, snow, sleet, hail, and condensates (dew and frost), either on land or on the oceans.

The precipitated water may be intercepted or transpired by plants, may run off over the land surface to streams (surface runoff), or may infiltrate the ground (Fig. 4–1). Much of the intercepted water and surface runoff (up to 80 per cent) is returned to the atmosphere by evaporation. Infiltrated water may be temporarily stored (average renewal time ca 280 days) as soil moisture and evapotranspired. Some of the water percolates to deeper zones to be stored as groundwater (average renewal time of 300 years). Groundwater is actively exchanged and may be used by plants, flow out as springs, or seep to streams as runoff. The runoff phase is exceedingly complex and variable because of the extensive involvement of biotic metabolism in its regulation, the extreme heterogeneity of substrate structure and composition, and variations in climate.

Runoff Flow Processes

The soil and rock substrate of the drainage basins of lake systems is of major importance in the regulation of the pathways and rates of hillslope runoff of water received as rain and meltwater (cf. review of Dunne, 1975). These pathways of runoff influence many characteristics of the landscape, the uses to which land can be put, and the requirements for effective land management.

When the rate of rainfall or meltwater influx exceeds the absorptive capacity of the soil, the unabsorbed excess water flows over the surface as overland flow. Overland flow is most common in arid and semiarid regions;

it also is found in humid areas where the original vegetation and soil structure have been disturbed, or in areas where normally porous soil contains a thin layer of concrete frost and consequently cannot absorb meltwater (Dunne and Black, 1970a, 1970b, 1971).

When precipitation is first absorbed by the soil, the water may be stored or it may move by gravity toward stream channels along several pathways (Dunne, 1975). If the soil or rock is deep and of relatively uniform permeability, the subsurface water moves vertically to the zone of saturation, and then follows a generally curving path to the nearest stream drainage channel. This simple pattern of groundwater flow often is disrupted by irregularities of the base geological structure (Davis and Dewiest, 1966; Leopold, Wolman, and Miller, 1964). Rates of groundwater flow are generally slow and the pathways long; hence much of this groundwater contributes to the sustained baseflow of streams between periods of precipitation. Drainage of water from storms is added to this more uniform groundwater discharge. Although the long route of groundwater flow generally dominates the baseflow of streams, the rate of flow can be more rapid in very permeable rock formations such as limestones and jointed basalts, and contribute significantly to runoff from stormflows to recipient drainage streams.

When the permeability of surface soil and weathered rock is high but this material is shallow and underlaid with relatively impermeable soil horizons, percolating water will be diverted horizontally as subsurface stormflow. The shallow subsurface pathway of drainage to a stream channel or lake basin is much shorter than the pathway of groundwater flow, and generally occurs in soil of high permeability. Therefore, subsurface stormflow can be volumetrically dominant to runoff, particularly among steep gradients of narrow drainage valleys.

In the progression of subsurface flow down the gradient of a hillslope, vertical and horizontal percolation can saturate the soil completely. Shallow subsurface flow encountering these saturated areas emerges from the soil surface as return flow, and continues to the recipient channel or basin as overland flow. Similarly, direct precipitation onto saturated areas flows over the surface of the soil. These processes assume greater significance in gently sloping, wide valleys with thin soils.

Each of these processes of runoff of rainfall or meltwater responds differently to variations in topography, soil, and characteristics of precipitation, and indirectly to variations in climate, vegetation, and land use. Therefore runoff flow processes govern the volume, periodicity, and chemical characteristics of the contributions to receiving streams and lake basins.

GLOBAL WATER BALANCE

The global water balance reflects the fact that more water evaporates from the oceans than is returned via precipitation (Fig. 4–2), whereas on the land more water is received via precipitation than is lost by evaporation. A majority of continental water income is from evaporation of the oceans. The water of land masses is not uniformly distributed over the major continents

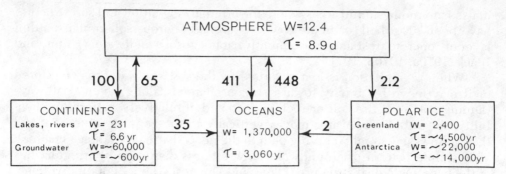

Figure 4–2 Global water balance. W = water content in 10^3 km^3, values on arrows = transport in 10^3 km^3 yr^{-1}, and τ = retention time. Estimate of groundwater is to a depth of 5 km in the earth's crust; much of this water is not actively exchanged. (Modified from Flöhn, 1973, after Lvovitch.)

(Table 4–1). The total and groundwater runoff are greatest in South America, nearly twice that per area of other continents.

Prior to recent times, the global water balance fluctuated very little. It is now apparent, however, that man has introduced regional fluctuations by extensive environmental modifications for irrigation, industrial, and domestic uses, such as land clearing to obtain more arable land, changes in drainage patterns, and exploitation of groundwater reserves (Flohn, 1973). Water demand for agricultural and industrial purposes is projected to increase the man-made fraction of continental evaporation from its present nearly 3 per cent to 10 per cent in ca 30 years, and to 50 per cent in about 70 years. The result will be an accelerated rate of continental freshwater turnover. Since freshwater supplies are inadequate where demand is highest, large-scale desalinization of seawater, at great energy expense, is the only practical contemporary alternative for continued expansion of fresh waters at sites of high demand. Manipulation of surface waters on a massive scale will inevitably lead to irreversible modifications of climate. Modifications of regional climatic

TABLE 4–1 Water Resources and Annual Water Balance of the Continents of the World[a]

	Europe[b]	Asia	Africa	N. America[c]	S. America	Australia[d]	Total
Area (10^6 km^2)	9.8	45.0	30.3	20.7	17.8	8.7	132.3
Precipitation (km^3)	7165	32690	20780	13910	29355	6405	110305
River runoff (km^3)							
Total	3110	13190	4225	5960	10380	1965	38830
Underground	1065	3410	1465	1740	3740	465	11885
Surface	2045	9780	2760	4220	6640	1500	26945
Total soil moistening (infiltration and renewal of soil moisture)	5120	22910	18020	9690	22715	4905	83360
Evaporation	4055	19500	16555	7950	18975	4440	71475
Underground runoff (% of total)	34	26	35	32	36	24	31

[a]After data from Lvovitch, 1973.
[b]Includes Iceland.
[c]Includes Central America but not the Canadian archipelago.
[d]Includes New Zealand, Tasmania, and New Guinea.

conditions have already occurred as a result of extensive alterations of large river systems in which surface waters and evaporation are greatly increased.

Among the continental land masses, three hydrological regions have been recognized (Hutchinson, 1957). The distribution of lakes is related partly to distribution of lake basins and partly to that of water. *Exorheic regions*, within which rivers originate and from which they flow to the sea, contain the major lake districts of the world and most of the lakes. *Endorheic regions*, within which rivers arise but never reach the sea, occur between subtropical deserts and the tropical and temperate humid regions. *Arheic regions*, within which no rivers arise, are desert areas that occur in the latitudes of the trade winds, and between which lies the zone of equatorial rains. Endorheic regions, transitional in nature between the other two, can shift to exorheic or arheic characteristics with relatively small changes in climate.

Water Balance in Lake Basins

The water balance of a lake is evaluated by the basic hydrological equation in which the change in storage of the volume of water in or on the given area per time is equal to the rate of inflow from all sources less the rate of water loss. Water income to a lake includes several sources:

(a) Precipitation directly on the lake surface. Although most lakes, largely in exorheic regions, receive a relatively small proportion of their total water income from direct precipitation, this percentage increases in very large lakes. Extreme examples include equatorial Lake Victoria, which receives a majority (>70 per cent) of its water from precipitation on its surface, and the endorheic Dead Sea, which receives practically no water from this source.

(b) Water from surface influents of the drainage basin[1]. The amount of the total water income to a lake from surface influents is highly variable. Lakes of endorheic regions receive nearly all their water from surface income. The rate of runoff from the drainage basin and corresponding changes in lake level are strongly influenced by the nature of the soil and vegetation cover of the drainage basin. One of the best examples of this effect resulted from the experimental forest cutting and use of herbicides to prevent vegetation regrowth in the Hubbard Brook drainage in New Hampshire (Likens, et al., 1967, 1970). Annual stream-flow increased 39 per cent the first year and 28 per cent the second year above the values when the drainage area was not selectively deforested.

(c) Groundwater seepage below the surface of the lake. Seepage of groundwater is commonly a major source of water for lakes in rock basins and lake basins in glacial till that extend well below the water table. Sublacustrine groundwater seepage forms the major source of water flow into, and from, karst and doline lakes of limestone regions.

---

[1]Reference is to the drainage or catchment area (*Einzugsgebiet; bassin versant*) which is, in American usage, equivalent to watershed, the region or area drained by a river. Watershed, as used in the United Kingdom, refers to the ridge or crest line dividing two drainage areas, and is defined similarly in German (*Wasserscheide*) and French (*ligne de partage des eaux*).

TABLE 4–2   Annual Water Budget for Lawrence Lake, Michigan, 1971[a]

Source/Loss	10^3 m^3	Per Cent
Inputs		
Inlet 1	146.6	32.1
Inlet 2	87.5	19.1
Groundwater	178.1	39.0
Precipitation	44.6	9.8
Total inputs	456.8	100.
Outputs		
Outlet	436.5	90.4
Evapotranspiration	46.2	9.6
Seepage losses[b]	0.	0.
Total outputs	482.7	100.

[a]After Wetzel and Otsuki, 1974.
[b]Indirect evidence indicated that seepage losses were negligible.

(d) Groundwater entering lakes as discrete springs. Sublacustrine springs from groundwater occur frequently in hard-water lakes of calcareous drift regions, where the basin is effectively sealed from groundwater seepage by deposits within the basin. For example, groundwater largely from springs contributed nearly 40 per cent of the annual water income to calcareous Lawrence Lake, Michigan, in comparison to 10 per cent from precipitation, and the remainder was from surface runoff in two streams (Table 4–2). Groundwater inputs were closely related to rates of precipitation and evapotranspiration each month.

In *drainage lakes*, loss of water occurs by flow from an outlet (Birge and Juday, 1934), and in *seepage lakes* by seepage into the groundwater through the basin walls. Deposition of clays and silts commonly forms a very effective seal in drainage lakes, from which most or all of the outflow leaves by the outlet (e.g., Table 4–2). In seepage lakes the lake seal is also likely to be effective over much of the deeper portions of the basin; losses to groundwater egress from the upper portions of the basin.

Further losses of water come about directly by evaporation, or by evapotranspiration from emergent and floating-leaved aquatic macrophytes. The extent and rates of evaporative losses are of course highly variable according to season and latitude, being greatest in endorheic regions. Lakes of semi-arid regions commonly have no outflow effluents and lose water only by evaporation. Such lakes are termed *closed* in contrast to *open lakes* that have outflow by an outlet or seepage.

Evaporative losses are greatly modified by the transpiration of emergent and floating-leaved aquatic plants, a subject treated in detail by Gessner (1959). Rates of transpiration and evaporative losses to the atmosphere vary greatly with an array of physical (e.g., wind velocity, humidity, temperature) and metabolic parameters, as well as with species. Further, many of the aquatic macrophytes are annuals and are predominantly seasonal in lakes of exorheic regions; in tropical lakes, many of the large hydrophytes are peren-

TABLE 4–3   Comparison of Water Loss from a Stand of the Emergent
Aquatic Macrophyte *Phragmites communis* to that of Open Water,
Berlin, Germany, 1950[a]

Date	Evapotranspiration $kg \ m^{-2} \ day^{-1}$	Evaporation $kg \ m^{-2} \ day^{-1}$	Ratio of Transpiration: Evaporation
11 May 1950	3.20	3.24	1.0
25 May 1950	2.50	1.44	1.6
27 July 1950	9.82	2.24	4.4
22 August 1950	16.01	2.29	7.0
17 October 1950	2.79	0.79	3.9

[a]Modified from Gessner, 1959, after Kiendl.

nials and grow more or less continually. In a majority of situations, transport of water from the lake to the air is greatly increased by a dense stand of actively growing littoral vegetation, as compared to evaporation rates from open water (Table 4–3). Since a majority of lakes are small and often possess well-developed littoral flora, these communities contribute significantly to the water balance of many lakes.

Analytical techniques for a detailed evaluation of the water balance of a lake and its drainage basin are complex and require a great deal of work and effort. Because of numerous local fluctuations in climate from year to year, analyses should extend over several years. Specific methodology varies greatly in relation to the objectives, and is beyond the scope of this work. Those interested in such analyses are referred to specific hydrological and isotopic compilations of methods such as those of Chow (1964), Stout (1967), and Grey (1970).

chapter 5

light in lakes

Solar radiation is of fundamental importance to the entire dynamics of freshwater ecosystems. Almost all energy that drives and controls the metabolism of lakes is derived directly from the solar energy utilized in photosynthesis, either autochthonously (within the lake) or allochthonously (within the catchment basin and brought to the lake in various forms of organic matter). The utilization of this energy received by the lake and its watershed, and factors that influence the lake's efficiency of conversion of solar energy into potential chemical energy, are basic to its productivity and to the extent of heterotrophic development of the microflora and fauna of the lake system.

In addition to these direct effects, the absorption of solar energy and its dissipation as heat have profound effects on the thermal structure, water mass stratification, and circulation patterns of lakes. The array of attendant effects on nutrient cycling, distribution of dissolved gases and biota, behavioral adaptations of organisms, and many others exert major controls on the environmental milieu. Therefore, the optical properties of lakes and reservoirs are important regulatory parameters in the physiology and behavior of aquatic organisms. They deserve detailed scrutiny.

LIGHT AS AN ENTITY

The term light is often confusing, in part because of different usages in absolute physical terms, reactions of visual receptors to light, and responses of plants to light energy. For the purposes of this discussion, it is essential that it be viewed physically, as part of the radiant energy of the electromagnetic spectrum. Light is energy, that is, something that is capable of doing work and of being transformed from one form into another, but it can neither be created nor destroyed. Radiant energy is transformed into potential energy by biochemical reactions, such as photosynthesis, or to heat. Energy transformations are far from 100 per cent efficient in a system such as a lake, and most of the energy is lost as heat.

Electromagnetic Spectrum

The electromagnetic spectrum is divided into units of frequency and wavelength. At one extreme are cosmic rays of very high frequency (10^{24} cps), and short wavelengths (10^{-13} cm), and at the other are radio and power transmission waves of low frequency (1 cps) and long wavelengths (to 10^{10} cm). For all practical purposes, solar radiation constitutes the total significant radiant energy input to aquatic systems. This solar flux of energy extends from wavelengths of 100 to >3,000 nm (1000 to >30,000 Å), from the ultraviolet (UV) to infrared radiation, as it is received extraterrestrially (Fig. 5–1). As solar radiation penetrates and diffuses in the atmosphere of the earth, the energy of certain wavelengths is strongly absorbed and attenuated by scattering. The visible portion of the spectrum, with maximum energy in the green (480 nm) portion of the visible range, is only a small portion of the total energy radiated by the sun (Fig. 5–1). UV energy is strongly absorbed by ozone and oxygen and infrared by water vapor, ozone, and carbon dioxide.

With respect to the mechanisms by which life receives light energy, it is important to view light as the radiation of packets of energy termed *quanta* or *photons*. A photon is a pulse of electromagnetic energy; as this energy propagates it has an electric (E) and magnetic (H) field, with respect to direction of flux and wave characteristics of wavelengths (λ) and amplitude (A) (Fig. 5–2). Hence, light is effectively a transverse wave of energy that behaves as a movement of particles with a defined mass. The photon carries energy in a wave conformation which, of course, carries no energy.

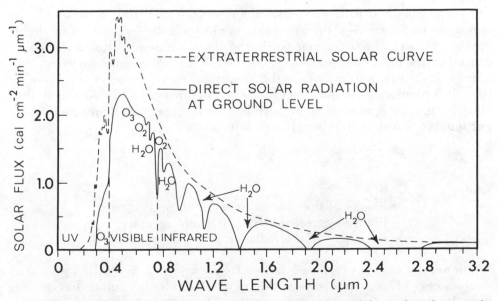

Figure 5–1 Extraterrestrial solar flux and that at the surface of the earth with major absorption bands from atmospheric O_2, O_3, and water vapor. (Modified from Gates, 1962.)

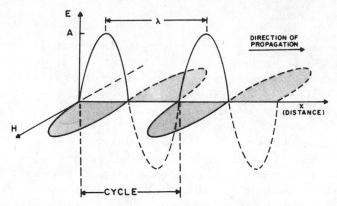

Figure 5–2   Instantaneous electric (*E*) and magnetic (*H*) field strength vectors of a light wave as a function of position along the axis of propagation (*x*), showing the amplitude (*A*), wavelength (λ), cycle, and direction of propagation. (From Bickford, E. D., and Dunn, S.: Lighting for Plant Growth. Kent, Ohio, Kent State University Press, 1972.)

Absorption of Light

Absorption of light by atoms and molecules is the result of resonation of the electrons of the atoms and molecules at definite frequencies which correspond to the quantum energy state. In the collision of an electron and a photon, the electron gains the quantum of energy lost by the photon. It is important to keep this basic photochemical relationship in mind, since the imparted quantum energy of the photon functions in relation to frequency, and each molecular or atomic species has a unique set of absorption characteristics or bands. Life responds to quantum energy of the photon at specific frequencies.

If the energy distribution of solar flux is plotted against wavelength, as was done in Figure 5–1, the maximum monochromatic intensity of sunlight appears to occur in the green portion of the visible spectrum, an illusion caused by the manner of presenting the data. It is more meaningful to express energy against frequency, where the area under any portion of the curve is directly proportional to energy (Fig. 5–3). The energy of the photon in the electromagnetic spectrum is proportional to frequency and inversely proportional to wavelength to accordance with Planck's Law:

$$\epsilon = h\nu$$

where:

ϵ = energy of the photon, ergs per second
h = Planck's constant, $6.62 \cdot 10^{-27}$ erg second
ν = frequency of the radiation in cycles per second.

When energy is thus expressed against frequency (Fig. 5–3), the true maximum energy of solar irradiance is shifted into the infrared, somewhat greater than 1000 nm or 1μm. The median value of irradiance occurs in the near infrared at 710 nm, slightly above the visible range. A major portion (29 per

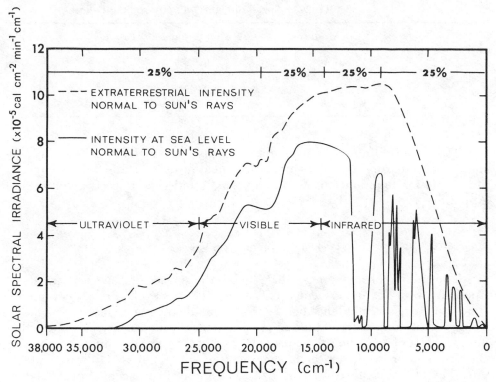

Figure 5–3   Solar spectrum at the mean solar distance from the earth as received outside of the earth's atmosphere and at sea level on a surface perpendicular to the solar rays (solar constant of 2.00 cal cm^{-2} min^{-1} cm^{-1}). (Modified from Gates, 1962.)

cent) of the incoming radiation occurs at wavelengths greater than 1,000 nm and 50 per cent beyond the red portion of the visible range. This distribution shifts somewhat as the solar radiation passes through the earth's atmosphere (Fig. 5–3). The point to be made, however, is that major portions of irradiance impinging on the surface of a lake are in the infrared portion of the solar spectrum and have major thermal effects on the aquatic system.

Wavelength (λ) is a quantitative parameter of any periodic wave motion, not only of water movements, as we will discuss later on, but also of light. It is defined simply as the linear distance between the crests of waves, and is equal in cm to the speed of light (c, 2.998×10^{10} cm sec^{-1}) divided by the frequency (v) in cycles per second (cps):

$$\lambda = c/v$$

Wavelength is also often expressed as the wave number (v^1) [or k], which is the number of wavelengths per cm, or the reciprocal of the wavelength:

$$v^1 = 1/\lambda$$

TABLE 5–1   Speed of Light (589 nm, sodium D-lines)

Medium	Speed (cm sec^{-1})
Vacuum	2.9979×10^{10}
Air (760 nm, 0°C)	2.9972×10^{10}
Water	2.2492×10^{10}
Glass	1.9822×10^{10}

The speed of light for transparent materials is reduced as it passes through media of increasing density in a roughly linear fashion (Table 5–1). A conversion table for commonly used units of length and irradiance is given in the Appendix.

Light Impinging on Lakes

The amount of solar energy that reaches the surface of a lake is dependent upon an array of dynamic factors. The amount of direct solar energy per unit of time from the sun, incident upon a surface outside the atmosphere perpendicular (normal) to the rays of the sun at an average distance of the earth from the sun, is referred to as the solar constant[1]. As a function of the angular height of the sun incident to the earth, the amount of energy received is strongly influenced by latitude and season (Fig. 5–4). The angle of contact of light rays has a marked effect upon the productivity of lakes, as will be illustrated repeatedly in subsequent chapters. In equatorial regions, where there is a vertical sun, the relative constancy of the energy regime [inputs] contrasts strongly with that of the temperate and polar areas, where the sun's angle changes with the sequence of the seasons. The time of day is another factor that strongly influences the solar flux reaching the surface of a lake, since it also influences the position of the sun and the distance of the path the light must travel through the absorbing atmosphere. In the polar extremes, for example, direct solar energy

---

[1]Solar constants are difficult to measure but recent evidence from satellite instrumentation indicates a value of 1.94 cal cm^{-2} min^{-1} (Drummond, 1971; Gates, personal communication).

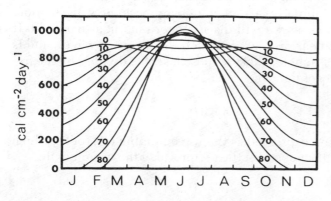

Figure 5–4   Daily totals of the undepleted solar radiation received on a horizontal surface for different geographical latitudes as a function of the time of year (solar constant 1.94 cal cm^{-2} min^{-1}). (After Gates, 1962.)

decreases to zero for over one-third of the year, and these waters receive only thermal radiation from indirect sources.

The absorptive capacities of the atmosphere for solar radiation are governed largely by the content of oxygen, ozone, carbon dioxide, and water vapor, as discussed previously. Additionally, atmospheric transparency can be modified strongly at regional levels by both industrial- and urban-derived contaminants. Scattering and absorption also increase in moist air, which is more common on the leeward side of large water bodies. The elevation of a lake and the angular height of the sun are both related to the quantity of atmosphere through which the radiation must travel. In sum, the amount and spectral composition of *direct* solar radiation reaching the surface of a water body vary markedly with latitude, season, time of day, altitude, and meteorological conditions, all of which are a function of the shifting dynamics of the angular height of the radiation, the molecular transparency of the atmosphere, and the distance the light must traverse.

Indirect solar radiation from the sky is largely the result of molecular scattering of light as it passes through the atmosphere. The extent of scattering is a function of the fourth power of the frequency, and therefore the UV and shorter wavelength radiation of high frequency is reduced by about one-fourth as a result of scattering. The result is the blue sky we see directly overhead on clear days. The factors influencing direct solar radiation also influence scattering, particularly solar height and the atmospheric distance through which the light must pass. The percentage of indirect radiation increases significantly as deviation of the rays from the perpendicular increase, e.g., 20 to 40 per cent at a sun elevation of 10° in contrast to 8 to 20 per cent at a sun elevation of 40°.

Distribution of Radiation Impinging on Lakes

Solar radiation impinging upon the surface of an inland water body does not penetrate the water totally. A significant portion is *reflected* from the surface, and is essentially lost from the system unless it is indirectly backscattered from the atmosphere or surrounding topography to the lake.

The extent of reflectivity of solar radiation directly and indirectly from the sky varies greatly with the angle of incidence of incoming energy, the surface characteristics of the water, the surrounding topography, and meteorological conditions. The reflection (R) of unpolarized direct sunlight as a fraction of the incident light is a function of Fresnel's law:

$$R = \frac{1}{2} \left[\frac{\sin^2 (i - r)}{\sin^2 (i + r)} + \frac{\tan^2 (i - r)}{\tan^2 (i + r)} \right]$$

where i = angle of incidence of light, and r = angle of refraction. This function states simply that the reflectivity is dependent upon the solar height from the zenith (Fig. 5–5A), i.e., the greater the departure of the angle of the sun from the perpendicular, the greater the reflection. Indirect radiation from the sky is also reflected from a water surface, but is less affected by solar height (Fig.

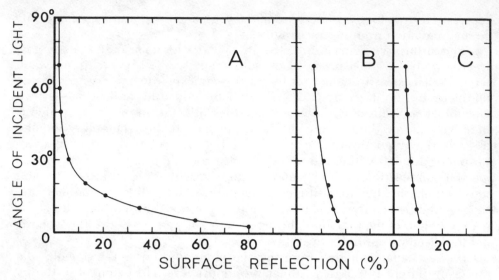

Figure 5–5 Surface reflection and backscattering as a percentage of total solar radiation at varying angles of incidence. A, clear, cloudless conditions; B, reflection of diffuse light under moderate cloud cover; C, heavily overcast conditions. (Generated from data in Steleanu, 1961, and from Sauberer, 1962.)

5–5B). Under an overcast sky the amount of indirect light that is reflected decreases (Fig. 5–5C). An average value of 6.5 per cent is common, although it is reduced further if surrounding topography such as mountains moderates low-angle radiation (Sauberer, 1962).

As the surface of the water is disturbed by wave action, reflection increases somewhat by about 20 per cent at low angles of incident light (ca 5°) to approximately a 10 per cent increase at higher angles (5 to 15°). The difference is small at angles of greater than 15°. Reflection may decrease slightly when the waves are very large and the light is exposed to the water surfaces at angles more closely approaching the perpendicular. Ice and especially snow cover markedly affect the reflectivity of light from the surfaces of lakes. Although data are very meager, clear, smooth ice acts with reflection characteristics similar to those of the undisturbed liquid phase. Changes in texture of the ice generally result in increases in reflectivity. Reflection increases markedly with the greatly increased angles and quantity of surface planes of granular ice in the form of snow cover. About 75 per cent of incident light striking snow is reflected.

Of the total incident light impinging upon the surface of a lake, a reasonable average amount that is reflected on a clear, summer day is 5 to 6 per cent. This mean value increases to about 10 per cent during winter. Qualitatively, light in the red portion of the spectrum is reflected to a slightly greater extent than light of higher frequencies, particularly at low angles of incidence. About one-half of the total quantity of light leaving the lake is by reflection, and half by scattering of light.

Scattering

Scattering of light from the water results in the loss of large amounts of light from the lake. This phenomenon is apparent to anyone who has looked down into relatively clear waters where the surface reflection is eliminated. Of the total light energy entering the water, portions are absorbed by the water and its suspensoids, as will be discussed in detail further on, and a significant portion is scattered. The scattering of light is the result of deflection of light energy by the molecular components of the water and its solutes but also, to a large extent, by particulate materials suspended in the water.

The scattering of light energy can be viewed in a simple way as a composite of reflection at a massive array of angles internally within the lake. The energy scattered in all directions within a volume of water varies greatly with the quantity of suspended particulate matter and its composition and optical properties. Volcanic siliceous materials, for example, scatter light much less than less transparent suspended particulate matter.

The scattering of light can change significantly with depth, season, and location in the lake and in response to variations in the distribution of particulate matter. When particulate matter is concentrated in the metalimnion of a thermally stratified lake, scattering of light can increase strongly either as a result of reduced rates of sinking as the particles encounter increased water densities, or as a result of the development of large populations of plankton in certain strata. Scattering can also increase markedly in areas of the lake where wind-induced currents and wave action agitate and temporarily suspend littoral and shore deposits of particulate matter which are carried out to the open water (Tyler, 1961b). When dimictic or amictic lakes (see next chapter) undergo complete circulation, a significant portion of the sediments of the lake basin are brought into resuspension (Davis, 1973; Wetzel, et al., 1972) and can affect the scattering properties of the lake for an extensive period (weeks). Similarly, the variable influxes of suspended inorganic and organic matter from stream inflows to reservoirs can radically increase the scattering of light energy in a nonuniform way within the lake basin. These alterations can be very temporary in relation to the composition and density of the material and the density characteristics of the recipient water at that particular time.

A significant amount of light can be reflected and scattered from the sediments both in the littoral zone and in the shallow areas of the lake, as well as from the bottom of moderately deep lakes (Fig. 5–6). The amount of light returned to the water is dependent upon the composition of the sediments; sand sediments or those rich in $CaCO_3$ (calcium carbonate, marl) reflect considerably more light than sediments of high organic content and detritus.

Differential scattering of light also depends on scattering coefficients for different wavelengths, as well as on absorption characteristics for different wavelengths. In very clear water, scattering is predominantly in the blue portion of the visible spectrum. Where the quantity and size of suspensoids increase, radiation of longer wavelengths is scattered to a greater extent, with greater absorption of the light of high frequencies. Hard-water lakes with high suspensions of $CaCO_3$ characteristically backscatter light that is predomi-

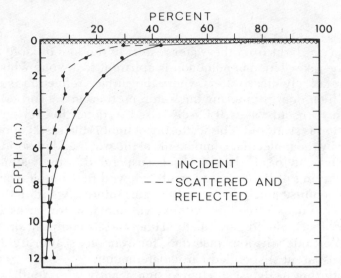

Figure 5–6 Comparison of incident light penetration with depth to that backscattered from concentrations of plankton especially at 3 m and from calcareous sediments. Lawrence Lake, Michigan; 17 March 1972; 23 cm cloudy ice.

nantly blue-green; those rich in suspended organic materials are commonly more green or yellow in appearance.

The diffuse light from scattering and reflecting sources is of obvious importance to organisms that utilize it directly in photosynthesis or indirectly in behavioral responses. In lake systems where direct light enters the environment unidirectionally from above, diffuse light can form a major supplementary source of energy. The percentage of light scattering can easily be one-fourth of that light absorbed by the water. The values are likely to be even higher, since most of the existing data on scattering is based on unidirectional (2π) instruments. When this type of instrument is contrasted with those approaching 4π geometry (e.g., Rich and Wetzel, 1969), higher values are commonly found. A portion of the scattered light is returned to the surface of the lake, and much (80 to 90 per cent) is lost to the atmosphere. Because of differences in light wave refraction properties at different frequencies, somewhat less of the red scattered light reaches the surface than the blue.

Total underwater light received by a receptor system, such as that received by an algal cell from all angles, is the optimum measure of radiant energy available for photosynthesis. This value is the photon scalar irradiance, which is defined as the total number of photons per unit time and area arriving at a point from all directions about the point when all directions are weighted evenly (Smith and Wilson, 1972). Techniques for measurement of photon irradiance are available, but unfortunately are complex and in very limited use.

Photosynthetic and phototropic responses of organisms are related to the number of quanta of light of specific frequencies (wavelengths) impinging upon biochemical receptor systems. The attenuation of illuminance underwater between a spectral range of from 350 to 700 nm is not the same for energy units as for quanta (Steemann Nielsen and Willemoës, 1971; Lewis,

1974). While divergence between absorption rates of energy and of quanta is small on an absolute scale, energy is absorbed at a higher rate than quanta in waters that are heavily stained with dissolved organic matter. Divergence in penetration of quanta and energy can increase in certain specific portions of the spectral range and thus affect the utilization of action spectra by organisms. For example, small differences in the utilization of illuminance per rate of photosynthesis between diatoms, green algae, and blue-green algae with differing action spectra of photosynthetic pigments are known when the rate of illumination is measured in energy units. The ratio of total quanta to total irradiance energy within the spectral region of photosynthetic activity, however, varies by no more than $\pm$ 10 per cent and is $2.5 \pm 0.25 \times 10^{18}$ quanta sec^{-1} watt^{-1} within a number of differing optical water types (Morel and Smith, 1974). Within this range, the ratio can be used to determine accurately the total quanta available for photosynthesis from measurements of total energy; the converse is also true.

Thermal Radiation in Lake Water

The absorption and back-radiation of thermal radiation of low frequencies ($<$14,000 cm) and long wavelengths (750 to $>$12,500 nm) by lake water operates much the same as a blackbody[2] which absorbs all of the radiant energy incident upon it. Because of surface reflection, about 97 per cent of this radiation is emitted into the atmosphere. The atmosphere does not function as a blackbody, but thermal radiation from it is influenced by water vapor pressure and the degree of cloud cover. The net emission of thermal radiation approximates [11 $\cdot$($^{\circ}$K of water $-$ $^{\circ}$K of air)] or simply [11 $\cdot$ (temperature of the water $-$ that of the air)] (cf. discussion in Hutchinson, 1957).

Net Radiation

The net amount of solar radiation affecting a lake then can be referred to as the net radiation surplus (Q_B):

$$Q_B = Q_S + Q_H + Q_A - Q_R - Q_U - Q_W,$$

where:

    Q_S = direct solar radiation
    Q_H = indirect scattered and reflected radiation from sky and clouds
    Q_A = long-wave thermal radiation from the atmosphere and from surrounding topography (the latter is usually insignificant)
    Q_R = radiation reflected from the lake

---

[2]Proportional to the fourth power of the absolute temperature (K), the thermal radiation $Q =$ 8.26×10^{-9} (Boltzman's constant) in cal cm^{-2} min^{-1} K^{-4} (degrees^{-4}) times (K)^4.

Q_U = radiation scattered upward and lost

Q_W = emission of long-wave radiation.

At night most components are negligible, and therefore the net radiation surplus is equal to the long-wave thermal radiation of the atmosphere minus that emitted from the water

$$Q_B = Q_A - Q_W,$$

or, approximately, $Q_B = -11$ (temperature of the water − the temperature of the air) in cal cm^{-2} day^{-1}. We will return to these relationships when discussing heat budgets in the following chapter. It should be emphasized, however, that even though the mean value of Q_B may be positive, the lake could be losing heat through evaporative and convective heating of the air (Hutchinson, 1957).

The collective value for the inputs $(Q_S + Q_H + Q_A)$ can be measured directly and rather effectively by sensitive Moll thermopile pyranometers. The subject of radiation measurement is treated excellently by Latimer (1972), Bickford and Dunn (1972), and Šesták, Čatský, and Jarvis (1971).

Transmission and Absorption of Light by Water

The quality and quantity of light energy penetrating the water is dispersed by the mechanisms discussed above, and absorbed. The diminution of radiant energy with depth, by both scattering and absorption mechanisms, is referred to as light *attenuation,* whereas *absorption* is defined as diminution of light energy with depth by transformation to heat (cf. Westlake, 1965). It is important to understand the selective absorptive properties of water, first in pure water and then in lake waters of differing optical characteristics.

The transmission and absorption of light in water can be approached in several ways. Perhaps the most direct way is to look first at the percentage of transmission or absorption of monochromatic light through given depths of pure water. This percentile absorption, or Birgean percentile absorption (after E. A. Birge who used the relationship extensively), is based on the expression

$$\frac{100 \,(I_o - I_z)}{I_o}$$

where

I_o = irradiance at the lake surface

I_z = irradiance at depth z, in this case taken as 1 m.

In distilled water, the percentile absorption is very high in the infrared region of the spectrum, decreases rapidly in the lower wavelengths to a minimum absorption in the blue, and then increases again in the violet and especially UV wavelengths (Table 5–2). These absorption relationships

TABLE 5–2   Percentile Absorption and Extinction Coefficients of
Monochromatic Light of Various Wavelengths by a One-Meter Column of
Distilled Water[a]

Wavelength (nm)	Percentile Absorption	Extinction Coefficient (η)
820 (infrared)	91.1	2.42
800	89.4	2.24
780	90.1	2.31
760	91.4	2.45
740	88.5	2.16
720	64.5	1.04
700	45.0	0.598
680 (red)	36.6	0.455
660	31.0	0.370
640	26.6	0.310
620 (orange)	23.5	0.273
600	19.0	0.210
580 (yellow)	7.0	0.078
560	3.9	0.040
540	3.0	0.030
520 (green)	1.6	0.016
500	0.77	0.0075
480	0.52	0.0050
460 (blue)	0.52	0.0054
440	0.70	0.0078
420	0.92	0.0088
400 (violet)	1.63	0.0134
380 (UV)	2.10	0.0255

[a] Modified from Hutchinson, 1957.

usually are expressed graphically in linear or, even better, in logarithmic forms to demonstrate these fundamental properties of light transmission (Fig. 5–7). Generally about 53 per cent of the total light energy is transformed into heat in the first meter.

The light intensity or irradiance, I_z, at depth z is a function, then, of intensity at the surface (I_o) to the log base of the negative extinction coefficient (η) at the depth distance, z, in meters:

$$I_z = I_o e^{-\eta z}$$

or $\ln I_o - \ln I_z = \eta z$.

The extinction coefficient (η) is a constant for a given wavelength; approximate values for pure water are given in Table 5-2. This relationship is imperfect in nature because sunlight is not monochromatic, but is a composite of many wavelengths.

Direct sunlight rarely enters the water at these idealized conditions of right angles to the surface, and indirect irradiance is not perpendicular to the surface at any time. Moreover, the natural total extinction coefficient (η_t) is influenced not only by that of the water itself (η_w), but also by absorption of particles suspended in the water (η_p), and particularly by dissolved com-

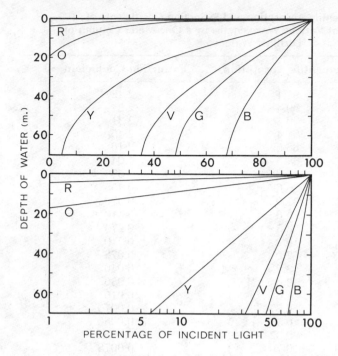

Figure 5–7 Transmission of light by distilled water at six wavelengths (R–720, O–620, Y–560, G–510, B–460, V–390 nm). Percentage of incident light that would remain after passing through the indicated depths of water expressed on a linear (*upper*) and a logarithmic (*lower*) scale. (After Clarke, 1939.)

pounds or color (η_c). Thus the in situ extinction coefficient (η_t) is a composite of these components (Åberg and Rodhe, 1942):

$$\eta_t = \eta_w + \eta_p + \eta_c.$$

At low concentrations, the particulate suspensoids have relatively little effect on absorption. With high turbidity, however, the effect is quite significant, particularly at lower wavelengths of the visible spectrum. In detailed analyses of the absorption of lake water and its dissolved components, the particulate fraction is commonly removed by filtration or centrifugation.

Effects of Organic Compounds

The effects of dissolved organic compounds or color on the absorption of light energy are very marked, and are best introduced by examples taken from the extensive work of James with Birge (1938). In comparison to distilled water, the percentile absorption of lake water with increasing concentrations of dissolved organic compounds, particularly humic acids, not only drastically reduces the transmission of light, but shifts the absorption selectively (Table 5–3). Common to all waters is a very high absorption in the infrared and red wavelengths, which results in the most significant heating effects in the first meter of water. At the other extreme, while distilled water absorbs relatively little UV light, even very low concentrations of dissolved organic compounds increase such absorption greatly. In lakes highly stained

TABLE 5-3   Percentile Absorption of Light of Different Wavelengths by
One Meter of Lake Water, Settled of Particulate Matter, of Several
Wisconsin Lakes of Progressively Greater Concentrations of Organic Color[a]

Wavelength (nm)	Distilled Water	Crystal Lake	Lake Mendota	Alelaide Lake	Mary Lake	Helmet Lake
800	88.9	89.9	90.5	92.4	91.7	93.2
780	90.2	91.3	91.9	93.5	93.0	94.5
760	91.4	93.5	92.6	94.5	94.8	96.0
740	88.5	89.3	91.5	92.7	93.0	96.2
720	64.5	67.6	71.0	78.0	78.0	86.9
700	45.0	50.4	49.7	66.3	70.7	82.5
685	38.0	45.2	42.2	65.7	71.7	86.6
668	33.0	40.3	36.8	65.0	72.3	88.0
648	28.0	37.0	31.9	64.5	75.2	91.2
630	25.0	34.4	28.9	65.8	77.8	94.0
612.5	22.4	32.1	26.3	66.8	80.3	96.0
597	17.8	27.5	22.5	67.0	83.2	97.6
584	9.8	22.0	17.6	67.1	85.7	98.2
568.5	6.0	19.3	14.0	67.6	88.5	98.6
546	4.0	19.2	13.5	70.9	91.6	99.3
525	3.0	19.8	14.1	74.5	94.8	—
504	1.1	20.7	15.2	81.0	97.4	—
473	1.5	21.7	21.7	88.6	99.4	—
448	1.7	23.8	27.8	92.2	—	—
435.9	1.7	24.4	31.0	95.2	—	—
407.8	2.1	28.1	44.3	99.0	—	—
365	3.6	40.0	80.0	—	—	—
Color Scale (Pt units)	0	0	6	28	101	264

[a]Selected data from James and Birge, 1938.

with humic compounds, such as Helmet Lake, absorption of UV, blue, and
green wavelengths is complete in much less than a depth of one meter. This
relationship of intense absorption of UV light by dissolved organic compounds has been used extensively as a relative assay of their concentrations,
as will be discussed in a subsequent chapter on organic matter.

The major effects of the dissolved organic matter in relation to particulate
suspensoids on absorption of light at varying wavelengths are illustrated
graphically in Figure 5-8. The total absorption characteristics in the spectrum
(T) are compared to the percentile absorption in one meter of distilled water
(W), absorption attributable to particulate suspensoids (P), and absorption
by water that has been filtered to remove particles >1 μm (C), termed color
absorption. Absorption by dissolved organic compounds, 'dissolved color,' is
highly selective and marked in the UV, blue, and green wavelengths. The
absorption by dissolved color at the red end of the spectrum is less selective,
and most likely is unrelated to those organic compounds absorbing at lower
wavelengths. The extinction coefficients of dissolved color (η_c) increase
directly with the color units of the water, which are measured by the relative

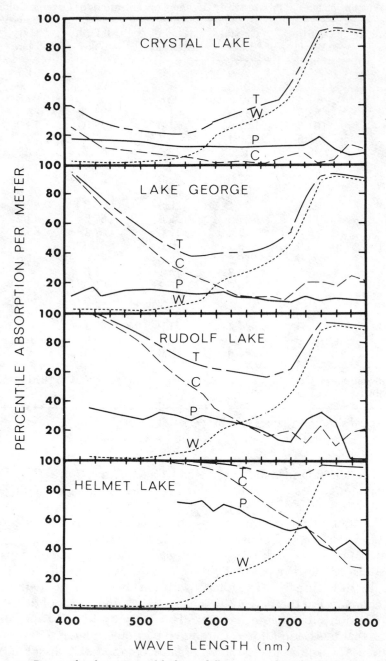

Figure 5–8 Percentile absorption of light at different wavelengths of 1 meter of water of 4 lakes of northern Wisconsin of increasing concentrations of dissolved organic matter. T = total absorption; C = absorption by dissolved organic color; P = absorption by suspended particulate matter; and W = absorption by pure water. (Modified from James and Birge, 1938.)

visual comparisons of the color of the filtered lake water under standard conditions to the color of a specific mixture of platinum-cobalt compounds in serial dilution (discussed further on). In the examples given in Figure 5–8, the water of Crystal Lake indicated zero Pt units, Lake George 24, Rudolf Lake 50, and that of Helmet Lake 236 Pt units or the color of weak tea.

Also apparent from this classical work is the relationship that absorption resulting from particulate suspensoids (P) is relatively unselective at different wavelengths, particularly at lower concentrations. The η_p functions essentially independently of the η_c but, along with the absorption of water (η_w), η_p values are additive for a particular lake at depth at a given time of year (cf. Åberg and Rodhe, 1942).

Analysis of Light Transmission

The transmission or absorption within a lake of the total white light from direct insolation and indirectly from the sky has been analyzed in many

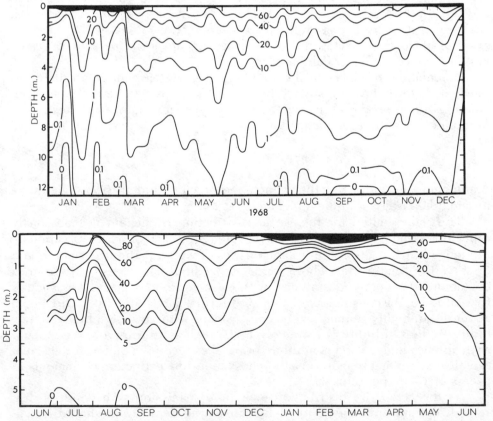

Figure 5–9 Isopleths of the percentage transmission of light at the surface with depth and time in unproductive Lawrence Lake (*upper*), and highly productive Wintergreen Lake (*lower*), southwestern Michigan. (From Wetzel, unpublished observations.)

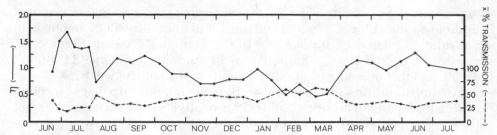

Figure 5–10 Average extinction coefficient (η_t) per meter and percentage transmission of light per meter of the water column, Wintergreen Lake, Michigan, 1971–72. (From Wetzel, unpublished data.)

ways. The vertical extinction coefficient is most commonly determined from the percentile absorption of light at the surface with depth (Fig. 5–9). The isopleths of these examples of depth distribution of light indicate some of the marked fluctuations that are found in natural waters, seasonally and vertically. The composite mean η_t of all depths in Lawrence Lake, an unproductive hardwater lake with rather high concentrations of particulate and colloidal $CaCO_3$ suspensoids, was 0.39 (N = 1,746), within an annual range of 0.05 to 1.02. The same value for the highly productive Wintergreen Lake was 1.00 (range 0.46 to 1.68). In the former case, the η_t is largely constant over an annual period, whereas in the latter situation the marked fluctuations in particulate suspensoids of algae are reflected in the mean η_t m^{-1} and mean percentage transmission m^{-1} of the water column (Fig. 5–10).

Calculations of the vertical extinction coefficient are not very reliable in the first meter below the surface because of surface agitation. Average calculations often exclude this region. Direct calculations are made using the formula given earlier, or changed to

$$\eta z = \ln I_o - \ln I_z.$$

A nomogram for estimating the extinction coefficient directly from light transmission data is given in the appendices.

The values cited for the vertical extinction coefficient (η_t) obviously represent a composite for all of the wavelengths, each of which is variously influenced by the absorption characteristics of water, particulate matter, and dissolved organic matter. The η_t values for natural lake waters vary from ca 0.2 (about 80 per cent transmission m^{-1}) per meter in very clear lakes, such as Crystal Lake, Wisconsin, Lake Tahoe, California, and Crater Lake, Oregon, to about 4.0 in highly stained lake waters or lakes with very high biogenic turbidity. Where turbidity is extremely high, such as in reservoirs near major river inflows under flood conditions, or in lakes receiving fine materials such as volcanic ash that remains in colloidal suspension, extinction coefficients in excess of 10 are not unusual.

As the composite distribution of light transmission with depth is dissected, the spectral selectivity of light reaching deep water should be apparent from the foregoing discussion. It is rather common for the green portion of the spectrum to penetrate most deeply (Fig. 5–11). In very clear

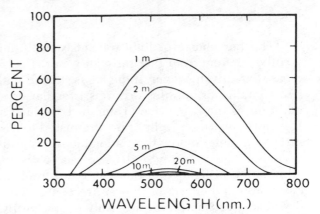

Figure 5–11 Average intensity of light radiation reaching different depths as a percentage of that at the surface in Lunzer Untersee, Austria. (After Sauberer, 1939.)

waters, the deepest penetration of light is in the blue; in Crater Lake it is at ca 469 nm, and in Lake Tahoe at ca 475 nm (Tyler and Smith, 1970; Smith, et al., 1973). In lakes that are highly stained with dissolved organic matter, practically no light of a wavelength below 600 nm penetrates below one meter.

The rapid attenuation of light transmission by dense population development of algae and bacteria at certain depths in stratified lakes is common. For example, in the lower depths where certain microorganisms are ideally adapted to low light intensities, their densities have a marked effect on the extinction of light (Fig. 5–12). Similarly, abiogenic suspensoids, brought into a water basin in inlet sources at temperatures colder than the surface strata, will penetrate to depth strata of comparable densities and can markedly influence the differential vertical extinction of light.

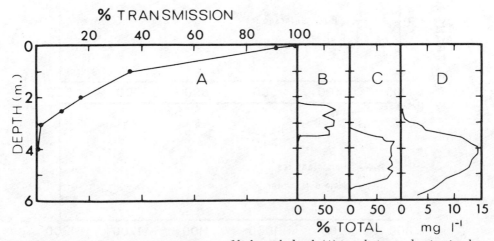

Figure 5–12 Percentage transmission of light with depth (A) in relation to dominating bacterial plate of *Thiopedia* (B) at 2.5 to 3.5 meters and *Clathrochloris* (C) at 3.5 to 5.5 meters as a percentage of total plankton and bacteriochlorophyll *d*, (D) in Wintergreen Lake, Michigan, 4 July 1971. (From Wetzel, unpublished data, and Caldwell, *et al.*, 1975.)

Transmission Through Ice and Snow

The percentage of light transmission through clear, colorless ice is not greatly different from that through water (Fig. 5–13). Absorption of light increases greatly, however, if the ice is stained with organic matter or is cloudy, i.e., contains air bubbles or forms irregular crystals upon freezing. With the addition of snow over ice (Table 5–4), reduction of light is extremely rapid. It is not unusual for light to be attenuated to essentially zero at a depth of as little as a meter below the underside of ice and wet heavy snow. Prolonged periods of heavy snow and ice cover, with attendant severe reduction or elimination of light from the water and from photosynthetic organisms, can have profound effects on the entire metabolism of the lake. In very productive lakes, consumption of dissolved oxygen by catabolic processes can exceed augmentation by photosynthesis, and lead to severe reductions in dissolved oxygen or even to total anoxia. This phenomenon is termed 'winterkill,' and is not uncommon in shallow, productive lakes and ponds in temperate latitudes (cf. Chapter 8).

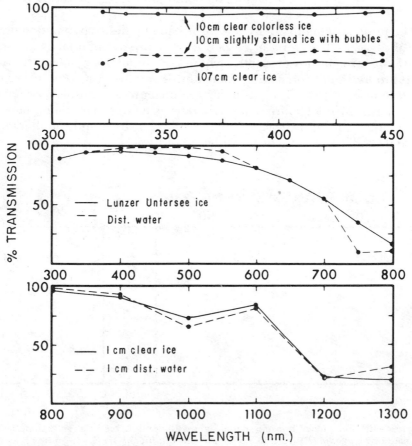

Figure 5–13 Percentage spectral transmission of light through ice in comparison to water. (From data of various sources cited in Sauberer, 1950.)

TABLE 5–4   Light Penetration through Ice and Snow Cover of Lakes under Different Conditions[a]

Ice-Snow Conditions	Thickness (cm)	Percentage Transmission of Surface Insolation
Clear ice	43	72
Clear ice with vestige snow	39	53
Milky ice with bubbles	29	54
Wet ice with bubbles	39	41
Translucent ice ('snow ice')	25	11–18
Ice with irregular surface	29	58
New snow	0.5	34
	5.0	20
	10.	9
	17–20	8.8–6.7
Compacted old snow	17–20	5–1

[a]Selected data from Albrecht, 1964.

The incoming light characteristics and amount of light penetrating the littoral zone of lakes are greatly altered by the type and extent of development of floating-leaved and emergent higher aquatic plants, as well as by reflection and scattering from the substratum. Reduction of over one-half of the incident light by dense vegetative stands is not uncommon (Szumiec, 1961; Gessner, 1955). These changes are dynamic, since the structure of the macroflora changes seasonally.

Color of Natural Waters

The observed color of lake water is functionally the result of light scattered upward from the lake after it has passed through the water to various depths and undergone selective absorption en route. Since molecular scattering of water is a function of the fourth power of the frequency, observed light is greater for shorter than for longer wavelengths, and blue dominates. Scattering of light from particulate suspensoids, however, is increasingly less selective with increasing particle size (Hutchinson, 1957). Colloidal $CaCO_3$, common to very hardwater lakes, scatters light in the greens and blues and gives these waters a very characteristic color appearance. Most of the color of lake waters results from dissolved organic matter and its rapid, selective absorption of the shorter wavelengths of the visible spectrum. The result is a dominance of emitted scattered light in the green portion of the spectrum and, with increasing concentrations of organic matter, especially humic compounds, an increase in yellows and reds.

When the density of particulate matter suspended in the water, collectively termed seston, becomes great, a seston color can be imparted to the water in spite of the relatively nonselective scattering properties of these materials. Suspension of large amounts of inorganic materials, such as clays and volcanic ash, can yield a yellow to brownish-red coloration. Seston color, however, is usually associated with large concentrations of suspended algae and, less frequently, with pigmented bacteria and microcrustaceans. Where blue-green algae or diatoms occur in great profusion and often accumulate in the surface waters, they may produce blue-green or yellowish-brown colors, respectively. Blood-red color is not an infrequent occurrence in lakes where conditions are temporarily ideal for massive development of populations of algae, such as *Glenodinium*. At low concentrations of suspended algae, as in the open ocean, the chlorophyll bands of algal pigments have no appreciable influence on water color (Yentsch, 1960).

Color Scales

Any color (shade or tint) always has two decisive characteristics: color intensity (brightness) and light intensity (lightness) (Albers, 1963). This duality in color intervals results in a highly subjective ability to discriminate colors in individuals. Moreover, visual memory is very poor in comparison with auditory memory. Therefore, the psychophysical nature of reactions of visual organs to light and color has led to several attempts to standardize observations by means of various color scales.

Several color scales have been devised to compare empirically the true color of lake water, after filtration to remove suspensoids, to various combinations of inorganic compounds in serial dilutions. Platinum units[1] is the most widely used comparative scale in the United States. Very clear lake water would yield a value of zero Pt units, and heavily stained bog water about 300 (cf. Table 5–3). In Europe the Forel-Ule color scale, involving comparisons to alkaline solutions of cupric sulfate ($CuSO_4$), potassium chromate (K_2CrO_4), and cobaltus sulfate ($CoSO_4$), is commonly used. A strong correlation exists between the brown organic color which is derived chiefly from peat and marsh detritus, and the amount of dissolved organic carbon in the surface waters (Juday and Birge, 1933). It is not infrequent for the color units to increase with depth in strongly stratified lakes; this is most likely related to increased concentrations of dissolved organic matter and ferric compounds near the sediments. Subjectivity of color evaluations can be reduced greatly by rather elaborate optical analyses and comparisons with standardized chromaticity coordinates (e.g., Smith, et al., 1973).

Transparency of Water to Light

The use of modern instrumentation for the in situ evaluation of the vertical extinction and spectral characteristics of light in lakes is now common.

---

[1]1000 Pt units = the color resulting from 2.492 g potassium hexachloroplatinate (K_2PtCl_6), 2 g cobaltic chloride hexahydrate ($CoCl_2 \cdot 6H_2O$), 200 ml concentrated hydrochloric acid (HCl), and 800 ml water.

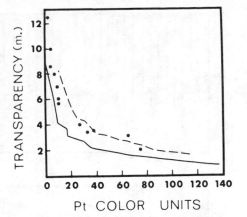

Figure 5–14 Relationship between Secchi disk transparency and dissolved color from Swedish lakes (points and dashed line) and 470 Wisconsin lakes (solid line). (From data of Åberg and Rodhe, 1942.)

An approximate evaluation of the transparency of water to light was devised by an Italian scientist, Secchi, who observed the point at which a white disk lowered into the water was no longer visible. This method continues to be widely used owing to its simplicity. The Secchi disk transparency is the mean depth of the point where a weighted white disk, 20 cm in diameter, disappears when viewed from the shaded side of a vessel, and that point where it reappears upon raising it after it has been lowered beyond visibility.

The Secchi disk transparency is essentially a function of the reflection of light from its surface, and is therefore influenced by both the absorption characteristics of the water and of its dissolved and particulate matter. Within general limits, a parabolic relationship exists between dissolved organic matter and transparency (Fig. 5–14). However, both theoretical analyses and a large number of empirical observations have shown that the reduction in light transmission in relation to Secchi transparency measurements is associated to a greater extent with increased scattering by particulate matter sus-

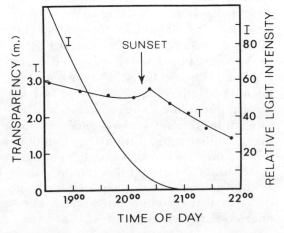

Figure 5–15 Changes in the Secchi disk transparency (T) in relation to light intensity (I) at the surface. I_0 at 2130 hours was 1.1×10^{-6} that at 1845 hours. (Slightly modified from Åberg and Rodhe, 1942.)

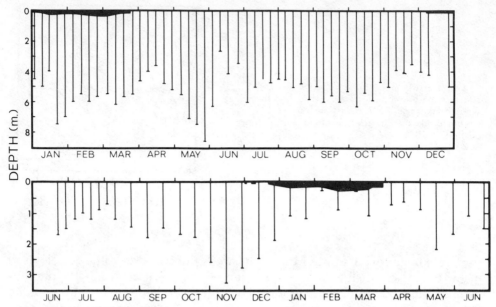

Figure 5–16   Annual variations in the Secchi disk transparencies of Lawrence Lake, 1968 (*upper*), and Wintergreen Lake, 1971–72 (*lower*), southwestern Michigan. (From Wetzel, unpublished data.)

pensoids (Štepánek, 1959; Szczepanski, 1968). This is particularly true in very productive lakes and in a very generalized way has been used to estimate the approximate density of phytoplankton populations. Although the transparency depth is independent of surface light intensity to a significant extent, results become erratic near dawn and dusk (Fig. 5–15), and the Secchi disk preferably should be used at midday.

Observed Secchi disk transparencies range from a few centimeters to over 40 m in a few rare clear lakes. Attention has been given to what percentage of surface light intensity occurs at the Secchi depth. The Secchi disk transparency correlates closely with percentage transmission. At the extremes, the depths of observation of the Secchi disk transparency can represent from 1 to 15 per cent transmission (Stepánek, 1959; Beeton, 1958; Tyler, 1968; and others). The variation is related to differences in the sensitivity of underwater photometers. Seasonal variations, such as depicted in Figure 5–16, are common for lakes of the temperate zone. The examples cited include an unproductive hardwater lake, and a highly productive lake with dense algal populations throughout much of the year, especially in the period of late winter through early summer. Even in the clearest lakes, such as alpine lakes, similar general trends are found on an annual basis.

Colored Secchi disks have been employed in estimations of the spectral distribution of light with depth (Štepánek, 1959; Elster and Štepánek, 1967). Within general limits, comparison of transparencies with a series of colored disks to that of white provides an approximate evaluation of the spectral characteristics of lake waters.

Utilization of Solar Radiation

Solar radiation is the major energy source driving the productivity of aquatic ecosystems, whether it is incorporated into potential energy by biochemical conversions within the lake directly by its flora, or by terrestrial components within the watershed and imported as organic matter to the lake. Algae and macrophytes use between 4 to ca 9 quanta of light energy per molecule of carbon dioxide (CO_2) reduced. The photosynthetic receptor is the chloroplast, which consists of lamellar structure from 1 to 10 μm in diameter and ca 0.025 μm in thickness. Absorption of light energy is specific for different chlorophyllous pigments: the absorption peaks of the primary pigment chlorophyll *a* are at 640 nm and 405 nm, and those of chlorophyll *b* at 620 nm and 440 nm. Certain plants, such as wheat, have broad responses to quanta of energy with relatively high efficiency of utilization in the green wavelengths, in addition to those of chlorophyll *a*. Other potential adaptations of plants to certain photic environments exist. For example, in the blue-green alga *Microcystis*, the action spectra indicate that the accessory blue-green algal pigment phycocyanin gains importance when this alga is grown in red light, and the cells lose much chlorophyll *a*. Other auxiliary pigments supplement the excitation of chlorophyll *a* and improve the efficiency of photosynthesis at long wavelengths.

In addition to the adaptations of photosynthetic organisms to the aquatic photic environment, utilization of light is also highly specific in aquatic animals. The visual receptors of animals have a quantum efficiency of one, the maximum possible. Among the invertebrates, for example, the opossum shrimp *Mysis* possesses visual receptors with a peak at 515 nm, and is found only in deep portions of clear lakes, that is, in a photically blue environment. The water flea *Daphnia* migrates extensively from deep to surface waters daily, and is adapted for life in changing photic environments with four visual receptor peaks at 370, 435, 570, and 685 nm (Chapter 16). The dragon fly, which spends much of its life cycle as an immature aquatic nymph, has visual receptors with a peak only in the green (530 nm) and possibly the blue (420 nm) portion of the spectrum (Ruck, 1965). Upon its shifting to the aerial adult stage, an increase in the number of receptor peaks (380, 420, 518, 530, and 550 nm) occurs concomitant with its extension into the terrestrial photic environment. Similar situations are found among the fishes; those living only deep in a blue photic environment have maximum reception at ca 485 nm, whereas shallow living fishes possess a sensitivity of receptors to longer wavelengths.

Behavioral adaptations to the utilization of light are also common. For example, *Daphnia* uses the angular light distribution with depth as a cue in body orientation in swimming. The long axis of the body is oriented with the vector of maximum light energy, which affects directional swimming of these crustaceans during vertical migrations, and hence the distribution of their populations within the lake.

It is now essential that we consider, in addition to these direct physiological effects and utilization responses, the effects of absorption of solar energy and its dissipation as heat on the physical and chemical structure and dynamics of lake systems.

chapter 6

fate of heat

The absorption of solar energy by lake water is, as we have seen, influenced, by an array of physical, chemical, and, under certain conditions, biotic properties of the water. These characteristics are dynamic, changing seasonally and over geological time for individual lake systems.

Light energy absorbed by a solution increases exponentially with the distance of the light path through the solution. For light of a wavelength of 750 nm, 90 per cent is absorbed in one meter and only 1 per cent is transmitted through two meters. Absorption is increased markedly by dissolved organic matter. Since much of solar energy is of low frequency in the infrared portion of the spectrum at wavelengths greater than 750 nm, the upper two meters of lake water absorb over one-half of the sun's radiation and function in heating the water.

The unique properties of the high specific heat of water permit the accumulation of light energy as it is dissipated as heat. The retention of heat is coupled with factors that influence its distribution within the lake system: physical work of wind energy, currents and other water movements, morphometry of the basin, and water losses. The resulting patterns of thermal stratification influence in fundamental ways the physical and chemical cycles of lakes, which in turn strongly govern their production, utilization, and decomposition.

DISTRIBUTION OF HEAT

It is evident from previous discussion that the greatest source of heat to lakes is solar radiation, and that most is absorbed directly by the water. Some transfer of heat from the air and from the sediments does occur, but in lakes of moderate depth this input is small in comparison to direct absorption. In shallow waters, either for the entire lake or in its littoral regions, sediments can absorb significant quantities of solar radiation and this heat is transferred in part to the water. However, such terrestrial heat is generally very small in comparison to direct absorption of solar radiation by the water. Exceptions include lakes that receive significant percentages of their volume inputs from surface runoff in short periods of time, such as certain reservoirs that have short retention times. Indirect heating also can be significant in lakes that have high volume inputs from groundwater sources and springs, especially in the case of hot springs and certain volcanic lakes. Condensation of water vapor at the water surface also provides some heat input to the lake from the air.

Heat is lost from lakes primarily by the thermal radiation (cf. Chapter 5). However, this is predominantly a surface phenomenon, restricted to the first few centimeters of water, because the thermal conductivity of water is very low. Measurable amounts of heat are lost nonetheless by specific conduction both to the air and, to, a much lesser degree, to the sediments. Heat is also lost through evaporation, during which 539.6 cal g^{-1} are required, the latent heat of evaporation or vaporization. The rate of evaporation increases with higher temperatures, reduced vapor pressure, lower barometric pressure, increased air movement over the water or ice surface, and decreases with increasing salinity.

For water at 0°C to freeze, 79.7[1] cal g^{-1} are released, and conversely when ice at 0°C melts to water at 0°C, a similar amount of heat must be absorbed. This latent heat of fusion is among the highest of known compounds. Also as ice sublimes to water vapor, 679 cal g^{-1} must be absorbed.

A significant portion of the heat increments to lakes or especially reservoirs can be lost through outflow. Since it is obviously dependent upon retention time and flow characteristics, outflow removal usually consists of the surface waters that are relatively warmer than underlying water strata.

The inputs and outputs of heat are therefore largely surface phenomena. Solar influx would be expected to dominate in the warmer seasons of the year, when the thermal structure, vertically with increasing depth, would approximate the attenuation profile of solar radiation (Fig. 6–1). The warmer, less dense and very stable heated water would successively overlie cooler and more dense water. Such is not the case, however, and what is observed (Fig. 6–1) is a relatively uniformly mixed upper portion of the lake that is isothermal often well below the photic zone.

Some convection currents do occur when the surface waters are cooled during the night, become more dense, and sink. Similarly, surface waters can cool during shifts in local meteorological conditions, such as under cloudy conditions, increased evaporation rates, cold rain, etc., or after seasonal decreases in air temperatures. Surface influents also may cool surface waters and induce convection currents.

Under most conditions of typical lakes of temperate latitudes, mixing of the surface waters by convection currents is weak (up to 3 m) and insufficient to produce the thermal profiles observed during the heating period of thermal stratification.

Solar radiation by direct absorption amounts to only about 10 per cent of the observed distribution of heat (Birge, 1916). The energy that distributes heat in a lake is derived almost totally from wind energy. As air currents

---

[1]The correct value is 79.72. The often cited value of Weast (1970) and earlier editions is in error and is corrected in newer editions of this handbook (personal communication).

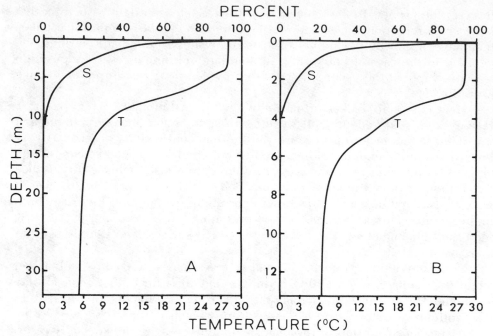

Figure 6–1 Vertical depth profiles of the penetration of solar radiation (S) and temperature (T) in Crooked Lake (A) and interconnected, adjacent Little Crooked Lake (B), Noble-Whitley counties, Indiana, July 18, 1964. (From Wetzel, unpublished data.)

move across the water interface, a frictional wind stress moves the surface waters and generates forced mixing and currents proportional to the intensity of the wind.

Thermal Stratification

It is instructive to begin a discussion of thermal stratification in lakes by examining typical conditions of lakes of the temperate zone that undergo strong contrasts in seasonal conditions. A majority of lakes are concentrated in the temperate and especially northern latitudes, and this permits some degree of generalization. Many exceptions to this general pattern can be found, and will be indicated when applicable.

Ice cover on the lake commonly deteriorates in the spring in a relatively slow, progressive way until it is permeated with air columns and saturated with water. Warm rains on ice often accelerate this process of erosion. Loss of ice cover is usually rapid, often taking place in a few hours, especially if associated with a strong wind that can move the central ice sheet to the lee-ward shore and cause it to break up in massive accumulations, under great stress. At this time, the water volume with depth is near the temperature of maximum density. Deviations from the temperature of the maximum density, either cooling below or warming above 4°C if weather conditions dictate, result in very small changes in density difference per change in temperature (cf. Chapter 2). As a result, there is relatively little thermal resistance to mixing, and only small amounts of wind energy are required to mix the water column. In a majority of lakes, the amount of wind energy impinging on the surface is adequate to circulate the entire water column. The circulation, aided to some extent by convection currents induced by cooling at night and by evaporation, can continue for varying periods of time. The duration of the spring turnover is governed by many factors. Lakes of small surface area, especially if protected from the wind by surrounding topography or vegetation, often circulate only briefly in the spring, usually for only a few days. Large lakes, in contrast, often circulate for a period of weeks, during which, weather conditions permitting, the temperature of the entire water mass can increase well above that of maximum density. During this period of mixing in spring turnover, before stratification occurs, the extent of heating is also a function of the volume of water that must be heated in relation to net solar income. In very deep, large lakes such as the Great Lakes and Lake Baikal, the heat income can be insufficient to increase the temperatures of the deep water significantly, even if the lake circulates completely. The circulation in shallow lakes, on the other hand, often may increase temperatures to well above 10°C.

As spring progresses, the surface waters of lakes of sufficient depth are heated more rapidly than the distribution of heat by mixing. Usually, the temperature distribution occurs during a warm, calm period of several days. As the surface waters heat and become less dense, the relative thermal resistance to mixing increases markedly. A difference of only a few degrees is sufficient to prevent further complete circulation of the entire water column (Fig. 6–2; another identical analysis is given for Lake Mendota by Birge, 1916). From that point onward, the water is divided into three regions of different temperatures, which are exceedingly resistant to mixing with each other. The initial temperature of the subsequent summer stratification is thus determined by the terminal temperatures of the spring turnover.

Figure 6–2 A summer temperature profile (single line) and relative thermal resistance to mixing (bars) for Little Round Lake, Ontario. The relative thermal resistance (R.T.R.) to mixing is given for columns of water 0.5 m long. One unit of R.T.R. $= 8 \times 10^{-5}$, i.e. the density difference between water at 5° and at 4°C. The R.T.R. of the lake water columns is expressed as the ratio of the density difference between water at the top and bottom of each column to the density difference between water at 5° and 4°C. (Modified from Vallentyne, 1957.)

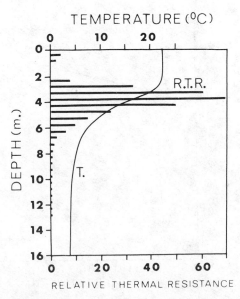

The period of summer stratification is characterized by an upper stratum of more or less uniformly warm, circulating, and fairly turbulent water, the *epilimnion* (Fig. 6–3). The epilimnion overlies a deep, cold, and relatively undisturbed region, the *hypolimnion*. The stratum between the epilimnion and the hypolimnion exhibits a marked thermal discontinuity, and is termed the *metalimnion*. The metalimnion is defined as the stratum of steep thermal gradient, and can be demarcated by the intersections of a plane along the gradient to the points of maximum curvature from the approximately homoio-thermal conditions of the epilimnion and the hypolimnion (graphically de-picted in Fig. 6–3). The term *thermocline* has been defined variously, but correctly refers to the plane or surface of maximum rate of decrease of tem-perature with respect to depth. An extensive discussion of these terms and their conceptual basis is given by Hutchinson (1957). Terms in wide use that are functionally synonymous with the above definition of the metalimnion include the German *Sprungschicht*, and discontinuity layer as used in the United Kingdom.

It should be pointed out that the greatest thermal discontinuity in the transition from spring turnover to summer stratification begins deep, often at the bottom of the lake, and progressively rises to the level at which the metalimnetic gradient eventually stabilizes. The rise in the thermal gradient, usually accepted as much more than 1°C per meter, its leveling, gradual de-pression, and loss in the autumn, is indicated by the dashed line in an exem-plary annual thermal cycle in Figure 6–4. This development of the metalim-netic gradient was pointed out by Ricker (1937), and recently emphasized with examples by Krokhin (1960).

It should also be noted that the hypolimnion gradually gains small

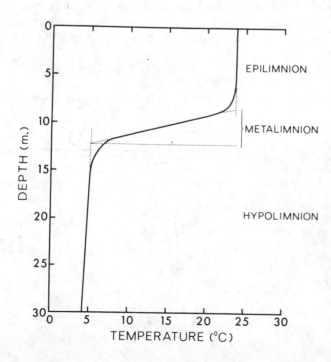

Figure 6–3   Typical thermal stratification of a lake into the epilimnetic, metalimnetic, and hypolimnetic water strata. Dashed lines indicate planes for determining the approximate boundaries of the metalimnion (see text).

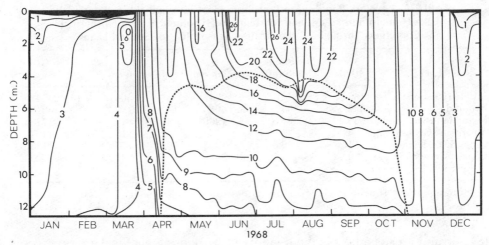

Figure 6–4   Depth-time diagram of isotherms (°C) in Lawrence Lake, Michigan, 1968. Dashed line indicates the upper metalimnetic-lower epilimnetic boundary. Ice-cover drawn to scale. (Modified from Wetzel, et al., 1972.)

amounts of heat during the period of thermal stratification (Fig. 6–4 and also particularly Fig. 6–8). It is clear from the elegant analyses of Hutchinson (1941) and subsequent works that the heating of the stratified lake is a combination of direct solar radiation, turbulent conduction, and density currents. Active turbulence of the epilimnion is carried into the metalimnion and hypolimnion. Conduction of heat by turbulence decreases as the stability of the stratification increases throughout the heating season, but within the hypolimnion this heat conduction varies little with increasing depth. Chemical density currents follow the contour of the basin along the sediments, and in movement to the deepest portions of the basin they can be effective in heat transport. Heat generated by biological oxidations of decomposition in the hypolimnion are entirely inadequate to account for the observed heating.

Solar heating of the hypolimnion can occur in lakes that are particularly transparent, such as in alpine situations. In a small clear mountain lake that is protected from much wind action, an estimated 65 to 85 per cent of the heating of the upper hypolimnion can be accounted for by direct solar heating (Bachmann and Goldman, 1965). As might be anticipated, the relative importance of direct solar heating and the transport of heat by turbulence from the upper water strata varies greatly with conditions of individual lakes. While wind-induced transport of heat to the hypolimnion apparently is of greater importance than direct solar heating in most lakes, particularly where light is rapidly attenuated with increasing depth, a spectrum of conditions can exist in a lake that will cause solar heating of the hypolimnion to dominate.

Stability of A Stratified Lake

The magnitude of thermal-density stratification and its resistance to mixing can be further indicated by estimations of the stability of a lake. Stability

TABLE 6-1   Work Required to Distribute Heat by Wind (B), to Render the
Lake Homoiothermal (S), and to Maintain Hypothetical Homoiothermal
Conditions Throughout the Heating Season (G) for Several Lakes

Lake	Area (ha)	Maximum Depth (m)	B (g − cm cm^{-2})	S (g − cm cm^{-2})	G (g − cm cm^{-2})
Atitlan (Guatemala)	136.9	341	3741	21500	25241
Pyramid (Nevada)	532	104	4055	10255	14310
Mendota (Wisconsin)	39.2	25.6	1209	514	1723
Amatitlan (Guatemala)	8.2	33.6	752	415	1167
Güija (El Salvador)	44.3	26.0	958	175	1133
Lunzer Untersee (Austria)	0.68	33.7	535	390	925

[a]Modified from Hutchinson, 1957; from various sources.

(S) per unit area of a lake is the quantity of work or mechanical energy in ergs required to mix the entire volume of water to a uniform temperature by the wind, without addition or subtraction of heat (Birge, 1915; Schmidt, 1915, 1928). Therefore during summer stratification, the stability expresses the amount of work needed to prevent the lake from developing thermal stratification. More importantly, it evaluates the resistance of that stratification to disruption by the wind, and therefore the extent to which the hypolimnion is isolated from epilimnetic and surface movements.

The stability of a lake is very strongly influenced by its size and morphometry (Table 6–1). The amount of work (B) contributed by the wind in distributing heat throughout the lake increases proportionally as the size and volume of the lakes decrease, whereas the stability (S) or amount of work per unit area to mix a lake to a uniform vertical density is greater in the larger lakes. The sum of B and S is the amount of work per unit area (G) needed by the wind to distribute the observed heat in a lake at the minimum winter temperature uniformly at all depths. It should be apparent that the metalimnetic thermal discontinuity represents an effective mixing barrier. While not a total barrier, it is nearly as effective as the sides of the lake basin. An immense amount of work is required to disrupt it.

Density differences that regulate stratification in lakes are not always thermally induced in inland waters. Differences in salinity can produce a similar stratification, often in combination with temperature differences. The stability so produced can exceed that resulting from thermal density differences, and lead to conditions where disruption of stratification is intermittent or never occurs (see discussion of meromictic lakes below).

Loss of Stratification

In late summer and fall, declining air temperatures result in a negative heat income to the lake, and loss of heat exceeds inputs from solar radiation. Surface waters thus cooled are more dense than underlying warmer epilimnetic water, and they sink, and mix by a combination of convection currents and wind-induced epilimnetic circulation. The penetration of the surface waters into the metalimnion continues as the lake continues to cool. A progressive erosion of the metalimnion from above can be observed as the stratum of thermal discontinuity is reduced and the homoiothermous epilimnion increases in thickness. On an annual basis, the result is an apparent lowering of the metalimnetic thermocline. The isopleth diagram of Figure 6–4 illustrates lines of equal quantities of a particular property, in this case lines of equal temperature or isotherms, over time against depth. These depth-time diagrams are particularly useful in limnology because they aggregate hundreds and often thousands of data points taken at different depths and times into an annual picture that indicates the seasonal dynamics of the physical, chemical, or biological properties of the lake. In the example from Lawrence Lake, it can be seen that stratification was maintained down to 11 m until late October. Finally, the entire volume of lake water was included in the circulation, and *fall turnover* was initiated. The transition from the final stages of weak summer stratification to autumnal circulation usually is dramatic, and can occur in a few hours, especially if associated with the high wind velocities of a storm.

Circulation continues with gradual cooling of the water column. The rate and duration of cooling are highly variable from lake to lake in relation to basin morphometry, and particularly in relation to the volume of water that must be cooled in response to prevailing meteorological conditions. For example, in Figure 6–4, Lawrence Lake, with a surface area of 5 ha and a volume of 0.292×10^6 m³, cooled down to the temperature of maximum density and lower by early December. Waters of nearby Gull Lake, with an area of 827 ha and a volume of 102.1×10^6 m³, usually circulate well into January before reaching comparable temperatures. Very large lakes, such as the Great Lakes, which are of similar latitudes to those just mentioned, circulate all winter without or rarely reaching conditions that permit the formation of an ice cover. In large, very deep lakes, for example Great Bear Lake of the Canadian Northwest Territories, late summer circulation often does not extend to the lower depths, e.g., below 200 m, because stability at lower levels is sufficiently strong to prevent mixing (Johnson, 1966). The temperature of the maximum density of the water below 200 m is 3.53°C, because of hydrostatic pressure. In cold years with slow spring heating, when the upper 200 m of water are cooled to 3.53°C in the autumn, it has then reached the same temperature of maximum density as the water strata below 200 m. Circulation then extends to the bottom, which is 450 m in Great Bear Lake.

Winter Stratification

As the temperature of the water reaches the point of maximum density—4°C—surface ice can form on a rapid loss of heat such as would occur on a calm, cold night. The difference in density between 4°C and 0°C is, as discussed earlier, very small, and results in only a minor density gradient just below the surface. Hence, this type of *inverse stratification,* in which colder water lies over warmer water, is easily disrupted by a small amount of wind energy. Under stormy weather conditions common at this time of year, it is not unusual for the lake to continue to circulate and cool to well below the temperature of maximum density. Temperatures of between 3 and 4°C of the circulating water mass are very common before conditions are such that ice cover can form, as for example is seen in Figure 6–4. Isothermy to well below 1°C has been observed frequently before the formation of ice cover, especially in large lakes subject to much wind action.

After the lake has frozen, the ice cover effectively seals it off from the effects of the wind. Immediately below the ice, water of 0°C is underlain with water of increasing density, to whatever the circulation temperature of the water prior to ice formation. This steep gradient near the surface is referred to as an inverse thermal stratification. The term inverse stagnation should not be used for this condition, because it implies that the water beneath the ice is stagnant and not subject to water movement. Most lakes exhibit considerable hydromechanical movement under ice (cf. Chapter 7).

Heating of the water under ice is characteristically found throughout winter to considerable depths (cf. Fig. 6–4). Clear ice is, as we have seen, quite transparent to solar radiation although this heat input decreases strongly, of course, with the type and amount of snow cover on the ice. From the few detailed studies on winter heating, it is clear that most (>75 per cent) is from solar radiation (Birge, et al., 1927). The relative contribution of heat stored in the sediments during the summer varies from lake to lake in relation to morphometry, summer heating conditions, and other factors (Hutchinson, 1957). Density currents, generated by heating of water through the ice in the shallow littoral areas, can flow towards the central portion of the basin along the sediments. Thermal-density instability is prevented by diffusion of solutes from the sediments, raising the density slightly but sufficiently, especially as winter progresses, to account for the commonly observed temperatures of above 4°C near the sediments.

It is not unusual to observe apparent thermal-density anomalies under the ice. For example, in Figure 6–4 a cell of water in excess of 6°C was found centered around 1 m in late March, in this extremely hardwater lake. The ice at this time was structurally poor and porous from heavy, warm rains on the lake the week prior to these measurements. Water very low in dissolved solutes that had entered through the ice and possibly from the margins of the lake, and was running under the ice was less dense than that of the normal water high in dissolved ions. By this means, solar heating through the ice was able to increase the temperatures in this dilute layer without producing instability. Most of these situations are quite transitory and variable (Koźmiński and Wisniewski, 1935), but may be related in part to the early onset of certain phytoplankton populations beneath the ice (cf. Chapter 14).

A detailed résumé on the types and characteristics of lake ice is given in Hutchinson (1957), based primarily on the extensive work of Wilson, Zumberge, and Marshall. Two recent books by Pivovarov (1973) and Ficke and Ficke (1974) summarize in great detail the diverse literature on the properties of ice and the thermal characteristics of lakes and rivers under differing conditions of ice cover.

Modifications in Stratification

The stratification picture just described, in which a typical temperate lake of moderate size undergoes two periods of mixing, one in the spring and one in the fall, and therefore is called *dimictic,* is not always so consistent. Many variations are found in relation to local or regional differences in climate, individual characteristics of the lake morphometry, and movement of the water masses.

A common divergence from the typical summer stratification pattern is the formation of secondary or even multiple discontinuity layers (Fig. 6–5). In these situations, which usually last only for a few days or weeks, secondary metalimnia are formed in the epilimnion of the initial stratification during the heating period. These conditions are common when intense heating alternates with periods of extensive mixing. The initial spring metalimnion forms deeply, and progessively stabilizes at a depth characteristic of the lake morphometry and prevailing meteorological conditions. This can be seen in the general limits of the epilimnetic-metalimnetic boundary as approximately indicated by the dashed line in the example given in Figure 6–4. The mixing of the entire epilimnion can cease during a calm, hot period, during which the surface waters can be intensely heated and perhaps mixed by light breezes for only a few meters of depth. In this way a secondary metalimnion

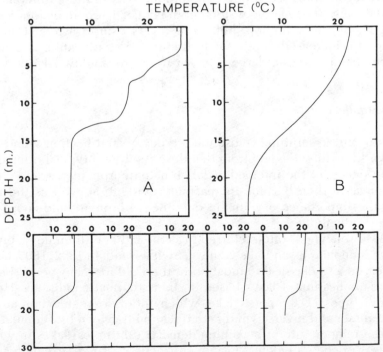

Figure 6–5 Variations in summer thermal stratification. *A,* Example of multiple thermal discontinuities (generalized from several sources); *B,* Example of thermal discontinuity greatly reduced by high through-flow volumes (generalized). *Lower series*: Profiles of thermal stratification in submersed depressions of Douglas Lake, Michigan, August 10–12, 1929. (From data of Welch and Eggleton, 1932.)

can be formed, overlying the first. The stability of these multiple metalimnia is usually not as great as the primary deeper one, and they are susceptible to disruption with ensuing periods of cooling and strong mixing of the epilimnion; on a small scale this process is analogous to normal autumnal erosion of the metalimnion and the ensuing circulation.

Thermal stratification can also be modified by inflow-outflow relationships, particularly if the volume of influent is large in relation to the volume of the epilimnion. For example, in reservoirs, high inflow from stream discharge, which is usually cooler than the water of the epilimnion, can cause much turbulence and reduce the thermal gradient appreciably (Fig. 6–5*B*). A similar phenomenon is observed frequently in alpine and northern lakes that receive large flows of glacial or snow meltwater during the later portions of the summer stratification.

Many other variations occur in relation to individual characteristics of lake heating and mixing. Individuality is a prominent characteristic of the observed patterns of thermal structure, and is governed strongly by climatic variations, volume of inflow and outflow in relation to the volume of the basin, basin configuration, surface area, position of the basin in relation to wind action, and other factors. Where the morphometry of the basin is complex, different stratification patterns can be observed within the same basin (e.g., Wetzel, 1966a). Lakes with many submersed depressions characteristically exhibit variations in both the position and stability of the metalimnion (Fig. 6–5), which can influence strongly the productivity of these different regions of such lakes. In the shallow water, depth is insufficient to maintain a typical epilimnion and hypolimnion, but these layers exhibit a reduction in temperature with increased depth. Such diminutive stratification is common in protected shallow areas of large lakes, or in entire shallow lakes.

Types of Stratification

It is useful to scrutinize closely the types of stratification found in lakes of the world as they deviate from that described for a typical dimictic situation. The effects of the interactions of climate, morphometry, and chemistry on stratification then become increasingly evident, as well as the implications of the importance of climatic influences on phyto- and zoogeography and lake productivity.

F. A. Forel, who is often referred to as the father of limnology, published extensive monographs on Lake Geneva, Switzerland, in 1892, 1895, and 1904, which served as a classical foundation in the field for many years. Forel proposed a classification of lakes based on their thermal conditions. He recognized three types: (1) temperate lakes which undergo a regular annual alternation of summer and inverse winter stratification between two circulation periods at the temperature of maximum density; (2) tropical lakes in which the water is never cooled below 4°C and is directly stratified, except for a single period of winter circulation; and (3) polar lakes in which temperatures never rise above 4°C, and the water is inversely stratified except for a single period of summer circulation. However, these terms also suggest differences in

latitude that are unfortunate, because many exceptions to the above classification can be found when, for example, an ideal 'tropical lake' occurs in the
northern temperate region, and so on.

Forel's terminology was modified in a number of ways, particularly by
Whipple (1898, 1927), Birge (1915), and Yoshimura (1936). Whipple's sub-
division of each of these categories, based on surface temperatures, into
orders based on temperatures of the deepest water, was widely adopted.
According to his scheme, in first-order lakes temperatures of bottom waters
remain at 4°C throughout the year. Circulation periods, such as occur in very
deep lakes, are often absent. In lakes of the second order, water at greatest
depth is near or above 4°C in the summer, and one or two complete circulation periods occur annually between the periods of thermal stratification.
Third-order lakes are those in which thermal stratification is never found, and
circulation is more or less continuous except when the lake is frozen.

As useful as these early categorizations were, it became apparent that
alternative groupings were necessary as more information on polar, high-
altitude, and tropical lakes became available. The following classification,
introduced by Hutchinson and Löffler (1956) and Hutchinson (1957), is now
generally accepted as most useful and as having the least ambiguity. The
definitions center on circulation patterns, as the roots of the names indicate,
and refer to lakes that are of sufficient depth to form a hypolimnion.

Amictic Lakes. Such lakes are sealed off perennially by ice from most
of the annual variations in temperature. Amictic, perennially ice-covered
lakes are rare and largely limited to Antarctica (Goldman, 1972), and under
special conditions can be found on very high mountains. Only a very few
cases of such lakes have been recorded in the Arctic, mostly from Greenland,
where in general conditions necessary for the formation of a permanent ice
cover are rare (Hobbie, 1973). It is clear that most of the heating of these lakes
is by means of direct insolation through the ice and conduction of heat from
the surrounding land through the sediments (Ragotzkie and Likens, 1964).

Cold Monomictic Lakes. These are lakes with water temperatures never
greater than 4°C, and with only one period of circulation in the summer at or
below 4°C. Cold monomictic lakes are largely restricted to the Arctic and to
mountain lakes which, although they may be ice-free for brief periods in the
summer, are in frequent contact with ice such as glaciers or permafrost.
Thermal cycles of Arctic lakes exhibit large variations in relation to their
location and changes in summer climate. Shallow lakes are extremely numerous in these areas, but usually lack sufficient depth to stratify except occasionally, and on a very temporary basis. Water temperatures often rise above 10°C,
but never exceed 15°C (Hobbie, 1973). Very deep lakes that become ice-free
for brief periods in the summer are truly cold monomictic lakes in which
temperatures never reach 4°C. Others, such as Char Lake, Cornwallis Island,
Canada (latitude 76°) warm to 4 to 5°C, and are therefore technically dimictic
(see later on), but do not always stratify. Lake Schrader in the Brooks Range,
Alaska, was dimictic and stratified in 1958, with the epilimnion at 10°C and
the hypolimnion at 4°C. The following year summer ice breakup occurred a
month later, and the lake did not warm above 4°C, i.e., it was cold monomictic
(Hobbie, 1961).

Dimictic Lakes. These lakes circulate freely twice a year in the spring and fall and are directly stratified in summer, inversely stratified in winter. Dimictic lakes represent the most common type of thermal stratification observed in most lakes of the cool temperate regions of the world. Their characteristics were described in detail earlier in this chapter. Such lakes also are found commonly at high elevations in subtropical latitudes.

Warm Monomictic Lakes. In these lakes temperatures do not drop below 4°C, they circulate freely in the winter at or above 4°C, and they stratify directly in the summer. Warm monomictic lakes are common to warm regions of the temperate zones, particularly those influenced by oceanic climates, and to mountainous areas of subtropical latitudes. Most lakes of the central and eastern portions of North America and the interior of Europe exhibit a distinct continental type of dimictic stratification, whereas a warm monomictic stratification is prevalent in many coastal regions of North America and Northern Europe.

Oligomictic Lakes. These lakes are generally tropical and have rare circulation periods at irregular intervals, and temperatures always well above 4°C. Lakes of small to moderate area or lakes of very great depth in the tropics often are observed to maintain stable stratification, even though only a small temperature difference may exist between the surface and bottom strata. These lakes are common in equatorial regions of high humidity, and may circulate only at irregular intervals during periods of abnormally cold weather. Circulation can be quite rare and occur between several years of continuous feeble stratification (Vollenweider, 1964).

Polymictic Lakes. These are lakes with frequent or continuous circulation. Polymictic lakes have been further divided (Ruttner, 1963). *Cold polymictic lakes* circulate continually at temperatures near or slightly above 4°C. These lakes commonly are of large area and moderate depth, and are found in equatorial regions of high wind and low humidity, where little seasonal change in air temperatures occurs. At very high altitudes in equatorial regions, cold polymictic lakes gain a significant amount of heat during the day, but nocturnal losses are sufficient to permit complete mixing during the night.

Warm polymictic lakes are usually tropical lakes that exhibit frequent periods of circulation at temperatures well above 4°C. Annual temperature variations are small in the equatorial tropics and result in repeated periods of circulation between short intervals of heating and weak stratification, followed by periods of rapid cooling. Under these circumstances convectional circulation is sufficient, in combination with wind, to disrupt stratification.

A useful schematic arrangement of these six types of lakes based on thermal and circulation characteristics is presented in Figure 6–6, in which generalizations are drawn on altitudinal and latitudinal distribution. Such generalizations are difficult to maintain; many exceptions exist, particularly at low altitudes, where there is a strong influence of oceanic ameliorations of climate. The diagram does, however, demonstrate the general observed geographical distribution.

The above discussion of thermal lake types refers to lakes with sufficient

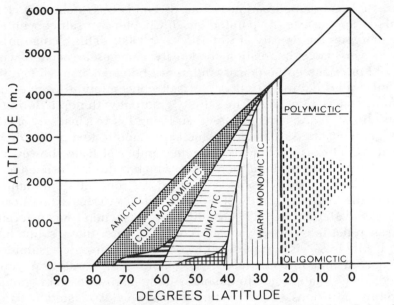

Figure 6–6 Schematic arrangement of thermal lake types with latitude and altitude. *Solid black*: cold monomictic; *black and white horizontal bars*: transitional regions; *horizontal lines*: dimictic; *crossed lines*: transitional regions; *vertical lines*: warm monomictic. The two equatorial types occupy the unshaded areas labelled oligomictic and polymictic, separated by a region of mixed types, mainly variants of the warm monomictic type (*broken vertical lines*). (Modified from Hutchinson and Löffler, 1956.)

depth to form a hypolimnion. Numerically speaking, when considering the thousands of small, relatively shallow lakes that literally cover the tundra and northern temperate region, it becomes evident that the majority of these lakes do not stratify during ice-free periods. The depth required to stratify thermally varies so greatly with surface area, basin orientation in relation to prevailing wind, depth-volume relations, protection by surrounding topography and vegetation, and other factors that generalizations in this regard are misleading.

Meromixis

The types of stratification discussed thus far are those in which circulation occurs throughout the entire water column, that is, *holomictic* lakes. A number of lakes do not undergo complete circulation, and the primary water mass does not mix with a lower portion. Such waters are termed *meromictic* lakes (Findenegg, 1935; Hutchinson, 1937). The water strata in such lakes are unique, and have been differentiated from those of normal holomictic lakes by three terms. The deeper stratum of water that is perennially stagnant is the *monimolimnion*; this underlies the upper *mixolimnion*, which periodically circulates. These two strata are separated by a steep salinity gradient which is called the *chemocline*.

Although the causality of such chemically-induced stratification has been treated and named variously, the divisions of Hutchinson (1937) are particularly appropriate. As pointed out in Chapter 2, a salt concentration of 1 g l^{-1} increases the density of water by ca 0.00008. This change in specific gravity is very large in relation to density changes associated with temperature. For example, the density difference between 4° and 5°C is 0.00008, and would only require 10 mg l^{-1} of salt concentration to give the same effect in resistance to mixing. The salinity gradient with depth during stratification is almost always much greater than 10 mg l^{-1}. In a majority of lakes, the salinity gradients are insufficient to increase stability to a point where wind energy does not cause holomixis. A large number of lakes, however, do exhibit temporary or permanent meromixis, which is directly related to salinity gradients which Berger (1955) has termed a "concentration stability," in contrast to the thermal stability of holomictic lakes discussed earlier.

Ectogenic Meromixis.   This condition results when some external event brings salt water into a freshwater lake or fresh water into a saline lake. The result in either case is a superficial layer of less dense, less saline stratum overlying a monimolimnion of denser, more saline water. As might be expected, such situations are found often along marine coastal regions where catastrophic intrusions of salt water from storms associated with unusual tidal activity are fairly common events. In estuarine lakes such events also occur routinely. More strictly defined ectogenic meromictic lakes are those that are isolated from routine marine influxes of water. Such isolation may be recent or old. An example of the latter is southern Norwegian Lake Tokke, which most likely has been permanently meromictic for about 6000 years; this isolation of the lake probably took place during a period when the former fjord depression and the surrounding land were elevated some 60 m above sea level (Strøm, 1955).

A most interesting variant of ectogenic meromixis is found in many saline lakes of arid regions or in small depressions that occasionally receive salt water intrusions (Hudec and Sonnenfeld, 1974). Often these saline lakes overlie large deposits of soluble salts such as magnesium sulfate, and are only a few meters in depth. Infusion of large amounts of fresh water, either naturally during wet periods or artificially as a result of increased irrigation of nearby land, creates a strongly meromictic stratification that may persist for many years. A striking example is Hot Lake, a shallow (3.5 m) saline body of water occupying a former epsom salt excavation in north central Washington (Anderson, 1958). Energy of solar radiation passing through the overlying mixolimnion of fresh water can accumulate as heat in the chemolimnion and monimolimnion, where circulation is absent. The resulting temperatures can be in excess of 50°C, and even when the surface of the lake is frozen, a temperature of nearly 30°C can be found at a depth of 2 m (Fig. 6–7). These dichothermic conditions in which the monimolimnion is considerably warmer than the overlying water of the chemolimnion are frequently observed in such lakes (Fig. 6–7).

Man's activities have been known to form meromictic lakes artificially. Creating a connecting channel between a freshwater lake and the sea for purposes of navigation has resulted in the intrusion of saline water into the lake and permanent meromixis. Another event that can form a large number of

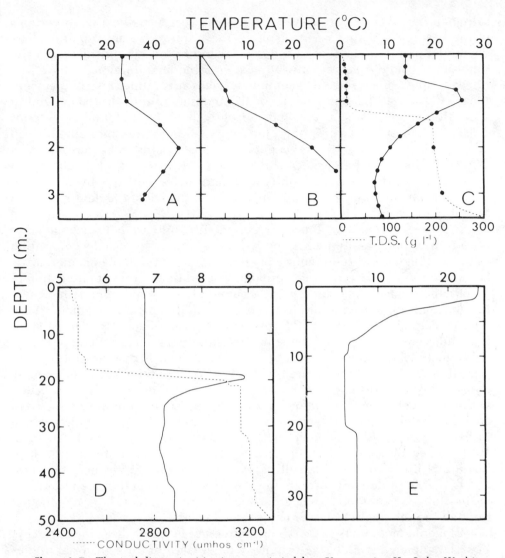

Figure 6–7 Thermal discontinuities in meromictic lakes. *Upper series*: Hot Lake, Washington (after Anderson, 1958): *A*, 23 July 1955; *B*, Heating under thin ice cover, 7 December 1954; *C*, Heating in a deeply colored layer, 0.5–1.0 m, 1 May 1956; *T.D.S.*=Total dissolved solids on evaporation. *Lower series*: *D*, Depth distribution of temperature and conductivity, Fayetteville Green Lake, New York, 16 December 1966 (from Brunskill and Ludlam, 1969); *E*, Thermal stratification in Ulmener Maar, Germany, 8 August 1911. (Modified from Ruttner, 1963, after Thienemann). Note variations in depth and temperature scales.

small meromictic lakes is the introduction of salt caused by runoff from street de-icing into lakes adjacent to roads (Judd, 1970). The excellent analyses of the paleolimnological record of Längsee, Austria, by Frey (1955) clearly show how the clearing of forests surrounding the lake in prehistoric times introduced ectogenic meromixis. The resulting silt-laden runoff flowed to the hypolimnion and initiated a monimolimnion, which was augmented subsequently by salinity inputs of biogenic origin (see further on).

    Crenogenic Meromixis. Crenogenic meromixis results from submerged saline springs that deliver dense water to the lake, usually at deep portions of the basin. The saline water displaces the water of the mixolimnion. The chemolimnion will stabilize at a depth related to the rate of influx, density differences, and the degree of wind mixing of the mixolimnion.

Examples of crenogenic meromictic lakes are numerous. In Ulmener Maar (Fig. 6–7), the temperature of the monimolimnion has stabilized at 7°C. Meromictic lakes of this type recently have been found in interior Alaska, when subsurface springs introduce saline water into small, deep lakes originating in the thawed craters of pingos (Likens and Johnson, 1966). Pingos are formed when water, rising by hydraulic pressure through gaps in the permafrost, freezes and uplifts a mound of ice covered by a layer of alluvium. The overburden usually ruptures and the ice of the depression can thaw to form small lakes.

Biogenic meromixis. Biogenic meromixis results from an accumulation of salts in the monimolimnion, which is usually liberated from decomposition in the sediments and sedimenting organic matter. Often biogenic meromixis is initiated when abnormal meteorological conditions prevent circulation of a normally dimictic lake and permit accumulation of hypolimnetic increases in biochemically derived salinity. These accumulations may be sufficient to permit meromixis to persist either temporarily or permanently. Temperature inversions are usually found in the monimolimnion, such as is seen in Figure 6–7D for Fayetteville Green Lake, New York. The high temperatures observed in the chemolimnion are the result of absorption of solar radiation by the dense bacterial plate that occurs at this level. The processes associated with the increasing temperatures in the lower monimolimnion which are consistently observed are less obvious. It is clear that the heat of metabolism in decomposition is inadequate to account for these increases (Hutchinson, 1941). When direct solar heating is not possible because of the depth, such heating is likely to be the result of warmer salt-laden water near the sediments that form density currents from overlying strata and descend into the basin along its contours. Heat from the surrounding land and conduction through the sediments is minor as a thermal source.

Biogenic meromixis increases in frequency in lakes that are very deep. For example, nearly all of the extremely deep lakes of the equatorial tropics are meromictic. But biogenic meromixis is common among lakes of small surface area, moderate depth, and that are sheltered, especially in continental regions that experience long, severe winters. *Partial* or *temporary meromixis,* when a normally dimictic lake skips a circulatory period, usually the spring period, is a result of dynamic processes of decomposition and sedimentary decomposition (Findenegg, 1937) in the lower levels of the lake after a combination of sheltered and unusual weather conditions. For example, after an unusually long and severe winter followed by a rapid thaw and warm conditions, Martin Lake did not undergo spring circulation (Fig. 6–8), and the chemically stabilized water of the temporary monimolimnion was carried over from the winter accumulation and augmented throughout the subsequent summer (Wetzel, 1973). Notable is the carryover of low temperatures from the circulation of the preceding fall which persisted throughout the summer and fall, prior to the ice-cover formation. Autumnal circulation in this protected lake did not occur until December, and then only briefly

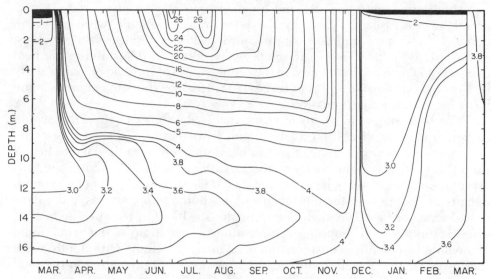

Figure 6–8   Isothermal variations (°C) in the major depression of Martin Lake, LaGrange County, Indiana, 1963–64. (From Wetzel, R.: Hydrobiological Studies, 3:91–143, 1973.)

before ice formation. The following spring a normal complete circulation was observed. This type of temporary, partial meromixis also was observed in two other lakes in northeastern Indiana, of the 13 investigated during the same period (Wetzel, 1966b). In his classical work on phosphorus cycles and enrichment effects in the small eutrophic Lake Schleinsee, Einsele (1941) found that complete spring turnover occurred on the average of only every other year. In Lawrence Lake, a small (4.9 ha), deep (12.6 m) hardwater lake of southern Michigan, temporary meromixis is observed about every third year (Wetzel, et al., 1972, and unpublished). Many other examples exist, but these are adequate to emphasize that this phenomenon of intermittent biogenic meromixis is quite common. It is not difficult to envision the combination of circumstances that can shift a lake from temporary to permanent meromixis. Both conditions have profound effects on the biota and productivity, as will be discussed later on.

Thermal Energy Content: Heat Budgets[2] of Lakes

Temperature and density stratification in lakes are dominant regulators of nearly all physico-chemical cycling, and consequently of lake metabolism and productivity. Therefore, it is important to understand the relative energy content of lakes, that is, what is a lake's thermal capacity both in the amount of heat energy required to develop stratification characteristics and the amount required to overcome resultant density differences.

---

[2]The use of the word 'budget' here is incorrect in terms of its normal definition. Reference actually is to the heat storage capacity of the lake, but the term budget is firmly entrenched in limnology.

Heating capacity and demand for energy to reach a given state is governed, in the case of a lake, by the heat inputs available within a given climate, the volume and accessibility of the water at depth to heat incomes, and similarly to heat losses. Although many means of determining the heat content of lakes were developed in the pioneer days of limnology, particularly by Forel, the extensive works of Birge (1915, 1916) on this subject provide the basis for the ensuing discussion.

The heat content of a body of water, that is, the amount of heat that would be released on cooling from its maximum temperature to 0°C, is simply the net amount of heat in calories that has been transferred to the lake to heat it from 0°C to its maximum temperatures at various depths. Of course, the lowest temperatures of many lakes are not near 0°C, or even its point of maximum density at 4°C. Hence, the minimum and maximum temperatures of the individual strata are used in such determinations. Although the specific heat of water is unity, which simplifies computations, calculations of the heat budget of a whole lake require consideration of changes in volume with depth, since the lower strata of smaller volumes make a smaller contribution to the total heat content than do the upper layers.

A Birgean heat budget for a typical dimictic lake can be defined as follows (Birge and Juday, 1914; Birge 1915; Hutchinson, 1957):

Summer Heat Income (Θ_s). This is the amount of heat necessary to raise the temperature of the lake from an isothermal condition at 4°C to the maximum observed summer heat content. This heat income is primarily a result of the heating of the surface layers by direct insolation, and the distribution of the heat by wind. It therefore has been called (Birge, 1915) the wind-distributed heat, even though not only wind energy is involved.

Winter Heat Income (Θ_w). This is the amount of heat necessary to raise the temperature from its minimum heat content to 4°C. The winter heat income is usually not considered as wind-distributed, even though wind is involved in loss of heat prior to the lake's reaching the minimum temperature in autumnal-early winter cooling. Accurate winter heat incomes must be corrected for the latent heat of fusion of ice, which can be sizeable when the ice cover is considerable. One cm thick ice is equivalent to 79.72 cal cm^{-2} or a correction of about 1000 cal cm^{-2} for 12.5 cm of ice.

Annual Heat Budget (Θ_a). This is the total amount of heat necessary to raise the water from the minimum temperature of winter to the maximum summer temperature. From the standpoint of the whole lake, this refers to the total amount of heat needed between the period of its lowest and its highest heat content.

The heat budget may be calculated by plotting the product $A_z(\Theta_{sz} - \Theta_{wz})$ against depth z, where Θ_{sz} and Θ_{wz} are the respective summer and winter temperatures at depth z. The area of the resulting curve is integrated by planimetry[3] and divided by the surface area of the lake (A_o). To separate the heat budget into summer or winter heat incomes, 4°C is substituted for the respective summer and winter temperatures at depth. A sample computation is given in Figure 6–9 for a small dimictic lake of moderate depth in south-

---

[3]The proper use of these ingeniously devised instruments is described in detail in Welch (1948).

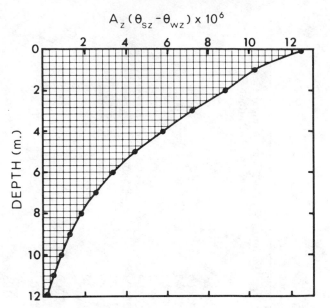

$$A_z(\theta_{sz} - \theta_{wz}) \times 10^6$$

Figure 6–9 Curve generated for calculation of the annual heat budget of Lawrence Lake, Michigan, 1968. Calculations resulted in a budget, including a correction for ice of 10,918 cal cm⁻². (From Wetzel, unpublished data.)

western Michigan, which indicates the decreasing contribution of decreasing volumes at greater depth of a rather typical basin.

A large number of annual heat budgets have been computed; about 100 are listed in Hutchinson (1957), and many have appeared in the literature since then. A few exemplary values are given in Table 6–2 to emphasize several points in regard to factors influencing the observed heat budgets.

Although a majority of the heat budgets have been derived for large, deep lakes, a few generalizations are possible. In temperate lakes that are sufficiently deep to maintain hypolimnetic temperatures at or near 4°C, annual heat budgets usually run between 30,000 to 40,000 cal cm⁻² year⁻¹, most of which (60 to 80 per cent) is gained in the spring-summer transition. Very few lakes have an annual heat budget in excess of 50,000 cal cm⁻² year⁻¹. Notable exceptions include Lake Baikal and Lake Michigan. Annual heat incomes of polar region lakes are reduced when θ_s values are in the range of 10,000 cal cm⁻² for large lakes. However, annual values are likely to be very much larger than this when incorporation of values of large latent heat of fusion for melting ice is included. The same is true for the apparently low values of lakes at high altitudes. In the temperate zone, annual heat budgets are strongly correlated with mean depth, area, and lake volumes, rising continuously with lake dimensions, though at a decreasing rate (Gorham, 1964). For lakes of a given volume, the deeper ones of lesser area take up slightly more heat than shallower ones of greater area. Dimictic lakes exhibit slightly higher heat budgets than warm monomictic lakes of similar size.

Heat budgets also vary considerably with the measurement of data over a period of time, especially in regard to determining the maximum heat content. Heat content can vary significantly (10 per cent) within 2 or 3 days; accurate determinations require frequent measurements over long periods (Stewart, 1973). Examples of this variation are given in Table 6–3.

The annual income and loss of heat in tropical lakes is very small in comparison to that of lakes at higher latitudes or mountain lakes. Also, the

TABLE 6–2 Heat Budgets of a Variety of Lakes, in cal cm⁻²yr⁻¹ [a]

Lake	Surface Area (A_o)	Maximum Depth (z_m)	Mean Depth $(\bar{z})$	Summer Heat Income Θ_s	Winter Heat Income Θ_w	Annual Heat Budget Θ_a
Baikal, USSR	31500	1741	730	42300	22800	65500
Michigan, USA	57850	282	77	40800	11600	52400
Washington, Washington	128	65	18	43000	[−2600]	43000
Geneva, Switzerland	581.5	310	154.4	36000	[−23200]	36000
Tahoe, California-Nevada	499	501	249	34800	0	34800
Green, Wisconsin	29.7	72.3	33.1	26200	7800	34000
Pyramid, Nevada	532	104	57	33600	[−7000]	33600
Tiberias, Israel	167	50	24	33500	[−26400]	33500
Ladoga, USSR	18150	223	56	18000	15300	33300
Mendota, Wisconsin	39.2	25.6	12.4	ca 18250	ca 5800	24073
Atitlan, Guatemala	136.9	341	183	22110	[−288300]	22110
Windermere, England						
North Basin	8.16	67	26.0	17500	[−2900]	17500
South Basin	10.5	44	17.7	15680	[−2450]	15680
Furesø, Denmark	9.9	36	12.3	14400	2700	17100
Lunzer Untersee, Austria	0.68	33.7	19.8	12300	1400	13700
Lawrence, Michigan	0.049	12.6	5.9	9168	1750	10918
Schrader, Alaska						
1958	13.2	57	33	9050	—	—
1961				8900	—	—
Chandler, Alaska	15	21	13.5	5760	—	—
Lanao, Philippines	357	112	60.3			
1970				7250	[−121300]	7250
1971				4500	[−121300]	4500
Ranu Klindungan, Java	2	134	90	ca 3410	[ca −189000]	ca 3410
Hula, Israel	14	4	1.7	2240	[−1600]	2240

[a] From numerous sources; area in km²; depth in m.

TABLE 6–3   Variations in the Annual Heat Budgets of Several Lakes[a]

Lake	Number of Annual Observations	Mean Depth $\bar{z}$	Mean Θ_a cal cm^{-2} yr^{-1}	Range cal cm^{-2} yr^{-1}
Green, Wisconsin	5	33.1	34200	32300–36400
Geneva, Switzerland	4	154.4	32325	22000–40200
Mendota, Wisconsin	5	12.4	24073	22308–25953
Menona, Wisconsin	3	7.7	17559	17256–18041
Waubesa, Wisconsin	3	4.6	11362	10948–11739

[a]From data of Birge, 1915, and Stewart, 1973.

mixing patterns are complex, often variable or oligomictic in nature. A recent example is given by Lewis (1973), who demonstrated the weak, variable stratification with the repeated formation of multiple, temporary metalimnia common to many tropical lakes of large size. The minimum yearly heat content, given as negative values in brackets in Table 6–2, represents the heat content above 4°C in those lakes that never cool to that temperature. In such lakes at northern latitudes, as in England, these values are very small; they are of course very high in equatorial regions.

The stratification patterns and thermal structure of reservoirs with high through-flow volumes are enormously complicated (Wunderlich, 1971). Where withdrawal of major volumes for generators occurs daily and intermittently, the patterns can be further disrupted and altered. Although draw-down usually occurs from the epilimnion, sometimes water is drawn from the hypolimnion. Influent waters also vary greatly in relation to surface runoff, heat content, and retention in series of reservoirs in tandem on major river systems. Normal Birgean heat dynamics must be modified considerably in order to be applicable to complex reservoir systems. Among the most detailed analysis is that of the Slapy Reservoir in Czechoslovakia, where seasonal thermal dynamics have been determined for over a decade (Hrbáček and Straškraba, 1966; Straškraba, Hrbáček, and Javornický, 1973).

Heat of Lake Sediments

The sediments of lakes absorb an appreciable amount of heat from the water during the warmer periods of the year, and transmit heat to the water during the winter period. Although few detailed studies are available, data from Lake Mendota illustrate this characteristic very well (Birge, Juday, and March, 1927). It will be noted from Figure 6–10 that the temperatures of the sediments at a water depth of 8 m follow the annual cycle of the epilimnetic water temperatures rather closely. This thermal variation is attenuated with depth in the sediments, so that at a depth of 5 m little seasonal oscillation is observed. A similar relationship has been observed in the sediments of the hypolimnion at a water depth of 23.5 m, but the amplitude and heat energy exchange are much less (Fig. 6–10). A lag in transmission of heat in the sediments of about one month per meter depth of sediments has been observed.

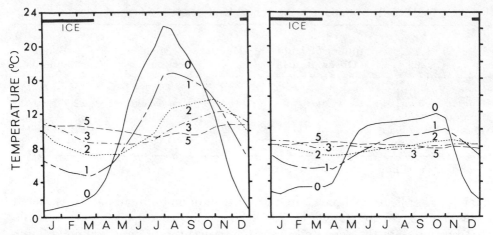

Figure 6–10 Annual temperature curves at 0, 1, 2, 3, and 5 m within the sediments of Lake Mendota, Wisconsin (means for 1918, 1919, and 1920). *Left*: Sediments at a water depth of 8 m; *Right*: Sediments at 23.5 m. (Modified from data of Birge, Juday, and March, 1927.)

Variations in temperatures spatially within the sediments at similar water depth have been found with regard to the type of adjacent land forms of the basin (Matveev, 1964). Sediments of bog lakes at the shoreline rim of the basin are considerably warmer than lake sediments, but seasonal variations are largely damped at a sediment depth of 4 m (Likens and Johnson, 1969).

Determinations of the heat budget of the sediments based on the differences between the minimum and maximum mean temperatures showed that in sediments at a lake depth of 8 m, this value was 2950 cal cm^{-2} yr^{-1}, at 12 m, 2200, and at the maximum of 23.5 m, 1100 cal cm^{-2} yr^{-1}. An estimated average heat budget of all of the lake sediments was ca 2000 cal cm^{-2} yr^{-1} for Lake Mendota. As lakes become more shallow, the heat budget of the sediments becomes a greater portion of the whole (Table 6–4). All of the examples of annual heat budgets of lakes cited above (Tables 6–2 and 6–3) are based on water temperatures, and ignore those of the sediments. While in large deep lakes, such an assumption of minor importance is perhaps justified, it should be noted that the sediments of shallow waters are significant to the thermal dynamics of lakes.

TABLE 6–4 Heat Budgets of Lakes in Which the Heat Budget of the Sediments Is Compared to that of the Water[a]

Lake	Mean Depth (m)	Heat Budget of Water Θ_a (cal cm^{-2} yr^{-1})	Heat Budget of Sediments Θ_m (cal cm^{-2} yr^{-1})	Heat Budget of Lake (cal cm^{-2} yr^{-1})	% Sediments of Total Heat Budget
Mendota, Wisconsin	12.1	23500	2000	25500	7.8
Stewart's Dark, Wisconsin	4.3	7000	730	7730	9.4
Beloie, USSR	4.15	8000	2500	10500	23.8
Tub, Wisconsin	3.6	8000	970	8970	10.8
Hula, Israel	1.7	2290	1400	3690	37.9

[a]Modified from Hutchinson, 1957, with data of Likens and Johnson, 1969.

water movements

THE HYDRODYNAMICS OF WATER MOVEMENTS IN LAKE BASINS

Lakes behave like large mechanical oscillators. They respond in numerous, complex ways to applications of force, and are frictionally damped by viscous forces associated with turbulence. Movements, forced by the transfer of wind energy to the water, give rise to a spectrum of rhythmic motions (oscillations), both of the water surface and internally deep within the basin. These oscillations, and their attendant currents, may be in phase or in opposition; their final fate is to break down into arrhythmic turbulent motions of a random chaotic nature. In addition, each basin has its own set of free modes of oscillation, both surface and internal, depending on gravity, on basin shape and size, and on the internal distribution of density. Which of these modes is brought into play during any particular disturbance or storm depends on the duration, the periodicity, and the spatial distribution of the applied force. In other words, the morphometry of the lake basin, its stratification structure, and its exposure to wind are important factors in water movement. The earth's rotation complicates these hydrodynamics as lakes increase in size and depth; but at the same time, the principal surface and internal water movements become correspondingly larger, making a study of these interrelationships somewhat easier.

The turbulence resulting from these water movements is of major significance for the biota and productivity of the lake. Limnological thought is still pervaded by the idea that summer and winter density stratification result in stagnation of large lake strata. There is little basis for this view. The hydrodynamics of water movements are integral components of the functional lake system, and consideration must be given to their effects on changes in temperature, on dissolved gases and nutrients, and on other chemical parameters. These movements influence not only the aggregation and distribution of nutrients and food, but also the distribution of microorganisms and plankton.

Flow of Water

Turbulent and Laminar Flow

At sufficiently slow speeds, flow of water in a smooth tube moves along the interface in an apparent orderly manner. If the velocity of movement is increased at the interface, in this case between the stationary walls of the tube and the water, a critical speed is reached above which the smooth and unidirectional laminar flow becomes disordered or turbulent. Such is also the case with two miscible fluids moving in opposing directions. An example would be opposing horizontal movements between two water strata of differing densities, such as the epilimnion and metalimnion of a lake.

Appreciable experimental work on turbulent shearing stresses (Reynolds) has demonstrated that, in lakes, the amount of velocity needed to shift from laminar flow to turbulent current is low. The critical velocity is a function of the fluid viscosities and densities, and decreases in proportion to increase in size of the channel or tube. In lakes, velocities of only a few mm sec^{-1} can induce turbulent flow, at which point disturbances arising on the interface between the layers will no longer be suppressed by buoyancy (gravity) forces which tend to keep layers of different densities separate. As a result, laminar flow will almost never be found in aquatic systems. Therefore it is important that the basic properties of turbulent flow and diffusion be understood in order to appreciate the magnitude of the resultant alterations in distribution of physico-chemical parameters and biota of lakes.

If a critical velocity difference across a given density interface is exceeded, disturbances grow in amplitude and break into vortices (Fig. 7–1)

Figure 7–1   Stages in vortex formation during shear instability on the interface of a stratified two-layer system. (From Mortimer, C. H.: Mitteilungen Int. Ver. Limnol., 20:131, 1974.)

(Mortimer, 1961, 1974). Such vortex formation increases mixing of the two layers by generating a transitional layer across which there is both a velocity gradient (shear) and a density gradient. This model represents in simplified form what is found when layers of nearly uniform velocity and density stream parallel to one another, as in the movements in the epi-, meta-, and hypolimnion discussed further on.

If the rate (a) of supply of energy to turbulent eddies[1] from the shear stresses exceeds the rate (b) at which they have to do work against gravity in disrupting the density stratification, then turbulence increases (Richardson, 1925; Mortimer, 1961). Rate a is proportional to the square of the shear, rate b is proportional to the density gradient, and the magnitude of the ratio b/a determines whether turbulence increases or decreases. In nondimensional form, that ratio is the Richardson number (R_i) expressed as:

$$R_i = \frac{g \; (d\rho/dz)}{\rho \; (du/dz)^2}$$

where g = acceleration of gravity,
    ρ = density
    u = velocity
    z = depth.

When R_i of a stratified fluid subjected to shearing flow falls below about one-quarter, flow becomes unstable. Or, put in another way, when R_i falls below 0.25, there is a sudden shift in the eddy spectrum from the molecular microturbulence of stable flow toward macroturbulence of the large vortices associated with unstable flow. The shift is accompanied by increases in friction and in mixing perpendicular to the direction to the current.

Thus in a lake with stratified shear layers, mixing and friction between the layers remains small as long as the flow remains stable ($R_i > 0.25$). When flow increases and becomes unstable ($R_i < 0.25$), the stirring action of vortex formation (macroturbulence) increases the interfacial area many times. Mixing is then rapidly completed by microturbulence and molecular diffusion. Soluble components (nutrients) and suspensoids (microalgae, microfauna) are carried along with this dispersion of water into the thus formed new, thicker layer in which the density and velocity gradients are less than those initially present.

The shear instability, resulting when R_i falls below 0.25, causes a sudden shift in turbulent energy with dissipation and collapse into a complex array of turbulent patches of temporary layers of nearly uniform density (Fig. 7–2). Exceedingly transient in nature, the statistical composite final product, after stability is restored, is what is most often observed. The thermal microstratification, commonly observed in detailed studies of the metalimnion (Whitney, 1937; Simpson and Woods, 1970), is possibly a result of dissipation of such shear instabilities in the thermocline region.

---

[1]The rotating region of fluid, an *eddy*, refers to an assemblage of shear waves of a spectrum of many lengths or 'eddy diameters' (Mortimer, 1974).

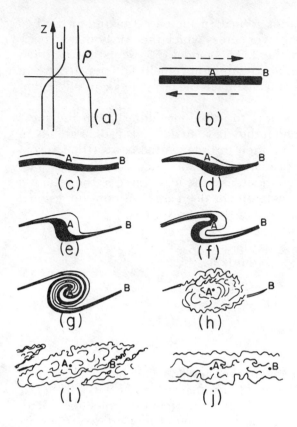

Figure 7–2 Growth of shear instability leading to turbulent mixing in a stratified fluid with the velocity and density distribution shown in *a*. A and B are fixed points, the arrows indicate direction of flow, and the lines represent surfaces of equal density. (From Mortimer, C. H.: Mitteilungen Int. Ver. Limnol., 20:134, 1974, after Thorpe.)

Eddy Diffusion and Conductivity

The distribution of turbulent motion must be viewed in a statistical model which considers the distribution of some fluid property such as heat or solute content as the composite of many small random movements. The diffusion models for fluids, developed early by Taylor and by Schmidt (Mortimer, 1974), are analogous to those that measure random movement of molecules in molecular diffusion, but are obviously on much larger scales.

Heat transfer across a thermal gradient within a layer in a lake may be used to estimate the extent of turbulent transport, because turbulent mixing and heat conduction are quite similar. In a hypothetical layer of water devoid of motion but with a temperature gradient, heat equilibrium is brought about only by conduction. If flow is applied across that gradient and it is turbulent ($R_i < 0.25$), the equalization of temperature is accelerated greatly over what would occur by conduction alone. Thus, heat (relatively warm water) is transported through a plane perpendicular to the gradient by eddy turbulence. The rate of transport (diffusivity or conductivity) across a fluid plane is the product of the gradient normal to the plane and an exchange coefficient (A). If expressed as a substance dissolved in the water across the gradient, it may be written as the gradient concentration per unit volume of water in terms of passage across unit area, as g cm^{-1} sec^{-1}.

Thus, the *coefficient of eddy diffusion* (A) is a measure of the rate or intensity of mixing (exchange) across the plane. It is not a constant but varies with average density, velocity of vertical motion, and mixing length. The horizontal dimensions of lakes are very much larger than the vertical dimensions of stratification; therefore, large differences exist between the magnitudes of horizontal and vertical eddy diffusion coefficients. The concept of an average A is subject to many variations related to distance in a spatial sense of turbulence, and is therefore an oversimplification. In reality, dispersion and mixing occur as a result of eddy movements over a wide range of spatial scales in which smaller eddies occur within larger ones. It is necessary to view the turbulent movements and consequent diffusion as taking place across a spectrum of turbulence (the eddy spectrum of Mortimer, 1974). Turbulent transport causes particles to move apart at right angles to the direction of flow. When particles are close together, their rate of separation is governed by the small eddies of turbulent motion. But as the distance of separation increases, the dispersion of the particles is influenced by eddies of increasing size. Over significant ranges of dispersion, the mean rate at which particles diffuse is proportional to the ⁴/₃ power of the distance between them (Stommel, 1949), a law that has been found to hold true for separation of particles over a range of from 10 to 10^8 cm (Olson and Ichige, 1959; Verduin, 1961). The most probable value of turbulent diffusion velocity is in the range of 0.3 to 1.0 cm sec^{-1}, with a single diffusion velocity of approximately 1.0 cm sec^{-1} being valid for all scales of diffusion processes (Noble, 1961).

The coefficient of eddy diffusion is, for practical purposes, identical to the coefficient of eddy conductivity which permits, in a general way, estimations of turbulence from changes in temperature. Estimations of eddy conductivity coefficients are useful in that they permit insight into the magnitude of current-induced turbulence in the hypolimnion. The magnitude of A and changes observed vertically with stratification, however, depend very much upon the methods of analysis and assumptions about the magnitude of direct solar heating of the upper hypolimnion, convective currents, lateral heating from the sediments, and accuracy of measurements of small temperature differences (Hutchinson, 1957; Dutton and Bryson, 1962; Lerman and Stiller, 1969; Hesslein and Quay, 1973).

Coefficients of eddy conductivity in the hypolimnion decrease with increasing stability. Therefore, in lakes of large area and depth that have longer fetch for wind exposure and supply of energy for turbulent transport, the coefficients of the hypolimnion are larger than in smaller shallow stratified lakes (Mortimer, 1941, 1942). Additionally, A decreases markedly from the turbulent epilimnion into the more stable metalimnion. This relationship holds seasonally as well: as the gradient of thermal density stratification increases from spring to summer, the coefficient of eddy conductivity decreases, i.e., A is inversely proportional to stability. In the deep hypolimnion A tends to increase somewhat, especially near the bottom where the effects of other water movements, to be discussed further on, complicate analyses of turbulent transport.

Surface Water Movements

Surface Waves (Progressive Waves)

The frictional movement of fluid air, wind blowing over water, sets the water surface into motion, producing a wind drift, and also sets the surface into oscillation, producing *travelling surface waves*. If these waves become large enough to break, then their momentum is transferred to the water. Although these surface waves are the most conspicuous periodic responses to an observer looking at a lake, their limnological importance is largely confined to shore action. In deeper water their motions are confined to surface layers, with little effect on displacement of the main deeper water masses.

These short surface waves cause the surface water particles to move in a circular path or orbit; in cross-section the path is *cycloid* (Fig. 7–3), with very little significant motion other than horizontal translocation (Smith and Sinclair, 1972). Water is displaced vertically and returned by gravity to an equilibrium state. Similarly, neighboring short surface waves are oscillating periodically in cycloid fashion. In synchrony, these movements result in a travelling wave whose path may be described as analogous to a point on the rim of a wheel moving along a plane where the hub is at the mean surface of the water (Fig. 7–3). Except at the shallow beach areas, the wavelengths (λ) of short surface waves are less than the water depth and, consequently, they are dispersive, i.e., they travel at speeds proportional to $\lambda^{1/2}$ deepwater gravity waves.

Of greater interest than the small horizontal movement of short surface waves is the influence such periodic oscillations have vertically, with depth. The amplitude (h) of this vertical oscillation is attenuated rapidly with depth (Fig. 7–3). The decrease of vertical motion with increasing depth can be approximately described as a halving of the cycloid diameter for every depth increase of $\lambda/9$. As we will see shortly, the amplitude or height of surface waves is not directly proportional to wavelength and is variable, although a ratio of about 1:20 of h:λ is a common average. If, for example, short surface waves had a $\lambda = 18$ m and h = 1 m, the vertical oscillation at a depth of 4 m ($2/9\lambda$) would be 25 cm, at 8 m or $4/9\lambda$, 6.25 cm, and at 18 m depth, 1.95 mm. In

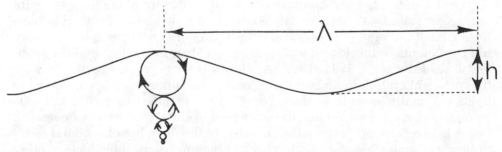

Figure 7–3 Surface wave indicating wavelength (λ), height (h), and attenuation of amplitude (h/2, the displacement positive or negative from the equilibrium in a sinusoidal wave) of the cycloid movement with depth.

other words, the amplitude at a depth equal to the wavelength will be $1/512$ of that at the surface.

Short surface waves with a wavelength greater than 6.28 cm (2π) are referred to as *gravity waves*. Waves of a length less than this minimum are called *ripples* or capillary waves. The ratio of amplitude (h) to wavelength (λ) is highly variable within a range of 1:100 to about 1:10. At great height the crests of the waves sharpen and become less stable. When the angle of the wave height increases to a h:λ ratio of about 1:10, collapse of the peak occurs and some apical water may be blown off, forming a whitecap.

For a given wind speed, wave height appears to be nearly independent of depth in small lakes, but in lakes of great area wave height and length increase with the depth (Hutchinson, 1957). The height of the highest waves observed on a lake appears, without good theoretical explanation, to be proportional to the square root of the fetch, or distance over which the wind has blown uninterrupted by land. Thus the maximum height (cm, crest to trough), where x = the distance in centimeters, is expressed as

$$h = 0.105 \sqrt{x}$$

from the edge of the lake to the point of measurement downwind. For example, the maximum wave height observed in Lake Superior was 6.9 m with a fetch of 482 km (4.82×10^7cm), which agrees well with a theoretical maximum of 7.3 m by this relationship.

The waves described in the foregoing paragraph are "deep water" or "short" waves, in which wavelength is much less than water depth. Where this condition no longer holds, and the wavelength becomes more than 20 times the water depth, the wave is transformed into a "shallow water" or "long" wave, and the cycloid motions are transformed into a to-and-fro sloshing which extends to the bottom of the water column. Shallow water waves are no longer dispersive, because their speed is proportional to the square root of the water depth, rather than being determined by wavelength. Therefore, as deep water waves enter the shallow water, their velocity decreases with the decrease in the square root of depth, and there is a simultaneous reduction in wavelength (Iverson, 1952; Mason, 1952; Hutchinson, 1957). Wave height first decreases slightly, then increases markedly and becomes asymmetrical and unstable. The collapse of water over the front of this now asymmetrical wave is termed a *breaker*, and occurs in a spectrum of forms between two extreme types (Fig. 7–4). In a *plunging breaker* the forward face of the wave becomes convex and the crest curls over, but collapses with insufficient depth to complete a vortex. The crest of a *spilling breaker* collapses forward, spilling downward over the front of the wave.

Because in shallow water the wave action extends right to the bottom, sedimentation can be prevented in sufficiently shallow water, and movement may be too severe for most aquatic plants to grow. In lakes with high rates of sedimentation in the littoral zone, such as in calcareous hardwater lakes where marl banks or benches extend out from the shoreline for many meters, recently sedimented material constantly is resuspended and removed to deeper areas. As a result, in suitably shallow water of about a meter, marl

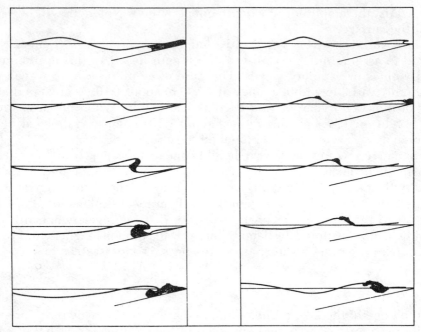

Figure 7–4 The breaking of waves on beaches. *Left*, Plunging breaker; *Right*, Spilling breaker. (From Hutchinson, G. E.: A Treatise on Limnology. Vol. 1, New York, John Wiley & Sons, Inc., 1957; after Iversen, 1952, and Mason, 1952.)

benches on the steep sides of these lakes do not increase in height. Thus, in a hardwater lake of northern Indiana, the surface sediments of a submersed lakemount island were found to be very old (Wetzel, 1970). The lakemount ceased accretion of sediments abruptly 1 m from the lake surface about 2740 years ago; more recent sediments were moved into the deeper parts of the lake.

Surface Currents

Currents are those nonperiodic water movements generated by external forces which include the frictional exertion of wind stress, changes in atmospheric pressure, horizontal density gradients caused by differential heating or by diffusion of dissolved materials from sediments, and the influx of water to a lake in relation to retention time and outflow. The latter subjects have been treated in other chapters, and therefore the following remarks are concerned primarily with wind-derived surface currents of open water masses.

Geostrophic effects of the deflecting (Coriolis) force of the earth's rotation are found in all currents of lakes of moderate to large size. Wind-driven circulation patterns tend to be cyclonic, and flow to the right or counterclockwise below a low pressure system in the Northern Hemisphere, and clockwise in the Southern Hemisphere. Beneath high pressure areas currents tend to be anticyclonic when currents flow clockwise in the Northern Hemisphere.

The combined effects of wind-induced acceleration of water and geostrophic deflection result in surface water movement downwind and to the right in the Northern Hemisphere. Under stratified conditions, there is also a general tendency for denser water to collect on the left side of the current and less dense water on the right side of the current.

The resulting wind-drift current is deflected 45° from the direction of the wind in a spiral manner (the Ekman spiral) in open waters of large, deep lakes. As the size and especially the depth of the lakes decrease, the magnitude of this angle also decreases until at lake depths below about 20 m, the angle of declination decreases and becomes insignificant in small lakes. However, in Lake Mendota, which is of moderate size (39.2 km²) and depth (25.6 m), a rotation with depth of wind-driven surface currents had an average angle of deflection of 20.6° to the right of the wind (Shulman and Bryson, 1961). The depth of frictional influence in Lake Mendota was found to be between 2 and 3.5 m. Stress exerted by the wind on the water surface was found to be a linear function of wind speed.

The wind factor, the ratio of water velocity of surface currents to wind velocity, is quite variable. In general, the velocity of wind-driven currents is about 2 per cent of the speed of the wind generating them, and is largely independent of the height of the surface waves. An average wind factor of 1.3 per cent for wind-driven currents in the upper half-meter of water was found in Lake Mendota at low wind velocities (Haines and Bryson, 1961). Water velocity in the surface layers increases with wind velocity until a critical wind speed is reached. Beyond the critical speed, surface velocity decreases, and the wind factor is nonlinear at higher velocities. In Lake Mendota the critical wind speed was 5.7 to 6.1 m sec^{-1}.

The surface currents of very large lakes tend to circulate in very large swirls or *gyrals*. For example, the normal summer surface circulation pattern of Lake Michigan, typically exposed to predominately westerly winds, is approximated in the measurements for the month of June shown in Figure 7-5. The large conspicuous gyrals are clearly influenced by geostrophic rotational forces (Noble, 1967), but are complicated by numerous other water movements related to standing water motions (see later discussion). Inertial currents on a smaller scale also occur in all of the Laurentian Great Lakes, not only at the surface, but at all depths and in all seasons (Verber, 1964, 1966). The geostrophic righthand acceleration of currents is also found under ice-cover. Many complex flow patterns have been observed: straight-line flow, sinusoidal or oscillatory, repeated crescent, circular or spiral, and rotary or screw flows. These gyrals of lakes the size of Lake Michigan are not simply the result of geostrophic forces; their direction is strongly modified by large long waves and especially by shifts in strong winds from another direction over an extended period.

Langmuir Circulation and Streaks

The turbulent energy of surface movement of the water and wave action is often sporadic and independent of wave direction, such as in the disper-

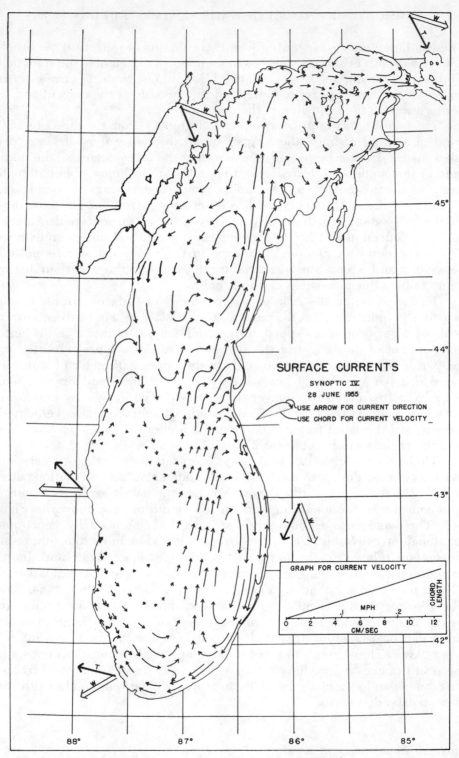

Figure 7–5 An example of typical summer surface current patterns of Lake Michigan (28 June 1955) with prevailing westerly winds. W = prevailing wind direction; T = direction of current water transport. (From Ayers, J. C., *et al.*: Currents and water masses of Lake Michigan. Publications Gt. Lakes Res. Inst. U. of Michigan, No. 3, 1958.)

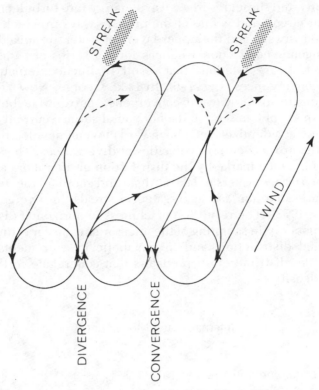

Figure 7–6 Diagrammatic representation of the helical flow of Langmuir currents in surface waters with aggregation of organic matter between streaks at the lines of divergence.

sion of wave energy in the mixing of the epilimnion when the lake is strati-fied, or in the mixing of the general water mass when the lake is not. Lang-muir (1938) demonstrated that under some circumstances the motions in-duced by turbulent transport are organized into vertical helical currents in the upper layers of lakes (Fig. 7–6). Convection from this vertical motion generates streaks, which are oriented approximately parallel to the direction of the wind and coincide with lines of surface convergence and downward movement. Between the streaks, which are marked by windrows of aggre-gated particulate and surface-active materials, are zones of upwelling. Such Langmuir circulation is seen on all waveswept lakes of any significant size. Convection from this source is active at wind speeds above 2 to 3 m sec^{-1}.

The mechanisms responsible for the generation of Langmuir circulation in lakes have been the subject of much debate (Stewart and Schmitt, 1968; Myers, 1969; Scott, et al., 1969; Faller, 1971). It is evident from Faller's (1969) work that the primary mechanism of Langmuir circulation most likely is associated with the vertical fluctuating oscillations produced by surface waves. When wind speed exceeds 2 to 3 m sec^{-1}, eddy pressure of the waves apparently resulting from the exponential decay of vertical oscillations that are maximum near the surface generates an instability that is immediately dispersed downward and begins a convection pattern. In this manner, wind

energy is converted through surface wave energy into turbulence. The resulting Langmuir circulation is one of the primary ways in which turbulence is
transported downward and the upper layers of water are mixed.

The Langmuir convectional helices form a series of parallel clockwise
and counterclockwise rotations that result in linear alternations of divergences and convergences (Fig. 7–6). In Lake George, New York, Langmuir
found the velocity of downward convergence currents to be 1.6 cm sec^{-1}
with a wind speed of 6 m sec^{-1} and a horizontal surface current of 15 cm sec^{-1}
and 10 cm sec^{-1} at a depth of 3 m. Downward movement velocities were about
three times the upward current velocities at divergences. These currents are
sufficient to influence markedly the distribution of microflora and fauna suspended in the surface waters of lakes. Algae and zooplankton, with limited or
no powers of locomotion, are aggregated in streaks in divergences (George
and Edwards, 1973). The result is a patchiness in horizontal distribution that
is important not only in sampling of the microorganisms for estimates of population size and distribution, and in metabolic measurements, but also in
determining the distribution of predators that congregate in these zones of
higher prey density.

Internal Water Movements

Metalimnetic Entrainment in Stratified Lakes

Transmission of epilimnetic turbulence of surface waves and Langmuir
circulation into the metalimnion and hypolimnion is suppressed by the stability barrier constituted by the density gradient in the metalimnion. As the
wind blows over the surface for a reasonable period of time, wind drift causes
water to pile up with a rise in surface level at the lee end of the lake. Movement continues downward by gravity and, when it encounters the more dense
water of the metalimnion, it flows back in the other direction, opposing that of
the prevailing wind (Fig. 7–7). The thermocline, or plane separating the two
layers of different density (termed pycnocline by Hutchinson, 1957), is tilted
in the opposite direction. We will return to the vertical movement of the entire water mass in the following section, after discussing the circulation pattern.

Progressive erosion of the lower layers occurs by *entrainment,* with a
corresponding lowering of the thermocline level (Mortimer, 1961). This
erosion is characteristic of the autumnal cooling phase, which begins when
the net heat flux across the water surface changes from downward to upward,
and which ends in the fall overturn. During the early stages (Fig. 7–7 A),
when the return current above the thermocline and the entrained layers
below it both are moving upwind, shear turbulence in the metalimnion is
low. As long as R_i does not fall below the critical value, the thermocline remains a slippery surface with little mixing across it. If the shear at the thermocline increases ($R_i < 0.25$), flow becomes unstable and large vortices appear
in the return current at the downwind end (Fig. 7–7B,C) and are carried

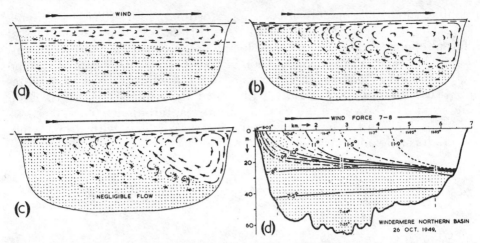

Figure 7-7   Stages of wind drift and circulation that led, after about 12 hours of strong wind, to the thermal situation depicted in (*d*), Lake Windermere, England, in late fall. Broken lines show equilibrium levels of water surface in (*a*), (*b*), and (*c*), and of the thermocline in (*a*). The initial layer below the thermocline is stippled; speed and direction of flow are roughly indicated by arrows. (From Mortimer, C. H.: Verhandlungen Int. Ver. Limnol., 14:81, 1961.)

upwind, where they become mixed into the surface drift. This process increases the density gradient at the downwind end to a point where it offsets the local shear to produce neutral turbulence stability, with R_i near 0.25. The resultant whole-basin disposition of isotherms is therefore fan-shaped (Fig. 7-7D), and corresponding large differences are produced in the shape of the metalimnetic density profile in different regions of the basin.

As the wind is reduced in a subsequent calm, the displaced layers slide over each other to redistribute into a new equilibrium. The oscillations so created may or may not be sufficient to create internal instability and waves. The resulting thermal profile retains much of its original form, but the mean metalimnion depth is pushed to a somewhat lower level. It is important to emphasize that this type of internal water motion is not continuous. The entrainment and metalimnetic erosion are progressive and forced by storms. Erosion can be counteracted by calm periods of heating, but in the fall when cooling exceeds heating, the erosion continues in step-wise progression with storms. A strong inverse relationship occurs between the magnitude of entrainment and the mean density gradient across the metalimnion, and a strong positive correlation exists between entrainment and basin dimensions, such as the mean depth of the basin (Blanton, 1973).

The entrainment rates are obviously of major importance in the control of the extent of heat intrusion into and nutrient return from the hypolimnion to the epilimnion in late summer and fall. A certain degree of compensatory motion occurs in the hypolimnion in opposing directions to that in the metalimnion. Turbulence in the hypolimnion is very much weaker than in the overlying layers, but increases as the basin dimensions increase where space is sufficient for their development.

The surface and epilimnetic movements and the effects of internal metalimnetic and hypolimnetic currents can be demonstrated very clearly in simple laboratory models of lakes (cf. Mortimer, 1951, 1954; Vallentyne, 1967). These models, stratified either thermally or by tripartite solutions of different chemical densities, are so effective in demonstrating these motions, seiches, and internal waves (discussion below) that their use is strongly urged in all introductory courses. For example, the upwelling areas of the epilimnion (Fig. 7–7) are seen clearly with tracer dyes in these models. Even more strikingly shown are the vortices of internal turbulence (curved arrows of Fig. 7–7B,C) on the upper downwind interface which, with strong wind, become unstable and break as waves.

Water Movements Affecting the Whole Lake

Long Standing Waves

Displacement of the water mass of the whole lake, as illustrated in Figure 7–7, gives rise to rhythmic motions of the whole basin involving both oscillations of the water surface and internal oscillations of isotherm depth. These motions take the form of very long waves, which have wavelengths of the same order as basin dimensions and which are reflected at the basin boundaries to combine into standing wave patterns. The water surface or the thermocline oscillates up and down like a seesaw about a line of no vertical motion (a node, Fig. 7–8), which coincides with regions of maximum to-and-

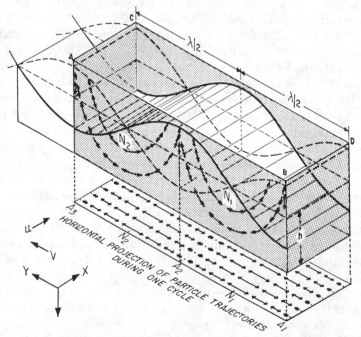

Figure 7–8 A long standing wave, without rotation, in a uniform depth (h) model. (From Mortimer, C. H.: Mitteilungen Int. Ver. Limnol., 20:157, 1974.)

fro motion of the water masses. Such standing waves, surface or internal, are referred to as *seiches*, a term originally referring to the periodic 'drying' exposure of shallow littoral zones. Figure 7–8 illustrates a constant depth model with two nodes, beneath which the horizontal to-and-fro sloshing of the whole water mass is at a maximum. Seiche motion, particularly the internal variety, has far-reaching effects on the vertical excursion of water masses, much larger than observed at the surface, and on their horizontal motions in the form of alternating currents.

The most common cause of seiches is the wind-induced tilting of the water surface and of the thermocline, as described in the previous section. When the wind stress is removed, the tilted water surface and the tilted thermocline swing back towards equilibrium. Momentum, however, is great and equilibrium is overshot (Fig. 7–8), resulting in a rocking motion about a *nodal point*. No vertical movement occurs at the nodal point, whereas maximum vertical movement takes place at the ends or *antinodes*. The nodal point is in reality a nodal line in this nonrotating rectangular situation running the width of the basin.

Surface Seiches.   The most conspicuous example of long standing waves is the *surface seiche*, which conforms to one or more resonant oscillatory frequencies (free modes) of a particular basin. The free mode of the surface seiche is a barotropic or surface wave, which affects the motion of the entire water mass of the lake, whether stratified or not, and attains its maximum amplitude at the surface. The surface seiche contrasts with the baroclinic or internal seiche, discussed further on, which is associated with the density gradient in stratified basins and attains its maximum amplitude at or near the thermocline.

The periodicity of vertical movement at the antinodes is a function of the length and depth of the basin. Particles at the node move back and forth horizontally with the same oscillatory rhythm as the vertical motion. This horizontal oscillation at the nodal line is large when the basin is very long in relation to depth, even though the vertical amplitude of the seiche is small at the antinodes. In a simple rectangular situation, analogous to many lakes in which the length greatly exceeds the mean depth, the period (t) of the uninodal surface oscillation is approximated by:

$$t = \frac{2l}{\sqrt{gh}}$$

where:

l = length of the basin at the surface
h = mean depth of the basin
g = acceleration of gravity (980.6 cm sec^{-1}).

Once the surface seiches are set into motion, frictional damping of the oscillations begins. The magnitude of frictional damping varies with water depth and complexity of basin shape. In deep lakes of uncomplicated shapes, for example, weakly damped oscillations may persist with slowly diminishing amplitudes long after the impact of the storm that set them into motion has

TABLE 7-1   Periodicity of Longitudinal Seiches[a]

		Observed Period, min				Observed Period, min
Loch Earn,				Lake Vetter,		
Scotland	t_1	14.52		Sweden	t_1	179.0
	t_2	8.09			t_2	97.5
	t_3	6.01			t_3	80.7
	t_4	3.99			t_4	57.9
	t_5	3.54			t_5	48.1
	t_6	2.88			t_6	42.6

Lake	Length (km)	$\bar{z}$ (m)	First Mode Period (t_1) (min)
Altausseer, Austria	2.8	34.6	5.3
Garry, Scotland	6	15.3	10.5
Mendota, Wisconsin			
NS axis	6.8	12.0	25.6
E axis	9.1	12.8	25.8
Ness, Scotland	38.6	132	31.5
Constance, Germany	66	90	55.8
Huron, USA-Canada	444	76	400
Erie, USA—Canada	400	21	786

[a]Compiled from data cited in Hutchinson, 1957, from various sources, and from Mortimer, personal communication.

passed. A few examples in Table 7-1 illustrate the periods of the surface seiche modes, which closely follow values calculated from hydrodynamic theory. In most lakes, the period of the second mode (t_2) is more than half of that of the first mode (t_1).

Transverse seiches are also found in long, narrow lakes with periodicities and amplitudes much less than those of longitudinal seiches, running the length of the longest axis—for example, the conspicuous transverse seiche of 132 min period in Lake Michigan (Mortimer, 1965). Where lake length differs little from breadth, complicated seiches can result from several wind directions and circular rotational movements.

Uninodal seiches, as described earlier, are common even in very large lakes. If pressure is exerted on the surface in the center of a basin and released, or periodically exerted and released, binodal surface seiches are generated. Multinodal seiches, with up to 17 nodes, have been observed.

The amplitudes of surface seiches are generally rather small in comparison to internal seiches. For example, in Lake Mendota, the amplitude is 1 to 2 mm with a periodicity of 25.8 minutes. In larger Lake Geneva, Switzerland, the maximum observed surface seiche amplitude was 1.87 m (t = 73 min), although the amplitude damped gradually in two weeks to less than 10 cm. In still larger lakes, such as Lake Erie, where amplitudes of surface seiches can exceed 2 m (t = 14 hours), the effects can have both destructive shoreline results as well as certain beneficial results, e.g., flushing of river delta and harbor areas. Surface seiches can occur under ice-cover in similar ways.

True lunar tides, the result of gravitational attraction of the moon and sun, have been poorly investigated in lakes. However, it is certain that true tidal movement, even in the largest lakes, is small (a few millimeters). The maximum observed amplitudes in Lake Baikal, USSR, are less than 15 mm and in Lake Superior 20 mm (Hutchinson, 1957). All observations are about one-half or less of the theoretical. The elastic yielding of the earth to the tide-generating forces presumably accounts for the difference (Mortimer, 1965).

Internal Seiches.   We have seen that the amplitude of surface seiches is relatively small, even in large lakes, and that their periodicity increases with basin dimensions. When the lake is stratified, the layers of differing density oscillate relative to one another. Most conspicuous is the successive oscillation of the metalimnion (Fig. 7–9), in which both the period and

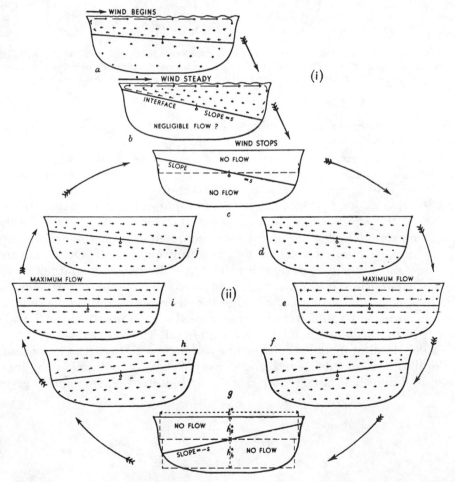

Figure 7–9   Movement caused by (*i*) wind stress and (*ii*) a subsequent internal seiche in a hypothetical two-layered lake, neglecting friction. Direction and velocity of flow are approximately indicated by arrows. δ = nodal section. (From Mortimer, C. H.: Proc. Royal Soc. London, Series B, 236:355, 1952.)

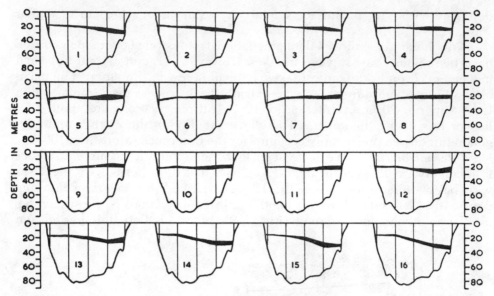

Figure 7–10 Successive hourly positions (0100 to 1600 hours, 9 August 1911) of the metalimnion bounded by the 9° and 11°C isotherms on a longitudinal section of Loch Earn, Scotland, (From Mortimer, C. H.: Schweiz. Zeitschrift f. Hydrologie, 15:94, 1953, after Wedderburn.)

amplitude of this internal standing wave or *internal seiche* are much greater than those of the surface seiche (Fig. 7–10).

In a simple rectangular, trough-like lake or lake model, without considering rotation for the time being, one observes the resultant horizontal flow depicted in Figure 7–9. Horizontal flow in this oversimplified stratified situation of only an epilimnion and hypolimnion of differing densities is maximal at the equilibrium point (node) of the oscillation; it ceases at the point of maximum deflection. Because of the much greater water movement associated with internal seiches, the resulting currents that rhythmically flow back and forth in opposing directions are the major deepwater movements of lakes. They lead to vertical and horizontal transport of heat and dissolved substances and significantly alter the distribution and productivity of phytoplankton and zooplankton, either directly or indirectly by means of changes in thermal and chemical stratification (cf. Thomas, 1951).

In the simplest model consisting of a rectangular basin with a homogeneous "epilimnion" of thickness z_e and density d_e and "hypolimnion" of thickness z_h and density d_h, the period (t) of the first internal seiche mode is given by

$$t = 2l \Big/ \sqrt{\frac{g\,(d_h - d_e)}{\dfrac{d_h}{z_h} + \dfrac{d_e}{z_e}}}$$

where:

$\quad$ l = length of the basin

$\quad$ g = acceleration due to gravity (980.6 cm sec^{-1}).

TABLE 7-2   Periodicity of Uninodal Metalimnetic Seiches of Several Lakes[a]

Lake	Dates	Length (km)	Calculated Period, Hours	Observed Period, Hours
Lunzer Untersee, Austria	Aug. 1927	1.51	4.0	3.7
Windermere, England				
Northern Basin	Aug.–Sept. 1951	6.5	14.4	12–14
Southern Basin	June–July 1951	8.9	23.1	23–25
Loch Earn, Scotland	Aug. 1911	9.6	17.2	16
Madüsee, Germany	July–Aug. 1910	13.8	27.3	25
Loch Ness, Scotland	Sept. 1904	37.0	62.4	60
Geneva, Switzerland	July–Aug. 1942	73	96	72–108
Baikal, USSR	Sept. 1914	675	1848	912

[a]From data cited in Mortimer, 1953, from various sources.

It will be noted that this equation is simply an extension of that for the periodicity of a surface seiche in which the density of the upper medium, air, was neglected as very small in comparison to that of water. Although more elaborate formulae have been developed for continuous density gradients in stratified lakes, the above formula yields reasonable estimates of periodicity, often very closely aligned with empirical observations in lakes and lake models.

The detailed analyses of internal seiches by Mortimer (1952, 1953) permit a few generalizations (Table 7–2). Of all the internal waves that are theoretically possible in stratified lakes, that most commonly set in motion is a uninodal seiche on the metalimnetic boundary (thermocline). In basins of fairly regular shape ranging (Table 7–2) from 1.5 km (Lunzer Untersee) to 74 km (Lake Geneva) in length, the uninodal internal seiche along the medial axis always appears as the main resonance. In basins possessing topographic features which may impede the uninodal internal seiche, such as constrictions, or shallows as in the southern end of Lake Windermere, the observed period is appreciably longer than that calculated. Multinodal internal seiches form a dominant type of resonance in very large lakes, especially when forced and damped by wind and other short-period disturbances.

Geostrophic Effects on Seiches

Coordinates which appear fixed to an earthbound observer are in fact rotating relative to coordinates in space (Mortimer, 1955). Inertia causes a moving water mass to attempt to follow a straight line in space, and the resultant track appears curved to the earthbound observer. He can account for the curvature by postulating a small force, the Coriolis force, directed at right angles to the line of motion. This force, which produces a deflection to the right in the Northern Hemisphere and to the left in the Southern Hemisphere, is zero at the Equator and maximum at the poles; it is a function of latitude and is proportional to the speed of the current. In basins sufficiently large for side constraint to be negligible, and in which friction is small, a water mass, once set in motion, follows a circular track, the "inertial circle." The time taken to complete the circle depends only on latitude. If the basin is stratified, this motion may be associated with internal waves, the period of which is referred to as the period of the inertia oscillation. For example, in Loch Ness, Scotland, at a latitude of 57° 16′ in which the velocity of a water mass was 10 cm sec^{-1}, the radius of the circle would be 810 m with a period of the inertia oscillation of 14.2 hours (Mortimer, 1955).

Motion in the inertia circle will only be encountered in basins whose width is many times the radius of the circle. In lakes whose width is of the same order or less than the radius, side constraints may be expected to modify the movement. If the lakes are stratified and conform to the simplified two-layered situation in Figure 7–9, then the flow depicted in that figure will be deflected by the Coriolis force onto the righthand shore (Northern Hemisphere) in the manner indicated in Figure 7–11. This rotating counterpart of the nonrotating conditions of Figure 7–9 has been demonstrated very clearly in Loch Ness (Mortimer, 1955), and subsequently in Lake Michigan and other large lakes. In addition to the morphometric parameters of the basin, important control factors include the timing of wind stresses, the time required to tilt the thermocline plane, and the period of the internal seiche.

The amplitude of internal seiches and the currents generated by them are of major importance. For example, in Lunzer Untersee with a fetch of 1.6 km, an internal seiche with a period of 4 hours had an antinodal amplitude of 1 m and generated maximal horizontal currents of 1 cm sec^{-1} at the node. Similarly, currents of 1 cm sec^{-1} were found in Lake Mendota. In larger lakes, for example at the ends of the long basin of Loch Ness and in Lake Michigan, amplitudes of > 10 m are common with horizontal currents of > 10 cm sec^{-1} near the nodes.

Internal Progressive Waves

The horizontal water movements associated with shearing flow at the metalimnetic interfaces can generate large *internal progressive waves* if the Richardson number falls below the 0.25 stability criterion (see Fig. 7–2). These internal waves on the thermocline are roughly an order of magnitude or more larger than waves found on the surface of large lakes. Examples include:

	Period	Mean Amplitude
Lake Mendota		
(Bryson and Ragotzkie, 1960)	1.5–7.9 min	0.15 m (max 1.0 m)
Lake Michigan	3–5 min	1.03 m (max 6 m)
(Mortimer, McNaught,	7 min	0.6 m
and Stewart, 1968)	10 min	0.3 m

These internal progressive waves propagate and break much the same as surface waves do (Bryson and Ragotzkie, 1960). Again, these wave movements can be demonstrated dramatically in simple lake models with dyed strata of differing densities (thermally or by salinity) (cf. for example the photographs

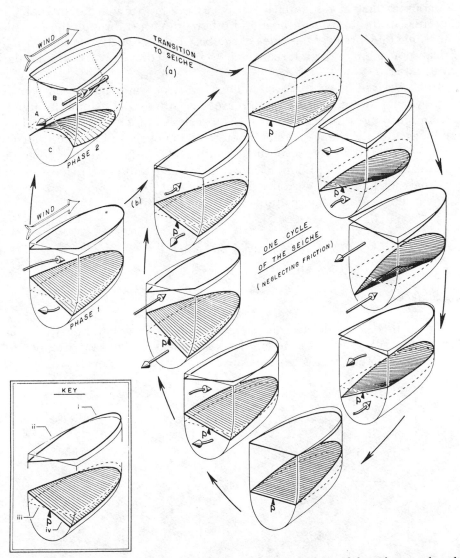

Figure 7–11 An internal seiche in a rotating two-layered lake model. In the inset key diagram, *i*, the oscillating lake surface is shown by a heavy line (this is the surface signature of the internal seiche mode, i.e., is *not* a surface seiche mode); *ii*, the equilibrium lake surface position is shown by a thin line; *iii*, the equilibrium interface position is shown by a broken line; and *iv*, the oscillating interface is shown by a shaded surface.

The two diagrams at upper left illustrate a hypothetical distribution of the layers during the application of the wind stress, which set the seiche in motion. *B* and *A* respectively indicate the wind-driven surface and return currents in the upper layer, both deflected to the right by the Coriolis force. *C* indicates the lower layer, the greater part of which has become displaced out of the half-basin shown.

Eight phases of one oscillation cycle of the first mode internal seiche are shown. Directions of flow in the upper and lower layers are shown by heavy arrows. The nodal point, *P*, the only point of zero elevation change, takes the place of the nodal line in the nonrotating model. Around this point, the internal seiche and its surface (out-of-phase) counterpart rotate counterclockwise. (From Mortimer, C. H.: Mitteilungen Int. Ver. Limnol., 20:169, 1974.)

of Mortimer, 1952). The turbulence associated with these internal waves, analogous to that at the surface but occurring on a much larger scale, are influential in the transfer of heat and other properties through the metalimnion. Since they break at the sides of the basin, their effects, coupled with the vertical movement of the internal seiche on which they move, are particularly significant.

In very large lakes, large circulatory cells of water movement can develop over the troughs of internal progressive waves and extend through the epilimnion to the surface. The result is a series of large, widely spaced convergences and divergences which operate analogously to Langmuir currents in aggregation and dispersion of suspensoids at the surface. Such aggregations are referred to as *slicks*, being of much larger size and more dispersed than the streaks of Langmuir currents. It is important to note, however, that the origins of these circulation patterns are quite different.

Other Water Movements

Long Surface and Internal Waves

Up to this point discussion has centered on long standing waves and short progressive waves. In the latter case, the wavelengths (λ), except at the beach, are less than water depth. Consequently, these short waves are dis-

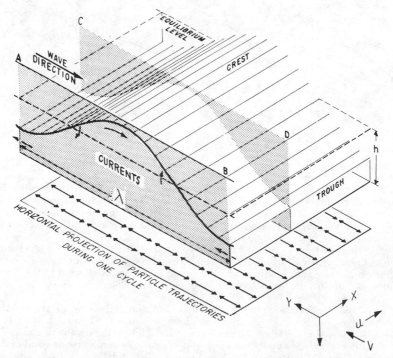

Figure 7–12 A long progressive wave, without rotation, in a uniform depth (h) model basin. (From Mortimer, C. H.: Mitteilungen Int. Ver. Limnol., 20:155, 1974.)

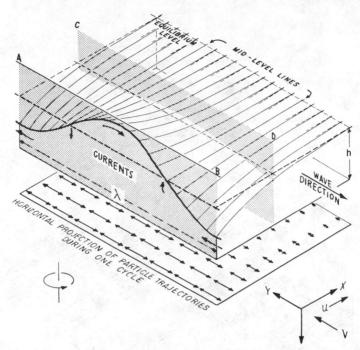

Figure 7–13 A long Kelvin wave in a uniform depth model basin rotating counterclockwise, as in the Northern Hemisphere, about a vertical axis. (From Mortimer, C. H.: Mitteilungen Int. Ver. Limnol., 20:167, 1974.)

persive, i.e., they travel at speeds in proportion to $\lambda^{1/2}$ (Mortimer, 1963, 1974). In contrast, *long waves,* with λ long in comparison to and much greater than the water depth, are nondispersive and travel at a speed independent of wavelength. A nonrotating model of such a wave, travelling in water of uniform depth, is illustrated in Figure 7–12. Its speed of progression is gh (in which g is the acceleration of gravity and h is the water depth), and the wave crests are horizontal. A combination of two such progressive waves, of equal amplitude and travelling in opposite directions, yields a model of the long standing waves (seiches) described earlier (Fig. 7–8).

The nonrotating models represented by Figures 7–8 and 7–12 provide simplified descriptions of events in small lakes. When the basin dimensions exceed about 15 km, the geostrophic effects of the earth's rotation must be introduced, and two separate model components emerge to provide the simplest interpretations of long surface and internal waves observed in large basins (Mortimer, 1963, 1974). Both model components, applicable to channels and basins of constant depth, were developed many years ago in connection with tidal theory and are named after the mathematicians concerned. The first component, associated with a shoreline at AB, is illustrated in Figure 7–13. In the Kelvin wave model, the rightward deflecting of the earth's rotation (Coriolis force) is everywhere exactly balanced by the components of gravity along a wave crest which slopes downward from right to left in the direction of progress (Northern Hemisphere), so that the righthand shoreline AB coincides with maximum wave amplitude. This condition of exact can-

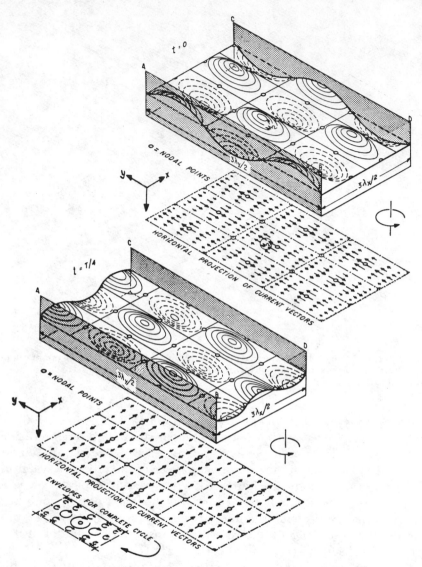

Figure 7–14 A standing Poincaré wave in which two phases, separated by ¼ cycle, of the oscillation are shown for the cross-channel trinodal case, with a ratio of long-channel to cross-channel wavelengths of 2:1. Current vectors rotate clockwise and attain their maximum values in the center of the cells. The two phases of the figure show the oscillations separated by a quarter period in which, in addition to the clockwise rotation of the vectors, the current directions of one cell remain approximately parallel and opposite in direction to those in neighboring cells. (From Mortimer, C. H.: Mitteilungen Int. Ver. Limnol., 20:179, 1974.)

cellation of the Coriolis force can only be met by an exponential decay of wave amplitude and wave-associated currents along a line normal to the shore. The result is that the currents are shore parallel everywhere, which satisfies the boundary condition of zero onshore current component along AB. That boundary condition is also met in the second model component, the Poincaré wave illustrated in Figure 7–14, but in that case, the wave amplitude does not decrease exponentially away from the shore. Rather it undulates in a standing wave pattern both across and along the model channel or basin, yielding a cellular pattern of alternating rising and falling hills and valleys with a corresponding cellular pattern of wave-associated currents, their direction rotating clockwise once every wave cycle. In very large basins, for example Lake Michigan, the period of observed internal Poincaré waves, equivalent to the Figure 7–14 model, is a little less than the period of motion in the inertial circle. This period depends on latitude, and is 17.5 hours for the central part of Lake Michigan.

The important difference between the Kelvin wave and Poincaré wave components is that the latter extends with undiminished amplitude right across the channel or basin, whereas the former decreases in amplitude away from the shore and is thereby 'trapped' or constrained to travel along the shore. In actual basins we must speak of Kelvin-type and Poincaré-type waves, although the mathematical models illustrated in Figures 7–13 and 7–14 (in which the wave surfaces can be visualized either as a water surface or a metalimnetic interface surface) are valid only for a constant-depth condition, not met in natural lakes. Nevertheless, even though they are oversimplified, the models do provide useful interpretations of what is observed in large basins (Mortimer, 1963, 1974).

When the Kelvin wave model is applied to typical conditions of summer stratification in a lake of 'medium' or 'large' width, for example Lac Léman and Lake Michigan, with respective widths of the order of 10 and 100 inertial circle radii, Table 7–3 predicts that an internal Kelvin wave travelling along one shore will have decreased to about one-sixth amplitude on the other shore in Lac Léman and to negligible amplitude on the other shore of Lake Michigan. In the open water of the latter lake, therefore, the Poincaré component will dominate the wave pattern, with upwellings and Kelvin-type wave responses restricted to nearshore bands of some 20 km width, as in fact has been observed (Fig. 7–15). We therefore must visualize a transition in the

TABLE 7–3   Parameters of Single Kelvin Waves in Lac Léman and Lake Michigan Representative of Nearshore Conditions in Late Summer[a]

Lake Conditions	Lac Léman Surface	Lac Léman Internal	Lake Michigan Surface	Lake Michigan Internal
Inertial period (t_p) in hr	16.6		17.4	
Thickness of upper layer (m)		15		15
Thickness of lower layer (m)		85		60
Mean °C of upper layer		19°		20°
Mean °C of lower layer		6°		6°
Density difference × 10^{-3}		1.54		1.74
Mean width (km)	7.9	7.4	102	100
Wave velocity (km hr^{-1})	113	1.59	104	1.60
Offshore distance intervals over which wave amplitude is successively halved	(207)	2.9	(190)	3.1
Per cent of wave amplitude still remaining at opposite shore	98%	17%	70%	$<1 \times 10^{-7}$%

[a]After data cited in Mortimer, 1963.

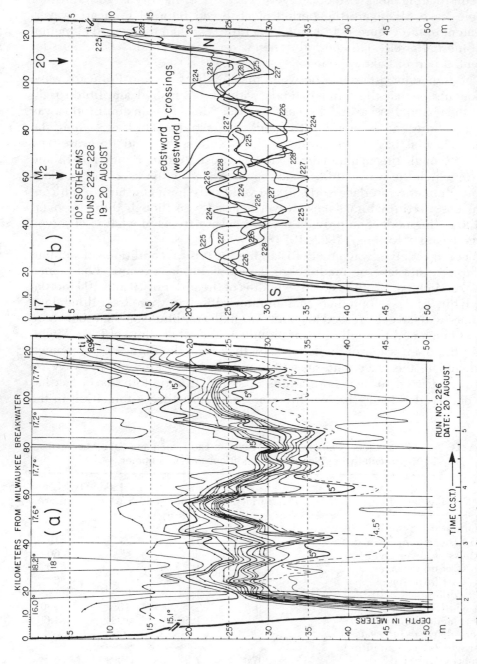

Figure 7–15 Lake Michigan, 19–22 August 1963. Distribution of the 10° isotherms, observed from the Milwaukee, Wisconsin-Muskegon, Michigan ferry, superimposed for 2 groups of 5 consecutive ferry crossings. (From Mortimer, C. H.: Mitteilungen Int. Ver. Limnol. 20:183, 1974.)

wave-induced current patterns from the cellular, clockwise-rotating, near-inertial periodicity of the offshore Poincaré-type waves in very large lakes, and the shore-parallel Kelvin-type currents which increase in amplitude as the shore is approached. This transition is sketched in Figure 7–16. In basins of Lac Léman size, on the other hand, it can be predicted that the internal Kelvin-type response will dominate. Metalimnetic oscillations with a counter-

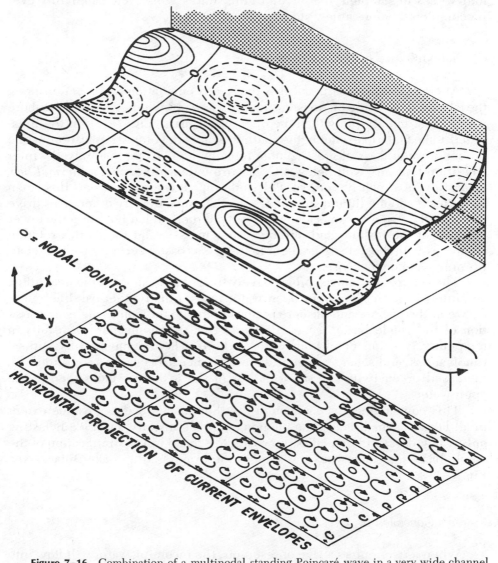

Figure 7–16 Combination of a multinodal standing Poincaré wave in a very wide channel model, of which only one side is shown, and a portion of a Kelvin wave travelling along the side. The elevation and current amplitudes associated with the Kelvin wave are at a maximum at the channel side (elevation amplitude, a), and decrease exponentially away from the side in the x-direction, falling to $1/e$ of the onshore value at a distance d_e from the side. The figure illustrates the transition, in current trajectories, from a nearshore pattern dominated by the Kelvin wave to a pattern dominated by the Poincaré wave at distances $>2d_e$ offshore. (From Mortimer, C. H.: Large-scale oscillatory motions and seasonal temperature changes in Lake Michigan and Lake Ohio. Special Report no. 12, Center for Great Lakes Studies, University of Wisconsin-Milwaukee, 1971.)

clockwise rotation of the internal wave motion in that lake can be interpreted as an internal Kelvin-type wave progressing counterclockwise around the basin in the manner illustrated in Figure 7–10, with a periodicity of a little over three days (Mortimer, 1963, 1974). As this usually coincides with the periodicity of storm passages, the oscillatory response of Lac Léman is strong. In Lake Michigan, by contrast, the internal Kelvin wave requires about four weeks to complete the circuit of the shores. This cycle can hardly ever be completed before a new storm disturbs its progress.

Circulation Caused by Thermal Bars

Another feature of large lakes is the steady circulation set in motion along the shore as a result of density gradients arising when shallow nearshore waters heat more rapidly than the open water mass (Fig. 7–17). In the illustrated Lake Ontario example, the shallow waters develop stratification, while the main water mass of great volume remains in its isothermal mixed winter condition typically well below 4°C. A narrow transition zone, a *thermal bar*, consisting of a nearly vertical 4°C isotherm, develops between this open water mass and the littoral stratified water. Thermal density differences drive downward flowing currents along the vertical thermal bar both into the lower portions of the inshore and offshore water masses, and the earth's rotation combines with the density gradient to induce a coastal current inside the bar, running counterclockwise.

The thermal bar moves progressively further from shore as the heat influx continues to warm the larger open water mass. Finally, thermal differences between the inshore and offshore regions lessen to the point where stratification of the whole basin occurs (Rodgers, 1966). This mechanism results in temporary isolation of inshore waters where, when augmented by surface runoff and river discharge, chemical enrichment can occur. As a consequence, increased productivity can occur earlier in the inshore regions than in the open water.

This convergence mechanism of the thermal bar occurs to some extent in all lakes. In small lakes, the phenomenon is transitory, most likely lasting only a few days. In large lakes, however, the transition to stratification of the whole basin may take weeks, as seen in the example from Lake Ontario (see Fig. 7–17).

Currents Generated by River Influents

When a river enters a lake or reservoir, the incoming water will flow into a density layer in the lake which is most similar to its density; this process is governed by temperature, dissolved material, and suspensoids. Depending on the density differences between the inflowing water and the lake water, three basic types of inflow water movements can result (Fig. 7–18). *Overflow* occurs when the inflow water density is less than the lake water density ($\rho_{in} < \rho$), and *underflow* results when the inflow water density is greater than

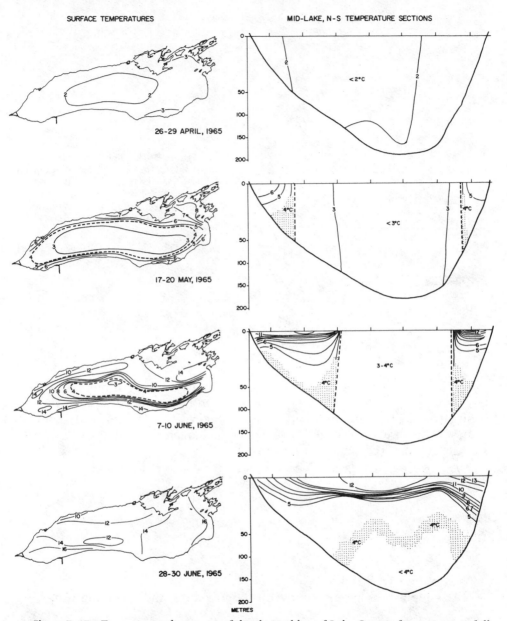

Figure 7–17 Formation and progress of the thermal bar of Lake Ontario from winter to full summer stratification. (From Rodgers, G. K.: Publications Gt. Lakes Res. Div. Univ. Mich., 15: 372, 1966.)

the lake water density ($\rho_{in} > \rho$). Where the density of inflow is greater than that of the epilimnion but less than that of the metalimnion or hypolimnion, flow enters in a plume at an intermediate depth, and *interflow* occurs ($\rho_{in} > \rho_1$, $\rho_{in} < \rho_2$).

A certain amount of turbulent entrainment and mixing nearly always, given adequate velocities, accompanies the influx of water into a lake. In

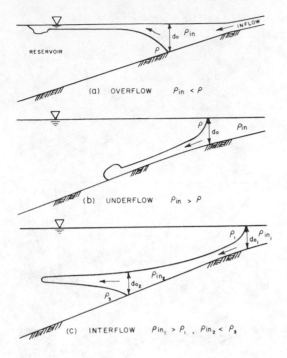

Figure 7-18 Types of inflow into lakes and reservoirs. (From Wunderlich, W. O.: The dynamics of density-stratified reservoirs. *In* Hall, G. E., ed., Reservoir Fisheries and Limnology. Washington, D.C., American Fisheries Society, 1971.)

many lakes, and especially in reservoirs, inflows enter through elongated bays which tend to inhibit lateral mixing. On the water's entrance into the basin, the depth of inflow increases and the velocity is gradually reduced until a critical section depth (d_o) is reached, characterized by the densimetric Froude number (F_o) (Wunderlich, 1971):

$$F_o = \frac{(v/A_o)}{\sqrt{g \,(\Delta \rho \,/\, \rho)\, d_o}}$$

where:
  v = flow rate
  A_o = cross sectional area at the critical section
  ρ = density of inflow
  $\Delta\rho$ = density difference between the incoming and receiving water.

Overflow or underflow initiates at this critical section, with a reduction of flow velocity that may be accompanied by deposition of suspended materials. A critical analysis of density differences and distribution of inflowing water in Lake Biwa, Japan, showed that the inflowing river waters do not mix readily if density differences are reasonably large, and flow occurs in complex overflow and interflow patterns (Morikawa, et al., 1959).

Variations in density differences between incoming water and those of the recipient water body are so great that generalizations are very difficult to make. Discharge also varies very widely seasonally, not only in volume but in the accompanying load of dissolved and suspended materials. For example, in alpine situations the density of river water is greater (cold, high dissolved load) than the recipient water and underflow is common, particularly if glacial erosion at the head of the river contributes to the suspended load. During the summer, although still cold, density is reduced by high-volume dilution of snow and ice melt runoff which interflows into the metalimnion. This situation has been documented frequently, but a particularly striking case is seen in the Bodensee, Germany, where light penetration is abruptly attenuated in the metalimnion over large areas of the lake by the intrusion of river water with a high suspended load (Lehn, 1965).

The extent of intrusion and current generation in the receiving lake is further a function of discharge volume of the river in relation to the volume of the lake. The theoretical retention time of a lake or reservoir, based on the relation of the total influent-outflow to the total volume of the lake, is realized only approximately in most lakes. Retention time varies with the dimensions and shape of the basin, seasonal rates of inflow, and stratification characteristics (Kajosaari, 1966). At high discharge rates, overflow and interflow of rivers may channel across or through the water mass of the stratified basin, whereas at lower discharge rates river water may penetrate more into the main water mass and may be mixed through more normal circulatory mechanisms. In larger lakes, geostrophic deflection of the incoming water intrusions, surface or interflows, is observed consistently.

Currents under Ice-Cover

It is clear that water of a lake under ice-cover, even lakes of closed basins with no appreciable inflow and outflow, exhibits considerable water movement by currents. Much of the evidence for the existence of these weak currents stems from careful direct measurements in which radioactive sodium (^{24}Na) was released at various points beneath the ice, and its short-term distribution was measured for several days. Slow horizontal currents, of velocities too low to be measured by mechanical current meters, have been observed in a small ice-covered lake in Wisconsin (Likens and Hasler, 1962) and in a large, permanently ice-covered lake in Antarctica (Ragotzkie and Likens, 1964). Similarly, radioactive tracers have been used to measure slow horizontal movements in a small lake of Nova Scotia (McCarter, et al., 1952), in a meromictic lake in Washington (Hutchinson, 1957), and in the monimolimnion of a small meromictic lake in Wisconsin (Likens and Hasler, 1960).

Horizontal displacements of radiosodium were found to be generally asymmetrical in a small ice-covered, closed basin bog lake (Fig. 7–19). Horizontal velocities of at least 10 m day^{-1} (42 cm hr^{-1}) were observed near the bottom at the deepest point of the basin, and maximum velocities of 15 to 20 m day^{-1}. The vertical component of this motion was between one and three orders of magnitude less than the horizontal component (Table 7–4). In further experiments Likens and Ragotzkie (1966) also found a definite rotary movement of horizontal currents, which was cyclonic near the center of the lake and anticyclonic near the perimeter (Fig. 7–20).

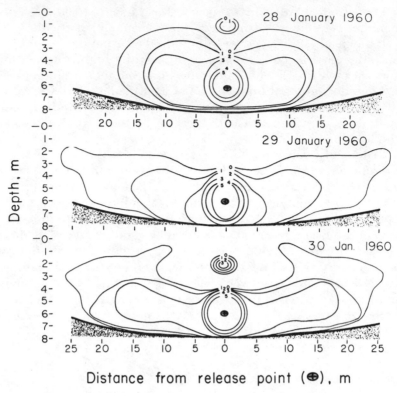

Distance from release point (⊕), m

Figure 7–19   Diagram of the horizontal and vertical dispersal of ^{24}Na in Tub Lake, Wisconsin, under ice-cover 24, 48, and 72 hours after release. Integers indicate the logarithmic value of the radioactivity concentration in counts min^{-1}. (From Likens, G. E., and Ragotzkie, R. A.: Journal of Geophysical Research, vol. 70, p. 2333, 1965. Copyright by American Geophysical Union.)

It is apparent from these detailed studies, as well as from studies of a larger lake in Swedish Lapland (Mortimer and Mackereth, 1958) that these movements are the result of convective cells of density currents generated primarily by relatively large quantities of heat flowing from the sediments. If the temperature of the water into which the heat flows is greater than 4°C, that flow will generate ascending buoyancy currents. If, however, the lake has cooled to well below 4°C before the ice-cover is established, the heat flux will increase the density of the water in contact with the sediments, causing it to flow as a density current down the sloping sides of the basin into the nearest depression. Chemical exchanges also occur between sediment and water, for example, oxygen uptake by the sediments. If the lake basin contains more than one topographically separated depression, the density flows may produce significantly different temperatures and chemical compositions in the bottom waters of these depressions (Mortimer and Mackereth, 1958).

Most of the heat of the sediments is gained during the previous summer period and fall turnover (cf. Chapter 5) and is dissipated slowly. Birge et al. (1927) found that about 650 cal cm^{-2} were conducted from the sediment to the water of Lake Mendota during the period of ice-cover (110 days), a heat gain

TABLE 7–4 Vertical Motion in Ice-Covered Tub Lake, Wisconsin, Calculated from Measurements of ^{24}Na Dispersal (Negative values of *w* indicate downward motion and positive values indicate upward movement.)[a]

Depth	Velocity (w) in cm sec^{-1} × 10^{-2}		
(m)	28–30 Jan 1960	26–27 Jan 1961	27–28 Jan 1961
0.5	0	0	0
1.5	0	+0.0132	+0.0378
2.5	+0.0033	−0.0057	+0.0047
3.5	−0.0041	+0.0007	−0.0158
4.5	−0.0233	−0.0498	−0.0508
5.5	−0.0066	−0.0497	−0.0219
6.5	+0.0381	−0.0583	−0.1180
7.5	+0.0661	−0.6990	−1.55

[a]From data given in Likens, G. E., and Ragotzkie, R. A.: Journal of Geophysical Research, 70:2333–2344, 1965. Copyright by American Geophysical Union.

of 5.9 cal cm^{-2} day^{-1}. Thermal gains of 3 to 4 cal cm^{-2} day^{-1} are common in small temperate lakes during the winter. That of Tub Lake was about 3 cal cm^{-2} day^{-1}, and of larger Lake Torneträsk, Swedish Lapland, about 1.0 cal cm^{-2} day^{-1}. This heat source is adequate to explain the observed currents. Biological oxidation in respiration in the sediments represents a caloric equivalent of about 0.04° cm^{-3}, which is totally inadequate as a heat source.

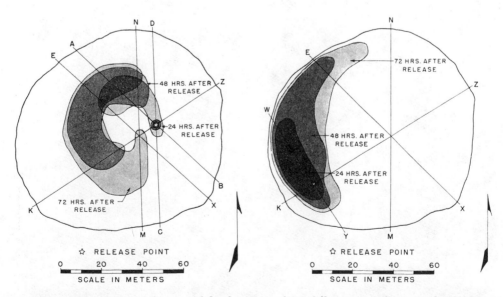

Figure 7–20 Maximum horizontal displacement of ^{24}Na following its release at a depth of 5 m (*right*), and of ^{131}I following its release at a depth of 3 m (*left*) in Tub Lake, Wisconsin, January 1962. (From Likens, G. E., and Ragotzkie, R. A.: Verhandlungen Int. Ver. Limnol., 16:126, 1966.)

Similarly, excess of solar radiation at the lake surface over heat losses is appreciable only at the surface and in late spring; it is absorbed primarily in melting of ice and snow. Geothermal heat flow (ca 1.23×10^{-6} cal cm^{-2} sec^{-1}) from the interior of the earth represents only about 0.1 cal cm^{-2} day^{-1} through the sediments of the lake. Groundwater that is relatively warm and circulating near the basin probably transmits some heat through the sediments to the lake water. In Wisconsin, this potential heat source was estimated to be about 1 cal cm^{-2} day^{-1}, a significant source of thermal conductivity in addition to that from heat storage from the previous summer and transfer in winter. Thus diffusion from the sediments of heat which had entered the lake during the previous summer is the major operator of thermally-induced convection currents under winter conditions.

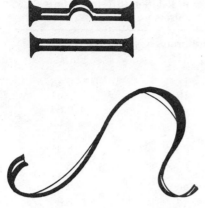

oxygen

THE OXYGEN CONTENT OF FRESH WATERS

Oxygen is the most fundamental parameter of lakes, aside from water itself. Dissolved oxygen is obviously essential to the metabolism of all aquatic organisms that possess aerobic respiratory biochemistry. Hence, the properties of solubility, and especially the dynamics of oxygen distribution in lakes, are basic to the understanding of the distribution, behavior, and physiological growth of aquatic organisms.

The control mechanisms of dissolved oxygen concentrations are important in relation to their availability for utilization by the lake biota. The dynamics of oxygen metabolism are, however, only one facet of the oxygen content of fresh waters. The sources of dissolved oxygen from the atmosphere and from photosynthetic inputs and the hydromechanical distribution of oxygen are counterbalanced by consumptive metabolism. The rates of oxygen utilization in relation to synthesis permit an effective evaluation of the metabolism of the lake as a whole.

The resulting distribution of oxygen affects strongly the solubility of many inorganic nutrients. The timing of changes of nutrient availability is governed by seasonal shifts from an aerobic to an anaerobic environment in regions of lakes. These changes in distribution of nutrients result in rapid growth of many organisms capable of taking advantage of changes in nutrient availability. Population responses may be temporary and transient. If, however, long-term changes in oxygen-regulated nutrient availability are sustained, the productivity of the entire lake can be radically altered and maintained. Although oxygen content is discussed here largely as a separate entity, its integration with the metabolism of all facets of biotic metabolism will become more apparent in subsequent discussions.

Solubility of Oxygen in Fresh Water

Air contains, by volume, about 20.95 per cent oxygen, the remainder being nitrogen, except for a very small percentage of other gases. Because oxygen is more soluble in water than nitrogen, the amount of oxygen dis-

TABLE 8–1   Solubility of Oxygen in Pure Water in Relation to Temperature
from Saturated Air at 760 mm Hg Pressure[a]

Temperature, °C	Oxygen mg 1^{-1}	Temperature, °C	Oxygen mg 1^{-1}
0	14.16	18	9.18
1	13.77	19	9.01
2	13.40	20	8.84
3	13.05	21	8.68
4	12.70	22	8.53
5	12.37	23	8.38
6	12.06	24	8.25
7	11.76	25	8.11
8	11.47	26	7.99
9	11.19	27	7.86
10	10.92	28	7.75
11	10.67	29	7.64
12	10.43	30	7.53
13	10.20	31	7.42
14	9.98	32	7.32
15	9.76	33	7.22
16	9.56	34	7.13
17	9.37	35	7.04

[a]From data of Truesdale, Downing, and Lowden, 1955.

solved in water from air is approximately 35 per cent and the remainder is largely nitrogen, even though their pressures are equal.

Solubility of oxygen is affected nonlinearly by temperature, and increases considerably in cold water. The general relationship of solubility to temperature is given in Table 8–1, although these values are subject to minor modifications as analytical assays improve (extensively discussed in Ohle, 1952; Mortimer, 1956; and Hutchinson, 1957). More detailed data are given in the references cited.

The solubility of gases in water is affected by pressure as well as by temperature. Therefore equilibrium of the oxygen of the atmosphere with the oxygen concentration in the water depends on the atmospheric partial pressure and hence the elevation of the surface of the lake. Usually, saturation is considered in relation to the pressure at the surface of the lake. The percentage saturation of oxygen in water has been extensively treated by Ricker (1934) and Mortimer (1956), both of whom have prepared nomograms to aid in such computations. The improved nomogram of Mortimer is included in this work (Appendix), although in routine analyses it is strongly recommended that the enlarged version, which is printed on material that does not change with humidity, be used.

The amount of a gas that will remain dissolved is influenced by the atmospheric pressure to which the lake is exposed at a given altitude, by meteorological conditions, and also by hydrostatic pressure exerted by the stratum of water overlying a particular depth. The amount of gas that can be held in water by the combined atmospheric and hydrostatic pressures at a particular depth is called the absolute saturation (Ricker, 1934). The actual pressure, P_z, at a given depth in atmospheres is equal to that at the surface, P_o, plus 0.0967 times the depth z in meters, or

$$P_z = P_o + 0.0967z.$$

The importance of this relationship lies in the amount of pressure required to prevent the formation of bubbles which rise and escape to the surface. Such bubbles are a composite of many gases, the most active of which is oxygen derived from photosynthesis and methane, resulting from anaerobic decomposition under certain conditions. The degree of supersaturation with oxygen necessary for bubble growth increases with depth. In still water, bubble growth may occur as a result of oxygen tension at depths above about one meter (Ramsey, 1962a). Below this depth, an unusually high degree of supersaturation is required, as well as an absence of turbulence or internal waves that may induce a depth displacement and change in pressure. Very small bubbles are stabilized by an organic adsorbed film, which, on rising to the surface, either rupture or accumulate on the surface as foam (Ramsey, 1962b). Below depths of 1 to 4 meters, depending upon the turbulence patterns, oxygen produced at these depths can remain dissolved because of the hydrostatic pressure. It can accumulate, several hundred per cent supersaturated relative to the pressure at the surface of the lake, and still be below its absolute saturation at depth.

Salinity also reduces the solubility of oxygen in water to an extent, and must be considered in analyses of oxygen in inland saline and brackish waters. Oxygen solubility declines exponentially with increases in salt content, and is reduced by about 20 per cent in normal sea water from the amount in fresh water. Revised tables and a nomogram for oxygen saturation in saline water are given in Green and Carritt (1967).

Although the methodology involved in the physical, chemical, and biological aspects of aquatic ecology is very complex and a full discussion would involve a work several times the length of this book, the importance of dissolved oxygen warrants a few comments. In spite of recent advances in electrode technology (Ohle, 1953; Golterman, 1969), the mainstay of oxygen analysis is the Winkler method, based on chemical fixation of the oxygen and colorimetric titration against reagents of known reaction with concentration. Experimenters should be aware of the numerous compounds interfering with precise measurements, such as iron, nitrates, and organic matter, that become significant in many waters, particularly in polluted waters. Modifications of the basic method are discussed in great detail in Welch (1948), *Standard Methods* (APHA, 1970), and Golterman (1969). In inland saline waters, other methods must be used (Walker, et al., 1970; Ellis and Kanamori, 1973).

Distribution of Dissolved Oxygen in Lakes

The diffusion of gases in water is a very slow process. For equilibrium of the oxygen of the atmosphere to be established within a reasonable period,

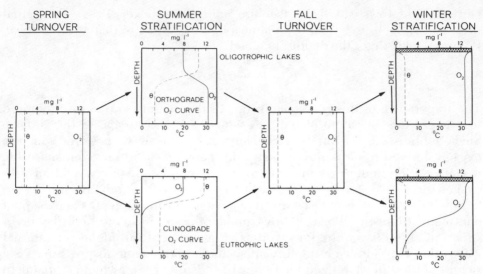

Figure 8–1 Idealized vertical distribution of oxygen concentrations and temperature during the four main seasonal phases of an oligotrophic and an eutrophic dimictic lake.

the water must circulate, such as occurs at periods of turnover or in the epilimnion of stratified lakes. If the dissolved oxygen concentrations at depth are not very far from saturation, equilibrium at prevailing temperatures and altitudinal pressure is established relatively quickly, usually in a matter of a few days. Very deep lakes, however, require longer periods for complete equilibrium saturation, which may or may not be achieved before thermal stratification effectively terminates circulation for a seasonal interval.

In an idealized lake, the oxygen concentration at vernal circulation is at or near 100 per cent saturation, which would be between 12 to 13 mg O_2 1^{-1} if occurring at about 4°C, and at an altitude near sea level (Fig. 8–1). This ideal concentration at 100 per cent saturation is based on physical control of diffusion, mixing, and saturation. Deviations are commonly found, most frequently in the form of slight supersaturation resulting from photosynthetic activity in excess of losses to the atmosphere, and subsequent reestablishment of equilibrium. Alternately, biochemical oxidations can result in slight undersaturation where consumption slightly exceeds the circulation equilibrium mechanisms.

Orthograde Oxygen Profile

Where the lake is very unproductive or *oligotrophic* (low in nutrients, with low organic production), the oxygen concentration with depth is regulated largely by physical means as summer stratification occurs. The oxygen concentration in the circulating epilimnion decreases as the temperature increases (Fig. 8–1); equilibrium is established with the atmosphere, and the solubility decreases with increasing temperature. With decreasing temperature in the metalimnion and hypolimnion, the oxygen concentrations increase, that is, they have remained in the hypolimnion essentially at saturation from the period of spring turnover just before summer stratification began. This oxygen profile has been termed (Åberg and Rodhe, 1942) *orthograde*. The important characteristic to note is that the percentage of saturation is more or less 100 per cent with increasing depth.

Clinograde Oxygen Profile

This idealized orthograde curve is found only in a few extremely unproductive lakes or in moderately oligotrophic lakes during the very early stages of summer stratification. Oxidative processes occur constantly in the hypolimnion, their intensity reflected by the amount of organic matter reaching the hypolimnion from the productive zones of the lake. As a result, the oxygen concentration of the hypolimnion becomes more reduced and undersaturated as the stratification season progresses. The oxygen content of the hypolimnion of very productive or *eutrophic* (high in nutrients, with high organic production) lakes is depleted rapidly by oxidative processes. The resulting curve, in which the hypolimnion is anaerobic, is termed *clinograde* (Fig. 8–1). The hypolimnetic oxygen content of highly eutrophic lakes is depleted often only a few weeks after summer stratification begins, and the hypolimnion remains anaerobic throughout this period.

The loss of oxygen from the hypolimnion results from the uptake during oxidation of organic matter, both in the water and especially at the sediment-water interface where bacterial decomposition increases greatly. Although plant and animal respiration can consume large, often catastrophic, amounts of dissolved oxygen, major consumption from the lake is associated with bacterial respiration in decomposition of sedimenting organic matter. Oxygen consumption in the free water by bacterial respiration is intensive at all levels, but in the hypolimnion it is not generally offset by renewal mechanisms of circulation and photosynthesis that occur in the epilimnion and metalimnion. The consumption is most intense at the sediment-water interface, at which accumulating organic matter and bacterial metabolism is greatest. The interface region becomes anaerobic during summer stratification very rapidly, and the hypolimnetic depletion usually is observed first at the bottom-most strata of the hypolimnion. Diffusion of oxygen into this depleted zone from overlying layers occurs slowly. Uniform horizontal distribution of the oxygen profile in the hypolimnion is assisted by vertical turbulence, horizontal translocations, and density currents that move along the basin sediments.

In addition to oxygen consumption by animals, plants, and especially by bacterial respiration in the open water, purely chemical oxidation of dissolved organic matter occurs. Lakes highly stained with humic organic compounds are frequently observed to be undersaturated even in epilimnetic strata. Although the mechanisms are not entirely clear, it is apparent that at least some of the oxygen uptake is the result of purely chemical oxidation or photochemical oxidation by ultraviolet light (Gjessing and Gjerdahl, 1970).

This chemical oxidation most likely is masked in very productive lakes by the intensive bacterial biochemical demands.

The relative importance of different mechanisms of hypolimnetic oxygen depletion varies with the individual lake system. In large, deep lakes bacterial respiration of organic matter of phytoplanktonic origin may dominate, and benthic decomposition then will play a minor role. In shallow lakes with relatively high inputs of organic matter from terrestrial and stream sources, benthic decomposition may dominate. And in highly stained, bog lakes that receive high inputs of dissolved humic compounds, chemical oxidation may assume greater significance. With the shift from an aerobic to anaerobic hypolimnion, a large, often major volume of the lake is excluded from habitation by most animals and many plants. Another major change is the necessary shift from aerobic to anaerobic bacterial metabolism, with a marked reduction in overall efficiency of decomposition of organic matter.

Fall overturn begins with the complete loss of summer stratification. At the terminal stages of stratification, with progressive deepening of the epilimnion, circulation of approximately oxygen-saturated water is carried deeper into the hypolimnion. When circulation is complete, oxygen concentration continues at saturation in accordance with solubility at existing temperatures (Fig. 8–1).

With the advent of ice formation, exchange of oxygen with the atmosphere ceases for all practical purposes. Based on solubility relationships, an oxygen concentration profile essentially constant at saturation in relation to temperature at depth would be expected. Such would be the case in an ultra-oligotrophic lake where biotic influences are minor (Fig. 8–1). These profiles are found very rarely in dimictic lakes. Much more frequently observed is a significant reduction in the oxygen concentration with depth, which is particularly acute near the sediment.

The mechanisms involved are several. In eutrophic lakes, synthesis of organic matter photosynthetically, although reduced, continues throughout the winter, and is often vigorous at the later stages of winter ice-cover. Light penetration is variable with changing conditions of ice- and snow-cover, but the photic zone generally is confined to the upper layers. Respiratory utilization and chemical oxidations increase with depth as during summer stratification, although at a lower rate because of the reduced temperatures. When the water of the main water mass under ice is colder than 4°C, as is commonly the case, water in the littoral areas can be heated slightly through the ice. This slightly more dense water will sink and flow in profile-bound density currents along the sediments to the deeper portions of the basin. Generally, movement is sufficiently slow so that en route, the oxygen of this water is reduced or depleted as it passes over the sediment-water interface. At the bottom of depressions it displaces water upward and results in a conspicuous depletion at the sediment-water interface, accentuating the decompositional utilization already occurring there.

The stages of the annual cycle discussed above can be collectively illustrated by isopleths of oxygen concentration and percentage saturation for a typical dimictic lake of moderate productivity, intermediate between extremes of oligotrophic and eutrophic conditions (Fig. 8–2). Noteworthy are

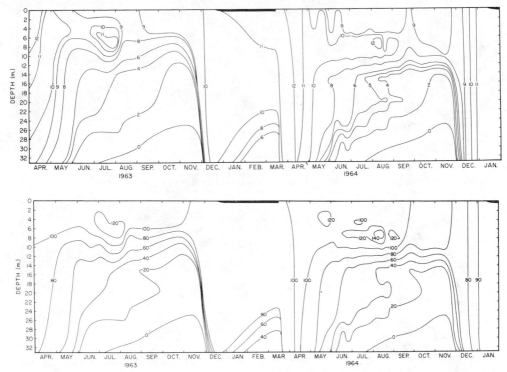

Figure 8–2 Depth-time diagram of isopleths of dissolved oxygen concentrations in mg l⁻¹ *(upper)* and percentage oxygen saturation *(lower)*, Crooked Lake, Noble-Whitley counties, northeastern Indiana. Ice-cover drawn to scale. (From Wetzel, unpublished data.)

the major points of high concentrations near 100 per cent saturation during the colder periods of spring and fall circulation. Epilimnetic concentrations of the warmer water are reduced, but at saturation. The high concentrations in the metalimnion are the result of active photosynthesis, and will be discussed in detail below. The slow, progressive reduction and final depletion of oxygen in the lower hypolimnion is seen in mid- and late summer. This same reduction, but not depletion, is seen under ice cover at the lower depths in Crooked Lake. More extreme changes are seen in the oxygen conditions of smaller eutrophic Little Crooked Lake (Fig. 8–3), in which the changes are similar to those of Crooked Lake but greater and more accelerated.

Variations in Oxygen Distributions

Metalimnetic Oxygen Maxima.

    Variations in the vertical and horizontal distributions of dissolved oxygen from those general conditions just discussed often are observed. The most common variation is an increase in oxygen in the metalimnion during stratification (Figs. 8–2, 8–3, 8–4); the metalimnetic maximum is referred to as

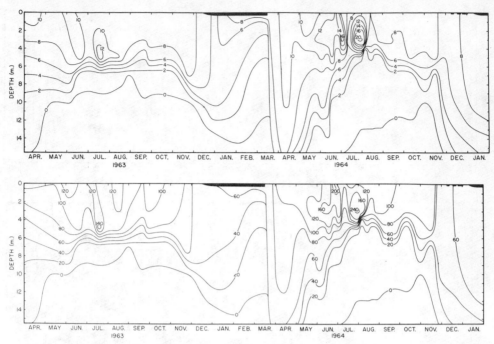

Figure 8–3 Depth-time diagram of isopleths of dissolved oxygen concentrations in mg l^{-1} *(upper)* and percentage oxygen saturation *(lower)*, Little Crooked Lake, Whitley County, northeastern Indiana. Ice-cover drawn to scale. (From Wetzel, unpublished data.)

a *positive heterograde curve* (Åberg and Rodhe, 1942). As the solubility of the epilimnion decreases with increasing summer temperatures, and oxygen consumption in the hypolimnion results in the typical clinograde reduction with depth, the result is an absolute oxygen maximum in the metalimnion which may be at or somewhat above saturation.

The metalimnetic oxygen maxima often are extremely pronounced, with supersaturated values above 200 per cent (Fig. 8–3). Concentrations of nearly 36 mg O_2 l^{-1} and 400 per cent saturation have been recorded (Birge and Juday, 1911). It is clear that the maxima are nearly always the result of oxygen produced by algal populations that develop more rapidly than they are lost from the zone of increased density by sinking (Fig. 8–4A). The algae are commonly stenothermal, adapted to growing well at low temperatures and light intensities, but have access to nutrient concentrations that are usually higher in the lower metalimnion than in the epilimnion. Blue-green algae, especially *Oscillatoria,* are often major contributors to this phenomenon (Eberly, 1959, 1963, 1964; Wetzel, 1966a, 1966b). The depth at which metalimnetic oxygen maxima occur is correlated directly with the transparency of the water (Thienemann, 1928; Yoshimura, 1935). In a majority of lakes they are found between 3 to 10 m, but in very clear lakes at depths of 50 m.

Extreme metalimnetic oxygen maxima are observed much more frequently in lakes that have a high stability of stratification. These lakes com-

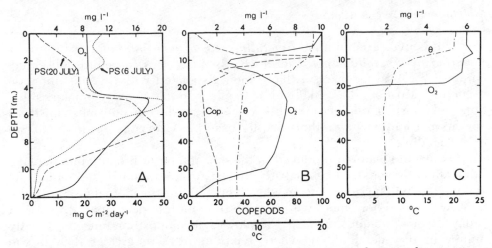

Figure 8–4   A, Metalimnetic oxygen maximum, showing a positive heterograde curve, in relation to temperature (Θ) and rates of phytoplanktonic photosynthesis (PS), Lawrence Lake, Michigan, 20 July 1971. (From Wetzel, et al., 1972.) B, Metalimnetic oxygen minimum, showing a negative heterograde curve, in relation to abundance of copepod microcrustaceans (Cop.), Lake Washington, Washington, 18 August 1958. (Drawn from data of Shapiro, 1960.) C, Oxygen concentrations in permanently meromictic Fayetteville Green Lake, New York, 3 September 1935. (Data of Eggleton, 1956.)

monly have high relative depths[1], i.e., their surface area is small relative to their maximum depth, and are protected from wind action by surrounding topography and vegetation. The average peak in oxygen concentration and relative depth of over 50 lakes that exhibited extreme metalimnetic maxima was 17.2 mg l^{-1} and 4.30 per cent, respectively (Eberly, 1964). Most lakes have a relative depth of less than 2 per cent.

Metalimnetic oxygen maxima are events that occur in a large number of lakes. When characteristics of morphometry and biota are suitable for their development, the maxima occur consistently from year to year at approximately the same time and depth. For example, in Lawrence Lake, Michigan, the metalimnetic maximum has occurred between 4 and 6 m, July through August for seven consecutive years of observation, and the data for four of these years are presented in Wetzel, et al. (1972). The maxima are superimposed exactly over maxima of photosynthesis by phytoplankton and summer growth maximum of dense beds of rooted aquatic plants of the steep slopes of the littoral zone. It is likely that water in the littoral zone, enriched with oxygen, dissipates into the metalimnion layer of equal density and contributes to the observed oxygen peak. Ruttner (1963) has reported a similar situation in a small clear lake with precipitous slopes, in which the metalimnetic oxygen maximum coincides seasonally with active growth of the macrophyte *Elodea*.

---

[1]Relative depth or z_r is an expression of the maximum depth (z_m) as a percentage of the mean diameter, and is determined by $z_r = \dfrac{50 \, z_m \, \sqrt{\pi}}{\sqrt{A_o}}$ where A_o is the surface area.

Metalimnetic Oxygen Minima

The converse condition, a *metalimnetic oxygen minimum* exhibiting a *negative heterograde curve* (Fig. 8–4B), is much less frequently observed. Numerous causal mechanisms have been associated with this metalimnetic reduction. Most or all are likely to be in effect simultaneously in many situations when such minima are observed. To single out one mechanism as operational and exclude others (as, for example Czeczuga did in 1959) can be misleading.

Oxidizable material produced in the epilimnion or brought into the lake from allochthonous sources outside the basin continuously sediments. Its sinking rate will be slowed down when it encounters the more dense metalimnetic water, and here there will be more time for it to decompose (Birge and Juday, 1911; Thienemann, 1928). Moreover, decomposition rates are usually higher in the metalimnion, where temperatures are greater than in the colder hypolimnion. As a result, more readily oxidizable organic matter is decomposed at this level, with concomitant consumption of oxygen by bacterial respiration; more resistant organic matter passes slowly into the hypolimnion (Kuznetsov and Karsinken, 1931; Vinberg, 1934; Åberg and Rodhe, 1942). In a review of the subject, Czeczuga (1959) presented some evidence from a lake in which, conversely, decomposition rates were greater in the epilimnion. However, renewal of epilimnetic oxygen and time of residence of sedimenting organic matter in the metalimnion were not clearly delineated. It is clear that a temporal balance between the transparency of the trophogenic zone and the depth of the metalimnion is important in relation to the depth at which photosynthetic maxima occur, and in relation to whether oxygen inputs are sufficient to offset decompositonal consumption (Ruttner, 1933). It is easy to conceive of a situation in which a pronounced metalimnetic maximum could shift to a minimum within a summer season.

In certain situations, concentrations of massive numbers of zooplankton microcrustacea in the metalimnion can contribute to a severe reduction of oxygen. Most likely, this respiratory consumption was the major cause of the metalimnetic minima observed in Zürichsee, Switzerland (Minder, 1923), and the same condition surely prevails in Lake Washington (Fig. 8–4B), where nonmigrating copepods develop in profusion in late summer (Shapiro, 1960).

The basin morphometry of a lake can also contribute to metalimnetic oxygen minima. In cases where the slope of the basin is gentle where it coincides with the prevailing metalimnion, a greater percentage of the sediments of high bacterial utilization of oxygen will be in contact with metalimnetic water than will the lower strata. Horizontal mixing and streaming of water from internal water movements is greatest in the metalimnion, where vertical density stability is greatest. As a result, reduction in oxygen content at the sediment-water interface is greater where the slope of the sediments is less. The reduction in the metalimnion may be sufficient to extend laterally over the entire lake or, as in the case of Skärshultsjön, Sweden (Alsterberg, 1927), it may occur more strongly in the metalimnion closer to the sediment-metalimnion interface.

There is evidence that biogenic oxidation of methane by methane-oxidizing bacteria can result in metalimnetic oxygen minima in certain productive lakes (Ohle, 1958; Kuznetsov, 1935, 1970). Methane of anaerobic fermentation in sediments rises from the sediments and is carried to water strata such as the metalimnion where, with warmer temperatures and greater oxygen, methane oxidation can occur rapidly. Severe oxygen reduction can result in a few days.

Distribution in Meromictic Lakes

The oxygen distribution of permanently meromictic lakes is one of extreme clinograde (Fig. 8–4C). Oxygen of the permanent monimolimnion receives constant inputs of organic matter or sulfates from saline water, which rapidly utilize any oxygen intrusions.

Temporary (partial) meromixis, discussed in Chapter 6, results in highly variable anomalous oxygen distributions. If a normally dimictic lake is quite productive, a strongly clinograde oxygen distribution both during summer and winter stratification would be anticipated, often with anaerobic deeper layers in both seasons. If spring or fall circulation and renewal of oxygen is incomplete, the reduced or anaerobic conditions of the lower strata will be carried over to the next stratified period and reinforced. Such was clearly the case in Little Crooked Lake in the fall of 1963 (Fig. 8–3). Mixing was incomplete in Lawrence Lake in the spring of 1968 (Fig. 8–5), which resulted in an accelerated reduction of the hypolimnetic oxygen the following summer, and anoxia below 11 m for three months. In normal years with complete spring circulation, as in the following four years (cf. Wetzel, et al., 1972), the hypolimnion did not become anaerobic or only very briefly just before fall turnover in a half-meter stratum immediately overlying the sedi-

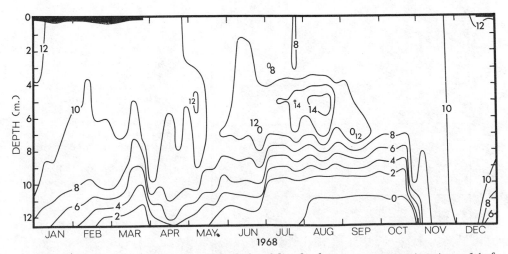

Figure 8–5 Depth-time diagram of isopleths of dissolved oxygen concentrations in mg l^{-1} of Lawrence Lake, Michigan, 1968. Ice-cover drawn to scale. (From Wetzel, R. G., et al.: Memorie Istituto Italiano Idrobiol., 29, Suppl., 185–243.)

ments. In 1973, incomplete mixing occurred again in the spring, resulting in an oxygen distribution pattern similar to that of 1968. The chemical nutrient and biotic implications of aerobic and anoxic hypolimnia are very important for the productivity of lakes, and will be pursued in detail further on.

Horizontal Variations and Diel Cycles

Horizontal variations in the distribution of dissolved oxygen of the open-water pelagic zone are generally rather small during the periods of spring and fall circulation. This is not always the case during the productive summer period and under ice-cover. Littoral aquatic plants often cover large areas of more productive lakes, to a depth of 3 to 10 m. Active photosynthesis of these macrophytes and of sessile algae growing attached to them and to the sediments generates large amounts of oxygen. Similarly, large quantities of oxygen are utilized in respiration at night and by littoral bacteria in a zone rich in dissolved and particulate organic matter. The oxygen regime of the littoral zone is usually totally different from that of the open water and undergoes marked diurnal cycles (Fig. 8–6). Moreover, the vertical distribution within the vascular plant beds can be highly stratified in accordance with the macrophyte portions that are actively photosynthesizing or those that are predominately respiring (Fig. 8–6).

The oxygen content of the littoral zone periodically can undergo very severe reductions. Such is the case when major populations of macrophytes (vascular hydrophytes and macroalgae) scenesce and die at the end of their

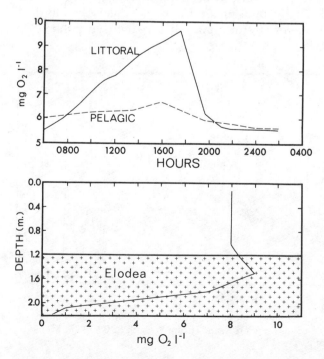

Figure 8–6 *Upper:* Changes in dissolved oxygen in the littoral and open water areas over a diurnal period in eutrophic Winona Lake, Indiana, 9 August 1922. (From data of Scott, 1924.) *Lower:* Vertical stratification of oxygen within the littoral zone of Parvin Lake, Colorado, 9 July 1955, in a luxuriant stand of the submersed macrophyte *Elodea*. (Generated from data of Buscemi, 1958.)

growing season. The intense decomposition can cause severe oxygen deficits for several months, and extend from the littoral zone for some distance into the open water (Thomas, 1960). In shallow lakes where submersed hydrophytes grow over the entire basin, decomposition of these stands at the end of the growing season, in late summer when temperatures are high, can be so intense that the oxygen content of the entire lake may be reduced severely to almost complete anoxia. Such a catastrophic event can result in massive die-offs of many species of fauna; often entire fish populations are lost in such a "summer-kill."

Where lakes are extremely eutrophic, the phytoplankton algae often occur in such profusion that the densities severely attenuate the light, and submersed macrophytes are limited to very shallow areas. Under such conditions of hypereutrophy, diurnal fluctuations in oxygen concentrations of the epilimnion are often just as marked as those of the littoral zone, as shown in Figure 8–6. Changes of 4 to as much as 6 mg O_2 1^{-1} are seen commonly between midafternoon and early morning darkness. Under clear ice-cover in such productive waters, where oxygen can accumulate in the surface waters, diurnal supersaturation is found. It is important to realize that oxygen analyses in surface waters of such productive lakes undergoing rapid diurnal changes, when taken only at one time of day, show only one stage of a much more complex situation.

Large horizontal variations also are found in lakes of complex basin morphometry. Where the lake consists of many bays and dendritic parts, the oxygen concentrations are very different in these areas which often operate as essentially individual lakes (Wetzel, 1966a).

In reservoirs, the structure of oxygen distribution becomes highly variable horizontally, vertically, and seasonally. The complex hydrodynamics are individual for each reservoir with regard to fluctuations in its influents, its morphometry, characteristics of drawdown and discharge, and other factors (see, for example, Straškraba, et al., 1973). Seasonal variations in influxes of natural organic matter from inflowing streams can place varying oxidative demands on the oxygen content and distribution of reservoirs.

Horizontal variations in oxygen are particularly conspicuous in the winter under ice-cover, especially in the surface strata. Fluctuations were pointed out earlier in response to seasonal changes in photosynthesis (Figs. 8–2, and 8–3). Photosynthesis by algae is strongly suppressed by snow-cover on ice, but often snow-cover is very spotty in accumulation on the surface of wind-swept ice. The effect is parallel variegated photosynthesis and oxygen production below the ice. Heavy snow-cover can plunge the lake into virtual darkness; if sustained for several weeks, heavy respiratory demands of decomposition can reduce the oxygen content to very low levels, or to anoxia (Greenbank, 1945). The resulting conditions are intolerable to many animals. For example, most fish cannot survive, even at low temperatures, at less than 2 mg O_2 1^{-1}. These "winter-kill" conditions are common in shallow, eutrophic lakes.

An example of the rapid seasonal changes in oxygen concentrations that can occur can be seen in Green Lake, a small reservoir lake in southern Michigan (Fig. 8–7). In early February, heavy ice and snow-cover had severely reduced the oxygen content of the lake, except in the area immediately adjacent to the small inlet. Three days later, following late winter rains on deteriorating ice, conditions had improved, although horizontal variations still were conspicuous. A week later, oxygen brought in primarily from increased flow of the inlet had virtually obliterated the former distribution.

Methane and hydrogen can be formed in anaerobic sediments of productive lakes in sufficient quantities to bubble up and rise to the surface. Rossolimo (1935) and Kuznetsov (1935) have shown that the methane (CH_4) and hydrogen (H_2) are effectively oxidized bacterially in the water, and under ice-cover can cause severe reductions in oxygen content (Fig. 8–8). Generation of these gases, and the oxygen utilization, is particularly great over the deeper depressions of the lake basin. The central depletion probably is accentuated by profile-bound density currents that slowly accumulate in the depression and force overlying water upward. Oxygen in these currents is reduced as they flow along the sediments to the deepest point.

Thus the oxygen regime of most dimictic lakes of the temperate zone is governed to a large degree by the quantities of oxidizable matter received during periods of stratification, either that produced within the lake autochthonously or that produced allochthonously within the drainage basin of the lake and brought to it. In lakes that tend toward permanent stratification, either meromictic variants or oligomictic lakes of low elevation in equatorial regions, the oxygen distribution is clinograde and is determined by the duration of the stratification in relation to the inputs of oxidizable organic matter. Lakes high in dissolved organic matter, such as bogs and bog lakes, have appreciable undersaturated oxygen concentrations.

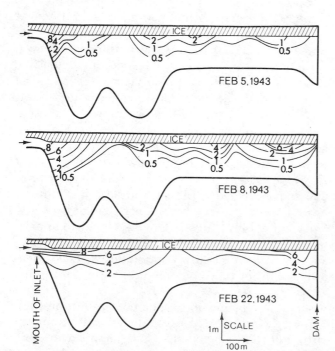

Figure 8–7 Horizontal isopleths of oxygen concentrations (mg 1^{-1}) during a two-week period in Green Lake, Michigan, winter 1943. (Modified from Greenbank, 1945.)

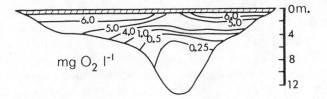

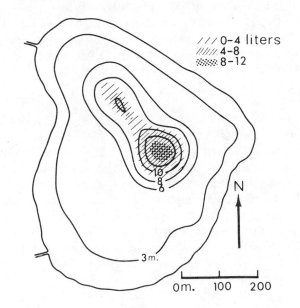

Figure 8–8 *Upper:* Isopleths of oxygen concentration in Lake Beloie, USSR, 15 March 1928; *lower:* quantity (*liters* m^{-2}) of gas, containing 20–24% CH$_4$, permeating the underside of the thawing ice in late winter. (Modified from Rossolimo, 1935.)

From the numerous examples of oxygen distribution and control in natural water bodies, the effects of artificial loading of lakes with pollutional sources of organic matter and other oxidizable materials should be readily apparent. Organic effluents from agricultural activities, sewage, and industry can very effectively and rapidly exceed the processing capacity of lakes. The resulting effects on aerobic organisms are direct and immediate. Indirect effects on the whole biogeochemical cycling and productivity are more gradual but, as we will see, more effective in producing generally undesirable conditions.

Oxygen Deficits

The amount of oxygen lost from the hypolimnion during summer stratification not only increases with depth, but becomes increasingly depleted as the stratification period progresses. The difference in amount of oxygen present at the beginning and at the end of stratification below a given depth is referred to as the *oxygen deficit*. In its simplest meaning, the oxygen deficit indicates the relationship of the metabolism in the *trophogenic zone,* the superficial stratum of a lake in which photosynthetic production occurs, and that in the underlying *tropholytic zone,* the aphotic deep stratum of heterotrophic decomposition of organic matter. Hence the amount of organic matter synthesized in the trophogenic zone that rains into and decomposes in the tropholytic zone is reflected in the rate of utilization of hypolimnetic oxygen. Although the dynamics of oxygen utilization are not that simple, the amount of oxygen consumption in the hypolimnion during stratification or oxygen deficit provides an indirect estimate of the productivity of the lake.

The recognition and development of the theory of oxygen deficits as an approximation of lake productivity has a long history since its foundations in the fundamental work of Birge and Juday (1911) and Thienemann (1928) shortly after the turn of the century. Since several of the older means of estimating oxygen deficit are still in use in spite of serious shortcomings, they should be briefly defined. All must be used cautiously, and in a general way.

Actual, Absolute, and Relative Deficits

The *actual oxygen deficit* of the water refers to the difference between the oxygen content observed at any point and the saturation value of that same quantity of water at its observed temperature at the pressure of the lake surface. A shortcoming of this measurement is that it assumes that the water at depth was saturated at the observed temperature during spring turnover, and that oxygen intrusion from the trophogenic zone during stratification by diffusion currents was negligible. The same criticism can be leveled at the *absolute oxygen deficit,* the difference between the observed oxygen concentration and the saturation value at 4°C at the pressure of the lake surface (Alsterberg, 1929). The assumption that the oxygen content during spring circulation is that of saturation at 4°C is not always borne out; indeed, more often this is not the case. Therefore, Strøm (1931) proposed and used the *relative oxygen deficit,* which is the difference between the oxygen content of the hypolimnion and that empirically determined at the end of spring turnover.

Use of relative oxygen deficit greatly improved the estimations of rates of consumption, because it took account of different oxygen deficits in layers below the hypolimnetic surface. Thus, consideration was given to the most basic property of changing volumes of the strata with depth, and the importance of hypolimnetic volumes among lakes. Thienemann (1926, 1928) emphasized this point clearly (Fig. 8–9). In comparing two lakes having identical volumes in the trophogenic zones, it would be assumed that the production and influx of organic matter in the trophogenic zone of each would be the same, and that similar amounts of oxidizable organic matter would sediment from the trophogenic layer into the tropholytic zones. However the volume of the first lake (Fig. 8–9A) is very much larger than that of the second lake (B), and it contains a much larger amount of oxygen to be utilized in decomposition. As a result, during the same interval of time, the oxygen content of the hypolimnion of the shallow lake (B) would be reduced more rapidly than that in the hypolimnion of the deep lake (A). One might anticipate a clinograde oxygen curve in the former case, and a more or less orthograde curve in the latter, deep lake. Much of the oxygen utilization in the shallow lake may

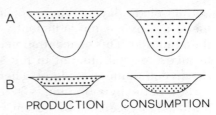

Figure 8–9 Diagrammatic representation of two lakes of equal production in the trophogenic zone but of differing volumes of their tropholytic zones, in which oxidative consumption occurs. (Modified from Thienemann, 1926, 1928.)

occur after the organic material has settled into the sediments. Turbulence and horizontal currents in the hypolimnion, although small, are usually sufficient to distribute much of the lower deoxygenated water into much of the hypolimnetic water. Thus, the relative oxygen deficit, which compensates for differences in hypolimnetic volume with depth, is an incorporation of the intensity of trophogenic production into the consumptive capacity of the hypolimnion or tropholytic zone.

This *relative areal deficit,* as introduced by Strøm (1931) and modified by Hutchinson (1938, 1957), is the mean deficit below one cm² of hypolimnetic surface of the individual deficits of a series of layers of decreasing volume with increasing depth. A computation from Gull Lake, southwestern Michigan, is illustrative of the general approach (Table 8–2). The data consist of measurements of oxygen concentration at several depths; a detailed hydrographic map is essential for accurate estimates of volume and area. Since the oxygen concentrations did not differ greatly from one depth to the next in Gull Lake at this time, it is possible to find the average concentration in any layer by taking the mean of the concentration at the top of the layer and that at the

TABLE 8–2 Sample Computation of the Relative Areal Oxygen Deficit of Gull Lake, Summer 1972[a]

Calculations for 6 June 1972:

Strata (m)	Volume (m³)	Mean O_2 (mg l^{-1})	Total O_2 in strata (metric t)[b]
15–20	12.910×10^6	9.42	121.612
20–25	5.221×10^6	8.90	46.464
25–30	0.203×10^6	8.68	1.762
30–33.5	0.0428×10^6	8.35	0.358
			170.196

metric t in hypolimnion

Calculations for
1 August 1972:   76.719
Difference: 93.477 t or 93.477×10^8 mg

The area of the surface of the hypolimnion at 15 m is 34.980×10^6 m², or 349.8×10^8 cm².

By division, the difference in oxygen or areal oxygen deficit was:
    2.67 mg cm⁻² 55 days⁻¹
or 0.0486 mg cm⁻² day⁻¹
or 1.46 mg cm⁻² month⁻¹

[a]Unpublished data of Tague, Lauff, and Wetzel.
[b]1 mg l^{-1} = 1000 metric tons km⁻³ = 1 metric t $(10^6$ m³$)^{-1}$

bottom of the layer. Where concentration gradients are steep, simple arithmetic means are not sufficiently accurate; a planimetric integration should then be used. This concentration is then multiplied by the volume of the layer to give the total amount in the stratum, and the amounts for each layer of the hypolimnion are added. Thermal and morphometric data show that the surface of the hypolimnion is 34.98 ha or 349.8×10^8 cm^2. By division the difference in oxygen is 2.67 mg cm^{-2} for the 55-day period between 6 June and 1 August, 0.0486 mg cm^{-2} day^{-1} or 1.46 mg cm^{-2} month^{-1}.

Many of the calculations of areal oxygen deficits are from older investigations that make direct comparisons of deficits with various indices of biological productivity difficult. Nonetheless, as would be expected, there is a strong tendency in lakes of moderate depth for the relative areal deficit to increase with increasing productivity. Based on limited data, Hutchinson (1938, 1957) set the ranges among the rather ambiguous general categories of unproductive to productive lakes as:

	Arbitrary Limit	Approximate Observed Range (mg cm^{-2} month^{-1})
Oligotrophic	<0.5	0.1–1.0
Mesotrophic		1.0–1.5
Eutrophic	>1.0	>1.5

Mortimer believes (Hutchinson, 1957) that the limits would be more realistic set at oligotrophy <0.75 and eutrophy >1.65 mg cm^{-2} month^{-1}.

Calculations of oxygen deficits are subject to numerous errors that detract from true representation of an index of trophogenic productivity. Significant inputs of allochthonous organic matter cause large errors, and productivity of moderately to highly stained waters (>10 Pt units) will be overestimated owing to oxygen consumption by dissolved organic matter. In both stratified shallow lakes and very deep, clear lakes, photosynthesis commonly occurs in the upper portion of the hypolimnion. On this basis, Hutchinson recommends use of hypolimnetic oxygen deficits only in dimictic lakes between approximately 20 and 75 m maximum depth.

There are a number of better methods for evaluating lake productivity. Autochthonous productivity can be measured directly; similarly, fluxes of particulate and dissolved organic matter can be evaluated in detail chemically and biotically. Additionally, in productive lakes where the hypolimnion becomes anaerobic during the period of stratification, the oxygen deficit cannot be used as an effective indicator. However, when detailed data on productivity are lacking, the oxygen deficit can be extremely informative about the general trophic status of the lake.

Changes in the hypolimnetic oxygen deficit rates over long periods of time can be indicative of overall changes in the productivity of the lake. A very good example of this is Douglas Lake, in the northern portion of the southern peninsula of Michigan. Although much is known on the character-

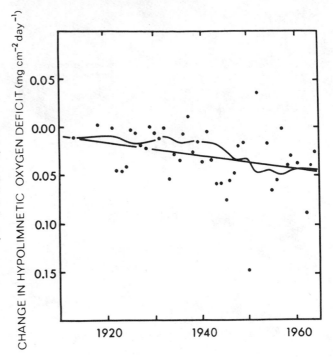

Figure 8–10 Change in estimated annual hypolimnetic oxygen deficits over a 53-year period in Douglas Lake, Michigan. The straight line represents the line of best fit by linear regression; the smooth curve represents an exponentially smoothed one-year forecast. (From data of Bazin and Saunders, 1971.)

istics and biota of this lake, which has been studied for many years, no good estimates of annual productivity exist. Analyses of the annual amount of oxygen below the metalimnion showed a progressive, slow decrease between 1911 and 1964 during each year considered except two (Fig. 8–10). Although considerable scatter in the annual change was found, the rate of change appears to be increasing in recent years (Bazin and Saunders, 1971), and reflects an accelerated nutrient input and eutrophication associated with human activity first as a result of deforestation, and second, as a result of the development of the area for recreational purposes. This pattern has been repeated many times in other lakes, but long-term data are available only rarely. The well-known rapid eutrophication of a much larger lake, Lake Erie, has been followed in a similar way, largely on the basis of losses of hypolimnetic oxygen and benthic fauna (Carr, 1962). Other indices of eutrophication will be discussed in the appropriate chapters.

chapter 9

salinity of inland waters

SALINITY IN SURFACE WATERS

The total salinity of inland waters usually is dominated completely by four major cations, calcium (Ca), magnesium (Mg), sodium (Na), and potassium (K), and the major anions, carbonate (CO_3), sulfate (SO_4), and chlorides (Cl). The salinity of surface waters has a world average concentration of about 120 mg l^{-1}, but varies appreciably among continents and with the lithology of the land masses. The salinity is governed by contributions from the rock runoff of the drainage basin, atmospheric precipitation, and the balances between evaporation and precipitation. A strong tendency exists for the proportions of major ions of surface waters of the world towards Ca>Mg≥Na>K and CO_3>SO_4>Cl. In soft waters and in lakes in coastal regions, Na and Cl often assume greater equivalent concentrations.

The ionic weathering of soil and rock is controlled by the processes of solution, oxidation-reduction, the action of hydrogen ions, and the formation of organic complexes. Salinity of atmospheric precipitation and fallout constitutes a major source of salinity to many dilute fresh waters and certain saline lakes of arid regions.

Concentrations of magnesium, sodium, and potassium and the major anion chloride are relatively conservative and undergo minor spatial and temporal fluctuations within a lake from biotic utilization or biotically mediated changes of the environment. Calcium, inorganic carbon, and sulfate are dynamic and the concentrations of these ions are strongly influenced by microbial metabolism.

The proportionate concentrations of major cations and the ratios of monovalent:divalent cations influence the metabolism of many organisms, especially certain algae and macrophytes, as much as absolute concentrations. Factors that influence availability of some cations disproportionately to others as, for example, organic complexing of calcium or biogenically-induced decalcification of the epilimnion, can indirectly affect seasonal population succession and productivity.

The relatively low salinity of fresh waters has influenced greatly the distribution of biota and the long evolutionary history of physiological adaptations for osmotic and ionic regulation in an extremely hypotonic environment. Although some groups of bacteria and algae are relatively homiosmotic, tolerating only a narrow range of salinity, most of the lower flora and fauna are euryhaline, i.e., adaptable to a wide range of salinity. Most higher freshwater animals originated from the sea or from land and adapted to freshwater secondarily. In comparison to marine forms, nearly all of these organisms have reduced osmotic pressures of body fluids, and have developed efficient mechanisms for active uptake of ions and renal mechanisms for ion retention.

Ionic Composition of Surface Waters

The ionic composition of fresh waters is dominated by dilute concentrations of alkalies and alkaline earth compounds of bicarbonates, carbonates, sulfates, and chlorides. The amounts of silicic acid, largely in undissociated form, are usually small but occasionally form a significant constituent of hardwater lakes. The concentrations of four major cations, Ca^{++}, Mg^{++}, Na^+, K^+, and four major anions, HCO_3^-, $CO_3^=$, $SO_4^=$, and Cl^-, usually constitute the total ionic *salinity* of the water for all practical purposes. The concentrations of ionized components of other elements such as nitrogen (N), phosphorus (P), and iron (Fe), and numerous minor elements are of immense biological importance, but from the standpoint of the composition of water they are small.

The term salinity is the correct nomen in chemical terminology for the ionic composition, expressed in mg l^{-1} or meq l^{-1} which are essentially equivalent as mass or volume in dilute solutions. In open lakes with an outlet effluent, the chemical composition is governed largely by the composition of influents from the drainage basin and the atmosphere. The water salinity of closed basins is greatly modified and concentrated by evaporation and precipitation of salts. *Soft waters* refer to waters containing low concentrations of salinity, and are usually derived from drainage of acidic igneous rocks (Hutchinson, 1957). *Hard waters* contain large concentrations of alkaline earths derived from drainage of calcareous deposits.

The salinity of fresh waters is best expressed as the sum of the ionic composition of the eight major cations and anions in mass or milliequivalents per liter. The quantity *total solids,* an estimation of inorganic materials dissolved in water by evaporation to dryness (105°C), is less satisfactory. Combustion of the residue at 550°C yields the *nonvolatile solids* per unit mass or volume and results in the loss of organic carbon and carbon dioxide (CO_2), largely from magnesium carbonate ($MgCO_3$), some alkali, and chloride. Therefore considerable underestimation can result from the CO_2 loss.

Salinity Distribution in World Surface Waters and Control Mechanisms

The salinity of the surface waters of the world is highly variable in relation to ionic influences of drainage and exchange from the surrounding land,

TABLE 9–1   Mean Composition of River Waters of the World (mg l⁻¹)

	Ca⁺⁺	Mg⁺⁺	Na⁺	K⁺	CO₃⁼ (HCO₃⁻)	SO₄⁼	Cl⁻	NO₃⁻	Fe (as Fe₂O₃)	SiO₂	Sum
North America	21.0	5.	9.	1.4	68.	20.	8.	1.	0.16	9.	142
South America	7.2	1.5	4.	2.	31.	4.8	4.9	0.7	1.4	11.9	69
Europe	31.1	5.6	5.4	1.7	95.	24.	6.9	3.7	0.8	7.5	182
Asia	18.4	5.6	5.5	3.8	79.	8.4	8.7	0.7	0.01	11.7	142
Africa	12.5	3.8	11.	–	43.	13.5	12.1	0.8	1.3	23.2	121
Australia[b]	3.9	2.7	2.9	1.4	31.6	2.6	10.	0.05	0.3	3.9	59
World	15.	4.1	6.3	2.3	58.4	11.2	7.8	1.	0.67	13.1	120
Cations (meq)	0.750	0.342	0.274	0.059							1.425
Anions	–	–	–	–	0.958	0.233	0.220	0.017	–	–	1.428

[a]Selected data from Livingstone, 1963, and Benoit, 1969.
[b]Values of calcium are likely less, on the average, than Na and Mg in Australian surface waters (Williams and Wan, 1972).

atmospheric sources derived from the land, ocean, and human activity, and equilibrium and exchange with sediments within the water body. The global mean composition given in Table 9–1 is based on a more recent evaluation than Clarke's 1924 monographic treatment of the subject. These data show a mean concentration of river water of the world of 120 mg l^{-1}, somewhat lower than Clarke's average but based on better data to compensate for the large amounts of dilute water draining tropical regions in major river systems. Principal differences among the continents are in the amounts of calcium and carbonate ions (all bicarbonate being converted to carbonate). Low values of these normally dominating ions are seen especially in South America and Australia.

The three major mechanisms controlling the salinity of world surface water are rock dominance, atmospheric precipitation, and the evaporation-precipitation process (Gibbs, 1970). Waters of the rock-dominated end of a spectrum are more or less in partial equilibrium with materials of their drainage basins. Their positions within this grouping depend on the relief and climate of each basin and the composition of rock material in the basin. Calcium and bicarbonate ions dominate (Fig. 9–1). The chemical composition of low-salinity waters, dominated by sodium and chloride, is influenced by

Figure 9–1 Diagrammatic representation of the general processes controlling the salinity of world surface waters. (Slightly modified from Gibbs, R. J.: Science, 170, 3962: 1088, 1970. Copyright 1970 by the American Association for the Advancement of Science.)

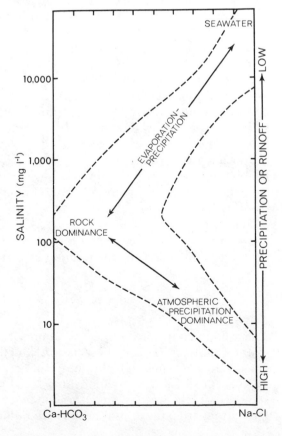

TABLE 9–2   Sources of Major Cations of Various Types of Rivers[a]

River	Contribution from Atmospheric Precipitation (%)	Contribution from Rock Deposits (%)
Rain-dominated Rio Tefé an Amazon tributary 1700 km inland from the Atlantic Ocean	81	19
Rock-dominated Ucayali River, an Amazon tributary of the Peruvian Andes	4.8	95.2
Evaporation-crystallization in Rio Grande, US-Mexico	0.1	99.9

[a]After data of Gibbs, 1970.

the dissolved salts derived from atmospheric precipitation which in turn is de-rived largely from the oceans. Further, the leached areas, mainly in tropical South America and Africa, generally are of low relief and high rainfall. The supply of dissolved salts from rocks is very low in proportion to the amount of rainfall. The third major process that influences salinity of surface waters is evaporation and fractional crystallization with precipitation. Fresh waters ex-tend in a series from calcium and bicarbonate-rich, to low salinity waters of rock dominance to sodium-dominated high-salinity waters (Fig. 9–1). Rivers and lakes at the extreme saline end of this series are generally located in hot and arid regions. Extremes in the series can be seen in the relative contribu-tions of the major cations from precipitation and rocks in three river systems given in Table 9–2.

The data discussed here represent average concentrations and general conditions of distribution on a global basis. The regional distribution of salinity of inland waters is highly variable and variegated in localized situa-tions. Hence some areas or lake districts can be found where, because of uniform rock formations, the ionic proportions are relatively consistent (Table 9–3). By contrast, in southwestern Michigan, extremely soft water seepage lakes may be found immediately adjacent to very hardwater cal-careous lakes, under conditions of salinity of lake water of complex, varying glacial till.

Over large regions of the temperate zone, the calcium and bicarbonate type of dominance prevails in open lake systems:

Cations: $Ca > Mg \geqslant Na > K$
Anions: $CO_3 > SO_4 > Cl$

The relationships are very consistent with those of the average world river water (Table 9–1), and there is a distinct tendency for the composition of lake waters to approach this uniform concentration (Rodhe, 1949; Gorham, 1955). These salinity relationships led Rodhe (1949) to propose a general standard of lake water concentration, given in Table 9–4 in abbreviated form. Deviations from these proportions are numerous, in general in the continents

TABLE 9-3 Mean Ionic Salinity in Equivalent Proportions of Several Natural Waters

Natural Waters	Ca^{++}	Mg^{++}	Na^+	K^+	$CO_3^=$	$SO_4^=$	Cl^-
Wisconsin soft waters (Juday, et al., 1938; Lohuis, et al., 1938)	46.9	37.7	10.9	4.8	69.9	20.5	9.9
N. German soft waters (Ohle 1955a)	36.0	14.3	43.0	6.7	42.4	14.1	43.5
Water from igneous rock (Conway, 1942)	48.3	14.2	30.6	6.9	73.3	14.1	12.6
Mean sedimentary source material (Hutchinson, 1957, after Clarke, 1924)	53.2	34.0	8.0	4.8	93.8	6.2	—
Upland Swedish sources (Rodhe, 1948, after Lohammar)	67.3	16.9	13.6	2.2	74.3	16.2	9.5
Hardwater lakes of N.E. Indiana (Wetzel, 1966b)	79.3	14.4	3.3	3.0	60.0	37.9	2.1
Hypereutrophic Sylvan Lake, Indiana[a] (Wetzel, 1966a)	57.3	11.8	26.9	3.4	48.7	34.7	16.6
Average river content of world (Livingstone, 1963)	52.6	24.0	19.3	4.1	67.1	16.3	15.4

[a]A hardwater lake that received large amounts of domestic effluents that had high sodium content, most likely from detergents.

of South America and Australia, and more specifically in drainage basins of igneous source materials and soft waters (see Table 9–3). Drainage from igneous rocks commonly has a salinity of less than 50 mg 1^{-1} and cationic proportions of:

$$Ca > Na > Mg > K$$

The anionic proportions of softwater systems shift toward decreasing carbonates and increasing halide concentrations:

$$Cl \geqslant SO_4 > CO_3$$

TABLE 9-4 Composition of Swedish Bicarbonate Fresh Waters (in mg 1^{-1}) in Relation to Specific Conductance[a]

Salinity	Ca	Mg	Na	K	CO_3	SO_4	Cl	μmhos cm^{-1} (20° C)
10.5	2.5	0.4	0.7	0.3	4.4	1.5	0.7	20
32.9	7.9	1.3	2.2	0.8	13.8	4.7	2.2	60
56.5	13.5	2.3	3.8	1.4	23.7	8.0	3.8	100
92.1	22.0	3.7	6.2	2.3	38.6	13.1	6.2	160
117.3	28.1	4.7	7.9	2.9	49.2	16.6	7.9	200
155.1	37.1	6.2	10.4	3.8	65.2	22.0	10.4	260
180.8	43.3	7.3	12.2	4.4	75.8	25.6	12.2	300
219.7	52.6	8.8	14.8	5.4	92.1	31.2	14.8	360
246.2	59.0	9.9	16.6	6.0	103.2	34.9	16.6	400

[a]From data of Rodhe, 1949, and Hutchinson, 1957.

The specific conductance of lake water is a measure of the resistance of a solution to electrical flow. The resistance of the water solution to electrical current or electron flow is reduced with increasing content of ionized salts. Hence, the purer the water is, i.e., the lower its salinity, the greater its resistance to electrical flow will be. By definition, specific conductance of an electrolyte is the reciprocal of the specific resistance of a solution measured between two electrodes 1 cm^2 in area and 1 cm apart. The resistance is expressed in μmhos cm^{-1} (reciprocal of ohms).[1] The temperature of the electrolyte affects the ionic velocities; conductance increases about 2 per cent per °C. The international chemical standard reference of 25°C is recommended in all cases, even though 18°C has been used widely in past limnological studies.

The specific conductance of the common bicarbonate-type of lake water is closely proportional (see Table 9–4) to concentrations of the major ions (Juday and Birge, 1933; Rodhe, 1949). Once the concentrations of the major salinity ions are known, changes in specific conductance reflect changes in ionic concentrations in a proportional way (Otsuki and Wetzel, 1974). This relationship is not true, however, for minor constituents (N, Fe, Mn [manganese], Sr [strontium], or especially P) of lake waters (Rodhe, 1951). As would be anticipated, there is a direct correlation between conductance and pH in the intermediate pH range of bicarbonate fresh waters, but this relationship deteriorates among lakes of low salinity and high dissolved organic matter content (Strøm, 1947).

The relative concentrations of the eight major ions of salinity are shown graphically in comparative studies among lakes in ionic polygonic diagrams (Maucha, 1932). A 16-sided regular polygon of standard area is divided into eight equal sectors, the four to the right of the vertical diameter representing the cations and the four to the left the anions (Fig. 9–2). Lines are drawn from the center bisecting each sector. The length of each radius bisecting a sector is proportional to the concentration of a particular ion in equivalent percentage of the total cations or anions.[2] Lines from the radii of the ionic sections are joined to the polygon and shaded. Thus, by incorporating a scale of polygonic area to equivalents per liter, expressions of quadrilaterals in which the area is proportional to concentrations of each ion were extended to include differences in actual concentrations as well (Broch and Yake, 1969).

---

[1] Specific conductance (μmhos cm^{-1} = $\dfrac{\text{R of } 0.00702 \text{ N KCl}}{\text{R of sample}} \times 1000$) where R is the resistance in ohms (cf. Rainwater and Thatcher, 1960; Golterman, 1969).

[2] The radius (R) within the sectors of a 16-sided polygon is

$$\frac{R^2 (\sin 22.5)}{2} = A/16$$

where A represents the area of the polygon, for example 200 mm^2. Thus R = 25 mm/0.38268, or R = 8.082 mm. The total equivalent per cent for each of the four cations or four anions is equal to 100 per cent, or in this example; 100 mm^2 (1 mm^2 = 1 equivalent %). Details of calculation are given in Gessner, 1959, and in Broch and Yake, 1969.

Figure 9–2 Ionic polygonic diagrams in which the relative concentrations in equivalent percentage of the major cations and anions are indicated for average concentrations of Pretty, Sylvan, and Crooked lakes, northeastern Indiana. The length of each radius bisecting a sector is proportional to the concentration of a particular ion in equivalent percentage of the total cations or total anions. (After Wetzel, 1966b.)

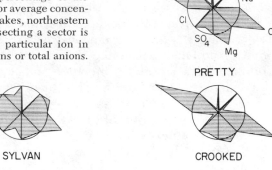

PRETTY

SYLVAN

CROOKED

SOURCES OF SALINITY

Weathering of Soil and Rock

The composition of soil and rock and their exchange capacity influence the rates of weathering and ion supply to the water of percolation and runoff. The adsorption of ions is dependent upon the cationic availability, concentrations, and proportions in a soil solution or leaching water, the nature and number of exchange sites on the exchange complex of soils, and the volume of water in contact with the exchange complex. Four general processes of weathering control ion supply: solution, oxidation-reduction, the action of hydrogen ions, and the formation of complexes (reviewed in detail by Gorham, 1961, and Carroll, 1962).

Ordinary solution not involving hydrolysis or acid weathering is important, primarily in sedimentary deposits rich in soluble salts. Leaching of marine salt deposits results in an enrichment of sodium and potassium chlorides relative to other ions in recipient lake waters. Relative leaching rates of different ions are time-dependent, and thus marked fluctuations in ionic proportions of runoff water are found during periods of high (flooding) and low rates of runoff because of differences in solubility.

Oxidation and reduction processes primarily affect iron, manganese, sulfur, nitrogen, phosphorus, and carbon compounds in soils. Iron sulfides are common constituents of rocks and water-saturated soils. Oxidation of these sulfides can be a major source of sulfate for natural waters, usually in the form of dilute sulfuric acid which solubilizes other rock and soil constituents as well. Microbial decomposition of organic compounds containing sulfur in soils, especially in woodland soils and peaty bog waters, adds sulfate to natural waters usually as sulfuric acid. Within the lake, oxidation-reduction processes play dominant roles in the sulfur cycle (see Chapter 13).

The action of hydrogen ions from the dissociation of carbonic acid, discussed at length in the following chapter, is of major importance in the weathering of soils and rocks. Concentrations of CO_2 and H^+ ions increase in soils undergoing intensive microbial decomposition of organic matter that is of plant origin and in close proximity to root systems. The importance of carbonic acid in weathering is illustrated by the high proportion of bicarbonate ions in most river waters. The proportion is high even in waters of low salinity (<50 mg 1^{-1}) draining igneous rocks. Strong acids in rainfall originating from air pollution accelerate the rate of weathering (Likens, et al., 1972). Colloidal acids of humic compounds and acid clays can supply large amounts of H^+ ions for weathering. The 'black' waters of tropical rivers draining acidic podzols are colored brown by humic substances originating from incomplete decomposition of plants of forests and swamps. The low pH (3.5 to 5) of these waters results largely from dissociation of humic acids, and the fact that the soils of lowland tropics contain very low levels of bases and of calcium.

Certain organic molecules can chelate or complex ions by bonding, thereby preventing these ions from reacting with others, and maintaining solubility of the complex. The magnitude of this process in weathering and transport of ions probably is quite significant, but as yet it has not been studied adequately.

The exchange of ions between soils and the soil solution is governed by exchange equilibria between hydrogen ions and ions attached to the soil minerals. Soil colloids have a finite absorptive capacity, and the exchange is highly variable. In humic, well-drained regions, water selectively removes cations from weathering rocks and soils. In areas of limited rainfall and stream activity, an accumulation of sodium and chloride deposited by rain in the soil water can result. The proportion of divalent to monovalent cations in the exchange positions in soils is a function of their concentrations and ratio in the soil water. Sodium ion adsorption to clay particles decreases with increasing concentrations of Ca^{++} and Mg^{++} in soil solution. Calcium and magnesium are the dominant exchangeable cations in most neutral or alkaline soils. In well-drained situations, the cations added by rainfall will have little effect on exchange leaching from soils or weathering rock. In contrast, when the soils are saturated, cations in exchange positions are in equilibrium with those in the soil water and groundwater. Cations added by rainwater to the surrounding water of clay minerals will establish equilibrium with those in exchange positions, and little weathering or removal of cations will occur. Under arid conditions, salts of soil solutions accumulate with no removal of cations derived from rainfall. Exchange sites contain predominately Ca^{++} and Mg^{++}; as the soil water becomes saturated with Na^+, the Ca^{++} and Mg^{++} are gradually replaced by Na^+ until all sites are filled. This Na^+ is released again during periods of rainfall, resulting in a much higher concentration of Na than the divalent cations. The clay mineral is affected very little by these exchanges and does not weather appreciably.

The salinity of natural waters is influenced further by the depth and mode of water percolation. Seepage lakes receiving largely surface runoff are usually very dilute in comparison to open lakes draining deeper soil horizons. The salinity of groundwater is generally much higher than that of surface runoff.

Atmospheric Precipitation and Fallout

A significant source of salinity for many dilute fresh waters and for some saline lakes of arid regions is from the atmosphere. Rainfall carries much of the atmospheric salt to lake and river waters (cf. reviews of Gorham, 1961, and Carroll, 1962). In less humid regions, dry fallout of salt particles occurs in significant quantities, often exceeding those washed down with rainfall. Although snow is less efficient in removing atmospheric salts than rain, at high latitudes contaminants form an appreciable supply of salinity. All of the major anions in natural waters are cycled in part through the atmosphere as gases as well as in dissolved and particulate form.

The sea is a major source of atmospheric supply of sodium, chloride, magnesium, and sulfate. Sea spray carries large amounts of sea water into the atmosphere which, on evaporation, forms salt particles that are capable of acting as nuclei for cloud and raindrop condensation. Atmospheric salinity can be carried for great distances inland over continents, although a majority of it is precipitated with rainfall in the coastal regions, and decreases with distance inland. Continental rain generally contains more sulfate than chloride; the chloride ion concentrations usually increase with proximity to the sea. In a similar fashion, sodium is the dominant cation of rain, and its concentration decreases in relation to amounts of magnesium and calcium with increasing distance from the sea. The effects of this atmospheric transport can be seen in lakes enriched with Na^+ and Cl^- in coastal maritime regions. This enrichment is particularly evident in coastal region mountains with igneous lithography, in which many of the maritime lakes of Japan can be found (Sugawara, 1961).

Windblown dust from the soil often contributes salts, especially of calcium and potassium, to rain and snow. Such is particularly the case in calcareous regions such as Sweden. Wind-transported salts from salt pans of lakes in endorheic regions, such as in western Australia or the United States, can be moved large distances to drainage basins of other lake systems.

An additional major source of atmospheric salinity is industrial and domestic air pollution. Although numerous ions and particles are emitted into the air, chlorides, calcium, and especially sulfates are major contaminants. The magnitude of the consequences of air pollution on the chemistry of surface waters now has reached such major proportions that global action is required to curtail its effects. Although much of the atmospheric pollution returns to the ground in the areas immediately adjacent to the industrial concentrations, sufficient sulfuric, nitric, and hydrochloric acids enter the atmosphere to influence the water of precipitation and surface waters of entire countries or portions of continents. The most infamous example is in Scandinavia, which receives air currents from southwestern Europe for much of the year. Contaminants from heavily industrialized regions of the United Kingdom, the Ruhr Valley, and elsewhere have caused a 200-fold increase in acidity of rain in a decade (Likens, et al., 1972). Values of pH less than 3 have been recorded, and over large areas of northern Europe the pH is less than 4.

An analogous situation has developed in eastern New York and smaller eastern states which can be traced to industrial origins in the central states. In addition to direct effects on the leaching rates of nutrients from plant foliage and plant metabolism, and leaching rates of soil nutrients, major nutrient inputs to and acidification of surface waters can result in changes in entire drainage basins. In the Hubbard Brook drainage basin of the White Mountains of New Hampshire, sulfate was found to be the principal ion in precipitation (Fisher, et al., 1968), and supplied most of the sulfate that was discharged by the streams. The input of ammonium and nitrate exceeded the discharge of these constituents. The annual deposition of hydrogen ion exceeded that of sulfate, and is a major determinant of pH of the waters in that region.

Environmental Influences on Salinity

Climate has a marked effect on the balance between precipitation and evaporation, and on the salinity concentrations of surface waters. This effect already has been discussed in global terms. From a more localized standpoint, climatic effects are manifested in a general increase in salinity with decreasing elevation of lakes. This correlation is related to the fact that much of the salinity of rain and fallout is deposited at lower elevations.

The salinity of lake waters of closed drainage basins is governed not only by inputs of dissolved ions of influents from runoff, but by the fate of these materials in evaporation (Hutchinson, 1957). A majority of closed lakes occur in regions with fluctuating long-term (a period of several years) climate, and are often exposed to periods of severe aridity. Commonly very shallow, these lakes may evaporate completely, or enough to expose large expanses of sediments. Loss of salts then can occur by wind deflation.

Saline lakes are generally distinguished on the basis of dominating anionic concentrations into carbonate, chloride, and sulfate waters. A spectrum of intermediate combinations may occur. The range of salinity in these lakes is extraordinary, from several hundred to over 200,000 mg l^{-1} in the Great Salt Lake of Utah and the Dead Sea. Borax Lake of northern California, one of the few shallow saline lakes that has been studied in detail over an annual period, exhibited a nearly four-fold decrease in volume and over a two-fold increase in salinity from less than 28,000 mg l^{-1} to nearly 60,000 mg l^{-1} (Wetzel, 1964). An analogous condition was recorded in Lake Chilwa, Malawi, which is typical of many shallow lakes of endorheic regions (Moss and Moss, 1969).

Other significant climatic factors in salinity are temperature and wind. Temperature influences the rate of rock weathering. Tropical waters, for example, which drain strongly weathered soils, are usually poor in electrolytes, and a large part of their total composition is constituted by silica. Wind direction and speed may affect strongly the chemical composition of atmospheric precipitation by the amount of incorporated sea salinity and sites of inland deposition. Losses of atmospheric salinity are greater in low elevation turbulent air masses. The type of vegetation growing on the drainage basin and its requirements for major ions are also influenced by climate. In rates of utilization and leaching of soil nutrients, the cycling of minerals in tropical perennial forests of dense vegetative cover differs greatly from cycling in deciduous vegetation of temperate regions.

DISTRIBUTION OF MAJOR IONS IN FRESH WATERS

The spatial and temporal distribution of the major cations and anions of salinity are separable into (1) conservative ions whose concentrations within a lake undergo relatively minor changes from biotic utilization or biotically mediated changes in the environment, and (2) dynamic ions whose concentrations can be influenced strongly by metabolism. Of the major cations, magnesium, sodium, and potassium ions are relatively conservative both in their chemical reactivity under typical freshwater conditions and their small biotic requirements. Calcium is more reactive and can exhibit marked seasonal and spatial dynamics. Of the major anions, inorganic carbon is so basic to the metabolism of fresh waters that it is treated in a separate chapter (Chapter 10), and later coupled with organic cycling (Chapter 17). Similarly, sulfate is greatly influenced by microbial cycling and the chemical milieu, and is treated separately (Chapter 13), along with iron and silica dynamics. Chloride is relatively conservative. The total ionic salinity, composed almost entirely of the eight major ions, is of major importance in osmotic regulation of metabolism and biotic distribution, and is briefly discussed in the concluding section of this chapter.

Calcium

Calcium has been implicated in numerous ways in the growth and population dynamics of freshwater flora and fauna. Calcium is required as a nutrient of normal metabolism of higher plants. A universal requirement for calcium has not been demonstrated for algae, but most likely it is required by the green algae, and is considered a basic inorganic element of algae. Where essential, calcium is usually needed as a micronutrient. Substitution of calcium by a closely related element, strontium, is readily acceptable in some algae (e.g., *Chlorella, Scenedesmus*) which require calcium and are indifferent to strontium; in other algae calcium utilization is inhibited strongly by strontium.

The two principal membranes of plant cells, the outer cytoplasmic plasmalemma and the inner cytoplasmic tonoplast, operate in ion transport, one of the basic biological functions of cell membranes (see reviews of Eppley, 1962, and Epstein, 1965). The central vacuole is the main repository of accumulated ions, and both cations and anions may be absorbed from dilute solutions until the vacuolar concentrations are vastly in excess of their external concentrations. The membranes of higher plants have a very low permeability to free ions; ionic exchange is more common among algae.

Active transport mechanisms presumably involve high-energy phosphate compounds of phosphorylation. The ion transporting mechanisms possess selectivity for ions, which varies among species, and the external concentrations of ions and ratios of divalent to monovalent cations can affect transport mechanisms markedly. Calcium is essential for maintenance of the structural and functional integrity of cell membranes in ion absorption and retention.

The distribution of certain algae has been correlated to differing concentrations of calcium. While causality cannot be established definitely for such relationships by singling out one affecting chemical species from an array of simultaneously interacting parameters regulating growth and distribution, it is clear that calcium is involved indirectly in the metabolism of certain organisms. The desmidium are one of the few large algal groups that are found in large part in waters of low salinity, especially waters of low calcium content, and those with a low pH of from 5 to 6. Of the few species of desmids studied, many exhibit high calcium requirements and sensitivity, and their restriction to soft waters is by no means universal. Species of the larger genera are divisible into those adapted to acidic (pH<6), calcium-deficient (<10 mg Ca l^{-1}) waters, through a series of those adapted to increasingly alkaline, calcium-rich waters. The detailed studies of Höll on this subject have been verified and are reviewed by Hutchinson (1967). Again, causality between calcium concentrations and the metabolism of these algae has not been established through experimental studies, even though the evidence for such an interaction is very strong.

Hardly a group of freshwater animals exists in which the distribution of some species has not been related to calcium concentrations (cf. Macan, 1961). Like algae and certain macrophytes, in some groups most of the species are found in calcareous waters, and their numbers decrease as the concentration of calcium lessens. Other species, which frequently are closely related to those in hard waters, are characteristic of waters poor in calcium. In particular, mollusks, leeches, and tricladian flatworms are divisible into hard-water species (ca>20 mg Ca l^{-1}), and those that can tolerate less than this concentration. Analogous correlations occur among the crustacean invertebrates, but less clearly; and with much variability. Although the requirements for calcium in invertebrates are known to some extent (cf. Robertson, 1941), little is known about how small differences in calcium concentrations affect distribution. Attempts to explain the metabolism and distribution of animal species on the basis of single chemical differences have been numerous but largely unrewarding, except where conditions are extreme. There are exceptions to this type of oversimplied study of the complexity of several interacting, simultaneously operating ions within a chemical spectrum of fresh waters. For example, calcium concentrations have been implicated in the aging of rotifers, a process which affects longevity and morphology by accumulation of calcium (Edmondson, 1948). These laboratory findings fit remarkably well into the observed distribution and population dynamics of rotifers in lakes of varying calcium concentrations where longevity and population success increase in waters of lesser calcium content.

The calcium content of softwater lakes remains well below saturation

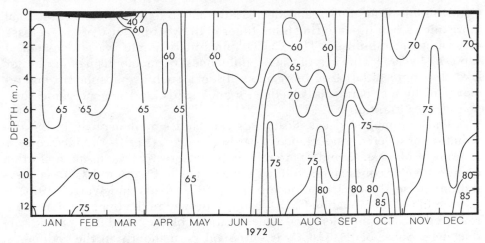

Figure 9–3 Depth-time distribution of isopleths of calcium concentrations (mg Ca^{++} l^{-1}) of oligotrophic, hardwater Lawrence Lake, Michigan, 1972. Opaque areas = ice-cover to scale. (From Wetzel, unpublished data.)

levels, and these concentrations exhibit minor seasonal variations with depth. Usually, the amount of calcium utilized by the biota is so small in comparison to existing levels that these effects cannot be seen in normal analyses. Some accumulation may occur from decomposition in the hypolimnion of productive softwater lakes during stratification.

The calcium concentrations of hardwater lakes, however, undergo marked seasonal dynamics. The changes depicted in Figure 9–3 for a hardwater lake in southern Michigan are quite typical. Between rather uniform concentrations during spring and fall periods of circulation, a conspicuous stratification that is repeated annually with minor variations may be seen. Both the calcium levels and total alkalinity (Fig. 10–3) decreased markedly as a result of precipitation of CaCO$_3$ during the summer months from May through September. Similar losses were seen during the period of winter stratification. In late winter the decrease just beneath the ice is associated in part with dilution by rains permeating the ice, as well as with increases in photosynthesis at that time just before ice loss. Decreases of concentrations and of inorganic carbon in the epilimnion and metalimnion are associated directly with rapid increases in the rates of photosynthesis by phytoplankton and littoral flora (Otsuki and Wetzel, 1974), indicating the major role of photosynthetic induction in epilimnetic decalcification.

Detailed analyses of the calcium and bicarbonate budgets of this hardwater lake were determined for the influxes, dynamics within the lake, and outflow over an annual period. Calcium inputs of two spring-fed inlet streams were uniform and contributed from 93 to 98 percent of the total monthly inputs to the basin during the summer months. After compensation for potential evapotranspiration and precipitation, inputs of calcium from groundwater influxes varied from 2 to 58 per cent of total monthly inputs. Inputs of bicarbonate of the two inlet streams were also uniform annually, and constituted from 82 to 97 per cent of the total monthly influx during the summer period.

Contributions from the streams decreased from 35 to 67 per cent of the total inorganic carbon input during other times of the year because of the increase in groundwater influxes. The $CaCO_3$ precipitation rate was calculated at 446 g m^{-2} year.$^{-1}$ This loss of $CaCO_3$ influences the metabolism of hardwater lakes by coprecipitation of inorganic nutrients, such as phosphorus, and removal of humic, particularly yellow organic acids, and other organic compounds by adsorption.

The epilimnetic decalcification is reflected simultaneously in the distribution of specific conductance of hardwater lakes (Fig. 9–4). Since concentrations of Mg, Na, K, and Cl are relatively conservative, as discussed further on, the specific conductance follows the changes in Ca^{++} and HCO_3^- concentrations in nearly a 1:1 relationship ($r = 0.997$; Otsuki and Wetzel, 1974). A portion of the precipitating $CaCO_3$ is resolubilized in the hypolimnion and is reflected in both the increased concentrations of Ca^{++} and the specific conductance. Some of the $CaCO_3$ is entrained permanently in the sediments, and commonly constitutes > 30 per cent of the sediments by weight in moderately hardwater lakes (Wetzel, 1970). Adsorption of organic compounds to $CaCO_3$ lowers the rates of dissolution in hypolimnetic strata of reduced pH. Further, Ohle (1955a) has demonstrated that calcium complexes with humic acids, especially in the sediments, which alters exchange equilibria in the hypolimnion.

The biogenically-induced decalcification of the epilimnion reaches an extreme in very productive hard waters. For example, the calcium concentrations of the epilimnion of Wintergreen Lake, Michigan, were reduced from nearly 50 mg l^{-1} to below analytical detectability in a few weeks (Fig. 9–5). The magnitude of this loss from the trophogenic zone and the resulting increase in monovalent:divalent cation ratios should be recalled in subsequent discussion of the effects of the calcium content and of cation ratios on species distribution.

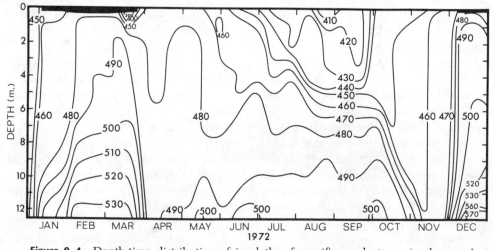

Figure 9–4 Depth-time distribution of isopleths of specific conductance (μmhos cm^{-1} at 25°C) of hardwater Lawrence Lake, Michigan, 1972. Opaque areas = ice-cover to scale. (From Wetzel, unpublished data.)

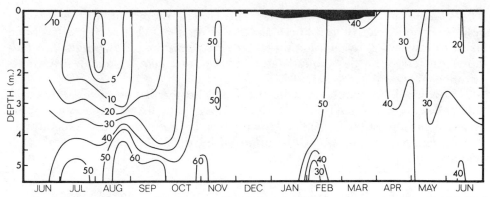

Figure 9–5 Depth-time distribution of isopleths of calcium concentrations (mg Ca^{++} l^{-1}) of hypereutrophic Wintergreen Lake, Kalamazoo County, Michigan, 1971–72. Opaque area = ice-cover to scale. (From Wetzel, et al., unpublished data.)

Magnesium

Magnesium is required universally by chlorophyllous plants as the magnesium porphyrin component of the chlorophyll molecules, and as a micronutrient in enzymatic transformations of organisms, especially in transphosphorylations of algae, fungi, and bacteria. The demands for magnesium in metabolism are minor in comparison to quantities generally available in fresh waters. Magnesium compounds, moreover, are in general much more soluble than their calcium counterparts. As a result, significant amounts of magnesium rarely are precipitated. The monocarbonates of hard waters are usually > 95 per cent CaCO$_3$ under ordinary CO$_2$ pressures. MgCO$_3$ and hydroxide precipitate significantly only at very high pH values (> 10) under most natural conditions. The concentrations of magnesium are extremely high in certain closed saline lakes.

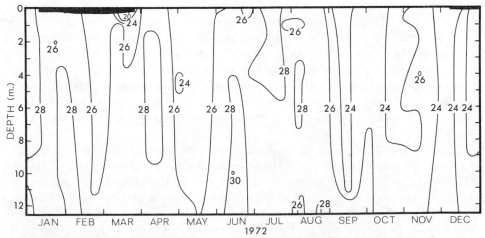

Figure 9–6 Depth-time distribution of isopleths of magnesium concentration (mg Mg^{++} l^{-1}) of hardwater Lawrence Lake, Michigan, 1972. Opaque areas = ice-cover to scale. (From Wetzel, unpublished data.)

As a result of these properties, the concentrations of magnesium are relatively conservative and fluctuate little (Fig. 9–6). This attribute has been used to advantage by employing the magnesium budget of inputs and out-flows to determine the groundwater influxes to a lake by magnesium mass balance (Wetzel and Otsuki, 1974).

Low available magnesium has been implicated as one of several factors influencing phytoplanktonic productivity in an extremely oligotrophic Alaskan lake (Goldman, 1960). However, such conditions are exceedingly rare in comparison to limitations imposed by restricted availability of other major nutrients.

Sodium, Potassium, and Minor Cations

The monovalent cations sodium and potassium are involved primarily in ion transport and exchange. An absolute sodium requirement has been demonstrated in only a few plants. The sodium requirements are particu-larly high in some species of the blue-green algae (Allen, 1952; Gerloff, et al., 1952; Allen and Arnon, 1955). Potassium and other elements of this series— lithium (Li), rubidium (Rb), and cesium (Cs)—cannot substitute for Na. A threshold level of 4 mg Na 1^{-1} is required for near optimal growth of several species (Kratz and Myers, 1954), a concentration that is about the mean for numerous hardwater lakes and presumably influences, among many other factors, the development of large populations of this algal group. Maximal growth of several blue-green algae was found at 40 mg Na 1^{-1}. The enrich-ment of waters with high levels of sodium and phosphorus, as is the case, for example, in domestic effluents with very high concentrations from synthetic detergents, was indicated as a potential contributor to effective competition among the blue-green algae under bloom conditions (Provasoli, 1958; Wetzel, 1966a, Ward and Wetzel, 1975). Both sodium and potassium occur in such abundance as highly soluble cations of numerous salts that alteration of the content of natural waters is not common.

The spatial and temporal distribution of sodium and potassium in lakes is uniform with very little seasonal variation, an indication of the conservative nature of these ions (Fig. 9–7; see also Stangenberg-Oporowska, 1967). Moderate epilimnetic reduction in potassium has been observed in extremely productive lakes; this is presumably related to potassium utilization by the massive algal populations, and to biotic sedimentation in the hypolimnion. An analogous situation has been found in fertilized, productive farm ponds (Barrett, 1957).

The rarer alkaline earth and alkali cations of natural waters vary con-siderably in concentration in relation to the lithology of the drainage basins. Their distribution is discussed by Durum and Haffty (1961) and Livingstone (1963) in general, and in closed-basin lakes in particular by Whitehead and Feth (1961). The general ratios to major cations are given in Table 9–5. Nutritional requirements of these elements have not been demonstrated, although they frequently can be substituted for the major cations in metabolic pathways.

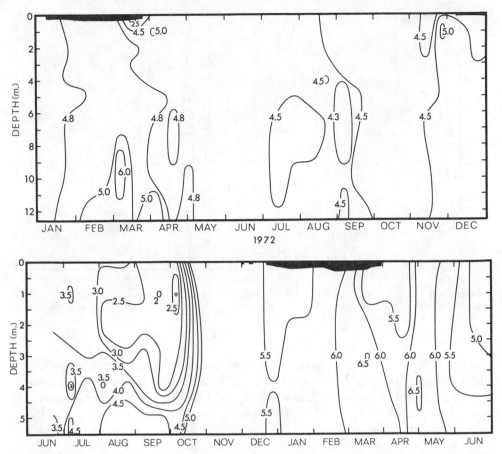

Figure 9–7 Depth-time distribution of isopleths of sodium concentrations (mg Na$^+$ l^{-1}) of Lawrence Lake, 1972 (*upper*) and potassium concentrations (mg K$^+$ l^{-1}) of hypereutrophic Wintergreen Lake, Michigan, 1971–72 (*lower*). Opaque areas = ice-cover to scale.

Chloride and Other Anions

In our prior discussion of the general distribution of chloride ions in lakes, it was emphasized that this anion is usually not dominant in open lake systems. Lakes of maritime regions often receive significant inputs of chlorides from atmospheric transport from the sea. Pollutional sources of chlorides

TABLE 9–5 Approximate Ratios of Minor Alkaline Earths and Alkalies to the Major Cations[a]

Na/Li 1500	Ca/Ba 400
Na/Rb 3600	Ca/Sr 3000–4500
Na/Cs 31900	Ca/Be ca 40000
	Ca/Ra 5 × 10^{10}

[a]From data of Livingstone, 1963.

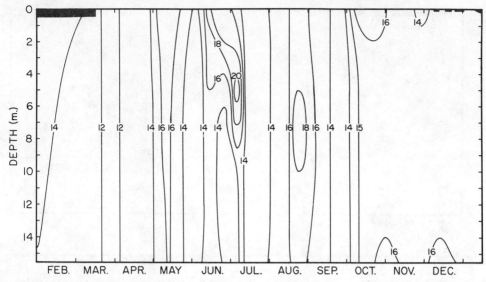

Figure 9–8 Depth-time distribution of isopleths of chloride concentrations (mg Cl⁻ l⁻¹) of eutrophic Little Crooked Lake, Whitley County, Indiana, 1964. Opaque areas = ice-cover to scale. (From Wetzel, unpublished data.)

can modify natural concentrations greatly. For example, the concentration of chloride in Lake Erie increased threefold in 50 years, being derived largely (ca 70 percent) from industrial sources, road salting, and municipal waste-waters (Ownbey and Kee, 1967). Chloride is influential in general osmotic salinity balance and ion exchange, but metabolic utilization does not cause significant variations in the spatial and seasonal distribution within a lake (Fig. 9–8). Variations observed were associated with the hydrology of the basin.

The concentrations of minor halides (Table 9–6) vary somewhat with the drainage lithology of the basin, and higher concentrations often occur in lakes in proximity to marine regions or in those that possess marine rock formations within their drainage basins. Boron is of greater limnological interest because it is a required micronutrient of many algae and other organisms.

TABLE 9–6 Approximate Average Concentrations of Halides and Boron in Natural Fresh Waters[a]

Element	Average Concentration (mg l⁻¹)	Chloride Ratio
Chloride	8.3	
Fluoride	0.26	Cl/F 32
Bromine	0.006	Cl/Br 1400
Iodine	0.0018	Cl/I 4600
Boron	ca 0.01	

[a]From data of Livingstone, 1963.

Concentrations in natural waters are relatively high in comparison to other minor elements. The average concentration in rain and snow is high (4.7 μg B l^{-1}) and apparently results largely from terrestrial continental sources rather than from the sea (Nishimura, et al., 1973). Concentrations reach exceedingly high levels in certain closed, saline lakes, approaching a gram per liter. As is the case with most micronutrients, high concentrations are generally quite toxic to most organisms, as was exemplified by the well-known excessive use of the micronutrient copper as a herbicide. Those organisms adapted to high concentrations of boron, to > 800 mg l^{-1}, were affected little by additions of further borate (Wetzel, 1964). On the other hand, aquatic angiosperms normally living in waters of very low boron content show a high resistance to large concentrations of boron. Such additions were found to be stimulatory to photosynthesis to levels of 100 mg l^{-1}, beyond which inhibitory responses occurred (Baumeister, 1943).

Cation Ratios

The question of the importance of monovalent to divalent cation ratios is particularly interesting in regard to the distribution and dynamics of algae and larger aquatic plants in fresh waters. Three major genera of diatoms common to oligotrophic waters, *Fragilaria*, *Asterionella*, and *Tabellaria*, are stimulated by high levels of calcium (Chu, 1942; Vollenweider, 1950). Increasing levels of potassium permit increased tolerance of these species to high concentrations of Ca and Mg. Numerous studies, particularly by Provasoli and coworkers (Provasoli, 1958), indicate that the ratio of monovalent to divalent cations (M:D) is significant to the observed growth of these algae. The concentrations of Ca and Mg can be manipulated over a wide range as long as the M:D ratio is maintained within reasonably narrow limits for different species. Calcium and Mg are widely interchangeable, and many species are quite tolerant to different Ca:Mg ratios. These experimental results provide excellent confirmation of the much earlier perceptive ecological observations of Pearsall (1921; 1932), who considered a M:D ratio below 1.5 favorable to diatoms and much higher ratios favorable to desmid algae. Diatoms dominate the algal flora of very hardwater lakes with M:D ratios much less than 1.5. As the epilimnion undergoes biogenically-induced decalcification in early summer in these calcareous waters, the concentration of the spring diatom maximum is often halved. It is unknown whether, along with other factors, this shift in M:D ratio, concomitant with the shift to mixed populations of predominately green algae with diatoms, is influential in phytoplankton succession.

The effects of cations on the photosynthesis and release of dissolved organic matter were studied on a submersed angiosperm, *Najas flexilis*, a macrophyte that occurs in many lakes but grows poorly in extremely calcareous hardwater lakes (Wetzel, 1969; Wetzel and McGregor, 1968). Concentrations of Ca over 30 mg l^{-1} and Mg over 10 mg l^{-1}, both exceeded by far in hardwater lakes, suppressed the rates of carbon fixation and altered the secretion rates of dissolved organic matter. Increasing the concentrations of

Na above levels commonly found in hardwater lakes increased both the rates of fixation and secretion. The overall effects observed were decreased rates of photosynthetic carbon fixation with decreasing M:D ratios.

The potential of dissolved organic compounds of fresh waters in regulation of M:D ratios and in indirectly influencing rates of photosynthesis was emphasized by Wetzel (1968). Sequestering of divalent cations by true chelation by amino substances and peptides is well-known. Complex formation is also possible by pyrophosphates, binding with macromolecules such as proteins, and the formation of peptized metal hydroxides of yellow organic compounds of humic acids. The mechanisms have not been adequately investigated in relation to their control over algal succession and dynamics, but it is clear that changes in dissolved organic matter can exert a major influence on productivity. This subject is discussed in greater detail in Chapter 14.

SALINITY, OSMOREGULATION, AND DISTRIBUTION OF BIOTA

Origins and Distribution of Freshwater Biota

The salinity of inland waters is generally very low in comparison to the sea, even though in semi-arid regions salinity of closed-basin lakes occasionally exceeds that of sea water by several times. The distribution of biota in fresh waters has been influenced by a long evolutionary history of physiological adaptations to a wide range of salinities or mechanisms for osmotic regulation. These changes have developed against a background of large differences in the salinity of the environment and that of the cytoplasm or body fluids. The general distribution of the freshwater biota and their tentative origins in terrestrial, freshwater, or marine sources are summarized in the detailed review of Hutchinson (1967). That summary, coupled with several major reviews of the physiological mechanisms and adaptation of osmotic regulation (Krogh, 1939; Beadle, 1943, 1957, 1959; Gessner, 1959; Robertson, 1960; and Potts and Parry, 1964), provides an introduction to the adaptations to fresh water life. The distribution of biota in the brackish water interface regions between marine and fresh waters is summarized in great detail in the symposium on brackish waters (1959), and especially in Remane and Schlieper (1971).

The number of species living in brackish water is very much smaller than that living in marine regions with similar habitats, and much smaller than the number of species in fresh water (Fig. 9–9). However, it should be pointed out immediately that, although there is a paucity of species in this region, this in no way implies a low productivity; productivity by adapted organisms can be exceedingly high. The lowest number of species is displaced close to fresh water at about 5 to 7‰. The salinity gradient seen here, with its associated osmotic and ionic properties, is the predominant factor influencing the distribution of biota in brackish waters. In these transitional waters, as in the spectrum of salinities of inland waters, the salinity range occupied by a species depends on the efficiency of the physiological mechanisms by which it is adapted to changes in salinity in the environment.

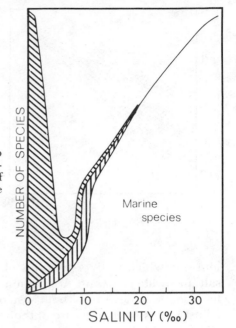

Figure 9–9 Number of species in relation to salinity. Diagonal hatching: proportion of fresh-water species; vertical hatching: proportion of brackish-water species; lower open area: marine species. (After Remane and Schlieper, 1971.)

*Osmotic Adaptations of Simple and Higher
Aquatic Plants and Animals*

Osmotic regulation functions primarily in the maintenance of a differ-ence in concentrations of ions inside and outside of the cells at appropriate operational physiological levels. The aquatic bacteria and blue-green algae (Monera: lacking mitochondria or chromoplastids) and the more primitive Protista (algae, fungi, and protozoa with mitochondria and, if they are photo-synthetic, with chromoplastids) demonstrate high evolutionary adaptability to changes in salinity (cf. Hutchinson's 1967 description of evolutionary euryhalinity) by means of relatively small genetic changes. Most freshwater bacteria and blue-green algae are relatively homoiosmotic, tolerating only a narrow range of salinity, but adapt to increasing salinity relatively rapidly by means of genetic change. Much adaptive radiation is seen among these groups. The flora and animals of the Protista, that are largely single-celled, retain considerable evolutionary euryhalinity and are widely distributed with respect to salinity, although some groups, especially among the green algae, are restricted to fresh water. The contractile vacuole is the primary osmoregu-latory organelle among the Protista.

Higher aquatic plants have developed adaptations to fresh water sec-ondarily from terrestrial origins. Only several major groups of angiosperms have developed extensively in fresh waters, and very few groups extend into saline waters of brackish and marine areas or hypersaline closed-basin lakes.

Nearly all of the higher freshwater animals originated from the sea or, especially among the insects, are terrestrial in origin. Both in terms of evolu-tionary and contemporary life in fresh water, osmoregulation is a major prob-

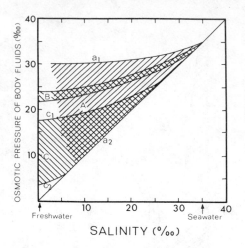

Figure 9–10 Relationship between the osmotic pressure, expressed as salinity, of body fluids of brackish-water organisms (A) and freshwater animals (B, C) and the salinity of external water. (After Beadle, 1959.) Relationships of the curves are detailed in the text.

lem, for which a diverse series of mechanisms have developed to regulate salt and water content. Adaptation to low salinities by some marine animals has been achieved without osmoregulation. Such *poikilosmotic* animals adjust the osmotic pressure of their body fluids such that they are more or less isotonic with the salinity of the medium. In contrast, a *homoiosmotic* animal will tend to retain its initial internal osmotic concentration upon being exposed to modest changes of salinity of the medium. The general relations between the osmotic pressure among different types of animals, expressed as salinity of the blood, and the salinity of the external water can be visualized from Figure 9–10. The range depicted by area A extends over a wide variation in osmotic pressure of body fluids that is found in brackish water animals, which tend to be more poikilosmotic at high concentrations and more homoiosmotic at lower salinities. The range of the osmotic pressure curve extends from the most homoiosmotic (a_1) to the most poikilosmotic species (a_2). The lower limits are very variable, represented by the undefined lefthand edge of area A, but all of these species have failed to colonize in fresh water.

A few species have succeeded in penetrating fresh water without a renal osmoregulatory mechanism. These brackish water animals are partially homoiosmotic (area B, Fig. 9–10), and maintain a very high osmotic pressure in hypertonic body fluids by the active uptake of ions from the water. The excretory organs are not involved, since the urine produced is isotonic with the blood.

In the majority of freshwater animals, however, the osmotic pressures of the body fluids have decreased to levels equivalent to 5 to 15‰ salinity, at which the osmotic gradients are reduced. These organisms have developed excretory organs that effectively recover ions and produce urine hypotonic to the body fluids. The majority of freshwater animals therefore effect osmoregulation by active uptake of ions and by a renal mechanism of ion retention. Extremes in osmotic pressures of blood delineate area C of Figure 9–10 by curves c_1 and c_2. The isotonic line along the righthand edge of area C indicates the upper salinity tolerance limit of most freshwater animals, and that they are incapable of hypotonic regulation. While most freshwater animals are capable of living in water of low salinity, the adaptation of body fluids of low osmotic pressure is apparently irreversible, and with few exceptions they are restricted to salinities of < 10 ‰.

Because of the slow diffusion of oxygen in water in comparison to that in air, movement of water over permeable membranes or tissue surfaces for respiratory needs is almost universal among aquatic animals. The pumping process places high energetic demands on the animals, and additionally exposes cellular surfaces to osmotic gradients. Mechanisms for taking up salts against a concentration gradient vary greatly among freshwater animals. In a few organisms, incorporation of salts with the food may be adequate, but usually some active organs have developed for this purpose. Aside from resorption mechanisms of the excretory organs, which are advantageous energetically, active uptake mechanisms for ions, especially sodium and chloride, are often associated with respiratory organs, or the gills of many invertebrates and vertebrates.

The fauna of extremely saline inland waters are relatively insensitive to the high salinity of these lakes and to large fluctuations in the chemical composition of the water (Beadle, 1969). Most of the animals of saline lakes are of freshwater origin, and include particularly representatives of the aquatic insects, phyllopods, copepod and cladoceran crustacea, and rotifers, all of which belong to predominantly freshwater groups. The blood of these saline-inhabiting animals maintains osmotic pressures at levels characteristic of those of fresh waters. The body surface of these animals exhibits very low permeability, and they possess effective excretory mechanisms for maintaining the body fluids strongly hypotonic to the external medium. Further, the water balance of saline inland waters frequently fluctuates widely; freshwater animals with resting stages capable of withstanding desiccation have a distinct advantage over marine animals which lack strong development of this characteristic.

chapter 10

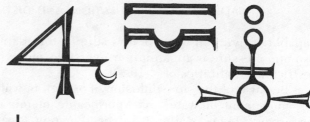

inorganic carbon

THE OCCURRENCE OF CARBON IN FRESHWATER SYSTEMS

A majority of the carbon in freshwater systems occurs as equilibrium products of carbonic acid. A smaller amount of carbon occurs in organic compounds as dissolved and particulate detrital carbon, and a small fraction occurs as carbon of living biota.

The complex equilibrium reactions of inorganic carbon and the distribution of species of total CO_2 ($\Sigma CO_2 = CO_2 + HCO_3^- + CO_3^=$) are understood in considerable detail. As atmospheric CO_2 dissolves in water containing bicarbonate, primarily associated with the cation calcium, dissociation kinetics of the inorganic carbon of fresh waters closely follow those of pure dilute solution chemistry below a pH of 8.5, and consist of: (a) free CO_2; (b) carbonic acid that is partly undissociated and partly dissociated (H_2CO_3, $CO_3^=$, H^+); (c) dissociated calcium bicarbonate (Ca^{++}, HCO_3^-, $CO_3^=$); and (d) hydroxyl ions (OH^-) formed by the hydrolysis of carbonic acid. A dilute solution of bicarbonate is weakly alkaline because slightly greater concentrations of OH^- ions than H^+ ions result from the dissociation of H_2CO_3. Above pH 8.5, total inorganic carbon dioxide concentrations are less than those predicted on the basis of apparent dissociation constants of carbonic acid. The pH of most fresh waters results from the H^+ ions of the dissociation of H_2CO_3 and OH^- ions of the hydrolysis of bicarbonate ions. An appreciable amount of precipitated particulate and colloidal $CaCO_3$, induced largely by photosynthetic utilization of CO_2, constitutes a major sink for inorganic carbon and organic carbon adsorbed to the calcium carbonate ($CaCO_3$) in hardwater lakes.

Although the exchange of CO_2 of the water with that of the atmosphere is rapid and relatively complete in aerated open water of lakes, the spatial and temporal distribution of ΣCO_2 and pH is altered by microbial metabolism in stratified zones of lakes and by photosynthesis in stagnated areas of the littoral zone. Lakes exhibiting an orthograde oxygen curve generally have an orthograde ΣCO_2 curve. In productive waters exhibiting a clinograde oxygen curve, an inverse clinograde ΣCO_2 curve is generally found. The ΣCO_2 distribution of hardwater calcareous lakes is modified greatly by biogenically-induced decalcification of the epilimnion. The accumulation of CO_2 in the hypolimnion

166

clearly has been related to rates of organic production in the trophogenic zone of small lakes with minimal inputs of organic matter from outside the basin.

Inorganic carbon constitutes a major nutrient of photosynthetic metabolism by algae and submersed macrophytes. Only rarely is inorganic carbon limiting to photosynthesis in natural situations. Assimilation of bicarbonate is known and can be induced among plants that have an affinity for both CO_2 and HCO_3^- ions under conditions of low free CO_2 and high bicarbonate concentrations.

Carbon Dioxide and Its Solution in Water

The carbon dioxide (CO_2) content of the atmosphere varies with locality and potential enrichment from industrial pollution. The global average is approximately 0.032 per cent by volume as of 1970, and is increasing progressively (Machta, 1973). Carbon dioxide is very soluble in water, some 200 times greater than oxygen, and obeys normal solubility laws within the conditions of temperature and pressure encountered in lakes. The amount of CO_2 dissolved in water from atmospheric concentrations is about 1.1 mg, 1^{-1} at 0°C, 0.6 at 15°C, and 0.4 mg 1^{-1} at 30°C.

As CO_2 dissolves in water, the solution contains unhydrated CO_2 at about the same concentration by volume (ca 10 μM) as in the atmosphere (reviewed extensively by Hutchinson, 1957; Kern, 1960; Stumm and Morgan, 1970; J. C. Goldman, et al., 1972):

$$CO_2 \text{ (air)} \rightleftharpoons CO_2 \text{ (dissolved)} + H_2O$$

The CO_2 of the water hydrates by the slow reaction with a half-time of ca. 15 seconds to yield carbonic acid:

$$CO_2 + H_2O \rightleftharpoons H_2CO_3$$

This reaction predominates at a pH of less than 8, and the concentration of H_2CO_3 is only about 1/400 that of the unhydrated CO_2 at equilibrium. Above a pH of 10, $CO_2 + OH^- \rightleftharpoons HCO_3^-$ is the dominant reaction.

H_2CO_3 is a fairly weak acid that dissociates rapidly in comparison to the hydration reaction:

$$H_2CO_3 \rightleftharpoons H^+ + HCO_3^-$$
$$HCO_3^- \rightleftharpoons H^+ + CO_3^=$$

The pK_i dissociation value of the first reaction, including both hydrated and unhydrated CO_2 as the undissociated molecule, is 6.43 at 15°C. The pK_2 of the second reaction is 10.43 (15°C).

The bicarbonate and carbonate ions dissociate after equilibrium is established:

$$HCO_3^- + H_2O \rightleftharpoons H_2CO_3 + OH^-$$
$$CO_3^= + H_2O \rightleftharpoons HCO_3^- + OH^-$$
$$H_2CO_3 \rightleftharpoons H_2O + CO_2$$

The hydroxyl OH^- ions formed by the first two of the above reactions result in alkaline waters above a pH of 7 in lakes that have a naturally high content of carbonates derived from surface and groundwater of the drainage basin. As water enters the soil, it becomes enriched with CO_2 from plant and microbial respiration. The carbonic acid effectively solubilizes limestone of calcium enriched rock formations, forming calcium bicarbonate ($Ca(HCO_3)_2$), which is relatively soluble in water, and increases the amount of ionized Ca^{++} and HCO_3^- of the water. With increasing content of HCO_3^- and $CO_3^=$ of hard waters of calcareous regions, common to much of the glaciated temperate region, the pH is increased by these reactions. In normal hard waters of the midwestern United States, for example, HCO_3^- predominates (ca. 60 per cent) among the anions with a shift of pH to > 8 and $[HCO_3^-] > 100$ mg l^{-1}. Saline lakes of endorheic regions often have a carbonate dominance of $CO_3^= > 10,000$ mg l^{-1} and $HCO_3^- > 1000$ mg l^{-1} with pH > 9.5. Such highly saline carbonate brines, in which total inorganic carbon exceeds several moles per liter, are rare.

Photosynthesis and respiration are two major factors that influence the amounts of CO_2 in water. However, the equilibria of the equations given above explain the buffering action of alkaline waters containing appreciable amounts of bicarbonate. The water tends to resist change in pH as long as these equilibria are operational. The addition of hydrogen ions neutralizes hydroxyl ions formed by the dissociation of HCO_3^- and $CO_3^=$, but more are formed immediately by reaction of the carbonate with water as long as the reservoir of carbonate ion is present. The pH remains essentially the same as before the addition, until the supply of carbonate or bicarbonate ions is exhausted. Similarly, when hydroxyl ion is added, it reacts with bicarbonate ion:

$$HCO_3^- + OH^- \rightleftharpoons CO_3^= + H_2O$$

If the pH of a solution is held constant with buffer reactions, and it is permitted to equilibrate with gaseous carbon dioxide, the total hydrated and unhydrated CO_2 in solution is independent of pH, while the bicarbonate and carbonate concentrations increase with pH in accordance with the pK values. These equilibria are influenced by temperature and by salt concentration. At the salinity of seawater, the pK_i for HCO_3^- is about 0.5 unit, and the pK_2 for

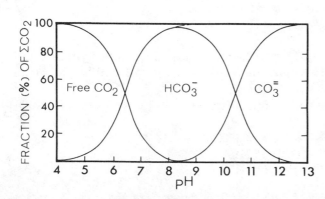

Figure 10–1 Relation between pH and the relative proportions of inorganic carbon species of CO_2 (+H_2CO_3), HCO_3^-, and $CO_3^=$ in solution. (Slightly modified from Golterman, H. L. (ed.), Methods for Chemical Analysis of Fresh Waters. IBP Handbook No. 8. Oxford England, Blackwell Scientific Publications, 1969.)

TABLE 10–1   Proportions of CO_2, HCO_3^-, and $CO_3^=$ in Water at Various pH Values[a]

pH	Total Free CO_2	HCO_3^-	$CO_3^=$
4	0.996	0.004	1.25×10^{-9}
5	0.962	0.038	1.20×10^{-7}
6	0.725	0.275	0.91×10^{-5}
7	0.208	0.792	2.6×10^{-4}
8	0.025	0.972	3.2×10^{-3}
9	0.003	0.966	0.031
10	0.0002	0.757	0.243

[a]From Hutchinson, G. E.: A Treatise on Limnology. New York, John Wiley & Sons, Inc., 1957, p. 657.

$CO_3^=$ is about 1 unit lower than in fresh water. Both oceanic and fresh waters are close to equilibrium with atmospheric CO_2. In the marine habitat, the inorganic carbon pool contains about 2 millimoles C 1^{-1}, largely as HCO_3^-, a reservoir some 50 times that of the atmosphere. In fresh waters, the total inorganic carbon (ΣCO_2) is much more variable, within a typical range of 50 micromoles to 10 millimoles 1^{-1}, and more pH-dependent in relation to $[HCO_3^-]$ and $[CO_3^=]$. From these dissociation relationships, the proportions of CO_2, HCO_3^-, and $CO_3^=$ at various pH values can be evaluated (Fig. 10–1). Free CO_2 dominates in water at pH 5 and below, and above pH 9.5 $CO_3^=$ is quantitatively significant (Table 10–1). Between pH 7 and 9, HCO_3^- predominates.

CO_2 Exchange Between the Atmosphere and Water

The diffusion of CO_2 from the atmosphere and the dissociation kinetics of dissolved carbonates are obviously of major importance to photosynthetic organisms dependent on the availability of inorganic carbon. The magnitude of CO_2 exchange between the atmosphere and water cannot be determined by partial pressure differences alone. Many lakes with surface waters near neutrality are slightly supersaturated with CO_2 relative to the atmospheric pressure of CO_2. Other waters, especially in alkaline bicarbonate lakes containing large amounts of carbonate, are apparently not in equilibirum with the CO_2 of the atmosphere, although they may be with other gases such as oxygen. Diffusion of atmospheric CO_2 has been elaborated by Broecker and coworkers (1965, 1968, 1971, 1973), who have utilized techniques of radium-226 and flux of radon-222 to determine gas transfer between the atmosphere and water. When applied to a softwater lake of the Canadian Shield of very low ΣCO_2, atmospheric invasion of CO_2 was adequate (0.12 ± 0.06 g C m^{-2} day^{-1}) to account for 30 to 90 per cent of carbon fixation by phytoplankton (Schindler, et al., 1972). In this relatively sheltered lake, an invasion rate of CO_2 from the air of 17 ± 8 mol CO_2 m^{-2} day^{-1} was estimated through a "stagnant boundary layer" at the surface of about 300 μm in thickness (Emerson, et al., 1973). This layer decreases in thickness with increased exchange as wind velocities increase, especially above 1.5 m sec^{-1}. Organic substances dissolved in or present on the water surface also decrease exchange rates of gas and evaporation. At the opposite extreme, in a hardwater lake in Michigan that receives high inputs of inorganic carbon, Otsuki and Wetzel (1974) determined from lake budgetary evaluations of calcium and total alkalinity that the estimated partial pressure in two inlet waters and groundwater was about 30 times higher than that in the atmosphere, and that in the outlet water was 7 times higher during the spring period, and 3 times higher even during the summer period of maximum utilization and losses by precipitation of $CaCO_3$. These estimates indicate that CO_2 in this lake was being released into the atmosphere throughout the year.

Proportions of Carbonates in Fresh Water

The total inorganic carbon concentration in fresh water depends on the pH, which is governed largely by the buffering reactions of carbonic acid, and the amount of bicarbonate and carbonate derived from the weathering of rocks. Carbonates exist as a number of polymorphic and hydrated forms. The most important carbonate of aquatic systems is $CaCO_3$, which occurs in natural waters principally as calcite and the metastable polymorph aragonite.

The solubility of CO_2 in water increases markedly in water that contains carbonate. A definite amount of CO_2 will remain free in solution after equilibrium is reached between calcium, bicarbonate, carbonate, and undissociated carbonate. The amount of excess CO_2 required to maintain stability of $Ca(HCO_3)_2$ in solution increases very rapidly with increasing content of bicarbonate in the water being derived from carbonates. If the amount of free CO_2 is increased above that required to maintain a given amount of $CaCO_3$ in solution at equilibrium as $Ca(HCO_3)_2$, this aggressive CO_2, as it is termed, will dissolve more $CaCO_3$.

If a solution of calcium bicarbonate in equilibrium with CO_2, H_2CO_3, and $CO_3^=$ loses a portion of the CO_2 required to maintain the equilibrium, $CaCO_3$ will precipitate until the equilibrium is reestablished by the formation of CO_2:

$$Ca(HCO_3)_2 \rightleftharpoons CaCO_3 \downarrow + H_2O + CO_2$$

CO_2 in excess, required to maintain large amounts of HCO_3^- in solution and equilibrium, can be lost in several ways, resulting in massive precipitation of $CaCO_3$. Groundwater of limestone regions, heavily enriched with CO_2 of terrestrial origin from decomposition, can release much CO_2 into the atmosphere when it flows to the surface, with resulting precipitation of $CaCO_3$. When such spring water, rich in bicarbonate, surfaces in streams or lakes, all substrata are covered with a dense encrustation of $CaCO_3$. A major cause of the loss of aggressive CO_2 in lakes and certain streams is photosynthetic utilization of CO_2 by littoral flora and phytoplankton. Hardwater lakes rich in bicarbonate undergo massive epilimnetic decalcification during the summer stratification period of active photosynthesis, a subject discussed later.

In the littoral zone, the submersed macrophytic vegetation of these lakes is encrusted densely with $CaCO_3$ precipitation induced by photosynthetic utilization of CO_2 by the macrophytes and epiphytic algae. The marl encrustations frequently are massive, and exceed the weight of the plant (Wetzel, 1960). Blue-green algae growing attached to substrata in the littoral of lakes and in streams also are active precipitators of large deposits of carbonates (Golubić, 1973).

In addition to highly dynamic demands for CO_2 from and inputs of CO_2 to fresh water, complex shifts in precipitation and dissolution reactions of carbonate occur spatially and temporally. In alkaline hardwater lakes, often much larger (2 times) perennial concentrations of calcium and bicarbonate are found than would be expected on the basis of equilibrium with pressures of CO_2 normally found in the atmosphere (Ohle, 1934, 1952; Wetzel, 1966b, 1972; Otsuki and Wetzel, 1974). The solubility product of $CaCO_3$ is low (0.48×10^{-8}), and $CaCO_3$ can start precipitating from calcareous waters when the pH is sufficiently high in a uniformly buffered system, or where equilibria are shifted in microzones associated with active photosynthesis of plant cells or surfaces. However, large amounts of inorganic carbon can exist as carbonate and $CaCO_3$ in metastable conditions. Its rate of precipitation is slow, unless induced metabolically, as by photosynthesis. The result is a supersaturation with respect to both Ca^{++} and HCO_3^- at concentrations often 2 to 3 times that expected from equilibria predictions. There is strong evidence that appreciable $CaCO_3$ occurs in stable colloidal form in markedly hardwater lakes. The trophogenic zone in moderately hardwater Lake Mendota was found to be supersaturated with respect to Ca^{++} and HCO_3^- in all seasons except winter (Morton and Lee, 1968). Extremely hardwater Lawrence Lake was found to be supersaturated continually (Otsuki and Wetzel, 1974), a situation found in numerous situations where dissolved CO_2 is not in equilibrium with the atmosphere.

The importance of colloidal $CaCO_3$, in addition to larger particulate $CaCO_3$, is just beginning to be appreciated in relation to indirect effects upon metabolism and flux rates of organic carbon (Wetzel and Rich, 1972). Organic compounds (amino acids, fatty acids, humic acids) adsorb strongly to particulate and colloidal $CaCO_3$ (Chave, 1965; Chave and Suess, 1970; Suess, 1968, 1970; Meyers and Quinn, 1971a; Wetzel and Allen, 1970; Otsuki and Wetzel, 1973; Orlov, et al., 1973). Although such adsorption could be viewed as a scavenging and concentrating process of labile dissolved organic carbon from dilute solution for a more ready utilization by bacteria, empirical evidence indicates, rather, a chemical competition with the bacteria for the organic substrates. During photosynthetic removal of CO_2, a large fraction of $CaCO_3$ is precipitated by algae and macrophytic vegetation. Frequently, the plant cells serve as a nucleus for particulate $CaCO_3$ formation, which occurs at the site of simultaneous secretion of organic compounds. This association of dissolved organic detrital carbon with $CaCO_3$ is a component of certain freshwater systems which affects the chemical milieu without clearly defined energetic transformations (Wetzel, et al., 1972). The organic coatings also reduce the rate of dissolution of sedimenting $CaCO_3$ in lakes, and form a major sink for inorganic and organic detrital carbon (Wetzel, 1970, 1972).

Alkalinity and Acidity of Natural Waters

Natural waters exhibit wide variations in relative acidity and alkalinity, not only in actual pH values, but also in the amount of dissolved material producing the acidity or alkalinity. The concentration of these compounds and the ratio of one to another determine the actual pH and the buffering capacity of a given water. Since the lethal effects of most acids begin to appear near pH 4.5 and of most alkalis near pH 9.5, it is evident that buffering can be of major importance in the maintenance of life.

Alkalinity of waters, as usually interpreted, refers to the quantity and kinds of compounds present which collectively shift the pH to the alkaline side of neutrality. The property of alkalinity is usually imparted by the presence of bicarbonates, carbonates, and hydroxides, and less frequently in inland waters by borate, silicate, and phosphates. The CO_2–HCO_3^-–$CO_3^=$ equilibrium system is the major buffering mechanism in fresh waters.

The terms alkalinity, carbonate alkalinity, alkaline reserve, titratable base, or acid-binding capacity are frequently used to express the total quantity of base, usually in equilibrium with carbonate or bicarbonate, that can be determined by titration with a strong acid (Hutchinson, 1957). The milliequivalents of acid necessary to neutralize the hydroxyl, carbonate, and bicarbonate ions in a liter of water are known as the total alkalinity. Alkalinity is numerically the equivalent concentration of titratable base, and is determined by titration with a standard solution of a strong acid to equivalency points dictated by pH values at which the contributions of hydroxide, carbonate, and bicarbonate are neutralized.

The least ambiguous usage of alkalinity is to express values as mass per unit volume; i.e., milliequivalents per liter, in which alkalinity remains unaffected by CO_2 added by conversion of carbonate to bicarbonate. In effect, alkalinity measures the proton deficiency with respect to the reference proton level CO_2–H_2O (this topic is reviewed at length by Stumm and Morgan, 1970). Alkalinity is often expressed in milligrams per liter (or parts per million) as $CaCO_3$. This expression assumes that alkalinity results only from calcium carbonate and bicarbonate, which in some lakes, e.g., closed alkaline lakes, implies much greater calcium than actually is present. In moderately hard waters, nearly all of the base is present as bicarbonate, and the term bicarbonate alkalinity, as mg HCO_3^- 1^{-1}, is sometimes used. The clearest expression, however, is milliequivalents per liter (= 50 mg 1^{-1} as $CaCO_3$).

The term *hardness* is frequently used as an assessment of the quality of water supplies. The hardness of a water is governed by the content of calcium and magnesium salts, largely combined with bicarbonate and carbonate (temporary hardness) and with sulfates, chlorides, and other anions of mineral acids (permanent hardness). The carbonate hardness can be removed by boiling which causes precipitation of $CaCO_3$. The fraction of calcium and magnesium remaining in solution as sulfates, chlorides, and nitrates after boiling constitutes the residual noncarbonate hardness. The extent of hardness has been expressed numerically in degrees of hardness in a remarkably heterogeneous system of scales among different countries (equivalents are given in Table 10–2). For example, one degree of hardness (H°) in the United States equals

TABLE 10–2   Various Scales of Hardness of Water[a]

1 German degree of hardness, dH°	$\begin{cases} = 10 \text{ mg CaO } l^{-1} \\ = 7.14 \text{ mg Ca } l^{-1} \\ = 17.9 \text{ mg Ca(HCO}_3)_2 \text{ } l^{-1} \end{cases}$
1 French degree of hardness, French H°	$= 10 \text{ mg CaCO}_3 \text{ } l^{-1}$
1 English degree of hardness, English H°	$\begin{cases} = 10 \text{ mg CaCO}_3 \text{ } 0.7 \text{ } l^{-1} \\ = 0.8° \text{ dh} \end{cases}$
1° dH	$\begin{cases} = 1.25° \text{ English H}° \\ = 1.79 \text{ French H}° \end{cases}$
1 French H°	$\begin{cases} = 0.56 \text{ German dH}° \\ = 0.7 \text{ English H}° \end{cases}$
1 American degree of hardness	$\begin{cases} = 1 \text{ mg CaCO}_3 \text{ } l^{-1} \\ = 0.056° \text{ German dH}° \end{cases}$
International degree of hardness, mval	$\begin{cases} = 1 \text{ meq } l^{-1} \\ = \dfrac{\text{German dH}°}{2.8} \end{cases}$

[a]After Höll, K.: Water: Examination, Assessment, Conditioning, Chemistry, Bacteriology, Biology. Berlin, Walter de Gruyter, 1972.

1 mg $CaCO_3$ l^{-1}; one German H° corresponds to a concentration of 10 mg lime (CaO) l^{-1}. Höll (1972) has proposed that the international unit be expressed in mval; 1 mval = 1 milliequivalent per liter of the material concerned.

Acidity is used infrequently as a parameter in limnological investigations, but it still is found in aquatic literature. Uncombined carbon dioxide, organic acids such as tannic, humic, and uronic acids, mineral acids, and salts of strong acids and weak bases are usually responsible for the acidity of natural waters. Free CO_2 of most waters is seldom present in large quantities, because of its reactions in equilibria of the carbonate complex and exchange with the atmosphere, discussed earlier.

In practice, acidity is a measure of the quantity of strong base per liter required to attain a pH equal to that of a molar solution of sodium carbonate (Na_2CO_3) equivalent to the total inorganic carbon; i.e., it is a measure of the active or free CO_2 expressed as meq or mg l^{-1} $CaCO_3$ rather than as a concentration of CO_2. Mineral acid acidity measures those materials present, other than CO_2, which result in the pH value of water below 4.5. Details on methodology for evaluation of alkalinity and acidity are given in the American Public Health Association publication (1971), and especially in Golterman (1969).

Hydrogen Ion Activity

Pure water dissociates weakly to H^+ and OH^- ions. The dissociation constant is very small (10^{-14}), however, and the amounts of H^+ and OH^- present are 10^{-7} g–ions per liter. Natural waters are, of course, not pure, and salts, acids, and bases contribute to the H^+ and OH^- ions in varying ways, depending on the individual circumstances. Since the dissociation constant of water is fixed, addition of one ion will result in a decrease of the other. The pH usually is defined as the logarithm of the reciprocal of the concentration of free hydrogen ions[1]. The 'p' of pH refers to the power (puissance) of the hydrogen ion activity. Therefore, more H^+ activity in an acid reaction increases the power from neutrality (10^{-7} or pH 7) to say 10^{-4} (pH 4). In more alkaline reactions, H^+ ion activity is decreased from neutrality to, for example, 10^{-10} (pH 10). By definition, pH values cannot be averaged arithmetically, but must be estimated by the logarithm of the reciprocals.

The pH of natural waters is governed to a large extent by the interaction of H^+ ions arising from the dissociation of H_2CO_3 and from OH^- ions, resulting from the hydrolysis of bicarbonate. The range of pH found in natural waters, however, extends between the extremes of < 2 to 12. Nearly all waters with pH values less than 4 occur in volcanic regions that receive strong mineral acids, particularly sulfuric acid. The oxidation of pyrite of rocks and clays in drainage basins can result in sulfuric acid and very acidic drainage to lakes.

Low pH values are found in natural waters rich in dissolved organic matter, especially in bogs and bog lakes that are dominated in the littoral mat by the moss *Sphagnum*. The pH of *Sphagnum* bogs is usually in the range of 3.3 to 4.5 (0.5 to 0.03 meq H^+ l^{-1}). The sources of the H^+ ion activity are several (Clymo, 1963, 1964). Although precipitation can be very acidic as a result of industrial pollution, and influence the pH of poorly buffered waters significantly (Likens, et al., 1972), in bog areas with no great pollution, the supply of H^+ in rain is unlikely to exceed 30 per cent of the total accumulating in a given time. The metabolism of proteins and the reduction of $SO_4^=$ by sulfur-metabolizing bacteria may contribute some H^+ ions to these waters, but their contributions are likely to be small. Although live *Sphagnum* plants secrete organic acids, the concentrations of these acids are usually insufficient to account for the observed acidity. Most of the H^+ ion concentrations appear to result from the active cation exchange in the walls of *Sphagnum*, during which H^+ is released.

Very high pH values of lakes are usually found in endorheic regions, where lake water contains exceedingly high concentrations of soda.

The range of pH of a majority of open lakes is between 6 and 9. Most of these lakes are the 'bicarbonate type,' i.e., they contain varying amounts of carbonate and are regulated by the CO_2–HCO_3^-–$CO_3^=$ system of buffering capacity. Calcareous hardwater lakes commonly are buffered strongly at pH values above 8. Seepage lakes and lakes of igneous rock catchment areas are less well-buffered and more acidic, with pH values usually somewhat less than 7.

Spatial and Temporal Distribution of Total Inorganic
Carbon and pH in Lakes

The total inorganic carbon (ΣCO_2) is distributed uniformly with depth during periods of circulation, during the vernal and autumnal turnovers of

---

[1]It should be noted that measurements of pH involve not the concentration but the activity of the hydrogen ion. One measures the differences of hydrogen ion activity, rapidly being released and incorporated from proton-donor molecules, between unknown solutions and standard buffers of known pH values. Concentration of hydrogen ions is a stochastic measure of that movement, which differs completely from the movement of a stable ion such as sodium.

dimictic lakes, and in shallow lakes of insufficient depth for thermal stratification. The ΣCO_2 content of the water is derived by the equilibria established between atmospheric CO_2, the bicarbonate-carbonate system, contributions from metabolic respiration, and utilization in photosynthesis. Photosynthesis is markedly influenced by numerous parameters; in the spring, increasing light and temperature exert major controls on photosynthetic uptake of CO_2 and generation of CO_2 in rates of microbial decomposition of organic matter.

During the period of thermal stratification, several conspicuous changes occur in the vertical distribution of ΣCO_2. Oligotrophic lakes exhibiting an orthograde oxygen curve and not possessing high concentrations of bicarbonate and carbonate usually also have an orthograde ΣCO_2 curve (Fig. 10–2). Where the dissolved CO_2–H_2CO_3 component constitutes an appreciable part of the total CO_2, as in softwater lakes, the warmer epilimnetic waters are influenced principally by decreased solubility. Photosynthetic utilization may exceed replacement exchange if periodically, mixing of the epilimnion is relatively incomplete, but this phenomenon would be uncommon in the open water. A slight increase in the ΣCO_2 is often observed in the hypolimnetic water overlying the sediments. The vertical pH distribution exhibits approximately the inverse pattern of ΣCO_2.

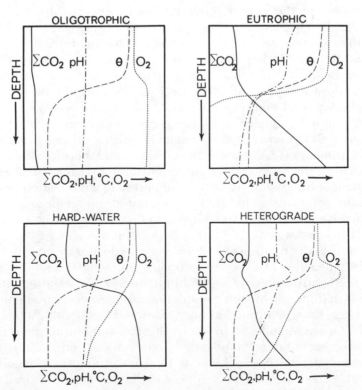

Figure 10–2 Generalized vertical distributions of ΣCO_2 and pH in stratified lakes of very low and high productivity, hardwater calcareous lakes exhibiting pronounced epilimnetic decalcification, and a lake with a distinct positive heterograde oxygen curve.

The vertical distribution of ΣCO_2 and pH is strongly influenced by various biologically-mediated reactions, each of which is treated separately elsewhere. Most conspicuous is the photosynthetic utilization of CO_2 in excess of respiration in the trophogenic zone, which tends to reduce CO_2 content and to increase pH, and the respiratory generation of CO_2 in the tropholytic zone and sediments, which decreases pH. In addition to heterotrophic degradation of organic matter, generation of CO_2 and reduction of pH are augmented by microbial methane fermentation, nitrification of ammonia, and sulfide oxidation. Further, denitrification of nitrate to molecular nitrogen and reduction of sulfate to sulfide result in a slight net increase in pH. The combination of decompositional processes results in an increase in ΣCO_2 of the hypolimnetic waters and a decrease in pH.

As the intensity of decomposition increases in the tropholytic zone, the amount of CO_2 and especially of HCO_3^-, increases markedly. The accumulation of ΣCO_2, both free and combined, is far in excess of oxygen consumed, and decomposition shifts from aerobic to anaerobic as the hypolimnion becomes anoxic. The origin of increasing concentrations of HCO_3^- in the hypolimnion, especially near the sediments, stems in part from bacterial production of ammonium bicarbonate in the sediments (Ohle, 1952). Ferrous and manganous ions are released as bicarbonates from the sediments when the redox potential is reduced sufficiently under anoxic conditions. In hardwater lakes, dissolution of part of the $CaCO_3$ precipitating in and sedimenting from the epilimnion to the hypolimnion, and sediments of lower pH, result in increases in HCO_3^- concentrations. It is clear, however, as discussed previously, that sorbed coatings of dissolved organic matter reduce the rates of $CaCO_3$ dissolution. The hypolimnetic increases of cations, especially calcium, usually are delayed somewhat from proportional increases in hypolimnetic HCO_3^-. Under oxidized conditions, Ca^{++} is complexed with the sediments. Under reducing conditions of anoxia, Ca^{++} is released nearly in proportion to bicarbonate, although some Ca^{++} is complexed with humic acids (Ohle, 1955). Hutchinson (1941) demonstrated conclusively that during stratification, the extent of sediment contact per volume of water (and therefore the morphology of the basin) in the metalimnion and hypolimnion is related directly to the development of increasing HCO_3^- as the period of stratification progresses.

Therefore, in eutrophic lakes possessing a clinograde oxygen curve, a marked inverse clinograde ΣCO_2 curve can be observed (Fig. 10–2). The pattern of pH decreases markedly in the hypolimnion.

The vertical distribution of ΣCO_2 of very hardwater calcareous lakes shifts seasonally when the ΣCO_2 of the trophogenic zone is reduced rapidly by photosynthetic utilization of CO_2 and induction of $CaCO_3$ precipitation (Fig. 10–3). This epilimnetic biogenic decalcification was described long ago (e.g., Minder, 1922; Pia, 1933), and has been observed frequently. While precipitation of $CaCO_3$ can be induced by many physical and biotic agents (increasing temperature, bacterial metabolism), photosynthetic utilization of CO_2 by algae and submersed macrophytes is by far the dominant mechanism (cf. Otsuki and Wetzel, 1974). The result is a marked decrease in the ΣCO_2 of the epilimnion by the end of summer stratification (see Fig. 10–2), and a slow progressive increase in the hypolimnion (Fig. 10–3). In the example shown,

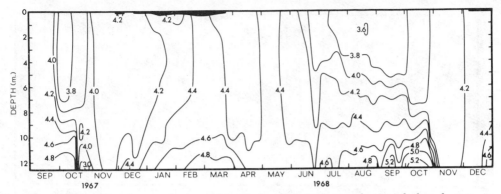

Figure 10–3 Depth-time diagram of isopleths of alkalinity, predominately bicarbonate, in meq 1⁻¹, Lawrence Lake, a calcareous hardwater lake of southwestern Michigan. Opaque area = ice-cover to scale. (From Wetzel, unpublished data.)

the hypolimnion became anoxic below a depth of 11 m in September and October, 1968, as reflected in the increasing HCO_3^- in this layer. The water of the trophogenic zone of unproductive calcareous lakes fluctuates very little seasonally in pH (Fig. 10–4). The bicarbonate buffering capacity is adequate to compensate rapidly for metabolic changes in ΣCO_2. The pH of the hypolimnion progressively decreases throughout the period of stratification. These changes also are observed under ice-cover, but they are less marked than during the summer period, since the rates of production and decomposition are reduced.

The rapidity with which the vertical changes in ΣCO_2 and pH occur is greatly increased in eutrophic lakes (Figs. 10–5 and 10–6). In the example given in the figures, the hypolimnion below 3 m becomes anoxic within a month of loss of the ice-cover and circulation, and becomes nearly anoxic

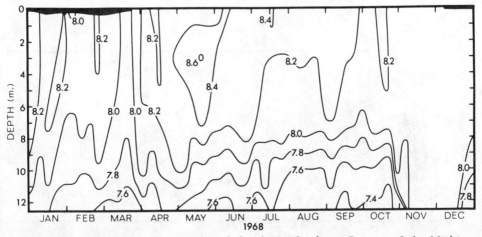

Figure 10–4 Depth-time diagram of isopleths of pH in hardwater Lawrence Lake, Michigan. Opaque area = ice-cover to scale. (From Wetzel, unpublished data.)

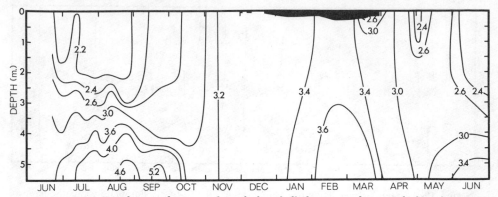

Figure 10–5 Depth-time diagram of isopleths of alkalinity, predominately bicarbonate, in meq l⁻¹, of moderately calcareous, hypereutrophic Wintergreen Lake, southwestern Michigan, 1971–72. Opaque area = ice-cover to scale. (From Wetzel, et al., unpublished.)

under ice-cover. The pH values of the epilimnion represent midmorning measurements. Under the conditions of intensive photosynthesis of phytoplankton, the pH undergoes appreciable diurnal fluctuations, often exceeding 10 in the late afternoon and decreasing below 8 during darkness. Biogenic decalcification of the epilimnion also is apparent in this moderately hardwater hypereutrophic lake (Fig. 10–5).

Where photosynthetic activity in a layer is particularly intense, as in the metalimnion, a positive heterograde oxygen curve is frequently found (see chapter 8). Commonly associated with this metabolism is a corresponding positive heterograde pH curve and the negative heterograde ΣCO_2 curve (Fig. 10–2, and at 3 m in May of Fig. 10–4). The opposite situation of a negative heterograde oxygen curve, with concomitant ΣCO_2 and pH curves, occasionally is found in strata of intensive respiration, e.g., containing plates of bacteria or aggregations of zooplankton (see Nagasawa, 1959).

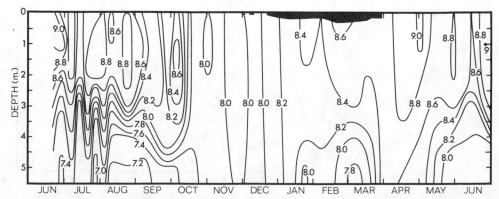

Figure 10–6 Depth-time diagram of isopleths of pH in hypereutrophic Wintergreen Lake, Michigan, 1971–72. Opaque area = ice-cover to scale. (From Wetzel, et al., unpublished.)

Hypolimnetic CO_2 Accumulation in Relation to Lake
Metabolism

As in the case of hypolimnetic oxygen deficits discussed earlier, changes in the concentrations of ΣCO_2 in the hypolimnion have been used to estimate indirectly the organic production of the trophogenic zone. The principle states that the accumulation of CO_2 in the hypolimnetic tropholytic zone from decomposition is proportional to the production of organic matter in the trophogenic zone that is transported to the hypolimnion. This approach to an estimate of lake metabolism is complicated by several other sources of organic matter that are not from the trophogenic zone, and by losses from the system that are not represented in CO_2 changes in the tropholytic zone. There is no question, however, that for many lake systems the measurement of hypolimnetic CO_2 accumulation provides a reasonable estimate of the intensity of lake metabolism.

The potential of production estimates from CO_2 accumulation was first suggested by Ruttner in his treatise on tropical lakes (1931). Einsele (1941) compared the CO_2 accumulation in the hypolimnion with several other estimates of production in his critical evaluation of the effects of phosphorus fertilization on productivity of algae, and found it to be in total agreement. By far the most comprehensive application of the hypolimnetic CO_2 accumulation principle to lakes, and its comparison with other measures of production, especially the oxygen deficit, was done by Ohle (1934, 1952, 1956). This approach has the advantage of being able to follow metabolically mediated changes under both aerobic and anaerobic conditions, whereas the oxygen deficit method is applicable only when the hypoliminion is oxic. More fundamentally, carbon, as the initial and endproduct of organic metabolism, is one of the best parameters by which to evaluate productivity.

The principles underlying calculations of hypolimnetic CO_2 accumulation make several assumptions. First, it is important that the water under consideration be relatively well-isolated from the atmosphere and from CO_2 exchange; these conditions may be found in the hypolimnia of small, sheltered lakes with steep thermal stratification gradients or when the entire lake is under ice-cover. It is assumed that the allochthonous inputs of organic matter from outside the basin are very small in comparison to autochthonous production within the basin, a situation that probably is not realized very often (cf. Chapter 17). Loss of organic production from the trophogenic zone via the outlet is assumed to be small, a condition which, again, is highly variable in open lakes. A major assumption is that most of the synthesized organic matter of the trophogenic zone decomposes after it has sedimented to the hypolimnetic tropholytic zone. In this case also, this assumption certainly causes underestimations, because much decomposition of algae, littoral macrophytes, and sessile algae occurs in the epi- and metalimnetic layers before reaching the relatively quiescent hypolimnion. The rates of decomposition in the trophogenic zone depend upon the proportion of production by the littoral flora of the total lake with changing species dominance, seasonal shifts occur in the proportion of phytoplankton algae reaching the hypolimnion, e.g., silicious diatoms sediment more rapidly than small algae, which are more neutrally buoyant. Finally, a significant amount of organic production will be incorporated permanently into the sediments without complete decomposition; the proportion permanently lost in this way generally increases in lakes of greater productivity. Despite these numerous limitations, the estimates of hypolimnetic CO_2 accumulation are correlated strongly with the general lake productivity and are directly proportional to mean depth of the basin.

Analysis of hypolimnetic CO_2 accumulation requires evaluation of the various components of total CO_2 and their origins. In bicarbonate-poor waters, the difference in CO_2 content at the beginning and after a period of stratification is determined. In bicarbonate-rich waters, the amount of free CO_2 is very small at the beginning of hypolimnetic stratification. As stratification continues, free CO_2 and bicarbonate accumulate from respiration of aerobic and anaerobic decomposition. In addition, a portion of the bicarbonate is present as 'volatile' ammonium carbonate of metabolic origin and as non-volatile bicarbonates, half of which are of metabolic origin. The other half is bound CO_2 of $CaCO_3$, which largely becomes incorporated into the sediment. The sum of these CO_2 inputs is assumed to be an estimate of lake metabolism, and a sample calculation is given in Table 10–3. If the CO_2 accumulation is calculated for the entire hypolimnion, and this is divided by the volume of

TABLE 10–3 Summary of Method of Evaluating the Hypolimnetic CO_2 Accumulation of a Lake with a Sample Calculation[a] (Example based on: 1.85 mg NH_4 1^{-1}; 13.1 mg CO_2 1^{-1}; 32.4 mg HCO_3 1^{-1}; 0.56 mg O_2 1^{-1}; °C = 6.7; period of stratification of 4 months.)

Calculations	Lake Example
(1) $NH_4:HCO_3 = 18.04:61.02$	(1) 1.85 mg NH_4 1^{-1} present as NH_4HCO_3
(2) $HCO_3 = \dfrac{(NH_4)\,(61.02)}{18.04} = (3.38)\,(NH_4) = \beta$	(2) $\beta = (1.85)\,(3.38) = 6.25$
(3) $HCO_3:CO_3 = 61.02 = 44.01$	(3) $b = (6.25)\,(0.721) = 4.51$
$\quad CO_2 = \dfrac{(HCO_3)\,(44.01)}{61.02} = (0.721)\,(HCO_3)$	
$\quad b = \beta\,(0.721)$	
(4) $\alpha = HCO_3;\; \alpha - \beta = x$	(4) $\alpha = 32.4$ mg HCO_3 1^{-1}
(5) $a = \dfrac{x(0.721)}{2} = x(0.361)$	$\quad x = 32.4 - 6.25 = 26.15$
	(5) $a = (26.15)\,(0.361) = 9.43$
(6) $c = CO_2$	(6) $c = 13.10$ mg CO_2 1^{-1}
(7) $\Sigma CO_2 = a + b + c$	(7) $\Sigma CO_2 = 9.43 + 4.51 + 13.10 = 27.04$
(8) $\delta O_2 = $ actual O_2 deficit	(8) mg O_2 $1^{-1} = 0.56$
$\quad \dfrac{CO_2}{O_2} = \dfrac{44}{32} = 1.375$	
$\quad \delta\,O_2\,(1.375) = \delta\,CO_2$	°C = 6.7 O_2 at turnover = 12.27 mg 1^{-1}
$\quad R.Q. = 0.85^b$	$\delta\,O_2 = 12.27 - 0.56 = 11.71$
$\quad \gamma = \delta\,CO_2\,(0.85)$	$\gamma = (11.71)\,(1.375)\,(0.85) = 13.69$
(9) $\Sigma CO_2 - \gamma = \Delta^1\,CO_2$	(9) $\Delta^1\,CO_2 = 27.04 - 13.69 = 13.35$
(10) $z = \Delta^1\,CO_2\,(2)$	(10) $z = (13.35)\,(2) = 26.70$
(11) $\mathcal{E} = z + \gamma$	(11) $\mathcal{E} = 26.70 + 13.69 = 40.39$
$\quad \dfrac{\mathcal{E}\,(30)}{\text{No. of days}} = \mathcal{E}$ per month	$\mathcal{E}$ per month $= \dfrac{(40.39)\,(30)}{120} = 10.1$ mg CO_2 1^{-1}

[a]After Ohle, 1952.
[b]A RQ value (CO_2/O_2) of 0.85 is a mean value based on a number of analytical analyses of plant and animal respiration, and agrees approximately with some whole closed lake evaluations (but see discussion of RQ variations in Chapter 17).

TABLE 10-4 Comparison of the Relative Assimilation Intensity (Hypolimnetic CO_2 Accumulation Expressed per Liter of the Epilimnion per Month) to Other Indices of Productivity in Lakes of Northern Germany[a]

	Relative Assimilation Intensity (mg CO_2 1^{-1} epilimnion month^{-1})	Chlorophyll of Epilimnion (μg 1^{-1})	Estimate of Photosynthetic Efficiency of. Solar Radiation (%)	Hypolimnetic O_2 Deficit (mg O_2 1^{-1})
Schaalsee	1.91	1.3	0.76	3.30
Schöhsee	2.13	1.4	0.56	6.96
Zanzen	2.33	1.6	0.96	–
Schmaler Lucin	2.70	1.8	0.80	3.16
Breiter Lucin	4.64	3.1	1.90	2.16
Techiner Binnensee	5.45	3.6	1.65	7.86
Tollensesee	7.04	4.1	2.09	5.49
Edebergsee	14.6	9.7	2.27	10.0

[a]After data of Ohle, 1952, 1955b.

the epilimnion, the quantity termed the "relative assimilation intensity" results. In the example given (Table 10-3):

$$(10.10 \text{ mg } CO_2 \ 1^{-1} \text{ month}^{-1}) \left[\frac{4.620 \times 10^6 \text{ m}^3}{7.442 \times 10^6 \text{ m}^3} \right] = 6.3 \text{ mg } CO_2 \ 1^{-1} \text{ month}^{-1}.$$

This estimate implies that approximately 6.3 mg CO_2 were combined per liter of water in photosynthesis to form carbohydrate in the epilimnion of this lake per month during the period of observation. The estimates of production by the CO_2 accumulation are usually always minimal or under, because of the confounding factors mentioned earlier. Further, some CO_2 loss from the upper hypolimnion occurs by turbulence, and some bicarbonate is released from the reduction of ferric hydroxide as redox potential decreases during stratification.

The hypolimnetic CO_2 accumulations in several lakes are compared in Table 10-4, along with several other indices of increasing productivity (Ohle,

TABLE 10-5 Estimations of Total Production in the Trophogenic Zone of Schleinsee, Germany, during Summer Stratification (May-September) in the Years Before and in the Year After Application of Phosphorus Fertilization[a]

Basis for Calculated Production	Kg Dry Weight		Ratio
	1935–1937	1939	
Biogenic decalcification	1900	5000	2.6
Ammonia N accumulation	2100	5900	2.8
Hypolimnetic CO_2 accumulation	3800	9600	2.5
Apparent total production	5600	17200	3.1
Related to surface area	43 g m^{-2}	117 g m^{-2}	2.7

[a]After data of Einsele, 1941.

1955b). The general correlations among various criteria hold. Einsele (1941) applied the CO_2 accumulation evaluation of productivity, along with several other measures, to the trophogenic zone of Schleinsee, Germany, before and after extensive fertilization of the lake with phosphorus (Table 10–5). Schleinsee lends itself well to such analyses because of its limited inflow and outflow. Although Einsele's estimations were quite approximate, again the comparison was good. Used in a relative manner, this method is of much value.

Utilization of Carbon by Algae and Macrophytes

Synthesis of organic matter, with a carbon content of approximately 50 per cent by dry weight, by photosynthesis of algae and submersed aquatic macrophytes requires an abundant and readily available source of carbon for high sustained growth. Abundant physiological evidence indicates that free CO_2 is most readily utilized by nearly all algae and larger aquatic plants. A number of algae and submersed macrophytes, particularly the mosses, can utilize only free CO_2. Many algae and aquatic vascular plants are capable of assimilating bicarbonate ions when free CO_2 is in very low supply and HCO_3^- is very abundant; a few species of algae require HCO_3^- and cannot grow with free CO_2 alone. There is no clear evidence that algae or higher aquatic plants assimilate $CO_3^=$ directly as a carbon source, although it has been implicated strongly in growth at very high pH values (see Felföldy, 1960). The most comprehensive review of the subject is given by Raven (1970).

Heterotrophic Augmentation

A large number of algae are heterotrophic, i.e., they can remain viable in bacterial-free culture by chemo-organotrophic uptake of dissolved organic compounds in the absence of light (cf. Danforth, 1962; Provasoli, 1963; Stanier, 1973). However, the rates of growth under the experimental conditions used are very low in comparison to normal light-mediated autotrophic growth, and the levels of substrate concentrations employed are usually exceedingly high, several orders of magnitude in excess of what would be found in natural waters. The range of organic compounds that support growth in the dark in blue-green algae is very narrow, because the pentose phosphate pathway is the energy-yielding dissimilatory pathway. Consequently, only exogenous substrates that are readily convertible to glucose-6-phosphate can support growth in the dark. Among obligately photoautotrophic algae, suitable enzymes, e.g., glucose permease, are absent. Investigations at naturally occurring substrate concentrations of the uptake of organic substrates by algae in lake water also containing bacteria that possess active permease enzyme mechanisms clearly indicate that the algae are ineffective competitors for the organic substrates (Wright and Hobbie, 1966; Wetzel, 1967; Hobbie and Wright, 1965; and others). Where active heterotrophic uptake of simple organic substrates is found under natural conditions, carbon assimilation is still low in comparison to the amount assimilated in photoautotrophy, even under very low light conditions. It might be anticipated that any significant augmentation of autotrophy by chemo-organotrophy would be confined largely to the blue-green algae, which have numerous morphological and physiological similarities to bacteria. This potential heterotrophic augmentation is seen in the blue-green alga *Oscillatoria*, which grows in massive densities in the microaerobic zone at the metalimnetic-hypolimnetic interface of some productive lakes (Saunders, 1972). The potential of photoheterotrophic growth in the light in the absence of exogenously supplied CO_2, which has been shown in two species of blue-green algae (Ingram, 1973a, 1973b) is unknown at present in natural conditions. Studies indicate that respired CO_2 from the substrate oxidation is assimilated by the photosynthetic reactions within the cells. All of the existing evidence for heterotrophic growth by algae indicates that under natural conditions, photoautotrophy by inorganic carbon uptake overwhelmingly dominates in freshwater systems. Supplementation of photosynthesis by algal heterotrophy is limited to a few specialized conditions and probably is insignificant for overall plant synthesis.

In most freshwater systems, dissociation rates of ionic CO_2 species and the maintenance of near-equilibrium conditions between atmospheric CO_2 and that of the water are sufficiently rapid that severe inorganic carbon limitation to algal photosynthesis is unlikely, even under conditions of low ΣCO_2. Below a pH of 8.5, the theoretical dissociation kinetics of the $CO_2-HCO_3^--CO_3^=$ complex, derived from pure solutions, agree very well with the concentrations of the ionic species found in natural waters (Talling, 1973). Above pH 8.5, however, the total inorganic carbon dioxide concentrations fall considerably below the values calculated on the basis of the apparent dissociation constants K_1' and K_2' of carbonic acid, as derived from measurements of alkalinity (Fig. 10–7). Although the explanation for this phenomenon is not completely clear, it appears to be associated with the presence of a noncarbonate, nonhydroxide alkalinity component resulting from ionized silicate, and can be corrected for (Talling, 1973).

Experimental evidence on freshwater photosynthetic utilization of carbon indicates a strong relationship between physiological availability and the forms of ΣCO_2 (Raven, 1970; Wetzel and Hough, 1973). In a majority of plants, the first stable major product of fixation of inorganic carbon is 3-phosphoglycerate. Up to 5 per cent of the inorganic carbon fixed is by β-carboxylation, a pathway involved in the formation of carbon skeletons or amino acid and porphyrin synthesis. In some tropical plants, all inorganic carbon fixed in photosynthesis occurs via the β-carboxylation pathway, but this mode of metabolism has not been found to occur to any significant extent in aquatic plants. The enzyme carboxydismutase is involved in the aquatic carboxylation reaction, which produces 3-phosphoglycerate as the first product of carbon fixation, and uses free, unhydrated CO_2 as its immediate substrate. Carbonic anhydrase, an enzyme that catalyzes the reversible hydration of carbon dioxide, is found universally in photosynthetic cells of plants. Where CO_2 is the carbon species entering the cell, there is no obvious biochemical role for carbonic anhydrase found in the plants, although some evidence indicates that this enzyme accelerates the diffusion of CO_2. Bicarbonate ions have been implicated as a critical factor in the evolution of oxygen in photosynthesis (Stemler and Govindjee, 1973). At high pH values and HCO_3^- concentrations, bicarbonate moves less effectively from binding sites of the chloroplast, a mechanism required for further photochemical reactions.

Utilization of Bicarbonate Ions

The ability to assimilate bicarbonate ions most likely is variable among planktonic algae, macroalgae, and submersed angiosperms (Raven, 1970). When this ability does occur, an additional reaction is needed for HCO_3^- assimilation, which is not required for CO_2 assimilation. Active bicarbonate transport with dehydration in the cytoplasm apparently is required, and is coupled with a similarly active stoichiometric excretion of hydroxyl ion from the cells. Carbonic anhydrase activity is found in plants that cannot use bicarbonate as well as in those that can, although such activity generally increases among the latter.

When aquatic plants have similar affinities for CO_2 and HCO_3^- ions, utilization of bicarbonate generally occurs when bicarbonate concentration exceeds that of CO_2 by more than 10 times. Free CO_2 concentrations (about 10 micromolar) of most fresh waters and of the sea are in approximate equilibrium with the atmosphere; however, many fresh waters contain bicarbonate concentrations far in excess of 10 times that quantity. Equilibrium-free CO_2, particularly in alkaline hardwater lakes with a pH > 8.5, is inade-

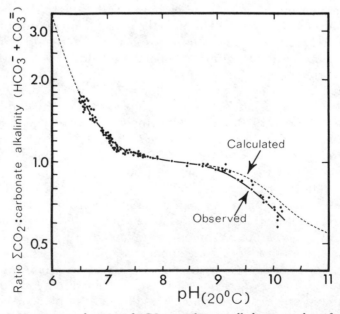

Figure 10–7 Variation in the ratio of ΣCO_2 to carbonate alkalinity resulting from HCO_3^- and $CO_3^=$ with pH for samples of water from the surface and bottom (14 m) layers of Esthwaite Water, England. The broken line indicates the relationship calculated from appropriate values of pK_1' (6.38) and pK_2' (10.32). (Slightly modified from Talling, 1973.)

quate to saturate photosynthesis in plants adapted to utilize bicarbonate. As these waters become more productive, and in densely populated littoral zones of less productive lakes, pH is rapidly modified by metabolism on a diurnal basis (pH ranges from 6 to 10 or more per 24 hours), and can be associated with reduced carbon fixation and bicarbonate assimilation. Under stagnant conditions, common to heavily colonized littoral zones of lakes, the shift to bicarbonate metabolism, as well as the increased pH, is often associated with severe reduction in CO_2.

Bicarbonate assimilation in media of high pH assumes greater significance in larger aquatic macrophytes that have morphologically long diffusion paths. During active transport of HCO_3^- from the dissociation of calcium bicarbonate and secretion of OH^- ions, $CaCO_3$ is precipitated. Many submersed angiosperms, but not all (Wetzel, 1969), and many algae can utilize bicarbonate. Aquatic mosses represent a group that only is able to utilize free CO_2, and is restricted to waters of relatively low pH and abundant CO_2, e.g., springs, mountain streams, bogs, or in shallow waters or the lower trophogenic zone where free CO_2 is higher than elsewhere in the lake. Moreover, many angiosperms with large intercellular gas lacunae refix CO_2 of respiration and photorespiration rather efficiently (Hough and Wetzel, 1972). The efficiency of refixation and photosynthetic efficiency of carbon fixation must be highly plastic, related in part to induced shifts to bicarbonate assimilation, and can affect rates of net primary production significantly.

In summary, evidence is available that, among the diversity of concentrations and states of inorganic carbon in fresh waters, there exists a large number of situations where free CO_2, in equilibrium with the atmosphere, may be inadequate for metabolism at high sustained levels or indirectly inadequate as a result of chemical losses from the system. Although it is highly doubtful that inorganic carbon per se is seriously limiting to photosynthetic metabolism under most natural situations, assimilation of bicarbonate at metabolic expense is known and can be induced among plants that have an affinity for both CO_2 and HCO_3^- ions under conditions of low free CO_2 and high bicarbonate concentrations. Possession of an affinity for bicarbonate is an adaptive advantage in a significant percentage of fresh waters, particularly for submersed angiosperms.

chapter 11

the nitrogen cycle

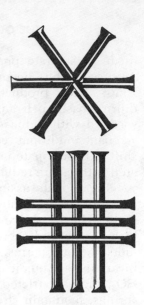

THE NITROGEN CYCLE

A major source of nitrogen of the biosphere originates from fixation of atmospheric molecular nitrogen. The nitrogen cycle is a biochemical process in which concentration of molecular nitrogen occurs by nitrogen fixation, assimilation, and denitrification in which nitrate is reduced to N_2. For all practical purposes, the nitrogen cycle of lakes is microbial in nature: bacterial oxidation and reduction of nitrogen compounds are coupled with photosynthetic assimilation and utilization by algae and larger aquatic plants. The role of animals in the nitrogen cycle of lakes is certainly very small, but under certain conditions it can influence periodic population responses of photosynthetic utilization rates of nitrogen compounds.

Microbial fixation of molecular N_2 in soils by bacteria is a major source of nitrogen compounds; in lakes, N_2 fixation by bacteria and certain blue-green algae is quantitatively less significant, except under certain conditions of severe depletion of inorganic nitrogen compounds from the trophogenic zone. Fixed nitrogen, or that nitrogen assimilated as nitrate or ammonia, is aminated into organic nitrogenous compounds within organisms. Most organic nitrogen is quantitatively bound and cycled in photosynthetic and microbial organisms. During normal metabolism of these organisms, especially at their death, it is liberated as ammonia and a variety of organic compounds of varying resistance to proteolytic deamination and ammonification by heterotrophic bacteria. In aerobic conditions, ammonia is largely oxidized first to nitrate, with a concomitant release of free energy that is utilized by nitrifying bacteria in decomposition of organic compounds. Many bacteria can utilize nitrite or nitrate as hydrogen acceptors, rather than oxygen. The reduction of these compounds, usually to molecular N_2 by denitrification, returns N_2 to the biosphere.

Nitrogen, as one of the major constituents of cellular protoplasm of organisms along with phosphorus, carbon, and hydrogen, forms a major nutrient that affects the productivity of fresh waters. Although the nitrogen cycle of freshwaters is understood qualitatively in appreciable detail, quantitative

186

transformation rates have not been delineated clearly, especially at the sediment-water interface zone of active bacterial metabolism. The nitrogen cycle of many lakes has been altered severely by increased loading of inorganic and organic nitrogenous compounds, and it is imperative that its dynamics be integrated with those of other nutrient cycles.

SOURCES AND TRANSFORMATIONS OF NITROGEN IN WATER

Nitrogen occurs in fresh waters in numerous forms: dissolved molecular N_2, a large number of organic compounds from amino acids, amines to proteins and refractory humic compounds of low nitrogen content, ammonia (NH_4^+), nitrite (NO_2^-), and nitrate (NO_3^-). Sources of nitrogen include: (a) precipitation on the lake surface, (b) nitrogen fixation both in the water and the sediments, and (c) inputs from surface and groundwater drainage. Losses of nitrogen occur by (a) effluent outflow from the basin, (b) reduction of NO_3^- to N_2 by bacterial denitrification with subsequent return of N_2 to the atmosphere, and (c) permanent loss of inorganic and organic nitrogen-containing compounds to the sediments.

Nitrogen of Precipitation and Fallout

The amount of influent of nitrogen to lakes and their drainage areas from atmospheric sources generally has been considered to be minor in comparison to that from direct terrestrial runoff. However, on closer inspection, and in view of exponentially increasing inputs of nitrogen from pollution to the atmosphere, the amount of nitrogen reaching lakes in this way is often significant for the nitrogen cycle and for productivity. For example, in relatively oligotrophic mountainous regions of granitic bedrock, nitrogen of precipitation forms a major source of many nutrients (Likens and Bormann, 1972) especially nitrogen that is contributory to the overall productivity of both fresh and marine waters (Menzel and Spaeth, 1962).

Atmospheric sources of nitrogen from precipitation and from direct bulk fallout are highly variable in relation to meteorological conditions and to the location of the lake with respect to technological outputs and prevailing atmospheric wind patterns. Nitrogen may enter a lake in many forms: dissolved N_2, nitric acid, NH_4^+, and NO_3^-, as NH_4^+ adsorbed to inorganic particulate matter, and as organic compounds, both in dissolved and particulate form. There is no direct proportional relationship between the volume of rainfall or snowfall and the quantity of nitrogen influx per area of land or water (Chapin and Uttormark, 1973). Up to ten times the nutrient content of rain falls as dry fallout and bulk precipitation from the air. Additionally, the nitrogen content of snow is often much higher than that of rain, and can contribute up to half of the total annual nitrogen influx, even though the snow generally comprises a small proportion of total precipitation. N_2-fixing bacteria occur in rainwater in low numbers (Visser, 1964a), but their contribution by in situ formation of NO_3^- most likely is small in comparison to other sources. Hutchinson (1944) discusses the reasonable evidence for combined nitrogen of the atmosphere being formed primarily from ammonia, which is released from microbial degradation of terrestrial organic matter that is partly oxidized to nitrate in the atmosphere.

The inputs of NO_3^- and NH_4^+ from atmospheric sources average about 0.1 g N m^{-2} year^{-1} over the continental United States, but are highly irregular in distribution. The north central and eastern regions, especially bordering the southern Great Lakes, receive the largest input of nitrogen from precipitation, on the order of 0.3 to 0.35 g N m^{-2} yr^{-1} (Chapin and Uttormark, 1973). Dry fallout sources increase these values by a factor of 3 to 4, so that in the Great Lakes region an atmospheric contribution of nitrogen occurs at a rate of ca 1 g N m^{-2} yr^{-1}. Neglecting other nutrients for the time being, this influx alone corresponds to the approximate amounts of nitrogen that are generally required to shift shallow lakes of a mean depth of <5 m from moderate to high productivity (Vollenweider, 1968). In this region of the country, which, moreover, receives very high inorganic inputs from runoff of high nitrogen-containing sedimentary rock formations, it would be very difficult to control productivity of lakes by means of nitrogen limitation because of large atmospheric contributions. We will return to this subject in later discussions.

Molecular Nitrogen and N_2 Fixation

Although N_2 is not particularly soluble in water, the N_2 concentrations usually are saturated with respect to the surface temperature and pressure at periods of circulation of lakes (Birge and Juday, 1911). Maximum concentrations are found in the winter period of increased solubility and colder temperatures (ca 15 to 20 ml l^{-1}). Under stratified conditions and heating of the epilimnetic waters, solubility is decreased (Fig. 11–1). The concentrations of the metalimnion and much of the hypolimnion may be slightly supersaturated, probably as a result of the effects of hydrostatic pressure maintaining excess gas from spring circulation, while the hypolimnetic temperatures increase somewhat during the summer. A decrease in the nitrogen content in the lower hypolimnion above the sediments has been observed

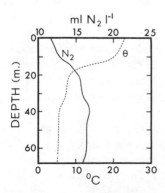
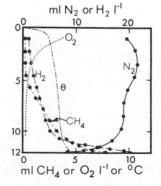

Figure 11–1 Vertical distribution of dissolved gases (*left*) in Green Lake Wisconsin, summer, 1906 (Birge and Juday, 1911) and (*right*) in Beloie Lake, USSR, March 1938. (After Kuznetsov and Khartulari, 1941.)

(Fig. 11–1), which presumably is related to bacterial fixation of N_2 in productive lakes (Kuznetsov and Khartulari, 1941). Conversely, late in the summer stratification of Lake Mendota, N_2 of the hypolimnion increased somewhat above what would be expected on the basis of solubility (Brezonik and Lee, 1971); this partly may be the result of active measured denitrification of nitrate in this lake.

Nitrogen Fixation: Blue-Green Algae

The importance of in situ nitrogen fixation to the productivity of lakes has been shown only within the last decade, even though blue-green algae were implicated in this process many years earlier (Burris, et al., 1943). An excellent comparative review is given by Hardy, et al. (1973). Thus far, the occurrence of nitrogen fixation in the open waters of lakes has been strictly correlated with the presence of blue-green algae that possess heterocysts (Fogg, 1971). Heterocysts are specialized cells that occur singly in most filamentous blue-green algae, except for the Oscillatoriaceae, and are the sole site of nitrogen fixation in aerobically grown, heterocyst-forming blue-green algae (Wolk, 1973). Although nitrogen fixation has been found to occur in some unicellular forms which do not produce heterocysts, among some species of blue-green algae, such as *Anabaena* spp., numbers of heterocysts correspond very approximately to observed nitrogen-fixing capacity (Horne and Goldman, 1972; Horne, et al., 1972).

In plankton of open water, nitrogen fixation is primarily light-dependent in that it requires reducing power and adenosine triphosphate (ATP), both of which are generated in photosynthesis. Nitrogen-fixing algae and some photosynthetic bacteria can fix only limited quantities of N_2 in the dark. Determinations in situ show a relationship of nitrogen fixation with depth similar to that of photosynthesis with depth. In full sunlight this process commonly is inhibited at the surface, reaches a maximum some depth below the surface, and involves a rapid, nearly exponential decrease with greater depth (Fig. 11–2; cf. Chapter 5) (Dugdale and Dugdale, 1962; Goering and Neess, 1964; Horne and Fogg, 1970).

Heterocyst formation and nitrogen fixation by blue-green algae are suppressed in the presence of a readily available source of combined nitrogen as nitrate or ammonia. Combined nitrogen suppresses synthesis of the nitrogenase complex rather than the activity of any existing enzyme (Fogg, 1971; Wolk, 1973), and this suppression of heterocysts by nitrate, even at very high concentrations, is often only partial (Ogawa and Carr, 1969; Ohmori and Hattori, 1972). Similarly, ammonia at low concentrations represses the formation of nitrogenase, but does not affect its activity. In general, then, one would expect an inverse relationship between the rate of nitrogen fixation by blue-green algae and the concentration of inorganic nitrogen in lake waters; this is often the observed situation. But because of the carryover of residual nitrogenase activity, the relationship is not always consistent. Thus N_2 fixation by blue-green algae sometimes may occur at greatly reduced levels in the presence of appreciable inorganic nitrogen in the water. The importance of the

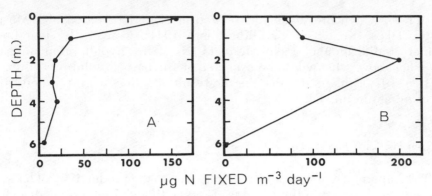

Figure 11–2   Variations in nitrogen fixation with depth (A) in Lake Windermere, and (B) in Esthwaite Water, England, 30 August 1966. (Modified from Horne and Fogg, 1970.)

microenvironment surrounding the cell also should be pointed out. Molecular N_2 is in higher concentrations and diffuses more readily than ammonium or nitrate ions. Within the massive mucilage sheaths surrounding many blue-green algae, a steep gradient could easily develop in which other inorganic nitrogen sources are greatly reduced in comparison to N_2 availability, regardless of the inorganic concentrations of the water.

Nitrogen fixation also has been positively correlated with concentrations of dissolved organic nitrogen occurring in the water (Horne and Fogg, 1970; Horne, et al., 1972). Algae secrete many simple and complex organic carbon and nitrogen compounds. There is no evidence at present to indicate a causative effect of the dissolved organic nitrogen compounds in situ on rates of nitrogen fixation. It appears, rather, that the secretion of the dissolved organic compounds reflects the growth of the blue-green algal populations and concurrent nitrogen fixation.

Diurnal rates of nitrogen fixation in open lake water are typically low in the early morning, reach a maximum midday at maximum insolation and photosynthesis, and then decline to low afternoon and evening rates (Rusness and Burris, 1970). The most commonly observed seasonal pattern in temperate lakes is for both the percentage and absolute rates of fixation to increase to maximum levels as heterocyst-bearing blue-green algal populations develop and sources of combined nitrogen are reduced or depleted. High concentrations of total phosphorus are necessary. Rates of fixation decline abruptly as the blue-green populations decrease (cf. Chapter 14 on algal successions). In winter, the rates of N_2 fixation are nonexistent or greatly reduced (Billaud, 1968; Horne and Fogg, 1970; Toetz, 1973). In productive tropical lakes where the periodicity of physico-chemical factors and algae is less marked, nitrogen fixation rates probably are more uniform throughout the year.

To determine the nitrogen cycle of a lake, estimates of the total nitrogen fixed per annum are required. Both horizontal variations and variations from year to year are large (Table 11–1); as a result, extensive measurements are required to account for the spatial and temporal heterogeneity of blue-green algal populations. For lack of data to the contrary, it had been assumed previ-

TABLE 11-1   Estimated Annual Rates of Nitrogen Fixation by Phytoplankton

Lake	Area (km^2)	N$_2$ Fixation (g N m^{-2})	Total N Fixed (megagrams = metric t)
Windermere, England			
North Basin 1966	8.2	0.037	0.30
South Basin 1965	6.7	0.287	1.92
1966	6.7	0.107	0.72
Esthwaite, England			
1965	1.01	0.127	0.13
1966	1.01	0.061	0.06
Clear Lake, California, 1970			
Upper Arm	127.0	0.352	361
Lower Arm	37.2	0.759	49
Oaks Arm	12.5	0.250	50
Total for Lake	176.7	0.384	460

[a]From data of Horne and Fogg, 1970, and Horne and Goldman, 1972.

ously that the magnitude of nitrogen income to lakes by N$_2$ fixation of the algal phytoplankton was very small or insignificant in comparison to other sources. It is now evident that a large spectrum of conditions exists, ranging from those lakes where N$_2$ fixation is insignificant, to those eutrophic lakes where extensive N$_2$ fixation permits higher rates of production to occur than would be possible otherwise. Direct measurements of N$_2$ fixation by ^{15}N$_2$ and acetylene reduction methods in oligotrophic lakes, or in lakes of moderate productivity but with high natural inorganic nitrogen sources, are consistently negative. In moderately productive Lake Windermere, the nitrogen fixation of benthic blue-green algae growing in abundance on littoral rocks and stones to a depth of 3.5 m was estimated conservatively to be 1 g N m^{-2} year^{-1} (Horne and Fogg, 1970). For a minimal rocky area of 2.86 km^2, the estimated total annual fixation by the benthic blue-green algae in Windermere is 0.8 megagrams. The littoral benthic contribution is therefore considerably larger than the nitrogen fixed by the phytoplankton (Table 11-1). Even though these values are large, the total N$_2$ fixation by blue-green algae contributes at most a small proportion (less than 1 per cent) of the total combined nitrogen income to this lake. An estimate (probably high) for small, eutrophic Chernoye Lake, USSR, indicated that the nitrogen fixation constituted about 13 per cent of the total nitrogen income (Kuznetsov, 1959). In eutrophic Clear Lake, California, biological N$_2$ fixation, largely associated with blue-green algal blooms, contributed at least 43 per cent to the nitrogen income, almost as important as NO$_3^-$ from river inflows as a nitrogen source.

A detailed study of the nitrogen fixation of a small subarctic lake of interior Alaska demonstrated that ammonia was the most important nitrogen source (Billaud, 1968). Of two main algal production periods, the first consisted largely of microflagellates under the ice, and depended on ammonia as a nitrogen source. Immediately after the ice melted from the lake, an algal population composed almost exclusively of the blue-green *Anabaena flos-aquae* developed. During the peak of the *Anabaena* bloom, molecular nitrogen constituted over 25 per cent of the nitrogen assimilated by the plankton. Nitrogen fixed by the plankton in the south basin of Lake Windermere in 1965 amounted to 72 per cent of the amount available as nitrate, although in 1966 and in other water bodies this proportion was less (1 to 48 per cent) (Horne and Fogg, 1970).

Nitrogen Fixation: Bacteria

Quantitative information on bacterial nitrogen fixation in lakes is very sparse. Heterotrophic nitrogen fixation commonly is disregarded, on the premise that nitrogen-fixing bacteria are limited by the availability of exogenous carbohydrate. One to 25 mg of nitrogen can be fixed per gram of carbohydrate utilized (Stewart, 1969). Such quantities of soluble carbohydrate rarely are available in natural waters, and there is an intense competition for these substrates by nonfixing heterotrophic bacteria.

The most common heterotrophic N_2 fixing bacteria comprise several species of *Azotobacter* and *Clostridium pasteurianum*, which are found in fair abundance living free in the water, epiphytically on submersed aquatic plants, and in the sediments (Kuznetsov, 1959, 1970). Their numbers generally are lowest in the open water, where soluble organic concentrations are low, and tend to increase, with *Azotobacter* being dominant, in bog lakes containing high concentrations of dissolved humic organic matter (Table 11–2). Based on very few data from eutrophic Chernoye Lake, USSR, numbers of open-water *Clostridium* increase in the spring just before and after ice-loss, decrease to a low in the summer, and increase in the autumn. This pattern is somewhat analogous to the lag observed in non-N_2 fixing bacteria populations that are slightly behind the productivity pulses of phytoplankton (cf. Chapter 14). In contrast, Niewolak (1972) observed a maximum of *Azotobacter* in Polish lakes in spring and summer.

TABLE 11–2 Occurrence of Nitrogen-fixing Bacteria in Water and Sediments of Lakes[a]

Lake Type	Number Lakes Studied	In Water (nos ml⁻¹)		In Sediment (nos g⁻¹ wet wt)	
		Azotobacter	Clostridium pasteurianum	Azotobacter	Clostridium pasteurianum
Oligotrophic	5	0	0	0–10	0–10
Mesotrophic	5	0–10	0–10	0–10	0–1000
Eutrophic	10	0–10	1–20	0–10	100–10000
Eutrophic with high humic content	2	10	1	—	—
Dystrophic with high humic content	3	1–10	0–1	—	—
Eutrophic reservoir	1	5–10	1–5	10–1000	1000–10000

[a]Modified from Kuznetsov, 1970.

Azotobacter is found in particular abundance growing epiphytically on submersed aquatic angiosperms and submersed portions of emergent macrophytes. A symbiotic relationship between the *Azotobacter* and the macrophytes is possible in that the larger plants secrete many dissolved organic compounds (Wetzel, 1969) that can serve as substrates for nitrogen-fixing bacteria, and the combined nitrogen of the bacteria may be utilized by the macrophytes. The planktonic *Azotobacter* populations of the littoral water are higher than those of the open water (J. Overbeck, personal communication). Although no quantitative data are available, it is known that the littoral zone may serve as a major site of nitrogen fixation by both heterotrophic bacteria and sessile blue-green algae. Further investigation is needed.

Large numbers of N_2-fixing bacteria occur in the sediments of lakes; most are concentrated in the upper 2 cm. Their numbers increase markedly in the transition from oligotrophic to productive lakes; this is particularly the case with *Clostridium* (Table 11–2). Seasonal changes in the numbers of *Clostridium* in the sediments of Chernoye Lake were similar to those observed in the open water: maximum numbers occurred in the fall, followed by low winter populations increasing to another peak in the spring. Lowest populations in the sediments were found in midsummer. Similar results were obtained for *Azotobacter* populations of sediments in Polish lakes (Niewolak, 1970, 1972).

Comparison of the intensity of N_2 fixation by *Azotobacter* and the blue-green *Anabaena* has indicated that fixation by the heterotrophic bacteria was several orders of magnitude less than that by the dominant alga (Table 11–3). In waters rich in dissolved humic organic compounds, such as in lakes Mary and Mize, where blue-green algae are rare, the nitrogen fixation rates are very low.

TABLE 11–3   Estimates of Fixation of Molecular Nitrogen by Heterotrophic Bacteria and Blue-Green Algae in Several Lakes[a]

Lake	Organisms	N_2 Fixation	N_2 Fixation in g, Whole Lake, Summer Stratification Period
Chernoye, USSR 1937	*Azotobacter*	1.6×10^{-11} mg cell^{-1} day^{-1}	0.14
	Anabaena scheremetievi	1.12×10^{-9}	13100
Mendota, Wisconsin 1960, 1967	Blue-green algal dominance	0.07–43 μg N l^{-1} hr^{-1}	
Sanctuary, Pennsylvania 1959	Blue-green algal dominance	0–6	
Wingra, Wisconsin 1961	Blue-green algal dominance	0.005–1	
Smith, Alaska 1963	*Anabaena* dominance	0–1	
Mary, Wisconsin June 1968	Heterotrophic bacteria	0.003–0.047	
Mize, Florida July 1968	Heterotrophic bacteria	0.083–0.308	
August 1968		0.000–0.083	

[a]After many sources cited in the text; also Brezonik and Harper, 1969.

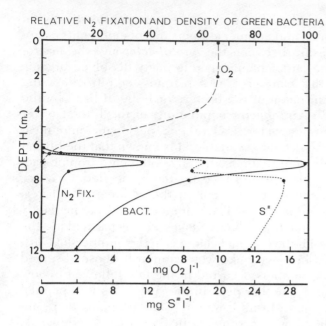

Figure 11–3 Nitrogen fixation in a Norwegian meromictic lake, June 1966, in relation to a dense layer of green photosynthetic bacteria, primarily *Pelodictyon*, on the surface of the moniomolimnion. (Drawn from data of Stewart, 1968.)

Unlike heterotrophic bacteria, whose capabilities for nitrogen fixation are limited to a few groups, nearly all photosynthetic bacteria are capable of fixing N_2. The photosynthetic bacteria include facultative aerobes and strict anaerobes. As will be discussed in more detail later on (Chapter 13), the photosynthetic bacteria commonly develop in great densities in highly structured depth strata at the interface regions between the aerobic epilimnion and the metalimnion and the anaerobic hypolimnion of eutrophic or meromictic lakes, if there is sufficient light to permit photosynthesis. N_2 fixation occurs only in the light, and intensive rates occur only under anaerobic conditions in the green and purple photosynthetic bacteria, including the facultative aerobes (Kondrat'eva, 1965). N_2 fixation by photosynthetic bacteria occurs simultaneously with the release of molecular H_2 by a noncyclic electron flux resulting from photophosphorylation. The source of these electrons is an exogenous H-donor, such as thiosulfate.

Very little quantitative information is available on the in situ rates of N_2 fixation by photosynthetic bacteria. In the example given in Figure 11–3, the nitrogen fixation is associated clearly with the green photosynthetic bacteria, mainly *Pelodictyon*, of the chemolimnion of this meromictic lake. This emphasizes the potential importance of bacteria to the nitrogen income of lakes under certain curcumstances.

Nitrogen Fixation: Wetland Sources

In addition to the inputs of nitrogen from drainage runoff from terrestrial sources, discussed further on, lakes and streams often are bordered by dense stands of shrublike trees that fix nitrogen from the atmosphere. The most common species of the genera *Alnus* and *Myrica* are nonleguminous angiosperms that form large nodules containing an actinomycetal fungal endophyte at or just below the soil surface. Nitrogen fixation by dense stands of *Alnus* can be as high as 225 kg N ha^{-1} year^{-1}, most of which enters the plant and subsequently the stream or lake water as leachate from direct leaf-fall or release during decomposition of foliage.

Alnus trees have been implicated as a significant nitrogen source for streams and lakes of the glaciated regions of Alaska that are particularly nitrogen deficient (Goldman, 1960; Dugdale and Dugdale, 1961). In Castle Lake, an alpine cirque lake of northern California, alder trees were found to be abundant on only one side (Goldman, 1961). *Alnus* leaves contained over four times as much nitrogen as those of other deciduous species. Soils, lake sediments, and spring waters draining to the lake from the alder side all contained higher nitrogen levels than those from the nonalder side of the lake. Bioassay of in situ rates of photosynthesis by the phytoplankton demonstrated the stimulatory effects of these nitrogen sources, and also that the primary productivity of the lake was significantly higher on the side on which the alders grew.

Inorganic and Organic Nitrogen

In addition to the nitrogen sources to fresh waters from the atmosphere in the form of precipitation, fallout, and fixation of N_2, discussed above, a major source of nitrogen income to lakes is from influents, both from surface land drainage and from groundwater sources that enter the basin by surface and subsurface springs. Inputs of nitrogen by groundwater can be a major part of the nitrogen in many lakes, especially in lakes of regions rich in limestone. In detailing the nitrogen cycle, the forms of nitrogen must be discussed individually, and integrated with biogeochemical utilization rates and transformations as well as nitrogen losses from the lake ecosystem. An obvious loss is via inorganic dissolved and organic (both dissolved and particulate) compounds in effluents flowing out of lake basins. Further losses occur as a result of permanent interment in the sediments of partially decomposed biota and inorganic and organic nitrogen compounds adsorbed to inorganic particulate matter. Nitrogen can be lost by diffusion of volatile compounds from the surface, such as ammonia at high pH, and as N_2 formed in microbial denitrification of nitrate.

Combined nitrogen occurs as ammonia (NH_4^+), hydroxylamine (NH_2OH), nitrite (NO_2^-), nitrate (NO_3^-), and dissolved and particulate organic nitrogen. NH_4-N can range from 0 to 5 mg l^{-1} in unpolluted surface waters, although levels are generally much lower than this extreme, and to well above 10 mg l^{-1} in anaerobic hypolimnetic waters of eutrophic lakes. The intermediate product NH_2OH usually is rapidly oxidized and occurs in very low concentrations. Similarly, NO_2-N levels of natural lake waters are generally very low, in the range of 0 to 0.01 mg l^{-1}, although up to 1 mg l^{-1} has been found in the interstitial waters of deep (>90 cm depth) sediments of Lake Mendota (Konrad, et al., 1970). Concentrations of NO_3-N range from 0 to nearly 10

TABLE 11–4 General Relationship of Lake Productivity to Average
Concentrations of Epilimnetic Nitrogen[a]

General Level of Lake Productivity	Change in Alkalinity in Epilimnion in Summer (mval 1^{-1})	Inorganic N (mg m^{-3})	Approximate Average Organic N (mg m^{-3})
Ultra-oligotrophic	<0.2	<200	<200
Oligo-mesotrophic	0.6	200–400	200–400
Meso-eutrophic	0.6–1.0	300–650	400–700
Eutrophic		500–1500	700–1200
Hypereutrophic	>1.0	>1500	>1200

[a]Modified from Vollenweider, R. H.: Scientific Fundamentals of the Eutrophication of Lakes and Flowing Waters, with Particular Reference to Nitrogen and Phosphorus as Factors in Eutrophication. OECD Report No. DAS/CSI 68.27, Paris, OECD, 1968, after data of E. A. Thomas and Lueschow, et al., 1970.

mg 1^{-1} in unpolluted fresh waters, but are highly variable seasonally and spatially. Organic nitrogen, much of which occurs in forms resistant to rapid bacterial degradation, commonly accounts for more than one-half of the total dissolved nitrogen.

Recently, much attention has been devoted to the importance of nitrogen concentrations of fresh waters in the regulation of algal productivity (for example, cf. Vollenweider, 1968). Although, as we will see in Chapter 12, phosphorus cycling rates are the most frequent regulating nutrient of high sustained productivity, the loading rates of nitrogen influents are of major importance to the maintenance of high flux rates within the nitrogen cycle. Though there are a number of exceptions, a direct correlation has been found between high sustained productivity of algal populations and average concentrations of inorganic and organic nitrogen (Table 11–4). Mean chemical mass, however, must be used with extreme caution when little consideration is given to the rates of mineralization, microbial transformation, and recycling.

In comparing the distribution of organic and inorganic forms of nitrogen in oligotrophic and eutrophic lake sediments, little correlation was found between amounts of total or organic forms of nitrogen in sediments and general productivity of the lake (Table 11–5) (Keeney, et al., 1970; Konrad, et al., 1970). The percentage of organic nitrogen as hexosamine-N decreased while amino acid-N increased with increasing lake fertility. NH_3-N of inter-

TABLE 11–5 Average Distribution of Nitrogen in Some Wisconsin Lake Sediments[a]

Lake Type	Sediment-N (mg kg^{-1})		Interstitial Water-N (mg 1^{-1})			Total Organic N (%)	Acid Hydrolyzable (% of Total N)
	Fixed NH_4	Exchangeable NH_4	Organic	NH_4	NO_3		
Softwater, oligotrophic	69	167	2.1	3.8	0.2	1.53	83.1
Softwater, eutrophic	44	415	2.2	9.6	0.3	3.26	80.4
Hardwater, oligotrophic	66	66	–	–	–	0.52	84.4
Hardwater, eutrophic	134	120	2.0	11.4	0.3	0.80	82.0

[a]Modified from Keeney, 1973.

stitial water was somewhat higher in sediments of eutrophic lakes than in those of oligotrophic lakes.

Ammonia

Ammonia is generated by heterotrophic bacteria as the primary end-product of decomposition of organic matter, either directly from proteins or from other nitrogenous organic compounds. Although intermediate nitrogen compounds are formed in the progressive degradation of organic material, these rarely accumulate and are deaminated rapidly by bacterial utilization. Although ammonia is a major excretory product of aquatic animals, this nitrogen source is quantitatively minor in comparison to that formed by decomposition.

Ammonia in water is present primarily as NH_4^+ and as undissociated NH_4OH, the latter being highly toxic to many organisms, especially fish (Trussell, 1972). The proportions of NH_4^+ to NH_4OH are dependent on the dissociation dynamics as governed by pH and temperature. The approximate ratios of NH_4 to NH_4OH are as follows (Hutchinson, 1957):

pH 6	3000:1
pH 7	300:1
pH 8	30:1
pH 9.5	1:1

Detailed dissociation relationships with pH and temperature are given in Trussell (1972). Ammonia is strongly sorbed to particulate and colloidal particles, especially in alkaline lakes containing high concentrations of humic dissolved organic matter.

Although ammonia would be a good source of nitrogen for plants, and

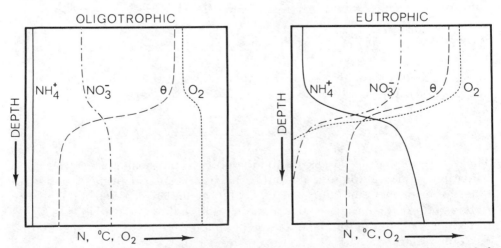

Figure 11-4   Generalized vertical distribution of ammonia and nitrate nitrogen in stratified lakes of very low and high productivity.

many plants can use it at alkaline pH values, most algae and macrophytes grow better with nitrate as their nitrogen source, even though nitrate must be reduced to ammonia. This relationship is influenced, in part, by the toxicity of NH_4OH at higher pH values that can prevail both in culture and in situ during periods of high daily photosynthesis in very eutrophic lakes (cf. Rodhe, 1948).

The distribution of ammonia in fresh waters is highly variable regionally, seasonally, and spatially within lakes in relationship to the level of productivity of the lake, and the extent of pollution from organic matter. While generalizations are difficult to make, the ammonia nitrogen (NH_3-N) of well-oxygenated waters is usually relatively low. Thus, results commonly show low concentrations of NH_3-N in unproductive oligotrophic waters, in the trophogenic zones of most lakes, and in most lakes after periods of circulation (Fig. 11–4). Where appreciable amounts of organic matter reach the hypolimnion of stratified lakes, NH_3-N tends to accumulate. The accumulation of NH_3-N greatly accelerates when the hypolimnion becomes anoxic. Under anaerobic conditions, bacterial nitrification, by which NH_4^+ is progressively oxidized through several intermediate compounds to NO_2^- and NO_3^-, ceases as the redox potential is reduced to below about 0.4 V. Moreover, with the loss of the oxidized microzone at the sediment-water interface under anoxic hypolimnetic conditions, the adsorptive capacity of the sediments is greatly reduced. The result is a marked release of NH_4^+ from the sediments.

Nitrification

Nitrification may be broadly defined as the biological conversion of organic and inorganic nitrogenous compounds from a reduced state to a more oxidized state (Alexander, 1965). Of the numerous oxidation and reduction stages outlined in Figure 11–5, initial nitrification by bacteria, fungi, and autotrophic organisms involves (Kuznetsov, 1970):

$$NH_4^+ + 1\frac{1}{2}\,O_2 \rightleftharpoons 2H^+ + NO_2^- + H_2O \ [\Delta F° = -66.0 \text{ kcal}],$$

which proceeds through a series of oxidation stages through hydroxylamine and pyruvic oxime to nitrous acid:

$$NH_4^+ \longrightarrow NH_2OH \longrightarrow H_2N_2O_2 \longrightarrow HNO_2$$

These intermediate products are highly labile to physical and heterotrophic oxidation, and are found only rarely in significant quantities relative to other forms of combined nitrogen (cf. Baxter, et al., 1973). Much of the energy (total exothermic energy -84.0 kcal mol^{-1}) released by the oxidations is used to reduce CO_2 in the formation of organic matter; detailed reactions are discussed by Alexander (1965a) and Kuznetsov (1970).

The nitrifying bacteria capable of the oxidation of $NH_4^+ \longrightarrow NO_2^-$

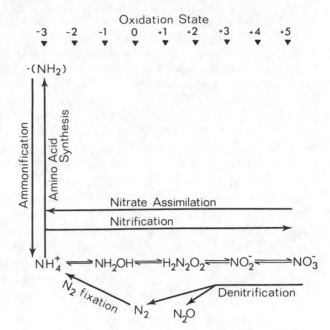

Figure 11–5 Biochemical reactions that influence the distribution of nitrogen compounds in water. (After Stadelmann, 1971, and Kuznetsov, 1970.)

are largely confined to *Nitrosomonas* (Nitrobacteriaceae, order Pseudomonadales), although several other genera are known to be capable of this process (Alexander, 1965). These bacteria are mesophilic, with a wide temperature tolerance range (1 to 37°C), and grow optimally at a pH near neutrality.

Oxidation of nitrite proceeds further to nitrate by:

$$NO_2^- + \tfrac{1}{2}O_2 \rightleftharpoons NO_3^- \; [\Delta F° = -18.0 \text{ kcal}]$$

Nitrobacter is the primary nitrifying bacterial genus involved in this oxidation. *Nitrobacter* is somewhat less tolerant of low temperatures and high pH, conditions that can lead to a slight accumulation of NO_2-N. The release of energy for synthesis of organic matter by the oxidation of nitrite is much lower at -18.0 kcal per mole than that of NH_4^+ to NO_2^-.

The overall nitrification reactions

$$NH_4^+ + 2O_2 \longrightarrow NO_3^- + H_2O + 2H^+$$

require two moles of oxygen for the oxidation of each of NH_4^+. Although conditions must be aerobic in order for nitrification to occur, these processes will continue to concentrations of 0.3 mg O_2 1^{-1} when the diffusion rates of oxygen to the bacteria become critical. In quiescent sediments where oxygen is very low or absent, nitrification is greatly reduced (Chen, et al., 1972), which indicates that the sediments do not contribute appreciable amounts of nitrate to the water by nitrification, except in the well-oxidized surficial layer such as in the littoral zone, or during periods of circulation (cf. Laurent and Badia, 1973; Landner and Larsson, 1973).

Oxidation of ammonia by autotrophic bacteria was found to be of relatively minor importance during summer stratification of eutrophic Pluss

See, Northern Germany (Gode and Overbeck, 1972). Heterotrophic nitrifying bacteria of several genera were found in much higher concentrations in both aerobic and anaerobic water strata. Experimental results suggested that these bacteria most likely were responsible for a majority of the nitrification that occurred in this lake.

Nitrification is inhibited severely by certain dissolved organic compounds, especially by tannins and tannin decompositional derivatives (Rice and Pancholy, 1972, 1973). Although this mechanism has been demonstrated only in soil systems, there is a strong possibility that an analogous situation exists in freshwater systems of high humic levels, where ammonia concentrations are often higher than those found in aerobic waters of other lakes. Therefore, one might expect reduced nitrification in neutral or alkaline waters containing high concentrations of dissolved humic organic matter. Such was apparently the case in several European waters (Nygaard, 1938; Karcher, 1939). Moreover, nitrification is reduced severely in acidic waters, such as in acid bogs and acidic bog lakes, where the pH is 5 or less. Nitrate produced in such lakes probably is utilized as rapidly as it is produced, so that most of the time only very low or undetectable quantities are found.

Nitrate Reduction and Denitrification

As nitrate is assimilated by algae and larger hydrophytes, it is reduced to ammonia. Molybdenum is required in the enzyme systems associated with this reduction. In a few lake regions, such as granitic mountainous regions, sources of molybdenum are extremely low. Rates of carbon fixation by in situ phytoplanktonic algae in lakes of these regions can be increased by additions of molybdenum (Goldman, 1960, 1972). These circumstances presumably are related to the rates of nitrate reduction.

The assimilation of nitrate and its reduction by green plants are clearly of major proportions in the trophogenic zone of lakes. Nitrate assimilation in oligotrophic lakes may be adequate to reduce the observed chemical mass of the trophogenic zone, as depicted in Figure 11–4, or may be counterbalanced by nitrification and inflow sources of NO_3-N. In eutrophic lakes, denitrification, discussed further on, is a major process influencing the vertical distribution of NO_3-N. Nitrate assimilation by photosynthesis can, however, greatly exceed sources of income and generation, in some cases to the point of reducing NO_3-N to below detectable concentrations.

The ratio of NO_3-N to NH_3-N in fresh waters is highly variable in relation to natural and pollutional sources of both forms of combined nitrogen. In regions draining calcareous sedimentary landforms, unpolluted lakes can have NO_3-N:NH_3-N ratios of 25:1. In other areas where natural sources of NO_3-N are low, the ratio approaches 1:1; where slight to moderate sewage contamination or agricultural applications of nitrogen fertilizers influence the lake, ratios in the range of 1:10 are common.

Denitrification by bacterial metabolism is the biochemical reduction of oxidized nitrogen anions, NO_3-N and NO_2-N, in the oxidation of organic matter. The general sequence of events of this process is:

$$NO_3^- \longrightarrow NO_2^- \longrightarrow N_2O \longrightarrow N_2,$$

which results in a significant reduction of combined nitrogen that can, in part, be lost from the system if it is not refixed.

Many facultative anaerobic bacteria, particularly of the genera *Pseudomonas, Achromobacter, Escherichia, Bacillus,* and *Micrococcus,* can utilize nitrate as an exogenous terminal H acceptor in the oxidation of organic substrates (Alexander, 1961). The denitrification reactions are associated with the enzyme nitrogen reductase and cofactors of iron and molybdenum, and operate similarly under both aerobic and anaerobic conditions (Bandurski, 1965).

An exemplary reaction of the oxidation of glucose and reduction of nitrate is (Hutchinson, 1957):

$$C_6H_{12}O_6 + 12\ NO_3^- = 12\ NO_2^- + 6\ CO_2 + 6\ H_2O\ [\Delta F^\circ = -460\ kcal\ mol^{-1}];$$

and for the reduction of nitrite to molecular nitrogen:

$$C_6H_{12}O_6 + 8\ NO_2^- = 4\ N_2 + 2\ CO_2 + 4\ CO_3^{--} + 6\ H_2O$$
$$[\Delta F^\circ = -720\ kcal\ mole^{-1}].$$

In both reactions, approximately as much energy is yielded as the aerobic oxidation of glucose by dissolved O_2 ($\Delta F^\circ = -699$ kcal mol^{-1}). The denitrification reactions occur intensely in anaerobic environments, such as in the hypolimnia of eutrophic lakes (Fig. 11–4) and in anoxic sediments, where oxidizable organic substrates are relatively abundant.

A specialized case, of much less general quantitative significance than the heterotrophic denitrification discussed above, is denitrification of nitrate concurrently with the oxidation of sulfur. The process is accomplished by denitrifying sulfur bacteria, particularly *Thiobacillus denitrificans,* that utilize S° or reduced sulfur compounds such as thiosulfate (Kuznetsov, 1970):

$$5S + 6KNO_3 + 2H_2O \longrightarrow 3N_2 + K_2SO_4 + 4KHSO_4$$

$$5Na_2S_2O_3 + 8\ KNO_3 + 2NaHCO_3 \longrightarrow 6Na_2SO_4 + 4K_2SO_4 + 2CO_2 + H_2O + 4N_2$$

Both processes occur chemosynthetically under dark, anaerobic conditions, and yield relatively small changes in free energy.

The rate of denitrification, as of nitrification, decreases in acidic waters (Keeney, 1973) and is very slow at low temperatures (ca 2°C), with optimum rates well above those temperatures of most natural fresh waters. At high temperatures the primary product is N_2, while at lower temperatures nitrous oxide (N_2O) predominates. However, N_2O is rapidly reduced to N_2, and has not been found in most lakes in appreciable quantities (Goering and Dugdale, 1966; Kuznetsov, 1970; Macgregor and Keeney, 1973).

Rates of denitrification by nitrate-reducing bacteria in eutrophic Pluss See, Germany, indicated marked seasonal and depth variations (Tan and Overbeck, 1973). Activity of nitrate reduction was particularly high, and was

TABLE 11–6   Approximate Rates of Bacterial Denitrification in Lake
Water and Sediments

| Lake | Rate of Denitrification (μg N 1^{-1} day^{-1}) | | Source |
	Water	Sediment	
Smith Lake, Alaska	15	90	Goering and Dugdale, 1966
Lake Mendota, Wisconsin	8–26	ca 580	Brezonik and Lee, 1968 Keeney, et al., 1971

correlated with cell numbers in the early portion of summer stratification before hypolimnetic nitrate concentrations were greatly reduced. High concentrations of oxygen and low levels of nitrate depressed rates of nitrate reduction, but had relatively little effect on cell numbers, especially in the winter.

Nitrification and denitrification can occur simultaneously. In lake sediments it has been found that denitrification of added NO_3-N, followed by $^{15}NO_3$, is rapid; within two hours up to 90 per cent of added NO_3-N was reduced, much to $^{15}N_2$ (Chen, et al., 1972). Much of the NO_3-N of lake sediments is incorporated into bacterial organic matter. Keeney, et al. (1971) found up to 37 per cent of 2 mg 1^{-1} NO_3-N on a volume basis incorporated into the organic fraction, and the remainder denitrified. Although there are few available data, it is known that denitrification rates of sediments are significantly greater than those of the overlying water (Table 11–6).

Dissolved and Particulate Organic Nitrogen

Much of the basic understanding of major nitrogenous fractions and their distribution in lakes has evolved from the classical studies of Wisconsin lakes, especially Lake Mendota (Peterson, et al., 1925; Domogalla, et al., 1925; Birge and Juday, 1926). Although analytical methodology has improved greatly since these early investigations, their results are still generally valid for many lakes of the temperate region.

The dissolved organic nitrogen (DON) of fresh waters often constitutes over 50 per cent of the total soluble nitrogen. Geographic variation is great, however, in relation to inputs of inorganic nitrogen from natural and artificial sources. Over one-half of the DON is in the form of amino nitrogen compounds, of which about two-thirds is in the form of polypeptides and complex organic compounds, and less than one-third occurs as free amino nitrogen (Table 11–7). The qualitative nature of the numerous nitrogen compounds is incompletely known (see reviews of Vallentyne, 1957, and Schnitzer and Khan, 1972). As will be discussed later, the simple amino acids are substrates that are readily utilized by bacteria; rates of decomposition are high, with resulting low instantaneous concentrations in free form in fresh waters.

As with organic carbon, the DON of lakes and streams is from 5 to 10 times greater than particulate organic nitrogen (PON) contained in the plank-

TABLE 11–7 Particulate Organic Nitrogen of the Seston and Dissolved
Organic Nitrogen of Several Lakes[a]

| | μg N l^{-1} | | | |
Lake	Particulate Organic N		Dissolved Organic N	
	O m	*20 m*	*O m*	*20 m*
Mendota, Wisconsin (means)	103	86		
Total soluble N			593	842
Amino N less free acids			170	177
Free amino N			88	88
Peptide N			181	173
Organic nonamino N			187	181
Furesø, Denmark	30–190		440–640	
Ysel, Netherlands	250–1400		590–1840	
Smith, Alaska	80–750			
Bodensee, Germany	10–160		50–150	
Lucerne, Switzerland (surface)	70–390		80–180	
Rotsee, Switzerland (surface)	180–1200		270–660	
Wintergreen, Michigan (surface)	50–2350		500–1320	
Augusta Creek, Michigan	ca 50–300		150–870	
Lawrence Lake, Michigan				
Pelagic	20–110		80–240	
Surface inlet streams			60–1550	
Groundwater			50–650	

[a]After several sources cited in the text.

ton and seston (Table 11–7) (Peterson, et al., 1925; Stadelmann, 1971; Barica, 1970; Manny, 1972a; Manny and Wetzel, 1973, 1974). The ratios of PON to DON decrease as the lakes become more eutrophic, are closer to 1:1 in the trophogenic zone, and increase in the tropholytic zones. More organic nitrogen apparently is synthesized by small phytoplanktonic algae ($<10\ \mu$m) per unit cell volume than by larger forms (Manny, 1972b). Algae, especially blue-green algae, also excrete polypeptides which are capable of forming complexes with metals such as iron and copper, and with phosphates, and of altering their solubility and physiological availability (Fogg and Westlake, 1955). Similar nitrogenous compounds have been found to be secreted by larger aquatic plants (Wetzel and Manny, 1972); in some situations where the littoral zone is extensively developed, these secretions can form a major source of organic nitrogen to the lake. Furthermore, as aquatic vascular vegetation decomposes, large quantities of organic nitrogen are released (Nichols and Keeney, 1973). Much of this organic nitrogen is absorbed by the sediments, in which decomposition can become rapidly limited by inorganic nitrogen, especially under anaerobic conditions.

Seasonal Distribution of Nitrogen

The seasonal changes in nitrogen vary greatly among lakes. Several general trends, however, do emerge from the seasonal patterns in the transition from oligotrophic to very productive lakes. Although many seasonal cycles of inorganic nitrogen have been studied, the cycles of the dissolved and the biota-bound particulate organic nitrogen are often overlooked. The ensuing discussion concerns five lakes: (a) Lawrence Lake, an oligotrophic, very hardwater lake of southern Michigan that receives high natural inputs of inorganic nitrogen; N_2 fixation is very low; (b) mesotrophic Vierwaldstattersee (Lake of Lucerne), Switzerland, a lake in which the hypolimnion remains aerobic throughout summer stratification; (c) moderately eutrophic Lake Mendota, Wisconsin, which was used to develop the fundamental studies of nitrogen dynamics; (d) highly eutrophic Wintergreen Lake, southern Michigan, a lake which undergoes normal dimictic circulation; and (e) Rotsee, Switzerland, an extremely eutrophic lake that circulates very poorly in spring and autumn.

The nitrogen inputs to Lawrence Lake, which is located in calcareous glacial till high in nitrates, occur largely as nitrate (Manny and Wetzel, 1974). Concentrations of two small spring-fed inlet streams varied from 1 to over 20 mg NO_3-N l^{-1}; groundwater inflow directly into the basin contained lower concentrations (mean 3 mg l^{-1}), but contributed about 25 per cent of the annual nitrogen income. Ammonia levels of the influents were much lower (between 0 to 190 μg NH_3-N l^{-1}), and constituted a relatively minor source

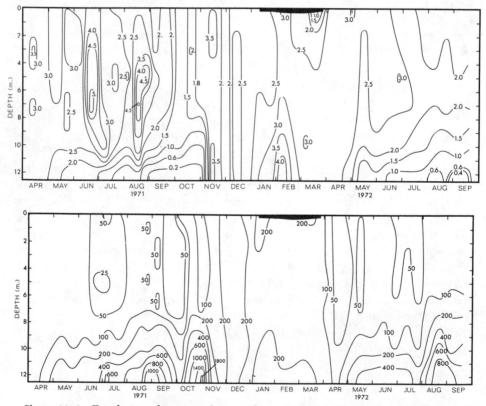

Figure 11–6 Depth-time diagrams of seasonal concentrations of NO_3-N + NO_2-N in mg l^{-1} (*upper*) and NH_3-N in μg l^{-1} (*lower*), Lawrence Lake, Michigan, 1971-72. Opaque areas = ice-cover to scale. (From Manny and Wetzel, unpublished data.)

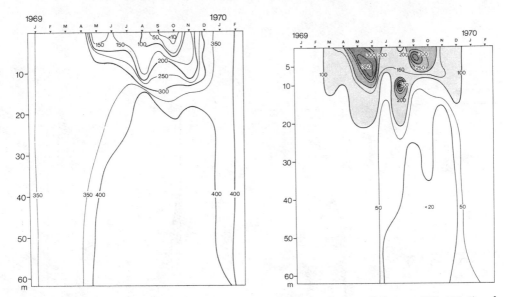

Figure 11–7 Depth-time diagrams of the concentrations of nitrate (*left*, in μg NO$_3$-N l^{-1}) and particulate organic nitrogen (*right*, in μg N l^{-1}). (After Stadelmann, P. Schweiz: Zeitschrift f. Hydrologie, 33:1-65, 1971.)

of nitrogen to this lake. Nitrite concentrations were very small. The total dissolved organic nitrogen (TDON) of the surface and groundwater influents, however, was a significant source and reached levels as high as 1.5 mg l^{-1}. This TDON was strongly correlated with the metabolic activity of the surrounding marsh vegetation through which the streams flowed.

Figure 11–6 contrasts the concentrations of NO$_3$-N + NO$_2$-N in mg l^{-1} (upper) and NH$_3$-N in μg l^{-1} (lower) in seasonal depth-time diagrams. NH$_3$-N levels of the epilimnion and metalimnion to 8 m were very low during stratification, and increased conspicuously only below 11 m, when anoxic conditions existed in late summer and autumn. This increase was accompanied by a simultaneous denitrification of NO$_3$ in the lower hypolimnion above the sediments. NO$_2$-N concentrations were always very low (<20 μg l^{-1}). During turnover in November, and throughout much of the winter period of ice-cover, NH$_3$ concentrations were distributed uniformly with depth at levels over 200μg l^{-1}. Nitrification is seen in the lower depths during the winter, as are marked decreases immediately below the ice, concordant with intensive growth of algal populations. After the spring period of uniformity during circulation, the pattern is repeated.

The dissolved organic nitrogen (DON) and particulate organic nitrogen (PON) of the open water of Lawrence Lake generally were found to be about equal in concentration (50 to 1000 μg l^{-1}). Concentrations of PON closely followed the dynamics of the biomass of the plankton; those of DON were maximal in the summer in the epilimnion and minimal in the winter in the hypolimnion.

The nitrogen budget of a large bay of Vierwaldstattersee was studied in detail by Stadelmann (1971). The nitrate concentrations (Fig. 11–7, *left*)

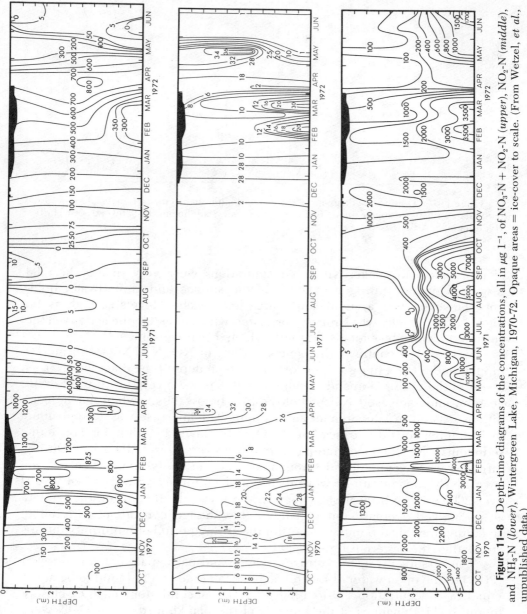

Figure 11–8 Depth-time diagrams of the concentrations, all in μg l⁻¹, of NO_3-N + NO_2-N (*upper*), NO_2-N (*middle*), and NH_3-N (*lower*), Wintergreen Lake, Michigan, 1970-72. Opaque areas = ice-cover to scale. (From Wetzel, *et al.*, unpublished data.)

of the epilimnion decreased sharply in the summer as growth of algae utilized major quantities, and this decrease is reflected in the amounts of planktonic particulate organic nitrogen (Fig. 11–7, *right*). NO_2-N concentrations followed a similar seasonal cycle, but never exceeded 11 μg NO_2-N 1^{-1}. Ammonia concentrations were always low, usually only in trace quantities; maximum levels (70 μg NH_3-N 1^{-1}) occurred in late summer near the sediments. During the period when epilimnetic concentrations of inorganic nitrogen approached zero, blue-green algal and heterocyst numbers reached their maximum, indicative of a heavy reliance on molecular N_2 fixation during this period. Aerobic hypolimnetic waters of this lake did not undergo significant denitrification.

As lakes become more eutrophic, these processes that create a balance between assimilation of nitrogen in the trophogenic zone and denitrification in the tropholytic zone intensify. Vertical gradients become extreme. As is exemplified in Wintergreen Lake (Fig. 11–8), nitrate and nitrite are reduced by algal assimilation and denitrification extremely rapidly, from over 1000 μg 1^{-1} to nil in slightly over a month (Wetzel, et al., 1975). Fixation of molecular N_2 was limited to the 0 to 3 m depths in Wintergreen Lake, and very high rates of fixation by *Anabaena* and *Aphanizomenon* were found only at these periods of greatly reduced combined inorganic nitrogen (Duong, 1972). Ammonification and denitrification are particularly intensive in the anaerobic tropholytic zone below a depth of 3 m. As the name of the lake (Wintergreen) implies, and as is common in hypereutrophic lakes (Wetzel, 1966a), production continues vigorously beneath the ice and the denitrification process is repeated, although reduced, during winter stratification. An analogous situation occurred in the extremely hypereutrophic Rotsee, Switzerland (Stadelmann, 1971). However, in this lake, which is protected from major wind-induced water movements by the surrounding topography, circulations in the spring and autumn were weak and incomplete. The hypolimnion was anaerobic all year long, except for a slight intrusion for a week in the spring; this condition is reflected in the high concentrations of ammonia throughout much of the lake most of the year (Fig. 11–9). Hypolimnetic concentrations of nearly 10 mg NH_3-N 1^{-1} are not uncommon near the sediments.

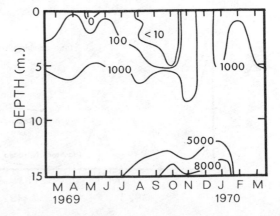

Figure 11–9 Depth-time diagram of the concentrations of ammonia (μg NH_3-N 1^{-1}), Rotsee, Switzerland, 1969-70. (Redrawn from Stadelmann, 1971.)

The seasonal distribution of the various forms of nitrogen in eutrophic Lake Mendota (Fig. 11–10) is not dissimilar from that commonly found in numerous lakes of temperate regions (Domogalla, et al., 1926; Domogalla and Fred, 1926; Barica, 1970). Autumnal minima have been observed which gradually increase in late fall to maxima in late winter and early summer. Looking at the seasonal processes involved in the surface and hypolimnetic strata over an annual period (Fig. 11–11), the net effects of nitrification and nitrate reduction-assimilation can be seen. In the oxygenated upper waters, nitrate reduction from the spring maximum is associated with increased rates of assimilation by the plankton and nitrate reduction. In the tropholytic zone, from the spring maximum of nitrate at turnover, nitrification decreases rapidly as the hypolimnion becomes anaerobic. Nitrification ceases and nitrate reduction increases markedly until stratification is disrupted in the fall.

Carbon to Nitrogen Ratios

Carbon of organic matter is on the average at least an order of magnitude greater in quantity than nitrogen. As the complex mixtures of organic compounds of biotic material of particulate and dissolved forms are decomposed and mineralized to inorganic carbon primarily as CO_2, and to inorganic nitrogen, the proteolytic metabolism of fungi and bacteria removes proportionately more nitrogen than carbon. As the rates of decomposition become slower with the greater resistance of the residual organic compounds of the organic matter, the net effect is an increasing C:N ratio.

If we look at the general distribution of organic carbon and nitrogen in many lakes, as Birge and Juday (1934) did in several hundred Wisconsin lakes, certain trends become apparent (Table 11–8). Despite the large seasonal and spatial variations in the particulate organic matter in lakes and streams, most of the organic matter is in the dissolved fraction which is much more constant

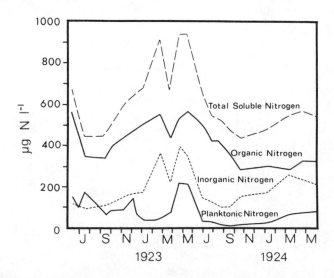

Figure 11–10 Average seasonal distribution of forms of nitrogen in Lake Mendota, Wisconsin, June 1922 through May 1924. (Redrawn from Hutchinson, 1957, after Domogalla, et al., 1926.)

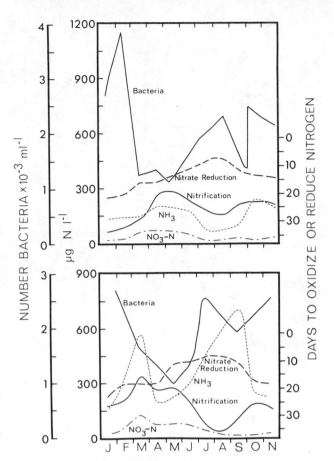

Figure 11–11 Rates of nitrification and denitrification, determined by the time required to oxidize or reduce added nitrogen, bacterial numbers, and the concentrations of ammonia and nitrate nitrogen in Lake Mendota, Wisconsin, 1925. (Redrawn from Domogalla, et al., 1926.) *Upper:* surface water of trophogenic zone; *lower:* near bottom water of tropholytic zone.

(cf. Wetzel, et al., 1972). Therefore, grouping of lakes into categories of increasing organic carbon and dissolved organic matter is reasonable. The reduction in organic nitrogen with increasing organic carbon content is clear, and results in increasing C:N ratios. This relationship suggests an increase in the refractory nature of the organic compounds, which become increasingly resistant to bacterial degradation under the conditions of the waters in which they occur. Such is indeed the case when waters with high concentrations of dissolved organic matter receive increasingly greater proportions of their organic content from organic plant material produced in terrestrial and littoral marsh areas surrounding the water. The allochthonous material becomes extremely significant as a major organic source in bog lakes and true bogs.

The organic matter of terrestrial and marsh sources undergoes varying degrees of decomposition prior to transport to the lake or stream water, during which much of the organic nitrogen is utilized. Therefore, in general (cf. Hutchinson, 1957), allochthonous organic matter contains about 6 per cent crude protein with a C:N ratio of 45 to 50:1. The dissolved organic matter contains a high percentage of humic acid compounds that are low in nitrogen

TABLE 11–8 Approximate Composition of Organic Matter of Water from Numerous Wisconsin Lakes[a]

Carbon of Total Particulate and Dissolved Organic Matter (mg l⁻¹)	Particulate Organic Matter (mg dry wt l⁻¹)	Dissolved Organic Matter[b]				
		Dry Weight (mg l⁻¹)	Crude Protein[c] (%)	Lipid Material[d] (%)	Carbohydrate (%)	C:N Ratio
1.0–1.9	0.62	3.1	24.3	2.3	73.4	12.2
5.0–5.9	1.27	10.3	19.4	1.3	79.0	15.1
10.0–10.9	1.89	20.5	14.4	0.4	85.2	20.1
15.0–15.9	2.32	31.3	12.9	0.2	86.9	22.4
20.0–25.8	2.22	48.1	9.9	0.2	89.9	29.0

[a]After data of Birge and Juday, 1934.
[b]Includes colloidal material; particulate matter removed by centrifugation.
[c]Total nitrogen content of organic fraction × 6.25.
[d]Ether extract.

content (Shapiro, 1956; Schnitzer and Khan, 1972) and impart a stained brown color to the water. In contrast, autochthonous organic matter produced by decomposition of plankton within the lake contains about 24 per cent crude protein with a C:N ratio of about 12:1.

Summary of the Nitrogen Cycle

A diagrammatic representation of the major nitrogen inputs, transformation pathways, and outputs to a general lake system is given in Figure 11–12. In spite of the numerous pathways presented, it is obviously a simplification of the complex mechanisms and processes. Numerous analogies in this cycle exist between lake and river systems.

The components of the cycle already have been discussed; only a few points should be emphasized. The processes indicated in the tropholytic zone represent a composite situation under aerobic conditions. Obviously, the obligately aerobic decomposition processes would cease in anaerobic hypolimnia of productive lakes. The metabolism of the littoral flora, including the vascular macrophytes and epiphytic associations of bacteria and algae, is not only a major or dominant source of organic nitrogen synthesis in many lakes, but can influence significantly the flux of nitrogen from the sediments to the water.

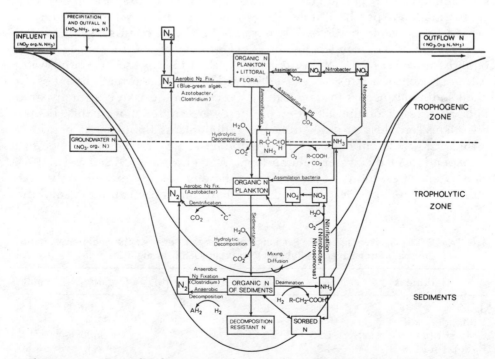

Figure 11–12 Generalized nitrogen cycle for fresh waters. PS = photosynthesis. (Greatly modified from Kuznetsov, 1970.)

TABLE 11–9 Approximate Nitrogen Content of Various Components of Bantam Lake, Connecticut[a]

Components	kg N per Lake	Per Cent
Lake water	10700	13
Algae	29200	33
Vascular aquatic plants	2500	3
Sediment (surface 1 cm)	44000	51
	86400	100

[a]Modified from estimates of Frink, 1967.

Many components and processes are highly variable seasonally and spatially. For example, inputs of nitrogen from guano of large migratory waterfowl that briefly reside on a lake in extraordinary densities (one per m^2) can represent a major input of nitrogen and phosphorus to certain lakes (Manny, et al., 1975; cf. also the extremely detailed review of Hutchinson, 1950). Sewage inputs of organic N and NH_3-N to rivers and recipient lakes are often highly pulsed. Similarly, agricultural applications of fertilizers play a very seasonal role in watershed retention and losses. Nitrogen fixation, normally a minor component of the total nitrogen income, can become a significant driving source of the system at certain times of the year.

The nitrogen dynamics of the sediments are poorly understood. Lake sediments typically contain on the order of 50 to 200 kg of nitrogen per 10 cm sediment depth per ha (Table 11–9), much of which is immobilized and sorbed to inorganic particles (Keeney, 1973). The interstitial water in sediments usually has a much higher concentration of soluble nitrogen compounds, mainly as NH_3-N and organic N, than that of the overlying water. Although diffusion rates are exceedingly slow, it is clear that mixing of surficial sediments occurs to some extent, even when the lake is strongly stratified, as a result of the activities of deep-water water movements, benthic organisms, and loss of gases from the sediments as bubbles. At periods of turnover, 5 to 20 cm of the surface sediment can be mixed, and much of it resuspended in the water column (e.g., Wetzel, et al., 1972; Davis, 1973). The extent of oxidation of the surface sediments usually extends only a few millimeters in depth, but the oxidized microzone is critical to the solubility

TABLE 11–10 Average Daily Exchange of Nitrogen Between the Sediments and Benthic Water Layers in Rybinsk Reservoir, USSR, in mg N m^{-2} day^{-1}[a]

Sediment	NO_3-N	NH_3-N	Albuminoid N	Organic N	Total N
Sand	+0.14	+ 1.14	+0.52	− 6.00	− 4.72
Unflooded soil	−0.44	+ 0.36	+0.41	+ 2.13	+ 2.03
Grey silt (fluvial areas)	−1.20	+21.43	+3.19	+10.98	+31.28
Redeposited peat	−4.3	+1.2	+1.3	+ 7.1	+ 4.0

[°]From Kuznetsov, S. L.: Limnol. Oceanogr., 13:211-224, 1968.

and sorption properties of the sediments for ammonia in particular, and for microbial transformations. The nitrogen exchange between sediments and water also varies greatly with sediment composition. For example, release of nitrogen from sediments of the Rybinsk Reservoir was greatest in silts high in organic matter (Table 11–10). Much loss occurred in the form of N_2 produced during anaerobic decomposition of sediment organic matter.

Nitrogen Budgets

Detailed evaluations of the nitrogen cycle that include close interval quantitative measurements of inputs, metabolic dynamics, and outputs are not available for freshwater systems. The cycle is obviously complex, and would require accurate analyses of the dynamics of all components for at least a year. At the present time, only a few approximate analyses of the nitrogen budgets of lakes are available. In these analyses the rates of bacterial metabolism are obtained indirectly, by assumptions of microbial processing or sedimentation by differences. Nonetheless, the calculated budgets are proximate and instructive.

The nitrogen budget of Lake Mendota, Wisconsin, demonstrates the roughly equivalent contributions from runoff, groundwater, and precipitation

TABLE 11–11   Estimated Nitrogen Budget of Lake Mendota, Wisconsin[a]

Sources	Nitrogen Income		Losses	Nitrogen Losses	
	$kg\ N\ year^{-1}$	%		$kg\ N\ year^{-1}$	%
Municipal and industrial wastewater	21200	10.4	Outflow	41300	20.4
			Denitrification	28100	13.9
Urban runoff	13700	6.8	Fish catch	11300	5.6
Rural runoff	23500	11.6	Weed removal	3250	1.6
Precipitation on lake surface	43900	21.6	Loss to groundwater	c	c
Groundwater Streams	35900	17.7	Sedimentation and other[d]	118850	58.6
Seepage	28500	14.1		202800	100
Nitrogen fixation	36100	17.8			
Marsh drainage	b	b			
	202800	100			

[a]Modified from Brezonik and Lee, 1968, with data improvements by Keeney, 1972.
[b]Considered significant, but data unavailable.
[c]Unknown; likely very small in this lake.
[d]By difference between total of other estimated losses and sum of income sources.

TABLE 11–12   Estimated Nitrogen Budget of the Rybinsk Reservoir, USSR, in Metric Tons N per Year[a]

	Factors Measured	Time		
		1 June 60– *1 June 61*	*1 April 61–* *1 April 62*	*1 June 62–* *1 June 63*
A	Measured nitrogen balance of water mass	+12519	+ 5399	+12170
B	Difference between inflow and outflow	− 180	− 7329	+ 1565
A − B = C	Increase in water mass	+12699	+12728	+10605
D	Precipitated in sediments (20-year average)	+12500	+12500	+12500
C − D = E	Theoretical balance in water mass	+ 199	+ 228	− 1895
A − E	Nitrogen input from atmosphere and fixation	+12320	+ 5171	+14065

[a]From Kuznetsov, S. I.: Limnol. Oceanogr., 13:211-224, 1968.

(Table 11–11). Major losses occur via sedimentation, denitrification, and outflow. Loss by seepage out of the basin probably is very small, since most lake basins are well-sealed (cf. Wetzel and Otsuki, 1974). Another type of nitrogen budget for a large reservoir of complex morphometry accounts for inflows from all river influents and losses by sedimentation and outflow (Table 11–12). Although the total nitrogen content of this reservoir decreased in 1962 because of low water levels, the inputs of atmospherically derived nitrogen amounted to roughly 10,000 tons per year. About 1300 metric tons entered with rain and snow; the remainder may be attributed largely to biological fixation.

General budgets of this type admittedly leave much to be desired because of the little insight they provide on the dynamics of internal processing and on metabolic control mechanisms. From an applied point of view, however, such budgetary information is important in relation to loading of aquatic systems with excessive influent sources of nitrogen, and their effects on increased productivity. We will return to this subject in the concluding discussion of eutrophication.

the phosphorus cycle

PHOSPHORUS IN FRESH WATERS

No other element in limnology has been as intensively studied as phosphorus. A great amount of quantitative data exists on the seasonal distribution of phosphorus in lakes. Ecological interest in phosphorus stems from its major role in biological metabolism, and the relatively small amounts of phosphorus in the hydrosphere. In comparison to the rich natural supply of other major nutritional and structural components of the biota (carbon, hydrogen, nitrogen, oxygen, sulfur), phosphorus is least abundant, and yet most commonly limits biological productivity.

Much of the quantitative information available on phosphorus distribution is open to question analytically. Only very recently have analyses of phosphorus dynamics of metabolism succeeded in explaining the complex rapid cycling of phosphorus, most of which is within the particulate phase of living biota, primarily algae. The secretion of highly labile compounds of low molecular weight from the particulate phosphorus fraction to a high molecular weight colloidal fraction is rapid and transitory. Slow losses of the colloidal fraction occur from the productive zones of lakes, and part is hydrolyzed to soluble orthophosphate, which is assimilated rapidly by the biota.

Losses of colloidal and particulate phosphorus fractions are replaced by regeneration of solubilized phosphorus fractions from decomposition at depth, by release of phosphorus from sediments, and by phosphorus contained in influents to the lake. The rates of biological productivity of a great number of lakes are governed to a large extent by the rate of phosphorus cycling in relation to the input loading of phosphorus from external sources. Influent loading of phosphorus to lakes has increased markedly in recent times as a result of man's accelerated use, and inefficient recovery, of phosphorus for agricultural nutrient, industrial, and detergent purposes, in addition to its release in domestic waste products. The recovery of phosphorus from many of these sources, coupled with substitution of nonphosphorus compounds in major products, is practical and economically feasible. Reduction of phosphorus inputs to many productive lakes results in the quantitative reduction of the phosphorus cycle, and a relatively rapid decline of the productive capacity of the lake system.

215

The Distribution of Organic and Inorganic Phosphorus in Lakes

Contrary to the numerous forms of nitrogen in lake systems, the only significant form of inorganic phosphorus is orthophosphate (PO_4^{---}). A very large proportion, greater than 90 per cent, of the phosphorus of lake water is bound organically in organic phosphates and cellular constituents in the living particulate matter of the seston or variously associated with or adsorbed to inorganic and dead particulate organic materials. It is instructive to discuss first the general aspects of the forms of phosphorus as they occur in fresh waters, and their distribution, before going on to analyze dynamics of exchange between the compartments.

The total inorganic and organic phosphorus has been separated in various ways in analyses; often, these fractions are related poorly to the metabolism, of phosphorus. The most important quantity, in view of the metabolic characteristics within a lake, is the total phosphorus content of unfiltered water, which consists of the phosphorus in suspension in particulate matter, and the phosphorus in 'dissolved' form (Juday, 1927; Ohle, 1938). Both compartments consist of several components. Particulate phosphorus includes: (1) Phosphorus in organisms as (a) relatively stable nucleic acids DNA and RNA, which are not involved in rapid cycling of phosphorus, and phosphoproteins, (b) low molecular weight esters of enzymes, vitamins, etc., and (c) nucleotide phosphates, such as adenosine diphosphate (ADP) and adenosine 5-triphosphate (ATP) of respiration and CO_2 assimilation. (2) Mineral phases of rock and soil, such as hydroxyapatite, are mixed phases in which phosphorus is adsorbed onto inorganic complexes such as clays, carbonates, and ferric hydroxides. (3) Phosphorus adsorbed onto dead particulate organic matter or in macro-organic aggregations. In contrast to the phosphorus of particulate matter, the dissolved inorganic phosphorus in the filtrate is composed of: (1) orthophosphate (PO_4^{---}), (2) polyphosphates, primarily of synthetic detergent origin, and (3) organic colloids or phosphorus combined with adsorptive colloids.

Because of the fundamental importance of phosphorus as a nutrient and major cellular constituent, much emphasis has been placed on analytical evaluation of its changes in concentrations with time. Chemical analyses all center around the reactivity of phosphorus with molybdate, and the changes in reactivity during enzymatic and acidic hydrolysis of complex forms of phosphorus compounds as they are converted to orthophosphate. Detailed analyses recognize eight forms of phosphorus, each of which is differentiated on the basis of its reactivity with molybdate, ease of hydrolysis, and particle size (Strickland and Parsons, 1972). Four operational categories result: (a) soluble reactive P, (b) soluble unreactive P, (c) particulate reactive P, and (d) particulate unreactive P. However, these operational methods do not necessarily correspond to the chemical species of phosphorus compounds, or to their role in the biotic cycling of phosphorus.

A majority of the phosphorus data for fresh waters is reported as total phosphorus and inorganic soluble phosphorus (orthophosphate). In more detailed studies, further differentiation into four general fractions has been done (Hutchinson, 1957), which are analogous to the four operational groups already cited: (a) soluble phosphate phosphorus, (b) acid-soluble suspended (sestonic) phosphorus, mainly ferric phosphate and calcium phosphate, (c) organic soluble and colloidal phosphorus, and (d) organic suspended (sestonic) phosphorus.

Total phosphate concentrations in nonpolluted natural waters extend over a very wide range from less than 1 μg 1^{-1} to extreme levels in closed saline lakes (>200 mg 1^{-1}). The total phosphorus concentrations of most uncontaminated surface waters are between 10 to 50 μg P 1^{-1}. Variation is high, however, in accordance with the geochemical structure of the region. Phosphorus levels are generally lowest in mountainous regions of crystalline geomorphology, and increase in lowland waters derived from sedimentary rock deposits. Lakes rich in organic matter such as bogs and bog lakes, tend to exhibit higher total phosphorus concentrations. A few sedimentary coastal areas, such as in the southeastern United States, are rich in phosphatic rock. Lakes with drainage from these deposits have abnormally high phosphorus levels.

In an extremely detailed treatment relating phosphorus and nitrogen to the productivity of lakes, Vollenweider (1968) demonstrated by several criteria that the amount of total phosphorus generally increases with lake productivity (Table 12–1). While there are a number of exceptions to this relationship, it demonstrates a general principle that is useful when dealing with applied questions of eutrophication.

Separation of the total phosphorus into inorganic and organic fractions in an appreciable number of lakes indicates that a large majority of the total phosphorus is in the organic phase (Table 12–2). Of the total organic phosphorus, about 70 per cent or more is within the particulate (sestonic) organic material, and the remainder is present as dissolved and colloidal organic phosphorus. Rigler (1964) demonstrated conclusively that data of former researchers who employed centrifugation and paper filtration methods of fractionation reduced the real significance of the sestonic organic phosphorus. Soluble organic phosphorus includes a major fraction of colloidal phosphorus. Inorganic soluble phosphorus is consistently very low, constituting

TABLE 12–1   General Relationship of Lake Productivity to Average
Concentrations of Epilimnetic Total Phosphorus[a]

General Level of Lake Productivity	Change in Alkalinity in Epilimnion During Summer (mval 1^{-1})	Total Phosphorus (μg 1^{-1})
Ultra-oligotrophic	<0.2	<5
Oligo-mesotrophic	0.6	5–10
Meso-eutrophic	0.6–1.0	10–30
Eutrophic		30–100
Hypereutrophic	>1.0	>100

[a]Modified from Vollenweider, R. A.: Scientific Fundamentals of the Eutrophication of Lakes and Flowing Waters, with Particular Reference to Nitrogen and Phosphorus as Factors in Eutrophication. OECD Report No. DAS/CSI/68.27, Paris, OECD, 1968, after numerous sources.

TABLE 12-2 Fractionation of Total Phosphorus in Lakes Employing Different Techniques of Separation

Lakes	Soluble Inorganic P		Soluble Organic P		Sestonic P		Total Organic P		Total P
	$\mu g\ l^{-1}$	%	$\mu g\ l^{-1}$	%	$\mu g\ l^{-1}$	%	$\mu g\ l^{-1}$	%	$\mu g\ l^{-1}$
Northern Wisconsin[a] (Juday and Birge, 1931)	3	13.0	14	60.9	6	26.1	20	87.0	23
Michigan Lakes[b] (Tucker, 1957)	1.5	12.0	5.7	46.9	5.0	41.1	10.7	88.0	12.2
Linsley Pond, Connecticut[c] (Hutchinson, 1957)	2	9.5	6	28.6	13	61.9	19	90.5	21
Ontario Lakes[c] (Rigler, 1964)	—	5.9	—	28.7	—	65.4	—	94.1	—

[a]Centrifugation techniques.
[b]Paper (No. 44 Whatman) filtration.
[c]Membrane filtration (0.5 μm).

only a few percent of total phosphorus, and as will be seen further on, it is cycled very rapidly in the epilimnetic zones of utilization. The ratio of inorganic soluble phosphorus to other forms of phosphorus of ca 1:20 or <5 per cent as inorganic phosphate phosphorus is remarkably constant in a large variety of lakes of the temperate zone. The percentage of total phosphorus occurring as truly ionic orthophosphate probably is considerably less than 5 per cent in most natural waters.

The phosphorus distribution within the fractions just discussed is the picture generally observed in the trophogenic zones of lakes. Phosphate, pyrophosphate, triphosphate, and higher polyphosphate anions are known to form complexes, chelates, and insoluble salts with a number of metal ions (Stumm and Morgan, 1970). The extent of complexing and chelation between various phosphates and metal ions in natural waters depends upon the relative concentrations of the phosphates and the metal ions, the pH, and the presence of other ligands (sulfate, carbonate, fluoride, organic species) in the water. Because phosphate concentrations are generally low, complex formations involving these major cations and various phosphate anions will have little effect on the metal ion distribution, but may have marked effects on the phosphate distribution (cf. Golachowska, 1971). Metal ions, such as those of ferric iron, manganous manganese, zinc, copper, etc., are present in concentrations comparable to or lower than those of phosphates. For these ions, complex formation can affect significantly the distribution of the metal ion, the phosphates, or both. For example, the solubility of aluminum phosphate ($AlPO_4$) is minimal at pH 6 and increases with higher pH. Ferric phosphate ($FePO_4$) behaves similarly, although it is more soluble than $AlPO_4$. Calcium concentration influences the formation of hydroxylapatite [$Ca_5(OH)(PO_4)_3$]. In a solution system without other compounds, a calcium concentration of 40 mg l^{-1} at a pH of 7 limits the solubility of phosphate to ca 10 $\mu g\ l^{-1}$. A calcium level of 100 mg l^{-1} lowers the maximum equilibrium of phosphate to 1 $\mu g\ l^{-1}$. Elevation of the pH of waters containing typical concentrations of calcium should lead to apatite formation. Moreover, increasing pH leads to formation of calcium carbonate, which coprecipitates phosphate with the carbonate (Otsuki and Wetzel, 1972). Sorption of phosphates and polyphosphates on surfaces is well known, particularly onto clay minerals (cf. Stumm and Morgan, 1970), by chemical bonding of the anions to positively charged edges of the clays and substitution of phosphates for silicate in the clay structure. In general, high phosphate adsorption by clays is favored by low pH levels (ca 5 to 6).

Following from these interactions, and the distribution of phosphorus in inorganic and organic fractions, the general tendency for unproductive lakes with orthograde oxygen curves to show little variation in phosphorus content with depth is reasonable (Fig. 12–1). Similarly, during periods of fall and spring circulation, distribution with depth is more or less uniform. Oxidized metals, such as iron, and major cations, particularly calcium, can induce precipitation of phosphorus.

Lakes exhibiting a strongly clinograde oxygen curve during the periods of stratification, however, possess much more variable vertical distributions of phosphorus. The most common observations show a marked increase in

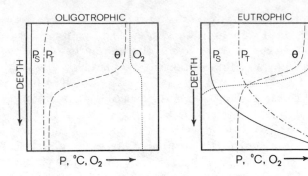

Figure 12–1 Generalized vertical distribution of soluble (P_s) and total (P_T) phosphorus in stratified lakes of very low and high productivity.

phosphorus content in the lower hypolimnion, especially during the latter phases of thermal stratification (Fig. 12–1). Much of the hypolimnetic increase is in soluble phosphorus near the sediments. The sestonic phosphorus fraction is highly variable with depth. Sestonic phosphorus in the epilimnion fluctuates widely with oscillations in plankton populations, whereas sestonic phosphorus in the metalimnion and hypolimnion varies with sedimentation of plankton, rates of decomposition with depth, and the development of bacterial and other plankton (e.g., euglenophyceans) populations at depth.

Phosphorus and the Sediments

The exchange of phosphorus between the sediments and overlying water is a major component of the phosphorus cycle in natural waters. Its importance rests in an apparent net movement of phosphorus into the sediments in most lakes. The effectiveness of the net phosphorus sink to the sediments and the rapidity of processes regenerating the phosphorus to the water relate to an array of physical, chemical, and metabolic factors. There is little correlation between the amount of phosphorus in the sediments and the productivity of the overlying water. The phosphorus content of the sediments can be several orders of magnitude greater than that of the water. The important factors are the ability of the sediments to retain phosphorus, and the conditions of the overlying water and the biota within the sediments that alter exchange equilibria and effect phosphorus transport back to the water.

Exchanges Across the Sediment Interface

Exchanges across the sediment interface are regulated by mechanisms associated with mineral-water equilibria, sorption processes (notably ion exchange), redox interactions dependent on oxygen supply, and the activities of bacteria, fungi, plankton, and invertebrates. The exchange rates depend on local diffusion coefficients and on environmental control of inorganic and organic, i.e., enzymatic, reactions. The sediment-water interface separates two very different domains. In all but the upper few millimeters of sediment, diffusion is controlled by motions on molecular scales with correspondingly low diffusion rates (Duursma, 1967), while in the water, exchange is regulated by much higher and extremely variable turbulent diffusion rates (Mortimer, 1971).

Oxygen Content of the Microzone

The most conspicuous regulatory feature of the sediment boundary is the mud-water interface and the oxygen content at this interface. The oxygen content at this microzone is influenced primarily by metabolism of bacteria, fungi, planktonic invertebrates that migrate to the interface, and sessile benthic invertebrates. Microbial degradation of dead particulate organic matter, sedimenting to the hypolimnion and sediments, is the primary consumptive process of oxygen in deepwater areas of lakes. In a large number of lakes, the amount of sedimenting organic material reaching the sediments reflects the intensity of photosynthetic productivity in the littoral and trophogenic zones. In reservoirs, organic loading from the drainage basin can become a significant source, in addition to the autotrophic production within the lake basin. The rate of oxygen depletion is governed by the rates of organic loading of the hypolimnia, and morphometric characteristics of volume (cf. Chapter 8). For example, it has been estimated that 88 per cent of the hypolimnetic oxygen consumption in the central basin of Lake Erie resulted from bacterial degradation of algal sedimentation from the trophogenic zone (Burns and Ross, 1971). Decomposition of more labile organic fractions, largely of plant origin, occurs en route to the sediments, depending on rates of input and sedimentation, and the sediments often receive organic residues that are more resistant to decomposition.

Sediment demand for oxygen from the water is high and is governed by the intensity of microbial and respiratory metabolism, a chemical demand arising from diffusion, and from the fact that inorganic elements, such as Fe^{++}, accumulate in reduced form when released into the sediment from biotic decomposition. Diffusion regulates transport and is essentially molecular in the sediments, unless the superficial sediments are disturbed by overlying water turbulence. Oxygen from well-aerated overlying water, such as in oligotrophic lakes or in more productive lakes at periods of complete circulation, will penetrate only a few centimeters into the sediments by diffusion, and is governed by the rate of supply to the sediments, turbulent mixing of superficial sediments, if any, and by the oxygen demand per unit volume of the sediment. The superb experimental and observational work of Mortimer (1941, 1942, 1971) has demonstrated the importance of an oxidized microzone to chemical exchanges, especially of phosphorus, from the sediments. The small difference of a few millimeters in depth of zero oxygen level is the critical factor regulating exchange between sediment and water. These relationships are exemplified by two lakes with organic sediments which we will describe; the first lake maintained oxygen concentrations at

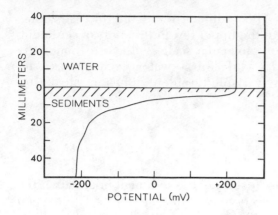

Figure 12–2 Diagrammatic profile of composite electrode potentials, not corrected for pH variations, across the sediment-water interface in undisturbed cores from the deepest portion of Lake Windermere before, during, and after stratification. (Based on data of Mortimer, 1971.)

the sediment interface greater than 8 mg l⁻¹ throughout summer stratification, while in the second oxygen levels at the interface decreased to <1 mg l⁻¹.

In the first situation, illustrated by Mortimer's studies of Lake Windermere, England, oxygen concentration at the sediment surface did not fall below 1 or 2 mg l⁻¹. Electrode potentials, an approximation of redox potentials evaluating composite redox reactants (cf. Chapter 13) — of the oxygenated overlying water and surficial sediments to a depth of ca 5 mm were uniformly high (+200 to 300 mv). Below 40 to 50 mm in the sediments the potentials were uniformly low (ca. −200 mv), indicative of extreme reducing conditions and total anoxia (Fig. 12–2). The conditions of oxidized sediment to a depth of about 5 mm, above the steep gradient of redox decline, were maintained throughout the period of summer stratification. Seasonal differences in sediment depth at which the transition from high to low potential occurred were observed, but the region of low potential never extended into the water. After five months of stratification, the point of zero mv increased towards the surface of the sediments to −5 mm from ca −12 mm at the time of spring turnover, and moved downwards to −10 mm during fall circulation. The point to be emphasized is that the integrity of the oxidized microzone was maintained in a thin but operationally very significant layer during stratification periods by diffusion and by turbulent displacement of the uppermost sediments to the overlying water during turnover periods (cf. Gorham, 1958). The effectiveness of the oxidized microzone in preventing significant release of soluble components from the interstitial waters of the sediments to the overlying water was demonstrated in experimental chambers for over five months (Fig. 12–3, *left*). Phosphorus, in particular, was prevented from migrating upward.

The ability of sediments to retain phosphorus beneath an oxidized microzone at the interface is related to several interacting factors. Much of the organic phosphorus reaching the sediments by sedimentation is decomposed and hydrolyzed (Sommers, et al., 1970). Most of the phosphorus is inorganic, such as apatite, derived from the watershed and phosphate adsorbed onto clays and ferric hydroxides. Additionally, phosphate coprecipitates with iron, manganese, and carbonates (Mackereth, 1966; Harter, 1968; Wentz and Lee, 1969). Work on Wisconsin lake sediments and the Great Lakes indicated

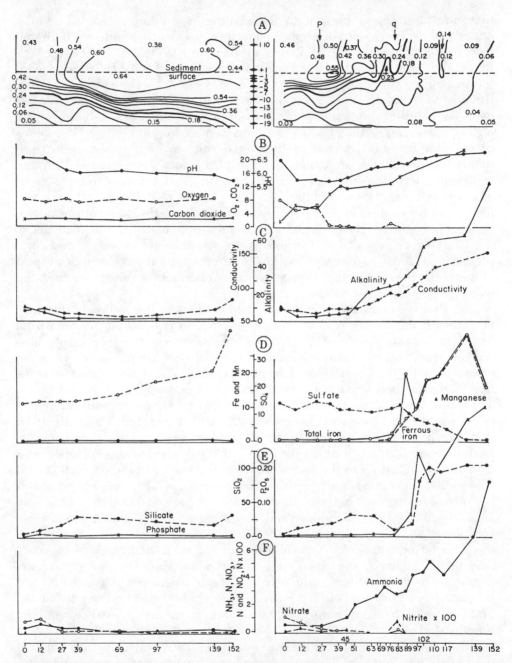

Figure 12–3 Variation in chemical composition of water overlying deep water Lake Winder-mere sediments over 152 days in experimental sediment-water tanks. *Lefthand series:* aerated chamber; *Righthand series:* anoxic chamber. A, Distribution of redox potential (E_7; Eh adjusted to pH 7) across the sediment-water interface in mm; B, pH, concentrations of O_2 and CO_2 in mg 1^{-1}; C, alkalinity expressed as mg $CaCO_3$ 1^{-1} and conductivity in μmhos cm^{-1} at 18°C; D, iron (total and ferrous as Fe) and SO_4 in mg 1^{-1}; E, phosphate as P_2O_5 and SiO_2 in mg 1^{-1}; F, nitrate, nitrite × 100, and ammonia, all as N, in mg 1^{-1}. (From Mortimer, C. H.: Limnol. Oceanogr., 16: 396, 1971, and J. Ecol., 29:280, 1941.)

that phosphorus was present in the sediments predominantly as apatite, organic phosphorus, and orthophosphate ions covalently bonded to hydrated iron oxides (Shukla, 1971; Williams, et al., 1970, 1971a, 1971b, 1971c; Williams and Mayer, 1972). In calcareous sediments of hardwater lakes containing 30 to 60 per cent $CaCO_3$ by weight, $CaCO_3$ levels were not related directly to inorganic and total phosphorus, and these sediments had a lower capacity to adsorb inorganic phosphorus than noncalcareous sediments. The observations imply that $CaCO_3$ sorption is less important than iron-phosphate complexes in controlling the concentrations of phosphorus in sediments. Although the exchange of adsorption and desorption within the sediments between sediment particles and interstitial water can be as rapid as a few minutes (Hayes and Phillips, 1958; Li, et al., 1972), the rate of transfer across the sediment-water interface depends on the state of the microzone. The oxidized layer forms an efficient trap for iron and manganese (cf. Chapter 13), as well as for phosphate, which is adsorbed on and complexed with ferric oxides and hydroxide, thereby greatly reducing transport of materials into the water while scavenging materials as phosphate from the water.

As the oxygen of the hypolimnion and water near the sediment interface is reduced, the barrier of the oxidized microzone is reduced. As seen from Mortimer's experiments (Fig. 12–3, *righthand series*), the release of phosphorus, iron, and manganese increased markedly as the redox potential decreased. With the reduction of ferric hydroxides and complexes, ferrous iron and adsorbed phosphate were mobilized and appeared in the water. The same general reactions were observed in the hypolimnetic water just overlying the sediments in eutrophic Esthwaite Water (Fig. 12–4), a pattern that has been observed repeatedly in productive dimictic lakes since its initial detailed description by Einsele (1936). A sudden release of ferrous iron and phosphate into the water occurs at the time when the +0.20 isovolt ($E_7 = +200$ mv) emerges above the interface surface. This event is preceded by nitrate reduction and the slow release of bases (alkalinity), CO_2, and ammonia. Manganese is reduced and mobilized at a higher potential than iron complexes.

The introduction of oxygen during autumnal circulation causes ferrous iron to be oxidized, and causes the simultaneous reduction of phosphate, in part as ferric phosphate, which is less soluble than ferric hydroxide, and in part by adsorption on ferric hydroxide and association with $CaCO_3$. Manganese is oxidized more slowly than iron, but nonetheless is effectively precipitated at the time of overturn. Ferrous iron released from the sediments is always in excess of phosphate, and when oxidized, it precipitates much of the phosphate. Some of the ferric phosphate in particulate form may hydrolyze slowly and restore some phosphate to the upper waters and littoral areas (Hutchinson, 1957). However, most phosphate is returned by eventual sedimentation.

In very productive lakes where hypolimnetic decomposition of sedimenting organic matter produces anoxic conditions and hydrogen sulfide, some ferrous sulfide (FeS) is precipitated. Ferrous sulfide, like many other metal sulfides, is exceedingly insoluble and forms at a redox potential of

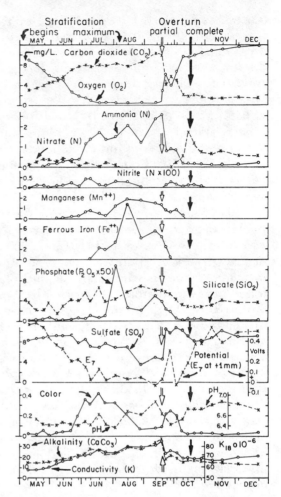

Figure 12–4 Seasonal distribution in composition (mg 1^{-1}) and properties of water within 30 cm of the sediments at 14 m in Esthwaite Water, England. Components as in Figure 12–3; color in arbitrary units. (From Mortimer, C. H.: Limnol. Oceanogr., 16:387, 1971.)

about +100 mv. If FeS precipitation is large, sufficient iron can be removed to permit some of the phosphate accumulated in the hypolimnion to remain in solution during autumnal circulation. The addition of sulfate to a lake to increase the bacterial production of hydrogen sulfide (H_2S) and to accelerate the loss of iron has been suggested as a method of fertilizing lakes by increasing the regeneration of phosphate from the sediments (Hasler and Einsele, 1948).

Phosphorus Release from the Sediments

Because of the obvious importance of phosphorus as a nutrient that often accelerates the productivity of fresh waters and generates attendant problems of excessive growth of plants, much interest has been devoted to the phosphorus content of sediments and its movement to the overlying water. Lake sediments contain much higher concentrations of phosphorus than the water

(Olsen, 1958, 1964; Holden, 1961; Hepher, 1966; and many others). Under aerobic conditions, it is clear that the exchange equilibria are largely uni-directional towards the sediments. Under anaerobic conditions, however, the interface inorganic exchange mechanism is strongly influenced by redox conditions. The depth of the sediment involved in active migration of phos-phorus to the water is considerable. In undisturbed anoxic sediments, given sufficient time (2 to 3 months), phosphorus moved upward readily from at least a depth of −10 cm to the overlying water, regardless of whether the sedi-ments were calcareous eutrophic muds or acidic and peaty in nature (Hynes and Greib, 1970). Comparison of movement in sterile sediments and sedi-ments with anaerobic bacteria showed no significant difference, and diffu-sion predominated.

Phosphorus-mobilizing bacteria, especially of the genera *Pseudomonas,* *Bacterium,* and *Chromobacterium,* were found to be abundant to at least 15 cm in reservoir sediments (Gak, 1959, 1963). Their abundance and ver-tical distribution varied with the type of sediments. Low numbers occurred in sandy sediments with small amounts of silt, and the bacteria were concen-trated near the interface. Their numbers increased with more uniform distri-bution with depth in sandy sediments with moderate amounts of organic matter and silts. The greatest numbers were found in silts with high organic matter. Hayes and Anthony (1958, 1959, and Anthony and Hayes, 1964) ex-amined the relationship of bacterial densities and organic content of sedi-ments in detail, and found only a weak correlation in a large number of lakes. Bacterial numbers in the sediments increased proportionally, especially at the interface, with several indices of increasing lake productivity in all clearwater lakes, i.e., lakes of neutral or alkaline pH and with low content of humic compounds resistant to rapid degradation. Biomass of bacteria in sediments of organically stained acidic bog lakes was also high, although these are lakes which by some criteria, but not all, are considered to be un-productive.

While bacteria are of major importance in the dynamics of phosphorus cycling in the water, as will be discussed in the following section, their role in expediting phosphorus exchange across the sediment interface is rela-tively minor in comparison to chemical equilibria (Hayes, 1964). Bacterial decomposition is proportional to bacterial densities of the interface and directly related to the general productivity of the lake (Hayes and MacAuley, 1959; Hargrave, 1972). The interface bacterial metabolism, however, has relatively little effect on biogenic fixation and removal of phosphate from the overlying water. Employing experimental conditions of sterilized and nat-ural sediments, it was found that microbial fixation and transport of phos-phorus from the water to the sediments amounted to less than 5 per cent of the total movement under anaerobic reducing conditions (Hayes, 1955; Olsen, 1958; Macpherson, et al., 1958; Pomeroy, et al., 1965). Under aerobic conditions, bacteria of the interface did increase the microbial transport of phosphorus to the sediments significantly (Hayes, 1955; Hayes and Phillips, 1958), and this is related to the input of microbial phosphorus sedimenting to the interface.

The rate of phosphorus release from the sediments increases markedly,

and in fact about doubles, if the sediments are disturbed by agitation from turbulence (Zicker, et al., 1965). Covering anaerobic sediments with sand or
polyethylene sheeting greatly impedes the loss of oxygen in the overlying
water, and decreases the rate of release of phosphorus, iron, and ammonium
from the sediment (Hynes and Greib, 1970).

Algae growing on sediments are able to utilize phosphorus of the sediments effectively (Golterman, et al., 1969). Moreover, algae suspended in
water with various particulate inorganic compounds of extremely low solubility were capable of extracting sufficient phosphorus for active growth;
without the phosphorus sources from the sediments, the phosphorus content
of the water was limiting to growth under experimental conditions. The presence or absence of bacteria had little effect on utilization. These results
stress the importance for growth of extractable phosphates of the sediments
if they are agitated into the water column, as in shallow lakes, even though
their solubilities may be extremely low.

Phosphorus Cycling by Aquatic Angiosperms and Benthic Organisms

The importance of submersed angiosperms, as well as floating-leaved
and emergent macrophytes, to the dynamics of the phosphorus cycle has
been implicated indirectly by several studies. Although phosphorus uptake
by roots of plants and cycling return of phosphorus, primarily by decomposition of returned organic matter, are well-known among terrestrial plants,
for many years this cycle was believed not to exist or not to be of importance
in aquatic habitats. The root and vascular systems of aquatic angiosperms are
greatly reduced because of their buoyancy and support of the water in comparison to air. Furthermore, the leaf morphology of these plants is simplified
with reduction or complete loss of cuticle on epidermal cell walls. Observing
these relationships, in addition to numerous physiological studies that indicated active uptake of nutrients by submersed leaves, led to the generalization that nutrients were absorbed primarily through the leaves, and that the
roots of these angiosperms functioned only in anchorage.

The idea that the roots of submersed angiosperms function in absorption
of nutrients evolved from numerous studies relating abundance and distribution of aquatic plants in lakes to the chemical composition of sediments.
Dominance of uptake by foliage absorption or by the root-rhizome system in
aquatic plants is highly variable among species. Evidence for shifts in functional dominance of uptake by leaves to that by roots during the life cycle is
discussed in more detail in Chapter 15 and by Sculthorpe (1967).

The importance of the littoral vegetation has been emphasized repeatedly
in early attempts to analyze the dynamics of phosphorus cycling in lakes by
the introduction of radioactive phosphorus to the surface waters (e.g., Hutchinson and Bowen, 1947, 1950; Coffin, et al., 1949; Hayes, et al., 1951, 1952;
Hayes and Phillips, 1958). These studies clearly indicated that when phosphorus was added to the water, rapid uptake by the littoral vegetation as well
as by the phytoplankton occurred. Moreover, it was observed that phosphorus
apparently was released slowly from the plants, but more rapidly from

epiphytic algae on the plants, to the water. With the decay of annual macro-phytes at the end of summer, releases of phosphorus from decay of the vege-tation were postulated.

A detailed investigation by Solski (1962) on the release of phosphorus from the leaves and roots of emergent, floating, and submersed macrophytes after death indicated the importance of the macrovegetation as a source of phosphorus to many aquatic systems in which the biomass is a major produc-tion component. Leaching of phosphorus from dead macrophytes under sterile conditions was rapid, and resulted in a loss of from 20 to 50 per cent of total phosphorus content in a few hours, and 65 to 85 per cent over longer periods. Leaching of phosphorus was more rapid from dried than from wet plants, but rates were affected very little by fragmentation of plant material or by different temperatures within the normal limnological range. Rates of leaching were found to be greater from roots than from leaves. Because the content of organically-bound phosphorus of the plant tissue changes during the vegetative period, generally decreasing with increasing age, the quantity of phosphorus leached to the water varies seasonally, but depends upon the total phosphorus content of the plants. Phosphorus released rapidly (within days) from decaying macrophytes is mineralized quickly, either utilized rapidly by bacterial and algal metabolism, or lost to the oxidized microzone of the sediments (Solski, 1962; Nichols and Keeney, 1973).

The presence of the macrophytes *Eriocaulon* or *Utricularia* in experi-mental sediment-water systems greatly increased the movement of radiophos-phorus from the water to the sediments (Hayes, 1955). The absorption and translocation of phosphorus were studied in bacteria-free cultures of the macroalga *Chara*, a common littoral macrophyte of hardwater lakes (Little-field and Forsberg, 1965). Results showed that all parts of the plants could absorb ^{32}P about equally, and that approximately the same proportion of absorbed phosphorus was translocated from the apices or rhizoids to other parts of the plants. Similar results were obtained for other freshwater plants (e.g., Schwoerbel and Tillmanns, 1964; DeMarte and Hartman, 1974).

Much insight into the role of aquatic macrophytes in phosphorus cycling has been provided by the studies of McRoy, et al. (1972). Rates of phosphorus uptake and excretion by both roots and leaves of eelgrass, *Zostera marina* L., were found to be dependent on the orthophosphate concentration of the medium. The interstitial inorganic phosphorus concentrations were nearly two orders of magnitude greater than those of the sediments. The plants absorbed 166 mg P m^{-2} day^{-1} from the sediments, assimilated 104 mg in the production of fresh plants, and excreted 62 mg into the water (Fig. 12–5). An amount equivalent to 41 per cent of the inorganic reactive phosphorus excreted was exported from the plant beds to the open water. Although this pathway of pumping reactive phosphorus from the sediment to water by aqua-tic vascular plants was demonstrated in a marine littoral area, there is no rea-son to suspect that quantitatively it is a less important pathway in freshwater systems. Indeed, much of the information on the quantitative cycling of the open water of lakes can be explained by the continual export of phosphorus from the sediments, coupled with the rapid cycling of phosphorus by the microflora of the phytoplankton.

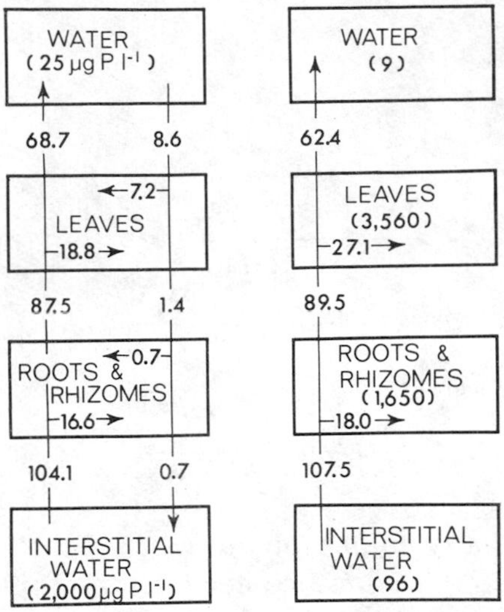

Figure 12–5 Phosphorus flux in a stand of the eelgrass *Zostera marina* L. *Left:* Phosphorus gradient, units in μg P (g plant)$^{-1}$ day^{-1}. *Right:* Net daily movement of phosphorus (mg m^{-2}). Amount of phosphorus (mg m^{-2}) in each compartment is in parentheses. (After McRoy, et al., 1972.)

Littoral Flora and the Sites of Phosphorus Flux

One way of distinguishing the major sites of phosphorus flux is by dividing them into three compartments: (a) that of the open water and organisms of the epilimnion, (b) that of the littoral organisms, and (c) that of the hypolimnion and sediments (Fig. 12–6). The lake has contact with the drainage basin via the epilimnion, and phosphorus enters with inflowing water and leaves with the outflowing water. In our division; the littoral flora, which is a major component in many if not a majority of lakes, is included correctly as part of the epilimnion and the trophogenic zone, both of which extend virtually to the same depth in many small lakes.

The application of compartmental analysis to the results of tracer experiments has demonstrated that phosphorus in the epilimnion is extremely mobile (Rigler, 1964), a subject treated at length in the following section. The turnover time of phosphorus in the epilimnion, that is, the time in which an amount of phosphorus equivalent to the total amount in a compartment leaves that compartment and a similar amount enters it, was found to be 3.6 days (Rigler, 1956). Within 20 minutes of the time it entered over 95 per cent of the added phosphorus was taken up by the plankton, and exhibited a turnover time of less than 20 minutes. In Toussaint Lake, a small acidic lake with a well-developed littoral zone of rooted aquatic plants, the littoral region was the most important contributor to the turnover of phosphorus in the epilimnion; phosphorus was lost to this compartment ten times faster than to the hypolimnion and sediments, and 50 times more rapidly than its loss through the outlet. Return of phosphorus from the littoral zone during the summer was some 20 per cent higher than loss. In comparable experiments in an acidic bog lake of Nova Scotia with extensive developments of littoral sphagnum moss, nearly all of the tracer phosphorus was taken up by the plankton and sphagnum (Coffin, et al., 1949; Hayes and Coffin, 1951). Essen-

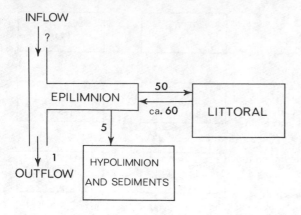

Figure 12–6 The three major compartments of phosphorus flux in a lake. The numbers indicate the relative fluxes of phosphorus between compartments in Toussaint Lake, Ontario, in midsummer. (After Rigler, 1964.)

tially no phosphorus reached the hypolimnion. When these studies are added to those on the effectiveness of the larger plants in assimilation and transport of phosphorus from the sediments, it becomes clear that the littoral flora can play a major role in the dynamics of the phosphorus cycle.

Although the rate constants of phosphorus loss from the epilimnion cannot be derived exactly from existing data, Rigler (1973) reanalyzed the results of several studies. He found that turnover time of phosphorus in the epilimnion ranged from 20 to 45 days (Table 12–3), and was correlated inversely with areas of the lakes and the estimates of development of the littoral vegetation.

Certain apparently conflicting conclusions on the importance of the littoral flora to the phosphorus circulation of lakes, based on summertime additions of tracer radioactive phosphorus, can be resolved by analyses of the differences among the communities of the lakes studied (critically reviewed by Confer, 1972). In experimental pond systems with regulated steady-state inputs of phosphorus to and outputs from the epilimnion, filamentous algae of the simulated littoral zone were a major site of phosphorus uptake before it reached the epilimnion. As the filamentous algae decayed, much phosphorus was released to the open water. In this relationship, the

TABLE 12–3 Calculated Rate Constants of Phosphorus Transport in Three Lakes of Differing Littoral Development[a]
(Lakes are ranked in order of decreasing amount of littoral vegetation, subjectively estimated.)

				Rate Constants	
Lake	Area (ha)	Littoral Vegetation (rank)	P Turnover Time (days)	k (out of epilimnion)	k (sedimentation from epilimnion)
Toussaint	4.7	1	20	0.05	0.01
Upper Bass	5.8	2	27	0.04	—
Linsley Pond	9.4	3	45	0.02	0.02

[a]Modified from Rigler, 1973.

equilibrium exchange system between phosphorus input to or release from the littoral flora and from or to the epilimnetic water varies with the amounts of phosphorus input from the drainage basin (Chamberlain, 1968). Also, its uptake by and release from the littoral macrophytes and attached algae will vary with the physical constraints of littoral development determined by the basin morphology. Moreover, this relationship will shift in proportion to seasonal changes in the active growth of the littoral flora and subsequent decay of annual plants in late summer. Among perennial submersed vascular plants or in tropical lakes where littoral growth is more or less continuous, a steady exchange of phosphorus between the littoral zone and the epilimnion generally is the rule.

Benthic Invertebrates and the Transport of Phosphorus

The effect of benthic invertebrates living on or in the sediments on the dynamics of phosphorus cycling between the sediments and the water is not completely understood. It is apparent that the burrowing activities of largely benthic invertebrates can alter significantly the surficial oxidized zone by physical penetration and permeation of the substrate. The effects of such disruption by dense populations of benthic fauna or bottom-feeding fishes (such as carp or perch) on phosphorus exchange as a result of physical manipulation of the interface are unclear. They are most likely quite small in comarison to overriding chemical-microbial regulation. In the development of populations of benthic invertebrates, phosphorus is incorporated into the fauna from the organic material fed upon in the sediments. Adsorption or direct assimilation of inorganic phosphorus is very low and insignificant, at least among the microcrustacea (Rigler, 1964). When the benthic invertebrates emerge as adults, they may emigrate from the sediments, thereby transporting phosphorus to other compartments of the system. For example, in streams a significant upstream migration of phosphorus by fish and invertebrates has been found (Ball, et al., 1963a, 1963b). However, this displacement of phosphorus by invertebrates plays only a small part in the overall quantitative cycling of phosphorus in the lake system.

The role of microinvertebrate activity of the sediment interface in relation to transport of phosphorus to the water also is unclear. Ciliates associated with the sediments are capable of hydrolyzing dissolved organic acids and of readily releasing inorganic phosphate to the water (Hooper and Elliot, 1953). Reduced oxygen, however, not only produces an unfavorable environment for the ciliates, but also inhibits the release of phosphate by the cells. Negatively phototactic cladoceran zooplankton, which migrate to the sediment interface region during daylight, presumably feed actively on the relatively rich microflora of that region. The extent of their transport of phosphorus to the epilimnion during subsequent night-time migration and release is unknown, but worthy of investigation. Although it was found that the experimental addition of snails and ostracod microcrustacea to the sediments altered plankton abundance in ponds, these benthic organisms produced no corresponding changes in the uptake rates of phosphorus in the open water (Confer, 1972).

Studies of a marine benthic amphipod demonstrated, however, that 43 μg P per gram animal (wet weight) per hour as dissolved inorganic phosphate, 25 μg P g^{-1} hr^{-1} as dissolved organic P, and 245 μg P g^{-1} hr^{-1} as particulate phosphorus were released into the water when the animal was feeding normally (Johannes, 1964a, 1964b). Most of the soluble phosphorus was utilized very rapidly by bacteria. There is no reason not to suspect that analogous situations exist in fresh waters. Rigler (1973) concludes that direct release of phosphorus from ultraplankton and excretion by zooplankton are equally important mechanisms regenerating phosphorus in the trophogenic zone of eutrophic lakes during summer stratification.

The Phosphorus Cycle within the Epilimnion

From the classical detailed studies of Einsele (1941) and many subsequent theoretical and applied analyses of the circulation of phosphorus that is added to the open water of the epilimnion, it is clear that phosphorus is incorporated very rapidly by the phytoplanktonic algae and bacteria. The specific rates of movement of phosphorus among its biologically important forms in lake water have been demonstrated only recently (Lean, 1973a, 1973b; Lean and Rigler, 1973; Rigler, 1973).

A steady state of exchange is attained rapidly between phosphorus and lake water in the summer, with the following composition (Fig. 12–7): (a) *particulate phosphorus*, the fraction removed when the water is filtered through a 0.45 μm porosity filter, which contains the bulk of the phosphorus (>95 per cent); (b) reactive inorganic *soluble orthophosphate* (PO_4^{-3}), with an extremely short turnover time; (c) a *low molecular weight* (ca. 250) *organic phosphorus compound* (XP); and (d) a soluble macromolecular *colloidal phosphorus* of a molecular weight >5,000,000. An exchange mechanism predom-

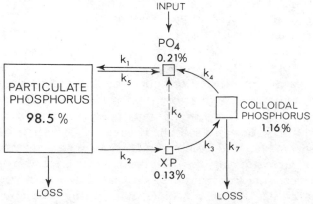

Figure 12–7   Phosphorus movement within the epilimnetic open water zone of lakes, elucidating the exchange mechanism between the phosphate and the particulate fractions. (Slightly modified from Lean, D.R.S.: Science, 179:678-680, 1973a, copyright 1973 by the American Association for the Advancement of Science; and J. Fish. Res. Bd. Canada, 30:1525-1536, 1973b.)

inates between the inorganic phosphate and the particulate fraction, but some phosphorus is excreted by the microorganisms in the form of the low molecular weight compound (XP). Polycondensation of the low molecular weight compound (XP) produces the high molecular weight colloidal compound. Both fractions, but primarily the latter, release phosphate in the soluble inorganic fraction which then becomes available for rapid uptake by the plankton.

The rate constants shown in Figure 12–7 demonstrate the rapidity of phosphorus dynamics in this algal-bacterial compartment for an exemplary lake:

(a) k_1, the uptake of $PO_4 = 0.9$ relative mass units min^{-1} or 0.002 μg liter^{-1} min^{-1} of total biologically active phosphorus.

(b) k_2, the release of XP, and k_3, the binding of XP by condensation to colloidal $P = 0.022$ min^{-1}. The excretion of extracellular organic phosphorus compounds is well-known among marine algae (Kuenzler, 1970; Lean, 1973a), and probably occurs among freshwater algae during periods of rapid growth and during senescence (Fogg, 1971).

(c) k_4, hydrolysis of colloidal P to $PO_4 = 0.0017$ min^{-1}. After the rapid formation of colloidal P via XP, in less than two minutes, subsequent additions of XP displace P from the colloid, again making the soluble inorganic PO_4 form available for uptake.

(d) k_5, the release of inorganic PO_4 by the particulate seston, was found to be very much greater than excretion of soluble organic phosphorus (XP). Mass balance calculations showed that k_5 is about 70 times k_2.

(e) k_6, the direct hydrolysis of XP to PO_4, was insignificant in comparison to release via the colloidal form.

(f) k_7, the rate constant for the loss of colloidal phosphorus, is small relative to the other rate constants. However, further aggregation of the colloidal fraction to particulate form or sorption to particulate matter followed by sedimentation represents a phosphorus sink from the epilimnion. Similarly, sedimentation of particulate phosphorus represents a slow but continuous loss from the epilimnion. Thus, during active growth of algae and bacterial heterotrophy, phosphorus must be replaced continuously through inputs by influents or from the littoral zone. In experimental systems, much of the colloidal fraction becomes biologically unavailable after one to five days. These rates of loss are congruous with those found in natural waters to which phosphorus additions are made.

From these insights on the cycling of phosphorus in the open water, it is apparent that the organic phosphorus of the seston must be separated into at least two fractions: (a) a rapidly cycled fraction is exchanged with the soluble forms in which PO_4 is transferred through the particulate phase to XP in less than 15 minutes; and (b) a fraction of sestonic phosphorus that is released much more slowly. The quantities of phosphorus in the different fractions and the kinetics of exchange have been investigated in a series of lakes ranging from oligotrophic to eutrophic, and in general apply to what we have discussed above (Rigler, 1964; Lean, 1973b).

The uptake and subsequent release of nutrients, especially phosphorus, by zooplankton microcrustacea have been suggested as significant regenerative pathways for nutrient cycling (Rigler, 1961, 1964, 1973; Johannes, 1964c; Hargrave and Geen, 1968). Although zooplankton take up inorganic phosphorus directly from solution, primarily by epizooic bacteria, this uptake is insignificant both in its effects on the phosphorus cycle and as a source of phosphorus for the zooplankton. Consumption of particulate phosphorus in the seston with subsequent release, however, is larger (e.g., 8.4 μg inorganic P per *Daphnia magna* per hour). The rate of excretion per unit weight increases as body weight decreases and varies with many factors that influence the metabolism and behavior of the animals. Hargrave and Geen (1968) estimated, by extrapolation of laboratory measurements of phosphorus excretion by crustaceans and rotifers to lake conditions, that the rates of excretion and abundance of zooplankton may, under some circumstances, be equal to the daily phosphorus requirements of the phytoplankton. Although after the removal of zooplankton Lean (1973b) could not demonstrate any effect on the rapid turnover of the phosphorus kinetics, the role of zooplankton in processing sestonic phosphorus can be potentially significant and needs much further study.

Algal Requirements for Phosphorus

Compounds containing phosphorus play major roles in nearly all phases of metabolism, particularly in energy transformation associated with phosphorylation reactions in photosynthesis. Phosphorus is required in the synthesis of nucleotides, phosphatides, sugar phosphates, and other phosphorylated intermediate compounds. Further, phosphate is bonded, usually as as ester, in a number of low molecular weight enzymes and vitamins essential for algal metabolism.

The importance of phosphorus in algal physiology is of special interest to the limnologist, because phosphorus is the least abundant element of the major nutrients required for algal growth in a large majority of fresh waters. Furthermore, the most important form of phosphorus for plant nutrition is ionized inorganic PO_4. Although phytoplanktonic algae are known to be able to utilize organic phosphate esters, such as glycerophosphates and pyrophosphates, this ability is highly variable among species, as is their ability to obtain phosphate groups enzymatically or by release of exoenzymes to the water for catalytic dissociation (see reviews of Krauss, 1958, Provasoli, 1958, and Overbeck, 1962a, 1963). A majority of the phosphorus released to the water during active growth of algae is inorganic soluble phosphate which is, in turn, very rapidly recycled. During lysis and decomposition, most of the algal phosphorus is in organic form and undergoes bacterial degradation (Kraus, 1964). Direct breakdown of dissolved organic phosphorus to dissolved inorganic phosphorus by phosphatase activity does not occur significantly (Watt and Hayes, 1963). Rather, the bacteria function as particulate phosphorus intermediaries in the degradation of dissolved organic phosphorus to dissolved inorganic phosphorus.

Phosphate Uptake and Light

The uptake of phosphate from the water is influenced by numerous external factors (Kahl, 1962; Vollenweider, 1968), the physiological importance of which has not been clearly established. The initial absorption of phosphate and subsequent uptake are greater in the light, especially under CO_2 limitation, among many algae studied in culture. The result is a reversible accumulation of cellular phosphate, much of which is released to the medium. Natural algal populations of the plankton are more or less synchronized in their productive and growth cycles, in which protein synthesis is most active in the initial portion of the daily light cycle (Soeder, 1965). Among populations of the green alga *Scenedesmus,* release of phosphate from the algae has been found to be greatest in the latter portion of the light period, prior to and during cell division in the dark (Overbeck, 1962).

Absorption of phosphate per cell in the light generally is dependent on the concentration in the medium, within a range that is rather specific for a given algal species, whereas uptake by phosphorus-deficient cells in the dark is independent of phosphate concentrations (Kahl, 1963). When phosphate is supplied in excess, the phosphorus content per cell remains constant; the amount in excess of physiological needs is incorporated and stored as polyphosphates, or it may be inhibitory. Phosphorus absorption is quite specific to groups of algae and often to species within a genus. Whereas nitrate absorption is independent of phosphate concentrations, optimal growth of many algae occurs at higher concentrations of phosphate when nitrate, rather than ammonia, is the nitrogen source.

Phosphate Uptake and pH

Many species of algae exhibit optimal growth and uptake of phosphorus within a distinct range of pH of the medium. The pH may alter the rate of phosphate absorption by direct effects on the activity of enzymes of the protoplasm, on the permeability of the cell membrane, or by changing the ionic form of the phosphate. For example, uptake of phosphorus by the ubiquitous diatom *Asterionella* is greatest at pH values between 6 and 7 (Mackereth, 1953). Uptake rates of phosphorus have been correlated directly to the presence of numerous ions and compounds of the water, such as potassium, availability of micronutrients, and organic compounds. The mechanisms involved are not clearly understood.

Phosphorus Concentrations Required for Growth

From an ecological standpoint, the growth of algae in both natural and laboratory cultures exhibits dependency on the amount of available phosphorus and the rate at which it is cycled in the trophogenic zone. Extensive investigations of minimal and maximal phosphorus concentrations, especially by Chu (1943) and Rodhe (1948), grouped freshwater algae into categories according to whether their tolerance ranges fell below, around, or above $20 \mu g$ PO_4–P l^{-1}:

(a) Species whose optimum growth and upper tolerance limit is below $20 \mu g$ PO_4-P l^{-1}, e.g., *Dinobryon, Uroglena*, and some species of the macro-alga *Chara*.

(b) Species whose optimum growth is below $20 \mu g$ PO_4-P l^{-1}, but whose tolerance limit is well above that level, e.g., *Asterionella* and other diatoms.

(c) Species whose optimal growth and upper tolerance limit is above $20 \mu g$ PO_4-P l^{-1}, e.g., green algae such as *Scenedesmus, Ankistrodesmus,* and many others.

Almost consistently in these studies of nutrient requirements, the phosphorus concentrations of the culture media required for optimal growth were higher than those of the water in natural habitats where the algae were growing. Many explanations were offered for this difference, such as the presence of unknown organic growth factors in the lake, but not the artificial media of the bacteria-free cultures. It is now apparent that the chemical mass of inorganic phosphorus of the water has relatively little relation to growth kinetics. Of overriding importance is the rapidity with which the phosphorus is cycled and exchanged between the particulate phosphorus and soluble inorganic and organic phases, as discussed in detail earlier. Secondly, it is of great ecological importance that many algae, when provided with a sufficient supply of phosphorus, can absorb phosphorus in quantities far in excess of their actual needs. A portion of this phosphorus is lost in normal active growth as both inorganic and organic compounds and recycled rapidly, at least in part. But a large majority (>95 per cent) of the phosphorus is bound in the particulate phase of algae and dead organic seston. These surplus amounts of phosphorus, often referred to as 'luxury' consumption, can provide a source of phosphorus within the cells as they are rapidly cycled in the epilimnion, even though the external concentration of phosphorus may be very low or depleted.

Of course, this recycling and utilization of stored organic phosphorus reserves cannot persist for very long. Losses continually occur from the colloidal phosphorus component, as well as from sedimentation from the particulate phosphorus component. Inputs of phosphorus to the system are needed continually, either from the littoral zone or externally, in order for high sustained growth to persist for extended periods. The amount of phosphorus inputs or loading, then, is important to sustenance of algal growth, as well as being one of many factors influencing the types of algae that are growing in a particular lake at a particular time of year.

It is therefore more relevant to the question of increasing algal productivity to view the phosphorus concentrations in terms of the total phosphorus, since most of the phosphorus is bound in the particulate component at any given time. From the few studies available in which common algal species of lakes of differing productivity were studied in relation to phosphorus requirements, the minimum phosphorus required per cell volume can be evaluated (Table 12–4). *Asterionella formosa* is found in oligotrophic waters, and has a low phosphorus requirement, reaching maximum densities at very low

TABLE 12–4 The Minimal Phosphorus Requirements Per Unit Cell Volume of Several Algae Common to Lakes of Progressively Increasing Productivity[a]

Algae	Minimum P Requirement, in μg mm^{-3} Cell Volume
Asterionella	<0.2
Fragilaria	0.2–0.35
Tabellaria	0.45–0.6
Scenedesmus	>0.5
Oscillatoria	>0.5
Microcystis	>0.5

[a]From data of Vollenweider, R. S.: Scientific Fundamentals of the Eutrophication of Lakes and Flowing Waters, with Particular Reference to Nitrogen and Phosphorus as Factors in Eutrophication. OECD Report No. DAS/CSI/62.27, Paris, OECD, 1968, after numerous sources.

phosphorus concentrations. The quantity of phytoplankton, expressed as cell volume, which may be produced by 1 μg P l^{-1}, is 2 to 5 mm^3 l^{-1} (Vollenweider, 1968), densities that are found commonly in lakes of low productivity. *Tabellaria* and *Fragilaria* are diatoms that reach their maximum densities at concentrations of ca 45 μg PO$_4$-P l^{-1}, while *Scenedesmus* needs higher concentrations of around 500 μg l^{-1}. The blue-green alga *Oscillatoria rubescens* does not reach its maximum phosphorus content until initial concentrations of the media used are about 3000 μg PO$_4$–P l^{-1}.

Studies on the kinetics of phosphate uptake and growth of *Scenedesmus* in continuous culture have shown that the uptake velocity is a function of both the internal cellular phosphorus compound and external substrate concentrations (Rhee, 1973). The apparent half-saturation constant (K_m) of phosphorus uptake was 0.6 μM (ca 18 μg P l^{-1}), whereas the apparent half-saturation constant for growth was less than K_m by an order of magnitude. Growth under these conditions of steady-state was a function of cellular phosphorus concentrations where the internal polyphosphates appeared to regulate growth rate, particularly in the process of cell division. The activity of alkaline phosphatase, the primary enzyme involved in the release of phosphates from polyphosphates, exhibited a relationship which was inversely related to growth rates and was correlated directly to both polyphosphates and internally stored surplus phosphorus. During phosphate limitations, the polyphosphates are mobilized for the synthesis of cellular macromolecules (protein, RNA, and DNA). Similar results were found for species of diatoms (Fuhs, et al., 1972).

Phosphorus Uptake of Algae vs. Bacteria

The minimum phosphorus content, the growth rates when phosphorus was nonlimiting, the Michaelis half-saturation constant of the uptake, and the maximum rate for uptake of phosphorus were compared for two diatoms and three bacteria (Table 12–5). These particularly illuminating findings by Fuhs and collaborators have far-reaching implications. The range of concentrations for limitation of growth rate by inorganic phosphate is less than

TABLE 12-5 Phosphorus-Dependent Growth and Phosphorus Uptake Kinetics of Two Diatoms and Three Bacterial Species[a]

Algae and Bacteria	Mean Cell Volume (μm^3)	Mean Cell Surface (μm^2)	a_o (10^{-15} g-atom P)	μ_m (doublings per day)	K_m (10^{-6} g-atom P liter^{-1})	V_m (10^{-15} g-atom P μm^{-2} day^{-1})
Diatoms:						
Cyclotella nana	77.5	103	0.95	1.6	0.6	2.0
Thalassiosira fluviatilis	1570	826	12.5	1.6	1.7	7.3
Bacteria:						
Corynebacterium bovis	0.71	6	0.19	4.8	6.7	7.7
Pseudomonas aeruginosa	0.41	4.2	0.10	48	12.2	17.9
Bacillus subtilis	0.39	3.9	0.15	12	11.3	12.5

[a]Modified from Fuhs, et al., Characterization of phosphorus-limited plankton algae. *In* Likens, G. E., ed., Nutrients and Eutrophication. Milwaukee, American Society of Limnology and Oceanography, 1972.

a_o = minimum P content per cell.

μ_m = growth rate during unrestricted growth at 22°C, other conditions at or near optimum.

K_m = Michaelis half-saturation constant of uptake mechanism.

V_m = maximum uptake rate for orthophosphate P per unit area of cell surface.

3 μg P l^{-1}, provided that depletion at the cell surface does not occur. Maximum growth rates were found to vary widely, but the phosphorus content depended on protoplasm volume. Maximum uptake rates of orthophosphate phosphorus per unit area of cell surface were similar among the organisms. Bacteria showed lower affinity for orthophosphate than did the algae, but potentially could outgrow the algae because of a more favorable surface area to volume ratio.

Studies on the competition for phosphate between the alga *Scenedesmus* and the bacterium *Pseudomonas* have demonstrated that algal growth is severely limited in the presence of the bacteria, but that the growth of the bacteria is affected little by the algae (Rhee, 1972). Competition, seen as the cessation of exponential algal growth, was observed some time after the external phosphate concentrations were exhausted. The concentrations of stored internal polyphosphates within the algae decreased to a critical value near zero at the time of suppressed algal and much faster bacterial growth rates.

MAN AND THE PHOSPHORUS CYCLE IN LAKES

Sources of Phosphorus

Precipitation

The phosphorus content of precipitation and fallout of particulate material of the atmosphere is highly variable in its contribution to the surface of lakes (Chapin and Uttormark, 1973). In general, the contribution of phosphorus from precipitation is less than that of nitrogen. In heavily fertilized agricultural regions, the phosphorus content of precipitation is much higher during the active growing season than in winter. Prevailing winds and water of storms emanating from oceans onto land are usually low in phosphorus content. The major source of phosphorus in precipitation is from dust generated over the land from soil erosion and from urban and industrial atmospheric contamination.

The phosphorus content of precipitation is generally low, at <30 μg P l^{-1} in nonpopulated regions over land. In the environs of urban-industrial aggregations, this content increases considerably to well over 100 mg l^{-1}. Based on relatively few data, the atmospheric contributions of phosphorus are approximately 0.01 to 0.1 g m^{-2} year^{-1} (0.1-1.0 kg ha^{-1} year^{-1}), with a majority of contemporary values in the lower portion of that range. This input from the atmosphere, however, is more significant than generally is believed when compared to the 0.07 g m^{-2} year $^{-1}$ value that Vollenweider (1968) estimated as a permissible phosphorus loading rate of lakes in general. Values of 0.13 g m^{-2} year^{-1} or higher are considered dangerous from the standpoint of eutrophicational control in lakes with a mean depth of less than 5m.

Groundwater

The phosphorus content of groundwater is generally low; average concentrations are 20 μg P l^{-1}, even in areas of soils containing relatively high phosphorus content. This is a result of the effectiveness of the high insolubility of phosphate-containing minerals, and the scavenging of surface phosphate concentrations by biotic removal and soil percolation.

Land Runoff and Flowing Waters

In general, the regional chemical characteristics of surface waters are closely related to the soil characteristics of their drainage basins (Keup, 1968; Vollenweider, 1968). The soils reflect the regional geological and climatic characteristics of the region, and surface drainage is often a major contributor of phosphorus to streams and lakes. The quantities of phosphorus entering surface drainage are influenced by the amount of phosphorus in soils, topography, vegetative cover, quantity and duration of runoff flow, land use, and pollution.

The parent rock material from which soil evolves by weathering is highly variable in phosphorus content, and this variability increases with the thickness and heterogeneity of the stratified soil layers. Basic igneous rock contains relatively little phosphorus as apatite; phosphorus percentages of other rocks are as follows: sandstone is ca 0.02, gneiss 0.04, unweathered loess 0.07, andesite 0.16, and diabase 0.03. Limestone containing ca 1.3 per cent P and rare deposits of rock phosphate (10 to 15 per cent P), in which biotic accumulations either from the organisms themselves or from guano were concentrated, are largely of sedimentary origin (cf. Hutchinson, 1950). Surface layers of soil are relatively rich in organic phosphorus from plant detritus in various stages of decomposition by soil fungi and bacteria. The exchange capacity of soils for phosphorus depends on the composition of the soil, and increases with greater quantities of organic and inorganic colloids. Phosphorus is most available and easily leached from soils with a pH of 6 to 7. At lower pH values phosphorus combines readily with aluminum, iron, and manganese, while at pH 6 and above progressively greater amounts of phosphate are associated with calcium as apatites and calcium phosphates.

The topography of the land in the catchment basin influences the extent of erosion and subsequent export of nutrients. Flat lands with little runoff and relatively high infiltration rates contribute less nutrient load to

TABLE 12–6   Concentrations of Nitrogen and Phosphorus in Runoff from Miami Silt Loam of Differing Gradients[a]

| | $g\ m^{-2}\ year^{-1}$ | |
Gradient (%)	Nitrogen	Phosphorus
8	2.0	0.06
20	4.25	0.2

[a]After data cited in Vollenweider, R. A.: Scientific Fundamentals of the Eutrophication of Lakes and Flowing Waters, with Particular Reference to Nitrogen and Phosphorus as Factors in Eutrophication. OECD Report no. DAS/CSI/68.27, Paris, OECD, 1968.

MAN AND THE PHOSPHORUS CYCLE IN LAKES **241**

TABLE 12–7   Relative Erosion from Soil in Relation to Vegetative Cover
and Land Use, Pacific Northwest of United States[a]

Crop or Practice	Relative Erosion
Forest duff	0.001–1
Pastures, humid region, or irrigated	0.001–1
Range or poor pasture	5–10
Grass/legume hayland	5
Lucerne	10
Orchards, vineyards with cover crops	20
Wheat, fallow, stubble not burned	60
Wheat, fallow, stubble burned	75
Orchards, vineyards, without cover crops	90
Row crops and fallow	100

[a] From data cited in Biggar and Corey, 1969, after Musgrave.

runoff than similar lands with increased gradients (Table 12–6). Further, the
relative erosion is influenced markedly by the type of vegetation and use to
which the land is put (Table 12–7). The literature on this subject is large, and
the reader is referred to the reviews examining phosphorus inputs from runoff
by Vollenweider (1968), Keup (1969), Biggar and Corey (1969), and Cooper
(1969). An attempt has been made to classify natural and agricultural areas on
the basis of exports of nitrogen and phosphorus in runoff (Table 12–8). Applications of fertilizers and land management practices, both in agriculture and
forestry, will modify and generally increase these values considerably.
Urbanization results in increases in phosphorus discharged to surface waters
in approximately direct proportion to population densities (Weibel, 1969).
Losses of phosphorus from heavy residential fertilization, storm sewer drainage, and leaves (Cowen and Lee,1973) increase phosphorus inputs to surface
drainage. For example, in the largest lake of southwestern Michigan, Gull
Lake, approximately 24 per cent of the total phosphorus entering this rapidly
eutrophicating lake was from fertilization of lakeside lawns, the soils of
75 per cent of which were saturated with phosphorus making additions unnecessary (Moss, 1972b).

Surface runoff waters receive very large quantities of phosphorus loading from domestic sewage. Of course, the extent of loading of these waters

TABLE 12–8   Grouping of Soils of Natural and Agricultural Areas on the Basis of Their
General Productivity and Export of Nitrogen and Phosphorus[a]

General Soil Productivity	Exports, g m^{-2} year^{-1}		
		Phosphorus	
	Inorganic Nitrogen	PO_4-P	Total
Low	<0.5	<0.01	<0.02
Medium	0.5-2.5	0.01-0.025	0.02-0.05
High	>2.5	>0.025	>0.05

[a] After Vollenweider, 1968.

TABLE 12–9   Surface Water Loadings of Nitrogen and Phosphorus per Unit Area from Human Excrement and Other Sources Based on an Average of 12 g N Caput^{-1} Day^{-1} and 2.25 g P Caput^{-1} Day^{-1}[a]

Population Density (caput km^{-2})	Nitrogen (g m^{-2} year^{-1})	Phosphorus (g m^{-2} year^{-1})
50	0.22	0.04
100	0.44	0.08
150	0.66	0.12
200	0.88	0.16
300	1.32	0.24
500	2.20	0.40
1000	4.40	0.80
2500	11.0	2.00
5000	22.0	4.05

[a]After Vollenweider, R. A.: Scientific Fundamentals of the Eutrophication of Lakes and Flowing Waters, with Particular Reference to Nitrogen and Phosphorus as Factors in Eutrophication. OECD Report no. DAS/CSI/68.27, Paris, OECD, 1968.

varies greatly with the density of population, the extent of treatment of the sewage for nutrient removal, and points of discharge of effluents. Average values are given in Table 12–9. Industrial inputs, especially those associated with food processing, are usually exceedingly high.

Somewhat ironically, cleaning detergents are one of the major sources of phosphorus that contributes so markedly to fertilization effects of many fresh waters. Synthetic detergents use phosphate builders as a major constituent, mainly sodium pyrophosphate and polyphosphates, to complex and inactivate cations of water supplies and permit more effective cleaning action. Until very recently, when their phosphorus content was reduced, from 7 to 12 per cent of the gross weight of detergents was phosphorus. The amounts of phosphorus used in the production of detergents are staggering; well over 2 million tons are produced annually in the United States alone. Although the percentage of phosphorus in detergents has been reduced somewhat, the phosphorus loading in water treatment facilities is still extremely high, and originates from a source that can be eliminated relatively easily with concerted technological effort. In a majority of existing treatment facilities, phosphorus has been reduced significantly, but not sufficiently to prevent accelerated productivity of many recipient lakes. The technology for nearly complete removal of phosphorus exists and is practical (cf. Vollenweider, 1968; Rohlich, 1969).

The amounts of phosphorus in flowing waters with only slight loading are generally less than 100 mg m^{-3} total phosphorus (Vollenweider, 1968; Keup, 1968). Phosphorus concentrations depend on waste water discharges, however, and levels of several hundred mg m^{-3} are not uncommon; serious cases exceed 1000 mg m^{-3}. Assimilation rates of phosphorus and uptake by stream biota are rapid, but a majority of the load is carried downstream in dissolved, particulate suspensoids, and scoured sediment deposits to receiving lakes or reservoirs of reduced flow velocities.

Effects of Phosphorus Concentrations on Lake Productivity

Most of the discussion of the interrelationships of phosphorus as a plant nutrient and its potential limitations to growth and seasonal succession of plant populations will be deferred to subsequent chapters on plant metabolism and dynamics. Several points, however, are germane to the study of phosphorus cycling in lakes.

Many definitions of lake eutrophication exist, based on an array of attendant conditions associated with increased productivity. The consensus of opinion among limnologists is that the term eutrophication is synonymous with increased growth rates of the biota of lakes, and that the rate of increasing productivity is accelerated over that rate that would have occurred in the absence of perturbations of the system. The most conspicuous, basic, and accurately measurable criterion of accelerated productivity is increasing rates of annual photosynthesis by algae and larger plants per given area.

The chemical composition of the biota delineates in an approximate way the requirements of the organisms that must be obtained from and supplied by the environment for sustained growth and reproduction with a suitable source of energy input. Oxygen and hydrogen exist in chemical abundance far in excess of requirements. The carbon:nitrogen:phosphorus ratio of plants is roughly 40C:7N:1P by weight. If the energy supply in light and in other factors such as micronutrients is adequate, as it is under a majority of conditions, then phosphorus will become the first of these three nutrients to become limiting. As the preceding chapters have shown, inorganic carbon exceeds nitrogen by an order of magnitude in most lakes, and exceeds phosphorus by 2 to 3 orders of magnitude.

Similarly, the nitrogen of most natural waters exceeds phosphorus by an order of magnitude, and usually by much more. Therefore, phosphorus is most rapidly and commonly the first limiting nutrient in a vast majority of lake systems. When lake systems are loaded with phosphorus, for example by sewage effluents that contain very high proportions of phosphorus — ca 6C:4N:1P — the phosphorus limitation is overcome. Subsequent increased levels of photosynthesis can continue until the next least abundant major nutrient, nitrogen, becomes limiting. Algae unable to fix nitrogen from the atmospheric pool are excluded; the blue-green algae with nitrogen-fixing capabilities can outcompete those without this ability very effectively. Only under the most extreme conditions of exceedingly high metabolism and stagnant conditions, where P and N are present far in excess of demands and carbon from the atmosphere is impeded, could carbon become limiting.

The amount of evidence for these relationships is so overwhelming, both from experimental and applied investigations, that it is difficult to appreciate how the bitter controversy on the importance of carbon rather than phosphorus as a limiting nutrient in fresh waters developed in Canada in the late 1960s. As a result of a combination of the magnification of a few findings from blue-green algal cultures and a few rare cases of discovered reduced carbon availability in natural habitats in which excessive amounts of nitrogen and phosphorus occurred, phosphorus was implied to be of less importance in the acceleration of eutrophication than carbon. So convincing was the mis-

interpretation of results by a few scientists, and so effective was the exploitation of the situation by industrial concerns of the detergent industry and
irresponsible journalists, that urgently needed legislation to reduce the effluent loading of surface waters with phosphorus was seriously impeded. The
matter finally was put to rest by the final rebuttal of Vallentyne (1970),
numerous subsequent investigations, and a national symposium on the subject (Likens, 1972). The effects of the controversy, however, continue to be
felt both in research on the subject and in the efforts to reduce the loading of
fresh waters with millions of tons of phosphorus annually so that ultimately
lakes will not become, in the words of J. R. Vallentyne, algal bowls.

When phosphorus is added as a pulse to unproductive lakes or ponds,
either experimentally, for purposes of intentional fertilization, or in effluents
resulting from man's activities, the usual response is a very rapid increase in
algal productivity (e. g., Einsele, 1941; Maciolek, 1954; Mortimer and Hickling, 1954; Vinberg and Liakhnovich, 1965; Vollenweider, 1968). The increased productivity is not sustained, however, and decreases rather rapidly
in a few weeks or months to levels approximate to those prior to the addition.
Losses in the colloidal fraction and from sedimentation of particulate phosphorus result in continuous losses from the trophogenic zone. Inputs to the
system must be maintained in a continuous or pulsed manner to sustain the
increased productivity. In other words, steady phosphorus loading of the
system is critical to sustaining increased productivity in a majority of lakes
of low or medium biological productivity.

Conversely, in order to reduce the productivity of a lake that is receiving a
continuous loading of nutrients, algal growth generally is decreased most
effectively by reduction of phosphorus inputs to below the level of losses
within the lake. A reduction of internal total phosphorus of the water is the
objective, since phosphorus is the major nutrient in greatest demand in relation to supply. Phosphorus is chemically reactive, technologically easier to
remove from water than nitrogen, and does not have major reservoirs in the
atmosphere.

Reduction of the productivity of lakes by decreasing phosphorus loading
can be very effective. Many examples are given in Vollenweider's (1968)
review. Among the most frequently discussed examples is Lake Washington,
Seattle, Washington (Edmondson, 1972). Lake Washington was enriched
with increasing volumes of effluent from secondary sewage treatment facilities during the period from 1941 to 1953. Production increased markedly,
and the abundance of algae greatly increased. Phosphate concentrations also
increased proportionally much more than those of nitrate or carbon dioxide.
Effluent was diverted away from the lake in 1963, and by 1969 phosphate had
decreased to 28 per cent of its 1963 value, and summer chlorophyll concentrations had decreased about as much, but nitrate and total CO_2 fluctuated
from year to year at relatively high values. Reduction in phytoplankton and
phosphorus is still continuing. Other examples are given by Edmondson
(1969).

The rate at which the productivity of a lake will revert towards conditions prior to increased loading is variable in relation to basin morphometry,
water chemistry, and whether the phosphorus sources are diffuse or concen-

TABLE 12–10 Provisional Permissible Loading Levels for Total Nitrogen and Total Phosphorus (Biochemically Active) in g m^{-2} Year^{-1} [a]

Mean Depth (m)	Permissible Loading		Dangerous Loading	
	N	P	N	P
5	1.0	0.07	2.0	0.13
10	1.5	0.10	3.0	0.20
50	4.0	0.25	8.0	0.50
100	6.0	0.40	12.0	0.80
150	7.5	0.50	15.0	1.00
200	9.0	0.60	18.0	1.20

[a]After Vollenweider, R. A.: Scientific Fundamentals of the Eutrophication of Lakes and Flowing Waters, with Particular Reference to Nitrogen and Phosphorus as Factors in Eutrophication. OECD Report no. DAS/CSI/68.27, Paris, OECD, 1968.

trated at point sources. The loading of lakes with nitrogen and phosphorus further is influenced by the ratio of the surface area of the drainage basin to that of the lake (Ohle, 1965). Depending on the percentage losses from the land, some of which may be under cultivation, inputs increase in a way roughly proportional to the 'surroundings' ratio.

In a series of particularly penetrating analyses of input-output models of substances in lakes, emphasizing phosphorus and nitrogen, Vollenweider (1968, 1969, 1972) attempted to evaluate the probable recovery time of a lake to attain a new steady-state under conditions of reduced inputs. The dynamic models he used incorporate average concentrations of phosphorus during early spring circulation, in relation to the annual inputs to the basin and to losses by outflow and sedimentation. Based on a combined consideration of numerous influencing conditions, very approximate provisional loading rates for nitrogen and phosphorus required for lakes to maintain steady-state can be estimated (Table 12–10). Rapid eutrophication is likely to occur as the loading is increased. Upon reduction of loading rates to or below those of the lake prior to accelerated inputs, for lakes of average size and average hydrological replenishment time, the period needed for recovery will be on the order of 2 to 10 years.

chapter 13

iron, sulfur, and silica cycles

BIOGEOCHEMICAL CYCLING OF ESSENTIAL MICRONUTRIENTS

The biogeochemical cycling of iron, and, to a large degree of manganese of similar reactivity, is regulated almost completely by spatial and seasonal variations in oxidation-reduction states, which are mediated by photosynthetic and bacterial metabolism. Concentrations of ionic iron are exceedingly low in aerated water; most iron occurs as ferric hydroxide in particulate and colloidal form and as complexes with organic, especially humic, compounds. The solubility of manganese is considerably higher than that of iron, but it reacts in an analogous manner. At low pH and redox, ferrous and manganous ions diffuse readily from the sediments and accumulate in anaerobic hypolimnia of productive lakes. At very low redox potentials, less than about 100 mv, sulfate is reduced to sulfide to form highly insoluble metallic sulfides, especially ferrous sulfide (FeS). Therefore in hypereutrophic lakes high concentrations of hydrogen sulfide from bacterial decomposition of sulfur-containing proteinaceous organic matter and from reduction can lead to significant decreases of iron but not the more soluble manganese as sulfides during the latter portion of stratification.

Iron and manganese are essential micronutrients of freshwater flora and fauna. Under certain conditions of restricted availability, photosynthetic productivity can be limited by these elements. Manganese clearly is involved in the seasonal succession of certain algal populations. Certain chemosynthetic bacteria can utilize the energy of inorganic oxidations of ferrous and manganous salts in relatively inefficient reactions of CO_2 fixation. Other autotrophic and heterotrophic iron-oxidizing bacteria deposit oxidized iron and manganese, and are restricted to zones of steep redox gradients between reduced metal ions and oxygenated water.

Quantitative information on the dynamics of other essential metallic micronutrients — zinc, copper, cobalt, and molybdenum — is limited. Micronutrient availability is governed by redox conditions, and reactions and extent of complexing with dissolved organic compounds and other inorganic ions. In most natural waters, micronutrient concentrations and availability usually

246

are adequate to meet metabolic requirements within other constraints of light, temperature, and macronutrient availability. A few clear cases of limitations to photosynthetic productivity by micronutrient deficiencies or physiological unavailability have been demonstrated. Dynamics of copper are strongly effected by redox conditions similar to that of iron; cobalt (Co), zinc (Zn), and molybdenum (Mo) dynamics are more closely related to organic microbial metabolism and transport and to complexing with organic compounds. Molybdenum exhibits greater mobility than the other micronutrient ions.

Sulfur is nearly always present in quantities adequate to meet high requirements for protein synthesis. The dynamics of sulfate and the hydrogen sulfide (H_2S) produced by decomposition of organic matter alter conditions in stratified waters that affect the cycling of other nutrients, productivity, and biotic distribution. Atmospheric transport of sulfur compounds in precipitation and dry fallout constitutes a major global source that in many natural waters exceeds inputs from rock and soil weathering and transport in surface runoff and groundwater.

Oxic waters largely contain sulfate; hydrogen sulfide tends to accumulate in anoxic zones of intensive decomposition where the redox potential is reduced below about 100 mv. Most of the sulfur of lakes is stored as sulfate ($SO_4^=$) and H_2S of the water, and as sulfide in the sediments; that fraction utilized by organisms does not influence the cycle materially. While oxygen is obtained from sulfate by sulfate-reducing bacteria, the H_2S generated readily oxidizes and utilizes oxygen on transport or intrusion to aerobic strata. Sulfur-oxidizing bacteria consist of two general types: chemosynthetic aerobes that oxidize reduced sulfur compounds and elemental sulfur to sulfate, and the photosynthetic sulfur bacteria that utilize light as an energy source and reduced sulfur compounds as electron donors in the photosynthetic reduction of CO_2. The redox requirements of these latter bacteria, especially those requiring light, are rather specific and distribution is often restricted to zones of steep gradients between anoxic and aerated strata. When conditions are optimal, the photosynthetic bacteria often develop in extreme profusion and contribute significantly to the annual productivity of lakes.

Silica occurs in relative abundance in natural waters as dissolved silicic acid and particulate silica. Diatom algae assimilate large quantities of silica and markedly modify the flux rates of silica in lakes and streams. Utilization of silica in the trophogenic zone of lakes by diatom populations often reduces epilimnetic concentrations and induces a seasonal progressive species succession of diatoms. Below a concentration of about 0.5 mg silica (SiO_2) l^{-1}, most diatoms cannot compete effectively with nonsiliceous algae, and are excluded until silica supplies are renewed, usually at autumnal circulation.

Oxidation-Reduction Potentials in Freshwater Systems

Purely inorganic chemical systems of oxidation-reduction (redox) reactions exhibit a flow of electrons between the oxidized and reduced states until an equilibrium is reached. There is a tendency for the reduced phase to lose electrons and transform to an oxidized state, e.g., $Fe^{+++} + e^- \rightleftarrows Fe^{++}$. Free electrons, however, usually inhibit this process and large quantities of free ions in reduced and oxidized states can exist together. If electrons are removed, as for example by an immersed platinum electrode, then a transformation of the reduced to oxidized states (Fe^{++} to Fe^{+++}) occurs, and a current of electrons passes through the electrodes. By applying a current (electrons) in the opposite direction, reduction can be induced; the reactions are reversible. Such potentiometric changes in electron flow can be accomplished with any solution containing ionic species, including that composite system of lake water, by immersion of calomel[1]-platinum electrodes and by drawing off electrons. The resulting measurement, taken against standard conditions, is an expression of the oxidizing or reducing intensity or condition of the solution. The current, referenced against the hydrogen electrode, is termed the redox potential or the electrode potential.

The Redox Potential

The redox potential is proportional to the equivalent free energy change per mole of electrons associated with a given reduction (Stumm, 1966; Morris and Stumm, 1967). Although aqueous solutions do not contain free protons and electrons, it is possible to define proton activity [pH = $-\log$ (H$^+$)] and electron activity [pE = $-\log$ (e$^-$)]. pE is large and positive in strongly oxidizing solutions (low electron activity), just as pH is high in strongly alkaline solutions (low proton activity). Thus both pH and pE are intensity factors of free energy levels and have nothing to do with capacity or condition (e.g., alkalinity, acidity).

When oxygen is dissolved in water, a redox potential is generated according to the reaction (cf. discussion in Stumm, 1966, and Stumm and Morgan, 1970):

$$H_2O \rightleftarrows \tfrac{1}{2} O_2 + 2H^+ + 2\ e^-$$

The pE of water in an equilibrium situation is relatively insensitive to change in oxygen concentration and extent of saturation, being influenced by the 4th root of the partial pressure of oxygen. If the oxygen changes by 99 per cent, the redox potential would be reduced by only about 30 mv. The activity of the hydroxyl ions, however, influences the activity of the hydrogen ions and therefore the redox potential is significantly changed by alterations of H$^+$ and is reflected in the pH. It is customary to express pE of redox reactions in natural waters at their activities in neutral water at pH 7 rather than at unity. Thus, the redox potential is expressed as E_h or as E_7, in which the correction is made for the change in redox at the pH of the sample to a pH of 7.[2] A rise of pH

---

[1] A calomel electrode consists of a chloride solution (usually KCl), solid Hg_2Cl_2 (calomel), and metallic mercury.

[2] E_h is compared to the hydrogen half-cell at any designated pH; usually it is pH O ([H$^+$] = 1). Thus, E_h is not necessarily corrected to pH of neutral water but is often so done.

of one unit is accompanied by a fall in redox potential of 58 mv. Hence, a common practice is to correct potentials to pH 7 by subtracting 58 mv for every pH unit on the acid side of neutrality, and by adding 58 mv for every pH unit on the alkaline side of neutrality. E_h is somewhat influenced by temperature, e.g., E_h of water $= 860$ mv at $0°C$, and 800 at $30°C$ at pH 7.

The preceding discussion reflects the ideal situation under conditions of equilibrium, in which the oxidation-reduction systems are reversible (Stumm, 1966). True redox equilibrium is not found in any natural aquatic system because of the extreme slowness of most redox reactions in the absence of appropriate biochemical catalysis and the continuous cyclic input of photosynthetic energy that disturbs the trend towards equilibrium conditions. In addition to the nonequilibrium redox conditions related to the highly dynamic state mediated by the activities of the biota, electrochemical measurements of redox depend on the nature and composite rates of reactions at the electrode surface, and in practice the redox measurements reflect electrochemical irreversible redox potentials. Thus in neutral, fully oxygenated water in equilibrium with air, redox potentials slightly greater than 500 mv are obtained, considerably less than the theoretical E_h of 800 mv.

Only a few elements (C, O, N, S, Fe, Mn) are predominant reactants in redox processes in natural waters. Photosynthesis, by conversion of energy into chemical bonds, produces reduced states (negative E_h) of high free energy and nonequilibrium concentrations of C, N, and S compounds (Stumm, 1966). Respiratory, fermentative, and other nonphotosynthetic reactions of organisms tend to restore equilibrium by catalytically decomposing, through energy-yielding redox reactions, the thermodynamically unstable products of photosynthesis and thereby obtain a source of free energy for their metabolic demands. The mean E_h is increased by these combined processes. It should be noted that organisms act as redox catalysts by mediating the reactions and electron transfer; the organisms themselves do not oxidize substrates or reduce compounds.

Reduced and oxidized iron (Fe^{++} and Fe^{+++}) are among the most electroactive ions of the redox reactants in natural water systems. Other than iron and to a certain extent manganese, the redox components of organic carbon, nitrogen, and sulfur of major importance in ecosystems are not electronegative and nonenzymatically do not yield reversible potentials. As a result, redox measurements in natural waters do not lend themselves to quantitative interpretation and comparison. Qualitative, relative comparisons of E_h, representing mixed composite potentials, however, can be extremely instructive, as was demonstrated by the monographic comparison of pH and E_h limits in different environments and among organisms (Baas Becking, et al., 1960).

From this discussion, one would anticipate little change in redox potential with increasing depth as long as the water contained some dissolved oxygen. Therefore, even though a distinctly clinograde oxygen curve may be found, as long as the water is not near anoxia, the E_h will remain fairly high and positive (300 to 500 mv). Such conditions have been found repeatedly in detailed studies of E_h in lake waters of widely diverse types (Kuznetsov, 1935; Hutchinson, et al., 1939; Allgeier, et al., 1941; Mortimer, 1942; Kjensmo, 1970). As the oxygen concentrations approach zero and anoxic con-

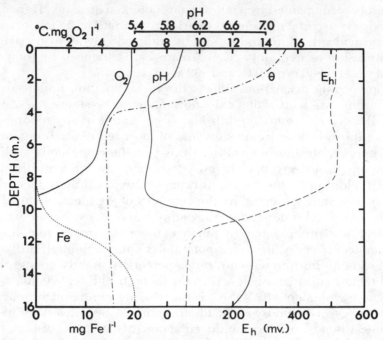

Figure 13–1 Vertical distribution of temperature, oxygen, pH, total iron, and redox potential E_h in permanently meromictic Lake Skjennungen, Norway, June 1967. (After Kjensmo, 1970.)

ditions appear, as in the lower hypolimnion and near the sediments, the E_h decreases precipitously. These relationships at the sediment-water inter-face were discussed in some detail in the previous chapter on phosphorus release from the sediments (cf. Figs. 12–2, 12–3, and 12–4). Within the sedi-ments, reducing conditions prevail and the E_h reaches approximately 0 mv or negative values within a few millimeters of the interface (Figs. 12–2, 12–3). Of the many reducing compounds that contribute to these reductions in E_h, ferrous iron of the sediments is the most important (Fig. 12–3).

Lower redox potentials generally are observed in lake systems containing relatively high concentrations of humic dissolved organic compounds (Allgeier, et al., 1941, Kjensmo, 1970). Humic acids, especially those de-rived from the moss *Sphagnum* of bogs and bog lakes, have a low E_h of about 350 mv (Visser, 1964b) and their reducing properties lead to a metal enrich-ment by complexing and adsorption to the acid molecules (Szilágyi, 1973). The redox potentials of the deep monimolimnia of meromictic lakes generally are extremely low, among the lowest found in natural waters (Fig. 13–1). Extremely high concentrations of soluble total iron frequently are found in the monimolimnia of these permanently stratified lakes of high dissolved organic matter content.

Iron and Manganese Cycling in Lakes

The similarities in chemical reactivity between iron and manganese permit us to discuss them together. While clear differences exist between the two metals, and much more is known about the dynamics and cycling of iron, they behave in similar fashion in freshwater systems. A strong interaction exists between the cycling of these metals, especially iron, and sulfur. The biogeochemical fluxes of iron and manganese reflect the combined spatial and temporal variations in physical chemistry of the lake systems which are controlled almost totally by the dynamic conditions regulating bacterial metabolism.

Chemical Equilibria and Forms of Iron and Manganese

Iron exists in solution as either the ferrous (Fe^{++}) or ferric (Fe^{+++}) state. (Detailed reviews are given in Hem and Cropper, 1959; Hem, 1960; Stumm and Lee, 1960; Doyle, 1968a; and Stumm and Morgan, 1970.) Amounts of iron in solution in natural water, and the rate of oxidation of Fe^{++} to Fe^{+++}, as occurs in oxygenated water, are dependent primarily on pH, E_h, and temperature. Ferrous constituents tend to be more soluble than ferric constituents.

Soluble ferrous iron occurs mainly as hydrated Fe^{++} and hydrated hydroxo ions, whose solubility is determined largely by the solubility of ferrous hydroxide [$Fe(OH)_2$], ferrous carbonate ($FeCO_3$), and ferrous sulfide (FeS). $Fe(OH)_2$ is exceedingly insoluble at the normal pH range of oxygenated natural waters. The solubility of ferrous iron generally is controlled by the solubility of $FeCO_3$ in which even at low pH, carbonate concentrations usually are sufficient to limit solubilization. For water that is in $CaCO_3$ equilibrium saturation, the solubility of $FeCO_3$ is about 200 times less than $CaCO_3$, unless very high pH values (>10) occur. Soft waters containing very low concentrations of bicarbonate usually contain somewhat higher concentrations of Fe^{++}, although most is oxidized to Fe^{+++}. For example, a water free of bicarbonate at pH 7 would contain more than 1000 times as much Fe^{++} as a water containing 2 meq 1^{-1} alkalinity, an average value for many waters. FeS also is exceedingly insoluble (Doyle, 1968b), and forms a stable slightly crystalline compound that darkens the color of anaerobic sediments. Bacterial reduction of sulfate to form sulfides is common in anaerobic hypolimnetic waters near the sediments, at low redox potentials that usually are found only in extremely productive or meromictic lakes.

Iron Complexes

The most common species of ferric iron in natural waters is hydrated ferric hydroxide, $Fe(OH)_3$. At equilibrium in the pH range of 5 to 8, $Fe(OH)_3$ is in the solid state, its solubility being very low (equilibrium constant ca 10^{-36} at 25°C). Other insoluble ferric salts are of less significance. For example, phosphate does not influence the solubility of Fe^{+++} when inorganic phosphorus concentration is less than 10^{-4} mol 1^{-1}, as is usually the case. It is apparent, therefore, that in the absence of organic matter and based only on solubility criteria as mediated by pH and E_h, a number of forms of inorganic iron can exist in natural waters. The simultaneous influence of hydro-

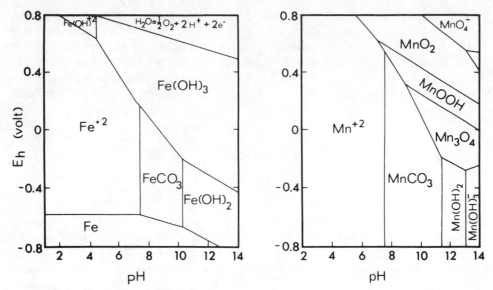

Figure 13–2   Approximate distribution of species of iron and manganese in relation to pH and redox potential E_h. Alkalinity is assumed to be equal to 2 meq 1^{-1}. Lines denote points at which the activities of soluble Fe and Mn are 10^{-5} mol 1^{-1}. (Modified from Stumm and Lee, 1960, and Stumm and Morgan, 1970.)

gen ions and electrons on the equilibria of aqueous iron is illustrated in Figure 13–2. All of the oxygenation reactions of Fe^{++} to Fe^{+++} are exergonic and capable of supplying energy; some of these reactions serve as an energy source for microorganisms.

Much of the iron of normal lake water is present as suspensions of ferric hydroxide in flocculent form, removable by filtration with membranes of a porosity of 0.5 μm (Hutchinson, 1957; Hem and Cropper, 1959; Hem and Skougstad, 1960). Under some conditions, a very finely divided precipitate of ferric hydroxide, which has the properties of a colloid (0.001 to 0.5 μm, charged), may form. Colloidal particles of $Fe(OH)_3$ commonly are positively charged, although a negatively charged sol can occur at high pH. Ions in solution, and negatively charged clay particles, organic colloids, and other suspended solids can neutralize the charges on the hydroxide colloidal particles. The uncharged aggregates join to form a rapidly settling precipitate. Metals, such as copper ions, can be adsorbed by and coprecipitated with the ferric hydroxide precipitate.

Although ferrous iron in solution occurs in extremely low concentrations in oxygenated water, apparently some occurs in particulate and colloidal form. Such suspended ferrous iron presumably is associated with soil and sediment particles. A portion of total iron is contained in the living plankton, but this quantity is a very small part of the whole. The iron content of higher aquatic plants averages 5 mg g^{-1} dry weight, about an order of magnitude greater than the content of terrestrial plants (Oborn, 1960). Plant roots contain a higher proportion of iron than stems or leaves.

Iron complexes with certain organic molecules, by which the solubility and availability of iron is greatly altered. Many organic bases form strong soluble iron complexes with ferrous (Gjessing, 1964) and ferric ions. An enrichment of iron commonly is found in surface waters with a high content of dissolved organic matter. These high concentrations of complexed soluble iron are associated with high levels of humic acids (Shapiro, 1957), tannic acids (Hem, 1960b), and other lignin derivatives. The intense yellow-brown color of bog waters is associated in part with these complexes. The use of natural and synthetic organic compounds, as for example citrate and ethylene-diamine tetraacetic acid (EDTA), of high chelating capacity for metals, is well-known in plant and especially algal nutrition as a means of maintaining iron available for assimilation. Shapiro (1964, 1966, 1969) extended his earlier work on the complexing of iron with humic derivatives, especially with yellow organic acids of low molecular weight. The primary mechanism apparently is peptization, in which iron is dispersed in a solubilized form of the $Fe(OH)_3$ precipitate as a result of adsorption of the organic acids onto the surfaces of the particles. Some iron is chelated with organic acids by weak chemical bonding, although chelation is not the primary complexing mechanism.

The typical range of total iron found in oxygenated surface waters of pH 5 to 8 is about 50 to 200 $\mu g\ l^{-1}$, almost none of which occurs in ionic form. These concentrations are found in hardwater lakes (Wetzel, 1972), and much higher levels occur in lakes heavily stained with dissolved limnohumic compounds, in acidic volcanic lakes, acidic bog waters, and certain alkaline, closed lakes rich in organic matter.

Manganese Complexes

The theoretical thermodynamic redox equilibria of manganese have been studied in some detail (Hem, 1963, 1964; Stumm and Morgan, 1970), permitting an evaluation of manganese species and equilibria under various conditions of oxidation-reduction found in surface natural waters. In general, the behavior of manganese in lakes follows these predictions.

Although Mn occurs in several valence states, Mn^{+3} is generally thermodynamically unstable in aqueous solutions under normal conditions and Mn^{+4} compounds are insoluble at most environmental pH values. Similar to iron, ionic divalent Mn^{+2} occurs at low redox potentials and pH (Fig. 13–2). Some form of oxidized Mn will be in equilibrium with Mn^{+2} under oxidizing conditions of high pH and E_h, and some form of Mn^{+2} may be in equilibrium with manganese carbonate under reducing conditions of low pH and E_h. The redox equilibrium E_h values are higher and the rates of oxidation slower than for iron; as a result, detectable quantities of Mn commonly are observed longer than comparable quantities of iron under lake conditions. Above a pH of 8.5, an intermediate oxide complex forms in which Mn^{+2} is adsorbed onto manganese oxides. The Mn^{+2} of these oxide complexes can react relatively rapidly with other anions and precipitate as manganous carbonate ($MnCO_3$), manganous sulfide (MnS), and manganous hydroxide ($Mn(OH)_2$). Manganese also is adsorbed to iron oxides and coprecipitates with ferric hydroxide when the pH is greater than 6 to 7.

Manganese has a solubility of about 1 mg l^{-1} in distilled water of E_h of

550 mv and pH 7; this decreases markedly with increases in E_h and pH. Manganese forms soluble complexes with bicarbonate and sulfate and increased bicarbonate activity decreases Mn solubility, but at high concentrations it has been found to reduce the oxidation rate. Analogous to iron, organic molecules can form stable complexes with Mn^{+2}, although their operation in aquatic systems is poorly understood. Manganese occurs in relative abundance in alkaline soils as hydrated oxides, and no doubt organic complexing plays a role in retention of Mn in a complexed dissolved form for transport once redox conditions are altered by microbial and plant metabolism. Igneous rock contains about 0.01 per cent manganese, which is similarly involved in biogeochemical cycling by weathering and subsequent mobilization. Drainage from forest litter, especially from more acidic coniferous forests, is often high in manganese content, which again points to the probability of the major role of organic complexes in the effective transport of easily oxidized metal ions.

Distribution of Iron and Manganese in Lakes

Under oxidized conditions, as in the epilimnia of lakes, very large amounts of iron are found only in acidic water (pH < 3), such as in lakes of volcanic origin (Yoshimura, 1936) or in drainage from acid strip mining operations. In both cases, organic content is relatively low, and acidity usually results from sulfuric acid. The range of total iron found in most typical neutral or alkaline lakes, 50 to 200 μg 1^{-1}, is constituted largely of $Fe(OH)_3$, organically complexed iron, and adsorbed iron of seston in particulate form. The world average distribution is variable among continents (Table 9–1), and higher than generally would be expected among most oxygenated surface waters of lakes. The range of manganese concentrations (ca 10 to 850 μg 1^{-1}) is also highly variable in relation to lithology and drainage of

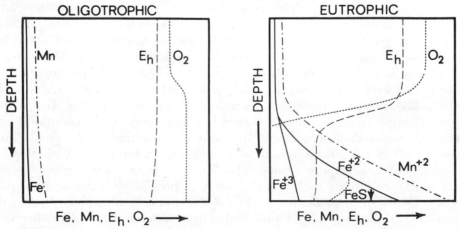

Figure 13–3 Generalized vertical distribution of iron, manganese, and redox potential (E_h) in stratified lakes of very low and high productivity.

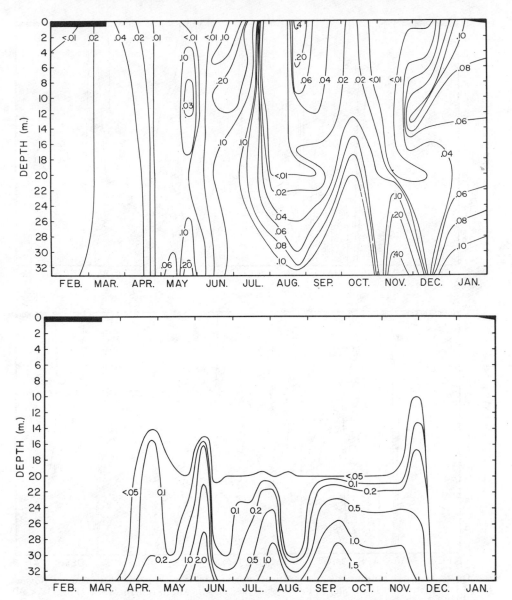

Figure 13–4 Depth-time diagrams of isopleths of total iron (*upper*) and manganese (*lower*) in mg l^{-1} of mesotrophic hardwater Crooked Lake, northeastern Indiana, 1963. (From Wetzel, unpublished data.)

the lake basins (Hutchinson, 1957; Livingstone, 1963); the average quantity is about 35 μg l^{-1}, somewhat less than that of iron. The ratio of Fe: Mn of the water is generally considerably lower than that of the lithosphere (50:1), indicative of the relative enrichment of Mn with respect to Fe, and in agreement with the reaction equilibria already discussed.

The vertical distribution of iron and manganese is reflected in the distribution of redox potentials. Ionic iron of oxygenated waters of oligotrophic lakes, epilimnia of more productive lakes, and of circulating waters is ex-

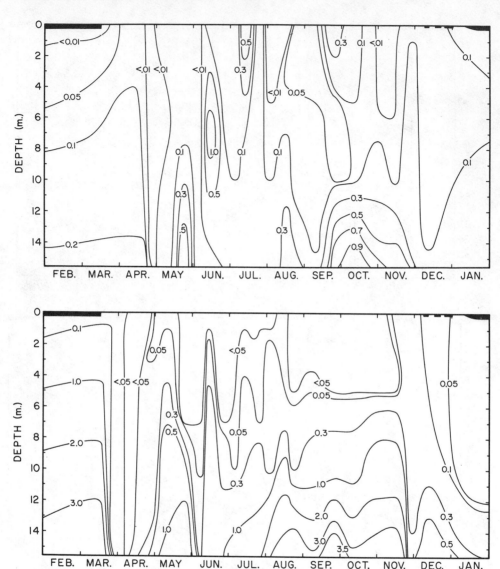

Figure 13–5 Depth-time diagrams of isopleths of total iron (*upper*) and manganese (*lower*) in mg 1^{-1} of eutrophic hardwater Little Crooked Lake, northeastern Indiana, 1963. (From Wetzel, unpublished data.)

ceedingly low. Manganese is somewhat more soluble (Fig. 13–3). Ferrous ions diffuse readily from the sediments when they are reduced to redox potentials of about 200 mv; migration of Mn^{+2} from the sediments occurs at somewhat greater redox potentials. These relationships were demonstrated very well in Mortimer's studies of the sediment interface, both experimentally (see Fig. 12–3) and in a eutrophic lake (see Fig. 12–4). Here it can be seen that the release of Mn precedes that of iron.

While the seasonal distribution of iron has been studied in some detail (reviewed in Hutchinson, 1957; McMahon, 1969), that of manganese has not been investigated as extensively (Delfino and Lee, 1968, 1971; Brezonik, et al., 1969). The general pattern of seasonal distribution is apparent from the examples given of a mesotrophic lake that undergoes hypolimnetic oxygen reduction in the later phases of summer stratification (Fig. 13–4), and an interconnected eutrophic lake (Fig. 13–5). The eutrophic lake, Little Crooked Lake, is a small, protected lake that frequently undergoes partial, temporary meromixis, as was the case in the year shown here when autumnal circulation was incomplete.

As the decomposition in the hypolimnion of very productive lakes continues throughout the period of stratification, the redox potential of the hypolimnetic waters can be further reduced to well below 100 mv. At an E_h below 100 mv, sulfate is reduced to hydrogen sulfide, and H_2S further is produced by bacterial decomposition of sulfur-containing organic compounds. Since ferrous iron is released in significant quantities from the sediments at a higher E_h of about 250 mv, much Fe^{+2} is present in the hypolimnion at the time of sulfide formation. The formation of FeS and other metal sulfides (cuprous sulfide, CuS, cadmium sulfide, CdS, etc.), all of which are very insoluble under normal lake conditions, can result in a significant reduction of iron towards the end of summer stratification. Manganous sulfide, on the other hand, is much more soluble and does not materially affect the Mn^{++} concentrations under normal lake conditions.

Iron concentrations of the hypolimnia of softwater lakes can reach very high levels under many conditions prevailing in small, deep basins, especially in bog waters receiving high concentrations of humic organic matter. Levels of sulfate are low in such waters, and sulfide concentrations seldom become sufficient to precipitate iron as FeS. Kjensmo (1962, 1967, 1968) demonstrated that the hypolimnetiç iron accumulations in protected lakes can reach such levels (> 250 mg 1^{-1}) that the salinity gradient becomes adequate to render the lakes permanently meromictic. The lack of sulfide precipitation of iron contributes to creating this condition.

These stages in the continuum of declining redox potentials can be divided into four phases of hypolimnetic conditions in stratified lakes of in-

TABLE 13–1 Changes in the Iron during the Continuum of Declining Redox Potentials in the Hypdimnia Stratified Lakes of Increasing Productivity[a]

Lake Status	[O_2]	E_h	Fe^{+2}	H_2S	PO_4^{-3}
Oligotrophic	High (orthograde)	400-500 mv	Absent	Absent	Very low
↓	↓	↓	↓	↓	↓
	Much reduced (clinograde)	400-500 mv	Absent	Absent	Very low
↓	↓	↓	↓	↓	↓
Eutrophic	Much reduced (clinograde)	ca 250 mv	High	Absent	High
↓	↓	↓	↓	↓	↓
Hypereutrophic	Much reduced (or absent)	<100 mv	Decreasing	High	Very high

[a] After discussion of Hutchinson, 1957.

creasing productivity, or within a very productive lake during the period of summer stratification (Table 13–1).

Utilization and Transformations of Iron and Manganese

The metabolic demands for iron and manganese are not generally such that the biota materially deplete the concentrations of these metals in the environment. Both iron and manganese, however, are essential micronutrients of microflora, plants, and animals (Oborn, 1960b; Wangersky, 1963; Coughlan, 1971) and are toxic at very high concentrations. Iron is required in the enzymatic pathways of chlorophyll and protein synthesis, and in the respiratory metabolism of all living protoplasm. The function of iron in cytochromes and as the basic component of hemoglobin in higher animals is well-known. Manganese is a functional component of nitrate assimilation, in the Hill reactions of photosynthesis, and among animals and bacteria it is an essential catalyst of numerous enzyme systems.

Although the requirements of these micronutrients are low, their reactivity, very low concentrations, and restricted availability (especially of iron) in the trophogenic zones of lakes and in streams suggest that under certain conditions the reduced availability of Fe and Mn, among other factors, may limit photosynthetic productivity. The mechanisms of assimilation of iron from the forms available in oxygenated natural waters are unclear. Some evidence, not completely satisfactory, indicates that ferric hydroxide is adsorbed onto algae and dead particles from which iron is assimilated. The peptizing and chelating properties of organic acids for iron, which have been found in a large number of lakes to varying degrees (Shapiro, 1966, 1969), appear to be the primary source of iron to algae, and demonstrate the importance of organic compounds in maintaining assimilable iron available to algae.

The available iron content of hardwater calcareous lakes is extremely low. For, example, the reactive iron of Lawrence Lake, Michigan, seldom exceeded 5 μg l^{-1} over an annual period (Wetzel, 1972). In this and similar lakes, it is evident that iron availability is so low that high sustained primary productivity is limited by an effective iron deficiency (Schelske, 1962; Schelske, et al., 1962; Wetzel, 1965a, 1966b, 1972). The addition of iron in complexed form, either by synthetic chelating or natural complexing organic compounds, resulted in immediate increases in photosynthetic rates. Additions of complexing organic compounds were less effective but presumably increased the availability of iron already present in the water. The effectiveness of natural organic compounds from the hypolimnion and amino compounds in maintaining solubility of iron in hardwater lakes also has been demonstrated (Wetzel, 1972).

High concentrations of manganese (> 1 mg l^{-1}) commonly are very inhibitory to blue-green and green algae (Gerloff and Skoog, 1957; Patrick, et al., 1969). An antagonistic response was demonstrated in which increasing calcium concentrations progressively reduced the inhibitory effects of high manganese. This relationship indicates that high levels of manganese, for example at the time of fall circulation, could be inhibitory to natural populations of blue-green and green algae. Manganese concentrations of $< 50 \ \mu g \ 1^{-1}$ were found to inhibit the development of green and blue-green algae in streams and to favor diatom growth strongly. If this response by diatoms is ubiquitous, then the moderate levels of Mn, combined with very high levels of Ca, of hardwater lakes may be contributory to the general dominance of diatoms and the stimulatory effects of primary productivity in response to small additions of chelated manganese found by Wetzel (1966b) in these lakes.

Bacterial Transformations of Iron and Manganese

The cycling of iron and manganese is dictated largely by the oxidation-reduction conditions of the lakes. While bacterial and photosynthetic metabolism greatly influence these controlling conditions, which indirectly regulate the states of these metals and their fluxes, the direct utilization of iron and manganese by bacteria in energetic transformations occurs but is minor in comparison to heterotrophic metabolism of organic substrates in most natural systems.

Chemosynthetic utilization of energy from inorganic oxidations in CO_2 fixation is relatively inefficient, especially in the case of the oxidation of iron and manganese. For example, in the oxidation of Fe^{+2} to Fe^{+3} by iron bacteria, an energy yield of only about 11 kcal per mole Fe is obtained. The cycling of iron and manganese is influenced by two processes (Kuznetsov, 1970). First, as pointed out earlier, reduction of the oxidized combined metal occurs under appropriate redox conditions as ferrous bicarbonate or is precipitated and sedimented as a sulfide. Example reactions are:

$$Fe_2O_3 + 3H_2S \rightleftharpoons 2FeS + 3H_2O + S$$

$$FeS + 2H_2CO_3 \rightleftharpoons Fe(HCO_3)_2 + H_2S$$

Second, sheathed and stalked bacteria, algae, protozoan flagellates, and specific true bacteria precipitate ferric and manganic oxides on their cells. The true iron bacteria occur in iron-rich waters of neutral or alkaline pH. Characteristic reactions of those few chemoautotrophic bacteria which deposit hydroxides and oxides are:

$$4Fe(HCO_3)_2 + O_2 + 6H_2O \longrightarrow 4Fe(OH)_3 + 4H_2CO_3 + 4CO_2 + 58 \ kcal$$

$$4MnCO_3 + O_2 \longrightarrow 2Mn_2O_3 + \ 4CO_2 + 76 \ kcal$$

Some species of *Leptothrix* are facultative iron bacteria that can oxidize both ferrous and manganous salts, whereas *Gallionella (Spirophyllum)* is restricted obligately to iron. Since at neutral pH and in the presence of oxygen Fe^{+2} is spontaneously oxidized, the iron-oxidizing bacteria are restricted to areas of steep redox gradients, in which they are competing effectively with oxygen for the reduced iron. Therefore, the iron bacteria usually are resticted to the interface regions of iron-bearing rock seeps, swamps, and bogs where the redox potential is sufficiently low for reduced iron to occur, and to upper hypolimnetic areas. For 0.5 g of cellular carbon to be produced, over 220 g of ferrous iron are required. As a result, much oxidized iron will be precipitated on the sheaths of the bacteria and extruded materials. While this iron may not be important from the standpoint of synthesis of organic carbon, it often is important economically in corrosion and clogging of pipes. These examples refer to strict autotrophic bacteria that satisfy their CO_2 requirements without organic matter; they obtain all energy required from the oxidation of some specific incompletely oxidized inorganic substance. Iron or manganese thereby enters the cells. There are certain facultative autotrophic bacteria (mixotrophic) that can develop in water containing only inorganic substances, but that are able to utilize organic substances as well.

Other groups of bacteria involved in these cycles are those which are heterotrophic. Certain filamentous forms (*Cladothrix*, some *Leptothrix*), in particular, deposit iron and manganese on the cell in sheaths in the process of metabolism of organic compounds. The colonial, coccoid cells or short rods of *Siderocapsa* of the Eubacteriales are a common form occurring at the oxic-anoxic interface zone of hypolimnion-metalimnion, especially in iron meromictic lakes (Dubinina, et al., 1973). *Siderocapsa* also is distributed widely in oxygenated zones of streams and lakes (Hardman and Henrici, 1939), and during periods of circulation (Sokalova, 1961). Species of this genus can mineralize humates and increase in numbers coincident with increases in iron humates during periods of high rainfall. *Spirothrix* is another dominant iron bacterium found in both aerobic and anoxic zones of Lake Glubok, USSR, whereas large populations of *Gallionella* developed only in the interface zone of the metalimnion. Over the period of an annual cycle in this lake, the iron bacterial numbers reached a very high percentage (19 per cent) of the total bacteria during the later portion of summer stratification. During other times of the year, the percentage was about 5 per cent of the total bacteria, and it is apparent that the iron bacteria have a significant function in the iron cycle of more productive lakes.

Another widely distributed iron-oxidizing bacterium is the autotrophic *Thiobacillus thiooxidans,* which oxidizes iron disulfides to ferric sulfate (ZoBell, 1973):

$$FeS_2 + 3\frac{1}{2}O_2 + H_2O \longrightarrow FeSO_4 + H_2SO_4$$

$$2FeSO_4 + \frac{1}{2}O_2 + H_2SO_4 \longrightarrow Fe_2(SO_4)_3 + H_2O$$

Closely related to *Thiobacillus* is *Ferrobacillus ferrooxidans*, which is abundant in very acidic mine waters where the pH is less than 3, in which ferrous iron is soluble and stable. *Ferrobacillus* oxidizes ferrous carbonate to ferric hydroxide with the formation of CO_2.

$$4\,FeCO_3 + O_2 + 6\,H_2O \longrightarrow 4\,Fe(OH)_3 + 4\,CO_2$$

A few heterotrophic species of iron-oxidizing bacteria of the genera *Sphaerotilis, Leptothrix, Clonothrix,* and *Siderobacter* deposit oxidized

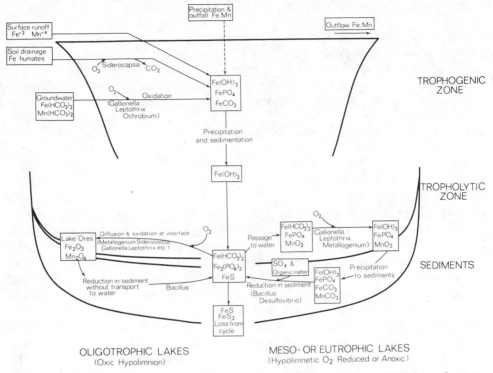

Figure 13–6 General iron and manganese cycles in lakes of low and high productivity emphasizing microbial interactions. (Modified from Kuznetsov, 1970.)

manganese along with iron in their sheaths on capsules. Some species of *Metallogenium* obtain part of their energy requirements from the oxidation of manganous oxide (MnO), manganous sulfate ($MnSO_4$), and manganous carbonate ($MnCO_3$) to manganese sesquioxide (Mn_2O_3) and manganese dioxide (MnO_2); others are heterotrophic, and along with several other manganese and iron oxidizing bacteria, contribute to the formation of manganese and iron oxides in lake sediments (Oborn, 1964; Perfil'ev and Gabe, 1969; Kuznetsov, 1970). *Metallogenium* is undoubtedly one of the dominant microorganisms involved in the deposition of manganese nodules in lakes (Sokalova, 1961; Sorokin, 1970).

A general summation of these inorganic and bacterial relationships is given diagrammatically in Figure 13–6. The primary distinction between the fluxes of iron and manganese in the hypolimnia of lakes of increasing productivity centers on the reduction of redox potential and pH. Such a cycle is qualitative, owing to the paucity of quantitative information of flux rates for both components of the systems as well as for whole lake systems. Very little is known about the role of the littoral flora in this cycling. Iron, for example, is translocated from the sediments to the leaves of submersed vegetation (Oborn and Hem, 1962; DeMarte and Hartman, 1974). The magnitude of this transport from the sediments to the water in organic-bound phases of plant material as the plants decay is unknown.

Minor Metals of Nutritional Value

Minor metallic elements, collectively referred to as micronutrients, include Fe, Mn, zinc (Zn), Cu, boron (B), cobalt (Co), Mo, and vanadium (V), all of which can be shown to be required for the nutrition of plants and in many cases of animals. The universality of an absolute requirement for all of these elements is not clear. In some cases, as with iron and manganese, the essentiality of the element is established; in others, such as vanadium, it is known that one element can substitute for another, e.g., vanadium can replace molybdenum.

Most of the information on requirements for micronutrients derives from studies of physiological deficiencies in cultures that present an array of experimental difficulties, particularly involving contamination (Weissner, 1962). Requirements further are complicated by observed differences that depend on the composition and concentration of other minerals, and antagonism among these elements. These effects, often found under the conditions of a controlled culture, become much more significant in the heterogeneous inorganic solution of natural waters undergoing constant seasonal and spatial variations. Hence, one finds a large number of analyses of concentrations of ionic and, less frequently, particulate micronutrients in water that may bear little relationship to actual metabolic requirements. The question of requirements concerns the *availability* of the micronutrient within the constraints of other ionic components of salinity, the extent of organic complexing of the micronutrient within a spectrum of degrees of ionic complexing by many ions, and the demands of varying species. In general, concentrations and availability of micronutrients are adequate in most natural waters for requirements of algal populations to sustain active population development within constraints of light, temperature, and macronutrient availability. There are, however, clear cases in which micronutrients can limit photosynthesis to a degree. A true deficiency of micronutrients is found in some oligotrophic aquatic systems of granitic alpine regions and volcanic areas, in which a paucity of these elements is well-known (reviewed by Goldman, 1972). In other situations, such as in hardwater calcareous lakes, the micronutrients are not deficient in the system but are in forms physiologically unavailable for assimilation (Wetzel, 1972). In both situations, the control of productivity can be quite transitory by being effective only on certain algal species and thereby indirectly influencing algal succession and productivity.

The importance of dissolved organic matter in regulating micronutrient availability and growth, although long known to be effective in cultures, has been emphasized only recently as a major controlling factor of productivity in lakes (Wetzel, 1968; Wetzel and Allen, 1970; Wetzel, et al., 1972). Although it is involved in lake metabolism in many ways (cf. Chapter 17), complexing of micronutrients by organic compounds increases the physiological availability of many micronutrients when the ratio of organic matter to micronutrient concentration is reasonably high. The effects of chelated iron and manganese on algal productivity, discussed earlier, provide only one example of this process.

Most metallic micronutrients are very toxic when in excess or when

TABLE 13–2   Average Concentrations (μg 1^{-1}) of Soluble Minor Metallic Elements of Natural Waters (Surface)

Water	Fe	Mn	Cu	Zn	Co	Mo	V
World surface lakes and rivers (Livingstone, 1963)	ca 40	35	10	10	0.9	0.8	ca 0.1
Alpine lakes, California (Bradford, et al., 1968)	1.3	0.3	1.2	1.5	<0.3	0.4	—
Northern German lakes (Groth, 1971)	31.5	28.6	2.9	6.6	0.05	0.39	—
Northeastern Indiana lakes (Wetzel, 1966b)	15.0	21.3	<2	12.9	<2	30.0	<3
Linsley Pond, Connecticut (Hutchinson, 1957)	350	140	53	—	0.05	—	—
South American lakes (Groth, 1971)	533	15.6	1.7	8.7	0.11	0.6	—

complexed organically to the point when their availability exceeds tolerance limits. Copper is a well-known metal that has been used repeatedly as a herbicide in high concentrations to control algal blooms. Fogg and Westlake (1955) have demonstrated that polypeptides secreted by blue-green algae can complex copper ions effectively and reduce toxic effects. This example is but one of many ways in which organic complexing can influence the availability of micronutrients. The effects of micronutrients and their availability to animals are less well-understood. It generally is assumed that requirements are met by means of uptake from food sources. Positive and negative correlations between fluctuations of micronutrient concentrations of the water and animal populations (see Parker and Hazelwood, 1962) yield little insight into this question, but suggest that food quality certainly can influence, in part, the succession and development of the fauna.

The general average concentrations of the soluble form of minor metallic micronutrients for a range of lakes and streams are given in Table 13–2. However, the ranges of concentration are extreme not only seasonally, but particularly in relation to contamination from industrial and other sources of pollution. Examples are discussed at length in the references given in Table 13–2, Chawla and Chau (1969), Robbins, et al. (1972), Mills and Oglesby (1971), and many others.

Distribution of the Minor Elements in Lakes

The distribution of the minor elements among ionic (soluble), organically complexed, and adsorbed fractions, as well as of that amount in living biota, is poorly understood. Quantitative rates of flux between the living and abiotic organic and inorganic phases are even less well-delineated; generally much is inferred from analogous chemical species, such as iron and manganese. Fragmentary evidence on the cycling of the minor elements indicates that these similarities are real and that their cycling is comparable.

The amount of Cu, Zn, Co, and Mo in ionic solution is generally very small in aerated surface waters. Transport of these trace metals in flowing waters can be partitioned analytically into (1) ionic form, (2) that complexed in organic materials, (3) that adsorbed and precipitated on solids, and (4) that incorporated in crystalline structures (Gibbs, 1973). The solubilities of the various elements vary somewhat (cf. Groth, 1971), but in general a large majority (> 70 per cent) of each is transported in crystalline solids and adsorbed solid phases. Much of the remainder is in organic complexes and dead seston, and very little in solution.

A summary of older literature on the cycling of Cu, Zn, Mo, and Co is given in Hutchinson (1957). The detailed analyses of Groth (1971) on Schöhsee in northern Germany highlight the basic phases of the dissolved and particulate fractions over an annual cycle. The total amounts of Co, Mo, and Zn, as well as of Fe and Mn, accumulated in hypolimnetic waters during summer stratification as shown in Table 13–3. Whereas the concentrations of Fe and Mn were found to be strongly related to redox conditions, the primary source of the hypolimnetic accumulation of Co, Mo, and Zn was from release and mineralization of sedimenting organic detritus. No significant vertical distribution of Cu was found in Schöhsee during summer stratification, although moderate increases in hypolimnetic sestonic Cu have been observed elsewhere in eutrophic lakes (Riley, 1939).

Co, Cu, and Zn form stable complexes with organic compounds, similar to Fe and Mn, so that losses of free ions by inorganic reactions forming insoluble hydroxides, sulfides, phosphates, and carbonates can be reduced appreciably (Groth, 1971). Molybdenum, however, exhibits greater mobility than the other ions (Mo > Cu > Zn > Co).

Phytoplankton tend to accumulate these minor metallic elements in the order: Fe > Zn > Cu > Co > Mn > Mo (Table 13–4). During decomposition and mineralization of plankton, release to a soluble phase occurs in the order of Mo > Co > Cu > Zn > Fe > Mn; this is reflected in the order of amounts of the elements found to be transported to the sediments in sedimenting organic detritus: Fe > Mn > Co > Zn > Cu > Mo. Coprecipitation of these trace elements with $Fe(OH)_3$ was greatest with Cu, in the order of Cu > Mo

TABLE 13–3 Average Concentrations ($\mu g \ l^{-1}$) of Minor Elements in the Epilimnion and Hypolimnion of Schöhsee, Germany[a]

Stratum	Mn	Fe	Cu	Zn	Co	Mo
Epilimnion (E)	4.5	15	1.0	1.8	0.03	0.21
Hypolimnion (H)	590	425	0.9	1.9	0.07	0.30
Enrichment ratio of H/E	130	28	0.9	1.1	2.3	1.4

[a]Data after Groth, 1971.

TABLE 13–4 Average Accumulation of Minor Metallic Elements in the Plankton and Sediment in Comparison to the Average Concentrations in the Epilimnion of Schöhsee, Germany, 1968–1969[a]

Element	Dissolved in Epilimnetic Water ($\mu g\ l^{-1}$)	In Plankton ($\mu g\ g^{-1}$)	In Sediment ($\mu g\ g^{-1}$)
IRON			
Concentration	15	950	58000
Enrichment factor	1	63×10^3	$3,900 \times 10^3$
MANGANESE			
Concentration	4.5	130	1600
Enrichment factor	1	29×10^3	355×10^3
COBALT			
Concentration	0.03	1.1	8.3
Enrichment factor	1	37×10^3	280×10^3
COPPER			
Concentration	1.0	60	95
Enrichment factor	1	60×10^3	95×10^3
ZINC			
Concentration	1.8	110	350
Enrichment factor	1	61×10^3	195×10^3
MOLYBDENUM			
Concentration	0.21	4.2	1.4
Enrichment factor	1	20×10^3	7×10^3

[a]Data after Groth, 1971.

> Co > Zn. Therefore, during summer stratification, the algal uptake and sedimenting detritus play a major role in the cycling of Cu, Co, and Zn, whereas concentrations of Fe and Mn are regulated largely by redox conditions. Although variations among lakes are great, the amounts of minor elements (especially copper) often are found to increase during fall circulation and during winter. Although ionic concentrations increase in winter somewhat, most of the peak is exhibited in the organic fractions (Riley, 1939; Kimball, 1973). In general, the activity of zinc (Bachmann, 1963) and cobalt (Parker and Hasler, 1969; Benoit, 1957) follows the conclusions summarized above. Molybdenum exhibits much greater mobility than the other minor elements discussed, and does not show analogous patterns of distribution (cf. Dumont, 1972).

The Sulfur Cycle

Sulfur in the form of both mineral and organic sulfates is utilized by all living organisms. Sulfate is reduced to sulfhydryl (−SH) groups in protein synthesis, with a concomitant production of oxygen that is utilized in oxidative metabolic reactions. Interest in the sulfur cycle of fresh waters, however, extends beyond nutritional demands of the biota, which almost always are met by the abundant distribution of this anion. Decomposition of organic matter containing proteinaceous sulfur and anaerobic reduction of sulfate in stratified waters contributes to altered conditions that markedly affect the cycling of other nutrients, productivity, and biotic distribution.

Forms and Sources of Sulfur

Sources of sulfur compounds to natural waters are rocks, fertilizers, and atmospheric transport in precipitation and dry deposition. At the present time, atmospheric sources, augmented greatly by combustion products of industry, dominate all other sources.

Geological processes of epigenesis of sulfate occur on rocks and soils containing either sulfides or free sulfur, which are oxidized in the presence of water to form sulfuric acid (ZoBell, 1973):

$$FeS_2 + 3\frac{1}{2}O_2 + H_2O \longrightarrow FeSO_4 + H_2SO_4$$
(pyrite)

$$2S + 3O_2 + 2H_2O \longrightarrow 2H_2SO_4.$$

These reactions tend to lower both the pH and E_h, which affects numerous other oxidative weathering reactions involving numerous minerals. Calcium sulfate, moderately soluble in water, is a common constituent of sedimentary rocks, and drainage from calcareous regions generally contains higher than average concentrations of sulfate. As will be discussed further on, bacteria contribute to the oxidation of sulfides and elementary sulfur both in soil and in water.

Reduced sulfur, as hydrogen sulfide, is added in large quantities to the atmosphere from volcanic gases, biogenic, and industrial sources (Kuznetsov, 1964; Kellogg, et al., 1972). H_2S undergoes a number of oxidative reactions to sulfur dioxide (SO_2), sulfur trioxide (SO_3), and sulfuric acid (H_2SO_4). Sulfur dioxide constitutes about 95 per cent of the sulfur compounds resulting from the burning of fossil fuels that contain sulfur. Although the oxidation of SO_2 to H_2SO_4 in air is slow (hours to days), when SO_2 is dissolved in atmospheric water, it is rapidly oxidized to sulfuric acid. The global cycling of sulfur compounds (Table 13–5) indicates clearly that in industrialized regions, man rapidly is approaching the point of exceeding inputs from natural sources. The removal processes over land are sufficiently slow (several days) to result in markedly increased concentrations in areas hundreds to thousands of kilometers downwind. In nonindustrial areas the primary source of sulfur as sulfate ($SO_4^=$) in rain and snow is atmospherically oxidized H_2S that is produced along coastal regions by anaerobic bacteria (Jensen and Nakai, 1961). The $SO_4^=$ derived from sea spray is largely returned to the ocean; this source over land is minor.

The relative contribution of sulfur compounds to natural waters of course varies with the regional lithology, agricultural application of sulfate-containing fertilizers, and atmospheric sources in relation to other sources. In calcareous areas of sedimentary rock, the atmospheric contributions are a

TABLE 13–5 Sources, Sinks, and Residence Times of Atmospheric Sulfur Compounds[a]
(Units are 10^6 tons as sulfate per year.)

Sources	Tons $\times 10^6$ yr^{-1}	Approximate Residence Time
Windblown sea salt $SO_4^=$ in precipitation, 10% of global total deposited on all land	130	
Bacterial and plant production of H_2S (SO_2)	268	0.5-6 days
Man-made production of SO_2 and $SO_4^=$ from fossil fuels (80% deposited on land; 93% of global production in the Northern Hemisphere)	150	0.5-6 days
Volcanic sources (H_2S, SO_2, $SO_4^=$)	2	<1 day
Deposition		
Rain over oceans (SO_2, $SO_4^=$)	217	
Rain over land (SO_2, $SO_4^=$)	258	
Plant uptake (SO_2, $SO_4^=$)	45	
Dry fallout deposition ($SO_4^=$)	30	

[a]From data of Kellogg, et al., 1972.

small portion of the total. By contrast, for example in crystalline rock areas of northeastern America, precipitation supplies nearly all of the sulfate of natural waters (Fisher, et al., 1968). Such acidic, sulfatic precipitation has equal or greater significance in geochemical processes than the carbonic acid system in these regions (Johnson, et al., 1972).

Distribution of Sulfur in Natural Waters

The sulfur content of biota ranges from 0.05 to nearly 5 per cent in a few bacteria; the average content is 0.2 per cent. The amounts in the biota and detritus, although significant, are generally small in comparison to the inorganic sulfur components of aquatic systems. The distribution and cycling therefore revolve about the chemical species under various conditions and the biotic influences in transitions among these forms, and sulfur transport in the system.

The lowest concentrations of sulfate in oxic waters, in the range of somewhat less than 1 mg l^{-1}, appear to be a common feature of numerous lakes of Africa in crystalline rock drainage basins (Talling and Talling, 1965). The other extreme is found in sulfate saline lakes (> 50 g l^{-1}). The usual range is within 5 to 30 mg l^{-1}, with an average of about 11 mg $SO_4^=$ l^{-1}. Low levels of $SO_4^=$ have been implicated in the suppression of algal productivity in Lake Victoria, Africa, although the evidence is not conclusive (Fish, 1956).

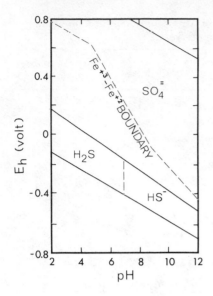

Figure 13–7 Approximate redox-pH fields of stability of sulfur species likely to occur in natural water. (Modified from Hem, 1960, with Chen and Morris, 1972.)

The predominant form of sulfur in water is the oxidized state as sulfate (Fig. 13–7). Nearly all assimilation of sulfur is as sulfate, but in decomposition of organic matter sulfur is released largely as hydrogen sulfide. Under oxic conditions H_2S is oxidized rather rapidly. Therefore little H_2S would be anticipated in aerated regions of aquatic systems. However, strict application of redox and pH to evaluation of reactions involving sulfur is difficult, because some are chemically slow and mediated by bacterial metabolism. Although $SO_4^=$ and H_2S dominate, HS^- and very small concentrations of $S^=$ occur in strongly alkaline solutions, since H_2S, which is very soluble in water, dissociates weakly (k_1 10^{-7}; k_2 10^{-15}) (Hutchinson, 1957; Hem, 1960c). Under certain conditions of low redox and pH, in which partial oxidation of sulfides occurs, free $S°$ may be formed.

Metal sulfides are exceedingly insoluble at neutral or alkaline pH values encountered in a majority of natural waters. The equilibrium ion activity product of FeS of anaerobic lake sediments is $10^{-17.7}$ (Doyle, 1968b). Therefore, Fe^{+2} released from sediments reacts vigorously with H_2S to form FeS. The water of, for example, an anaerobic hypolimnion must be somewhat acidic in order for appreciable H_2S to accumulate. If the water is alkaline, H_2S will accumulate only after most of the Fe^{+2} has been precipitated as FeS. The removal of sulfide by the release of ferrous ions will permit an increase in the migration of other metals from the sediments, such as Cu, Zn, and lead (Pb), which form even more insoluble sulfides than that of FeS.

The generalized vertical distribution of sulfate and hydrogen sulfide for stratified lakes is depicted in Figure 13–8, although much individual variation is found among lakes. Under oxic conditions, as is the case in many oligotrophic and mesotrophic lakes, and during periods of circulation, H_2S is absent and $SO_4^=$ concentrations change little with depth. Some release of $SO_4^=$ occurs from the sediments, and this increase in sulfate in the hypolimnion can become more pronounced in hypolimnia of mesotrophic or eutrophic

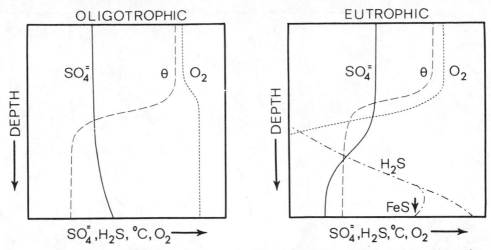

Figure 13-8 Generalized distribution of sulfate and hydrogen sulfide in lakes of very low and very high productivity.

lakes in the earlier phases of summer stratification (Fig. 13-9). Reduction of sulfate to H_2S occurs as the redox potential, because of decomposition, is lowered to less than about 100 mv. Some, and in many cases much, of the H_2S, especially near the sediments, reacts with Fe^{+2} ions released from the sediments to form insoluble FeS. In this way, considerable quantities of sulfur are lost to the sediments. Lakes receiving rich sources of sulfate from inflowing water, such as meromictic lakes of crenogenic formation, often contain immense concentrations of H_2S in their anoxic monimolimnia. Analogous situations occur in certain anoxic stretches of rivers that are polluted with sulfate-rich organic wastes. Effluents from paper-producing industries are a common source of such pollution. Horizontal variations in the distribution of sulfate and sulfides can be large and, especially in reservoirs, complicated by flow patterns (e. g., Hanušová, 1962).

As a result of the reduction of $SO_4^=$ to sulfide, some of which is lost to the sediments as insoluble metallic sulfides, and oxidation of H_2S to sulfate, these reactions play a significant role in the modification of conditions for numerous other nutrients, especially for the mobilization of phosphate. In an excellent discussion of these reactions and their mediation by various bacterial groups, Ohle (1954) characterizes sulfate as a 'catalyzer' of limnetic nutrient cycling.

The sulfur cycle in Linsley Pond, Connecticut, was studied during summer stratification in appreciable detail with ^{35}S labelled sulfuric acid (Stuiver, 1967). In this eutrophic lake, large quantities of sulfate were lost from the metalimnion and hypolimnion, especially at the sediment-water interface. The rates of reduction of sulfate in the metalimnion and hypolimnion differed little between regions that were anoxic or partly devoid of oxygen, but were about 10 times faster than in the fully aerobic epilimnion. Vertical diffusion of sulfate in the metalimnion and upper layers of the hypolimnion was very small during stratification. Horizontal diffusion at a given depth within the lower strata of water was sufficient to transport sulfates to the surrounding sediments, in which reduction took place. Transport of sul-

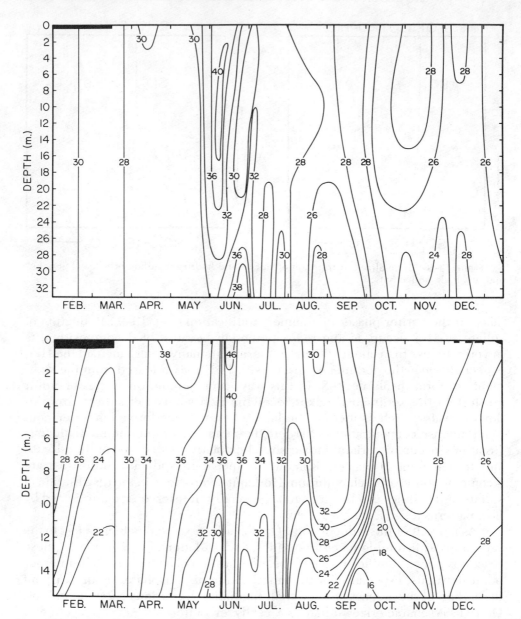

Figure 13–9 Depth-time diagrams of isopleths of sulfate concentrations (mg l⁻¹) of meso-trophic hardwater Crooked Lake (*upper*) and interconnected eutrophic Little Crooked Lake (*lower*), Noble-Whitley counties, northeastern Indiana. Opaque areas = ice cover to scale. (From Wetzel, unpublished data.)

fur by means of sedimentation of biological material was negligible and influenced the sulfur cycle of this lake only by providing organic substrates for bacterial metabolism in the hypolimnion and sediments.

This analysis permitted calculation of a total sulfur budget, estimated from the ^{35}S activities of the water and organic compounds (Fig. 13–10). Most of the sulfur was stored as sulfate and sulfide of the water, and as sulfide in the sediments. That fraction utilized by organisms was small and did not materi-

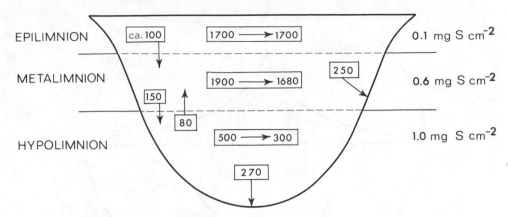

Figure 13–10 The total sulfate budget of Linsley Pond, Connecticut, estimated for an early stage of thermal stratification (*lefthand numbers in central squares*) and final stages (*righthand numbers in central squares*). The other numbers give the transfer rates of sulfate, in kg S, between the different strata. At the right is the total amount of sulfate reduced and stored per cm^2 of sediment during the 4-month period. The total amount of dissolved H$_2$S at the end of stratification was about 15 kg S in the hypolimnion. (After Stuiver, 1967.)

ally influence the total cycle. The H$_2$S of the hypolimnion was oxidized in the epilimnion; escape of H$_2$S to the atmosphere by diffusion or gas bubbles was very small.

The sulfur of black sediments of eutrophic lakes consists of sulfide dissolved in interstitial water, acid-soluble sulfide, elemental sulfur, organic sulfur, and sulfates (Doyle, 1968a, 1968b; Nriagu, 1968). At least 40 per cent of the acid-soluble sulfide of Linsley Pond was identical to tetragonal FeS (mackinawite). In Lake Mendota, about 45 per cent of the sulfur precipitated as sulfide was estimated to be derived from mineralization of organic matter, and the remainder from bacterial reduction of sulfates.

Bacterial Metabolism and the Sulfur Cycle

As has been discussed earlier, sulfate is utilized in protein synthesis in photosynthetic and animal metabolism in which SO$_4$ is reduced to sulfhydryl (−SH) form. Further reduction to H$_2$S occurs upon decomposition of this organic material by more typical heterotrophic bacterial metabolism (Fig. 13–11). The most important of a large number of protein decomposing bacteria belong to the genus *Proteus*, which are gram-negative, nonsporing rods, usually possessing large numbers of flagella (Butlin, 1953), and are particularly active in soil systems. The dominant bacteria involved in protein-aceous decomposition forming H$_2$S in lakes of varying productivity (Table 13–6) and the stages of degradation are discussed in Kuznetsov (1970). Bacterial densities in the water are very much lower (1 to 3 orders of magnitude) than in the surface sediments. In addition to this pathway, a number of bacteria reduce sulfate, sulfite, thiosulfate, hyposulfite, and basic sulfur to hydrogen sulfide. These *sulfate-reducing bacteria* are heterotrophic and an-aerobic, and use the sulfur compound as a hydrogen acceptor in the oxidative

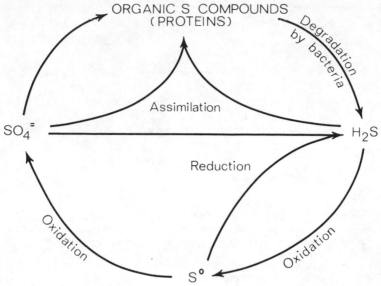

Figure 13–11   General sulfur cycle in nature. (Modified from Butlin, 1953.)

metabolism. Finally, several groups of bacteria oxidize sulfide to sulfur and sulfur to sulfate.

The *sulfur-reducing bacteria* of the genera *Desulfovibrio* and *Desulfo-tomaculum* are strictly anaerobic, and derive oxygen from sulfate for the oxidation of either organic matter or molecular hydrogen:

$$H_2SO_4 + 2(CH_2O) \longrightarrow 2CO_2 + 2H_2O + H_2S$$

$$H_2SO_4 + 4H_2 \longrightarrow 4H_2O + H_2S \ (\Delta F = -60 \text{ kcal mole}^{-1})$$

TABLE 13–6   Predominating Bacteria that Form Hydrogen Sulfide from the Decomposition of Proteinaceous Organic Matter in Different Types of Lakes[a]

Type of Lake	Dominant Bacteria
Oligotrophic	*Mycobacterium phlei, Mycobacterium filiforme*
Mesotrophic	*Bacterium nitrificans, Pseudomonas liquefaciens, Chromobacter aurantiacum*
Eutrophic	*Pseudomonas liquefaciens, Bacterium delicatum*
Bog lakes of high organic matter	*Pseudomonas fluorescens, Bacillus pituitans*
Saline lakes and estuaries	*Mycobacterium luteum, Micrococcus nitrificans, Achromobacter halophilum, Flavobacterium halophilum, Bacterium albo-luteum, Vibrio hydrosulfureus*

[a]After data of Kuznetsov, 1970.

While no oxygen is consumed directly, the H_2S generated by sulfate-reducing bacteria readily oxidizes and utilizes oxygen upon intrusion or transport to aerobic regions. The sulfate-reducing bacteria can reduce sulfite more rapidly, and thiosulfate less rapidly, than sulfate. Colloidal sulfur, but not pure noncolloidal sulfur, is reduced very slowly. In a concentration range of 20 to 130 mg SO_4 l^{-1}, the rate of production of H_2S by bacteria is roughly proportional to SO_4 concentration (Ohle, 1954). The biological oxygen demand of oxidizable organic matter theoretically can be satisfied by about 1.6 g SO_4 g^{-1} by this reduction.

The *sulfur-oxidizing bacteria* commonly are differentiated into two groups. The *chemosynthetic (colorless) sulfur-oxidizing bacteria* are mostly aerobic forms that oxidize H_2S, and are of two types. The first deposits sulfur *inside* the cell

$$H_2S + \frac{1}{2}O_2 \longrightarrow S° + H_2O \ (\Delta F = -41 \text{ kcal mole}^{-1}),$$

which accumulates as long as H_2S is available. As sulfide sources are depleted, the internally stored sulfur is oxidized with the release of sulfate:

$$S^0 + 1\frac{1}{2}O_2 + H_2O \longrightarrow H_2SO_4 \ (\Delta F = -118 \text{ kcal mole}^{-1})$$

Beggiatoa, a long, filamentous bacterium, and *Thiothrix,* are common bacteria that oxidize H_2S with deposition of sulfur intracellularly and occur in areas where H_2S is being formed, e.g., canals, swamps, and sulfur springs.

By similar reactions, a second type of chemosynthetic sulfur-oxidizing bacteria deposits sulfur *outside* of the cell. This assemblage is represented best by the genus *Thiobacillus,* which oxidizes sulfide, S°, and other reduced sulfur compounds such as thiosulfate:

$$2 \ Na_2S_2O_3 + O_2 \longrightarrow 2 \ S° + 2 \ Na_2SO_4$$

Of the many species of *Thiobacillus,* some (e.g., *T. thiooxidans*) are restricted to acidic waters (pH 1 to 5), while others such as *T. thioparus* grow optimally at neutral or alkaline pH values. The anaerobe *T. denitrificans* oxidizes thiosulfate in alkaline waters by reduction of nitrate to N_2:

$$5 \ S_2O_3^= + 8 \ NO_3^- + 2 \ HCO_3^- \longrightarrow 10 \ SO_4^= + 2 \ CO_2 + H_2O + 4 \ N_2$$

or,

$$5 \ S° + 6 \ NO_3^- + 2 \ CO_3^= \longrightarrow 5 \ SO_4^= + 2 \ CO_2 + 3 \ N_2 \ (\Delta F = -179 \text{ kcal mole}^{-1}).$$

These sulfur-oxidizing bacteria commonly adhere to sulfur granules, continuously utilizing a little at a time in the formation of sulfate.

The other major group of sulfur-oxidizing bacteria is the *photosynthetic (colored) sulfur bacteria,* anaerobes that can be divided conveniently into the *green sulfur bacteria* (Chlorobacteriaceae) and the *purple sulfur bacteria*

(Thiorhodaceae). Excellent detailed reviews of the photosynthetic bacteria and their metabolism are given by Gest, et al. (1963), Vernon (1964), Kondrat'eva (1965), and Pfennig (1967). The green sulfur bacteria require light as an energy source, and use sulfur of H_2S as an electron donor in the photosynthetic reduction of CO_2:

$$CO_2 + 2\ H_2S \xrightarrow{\text{light}} (CH_2O) + H_2O + 2S$$
$$2CO_2 + 2H_2O + H_2S \xrightarrow{\text{light}} 2(CH_2O) + H_2SO_4$$

A few species can utilize molecular hydrogen alone. The green sulfur bacteria, notably the genera *Chlorobium* and *Pelodictyon*, are generally unicellular, nonmotile, and produce sulfur granules outside of their cell membranes. At least four bacteriochlorophylls occur in the photosynthetic bacteria which differ from chlorophyll *a* in a primary absorption maximum at higher wavelengths (770 to 780 nm). The green sulfur bacteria tolerate fairly high concentrations of H_2S, whereas the purple sulfur bacteria are less tolerant of H_2S and grow optimally at high pH values (9.5).

The *purple sulfur bacteria* require light energy for the oxidation of H_2S and other reduced sulfur compounds, especially thiosulfate, to sulfate in the photosynthetic reduction of CO_2. Members of this group are generally large (5 to 10 μm), actively motile, and free $S°$ is deposited intracellularly. Sulfide is oxidized to S and $SO_4^=$ by the same reactions described for the green sulfur bacteria. In addition, some of the purple sulfur bacteria are able to grow photoautotrophically, with thiosulfate as the electron donor (*Thiopedia, Thiocapsa, Thiocystis, Rhodothece*). Other important genera (*Chromatium, Chlorobium, Thiospirillum*) are unable to utilize significant amounts of thiosulfate. Like the green sulfur bacteria, many purples can utilize hydrogen as the only electron acceptor, simultaneously with an assimilatory sulfate reduction system.

The photosynthetic Calvin cycle has been shown to be operational in all photosynthetic bacteria studied in this respect. Many strains can utilize low molecular weight organic substrates, especially fatty acids, as their carbon source, singly or in combination with CO_2. Nearly all green and purple sulfur bacteria require vitamin B_{12} from exogenous sources which, as will be discussed further on, is an organic micronutrient that influences photosynthetic productivity under some limnological conditions.

A third group, the *purple nonsulfur bacteria* (Athiorhodaceae), is included here among the other pigmented photosynthetic bacteria because of their many metabolic and distributional similarities. The nonsulfur purple bacteria, including *Rhodopseudomonas, Rhodospirillum,* and *Rhodomicrobium,* are facultative photoautotrophs that grow photosynthetically or heterotrophically either aerobically or anaerobically in the dark on organic substrates. Some *Rhodopseudomonas* can utilize thiosulfate anaerobically as a hydrogen donor:

$$2\ CO_2 + Na_2S_2O_3 + 3\ H_2O \xrightarrow{\text{light}} 2(CH_2O) + Na_2SO_4 + H_2SO_4$$

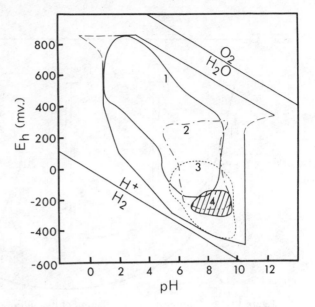

Figure 13–12 General E_h-pH environmental limits of *1,* chemosynthetic (colorless) sulfur-oxidizing bacteria, *2,* photosynthetic purple bacteria, *3,* sulfate-reducing bacteria, and *4,* green sulfur bacteria, all within the composite distributional field of E_h-pH measurements of habitats of organisms. (After data of Baas Becking, et al., 1960.)

Sulfur is not stored intracellularly. Hydrogen sulfide inhibits the growth of Athiorhodaceae.

The occurrence and distribution of the various sulfur-oxidizing reducing bacteria are restricted, therefore, by the redox and pH conditions in relation to oxygen and the state of sulfur compounds (Fig. 13–12). The reducing conditions for strict anaerobic photosynthetic sulfur bacteria, for example, must coincide with adequate light of high wavelength for these populations to develop extensively. Often these optimal conditions occur in stratified lakes in sharply defined layers with steep physical and chemical gradients, and result in narrow layering of the bacterial populations.

The microbial processes involved in the sulfur cycle of lakes are diagrammatically represented in Figure 13–13 in a composite manner. The processes on the lefthand side of the figure would be more characteristic of a lake with relatively high concentrations of sulfur in various forms. The gradient between the oxic upper strata and the lower H_2S-rich strata would be steep, but with a diffusion interface zone where both gases occurred. Those processes on the righthand side of the lake figure are more representative of lakes with low sulfate content but adequate amounts for high photosynthetic production.

The rather specific requirements of the sulfate-reducing bacteria, photosynthetic and colorless sulfur bacteria, and sulfur-oxidizing bacteria would lead one to expect development of massive populations in rather localized strata. Such is indeed the case in many situations, particularly in meromictic lakes where gradients within the chemolimnion are steep. Green sulfur bacteria commonly are found in profusion in a narrow layer immediately below a dense population of purple sulfur bacteria at the interface of the oxic-anoxic layer of very productive lakes. Light levels at this region are almost always low, usually less than 10 per cent of intensities at the surface. The

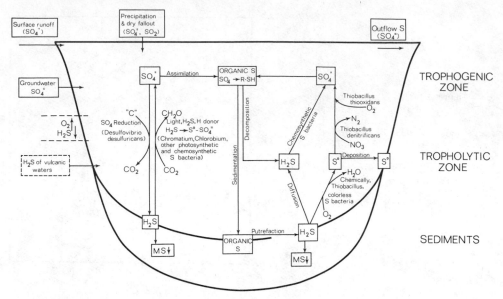

Figure 13–13 Composite representation of the sulfur cycle in a lake, with emphasis on the microbiological processes. MS = metallic, primarily iron, sulfides. (Greatly modified from Kuznetsov, 1970.)

seasonal development of optimal conditions for these specific groups can be transitory, so that their contribution to the total annual productivity of the lake may be short-lived. Among meromictic lakes, the bacterial plates can persist more or less continuously.

While descriptions of the distribution of the sulfur bacteria in aquatic systems are fairly common, and much is known about the physiology of these interesting organisms, little is known of their contribution to the total productivity of lakes. It is clear that at certain periods in productive dimictic and in meromictic lakes, bacterial photosynthesis easily can exceed that of algae and macrophytes (Table 13–7). Most of the data given in Table 13–7, however, are taken from periods when the photosynthesis of bacteria was maximal. Occasionally, when the algal photosynthesis is low, as in the example of Smith Hole Lake, which receives a high proportion of allochthonous organic matter, the brief but very productive photosynthetic bacterial development represents a major part of the lake's annual photosynthetic productivity (Wetzel, 1973). In most situations, however, the contribution of bacterial photosynthesis to the entire system over an annual period is small, even though spectacular localized populations may develop periodically.

The rates of bacterial sulfur metabolism in situ in aquatic systems have been studied in only a few cases, and largely by Sorokin (1964, 1966, 1970). In several systems, as illustrated in Figure 13–14 for Lake Gek Gel, an oligotrophic meromictic lake of crenogenic origin, and in the deep, meromictic Black Sea, the most intensive rates of bacterial reduction of sulfates were at the near sediment strata and at the water layers near the littoral slopes at the upper boundary of the anaerobic H_2S zone. The higher rates of sulfate reduction in these cases apparently are related to higher inputs of organic matter

TABLE 13–7 Comparison of Algal and Bacterial Photosynthetic Rates and Rates of Bacterial Chemosynthesis in Several Lakes (PS = photosynthesis.)

Lake	Units of Measurement	Algal PS	Bacterial PS	Bacterial PS/Algal PS × 100 (%)	Chemosynthesis	Chemosynthesis/Total PS × 100 (%)	Remarks
Hiruga, Japan 21 July 66 (Takahashi and Ichimura, 1968)	mg C m^{-2} hr^{-1}	29.7	0.9	3.0	4.3	14.1	
Waku-ike, Japan 4 Aug. 65	mg C m^{-2} hr^{-1}	19.8	1.8	9.1	—	—	
Kisaratsu, Japan 16 Aug. 66	mg C m^{-2} hr^{-1}	22.1	128	579	3.5	2.3	
2 Sept. 65		12.0	10.6	88	2.5	11.1	
Smith Hole, Indiana Annual period 1963–1964 (Wetzel, 1973)	annual mean mg C m^{-2} day^{-1}	194	91	47	6	<10	Massive late summer population of *Chromatium*
14 Aug. 63	mg C m^{-2} day^{-1}	540	5960	1104	—	—	
Wadolek, Poland June 66	mg C m^{-2} day^{-1}	32.8	55.3	169	—	—	*Chlorobium* in metalimnion
July 66		144.5	19.4	13	—	—	
Sept. 66 (Czeczuga, 1967, 1968a, 1968b)		24.6	19.1	78	—	—	
Muliczne, Poland 19 Aug. 67	mg C m^{-2} day^{-1}	478	157	32.8	—	—	*Thiopedia* in hypolimnion
16 Sept. 67		258	136	52.7	—	—	
20 Oct. 67 (Czeczuga, 1968c)		281	28	10.0	—	—	
Belovod, USSR July (Sorokin, 1970)	mg C m^{-2} day^{-1}	500	55	11	15	2.7	
13 m	mg C m^{-3} day^{-1}	0	210	—	74	35	
14 m		0	79	—	44	56	
18 m		0	2.7	—	16.4	607	
24 m		0	0	—	5.5	—	

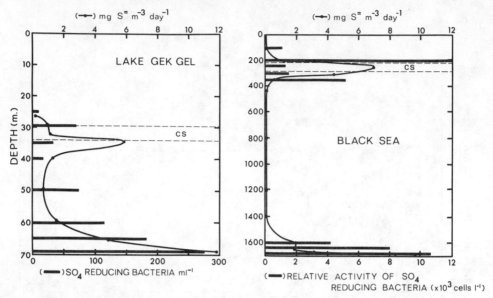

Figure 13–14   Rates of sulfate reduction (mg H₂S formed m⁻³ day⁻¹) and the biomass and activity of sulfate-reducing bacteria in meromictic Lake Gek Gel and Black Sea, USSR. CS = zone of active chemosynthetic bacterial metabolism. (After data of Sorokin, 1964, 1970.)

from the littoral zone brought in allochthonously from the drainage basin. Higher H₂S concentrations along these sediment interfaces are then dispersed by water currents, weak as they are, to the open water (Sorokin, 1970). Furthermore, high rates of sulfate reduction occur below the zone of most active bacterial chemosynthesis (Fig. 13–14), where again the concentrations of organic substrates presumably are higher just below the trophogenic zone.

A similar dichotomous distribution of sulfate-reducing bacteria below the zone of most active chemosynthesis is seen even more clearly from data of Lake Belovod (Fig. 13–15). The major bacteria in the zone of chemosynthesis were several species of *Thiobacillus,* overlain by a dense population of purple sulfur bacteria. Sorokin has provided evidence on the movements of zooplankton populations in response to changes in bacteria stratification, and studies of feeding which indicate that zooplankton, especially the cladoceran microcrustacea, actively feed on the dense populations of bacteria.

The Silica Cycle

Silica (SiO₂) generally occurs in moderate abundance in fresh waters and, although it is relatively unreactive chemically, it assumes major significance in the cycles of diatom algae. Diatoms assimilate large quantities of silicon in the synthesis of their frustrule cell structure. Silicon is a major component of algal production in many lakes, and diatom utilization greatly modifies the flux rates of silica in lakes and streams. Availability of silica can have a strong influence on succession and productivity.

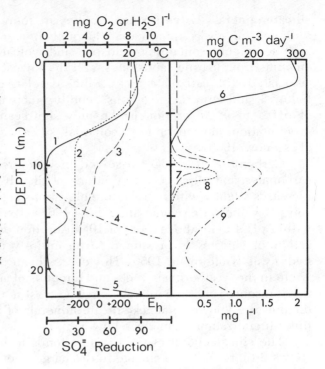

Figure 13–15 Distribution of midsummer characteristics and intensity of biological processes in the central depression of meromictic Lake Belovod, USSR. (After Sorokin, 1970.) *1*, oxygen, mg l^{-1}; *2*, °C; *3*, E_h in mv; *4*, H_2S, mg l^{-1}; *5*, rate of sulfate reduction, mg H_2S formed m^{-3} day^{-1}; *6*, photosynthesis of algae, mg C m^{-3} day^{-1}; *7*, chemosynthesis, mg C m^{-3} day^{-1}; *8*, photosynthesis of purple sulfur bacteria, mg C m^{-3} day^{-1}; *9*, biomass of bacteria, mg l^{-1}.

Forms and Sources of Silica

Silica occurs in fresh waters in the forms of: (1) *dissolved silicic acids,* which form stable solutions of H_4SiO_4 at much higher concentrations than are encountered in fresh waters (60 to 80 mg SiO_2 l^{-1} at 0°C to 100 to 140 mg SiO_2 l^{-1} at 25°C at commonly occurring pH values; Krauskopf, 1956). Unreactive silicon is not generally found; polymeric silicon is unstable and depolymerizes rather rapidly (within hours) (Burton, et al., 1970). Solution from various rock sources of silica is modified, however, by surface adsorption of silicic acid which reduces solubility and leads to a general situation in which nearly all natural waters are greatly undersaturated in silica content (Stöber, 1967; Tessenow, 1966). (2) *Particulate silica* is found in two forms, that in biotic material, in particular diatoms and a few other organisms that use large amounts of silica, and that adsorbed to inorganic particles or complexed organically. The formation of silicate complexes with iron and aluminum hydroxides decreases the solubility of silicates in sediments and these complexes increase in effectiveness with decreasing silicate solubility of interstitial waters at pH values above 7 (Ohle, 1964). Solubility is increased by humic compounds and the formation of iron and aluminum-silicate-humic complexes.

The silica content of drainage to natural waters is less variable than many of the other major inorganic constituents. The world average is about 13 mg SiO_2 l^{-1}, with relatively little variation among the continents, although the average of groundwater is somewhat higher than that of surface drainage (Davis, 1964). The major source of silica is from the degradation of aluminosilicate minerals. Largest concentrations are found in groundwater in contact with volcanic rocks, intermediate amounts in association with plutonic rocks and sediments containing feldspar and volcanic rock fragments, small amounts from marine sandstones, and the lowest amounts of silica in water associated with carbonate rocks. Silica is relatively immobile at pH values below 3, and forms aggregations; mobility increases somewhat in the range of pH 4 to 9. Adsorption is the only significant mechanism of inorganic precipitation; flocculation of colloidal silica could not be demonstrated (Tessenow, 1966).

Carbonic acid of CO_2 dissolved in water reacts with silicates to form carbonates and silica. Concentrations of dissolved silica in water of soils increases slightly with temperature elevations and decreases with increasing soil pH values (McKeague and Cline, 1963a, 1963b) as adsorption increases within pH 4 to 9. Above a pH of 10, adsorption decreases sharply. Mineralization of silicates is presumed to be entirely or largely a nonenzymatic hydrolysis (Golterman, 1960). However, silicate bacteria are known to play a role in the weathering of rocks and minerals of arid tropical regions without the presence of organic substrates (Savostin, 1972), and benthic-living diatoms are known to attack silicate minerals of sediments. Mechanisms of this mineralization are unclear, however.

The silica content of river waters tends to be remarkably uniform and shows little response to change in discharge rates (Edwards and Liss, 1973). This situation is in distinct contrast to other major constituents of river water that show a general inverse relationship between concentration and discharge rates. Although rapid diurnal changes in the silica content of river waters are known and can be associated with division rates and growth of diatoms (Miller-Haeckel, 1965), these and other biological factors are insufficient to explain the relative stability of content. An abiological buffering mechanism apparently is also operational by adsorption reactions between dissolved silica and solid phases in which silica is associated with hydrated oxides. Adsorption and desorption equilibria can respond to accommodate changes in concentration over a period of several days. The effects of dissolved silica and degraded silicates assume much more significance in the oceans, where reactions occur in which silica and alkali metal cations are fixed. Hydrogen ions are released and result in an effective buffering of pH and silica concentrations (Garrels, 1965; Mackenzie et al., 1965, 1967).

Distribution of Silica in Lakes

The concentrations of silica within lakes frequently exhibit marked variations in seasonal and spatial distribution. Even in oligotrophic waters, a conspicuous decrease in silica often is found in the epilimnetic strata during early winter, and in the spring during circulation and following thermal stratification (Fig. 13–16). In eutrophic lakes the trophogenic zone commonly is extremely reduced in silica content to near analytical undetectability. These reductions as well as metalimnetic reductions, resulting in a negative heterograde silica curve against depth (Figs. 13–16 and 13–17), clearly are

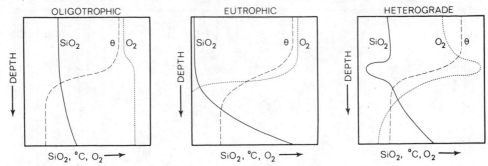

Figure 13–16 Generalized distribution of silica concentrations in unproductive and very productive lakes, and in a lake exhibiting a metalimnetic development of diatom algae and a negative heterograde silica curve.

associated with intensive assimilation of silica by diatom algae and sedimentation of the diatoms from the trophogenic zone more rapidly than silica is replaced by inputs to the system from surface and groundwater. Other sinks of silica are minor in comparison to assimilation by diatoms and subsequent sedimentation by diatoms. Abiogenic precipitation in the open water is relatively unimportant in the cycle of silica. Dilution by water low in silica can occur, as for example by rain percolating through decaying ice (March-April, Lawrence Lake, Fig. 13–17).

The seasonal cycles of silica, demonstrated in Figure 13–17, have been observed frequently. By far the most detailed work on experimental investigations of mechanisms controlling silica dynamics in lakes is that of Tessenow (1970), from which many of the following statements are drawn. Utilization of silica by diatoms occurs during photosynthesis and increases somewhat in darkness when living. Adsorption of SiO_2 to dead cells under some conditions can lead to reduction of silica content even below the trophogenic zone. Silica concentrations usually increase in the tropholytic zones of lakes during both summer and winter periods of stratification. The silica gradient becomes more steep in eutrophic lakes, exhibiting an anaerobic clinograde oxygen curve, and in most lakes in water immediately above the sediments.

The amorphous silica of diatoms is embodied in the sediments almost completely. This biochemical condensation of dissolved silica and sedimentation greatly exceeds inputs from abiogenic sources, and reaches the sediments with variable periodicity. Other diatom production is greatest in the spring and early winter, but often proceeds more or less continuously, particularly in oligotrophic lakes.

Interstitial solutions of sediments are enriched in dissolved silica at concentrations far in excess of those of water entering the lake (Tessenow, 1966; Harriss, 1967). Interstitial concentrations increase as the pH is reduced below 7, decreases between pH 7 to 9, and greatly increases above pH 9, and also increases at higher temperatures within the range of limnological interest. The dissolved silica of interstitial water is not equilibrated with amorphous silica, but rather with chemically bound or adsorbed silica. Liberation of silica to the overlying water in relatively isolated lake sediments is

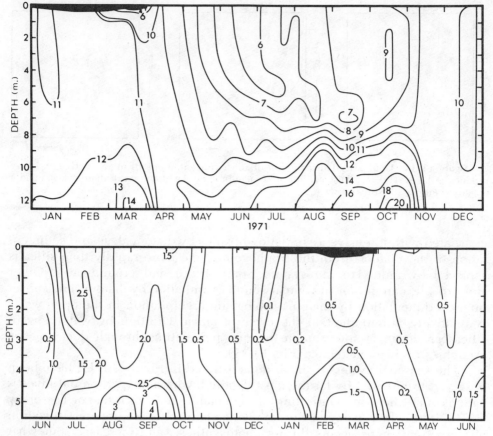

Figure 13–17 Depth-time diagram of isopleths of silica concentrations (mg SiO$_2$ l^{-1}) in oligotrophic hardwater Lawrence Lake, 1971 (*upper*), and hypereutrophic Wintergreen Lake, 1971–72 (*lower*), southwestern Michigan. Opaque areas = ice-cover to scale. (From Wetzel, unpublished data.)

governed by these equilibria, which are stable only when the solid phase for adsorbed silica is present. Exchange between sediment and water decreases the concentrations of interstitial water and results in greater redissolution from the sediments. The rate of release from the sediments is influenced by temperature and the difference in silica concentrations between the sediments and the overlying water. The tendency towards equilibrium is not attained, because of the slowness of diffusion (weeks). The difference between silica of interstitial water and of the overlying water is influenced by currents, movements produced by benthic organisms (e. g., larvae of chironormid insects; Tessenow, 1966), and by gas bubbles escaping from the sediments.

In lakes dominated by diatom algae, a common situation, large percentages of sedimenting diatom frustules can accumulate within the sediments and be lost permanently from the system. The extent of this permanent loss depends on the morphometry of the lake basin and what percentage of the sediments lie in the quiescent waters of the deep hypolimnion. The dissolution of suspended silica can be accelerated by consumption and fragmentation of diatom frustules by large zooplankton, a process that is apparently of greater significance in productive shallow ponds than in deep lakes. Silica sedimented from biogenic sources to shallower sediments is returned more rapidly to the overlying and circulating waters. Therefore, the metalimnetic and upper hypolimnetic waters that experience greater movement than the deeper waters commonly are enriched with silica in relation to strata above and below. A resulting positive heterograde silica curve often has been observed (numerous examples are given in Tessenow, 1966). The increased exchange induced by circulation over littoral sediments in shallow lakes can lead to silica concentrations that are sometimes greater than those of drainage inlet sources, and also result in silica losses from the system by outflow. The silica cycle and economy of most lakes are regulated largely by autochthonous metabolism within the lake, but also are balanced strongly by allochthonous inputs.

Diatom Utilization and Role of Silica

Silicified structures occur in many aquatic organisms, but none approaches the importance of the diatoms (Bacillariophyceae). All cells of diatoms are enclosed in a silica wall or frustule in which silicic acid dehydrated and polymerized to form silica particles (J. C. Lewin, 1962). The vegetative cells of some species of yellow-brown algae (Chrysophyceae) bear discrete siliceous scales and form cysts with silicified walls. These algae, as well as certain silicoflagellates, however, probably have only an insignificant impact on silica cycling in comparison to the active utilization and more extensive distribution of the diatoms. Similarly, utilization of silicon by certain aquatic macrophytes (*Equisetum*) and siliceous sponges, while of major importance to the organisms per se and their development and productivity, rarely is sufficient to alter the quantitative cycling of silica in lakes.

The succession and productivity of algal populations will be discussed in detail further on, but it is appropriate here to stress the importance of diatoms on the silica cycle and their effects on their own population dynamics. Of all of the aspects of chemical determination of succession and productivity, the relation between diatoms and silica concentrations is among the most apparent. The data, upheld by the detailed investigations of Lund (1949, 1950, 1954, 1955; Lund, et al., 1963; Heron, 1961), are irrefutable, and have been corroborated extensively by experimental work.

The development of phytoplankton algae in temperate waters usually undergoes a spring maximum. This population development may begin beneath the ice as light conditions improve, but is most conspicuous during and following spring circulation when the water is relatively rich in nutrients as the winter accumulations are mixed throughout the water column. One or several species usually dominate this exponential growth maximum for several weeks, and in a large number of lakes diatoms constitute the predominant algae of the spring maximum. Increasing light, and to a lesser extent a rise in temperatures, are major factors initiating the development of major diatom populations from small planktonic residual populations in the water. Circu-

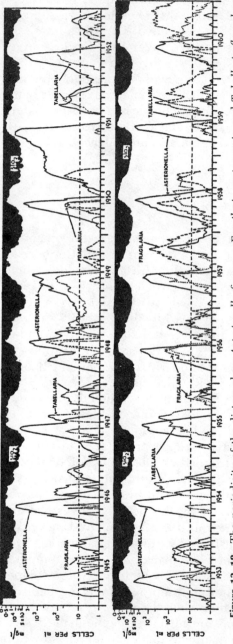

Figure 13–18 The periodicity of the diatom algae *Asterionella formosa*, *Fragilaria crotonensis*, and *Tabellaria flocculosa* in relation to fluctuations in the concentration of dissolved silica, 0–5 m in Lake Windermere, England, 1945–1960. (From Lund, J. W. G.: Verhandlungen Int. Ver. Limnol. *15:37*, 1964.)

latory turbulence is much higher in the spring than later in the season, and assists in retention of the relatively dense diatom cells in optimal light zones. Other factors influencing the development of succession and development of the algal populations will be treated later on (Chapter 14).

The diatom *Asterionella* commonly precedes other diatoms such as *Cyclotella, Fragelaria,* and *Tabellaria* in the spring maximum. Collectively, the spring maximum often declines abruptly as the silica concentrations are reduced rapidly by diatom utilization and sedimentation to levels below 0.5 mg l^{-1} (Fig. 13–18). The same pattern, which is illustrated for Lake Windermere, has been demonstrated continuously for over 20 years and observed in many other lakes. Factors such as light intensity, temperature, grazing by zooplankton, fungal.parasitism, and changes in other nutrients, especially nitrogen and phosphate, could not be demonstrated to be associated with the decline of the maximum. Experimental results on silica requirements showed that the silica reduction was clearly the major factor contributing to the decline of the diatoms.

To generalize from this simple situation would be misleading, since in most lakes, even within the English Lake District where this sequence occurs rather commonly, numerous other interacting factors and species

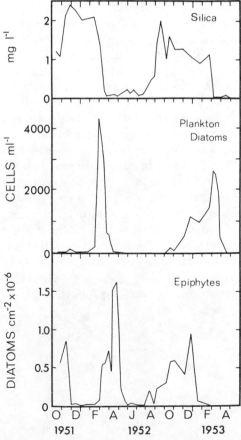

Figure 13–19 Variations in numbers of the dominant phytoplanktonic diatom (*Stephanodiscus hantzschii*), silica concentrations at the surface, and ephiphytic diatoms per cm^{-2} of *Phragmites* stems at a depth of 20 to 30 cm, Furesø, Denmark. (After data of Jørgensen, 1957).

requirements for effective competition are involved. Silica concentrations and their biogenic reduction from epilimnetic waters, however, are certainly major factors in many situations. In lakes where silica levels remain high, even though severely reduced during the summer productive period (e.g., Lawrence Lake, Fig. 13–17, upper), the spring diatom maximum persists longer into the summer and is overtaken gradually by a predominance of green algae. In very productive lakes, the maximum diatom peak can be found in the fall and early winter, as for example in Wintergreen Lake (Fig. 13–17, lower), during which time a competitive advantage is apparent until the silica levels are reduced to very low concentrations (< 100 μg l^{-1}). After spring overturn diatom growth is very short-lived in this lake, followed by a brief dominance of green algae until inorganic nitrogen sources are depleted. The nitrogen-fixing algae then quickly dominate and persist until the combined effects of increased inorganic nitrogen, reduced light and temperature, and renewed silica reoccur in late summer and early autumn.

The sequence of succession of different species of diatoms can be seen not only within a composite spring maximum, but also in the patterns of diatom periodicity as lakes become more productive. In lakes in which silica concentrations are moderate to low (e.g., < 5 mg l^{-1}), progressive long-term enrichment with phosphorus and nitrogen can lead to rapid biogenic reduction in silica levels whereby the diatoms cannot effectively compete and are replaced by nonsiliceous phytoplankton (Kilham, 1971). The current eutrophication of Lake Michigan is imposing such circumstances on the diatom populations, as a result of which they are being excluded gradually by green and blue-green algae during silica depletion in the summer period (Schelske and Stoermer, 1971).

A strong interaction can exist between the population level of littoral diatoms and the development of planktonic diatoms. For example, in Furesø the spring maximum of the planktonic diatoms, primarily *Stephanodiscus*, reduced the silica concentrations to < 40 μg l^{-1}, levels experimentally demonstrated to inhibit the growth of these algae (Jørgensen, 1957) (Fig. 13–19). An immediate increase in the diatoms epiphytic on the submersed portions of the emergent macrophyte *Phragmites* accompanied this decrease. The *Phragmites* stems were shown to possess large quantities of easily dissolved silica, and their silica content decreased during the development of the epiphytes on them during their maxima. Coupled with this diatom development in relation to silica availability, evidence was obtained that when plankton populations were high, growth-inhibiting organic substances were excreted by green algae and diatoms of the plankton that held back the development of the epiphytic diatoms (Fig. 13–19).

chapter 14

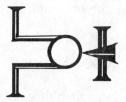

phytoplankton

The algae of the open water of lakes and large streams, the phytoplankton, consist of a diverse assemblage of nearly all major taxonomic groups. In spite of differing physiological requirements and variations in terms of limits of tolerance to physical and chemical environmental parameters, a number of species populations coexist in the phytoplankton. Knowledge of the basis underlying this quasi-balanced coexistence and of the factors that lead to seasonal succession of the species populations is fundamental to our understanding of both the phytoplanktonic[1] communities and their composite productivity.

Several major classes of environmental factors interact to regulate spatial and temporal growth of the phytoplankton. Within the basic physiological requirements of temperature and light, and concerning the means of remaining within the photic zone sufficiently long to complete growth and reproduction, a number of inorganic and organic nutrient factors play critical roles in succession of algal populations. The duration and characteristics of population development can be influenced further under certain conditions by herbivorous predation and parasitism.

The successional periodicity of phytoplanktonic biomass and productivity of unperturbed systems is quite constant from year to year. Seasonal variations are muted in tropical waters and increase greatly, to several thousand-fold, with increasing latitude and attendant large climatic changes each year. Although periodicity is observed in both phytoplanktonic biomass and productivity, the two are often quite out of phase as growth rates of smaller species with faster turnover times predominate.

As nutrient limitations of the phytoplankton of infertile waters are met increasingly by nutrient inputs to the systems, rates of algal production increase. The densities of phytoplankton populations progressively reduce the light available and the depth of the trophogenic zone. A point is rapidly reached at which self-shading inhibits further increases in productivity, regardless of nutrient loading and availability. Maximum photosynthetic efficiencies of light utilization are very low, usually less than one per cent of

---

[1]The word *planktonic*, while etymologically incorrect (see discussion by Rodhe, 1974, and Hutchinson, 1974), is so ingrained in aquatic ecology that change is not desirable.

radiation incident on the water, and considerably less than potential utilization by terrestrial systems.

COMPOSITION OF THE ALGAE OF PHYTOPLANKTONIC ASSOCIATIONS

The phytoplankton consist of the assemblage of small plants that have no or very limited powers of locomotion and are more or less subject to distribution by water movements. Certain planktonic algae have limited powers of locomotion; they move largely by means of flagella or various mechanisms which alter their distribution by changes in buoyancy. Most, however, are "freely floating," a term commonly used when referring to plankton in general, even though it is obviously somewhat of a misnomer in that most algae do not float. Most algae sink and sediment from the water, being slightly more dense than water. Phytoplankton largely are restricted to lentic ("standing") waters and larger rivers of reduced current velocity (cf. Wetzel, 1975b). Phytoplankton of streams of moderate flow, either having broken loose from the algal communities attached to substrata or having entered a stream from the outflow of lakes, usually are reduced rapidly because of the abrasive action of turbulence and of the substrata (see, for example, Chandler, 1937). Distinctly macroscopic algae with long filamentous forms usually inhabit parts of the littoral zone. Others, such as the primitive but morphologically complex stoneworts (Characeae), are large algae that are attached to the substratum. Both of these groups will be treated in the ensuing chapter on macrophytes and benthic algae. The present discussion is limited to the algal populations of the open-water, pelagic zone of lakes.

Morphological and Physiological Characteristics

Prior to summarizing the algal associations and population dynamics of phytoplanktonic communities of lakes, a few of the major morphological and physiological characteristics of the primary algal groups should be emphasized. The systematics of the large assemblage of different algae comprising the phytoplankton of lakes is a subject that has been treated well in a number of detailed works. Much of the systematics of algae is incomplete and is continually being revised and improved. The emphasis here is functional, from the standpoint of interactions of physiological characteristics with environmental variables, and their effects on primary productivity and resultant population successions within communities.

In this physiologically functional and ecological treatment, it is defeating to become engrossed in discussion of whether certain groups are truly algae in a strict sense. In many respects the blue-green algae (Myxophyceae) are physiologically more closely aligned to the bacteria. Many motile and colonial flagellates are classified as Protozoa on the basis of certain morphological and reproductive characteristics. The important characteristic, however, is that autotrophic photosynthesis is the primary mode of nutrition and synthesis of new organic matter of these microbiota. Therefore they are treated collectively with the true algae of the phytoplankton.

Pigments

A primary characteristic among the algal groups is the distribution of photosynthetic pigments—the chlorophylls, carotenoids, and biliproteins. Chlorophyll *a* is the primary photosynthetic pigment of all oxygen-evolving photosynthetic organisms and is present in all algae (Table 14–1) and photosynthetic organisms, other than the photosynthetic bacteria (discussed separately in Chapter 13). This pigment has 2 in vitro absorption bands, as discussed on page 65, in the red-light region at 660 to 665 nm and at lower wavelengths near 430 nm.

Chlorophyll *b*, although common to higher plants, is found only in the green algae and the euglenophytes (Table 14–1). This pigment functions as a light-gathering pigment in which absorbed light energy is transferred to chlorophyll *a* for primary photochemistry (reviewed in Govindjee and Brown, 1974). Maximum absorption bands are approximately 645 nm and 435 nm. Chlorophyll *c*, consisting of two spectrally distinct components, most likely functions as an accessory pigment to photosystem II. Maximum extraction absorption bands are at about 630 to 635 nm, and in the dominant blue portion of the spectrum at about 583 to 586 nm and 444 to 452 nm (Meeks, 1974). Chlorophyll *d*, of no known function, is a minor pigment component found only in certain red algae. A fifth chlorophyll (*e*) is suspected but as yet has not been verified as a specific compound.

Among the many carotenoids of algae, the carotenes are linear unsaturated hydrocarbons and the xanthophylls are oxygenated derivatives of carotenes (Goodwin, 1974). As is the case with chlorophyll *b*, light energy absorbed by carotenoids and biliproteins is transferred to chlorophyll *a*, leading to fluorescence and excitation of chlorophyll *a*. β-carotene is the most widely distributed of the carotenes, and is replaced by α-carotene only in certain green algae and in the Cryptophyceae. The biliproteins are a water-soluble, pigment-protein complex occurring primarily in the blue-green algae and to a lesser extent in certain cryptophytes and red algae.

Blue-Green Algae

Perhaps no other group of algae has received as much attention in the areas of physiology and ecology as the blue-green algae. This primitive group, the Cyanophyta or Myxophyceae, depending on the classification system used, is the only group of algae that, like the bacteria, is prokaryotic in cell structure. Prokaryotic cells are undifferentiated and exhibit a lack of mitochondria, chloroplasts, and internal membranes. A cell wall of mucopeptides is present. Reproduction is usually by binary fission, and nuclear division does not occur by mitosis, as in eukaryotic cells of other algae and higher organisms.

The blue-green algae are mostly filamentous in form but a number of important planktonic members are unicellular and usually occur in large colonies. All occur in mucilaginous sheaths either individually or in colonies, often of irregular and variable size and form. A majority of the planktonic

TABLE 14–1 Distribution of Photosynthetic Pigments Among the Algae[a]

Pigments	Cyanophyceae (Myxophyceae)	Chlorophyta	Xanthophyceae	Chrysophyceae	Bacillariophyceae	Cryptophyceae	Dinophyceae (Pyrrophyta)	Euglenophyceae	Phaeophyceae	Rhodophyceae
Chlorophylls:										
chlorophyll *a*	+	+	+	+	+	+	+	+	+	+
chlorophyll *b*	–	+	–	–	–	–	–	+	–	–
chlorophyll *c*	–	–	–	+	+	+	+	–	+	–
chlorophyll *d*	–	–	–	–	–	–	–	–	–	+
Carotenoids:										
Carotenes										
α-carotene	–	+	–	–	–	+	–	–	–	+
β-carotene	+	+	+	+	+	–	+	+	+	+
γ-carotene	–	+	–	–	–	–	–	–	–	–
ε-carotene	–	–	–	–	–	+	–	–	–	–
Xanthophylls										
lutein	+	+	–	–	–	–	–	–	–	+
violaxanthin	–	+	+	–	–	–	–	–	+	–
fucoxanthin	–	–	–	+	+	–	–	–	+	–
neoxanthin	–	+	–	–	–	–	–	+	–	–
astaxanthin	–	+	–	–	–	–	–	+	–	–
diatoxanthin	–	–	+	+	+	–	–	–	+	–
diadinoxanthin	–	–	+	+	+	–	+	+	+	–
peridinin	–	–	–	–	–	–	+	–	–	–
dinoxanthin	–	–	–	–	–	–	+	–	–	–
teraxanthin	–	–	–	–	–	–	–	–	–	–
antheraxanthin	–	–	–	–	–	–	–	–	–	+
myxoxanthin	+	–	–	–	–	–	–	–	–	–
myxoxanthophyll	+	–	–	–	–	–	–	–	–	–
oscilloxanthin	+	–	–	–	–	–	–	–	–	–
echinenone	+	–	–	–	–	–	–	–	–	–
Biliproteins:										
phycocyanin[b]	+	–	–	–	–	+	–	–	–	+
phycoerythrin[b]	+	–	–	–	–	+	–	–	–	+

[a] From Morris (1967), Meeks (1974), and Goodwin (1974).

[b] These chromoproteins consist of distinctly different types in the algal groups in which they occur. Differentiation is based on their absorption spectra (cf. Goodwin, 1974).

blue-green algae consists of members of the coccoid family Chroococcaceae (e.g., *Anacystis = Microcystis, Gomphosphaeria = Coelosphaerium,* and *Coccochloris*) and filamentous families Oscillatoriaceae, Nostocaceae, and Rivulariaceae (e.g., *Oscillatoria, Lyngbya, Aphanizomenon, Anabaena*).

The filament of filamentous blue-green algae consists of the mucilaginous sheath and a trichome of linearly arranged cells within the sheath. Although vegetative reproduction by fragmentation of trichomes is common, a number of cells are differentiated and specifically involved in reproduction and perennation. Spores or akinetes form in a number of species as enlarged, thick-walled cells that accumulate proteinaceous reserves in the form of cyanophycin granules (cf. Fogg, et al., 1973). Under favorable conditions, akinetes germinate directly into a trichome or into hormogonia. Hormogonia are short, slightly modified pieces of trichome that fragment from the parent trichome and move away by means of gliding motions to develop into a new filament. In a few species the hormogonia develop within the parent sheath as multiple trichomes. A few other spore structures develop in species that do not form hormogonia, in which development is direct without resting stages.

The unique differentiation of certain vegetative cells into heterocysts, common to all filamentous blue-green algae except the Oscillatoriaceae, has been discussed earlier in relation to nitrogen fixation (Chapter 11). These vegetative cells develop a thick envelope over the cell wall, except at the polar regions, in which the heterocyst is connected to adjacent vegetative cells by a pore channel. It is through this pore that the plasma membranes are adjoined and exchange of metabolic products occurs. In contrast to vegetative cells, heterocysts lack phycobilins, which are the primary light absorbers in blue-green algae, lack an oxygen-evolving photosystem, and show higher reducing activity. Carbon assimilated by vegetative cells in the light passes into the heterocysts in the dark and provides a source of reductant and energy for their metabolism in nitrogen fixation (Wolk, 1968, 1973). While nitrogen fixation is not limited solely to heterocysts, and occurs in some nonheterocystous blue-green algae, heterocysts certainly are the major sites of nitrogen fixation under aerobic conditions.

Green Algae

The Chlorophyta is an extremely large and morphologically diverse group of algae that is almost totally freshwater in distribution. A majority of the planktonic green algae belong to the orders Volvocales (e.g., *Chlamydomonas, Sphaerocystis, Eudorina, Volvox*) and Chlorococcales (e.g., *Scenedesmus, Ankistrodesmus, Selenastrum, Pediastrum*). Many members are flagellated (2 or 4, rarely more) at least in the gamete stages; in the desmids (Conjugales or Desmidiales), the gametes are not flagellated but rather amoeboid.

Asexual reproduction by vegetative division is common to most of the green algae, but is lacking in most of the Chlorococcales and Siphonales. Cell division very often is synchronized, with nuclear and cell division occurring at night. Cell division in colonial species results in enlargement of the colony; new colonies are formed only by fragmentation of the colony. Among filamen-

tous species, such as *Spirogyra,* fragmentation by weakened cells of the filament is common. Vegetative formation of zoospores, singly or in numbers
within a cell, also is common. When liberated from the parent cell, the flagellated zoospores are actively motile and positively phototactic for varying
periods of time before losing the flagellae and entering the resting spore stage.

Sexual reproduction in the green algae is diverse. In the simplest case of
isogamy, the flagellated male and female gametes are morphologically similar
in size and structure. In anisogamy, the flagellated female gamete of a fusing
pair is larger than the male gamete. Among more specialized algae, oogamy is
common, in which union occurs between a large, nonflagellated female
gamete and a small, flagellated male antherozoid. Some green algae are sexually monoecious (gametes are derived from the same cell), and others are
dioecious (male and female gametes are derived from different cells).

Whereas the planktonic Volvocales and Chlorococcales are rather ubiquitous in distribution among waters of differing salinity within the normal
limnological range, the distribution of desmids of the Conjugales largely is
correlated directly with low to very low concentrations of the divalent cations
calcium and magnesium. Although not totally restricted to waters of low
salinity, the desmids are most common and the species diversity is greatest
in soft waters draining granitic, igneous rock land forms, and especially in
waters with a high content of dissolved organic matter. Their abundance,
with an enormously large diversity of species, is often greatest in bog waters
that drain through deposits of the moss *Sphagnum.* Many desmids, however,
are distributed widely, but as a whole they are less cosmopolitan than most
unicellular algae (cf. Hutchinson, 1967).

Yellow-Green Algae

The Xanthophyceae are unicellular, colonial, and filamentous algae that
are characterized by the conspicuous, often excess, amount of carotenoids
over chlorophylls that result in their predominantly yellow-green coloration.
Nearly all of the motile cells possess two flagellae, one of which is longer than
the other (hence the older name Heterokontae). A cell wall is often absent;
when present, it contains large amounts of pectin and in many species is
silicified. Asexual reproduction is usually by fission, with the formation of
zoospores. Sexual reproduction is poorly understood in this group, but is
most often isogamous.

A majority of the xanthophycean algae are associated with substrata, and
many are epiphytic on larger aquatic plants. A few members are planktonic
and include common genera such as *Chlorobotrys, Gloeobotrys,* and *Gloeo-
chloris.*

Golden-Brown Algae

The chromatophores of the Chrysophyceae often yield a distinctive
golden-brown coloration because of the dominance of β-carotene and spe-

cific xanthophyll carotenoids, in addition to chlorophyll *a*. Most of the chryso-phycean algae are unicellular. A few are colonial; they are rarely filamentous. Flagellation is variable; most often cells are uniflagellate, but some possess 2 flagellae, usually of equal length. Many species lack a cell wall and are bounded only by a cytoplasmic membrane; others possess a cell surface covered by delicate siliceous or calcareous plates or scales. Vegetative repro-duction by longitudinal cell division is most common, especially among the motile unicellular species. Although it has rarely been observed, sexual reproduction is isogamous, forming a cyst-like zygote that undergoes reduc-tive division under certain conditions.

A number of chrysophycean algae form important components of the phytoplankton. The unicellular species with a single flagellum (e.g., *Chro-mulina, Chrysococcus, Mallomonas*) are usually very small algal constituents of the nannoplankton.[2] Larger colonial forms such as *Synura, Uroglena,* and particularly *Dinobryon* are distributed widely and are often major components of the phytoplankton under certain environmental conditions. The conspicu-ous, apparently obligate, nutritional requirements of low phosphorus concen-trations for certain species of *Dinobryon* and *Uroglena* has been mentioned earlier (page 236). In contrast, some species of *Dinobryon* and *Synura* have high phosphorus requirements.

Diatoms

By far the most important group of algae of the phytoplankton is the diatoms (Bacillariophyceae), even though a large majority of the species is sessile and associated with littoral substrata. Their primary characteristic, silicified cell walls, has been discussed in the preceding chapter. Both uni-cellular and colonial forms are common among the diatoms. The group com-monly is divided into the centric diatoms (Centrales), which have radial symmetry, and the pennate diatoms (Pennales), which exhibit essentially bilateral morphology. The cell wall or frustule of diatoms consists of 2 lid-like valves, one of which fits within the other; the overlapping area of the valves is connected by bands that constitute the girdle. The beautiful struc-

---

[2]The term nannoplankton was introduced long ago in reference to small plankton compo-nents (bioseston) of the total particulate material (seston) that pass through the older plankton nets of a mesh size of about 50 to 60 μm. The generally accepted size ranges, as commonly used (see, for example, Strickland, 1960), are:

Macroplankton	> 500 μm
Microplankton	ca 50 to 500 μm
(net plankton)	
Nannoplankton	10 to ca 50 μm
Ultraplankton	0.5 to 10 μm

Hutchinson (1967) discusses further separations within these divisions and gives a more complex terminology. Since so many phytoplankton algae are smaller than the common porosity of plank-ton nets (50 to 60 μm), nets are relatively useless for even qualitative studies. Plankton nets have been relegated to limnological museums for algal work, and are employed now only for macro-zooplankton studies.

tures of the siliceous cell walls of diatoms exhibit great variability and are used as taxonomic characteristics.

The arrangement of the valve structure of pennate diatoms exhibits various areas of cell thickenings and dilations. In some species a slit, termed a raphe, traverses all or part of the cell wall; in others, a depression in the cell wall, termed a pseudoraphe, is found in the axial areas of the cell wall. The 4 major groups of pennate diatoms are differentiated on the basis of these structures: (a) the Araphidineae, which possess a pseudoraphe (e.g., *Asterionella, Diatoma, Fragilaria, Synedra*); (b) the Raphidioidineae, in which a rudimentary raphe occurs at the cell ends (e.g., *Actinella, Eunotia*); (c) the Monoraphidineae, which have a raphe on one valve and a pseudoraphe on the other (e.g., *Achnanthes, Cocconeis*); and (d) the Biraphidineae, in which the raphe occurs on both valves (e.g., *Amphora, Cymbella, Gomphonema, Navicula, Nitzschia, Pinnularia, Surirella*). These divisions are of more than taxonomic interest in that, as will be pointed out subsequently, distinct nutritional requirements favor the growth of one group over another.

Vegetative reproduction by cell division, usually at night, is the most common mode of multiplication. Sexual reproduction occurs periodically when the size of the cells becomes reduced from asexual reproduction. Amoeboid gametes occur in isogamous reproduction in the Pennales, whereas in the Centrales reproduction is oogamous with flagellated spermatozoids. After fusion of the gametes, an auxospore develops and divides; half of the resulting cells give rise to vegetative cells that are of the largest dimensions of the species.

Cryptomonads

Of the cryptophycean algae, most are naked, unicellular, and motile. This class is very small and most of the planktonic members belong to the Cryptomonadineae (e.g., *Cryptomonas, Rhodomonas, Chroomonas*). All members are small and not a great deal is known about them. It is suspected that their significance in the metabolism of lakes is much greater than generally is believed, especially because dense populations of these algae often develop during cold periods of the year under relatively low light conditions.

The cryptomonads usually are flattened dorso-ventrally, with an anterior invagination that bears 2 generally equal flagella. One or usually two chromatophores include a variety of pigments (Table 14–1) that can yield a spectrum of apparent colors including olive green, blue, red, or brown. Reproduction is by longitudinal division; sexual reproduction is unknown in the group and apparently is lacking.

Dinoflagellates

The dinoflagellates (Dinophyceae of the Pyrrophyta) are unicellular flagellated algae, many of which are motile. Although a few species are naked or without a cell wall (e.g., *Gymnodinium* of the Gymnodiniales), most develop a conspicuous cell wall that usually is elaborated in sculptured patterns with the formation of large spines and cell wall processes (the Peridiniales, e.g., *Ceratium, Glenodinium, Peridinium*). In both types the cell surface has transverse and longitudinal furrows that connect and contain the flagella, whose movements create water currents sufficient for weak locomotion and disruption of chemical gradients at the cell surface. Although sexual reproduction occurs, asexual reproduction by the formation of aplanospores, in which a motile phase is omitted, is apparently predominant. These asexual resting stages or cysts undergo considerable periods of diapause, especially in the large *Ceratium*, in the autumn decline of summer populations of the temperate region. The distribution of dinoflagellates in relation to major chemical characteristics shows that, while some species are widely tolerant and ubiquitous, especially among species of *Ceratium* and *Peridinium*, most dinoflagellate species are quite specific in their range to calcium, pH, dissolved organic matter, and temperature.

Seasonal polymorphism or cyclomorphosis is a very conspicuous phenomenon among the zooplankton and is discussed in detail in Chapter 16. It is rare among the phytoplankton, however, and the most marked cyclomorphic development is found in the dinoflagellate *Ceratium* (Hutchinson, 1967). In *Ceratium*, the conspicuous separation of the cell wall by the transverse furrow into the upper part (epitheca) and lower part (hypotheca) is accentuated further by the development of extensions or horns. The epitheca always has 1 horn; the hypotheca usually has 1 to 3 horns. As *Ceratium* germinates from encystment in the spring, a number of seasonal changes in morphology can be observed. As the water warms, the cells of succeeding populations exhibit a decrease in the length of the horns, in the width and general size of the cell, and a lessening of the angle of divergence of the hypothecal horns (Fig. 14–1). In midsummer, with warmer temperatures, the development of a short, 4th horn is common. Field and experimental evidence indicate that these changes are induced primarily by increasing temperatures. Some experimental evi-

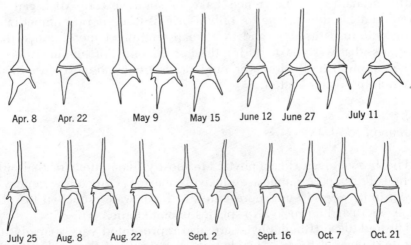

Figure 14–1 Seasonal polymorphism in *Ceratium hirundinella* in Kirchbergteich, Darmstadt, Germany. (From Hutchinson, G. E.: A Treatise on Limnology. Vol. 2, New York, John Wiley & Sons, Inc., 1967, after work of List.)

dence points to the fact that the sinking rate is reduced appreciably in less viscous warmer water by the reduced cell dimensions, increased divergence of the horns, and the development of the 4th horn. Presumably, these changes are of adaptive significance to *Ceratium* in that they reduce the rate of sinking out of the photic zone as the viscous resistance of the water decreases at summer temperatures.

Euglenoids

Although the euglenoid algae (Euglenophyceae) are a relatively large and diverse group, few are truly planktonic. Nonetheless, when conditions are favorable, the euglenoids can develop in great profusion. Almost all euglenoids are unicellular, lack a distinct cell wall, and possess 1, 2, or 3 flagella that arise from an invagination in the cell membrane. Reproduction occurs by longitudinal division of the motile cell; sexual reproduction has not yet been substantiated.

Although some euglenoids are unpigmented and are phagotrophic, or able to ingest solid particles, and are best treated as Protozoa, most are photosynthetic and facultatively heterotrophic. Even when phagotrophy constitutes the dominant mode of carbon assimilation among certain nonpigmented members of the Dinophyceae, Chrysophyceae, and especially the Euglenophyceae, none of these algae depend solely on phagotrophy for their major carbon requirements (Droop, 1974). Nutrition is supplemented by the uptake of dissolved organic compounds. Ammonia and dissolved organic nitrogen compounds are the dominant sources of nitrogen among most euglenoid algae. Their development in the phytoplankton is found most often in seasons, strata, or lake systems in which ammonia and especially dissolved organic matter concentrations are high. For example, detailed seasonal analyses of the phytoplanktonic populations at weekly intervals at each meter of depth for 16 months in hardwater Lawrence Lake of southwestern Michigan showed euglenoid algae developing for only a brief 2-week period immediately following autumnal turnover and in a small stratum, 1 meter above the deep sediments, that was anoxic during the last 2 months of summer stratification (Wetzel, unpublished). However, euglenoids are found most often in shallow water rich in organic matter.

Brown and Red Algae

The brown algae (Phaeophyta) are mostly filamentous or thalloid algae which, as a group, are almost exclusively marine. Only a few genera of this large, primitive group are represented in fresh water attached to substrata, such as encrusted on rocks. No species is planktonic.

The red algae (Rhodophyta) are also represented very sparsely in fresh water, and none of these is planktonic. A majority of the freshwater genera are found only in fresh water. The thalloid or filamentous species are nearly all restricted to fast-flowing streams of well-oxygenated, cool waters.

PHYTOPLANKTONIC COMMUNITIES

The outstanding feature of phytoplanktonic communities is a number of simultaneous species populations co-occurring in the lacustrine habitat. Each species has a *niche* based on physiological requirements in relation to the variations in all factors within the habitat. The combination of variations of these factors required by a species can be expressed theoretically in multi-dimensional space or hyperspace corresponding to the values of variables that permit a species to exist in the habitat (summarized in Hutchinson, 1967). Under the well-known Gausian principle of competitive exclusion, in a relatively uniform environment, such as the pelagial zone of lakes, in which a number of species are competing for the same resources with only slight differences, one would expect a tendency towards a unispecific equilibrium. Indeed, one species is often found to be in greater abundance than others, or there may be 2 or more dominant species of the phytoplanktonic assemblage. However, a number of rarer species always exist among the dominant or subdominant algae. This paradox of the plankton, as discussed by Hutchinson (1961), results in apparent conditions of multispecific equilibrium among what seem to be physically uniform conditions of the turbulent open water of lakes.

A number of explanations have been advanced for the coexistence of these conditions of multispecific equilibrium, in which species diversity is much higher than would be anticipated from theory and mathematical derivations. First, it must be assumed that in order for the relatively slow process of competitive exclusion to occur, the physical conditions must be uniform for a sufficient period of time. If conditions change sufficiently rapidly, the advantages gained by one species that is a better competitor may not exist long enough to result in the competitive exclusion of other species. Differences in the efficiencies of resource utilization among species may be too small for competitive exclusion to occur before conditions change. Expressed in another way, the niche hypervolumes of species often exhibit a large degree of overlap.

Competition can be negated partially if commensalism and symbiotic relations exist among species. As discussed further on, a number of organic compounds released extracellularly by one alga can influence the metabolism of another species, although most of these compounds are inhibitory and antibiotic in nature. Also, a majority of those algae requiring exogenous sources of vitamin micronutrients are motile and small (Chrysophyceae, Cryptophyceae, Euglenophyceae, and Dinophyceae). The differences in organic micronutrient requirements and the release of the excess by species capable of their synthesis would serve as a means of encouraging mixed equilibrium populations.

Predation on one species of algae to a greater extent than another would encourage coexistence, even if the species being preyed upon had a much greater competitive advantage under the conditions of its niche. Selective grazing, largely on the basis of size, by zooplankton is a very real phenomenon and is discussed in some detail in Chapter 16.

Some species of algae are truly planktonic in that, as far as is known, their population abundance oscillates temporally, dominating for a period and then becoming extremely rare but nonetheless still remaining planktonic (Hutchinson, 1964). Alternatively, some species enter resting stages and thereby leave the competitive arena for a period of time. Some of these species quiesce in the littoral zone sediments and later develop sufficiently to form a significant component of the phytoplankton. Although evidence for benthic algal populations of the littoral zone serving as a "seed" inoculum for some phytoplankton species is sparse (cf. Chapter 15), when this occurs diversity is increased in the phytoplankton. This mode of opportunistic expansion into the pelagial region certainly is more common in more shallow waters with extensive littoral development.

None of these hypotheses is mutually exclusive, and all probably operate simultaneously to a greater or lesser degree. Which one dominates at a particular time in a given water surely varies. There is some evidence that the distribution of phytoplankton is not necessarily as uniform as the concepts of competitive exclusion and multispecies equilibrium would indicate. For example, the phytoplankton of an alpine lake showed a high degree of patchiness for many species, which indicated that the rate of mixing was sufficiently slow relative to the reproductive rate of the algae that niches of many different species could exist simultaneously (Richerson, et al., 1970). Under these hypothetical conditions of "contemporaneous disequilibrium," at any given time many patches of water exist in which one species has a competitive advantage relative to the others, but the masses are obliterated frequently enough to prevent exclusive occupation by a single species.

Types of Algal Associations

A great deal of descriptive work has been devoted to the associations of algal species among differing fresh waters. The diversity of algae among the thousands of species is great, and many exhibit a very wide tolerance to environmental conditions found under natural limnological situations. Nonetheless, certain characteristic phytoplanktonic algal associations occur repeatedly among lakes of increasing nutrient enrichment and cycling from oligotrophy to eutrophy. Some of the commonly observed major associations are set out in Table 14–2, based on the detailed discussion of Hutchinson (1967). Such categorizations are not very satisfactory because of the wide spectrum of intergradations often observed and the shifts that occur seasonally from one type to another, especially among more productive waters. Even though such characterizations yield little insight into regulating environmental factors, they are useful from the standpoint of general correlations between qualitative and quantitative abundance and available nutrient supply.

A number of phytoplankton indices have been developed, particularly by Thunmark (1945) and by Nygaard (1955), in admirable attempts to quantify somewhat more precisely the relationships of rarely occurring and dominant algal species as indicators of lake productivity. Their use is revived periodically with limited and varying degrees of usefulness. The indices developed

TABLE 14–2   Characteristics of Common Major Algal Associations of the Phytoplankton in Relation to Increasing Lake Fertility

General Lake Trophy	Water Characteristics	Dominant Algae	Other Commonly Occurring Algae
Oligotrophic	Slightly acidic; very low salinity	Desmids *Staurodesmus, Staurastrum*	*Sphaerocystis, Gloeocystis, Rhizosolenia, Tabellaria*
Oligotrophic	Neutral to slightly alkaline; nutrient-poor lakes	Diatoms, especially *Cyclotella* and *Tabellaria*	Some *Asterionella* spp., some *Melosira* spp., *Dinobryon*
Oligotrophic	Neutral to slightly alkaline; nutrient-poor lakes or more productive lakes at seasons of nutrient reduction	Chrysophycean algae, especially *Dinobryon,* some *Mallomonas*	Other chrysophyceans, e.g., *Synura, Uroglena;* diatom *Tabellaria*
Oligotrophic	Neutral to slightly alkaline; nutrient-poor lakes	Chlorococcal *Oocystis* or chrysophycean *Botryoccocus*	Oligotrophic diatoms
Oligotrophic	Neutral to slightly alkaline; generally nutrient poor; common in shallow Arctic lakes	Dinoflagellates, especially some *Peridinium* and *Ceratium* spp.	Small chrysophytes, cryptophytes, and diatoms
Mesotrophic or Eutrophic	Neutral to slightly alkaline; annual dominants or in eutrophic lakes at certain seasons	Dinoflagellates, some *Peridinium* and *Ceratium* spp.	*Glenodinium* and many other algae
Eutrophic	Usually alkaline lakes with nutrient enrichment	Diatoms much of year, especially *Asterionella* spp., *Fragilaria crotonensis, Synedra, Stephanodiscus,* and *Melosira granulata.*	Many other algae, especially greens and blue-greens during warmer periods of year; desmids if dissolved organic matter is fairly high
Eutrophic	Usually alkaline; nutrient enriched; common in warmer periods of temperate lakes or perennially in enriched tropical lakes	Blue-green algae, especially *Anacystis* (=*Microcystis*), *Aphanizomenon, Anabaena*	Other blue-green algae; euglenophytes if organically enriched or polluted

by Nygaard (1949) are based on the ratios or quotients of number of species within general groups, such as those of blue-green algae to desmids, chlorococcalean algae to desmids, centric to pennate diatoms, and euglenophytes to blue-green and green algae. His compound index was the ratio of all species of blue-green, chlorococcalean green, centric diatoms, and euglenoid algae to the species of desmid algae. A general relationship was found between low compound index values of less than one and oligotrophic waters. Lowest values occurred in desmid-rich dystrophic lakes enriched by dissolved organic and humic compounds. The values of the compound index increased in more productive waters and approached 50 in highly nutrient enriched eutrophic lakes.

Comparisons of these indices with more modern measurements of algal productivity show weak positive correlations, but exceptions are many. Great variations are found among different regions of the world and among lake districts. It is apparent that these indices, while having some value in determining species relationships, are much too superficial in physiological foundation to be of significant use in evaluations of productivity among lakes and of causal mechanisms underlying the composite growth of algae.

Pleustonic Organisms

A number of specialized organisms, the *pleuston,* are adapted to the interface habitat between air and water. Although there are a number of macroscopic organisms such as many of the duckweeds (Lemnaceae) and adult insects (e.g., gerrid hemipterans) that are morphologically adapted to the interface, the bacteria and algae have several specialized representatives including diatoms, chrysophyceans, and xanthophyceans. The microscopic components of the pleuston are collectively termed *neuston,* and are separated into those organisms adapted to living on the upper surface of the interface film (the *epineuston*) and those living on the underside of the surface film (*hyponeuston*). A majority of the neustonic organisms live on the upper surface of the interface. A few examples of algal and protozoan neuston are illustrated in Figure 14–2.

Although the development of neuston is most extensive in sheltered, quiescent waters, it is by no means restricted to quiet waters, as attested to by their development in marine systems (cf. treatise by Zaitsev, 1970). Conspicu-

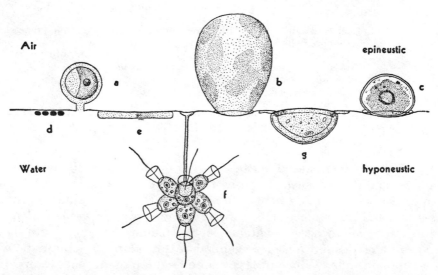

Figure 14–2 Exemplary organisms of the neuston at the surface film. Epineustonic: *a, Chromatophyton* (Chrysophyceae); *b, Botrydiopsis* (Xanthophyceae); *c, Nautococcus* (Chlorococcaceae). Hyponeustonic: *d, Lampropedia* (Coccaceae); *e, Navicula* (Bacillariophyceae); *f, Codonosiga* (Craspedomonadaceae); *g, Arcella* (Rhizopoda). (From Ruttner, F.: Fundamentals of Limnology. Translated by Frey, D. G., and Fry, F. E. J., 3rd ed., Toronto, University of Toronto Press, 1963.)

ous development of neuston, particularly of the epineuston, is common in small, quiet waters in the littoral or along the margins and backwater areas of streams. Often the density is so great that as light is reflected from the chromatophores, the water appears to be covered with a dry film of varying coloration. A few cladoceran zooplankters and insect larvae are adapted to feed on the neuston from the underside of the interface film.

Very little is known of the physiology and metabolism of the neuston of fresh waters in comparison to that of marine waters (cf. Zaitsev, 1970; Hardy, 1973). Although the neustonic productivity and contributions to the total metabolism of fresh waters probably are small in a majority of situations, their contributions to shallow lake and marsh systems deserve much further study.

Growth Characteristics of Phytoplankton

Analyses of the seasonal and spatial growth characteristics of phytoplanktonic algae must be viewed within a dynamic array of interacting environmental parameters and in relation to the physiological characteristics of the organisms. Scrutiny of the interacting factors regulating growth and succession is needed, particularly of (a) light and temperature, (b) means of remaining within the photic zone by alterations of sinking rates, (c) inorganic nutrient factors, (d) organic micronutrient factors and interactions of organic compounds with inorganic nutrient availability, and (e) biological factors of competition for available required resources and predation by other microorganisms. Each species of algae of the phytoplanktonic associations possesses a range of tolerance to extremes of these factors, and growth proceeds best at a given optimal combination of the interacting factors. The optimal combination of factors required for greatest growth and productivity may not be, and probably seldom is achieved under most natural conditions. Optimality and competitive advantage of one species over another are relative within the highly dynamic phasing of physical and biotic conditions affecting growth over time.

Light and Temperature

The ecological effects of light and temperature on the photosynthesis and growth of algae are inseparable because of the interrelationships in metabolism and light saturation. Qualitative factors of the selective attenuation of light energy with increasing depth have been discussed earlier (Chapter 5) in relation to photoreceptors and environmental parameters such as humic dissolved organic matter. Growth of algae and photosynthetic rates are directly related to quantitative light intensity. Response to light intensity, however, is variable among species and in many a considerable degree of adaptation to changing light intensities occurs.

One common change in rates of photosynthesis of planktonic algae is an increase in the rate at which light saturation occurs at higher temperatures (Fig. 14–3A). In this type of algae, exemplified by *Chlorella*, adaptation to

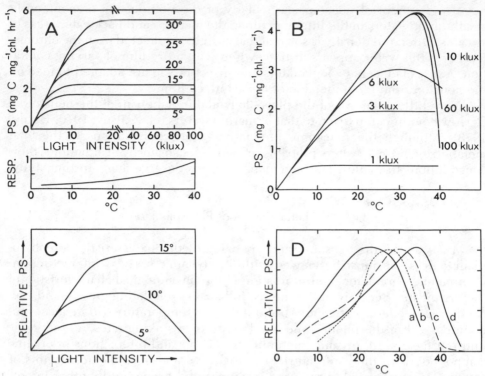

Figure 14–3 Interactions between photosynthesis (*PS*) and light intensity and temperature in plankton algae. *A*, Generalized photosynthetic rates (mg C fixed per mg chlorophyll per hour) in relation to increasing light intensities and rates of respiration at different temperatures; *B*, Commonly observed increasing inhibitory effects of high light intensities that are accentuated at higher temperatures; *C*, Effects of high light intensity on photosynthetic rates of algae adapted to cold temperatures, as in arctic lakes; and *D*, Optimum temperatures for photosynthesis of different algae: *a, Synedra, b, Anabaena, c, Chlorella*, and *d, Scenedesmus*. (After data of Aruga, 1965.)

higher or lower light intensity occurs mainly by changing the amount of pigment per cell (Steemann Nielsen and Jørgensen, 1968a; Jørgensen, 1969). Cells adapted to high light intensity have a lower chlorophyll *a* content per cell than those adapted to low light intensities. In *Chlorella pyrenoidosa*, for example, cells grown at 1 klux[3] have about 10 times more chlorophyll per cell than those grown at 21 klux. The actual rate of photosynthesis is not much greater at high light intensities than at low intensities within the range below light inhibition at very high intensities (Fig. 14–3B).

In other algae, such as in many diatoms, adaptation to changing light occurs only by changing the rate of light saturation of photosynthesis. The chlorophyll content is the same in cells grown at low and high light intensities. The light-saturated rate of photosynthesis is generally much higher at higher light intensities than at lower illuminance, and the actual rates of photosynthetic carbon fixation are higher (Fig. 14–3A and 14–3B). Some algae are intermediate between the extreme types discussed here.

---

[3]See Appendix for conversions among units of illuminance.

We therefore observe considerable variation in light saturation curves, as exemplified in Figure 14–3, in which the sloping portion is the result of the physiological photochemical response of pigments to increasing light. At light saturation, in the horizontal portions of the family of curves, rates of enzymatic processes are operational which are dependent on the concentrations of active enzymes and temperature. Some algae, such as *Skeletonema*, adapt to lower temperatures by increasing enzyme concentrations so that the same rate of photosynthesis is achieved per given light intensity at high and low temperatures (see, for example, Jørgensen, 1968). In *Skeletonema* adapted to low temperatures, the amount of protein per cell was twice as great at 7°C as at 20°C, and the size of cells was greater at low temperatures. While the rates of respiration are affected very little by changes in light intensity, they do increase with increasing temperatures (Fig. 14–3A). Photosynthetic rates of planktonic algae adapted to light and temperature conditions of arctic regions can be nearly as high as in temperate regions during summer conditions of relatively high light (see, for example, Steemann Nielsen and Jørgensen, 1968b).

Light of high intensity is distinctly detrimental to many algae (Fig. 14–3B, C). Ultraviolet light is particularly harmful to planktonic algae (Gessner and Diehl, 1952; Godward, 1962; van Baalen, 1968), but the effects are partly reversible if exposure is not too long, e.g., if plankton exposed to more intensive UV radiation at the immediate surface waters are circulated to deeper waters in which UV is attenuated rapidly (cf. Chapter 5). Effects of high light intensities are highly variable among species, and some adaptation occurs with time. On bright days it is very common to observe a distinct depression of rates of in situ photosynthesis near the surface. The decrease in photosynthetic rates is associated with photo-oxidative destruction of enzymes, and does not involve any destruction of the chlorophyll (Steemann Nielsen, 1962; Steemann Nielsen and Jørgensen, 1962). The depression of enzymatic processes at high light intensities takes some time to be effective. If exposure is not too long (several hours), reactivation can occur and permanent damage is avoided. As photosynthesis is depressed, rates of chlorophyll synthesis are also depressed and this process probably is partly involved in the depression of pigment concentrations often observed at midday on bright days. Some algae, such as the diatom *Cyclotella*, adapt rapidly (<24 hrs) to changing light intensities and are not inhibited at very high intensities (Jørgensen, 1964). Other algae, such as natural populations of phytoplankton of Antarctic lakes exposed to continuous light in summer or of Arctic lakes after loss of ice-cover in summer, are inhibited severely by high light intensities (Goldman, et al., 1963; Hobbie, 1964). In Arctic ponds the phytoplankton exhibit increasing susceptibility to photoinhibition at colder temperatures (Fig. 14–3C and Hobbie, personal communication). As the water temperatures increase during the brief 100-day period when the ponds are not frozen solid, adaptation to higher illuminance occurs.

Therefore, a vertical depth distribution of photosynthesis in which a zone of maximum photosynthetic rates at light saturation is underlain by a zone of near-exponential decline of rates with increasing depth frequently can be observed (Fig. 14–4A). When light intensities exceed the photoinhibition

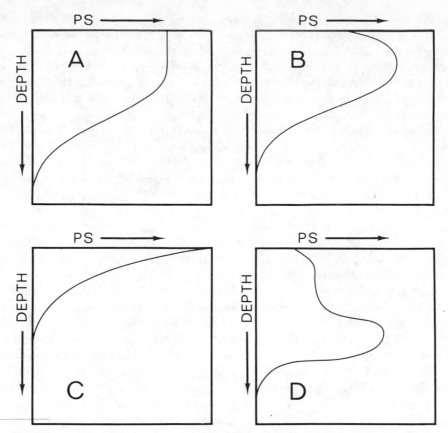

Figure 14–4   Generalized variations in vertical profiles of rates of photosynthesis (*PS*) among phytoplankton. *A*, Light saturation in surface waters, without inhibition, underlain with decreasing rates as light is reduced with depth; *B*, Profile when severe surface photoinhibition occurs; *C*, Photosynthetic rates when biogenic turbidity is very high; *D*, Photosynthetic maximum in the metalimnion.

threshold of the dominant producer algae of the composite phytoplankton, a surface photoinhibition very often occurs (Fig. 14–4*B*). The depth at which maximum rates of photosynthesis occur varies with the transparency conditions of the water, which are governed by the concentration of dissolved and particulate organic matter and abiotic turbidity. When the densities of phytoplankton, i.e., the biogenic turbidity, increase greatly, their self-shading effects can greatly reduce the depth of light penetration and the trophogenic zone (Fig. 14–4*C*). Although surface photoinhibitory effects occur, these are masked by rapid light attenuation by the self-shading of the algal populations. In some cases the algal populations adapted to relatively low light intensities of deeper layers develop greatly and have higher photosynthetic rates than those of the epilimnion (Fig. 14–4*D*; cf. also Fig. 8–4*A*).

There is great diversity in tolerances to variations in temperature among the algae (Fig. 14–3*D*). Aside from the adaptation of photosynthetic rates to temperatures discussed earlier, above maintenance levels the minimal temperatures for photosynthesis vary. For many diatoms, the critical temperature is about 5°C; in others, about 15°C (e.g., Rodhe, 1948). In others, especially many green and blue-green algae, the incipient temperature is higher for significant photosynthesis. The blue-green algae as a group are generally much more tolerant of higher temperatures than other algae. Many species of blue-green algae have higher temperature optima than eukaryotic algae of the same waters. Further, thermophilic algae with a maximum optimal growth at temperatures above 45°C are almost exclusively blue-greens, and photosynthesis is known to occur at constant temperatures as high as at least 74°C (Castenholz, 1969). Respiration and especially photorespiration increase at high temperatures (Döhler and Przybylla, 1973). In the ensuing discussion of seasonal succession of algal populations, we will see again the frequently observed preponderance of diatoms at colder temperatures and the increasing diversity of algae as the waters warm.

Flotation Mechanisms and Water Turbulence

The importance of water movements in the transport of particulate organic matter, in this case planktonic algae, has been emphasized at length earlier (Chapter 7). Not only are the water movements important in the physical movement of algae into or out of the photic zone, but they are critical in the vertical transport of mineralized matter from lower depths and littoral regions to the open water. The turbulence and movement are obviously critical in the basic regulation of algal periodicity and production.

The importance of basin morphometry to water movements again must be emphasized in relation to composition of planktonic algae aside from nutrient considerations. An excellent example is seen in the analyses of the spatial distribution of predominantly littoral desmid algae in several lakes of Minnesota (Bland and Brook, 1974). Many desmids occur throughout the year on and among submersed macrophytes of the littoral zone and overwinter there. During spring circulation, many desmid algae are dispersed, particularly to other areas of the littoral on newly growing annual macrophytes. In small lakes, circulation in the littoral is not as intense as in larger lakes. Larger desmids are carried out of the littoral less frequently in these smaller lakes. The more active circulation of larger lakes, however, transports a greater variety and larger desmids into the pelagic regions from the littoral zone, and this input increases the diversity of the planktonic populations.

The density of most freshwater planktonic organisms is 1.01 to 1.03 times that of the water. This slightly greater density causes them to sink when in undisturbed water. Spherical particles of a mean diameter of not more than 0.5 mm fall according to Stoke's law: the velocity of sinking varies inversely with the viscosity of the medium (cf. Chapter 2), directly with the excess density of the particle over that of the medium, and directly as the square of the diameter or some appropriate linear dimension (reviewed in Hutchinson, 1967). It is apparent that in fresh water in which light is attenuated approximately exponentially with increasing depth, sinking out of the photic zone is a distinct disadvantage for photosynthetic algae. Movement of a cell through water, either on its own by motility or by sinking, has the advantage of disrupting nutrient gradients around the cell and increasing the chances, because diffusion is so slow, of contact with nutrient molecules. Thus the disadvantage of sinking out of the light zone, which is usually fatal, is offset by movement and transport by water turbulence.

Various characteristics are found among planktonic algae that improve flotation or reduce sinking rates (Lund, 1959, 1965; Fogg, 1965; Hutchinson, 1967). (a) Buoyancy is improved by a reduction in sinking rate by increasing *surface-to-volume ratios* (Table 14–3). The "form resistance" is the decrease in sinking rate caused by the shape of the alga relative to that of a sphere of equal density and volume. While the rough surface texture of the cell wall or membrane has relatively little effect on sinking, the elongation into cylindrical or discoid shapes distinctly decreases the settling rates. Projections or more elaborate protrusions such as spines decrease the settling rate by frictional resistance, and shift the orientation of the cell upon descent. (b) Production of *mucilage* reduces the sinking rate. Such gelatinous sheaths occur in nearly all blue-green phytoplankton, some diatoms and green algae, and most desmids with rather elaborate projections and irregular cell morphology. The mucilage generally has a density less than that of the cell and closer or equal to that of the water, while increasing size and frictional resistance somewhat. The sheath, however, reduces the efficiency of nutrient uptake by creating an additional diffusional barrier. (c) *Gas vacuoles.* These vacuoles are gas-filled structures of a more or less permanent nature found in the protoplasm of living cells only among prokaryotic bacteria and blue-green algae (cf. reviews of Walsby, 1972, and Fogg, et al., 1974). Gas-vacuole membranes are freely permeable to gases such as nitrogen, oxygen, and carbon dioxide, and the gas

TABLE 14–3   Mean Sinking Rates of Several Marine Diatoms in Relation to Area:Volume Ratios of Cells[a]

Cell Type	Area:Volume ($\mu m^2 : \mu m^3$)	Cell Size (μm diameter or width × length)	Sinking Rate (m day^{-1})
Centric, unicellular			
Thalassiosira nana	0.88–1.20	4.3–5.2	0.10–0.28
Centric, elongate			
Rhizosolenia setigera			
Normal pre-auxospore	–	6 × 245	0.19–0.44
Spineless pre-auxospore	0.62–0.75	–	0.22–0.63
Post-auxospore	0.10–0.16	33 × 363	0.79–1.94
Total population	–	–	0.63–0.95
Centric, chain-forming			
Thalassiosira rotula	0.23–0.29	19–34	0.39–2.10
Skeletonema costatum	0.81–1.01	5 × 20	0.31–1.35
Bacteriastrum hyalinum	0.29–0.33	10 × 30	0.39–1.27
Chaetoceros lauderi	0.19–0.41	20 × 34	0.46–1.54
Pennate, chain-forming			
Nitzschia seriata	1.18–1.65	4 × 40	0.26–0.50

[a]After data of Smayda and Boleyn, 1965, 1966a, 1966b.

is of the same content and at partial pressures similar to those in the surrounding water. Water is excluded from the vacuoles by the surface tension of the membrane.

The gas vacuoles decrease the density of blue-green algae to below that of water, even though the volume of the vacuoles often occupies less than one per cent of the total cell volume. Blue-green algae possessing gas vacuoles float towards the surface and become independent of turbulence required by other algae to remain in the photic zone. Changes in the gas vacuole:cell volume ratio determine whether the alga sinks or floats. The ratio changes by growth when the vacuoles become proportionately less of the total volume in exponential growth phases. Gas vacuoles become more abundant when light is reduced and growth rate slows. Rises in the turgor pressure of the cells as a result of the accumulation of photosynthate cause a decrease in existing gas vesicles and a reduction in buoyancy (Dinsdale and Walsby, 1972). By these mechanisms, blue-green algae are able to regulate buoyancy and undergo limited vertical migration to poise themselves within vertical gradients of physical and chemical gradients favorable to growth (Fogg and Walsby, 1971). The population maximum, coupling population growth with movement downward, apparently often occurs as epilimnetic nutrient concentrations are depleted in summer. Movement to lower strata of low light and temperatures and increased nutrient availability occurs. Vertical movement has the added advantage of disrupting nutrient gradients surrounding actively growing cells. The additional effects of light scattering by gas vacuoles are of possible benefit to blue-green algae in that they shield the cells from excessively bright light. The mechanisms involved, if real, are unclear at this time. (d) The density of some algae is decreased by the *accumulation of fats*. Some algae, such as *Botryococcus*, can contain lipoids up to 30 to 40 per cent of dry weight and can float because of this fat accumulation (Fogg, 1965). Usually, however, fat accumulation occurs in senescent cells that are in various stages of cell breakdown.

Inorganic Nutrient Factors

The importance of inorganic macro- and micronutrients, particularly the two major elements phosphorus and nitrogen, to algal nutrition has been emphasized in preceding chapters on biogeochemical cycling. Reiteration is not necessary here. Facets of algal nutrition and nutrient interactions will be brought into further discussion in the sections on the periodicity and succession of phytoplanktonic populations.

The importance of the limiting nutrient concept, originally developed by Liebig in the middle of the last century as the "Law of the Minimum," should not be dismissed in spite of considerable misuse resulting largely from oversimplification. Simply paraphrased, the law states that yield of any organism will be determined by the abundance of the substance that, in relation to the needs of the organism, is least abundant in the environment (cf. Hutchinson, 1973). Since yield is a result of growth, rate of growth has been substituted for yield in many subsequent analyses, the most important of which is the well-known Monod model for nutrient limitation in the growth of microorganisms (cf. Droop, 1973):

$$\mu/\mu_m = s/(K_s + s),$$

when

μ = specific growth rate (increase in biomass per unit biomass per unit time),

μ_m = maximum specific growth rate at infinite external substrate concentration,

s = external substrate concentration (mass per unit volume),

K_s = saturation constant, when external substrate concentrations result in half the maximal rates of uptake.

Irrespective of the internal nutrient concentrations, the specific rate of uptake depends upon substrate concentration in a Michaelis function, that is, the rate of uptake increases with increasing substrate concentration to a specific substrate level beyond which no further change in rate of uptake occurs.

Under conditions of steady state with nutrient concentrations that are not limiting, the specific rate of nutrient uptake (u, mass per unit biomass per unit time) equals the product of the specific growth rate (μ) and the cell nutrient quota (Droop, 1973):

$$u = \mu q$$

when q is a demand coefficient or cell nutrient quota in the absence of nutrient excretion [(internal substrate concentration) (mass per unit biomass)]. The specific rate of uptake and the cell nutrient quota have a linear relationship:

$$\mu q = \mu_{m'}(q - k_q) \ or \ \mu/\mu_{m'} = 1 - k_q/q$$

when

$\mu_{m'}$ = maximum specific growth rate at infinite internal substrate concentration,

k_q = the subsistence quota, i.e., the q intercept for zero μ.

The importance of these models, which are confirmed by many metabolic data from planktonic algae, lies in their applicability to obtaining specific growth values from more easily determined uptake rates (Eppley and Thomas, 1969).

Furthermore, it is important to view nutrient uptake as it really takes place, since uptake of several nutrients occurs simultaneously. Any nutrient limitations of photosynthesis can be highly dynamic both spatially and temporally (Wetzel, 1972). Of importance among the many interacting potentially limiting nutrients is the relative position of limitations within an *intensity* spectrum which often undergoes constant change spatially and temporally. Limitations interact simultaneously in an activity series within a given phytoplanktonic community, and therefore are best considered simultaneously.

Taking the previous model relationship of Droop, if a limiting parameter is independently proportional to 2 or more functions, it is also proportional to their product. Therefore,

$$\mu/\mu_m' = (1 - k_{q_A}/q_A)\,(1 - k_{q_B}/q_B)\,(1 - k_{q_C}/q_C)\,(.....,)$$

is a simple polynomial in which subscripts A, B, C, ..., refer to various nutrients. The latter equation does not require any postulation as to which nutrients are limiting or whether any are more intensely limiting than others. The parameter μ_m' is an abstract proportionality constant in this case, whose value is determined by all factors (physical, nutritional, etc.) excluded from the equation. It is important to note, as will be discussed further among the bacteria (Chapter 17), that nutrient uptake can be influenced by the previous substrate (nutrient) concentration. If lack of a nutrient previously has limited growth severely, the uptake sites for that nutrient can be reduced. Replenishment of the limiting nutrient may not result in immediate uptake and growth response. This fact has important implications for short-term enrichment bioassays in which the effects of the added nutrient may not be realized in population growth for a considerable period, possibly longer than the duration of the experimental analyses.

In fresh waters of low nutrient concentrations and slow turnover rates, it is metabolically advantageous to be both small and motile. Small size increases the ratio of absorptive surface area to cell volume. Since growth rates are slow and may not be adequate to offset mortality by sedimentation and predation, small size is further advantageous in reducing settling rate and possibly predation by some grazers. Motility also is effective in breaking nutrient gradients surrounding cell membranes during active growth. These relationships suggest that one would anticipate a greater percentage of the algal populations of oligotrophic lakes to be small and motile. Such is indeed the case, with a few exceptions, in the majority of situations. Usually a much larger percentage of the total rates of primary productivity derives from nanno- and ultraplanktonic algae of oligotrophic lakes than from larger algae. In some extreme cases, the whole of photosynthetic phytoplanktonic productivity is by algae of a size less than 30 μm.

Organic Matter and Algal Growth

Dissolved organic matter is coupled intimately with the varying nutritional requirements of algae for organic growth factors and a number of inorganic conditions of fresh waters that influence the availability of these compounds. Furthermore, a number of algae can supplement primary photosynthetic autotrophy by the uptake and utilization of organic substrates. Therefore, algae capable of synthesis of required organic growth factors, e.g., vitamins, have a distinct competitive advantage over algae that cannot perform this function and must rely on exogenous sources synthesized by other microorganisms.

In addition to this direct utilization as a micronutrient or supplementary energy and carbon sources, organic compounds can function as nonessential accessory growth substances that may stimulate growth of the algae. Growth substances or nonessential organic micronutrients include a large group of hormones known to be produced by algae and effective in growth regulation and cell development (Bentley, 1958a, 1958b; Conrad and Saltman, 1962; Provasoli and Carlucci, 1974). Little is known of the physiological mechanisms of these substances in freshwater algae, although they probably are of some significance in the success and succession of certain more specialized algae under natural conditions.

Organic Micronutrient Requirements

Knowledge of the specific requirements of phytoplanktonic algae for growth factors or essential organic micronutrients, especially the vitamins, is extensive (cf. reviews of Provasoli, 1958; Hutner and Provasoli, 1964; Provasoli and Carlucci, 1974). Although most higher plants do not require vitamins, many algae do and these are termed *auxotrophic*. In the case of auxotrophic growth, the requirements for specific organic compounds are low and these compounds do not contribute significantly to the cell carbon.

Only 3 vitamins are known to be required by algae. Of the 3 primary water-soluble vitamins, vitamin B_{12} (cobalamine), thiamine, and biotin, vitamin B_{12} and thiamine are required alone or in combination by a majority of the auxotrophic algae, and B_{12} is required more often than thiamine. Biotin is known to be required by a few chrysomonads, dinoflagellates, and euglenoid algae.

The distribution among algae of vitamin requirements or auxotrophy, and the lack of it, does not allow any strict lines of separation. In general, however, the differences are sufficient to permit some generalizations. On the basis of relatively few analyses in pure cultures in relation to the total number of species, a significant number of blue-green algae, diatoms, green algae, and dinoflagellates require exogenous sources of vitamin B_{12} and are auxotrophic (Table 14–4). Only a few groups show any significant requirements for thiamine. Only the xanthophycean algae exhibit an apparent lack of need for these major water-soluble vitamin growth factors, although in several other groups the requirements are rare or only stimulatory in nature (Table 14–4). The Cyanophyceae, Chlorophyceae, Xanthophyceae, and Phaeophyceae exhibit the least number of species requiring vitamins. Most of the species in these groups are strictly autotrophic in metabolism. In contrast, a clear predominance of auxotrophic species occurs in the Chrysophyceae, Dinophyceae, Cryptophyceae, and Euglenophyceae.

When algae are unable to synthesize an organic micronutrient, the need for an exogenous source of the correct chemical moiety is obvious. Vitamins are synthesized by certain bacteria and algae, and some are released extracellularly during active growth and upon death. Appreciable concentrations of vitamins can enter aquatic systems from precipitation (Table 14–5); these may be derived from airborne soil particles, pollen, or active microorganisms (Parker and Wachel, 1971). The limited number of analyses of organic micro-

TABLE 14–4   General Requirements for Vitamins Among the Algae[a]

Algal Groups	Biotin	Thiamine	Vitamin B_{12}	Predominant Vitamin Requirements
Cyanophyceae	0	0	++	B_{12}
Rhodophyceae	0	0	++	B_{12}
Bacillariophyceae	0	+	++	B_{12}
Xanthophyceae	0	0	0	None
Phaeophyceae	0	0	+	None
Chlorophyceae	0	+	++	B_{12}
Chrysophyceae and Haptophyceae	–	++	+	Thiamine
Cryptophyceae	–	–	+	None
Dinophyceae	+	0	++	B_{12}
Euglenophyceae	–	–	+	None

++ = required in many species
+ = few species
– = requirement rare
0 = no known requirement
[a]After discussion of Provasoli and Carlucci, 1974.

nutrient concentrations in fresh waters indicates that very low concentrations exist, especially of vitamin B_{12}, which is in widespread demand (Table 14–5). These concentrations, however, yield little insight into rates of turnover between supply sources and demand. All assays of the absolute vitamin requirements of algae necessarily are based on bacterial-free cultures, and in most cases indicate that the requirements are much lower than observed concentrations in fresh waters.

Concentrations in natural waters are so low that they are determined by bioassay organisms of known growth responses to concentrations of the micronutrients. It must be emphasized that concentrations utilizable by the assay organism may not be equally utilized by algae. Moreover, some evidence indicates that some or much of the natural organic micronutrients may not be readily available for assimilation. Photosynthetic rates of carbon fixation and cell numbers of natural phytoplanktonic populations can often be enhanced markedly by additions of vitamin B_{12} or thiamine (see, for example, Wetzel, 1965a, 1966b, 1972; Hagedorn, 1971). Concentrations needed, especially in hardwater lakes, are often greatly in excess of those generally known to be required by algae in pure culture. White and Wetzel (1975) have shown experimentally that much vitamin B_{12} is adsorbed to particulate calcium carbonate and removed from the trophogenic zone by coprecipitation and sedimentation. Other losses are likely. It is doubtful whether the rates of synthesis generally are limiting to total growth and productivity of phytoplankton of lakes. However, large vertical seasonal fluctuations are known to occur (references cited in Table 14–5), and these organic micronutrients most likely play a significant role in the succession and competitive success of algal community populations.

TABLE 14–5 Range of Concentrations of Organic Micronutrients Found in Fresh Waters (in μg l^{-1} or mg m^{-3})

Lake	Vitamin B$_{12}$ (Cyanocobalamin)	Thiamine	Biotin	Niacin (Nicotinic Acidamide)	Pantothenic Acid	Folic Acid
Linsley Pond, Conn. (Hutchinson, 1967; Benoit, 1957)	0.06 –0.075	0.008–0.077	0.0001–0.004	0.15 –0.89	—	—
Northern German lakes (Hagedorn, 1971)	—	0.05 –12	—	—	—	—
Sagami, Japan (Ohwada and Taga, 1972)	0.005 –0.85	0.001–0.38	0.010 –0.068	—	—	—
Tsukui, Japan (surface) (Ohwada, et al., 1972)						
Dissolved	0.0005–0.0042	0.075–0.436	0.013 –0.058	—	—	—
Particulate	0.0019–0.0203	0.031–0.159	0.0005–0.0042	—	—	—
Kasumigaura, Japan (Kashiwada, et al., 1963)	0.005 –0.028	—	0.0021–0.050	0.30 –3.3	0.01–0.26	0.040–0.244
Small Swiss Ponds (Clémencon, 1963)	—	<0.001–1.0	<0.001 –0.004	<0.001–3.0	<0.01–0.034	<0.01 –0.48
Mean of Precipitation, Mo. (Parker and Wachtel, 1971)						
April-November	0.001	—	0.004	2.0	—	—
November-March	0.0004	—	0.0008	0.42	—	—

Heterotrophy of Organic Carbon by Algae

By far the most dominant mode of metabolism among the algae is photosynthesis, in which cell carbon is obtained by reduction of carbon dioxide when the inorganic reductant is water which is oxidized to oxygen. Thus, a majority of algae are obligate *photoautotrophs* that require light energy for these transformations, and suitable pigment receptor mechanisms.

In pure culture, an appreciable number of algae have been shown unequivocally to be able to assimilate and utilize dissolved organic compounds as a source of carbon and energy both in the dark and the light. *Heterotrophy* or chemo-organotrophy in algae implies the capacity for sustained growth and cell division in the dark, in which both energy and cell carbon are obtained from the metabolism of an organic substrate(s). Heterotrophy in algae occurs by means of aerobic dissimilation, and carbon dioxide may or may not be required. Excellent recent reviews on the physiology and biochemistry of heterotrophy by algae are given by Droop (1974) and Neilson and Lewin (1974).

A few algae are *mixotrophic*, assimilating carbon dioxide in small amounts simultaneously with organic compounds both in the light and especially in darkness. In other words, photoautotrophy in the light can be supplemented by the assimilation of organic compounds in the dark. True *photoheterotrophy*, in which the organic substrates serve as a significant source of cell carbon during growth, is more restrictive in that different metabolic pathways of cyclic photophosphorylation are involved.

In comparison to bacteria and fungi, heterotrophic algae can utilize only a few substrates, such as acetate and related compounds (pyruvate, ethanol, lactate, higher fatty acids), glycolate, hexose sugars, and amino acids. All algae known to grow heterotrophically cannot do so under anaerobic conditions. The ability to utilize organic substrates requires specific enzymes for transport across cell membranes, and many algae are deficient in these enzymes. In certain species that possess the enzymes, organic substrates still are not utilized because of an apparent impermeability of the cells or an inability to couple their dissimilation metabolically with the generation of adenosine 5'-triphosphate (ATP).

It must be emphasized that in a majority of cases in which heterotrophic utilization of organic substrates has been demonstrated among pigmented algae, the concentrations of substrates are very high, usually several orders of magnitude greater than those found in most natural waters. Furthermore, culture conditions have been bacteria-free, eliminating the competitive interactions of bacteria that possess much more efficient active uptake mechanisms. Nearly all analyses in situ among natural populations have led to the following conclusions: (a) At naturally occurring substrate concentrations, under most conditions the low affinity of algae for simple organic substrates results in heterotrophy being a relatively unimportant process in comparison to photoautotrophy. (b) Algae cannot compete effectively with bacteria for available substrates (e.g., Wright and Hobbie, 1966). This subject is discussed later on (Chapter 17).

While it is obvious that photoautrophic metabolism is the primary mechanism of synthesis and growth among algae in lake systems, it is incorrect to dismiss heterotrophic metabolism in natural populations of algae as unimportant. Indeed, there is sufficient evidence that heterotrophy and photoheterotrophy can be instrumental in subtle but important ways in the survival, competition, and succession of algal populations. Under conditions of light limitation, such as deep in the hypolimnion, under heavy snow and ice-cover, at extremely high latitudes at which total darkness exists for a period, or when very dense populations lead to shading, an ability to be facultatively heterotrophic (mixotrophy) can constitute a distinct competitive advantage over metabolism of obligate photoautotrophic algae. Observations of apparently viable and even reproducing phytoplanktonic algae under ice in northern Scandinavia during almost total darkness are incompletely explained (Rodhe, 1955). Some evidence exists that phytoplankton can compete with bacteria for acetate, but not glucose, under low light conditions of winter (Maeda and Ichimura, 1973). Certain deep-water blue-green algal populations, especially *Oscillatoria*, are facultatively heterotrophic under very low light conditions (Saunders, 1972). Earlier remarks should be recalled on the possible utilization by these populations of extremely low light intensities to generate ATP by photophosphorylation without concomitant CO_2 reduction, which could be potentially important in maintaining viability without significant growth.

When organic loading is very high, such as in sewage oxidation ponds, algae such as *Chlamydomonas* can assimilate most of the acetate generated by anaerobic bacteria by photoheterotrophy involving photosynthetic production of ATP and reducing power (Eppley and MaciasR, 1963). While photoheterotrophy is generally not of quantitative importance in most natural waters, recent evidence indicates that it can amount to at least 20 per cent of total inorganic carbon fixation at low light intensities in certain oligotrophic lakes (McKinley and Wetzel, unpublished). More importantly, microautoradiographic techniques have demonstrated selective utilization of organic substrates among the algal species of the populations.

Indirect Effects of Dissolved Organic Matter

Another function of dissolved organic compounds is as biotics, either in a stimulatory fashion as growth substances or more commonly as antibiotics. The extracellular production and release of antibiotics or growth inhibitors have been suggested in numerous cases from observations of the inhibitory effects of one alga on another from both mixed cultures and natural populations (Krauss, 1962; Lefèvre, 1964; Hellebust, 1974). The chemical composition of these compounds is poorly understood. Inhibitory compounds include peroxides of fatty acids and possibly polyphenolic substances. Stimulatory compounds include a number of weak organic acids, especially glycolic acid. The effectiveness of antibiotics is governed in part by species specificity, relative concentrations, and potential rates of bacterial degradation. Such substances may play important roles in the succession of species under natural conditions, and are a fertile field of investigation.

Other indirect effects of dissolved organic compounds include a number of ways in which the organic substances affect the availability of and capacity for assimilation of inorganic micronutrients (Saunders, 1957; Wetzel, 1968). Chelation of metal ions, an equilibrium reaction between a metal ion and an organic chelating agent resulting in the formation of a stable ring structure incorporating the metal ion, can function in several complex ways in accordance with environmental conditions. The extent to which chelation of metallic ions, better known in soil systems, occurs within a heterogeneous solution such as lake water is governed by the bonding characteristics of the organic compounds, the ratio of chelator to metal ions, and stability constants of chelates for different ions. Temperature, hydrogen ion concentration, and the concentration and ionic strength for various anions further influence the extent of complexing. Other organic and inorganic compounds present in fresh waters function as sequestering agents for metals and cations (cf. Chapters 9 and 13). Complex formation occurs by pyrophosphate, metal binding by the complex formation with macromolecules such as proteins (Povoledo, 1961), and the formation of peptidized metal hydroxides of yellow substances (Shapiro, 1964).

Natural functions of organic complexing of metal ions and major cations are only beginning to be appreciated as factors influencing the specific composition and succession of algal populations. Examples of some of the mechanisms involved include: (a) increases in the physiological availability of inorganically reactive ions, such as iron and manganese (Chapter 13); (b) modifications of membrane permeability and osmoregulation by complexing of cations and resulting in changes in monovalent:divalent cation ratios (Chapter 9); (c) if a trace metal is toxic to a particular organism, its availability may be effectively reduced to a concentration below the threshold of toxicity by excessive complexing of the metal ions by organic compounds; and (d) under certain circumstances, organic compounds could complex and effectively compete for a metal ion that is antagonistic to a toxicant, and thereby increase the relative concentration to the threshold of toxicity.

SEASONAL SUCCESSION OF PHYTOPLANKTON

In spite of a number of generalizations about the common seasonal succession of phytoplanktonic algal populations in freshwaters that pervade much of the literature, a great diversity of patterns emerges upon close inspection of existing data. Some of the disparity is related to the choice of study methods employed. For example, many of the older analyses were based solely on number of organisms, which is quite a biased indicator in comparison to biomass, because of great differences in size among algae. Furthermore, early analyses often used plankton nets of fairly large porosity, in which significant and variable portions of the algae were not retained.

It is also desirable to differentiate between growth changes in biomass, by whatever criteria used, and rates of in situ metabolism of the algae. Prior to discussing the seasonal succession of phytoplanktonic populations, the terminology employed in biotic evaluations of fresh waters and the criteria used should be discussed briefly.

Productivity

In any analysis of productivity, it is imperative that the values obtained be comparative, i.e., expressed in such a way that they are comparable effectively among different ecosystems, among different community components of the same system, or among different responses of the components to environmental dynamics, whether they are natural or perturbed. The long history of conceptual framework behind productivity has been characterized by numerous definitions largely based on agriculture and economics within society. Perhaps more so than in any other field, in aquatic ecology the questions of productivity have been addressed in detail, and an array of often incongruous terms and definitions has been proposed. These definitions, many of which are unnecessarily complex, have caused much confusion that has led to ambiguous thinking and expression.

The basic terminology and definitions of the concepts of productivity have been discussed in detail by MacFadyen (1948, 1950), Elster (1954a), Balogh (1958), Davis (1963), and Westlake (1963, 1965a). Much of the confusion emanates from early concepts that considered productivity as the maximum growth and development of organisms under optimal conditions (Thienemann, 1931), i.e., the potential production of organisms or organic matter per unit volume or surface area per unit time (Dussart, 1966). While one can consider the potential of organisms to produce and increase towards infinity as a useful conceptual framework, in the real world the constraints of an array of environmental factors regulate these increases. Optimal conditions of an organism, population, community, or ecosystem can, at best, only be approximated by extensive investigation. Even when the detailed physiological optima for an organism are determined, their applicability to theoretical maximum growth under the dynamics of a nearly infinite number of competitive interactions in natural systems is relegated to the abstract. Such information is much more useful in the interpretation of environmental abiotic and biotic control of observed growth under in situ natural or perturbed conditions.

It is therefore much more meaningful to define the terms production and productivity in relation to realized or actual production of organisms, a functional group of organisms, or an ecosystem. Changes in production are related to time and dynamics of environmental regulatory parameters. However, to generate separate terms for productivity under arbitrarily defined "natural conditions" and under perturbed conditions, as has been variously done, has introduced anthropocentricity and unnecessary complexity into the problems being addressed. The following definitions of terms represent common usage, all in basic agreement with the long history of theoretical discussions (reviewed in detail by Westlake, 1963; 1965a).

Terminology and Definitions

Standing crop (also referred to as standing stock) is the weight of organic material that can be sampled or harvested by normal methods at any one time from a given area. Standing crop does not necessarily include the whole plant population, because certain species or inaccessible parts of the sampled species may be omitted by the sampling procedure. The term derives from agricultural usage of the word *crop,* the total weight of organic material removed from a given area over a period of time in the course of normal harvesting practice. The crop is a much less variable measurement than the standing crop, which depends on the time of measurement. For example, a wheat crop is the annual maximum standing crop of above-ground foliage, an alfalfa crop is often the annual total of several standing crops of plant tops, and a sugar cane crop may be a standing crop of tops after nearly two years. Omissions of underground organs in measurements can constitute a large portion of the total plant mass, and there may be great differences in the standing crop according to the time of sampling.

Yield is the crop expressed as a rate.

Complications have resulted from the wide usage of the standing crop measurement in limnology. When applied to plankton, standing crop (or stock) is synonymous with biomass. *Biomass* is the weight of all living material in a unit area at a given instantaneous time. Ambiguities arise when standing crop, which refers only to the upper above-ground portions, is applied to aquatic macrophytes; biomass includes the entire plant. Biomass evaluations are essential in any analyses of the aquatic plant population or of productivity dynamics. Thus because of these inherent differences between the terms, the measurement of standing crop should be abandoned in limnology and only biomass used. If, for some reason, e.g., in herbivory, the foliage above ground is specifically of interest, it should be labeled as such: above-ground biomass.

Production is the weight of new organic material formed over a period of time, plus any losses during that period. Production thus refers to the increase in biomass observed over a period plus any losses by respiration, excretion, secretion, injury, death, or grazing. Thus primary production is the quantity of new organic matter created by photosynthesis (or chemosynthesis), or the stored energy which this material represents. If a photosynthetic organism also uses organic substrates (i.e., is mixotrophic), this energy flow is secondary production even though new organic material may be produced by transformation.

Productivity is the rate of production expressed as production, divided by the period of time. Productivity is usually an average of the instantaneous rates over some period such as a day or a year, since natural systems have so many factors causing rapid, frequent, and irregular changes in the instantaneous rates that only average rates can be determined in normal study. It is imperative, as we will stress repeatedly, that the evaluations be done in accordance with the regeneration times of the organisms under consideration. Weekly evaluations of the instantaneous production of bacteria, for example, are relatively meaningless because much of their population dynamics will be excluded; however, this sampling frequency may be adequate for certain animal components.

Losses of production at any ecosystem level occur as a result of nonpredatory losses (respiratory utilization to form CO_2 and heat, excretion and secretion of dissolved organic materials, and death or injury) and predatory losses (grazing by herbivory or by carnivory). *Gross productivity* (sometimes termed real productivity) refers to the observed change in biomass, plus all predatory and nonpredatory losses divided by the time interval. Thus gross primary productivity is the rate of production of new organic matter, or fixation of energy, including that subsequently used and lost during the time interval. *Net productivity* (occasionally termed apparent productivity) is the gross rate of accumulation or production of new organic matter, or stored energy, less losses divided by the time interval. Sometimes only respiratory losses are subtracted, particularly among evaluations of plant productivity in which losses from processes other than respiration are small. However, all losses should be considered in a true evaluation of net productivity.

Biomass and Productivity

Numerous criteria have been used in evaluations of the biomass and productivity of aquatic organisms including: enumeration, volume, wet (fresh) weight, dry weight, organic weight, content of carbon, pigments, energy as heat on combustion, and ATP, and the rates of exchange of oxygen and carbon dioxide. Often these criteria have been used uncritically, which makes comparability and interpretation difficult (cf. reviews of Lund and Talling, 1957; Strickland, 1960; Westlake, 1965b; Vollenweider, 1969b; Edmondson and Winberg, 1971; Stein, 1973).

Enumeration and Volume.   Enumeration of organisms per unit volume is a commonly used method among the microorganisms. While they afford great advantages in permitting qualitative differentiation among species and of organisms from detrital particles, numbers do not give a true evaluation of biomass because organisms differ greatly in size (Table 14–6). The number of individuals of each species therefore is expressed by the volume per volume of water or area of sediment (numbers times average cell volume determined from mean dimensions of the cells). Among larger organisms, such as zooplankton, benthic fauna, and fish, size and biomass characteristics for allometric evaluations are obtainable more readily among individual groups. Volume is a poor measure of freshwater macrophytes because the density is mainly a function of the proportion of internal air spaces; thus the relationship between volume and biomass criteria is highly variable.

Weight.   Fresh weight, the weight of the organism without any adherent water, with appropriate precautions, is essentially equivalent to wet weight. Owing to the highly variable water content of nearly all organisms, however, wet weight is a criterion to be avoided. If elaborate precautions are taken, wet weight analyses can be converted to dry weight on a species basis from a particular environment. Dry weight is variable at temperatures below 105°C; if loss of the small percentage of volatile organic constituents is germane to the analyses, lyophilization (freeze drying) is the preferred method.

Although dry weight is employed widely in production analyses, organic (ash-free) dry weight, reasonably determined in most situations by the loss in weight after ignition at 550°C, is the preferred general biomass criterion

TABLE 14–6 Calculated Volumes of Representative Species of Freshwater Plankton Organisms (in μm^3)[a]

Classification	Volume (μm^3)	Classification	Volume (μm^3)
Cyanophyta		*Bacillariophyceae*	
Anabaena flos-aquae (col.)	80,000	Amphiphora ornata	17,650
Aphanocapsa delicatissima	4	Asterionella formosa (Michigan)	350
Aphanothece clathrata	10	(Europe)	700
Aphanothece nidulans	5	Cyclotella bodanica	10,000
Chroococcus limneticus (col.)	400	Cyclotella comensis	400
Chroococcus turgidus (col.)	1000	Cymatopleura solea	80,000
Coelosphaerium naegelianum (col.)	15,000	Diatoma vulgare	4350
Dactylococcopsis smithii (col.)	1500	Fragilaria crotonensis (1 mm)	200,000
Gloeocapsa rupestris	18	Nitzschia gracilis	240
Gomphosphaeria lacustris (col.)	2000	Melosira granulata (1 mm)	60,000
Merismopedia tenuissima	8	Melosira islandica (1 mm)	80,000
Microcystis flos-aquae	50	Stephanodiscus astraea	2000
Microcystis aeruginosa (col.)	100,000	Stephanodiscus hantzschii	
Oscillatoria limnetica (1 mm)	17,500	var. pusillus	200
Oscillatoria rubescens (1 mm)	30,000	Stephanodiscus niagarea	5000
Synechococcus aeruginosus	350	Synedra acus	250
		Synedra acus angustissima	1000
Chlorophyta		Synedra capitata	950
Ankistrodesmus falcatus	250	Synedra delicatissima	300
Chlamydomonas subcompleta	250	Synedra ulna	50
Botryococcus braunii (col.)	10,000	Tabellaria fenestrata (Michigan)	3000
Chlorella vulgaris	200	(Europe)	4000
Closterium aciculare	4000		
Cosmarium phaseolus	3000		
Cosmarium reniforme	30,000	*Pyrrophyta*	
Gloeococcus shroeteri (col.)	5000	Ceratium hirundinella	4000
Pandorina morum (col.)	4000	Gymnodinium fuscum	10,000
Oocystis solitaria	400	Gymnodinium helveticum	20,000
Scenedesmus quadricauda	1000	Gymnodinium ordinatum	400
Staurastrum paradoxum	20,000	Peridinium cinctum	40,000
Tetraedron minimum	40	Peridinium willei	40,000
Ulothrix zonata (1 mm)	6000		
		Euglenophyta	
Cryptophyta		Trachelomonas hispida	4200
Chroomonas nordstedtii	35	Trachelomonas volvocina	1800
Cryptomonas erosa (Michigan)	1000		
(Europe)	2500		
Cryptomonas ovata	2500		
Rhodomonas lacustris (Michigan)	175		
(Europe)	200		
Rhodomonas minuta	200		
Chrysophyta			
Chromulina pyriformis	50		
Dinobryon borgei	1500		
Dinobryon divergens	800		
Dinobryon sociale	800		
Mallomonas caudata	12,000		
Mallomonas urnaformis	1200		
Rhizochrysis limnetica	1200		
Uroglena americana (col.)	90,000		

[a]After Nauwerck, 1963, Findenegg, unpublished in Vollenweider, et al., 1969, and Wetzel and Allen, unpublished.

among larger organisms. The difficulty of separating bacteria, algae, and other small microorganisms from detrital particulate organic matter limits its application to only larger organisms.

Cellular Constituents.   Perhaps the most satisfactory means of measuring the biomass of photosynthetic organisms is to oxidize the organic plant material back to carbon dioxide, from which it originated in photosynthetic reduction. The organic carbon content of plants is one of the least variable constituents, and falls nearly without exception between the range of 40 to 60 per cent of ash-free dry weight. The average carbon content among algae is 53 ± 5 per cent and among aquatic macrophytes 47 per cent of ash-free organic dry weight. Exclusion of extraneous organic detritus is not practical at present, and analyses of particulate organic carbon of the pelagic zone (cf. Chapter 17) include a very high percentage of nonliving organic material. Algal carbon content usually is estimated from average species content and extrapolated to natural populations by volume measurements of the populations, since the carbon to volume relationship has been shown to be allometric (Mullen, et al., 1966).

A number of other cellular constituents have been employed variously to estimate changes in population biomass. Except for carbon, however, measurement of biomass by other elements is so complicated, owing to the extreme variability of the composition in response to environmental variables, that their use is limited to specific physiological analyses rather than changes in population biomass. While the pigment content also varies appreciably with environmental parameters, being able to correct accurately for pigment degradation products in order to measure only the functional pigment content of plants (separate from particulate detritus) permits effective analyses of composite population dynamics among algae.

Several conversion factors have appeared in the literature, by which biomass estimates of one cellular component may be made from another. Even under the most favorable conditions, such factors must be used with utmost caution; under most conditions their use cannot be justified and is best not attempted.

Productivity.   Rates of production have been estimated by numerous techniques, most of which are best discussed in ensuing treatments of specific groups of organisms. Among aquatic macrophytes, and to a certain extent attached algae, primary productivity can be estimated rather well from changes in biomass over time (Chapter 15). Estimates of production rates by planktonic microflora from changes in biomass are much more difficult (Vollenweider, 1969b). A temporal set of biomass measurements results in minimal estimates or underestimations of net productivity because of losses by grazing, current transport, sedimentation, death, and decomposition. Rarely are these parameters sufficiently evaluated to permit analyses of production rates from changes in biomass; exceptions include the excellent investigations of Lund (1949, 1950, 1954), and especially Grim (1952).

Primary Productivity of Phytoplankton

The primary productivity of phytoplankton has received an extraordinary amount of attention in limnology, and has been measured in great detail in a number of aquatic systems. The reasons for this abundance of information on rates of primary production are manifold.

Fundamental, of course, is that phytoplanktonic productivity represents a major synthesis of organic matter of aquatic systems which can be summarized in the universally known equation:

$$6 \; CO_2 + 12 \; H_2O \; \xrightarrow[\substack{\text{pigment} \\ \text{receptor}}]{\text{light}} \; C_6H_{12}O_6 + 6 \; H_2O + 6 \; O_2.$$

This equation is a gross oversimplification of the complex Calvin-Benson metabolic pathway of photosynthesis, which is basically a redox reaction:

$$CO_2 + 2 \; AH_2 \; \xrightarrow{h \cdot v} \; (HCOH) + 2 \; A + H_2O,$$

in which AH_2 represents a hydrogen donor, normally water but, as discussed earlier, can include an array of reduced sulfur (e.g., H_2S) or organic carbon compounds among autotrophic and certain other bacteria. Particularly in large lakes, phytoplanktonic productivity often represents the dominant input of new organic matter and potential energy that drives the system.

Secondly, as is often the case in ecology, the development of reasonably accurate techniques for the measurement of in situ rates of primary production has led to their wide application, unfortunately often without any sound rationale as to why the measurements were being made. Only now, after many years of experimentation, is a reasonable understanding of the methodology being realized. The appeal of these techniques lies in their property of being able to measure direct rates of metabolism in situ, while most other evaluations of productivity are forced to use indirect methods.

In practice, the changes in oxygen production or rates of carbon uptake usually are measured on isolated samples of the natural communities which are incubated for brief periods at the points of collection or under simulated natural conditions aboard ship. Certain environmental factors, e.g., temperature and light, are simulated closely; other factors such as turbulence, nutrient replenishment, and grazing, can differ to varying degrees in the isolated samples. Alternately, productivity estimates can be made from measurements of changes in oxygen, pH, carbon, or conductivity over short intervals directly in the natural environment on nonisolated communities. This approach, in addition to possessing numerous complications, yields estimates of community metabolism.

The light and dark bottle techniques for estimating primary productivity have received wide application. In the oxygen method, samples of phytoplankton populations are incubated in a depth profile in clear and opaqued bottles. The initial concentration of dissolved oxygen (c_1) can be expected to be reduced to a lower value (c_2) by respiration in the darkened bottles, and increased to higher concentrations (c_3) in the clear bottles according to the difference between photosynthetic production and respiratory consumption. The difference ($c_1 - c_2$) represents the respiratory activity per unit volume over the time interval of incubation, the difference $(c_3 - c_1)$ the net photo-synthetic activity, and the sum $(c_3 - c_1) + (c_1 - c_2) = (c_3 - c_2)$ the gross photo-synthetic activity. Numerous assumptions are made in the method that can alter the photosynthetic measurements appreciably, e.g., respiration rates are not necessarily the same in light and dark, since photorespiration clearly oc-curs in algae, other processes such as photo-oxidative consumption utilize oxygen separately from apparent respiratory uptake, nonphotolysis of water by bacterial photosynthesis, etc. Under many circumstances these errors are small, but the technique can be considered only as a reasonable estimate. It is probable, however, that these and analytical errors in determination of oxygen concentrations are appreciably less than sampling errors of heter-ogeneous plankton populations in the lake for the analyses. Large portions of recent methodological works are devoted to detailed discussions of this and the following ^{14}C techniques, and should be used critically when applying the methods (cf. especially Strickland, 1960; Vinberg, 1960; Vollenweider, et al., 1969b; Strickland and Parsons, 1972).

The incorporation of ^{14}C tracer into the organic matter of phytoplankton during photosynthesis has been used as a highly sensitive measure of the rate of primary production. If the content of total CO_2 of the experimental water is known, and if a definite amount of $^{14}CO_2$ is added to the water, then determination of the content of labelled carbon in the phytoplankton after incubation permits calculation of the total amount of carbon assimilated. Numerous methodological and physiological problems confront application of the ^{14}C light and dark technique; however, with care, most technical problems can be successfully approached and errors evaluated, e.g., respira-tory losses of CO_2, and secretion rates of soluble organic products of photo-synthesis. Rates of respiration are difficult to evaluate directly by this tech-nique. Under many situations of application, the ^{14}C method yields a measure close to net photosynthetic rates. Comparison of the oxygen and ^{14}C methods under optimal conditions shows close agreement with a photosynthetic quotient ($PQ = \Delta O_2 / - \Delta CO_2$, by volume) somewhat greater than unity (Fogg, 1963). The photosynthetic quotient varies from near unity when carbo-hydrates are the principal photosynthetic products to as high as 3.0 during fat synthesis.

Assuming a photosynthetic quotient of 1.2 and a statistical probability at the 0.05 limit, the smallest amount of photosynthesis that the oxygen light and dark technique can measure under ideal conditions is about 20 mg C m^{-3} day^{-1} and the potential error range is similarly $\pm$ 20 mg C m^{-3} (Strickland, 1960). The limit of sensitivity of the ^{14}C method is some 50 to 100 times greater, on the order of 0.1 to 1 mg C m^{-3} day^{-1}.

Seasonal Patterns and Periodicity

A distinct periodicity in the biomass of phytoplankton is observed in polar and temperate fresh waters. Growth is greatly reduced or negligible during the winter period of low light and temperatures. Phytoplankton num-bers and biomass normally increase greatly in the spring under improved light conditions, building up to a spring maximum. The spring maximum can begin under the ice in late winter, and often consists predominantly of diatoms adapted to low temperatures. In many dimictic lakes, the spring maximum does not develop fully until after the spring circulation and the period of summer stratification have begun.

The spring maximum of phytoplanktonic biomass generally is short-lived, usually less than 3 months in duration. This maximum often is followed by a period of low numbers and biomass that may extend throughout the summer. Among more eutrophic lakes of the temperate region, the summer minimum is often brief and phases into a late summer profusion of blue-green algae that persists into the autumn until the disruption of thermal stratification begins. The summer populations of phytoplankton are often low throughout the summer in temperate oligotrophic lakes and develop a second maximum in the autumn period, again usually predominantly of diatoms. This second maximum of the autumn generally is not as strongly developed as that of the spring period. Decline of the populations into the winter minimum frequently also is more rapid and irregular in the autumn.

The limited growing season of lakes of high latitudes and polar regions often is reflected in a conspicuous single summer maximum of phytoplanktonic biomass. By contrast, the maximum in tropical lakes often is observed in the winter.

Generalizations are difficult to make because of the great variability observed among phytoplanktonic numbers and biomass from lake to lake. Several points are reasonably consistent, however: (a) The successional seasonal periodicity of phytoplanktonic biomass is reasonably constant from year to year. If the freshwater system is not perturbed by outside influences, such as the activities of man in modifications of the watershed, nutrient loading, etc., the characteristic seasonal changes in the phytoplanktonic populations are very repetitious from year to year on a short-term time basis. (b) The seasonal amplitude of changes in phytoplanktonic numbers and biomass is usually very great, on the order of a thousandfold in temperate and polar fresh waters. In keeping with the relatively constant environmental conditions, the seasonal variation in tropical waters is much lower, often as little as fivefold (Fogg, 1965). (c) The maxima and minima observed in numbers and biomass of phytoplankton often are quite out of phase with measured periodicity of rates of primary production. Primary productivity usually follows the annual cycle of incident solar radiation in temperate, less productive lakes more closely (see following discussion). The spring maximum, if conspicuously developed, often is composed of larger algae such as diatoms with slower turnover rates than summer algae which occur in warmer and more favorable light conditions.

Within these general quantitative changes in seasonal total biomass, a periodicity of species composition occurs continuously. The species composition fluctuates in a regular way from year to year, as long as the system is not perturbed seriously. The seasonal community of algal populations consists of a composite of perennial species (holoplanktonic) that are present throughout the year, and an intermittent species (meroplanktonic) that enters some type of diapause in resting stages. In both cases, but much more markedly so in the latter type, an interplay of environmental conditions results in fluctuations in growth and competition among other species. Among the perennial species, population numbers decrease to extremely low levels but are present in the water in sufficient numbers to reinoculate the community when growth conditions improve.

Winter Populations and the Spring Maximum

Both the biomass and productivity of phytoplanktonic algae of the winter season in temperate and polar lakes usually are low. If nutrients are adequate, growth is limited to species that are adapted to low temperatures and low light irradiance. Population accrual by slow growth often is offset by respiratory losses, secretion of organic compounds, and sedimentation of cells from the limited photic zone under ice-cover. In polar regions, inappreciable or zero growth exists in extremes under conditions of midwinter near-total darkness and very heavy ice and snow-cover. At lower latitudes, within the temperate zone or at latitudes influenced by maritime amelioration of winter conditions, snow and ice is less thick and opaque, allowing more penetration of the low available irradiance.

The population of winter algae beneath ice usually is dominated by small and often motile algae. Cryptophyceans, such as *Rhodomonas* and *Cryptomonas*, are particularly common, as well as pyrrophyceans (*Gymnodinium*), small green algae (*Chlamydomonas*), chrysophyceans (*Dinobryon*, *Mallomonas*, *Chrysococcus*, *Synura*), some diatoms (*Synedra*, *Tabellaria*, *Fragilaria*), and euglenophyceans (*Trachelomonas*) (Wright, 1964; Rodhe, 1955; Verduin, 1959; Tilzer, 1972; and Maeda and Ichimura, 1973, among many others). The cold-water, low light-adapted species are evidently quite species-specific in their preferred tolerances to light quantity and quality, and distribute in narrow strata in the photic zone beneath the ice-cover. For example, the layers of the water column of an ice-covered lake of eastern Massachusetts receiving 0.5 to 20 per cent of incident radiation accounted for over 95 per cent of the phytoplankton (Wright, 1964). As discussed earlier, there is some evidence for mixotrophic growth in which photoautotrophy is supplemented somewhat by heterotrophic assimilation of simple organic substrates under darkness or very low irradiance. However, photoautotrophy clearly dominates the metabolism of these algae, even under exceedingly low light conditions. Changes in light, as occur with snowfall on ice, result in a rapid vertical shift in the distribution and rates of photosynthesis (Fig. 14–5).

Rates of primary production under ice-cover can constitute a very significant portion of the total annual primary productivity of the phytoplankton. Nearly one-quarter of the annual phytoplanktonic productivity of hypereutrophic Sylvan Lake in northeastern Indiana occurred under the three-month period of ice-cover (Wetzel, 1966a). Similar values have been found since in numerous lakes in the temperate zone, emphasizing that the widely held assumption that winter productivity is insignificant is not valid universally.

The spring circulation, with the concomitant loss of ice-cover and the

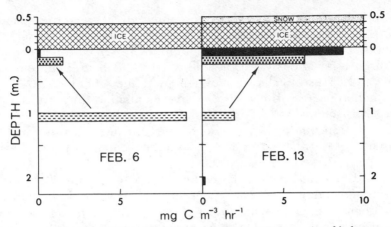

Figure 14–5 Rates of primary production beneath ice (6 February) and below ice and snow (13 February). Rectangles with dots are rates obtained when samples taken from 1 m were incubated alongside the surface samples. (Modified from Wright, 1964.)

season progression, results in mixing of nutrient-laden water from the lower depths of winter stratification with surface strata of increasing light. In smaller lakes in which the light penetrates throughout much of the water column, the spring algal maximum, often initiated by cold-water, polyphotic[4] species under ice in late winter, flourishes and develops throughout the period of circulation and continues for a time after summer stratification has set in. Circulation of small lakes is commonly weaker and of shorter duration than that of larger lakes. Among large lakes, net growth of phytoplankton may not occur during circulation if mixing occurs at such a rate and to such a depth that the algae are carried out of the photic zone faster than they can multiply. The period of spring circulation is much longer in large than in small lakes, and can suppress net growth significantly. The effects of circulation are accentuated in regions, such as much of the United Kingdom, in which lakes rarely freeze or do so for only brief periods. In these areas circulation can continue practically all winter, with a significant loss of algal accrual occurring as a result of outflow and other mechanisms (e.g., Lund, et al., 1963).

There is little question that increasing light in the spring is the dominant factor contributing to the development of the spring "outburst," even though water temperatures are still low. The spring maximum frequently is dominated by one species, often diatoms, such as *Asterionella* discussed earlier (Fig. 13–18), *Cyclotella*, or *Stephanodiscus* (see, for example, Pechlaner, 1970). While a lag phase often is not seen before the dominants reach exponential growth phases, it obviously exists and is observed upon frequent sampling (e.g., Lund, 1950). In Lake Erken, Sweden, some lag in the response of the algae of the spring maximum to the increase in radiation was caused by the necessity for light adaptation, even when the intensities of photosynthetically usable energy were still low (Pechlaner, 1970).

---

[4]Adapted to a wide range of light intensities.

The relative rates of increase[5] of the dominant algae of the spring maximum during exponential growth are usually much less than those observed under "optimal" conditions of light and temperatures of cultures (Fogg, 1965). This difference is variously attributable to restrictions of light, temperature, nutrients, and losses of cells by sedimentation and other causes. The larger species, such as major spring diatoms, have generation times[5] of about 4 to 8 days, which is somewhat longer than those of smaller species under natural conditions. The specific growth rates of the dominant species are considerably higher than those of the entire algal assemblage combined (e.g., Pechlaner, 1970).

It should be noted that cold, low-light adapted species of algae that are well-developed under winter conditions are limited severely in the spring, or move to deeper strata in which temperatures and irradiance remain low. Movement may consist of actual migratory movement, as occurs among the dinoflagellates of alpine lakes, or the development of sedimenting populations in lower strata, as is the case among some ubiquitous blue-green algae, e.g., *Oscillatoria rubescens*. Findenegg (1943), and others since, have demonstrated the characteristic weak development of *Oscillatoria* in upper layers under winter conditions. Before, during, and after spring circulation the *Oscillatoria* population weakens in the upper strata and continues to increase in the deep, cold layers of the lower metalimnion and upper hypolimnion.

In high-mountain lakes above the tree line, the nannoplankton constitute total dominants of the phytoplankton, and are autotrophic throughout the long winter conditions (Pechlaner, 1971). These algae are adapted to the low light intensities that enter the water despite heavy ice- and snow-cover. Both the concentration of algae and rates of in situ photosynthesis shift from the surface layers in winter to deep water strata during spring and summer. Immediately before ice breakup, the increasing light induces a downward migration of predominating dinoflagellates.

Spring Decline and Summer Populations

The decline of the spring maximum of phytoplankton in temperate lakes also is associated with an interaction of physical and biotic parameters. It is clear that in many straightforward cases, reduction of nutrients in the photic zone of the epilimnion slows the growth of populations of the dominant as well as rarer algae. Since diatoms are often the dominant component of the phytoplankton and the spring maximum, especially in temperate fresh waters,

---

[5]The generation time (G) is the mean doubling time if the cells divide by two or

$$G = \frac{\log 2}{k'} = \frac{0.301}{k'}, \text{ when}$$

k' = the relative growth constant under exponential conditions, or $k' = \frac{\log_{10} N - \log_{10} N_0}{t}$ when N_0 is the initial concentration of cells, N is the final cell concentration, and t the time interval.

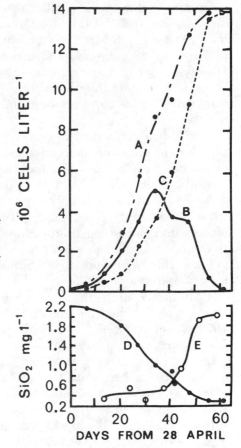

Figure 14–6 Production and loss of *Asterionella formosa* in the epilimnion of Lake Windermere, England, from 28 April to 30 June 1947. *A*, Cumulative total of cell production computed from silicate uptake; *B*, Mean concentration of cells in epilimnion; *C*, Cumulative loss of cells from the epilimnion (*A* minus *B*); *D*, Epilimnetic silicate concentration; and *E*, Relative rates of loss of cells from the population. (Modified from Lund, et al., 1963.)

after excessive growth, silica availability is a prime candidate for suppression. The reduction of silicate concentrations to limiting levels (<0.5 mg l^{-1}) has been shown repeatedly in many fresh waters, but is especially known to occur in the northern British lakes, in which maximum concentrations are low (2 to 3 mg l^{-1}) (Fig. 14–6; see also Fig. 13–18). Reduction of silicate to limiting concentrations occurs in less than 2 months in conjunction with reduced turbulence and increased sedimentation (Fig. 14–6).

In calcareous lakes in which silica concentrations are naturally high (>10 mg l^{-1}), their levels are reduced appreciably during the spring and summer in the epilimnion (e.g., Fig. 13–17, *upper*). Utilization by diatoms continues throughout the period of the summer stratification. As the hardwater lakes become more productive, the rapidity of the reduction increases. For example, in extremely eutrophic Wintergreen Lake (Fig. 13–17, *lower*), concentrations are reduced to and below generally accepted inhibiting levels in both winter and spring periods.

So distinct is the correlation between the observed decline of the diatom maximum with regressing silica concentrations and the experimental results of bioassays, that the relationship appears to be predominantly causal. Factors of light intensity, temperature, nitrogen, phosphorus, zooplankton grazing, and fungal parasitism do not appear to be instrumental in the decline. However, much more complex interactions of both physico-chemical and biotic factors undoubtedly exist in the seasonal succession of diatom populations when silica concentrations are not reduced to limiting levels for exponential growth.

As silica concentrations are reduced in productive lakes, the diatom populations are often succeeded by a preponderance of first green algae and later blue-green algae. Growth in these eutrophic lakes can be so intense that combined nitrogen (NO_3, NH_4^+) sources are reduced to below detectable concentrations in the trophogenic zone. When this happens, often by midsummer when the warmest epilimnetic temperatures occur, blue-green algae with efficient capabilities for fixing molecular nitrogen have a competitive advantage and predominate (cf. Chapter 11). These lakes require, as a general rule, a reasonably sustained and heavy loading of phosphorus (cf. for example, Moss, 1973c).

Similar relationships have been demonstrated to be in effect on a much larger scale in the Laurentian Great Lakes (Schelske and Stoermer, 1971, 1972; Schelske, et al., 1972). Phytoplanktonic algae in the upper Great Lakes are distinctly phosphate limited. As the loading inputs of phosphorus to the lakes are progressively increased, diatom metabolism of available silica concentrations (ca 2 to 3 mg 1^{-1}) is more and more rapidly reduced to limiting levels. The diatoms then are replaced rapidly by green and blue-green algae. Increased loading with trace metals, vitamins, and dissolved organic matter further accelerates the algal succession.

A series of important papers by Moss (1972c, 1973a, 1973b, 1973c) is instructive in relation to both the seasonal succession and the distribution of phytoplankton. Shifting concentrations of necessary nutrients are just as important to seasonal periodicity within a lake as different nutrient concentrations among lakes are important in determining algal distribution. Working with 16 species of diverse but widely distributed phytoplanktonic algae, Moss found little effect of calcium ions above a concentration of 1 mg 1^{-1} among both oligotrophic and eutrophic species. No evidence was found to support the contention that oligotrophic desmids are calciphobic. Certain oligotrophic algae species were unable to grow at pH values greater than 8.6 common to many hardwater lakes, in which free CO_2 occurs in very low concentrations, whereas eutrophic species able to utilize both CO_2 and bicarbonate grew well at pH values above 9. Temperature optima for good growth were similar to those discussed earlier: diatoms have an extended range into lower temperatures, whereas green and blue-green algae grow well at temperatures above 15°C. Temperatures above 15°C are common in many temperate waters throughout much of the summer and in tropical waters for the entire year. These and other nutrient interactions generally could be verified in experimental bioassays using a variety of natural lake waters ranging from oligotrophic to extremely eutrophic.

Parasitism and Grazing

Parasitism of phytoplanktonic algae has been shown to be significant only recently. Although the best-known organisms are chytridiaceous or bi-flagellated phycomycetous fungi, viral and bacterial infection of algae surely occurs under natural conditions (Shilo, 1971). Viral lysis of specific blue-green algae has been demonstrated (Saffermann and Morris, 1963; Shilo, 1971), but attempts to exploit this specificity for the control of blue-green algal blooms have not been very successful (V. Sládeček, personal communication).

The best studies of chytrid fungal parasitism, centered in the English Lake District, indicate that infection of desmids, diatoms, and blue-green algae is fairly common. Even though the fungal percentage infection can be very high (>70 per cent) and can parasitize healthy, rapidly growing cells as well as declining senescent algae, parasitism usually does not greatly alter the overall seasonal pattern of periodicity of diatoms and desmids (Canter and Lund, 1948, 1969). Fungal parasitism, however, can have a significant effect on interspecific competition of algae because of the degree of specificity in the algae attacked. The reduction of dominant desmids apparently is influenced to a greater extent by parasitism, whereas the diatom decline, particularly of *Asterionella*, results from a combination of mainly nutrient and parasite interactions. Parasitism probably increases in eutrophic waters, but little is known of its quantitative significance. Much further study is badly needed.

Predation of phytoplankton by animals, particularly by the microcrustacea, often has been invoked as a significant factor in the decline of the algal populations, and as contributory to seasonal succession. Although most of this discussion will be deferred until the subsequent chapter on zooplankton, a few points are appropriate here.

Evidence for the induction of quantitative changes in phytoplanktonic abundance and species or size specificity is conflicting. As temperatures of the water increase in temperate waters, the reproduction and feeding of zooplankton increase markedly. In some cases the population maxima of zooplankton coincide with the decline of algal maxima; in other cases inverse correlations between the two are poor, or changes in zooplankton population are correlated better with those of bacterial or detrital concentrations. Although certain recent studies indicate that total grazing of the trophogenic zone volume per day by rotifers and microcrustacea may occur at certain times of the year (cf. Chapter 16), much lower rates generally occur. A degree of size and species specificity also exists among the grazers. Such selectivity can lead to a competitive advantage by less effectively grazed species, and influences seasonal succession of algae within prevailing physical and nutrient constraints. In this way higher invertebrates and fish possessing size-selective feeding habits with respect to zooplankton can influence their grazing effectiveness and in turn the algal succession.

Size-selective feeding is particularly conspicuous among the protozoans. Certain herbivorous protozoans feeding on small colonial algae have been shown to decimate over 99 per cent of certain chlorophycean planktonic algal populations in short periods of from 7 to 14 days (Canter and Lund, 1968).

Some of these protozoans, such as *Pseudospora,* have been shown to consume only certain species of algae of similar size. The effects on seasonal succession of algal populations are obvious.

Competitive Interactions and Successional Diversity

The phytoplanktonic community consists of a diverse assemblage of species populations. Each of these species components and their growth dynamics are influenced by an array of environmental parameters (physical, chemical, and biotic) that undergo constant temporal variations. Some of the parameters, such as temperature, conservative nutrients, etc., undergo slow, periodic oscillations over an annual period. Other parameters, such as light intensity and quality with depth, nutrients in rapid flux, etc., exhibit very rapid temporal variations, physiologically important changes in terms of minutes and hours. The intensities of these forces greatly affect the resultant algal physiology and growth, some with regularity and others in a highly irregular fashion. Therefore, in attempting to characterize the multitude of dynamic variables defining a species niche, they can only be viewed as a multidirectional hypervolume that varies quantitatively between extremes and with time, some variables rapidly and some slowly.

A crude analogy might be drawn to the irregular variations in length and size of contracting and expanding pseudopodia of an *Amoeba* protozoan. If each of say 30 to 60 pseudopodia represented a dynamic environmental parameter affecting growth, some would extend in length (range of tolerance) and size (magnitude of parameter) only slightly, while others would extend greatly. The length (tolerance) might increase or decrease over the duration of the algal population, depending upon the adaptation to changes. The size of the leg of the amoeboid hypervolume may change seasonally as a result of external forces (e.g., climate-induced changes of nutrient input). The pseudopodium would contract and enlarge with time in response to the daily or seasonal variations in periods and amplitudes of the parameters. The oscillation might be very regular for a given time (e.g., daily insolation on a clear day) or highly irregular (e.g., the rapidly changing light of a moving broken cloud cover). Thus we may envision a species hypervolume as a cloud of pseudopodia, some short and thick, others long and narrow, each oscillating in magnitude regularly or irregularly in response to temporal changes in parameter variables within tolerance ranges.

It is doubtful that an ideal species hypervolume is ever met under natural conditions. The fit of a species within an array of competing species is relative and constantly changing. The ability of an alga to store a critical nutrient imposes another time factor interaction that stresses the importance of past environmental conditions as well as contemporary ones. On the basis of the number of interacting parameters and ranges of variations tolerated by algae, it is not too surprising that (a) a species may be able to survive indefinitely provided that a suitable combination occurs with sufficient frequency and duration, and (b) the interactions are so complex and variable in combination that, in spite of species overlap in their niche hypervolumes, coexistence can occur even though competitive interactions are in effect. The beginnings of a mathematical derivation of these competitive interactions are set forth in the stimulating theoretical discussion of Grenney, Bella, and Curl (1973).

Diversity of phytoplanktonic algal species has been determined in a number of ways because of observed species changes as lakes become more eutrophic. Possibly among the best indices of species are those that are largely independent of sample size.[6] Diversity indices often are determined primarily by the proportions of the more common species (equitability), and only secondarily by the number of species according to derivatives of the Shannon formula[7] (Sager and Hasler, 1969; Moss, 1973d). Predominance of one or two species results in low diversity values; high values occur when populations of several species each form moderate proportions of the whole.

Although data are far from satisfactory because of the variations found, there is a general tendency for species diversity to decrease with increasing fertility of the water. Presumably, the slower growth rates attainable in oligotrophic lakes permit a greater number of species with reasonably similar requirements (high degree of niche overlap) to coexist within the temporal variations of regulating parameters than would be found in more eutrophic waters.

On a seasonal basis, particularly in eutrophic temperate waters, diversity tends to increase in summer and be low in winter (Moss, 1973d). Again much of the observed relationship is believed to be related to nutrient availability, which is high at and after spring circulation during a period of increasing illuminance. At this time, fewer species compete successfully with the fast growing dominants. Obviously, other factors such as parasitism and predation are involved, although nutrients most likely predominate in most cases.

Congruous with these findings are the results of a measure of succession rate, or a seasonal rate of change of species composition. Jassby and Goldman (1974), employing a mathematically more involved but more realistic expression, showed that the rate of change in species composition per time decreases precipitously following the spring maximum and midsummer plateaus. Thus, when the community is disturbed by the rapid changes of spring circulation as in this case, or is perturbed by some catastrophic event (e.g., fertilization by man's activities), rapid changes in species composition and increased successional rates would be expected.

PERIODICITY AND DISTRIBUTION
OF PRODUCTIVITY

The distribution of phytoplanktonic biomass and in situ rates of primary productivity has received a great deal of attention and represents perhaps the

---

[6]Increasing sample size tends to produce a greater number of species, as more and more of the rarer species are encountered. Many of the rarer species are of littoral zone origin, in which diversity of species is much greater, and not true members of the phytoplankton (cf. Moss, 1973d; Symons, 1972).

[7]Diversity $(H') = -\sum_{i=1}^{s} p_i \log_2 p_i$ = units of bits per individual per liter, when the p_i is estimated from n_i/N as the proportion of the total population of individuals (N) belonging to the ith species (n_i) using logarithms to the base 2.

most thoroughly investigated aspect of the biota of lakes. In contrast, the growth and productivity of phytoplankton of streams and rivers, although of lesser importance, is poorly understood (cf. reviews of Hynes, 1970 and Wetzel, 1975b).

Throughout this entire book, concerted attempts have been made to point out the influence of major interactions of environmental control on productivity. As we now view examples from the huge spectrum of variability in the composite biomass and productivity of the phytoplankton, the complexity of the multipopulations of species composing this community again underscores the difficulties of understanding which factors dominate the control of photosynthesis at any given time and depth.

Individual patterns from lake to lake and within lakes from one time interval to another are the dominant feature of phytoplanktonic productivity. However, some patterns of common distribution are found. Moreover, there are definite limits to the productivity of phytoplankton, beyond which the productive capacity cannot be increased.

Vertical Distribution

The vertical depth distribution of phytoplanktonic biomass varies greatly from season to season with the shifts in species composition. If the depth-time distribution of phytoplanktonic pigments of an oligotrophic hardwater lake of southern Michigan is taken as an example, conspicuous temporal changes are seen. In the seasonal depth distribution of chlorophyll a concentrations (Fig. 14–7), corrected for degradation products, the biomass is low in winter, increases conspicuously in the spring maximum, and then increases at different strata during the period of summer stratification. Detailed analyses of the phytoplankton composition of some 150 major species showed cryptomonads

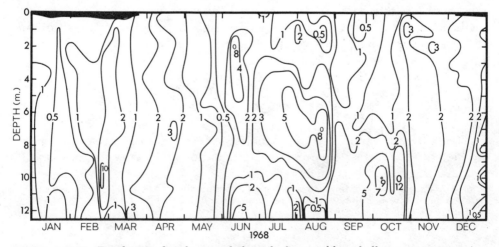

Figure 14–7 Depth-time distribution of phytoplanktonic chlorophyll a concentrations (mg m^{-3}), Lawrence Lake, Michigan, 1968. Opaque areas = ice cover to scale. Sampled at strict 7-day intervals at each meter of depth. (From Wetzel, unpublished data.)

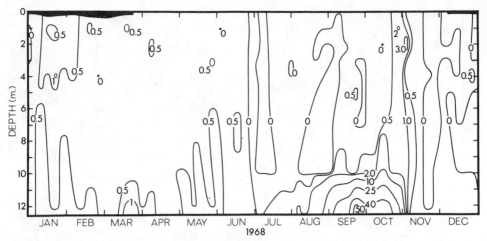

Figure 14–8 Depth-time distribution of phytoplanktonic chlorophyll *b* concentrations (mg m^{-3}) Lawrence Lake, Michigan, 1968. Sampled at each meter of depth at strict 7-day intervals; opaque areas = ice cover drawn to scale. (From Wetzel, unpublished data.)

and small green algae predominating in the winter, shifting to predominately diatoms in the spring, and green algae in June and late August during this particular year. The metalimnetic maximum in summer was related largely to green and nonnitrogen-fixing blue-green algal populations with lesser percentages of diatoms. The development deep (10 to 11 m) in the later portion of summer stratification was related to the very dense populations of euglenoid algae just at the aerobic-anaerobic interface one meter above the sediments. Concentrations of chlorophyll *b* were very much lower throughout the year and irregularly distributed, except in the lower hypolimnion in late summer and autumn (Fig. 14–8). Chlorophyll *c* concentrations, requiring more elabo-

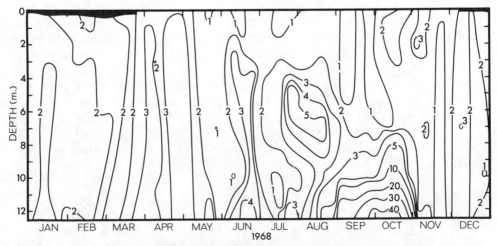

Figure 14–9 Depth-time distribution of phytoplanktonic total plant carotenoids (ca mg m^{-3}), Lawrence Lake, Michigan, 1968. Sampled at meter depth intervals each 7 days; opaqued areas = ice-cover to scale. (From Wetzel, unpublished data.)

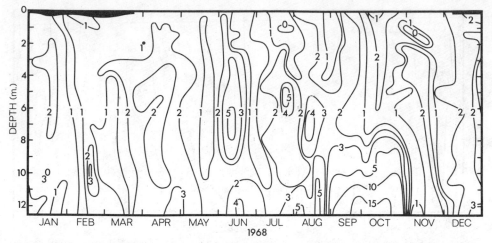

Figure 14–10 Depth-time distribution of phytoplanktonic phaeopigment concentrations (mg m^{-3}), Lawrence Lake, Michigan, 1968. Sampled at meter depths at strict 7-day intervals; opaqued areas = ice-cover to scale. (From Wetzel, unpublished data.)

rate separation procedures, were most abundant in algal populations dominated by diatoms. The total composite plant carotenoids, somewhat more stable than the chlorophyllous pigments, exhibited a vertical depth distribution seasonally similar to that of chlorophyll a (Fig. 14–9).

In using pigments as estimates of phytoplanktonic biomass, it is important to emphasize that much of algae is nonviable, i.e., is particulate detritus in various stages of decomposition. The chlorophyllous pigments can be corrected for degradation products. Concentrations of these heterogeneous phaeopigments give some insight into the magnitude of nonfunctional pigments in the particulate fractions. Taking the same data for Lawrence Lake, Figure 14–10 sets forth the concentrations of phaeopigments with depth over the year. It will be noted that at many times of the year the phaeopigment concentrations equal or exceed those of chlorophyll a. Moreover, the phaeopigments are often displaced in concentration at slightly greater depths and at slightly later times than those of the functional pigments. The importance of this detrital particulate matter of algal origin to lake metabolism is discussed at length later (Chapter 17). Here, it is important to realize that cell integrity and pigment concentrations can persist for several days after the cell is no longer viable (e.g., Gusev and Nikitina, 1974). Chlorophyll a degrades most rapidly upon death, and when corrected for more stable phaeopigment degradation products is a reasonable estimate of biomass.

Vertical Distribution and Maximum Growth

The above discussed examples of pigment biomass are very low, typical of a rather oligotrophic lake. Considerably lower concentrations, about one mg chlorophyll a m^{-3}, are not unusual among pristine arctic and alpine oligotrophic lakes. It is now necessary to examine the other extreme of maximum growth and pigment concentrations. Again light limitations become restrictive of productivity, and the biomass of the algae themselves imposes controls upon further growth by self-shading.

Photosynthetic yield can continue to increase, given adequate nutrient availability, to a theoretical point at which total absorption of photosynthetically active incident radiation by photosynthetic pigments exists. This absorption is distinct from that absorbed by the water itself, by dissolved organic matter, and by abiogenic particulate matter (cf. Talling, 1960, 1971). The theoretical maximum, as indicated by culture analyses, is somewhere in the range of about 1 g of chlorophyll per m^3 (1 mg l^{-1}), but varies among algal types, being higher for blue-green algae and less for green algae and diatoms

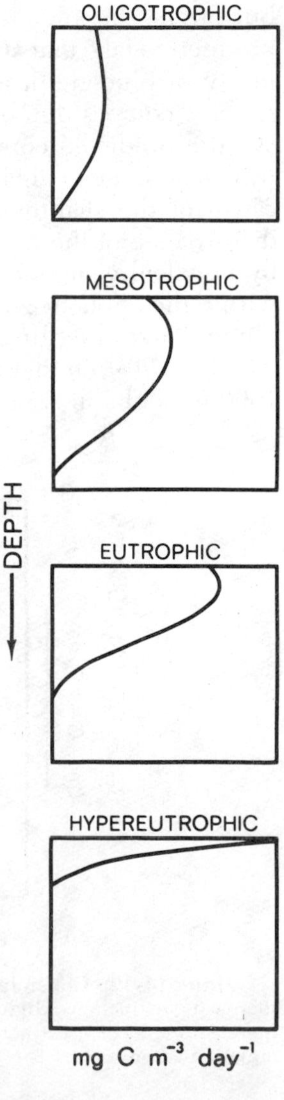

Figure 14-11 Generalized increases in productivity of phytoplankton per unit volume of water, with simultaneous reduction of the thickness of the trophogenic zone, in a series of lakes of increasingly greater fertility.

OLIGOTROPHIC

MESOTROPHIC

DEPTH

EUTROPHIC

HYPEREUTROPHIC

mg C m^{-3} day^{-1}

(e.g., Steemann Nielsen 1962b). As the biogenic turbidity of the algal populations increases in progressively more fertile lakes, the productivity per
volume of water increases greatly (Fig. 14–11). However, the productivity per
square meter of water column does not increase proportionately, because the
biogenic turbidity of the algae is reducing light penetration and the effective
trophogenic zone.

Therefore, as the composite extinction coefficient of light from biogenic
sources increases, the productivity per square meter of water column increases to the maximum imposed by self-shading effects of the algal densities
(Fig. 14–12). Increasing the extinction coefficient further, such as by high
concentrations of humic dissolved compounds or inorganic particulate and
colloidal turbidity, results in a progressive depression of maximum possible
productivity.

The highest recorded concentrations of chlorophyll of phytoplanktonic
origin emanate from tropical, eutrophic lakes. For example, shallow (2.5 m)
but large (250 km^2) equatorial Lake George of western Uganda is a highly
productive lake that supports very dense phytoplanktonic populations, particularly of blue-green algae, all year round with very little change in either
species composition or densities. Based on in situ estimates of photosynthesis, the predicted concentrations of chlorophyll a in this lake stabilized approximately at an algal density of about 500 mg chl a m^{-2} (Ganf, 1974a). Because of the shallow nature of this large lake, frequent daily mixing, and
disturbance of the flocculent sediments, light penetration is reduced further
by nonalgal components. Therefore the maximum quantity of chlorophyll a
within the euphotic zone is between 230 to 310 mg m^{-2}. However, many of the
phytoplankton settling to the sediments are viable and resuspended each day
(Ganf, 1974b), so that a more realistic mean algal biomass may be as high as
1000 mg chl a m^{-2}.

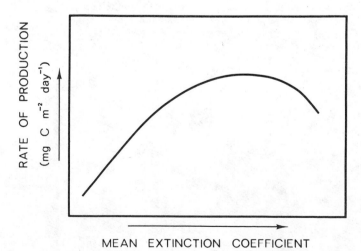

Figure 14–12   Generalized maximum possible productivity of phytoplankton in relation to
increasing extinction of light, resulting from biogenic turbidity initially, and then combined with
increasing effects of extinction by dissolved organic matter and/or particulate and colloidal inorganic matter.

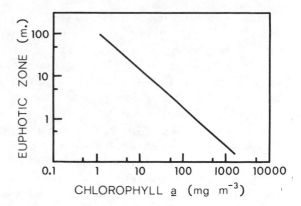

Figure 14–13 Generalized relationship of the depth of the euphotic zone in relation to the maximum concentrations of chlorophyll *a* per volume of water. (After data of several authors and Talling, et al., 1973.)

The maximum recorded values of algal biomass on the basis of chlorophyll content appear to be from two Ethiopian soda lakes (Talling, et al., 1973). In these extremely productive lakes, favored by high temperatures, abundant light, phosphorus, carbon, and other nutrients, the euphotic zone was less than 0.6 m in thickness and the algal populations were limited severely by self-shading. Although concentrations of chlorophyll *a* exceeded 2000 mg m^{-3} in one lake, which appears to be about the maximum possible under natural planktonic conditions (Fig. 14–13), the content of chlorophyll *a* per unit area of the euphotic trophogenic zone for these lakes was in the range of 180 to 325 mg m^{-2}. These values are similar to those maxima (180 to 300 mg chl *a* m^{-2}) that have been estimated indirectly or directly to occur in nature.

Seasonal Rates of Photosynthesis

Consideration of the rates of in situ photosynthesis of the phytoplankton necessitates evaluation over an annual period to compensate for major differences in the length of active growing season with increasing latitude and altitude. Although summer productivity of an arctic or high mountain lake or pond may have just as high or higher rates of photosynthesis per given volume of the euphotic zone as water at a much lower latitude or elevation, the length of the growing season is greatly constricted. Since composition of phytoplanktonic productivity must be made on an annual basis, it is necessary to look first at the magnitude of these seasonal variations.

Further, it is important to keep in mind the ecological significance of gross versus net productivity. As discussed earlier, the rates of the opposing processes of photosynthesis and respiration are very difficult to evaluate in heterogeneous planktonic populations of composite phytoplankton, bacteria, and zooplankton. The widely used ^{14}C light and dark bottle generally results in values close to net productivity under a majority of in situ conditions. This method permits a further evaluation of the extent of extracellular release of soluble organic compounds during growth of the algae. Respiration cannot, however, be evaluated by the ^{14}C method. Therefore, the sensitive ^{14}C method permits an estimation of the in situ rates of photosynthesis near to net productivity, which is realistic in the sense that it is this particulate matter that is potentially available for consumption by higher organisms. More elaborate methods, however, are required to ascertain the fate of the total productivity, since a majority of the synthesized organic matter is not consumed by animals but enters the detrital food chain of degradation and utilization (cf. Chapter 17).

The less sensitive oxygen change techniques measure community metabolism, although the oxygen production over time is clearly most often the result of the phytoplanktonic algae. The importance of the effects of light-mediated photorespiration on the assumption that the dark bottle estimates of respiration are similar to those in the light have not been evaluated satisfactorily. Similarly, bacterial respiration is assumed to be small in relation to that of the planktonic algae, and the same in the light with different rates of soluble organic matter inputs as it is in the dark.

It can only be concluded that at best, the contemporary methods provide estimates of in situ rates of photosynthesis that are comparable only in a general manner. Nonetheless, the advantages of being able to measure rates directly in situ are great, and are exceedingly valuable if kept in their correct perspective of approximations of productivity.

Several examples of the seasonal primary productivity illustrate the types of variations that can be encountered. The changes in productivity of the phytoplankton at depth over the season set out in Figure 14–14 for a hardwater lake of low productivity in Michigan show a common pattern. Rates are low in the winter with periodic minor surges of near-surface populations under ice, particularly towards the end of winter. Rates often decrease during spring circulation, which in this lake for this year was very brief (until early April). Rates of production increased during the spring maximum of diatoms, entered a low period in May, and then succeeded through a series of pulses through

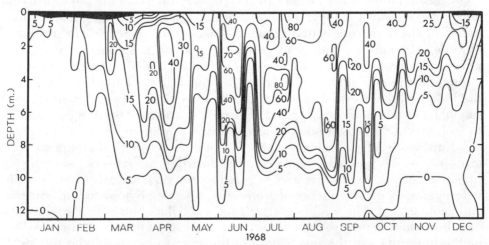

Figure 14–14   Depth-time distribution of the in situ rates of production of phytoplankton in mg C m^{-3} day^{-1}, Lawrence Lake, Michigan, 1968. Measured at meter depth intervals every 7 days; opaqued areas = ice-cover to scale.

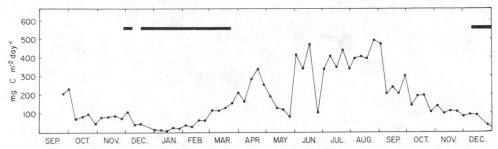

Figure 14–15 Integrated areal primary productivity per square meter of the pelagial zone of Lawrence Lake, Michigan, 1967-68. Bars indicate periods of ice-cover. (From Wetzel, unpublished data.)

the summer at higher values than during the spring. Metalimnetic maxima are conspicuous in July and late August prior to a general slow decline in the autumn. These rates of production of phytoplankton are composite values for the community of many species populations, which show a great regularity in periodicity from year to year (monitored continuously for nearly a decade).

When these data at each depth interval are integrated over a square meter of water column, the rates of in situ productivity per surface area of the pelagial zone emerge (Fig. 14–15). It is usual to integrate the areal productivity under such curves for an annual period to obtain the estimate of the total annual productivity by phytoplankton for a year. Division by 365 results in an estimate of the mean daily productivity for the lake. Obviously, the conically shaped lake basins have much greater volume for phytoplanktonic productivity in upper strata than at greater depth within the trophogenic zone. Therefore, it is much better to determine the annual productivity for each stratum, say at meter intervals, over the year and integrate these values prior to summation of the productivity of each stratum for the whole. In this way the changes in volume available for productivity are taken into account.

A feature of the total phytoplanktonic productivity and the mean daily productivity for the year is their stability from year to year as long as the system is not perturbed. In spite of internal variations in the seasonal rates of observed productivity and numerous sources of potential error in measurement and sampling, the values are reasonably constant. Examples of this consistency can be seen in Table 14–7. Some variations are to be expected, in this case, for example, in 1972, which was a particularly variant year with abnormally high cloud cover and rainfall. Within this type of variation, the annual primary productivity values serve as one of the best available criteria for following changes in the metabolism of lake systems in response to alterations by human activities in the short-term, and those changes involved in the recovery of perturbated systems.

Looking at the primary productivity of phytoplankton of more eutrophic lakes of the same latitude as the previous example, three points are conspicuous (Fig. 14–16). First, the thickness of the trophogenic zone is reduced markedly to the point at which a large majority of the productivity is occurring in the first 2 meters of water. Secondly, the succession of productivity seasonally is much more irregular and exhibits marked fluctuations in comparison to less productive lakes. The effects of algal utilization and reduction of

TABLE 14–7   Rates of In Situ Production of Phytoplankton of
Lawrence Lake, Michigan, over Several Years

Year	g C m^{-2} Year^{-1}	Mean mg C m^{-2} Day^{-1}
1968	46.57	127.4
1969	43.39	119.0
1970	45.49	124.6
1971	38.20	104.7
1972	30.46	83.5
1973	42.51	116.5
1974	41.01	112.4
Mean 7 years	41.09	112.6

critical nutrients to levels limiting for certain algal groups, discussed earlier, progress very rapidly and contribute to the marked observed oscillations in composite photosynthetic rates. Finally, the data of Figure 14–16 again emphasize the importance of primary productivity under ice and winter conditions and should not be ignored in contemporary studies.

Because the primary productivity of this hypereutrophic Wintergreen Lake is attenuated so rapidly with increasing depth by the self-shading effects of the algae, the majority of the productivity is restricted to the largest volume strata of the lake basin. Therefore evaluation of the productivity of the lake on a per square meter of water column basis throughout the trophogenic zone is similar to values for the entire lake, in which compensation for the decreasing volume of the lower strata is taken into account (Fig. 14–17).

As the seasonal climatic changes lessen in tropical regions, the length of the active growing season is extended to the entire year. The annual rates of primary production of phytoplankton increase accordingly and when nutrient limitations are not severe, the equatorial lakes exhibit the highest

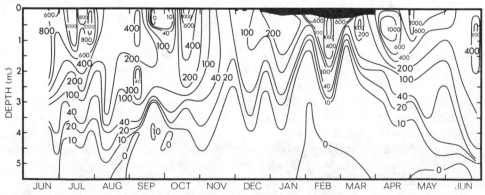

Figure 14–16   Depth-time distribution of the in situ rates of production of phytoplankton in mg C m^{-3} day^{-1}, Wintergreen Lake Michigan, 1971-72. Opaqued area = ice-cover to scale. (From Wetzel, et al., unpublished.)

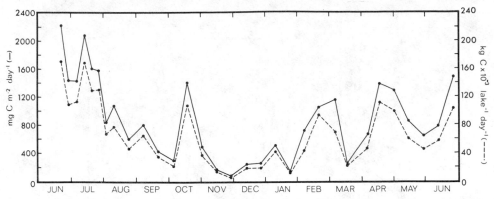

Figure 14–17 Integrated primary productivity of the phytoplankton of hypereutrophic Wintergreen Lake, Michigan, 1971-72, per square meter of the trophogenic zone and for the lake, compensating for volumetric changes. (From Wetzel, et al., unpublished.)

values of phytoplanktonic productivity recorded in aquatic systems (for example, Talling, 1965; Lewis, 1974b). In spite of rather continuous conditions nearly optimal for sustained high growth rates, definite seasonality persists even though it is not nearly as marked as that found at higher latitudes. In tropical lakes which are shallow, without any stratification and with high nutrient inputs, the productivity values are not only very high, but are relatively uniform throughout the year. Such is the case with Lake George on the equator in Uganda (Talling, 1965; Ganf, 1974). Deeper tropical lakes that stratify thermally, even though weakly so, at certain periods of the year tend to exhibit some periodicity in productivity. For example, in equatorial Lake Victoria of Africa, somewhat higher rates occur in the early months of the year and in June and July which are associated with the breakdown of stratification. The net primary productivity of weakly monomictic Lake Lanao (8°N) in the southern Philippines was relatively high throughout the period of stratification (May-November) and decreased precipitously during the winter period of minimum water temperatures and deepest seasonal mixing (Fig. 14–18). Respiration rates as a percentage of gross photosynthesis generally are higher (towards 50 per cent) than the average for temperate waters (20 to 30 per cent of total fixed carbon).

Therefore, when viewing the annual cycles of primary productivity in lakes in relation to latitude, the variable input of solar radiation at higher latitudes is the general controlling mechanism. Near the equator this seasonal differentiation is muted, and the incidence of vertical mixing is important in any observed periodicity (Talling, 1969). In some tropical lakes, such as in Lake Victoria, seasonal cooling and mixing, regulated chiefly by atmospheric factors like wind and humidity, are decisive events. Nutrient supply has been implicated as a dominant controlling factor of primary productivity of Lake Lanao during stratification (Lewis, 1974b). Nutrient depletion is relieved at frequent intervals by changes in the depth of mixing associated with irregular storms.

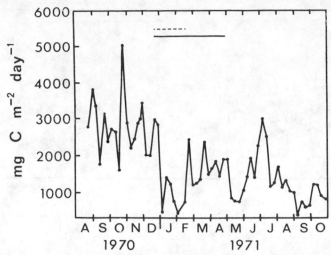

Figure 14–18 Net rates of primary production of the phytoplankton of Lake Lanao, Philippines, 1971-72. The dotted line marks the period of minimum water temperature and deepest seasonal mixing; the solid line marks the period when the lake lacked stable stratification. (Modified from Lewis, 1974b.)

Efficiency of Light Utilization

The productivity efficiency of algae in terms of light utilization can be estimated in several ways. All methods are related to the amount of irradiance available at depth and to the production rates of transformation of light energy to chemical energy by photosynthesis. Existing values can only be viewed as estimates because of the difficulties in measuring the caloric equivalents of photosynthesis per unit volume at depth, or integrated for the water column and the variable light inputs of photosynthetically active radiation at depth. In spite of these limitations, it is of great interest to see the range of values encountered in fresh waters because of the desires of some to utilize aquatic systems as potential food sources.

It is well-known that the efficiency of utilization of light energy by phytoplankton in aquatic systems is generally much lower than photosynthetic efficiencies of terrestrial systems. Nearly all of the estimated efficiencies are less than 1 per cent, and the highest values reported for phytoplankton are from tropical areas in the range of 2 to 3 per cent. It should be emphasized that the productivity of the littoral zone normally is ignored as a component of aquatic systems. While the littoral productivity may be small on a worldwide basis (its contributions scarcely have been evaluated), it will be emphasized in the following chapter that it is high and constitutes a significant component in most lakes.

Photosynthetic efficiencies, based on the ratio between the caloric equivalent of integral photosynthesis and radiation inputs, range from <0.01 to about 3 per cent at the maximum. In the surface waters, especially in transparent waters, efficiencies are relatively low if light inhibition is severe (Fig. 14–19; cf. also Tilzer, et al., 1975). At low light intensities, as are found at the

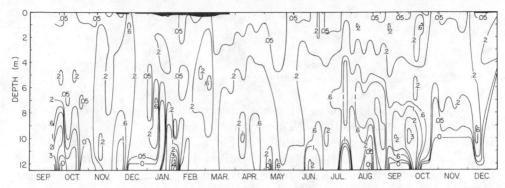

Figure 14-19 Depth-time distribution of estimates of the photosynthetic efficiency of utilization of photosynthetically active light income by phytoplankton of Lawrence Lake, Michigan, 1967-68. Values expressed as per cent, most of which are much less than one per cent. (From Wetzel, unpublished data.)

lower limit of the trophogenic zone, adaptation to low light intensities can occur, which increases efficiencies somewhat. The photosynthetic values in these strata, however, are usually so low that they increase the overall efficiency little.

As the concentrations of phytoplanktonic biomass increase, the integral photosynthetic efficiencies generally increase until the maximum levels are restricted by light limitations imposed by self-shading. Therefore, the maximum values are found in very productive waters with dense algal populations.

TABLE 14-8 Estimates of Integrated Photosynthetic Efficiencies of Utilization of Photosynthetically Active Radiation by Phytoplankton[a]

Lake	Percentage Efficiency
Tahoe, Calif.-Nev.	0.035
Castle, Calif.[b]	0.040
Finstertaler, Austria[b]	0.068
Oliver, Ind.	0.19
Olin, Ind.	0.23
Walters, Ind. (mean of 4 basins)	0.26
Pretty, Ind.	0.26
Chad, Chad, Africa	0.26
Crooked, Ind.	0.32
Little Crooked, Ind.	0.33
Martin, Ind.	0.34
Smith Hole, Ind.	0.38
Sammamish, Wash.	0.42
Wingra, Wisc.[b]	0.45
Goose, Ind.	0.57
Sylvan, Ind. (mean of 3 interconnected basins)	0.98
Leven, Scotland	1.76

[a]After data of Wetzel, 1966b, and Tilzer, et al., 1975. Estimates are based on different criteria of conversion; values are only approximately comparable but should be within about 30 per cent.
[b]Ice-free period only.

Exemplary ranges encountered in lakes of increasing productivity are set out in Table 14–8.

Extracellular Release of Organic Compounds

Algae now are known to release a large number of organic compounds into the water extracellularly. These soluble compounds include glycolic acid, carbohydrates, polysaccharides, amino acids, peptides, organic phosphates, volatile substances, enzymes, vitamins, hormonal substances, inhibitors, and toxins. Recent literature on this subject is large and is reviewed excellently by Fogg (1971) and Hellebust (1974).

The release of organic compounds extracellularly represents first a significant loss of carbon fixed in photosynthesis. Second, the release of such organic compounds undoubtedly is of much greater importance than is realized in a number of subtle but significant ways in modifying growth, behavior of organisms, and the successional dynamics of algal populations.

Two types of extracellular products generally are recognized (Fogg, 1971). (a) Metabolic intermediate compounds of low molecular weight. Glycolic acid, as an intermediate compound in photosynthesis, is known to be released under physiological stress conditions, especially under oligotrophic nutrient conditions and when photosynthesis is light inhibited. Glycolic acid and polysaccharides released during active growth are utilized readily by bacteria (cf. Chapter 17). Intermediate compounds of respiration that are released include organic acids, organic phosphates, and, to a lesser extent, amino acids and peptides. (b) End products of metabolism, usually of high molecular weight, the liberation of which does not depend on equilibrium and which are approximately proportional to the amount of growth. Included in this miscellaneous group are carbohydrates, peptides, volatile compounds such as aldehydes and ketones, enzymes, and a number of growth promoting and inhibiting substances.

The release of simple compounds, e.g., sugars, amino acids, and organic acids, by actively growing cells probably occurs mainly by diffusion through the plasmalemma (Hellebust, 1974). The rates of release depend on the concentration gradient of the substance across the membrane, and the permeability constant of the membrane for the compound. Although active excretion of small compounds is also possible, no convincing evidence for such processes in algae exists. Large molecules, such as polysaccharides, proteins, and polyphenolic substances, probably are excreted by means of more complex processes such as fusion of intracellular vesicles containing the compounds with the plasmalemma. The rates of extracellular release depend on the physiological and environmental factors affecting membrane permeability, and on intracellular concentrations.

The rates of extracellular release under in situ conditions are quite variable. Although rates have been reported ranging from those equal to the rates of carbon fixation into cellular constituents down to less than 1 per cent of rates of carbon fixation, on the average most values are less than 20 per cent. For example, in the unproductive hardwater Lawrence Lake, in situ rates of

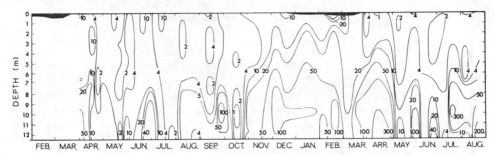

Figure 14–20 Depth-time isopleths of extracellular release of organic carbon by phytoplankton as a percentage of rates of carbon fixed photosynthetically, Lawrence Lake, Michigan, 1968-69. (From Wetzel, et al., 1972.)

release of organic compounds by phytoplankton monitored continuously over nearly 2 years ranged from 0.0 to 22.5 mg C m^{-2} day^{-1} with a mean of 7.3 mg C m^{-2} day^{-1}. The annual mean percentage secretion was 5.7 per cent of net phytoplanktonic primary productivity (Wetzel, et al., 1972). Rates of release were greatest in April when production rates were low but increasing (Fig. 14–20). Secretion reached another maximum during the latter portion of summer stratification as primary productivity began to decrease while particulate organic carbon remained high.

The mean annual amount of released organic carbon in Lawrence Lake was highest at 1 m and decreased with increasing depth except below 10 m, at the upper level of the hypolimnion (Fig. 14–20). In these lower strata of very low light and low rates of photosynthesis, secretion, although extremely low in absolute value (cf. Fig. 11 of Wetzel, et al., 1972), was relatively high in relation to rates of primary production. Therefore, expressed as the mean percentage of extracellular release of all dates and samples, 23.5 per cent of the phytoplanktonic particulate productivity was secreted.

As lakes become more productive, the absolute amounts of organic carbon released increase markedly. For example, the annual range in hypereutrophic Wintergreen Lake, Michigan, was between 2 and 100 mg C m^{-3} day^{-1}, with the highest values in the epilimnion (Fig. 14–21). However, because of the

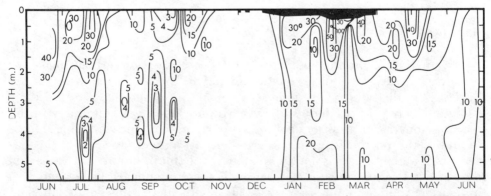

Figure 14–21 Depth-time isopleths of the rates of extracellular release of organic carbon by phytoplankton (mg C m^{-3} day^{-1}), hypereutrophic Wintergreen Lake, Michigan, 1971-72. (From Wetzel, et al., unpublished.)

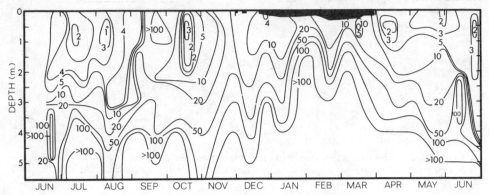

Figure 14–22 Depth-time isopleths of extracellular release of organic carbon by phytoplankton as a percentage of rates of carbon fixation, Wintergreen Lake, Michigan, 1971-72. (From Wetzel, et al., unpublished.)

extremely high rates of photosynthetic carbon fixation and their rapid seasonal oscillations in the compressed trophogenic zone, the percentage release values are generally lower than in oligotrophic waters (Fig. 14–22).

High rates of release of extracellular products of algal photosynthesis have been shown to be a function of CO_2 limitation (e.g., by high pH), high population densities, inhibiting light intensities, low light intensities, and low cell densities and growth rates. The high percentage released at both extremes of the light and pH continuum indicates that the release of a high proportion of photoassimilates is favored by any environmental condition which inhibits cell multiplication but still permits photoassimilation to occur. Membrane permeability or damage is probably the case in situations of high light intensities and other distinctly detrimental conditions.

It should be noted that bacterial utilization of excreted organic compounds is extremely rapid (e.g., Nalewajko and Lean, 1972). As a result, the simpler low molecular weight compounds are utilized and degraded rapidly and large molecular weight substances dominate in time because of their much slower rates of degradation by bacteria (cf. Chapter 17).

Diurnal Changes in Phytoplanktonic Production Rates

Evaluation of the methodology of oxygen change or carbon uptake for measuring in situ rates of production of phytoplankton has shown that the diurnal periodicity of photosynthesis often is not proportional to the daily insolation curve. Most studies show maximum photosynthetic rates in the morning hours, with a subsequent reduction at midday, and sometimes recovery in the afternoon. This diurnal periodicity is particularly acute at the surface, and undoubtedly is partly related to photoinhibition and increased rates of extracellular release of organic photoassimilates. However, much of this surface midday depression dissipates with increasing depth, often in less than 1 meter, as light is attenuated and photosynthesis is undersaturated with light (Vollenweider, 1965).

Diurnal synchrony of cell division and growth is well-established in cultures of algae, and there is every reason to expect it to take place in the same way in natural populations under certain environmental conditions (Soeder, 1965). As a result, the composite phytoplanktonic populations consist of various combinations of algae in (a) active, strongly anabolic growth stages in which cell numbers are increasing and overriding losses by sinking, death, and grazing; (b) "neutrally" active stages that are slightly anabolic, in which cell numbers do not lead to significant positive or negative changes in biomass; and (c) inactive, catabolically metabolizing stages in which cells in resting, decaying, or degenerating stages are largely decreasing. Thus, depending on the combination of algae in various physiological stages of population development, and the dominant stage within the composite phytoplanktonic community, diurnal periodicity of active early morning photosynthetic rates from cells dividing before dawn may be more acute in the early part of the day than at other times. It might also be mentioned that photorespiration increases at midday in submersed angiosperms (Chapter 15). Photorespiration certainly occurs in algae as well, but little is known as yet of its magnitude and diurnal periodicity under natural conditions.

Vertical migration of phytoflagellates can contribute significantly to observed diurnal periodicity in stratification and photosynthesis of phytoplankton (e.g., Nauwerck, 1963; Tilzer, 1973). In high mountain lakes, for example, the dominant flagellates (*Gymnodinium*) usually ascend in the afternoon and evening and migrate downward with increasing light intensities (maximum rates ca 1 m hr^{-1}). Photosynthesis is reduced near the surface during the midday hours as a result of light saturation, photoinhibition, and downward migration, but at greater depths utilization of light energy usually increases at this time.

Horizontal Variations in Productivity

It is commonly assumed in measurements of primary productivity of phytoplankton that horizontal variations within the pelagial zone are relatively minor. Such is the case in many small to moderate sized lakes in which variations in different parts of the lakes are relatively small (<10 per cent), and may often be within the sampling and experimental errors of measurement. Horizontal variations become much more significant, however, in very small lakes, extremely large lakes such as the Great Lakes, in proximity to the littoral zone of most lakes, and as the morphometric complexity of the lake basins increases.

Even in very productive tropical lakes, phytoplanktonic productivity increases more in the pelagial zone closer to the perimeter of the lake than in the central portion of the pelagial (e.g., Talling, 1965). In very large lakes, this horizontal disparity becomes more acute (Glooschenko, et al., 1973, 1974a, 1974b). Nutrient loading from specific regions of the drainage basins of such lakes is sufficient to cause significantly higher, often several-fold, rates of primary productivity in near-shore areas or sectors of the pelagial

TABLE 14–9 Variations in the Annual Primary Productivity of Phytoplankton of Different Sections of Sylvan Lake, Indiana, a Basin of Complex Morphometry[a]

Region of Lake	Annual Mean mg C m^{-2} Day^{-1}	Annual Productivity kg C ha^{-1}
Shallow arm near inlet	1475	5383
Large, isolated bay	1509	5509
Central main basin	1583	5777
Lower arm near outlet	1691	6171
Mean for lake	1564	5710

[a]After data cited in Wetzel (1966a).

zone. Phytoplanktonic productivity within the littoral zone or in the pelagial immediately adjacent to it generally is reduced significantly; the causes of this are discussed in some detail in the following chapter.

In lakes of complex morphometry, particularly in dendritic valleys with numerous large, long bays, primary productivity within these relatively isolated bays can be quite different from that of the major basin (Table 14–9). Often the productivity of these isolated arms is higher than that of the primary basin, and they receive proportionally higher loading of nutrients per unit volume.

The primary productivity of phytoplankton of reservoirs is very variable seasonally, and can be so from year to year. During periods of high runoff from the drainage basin, abiogenic turbidity can be very high in the section of the basin that receives the primary influents. In this region, primary production rates often are greatly reduced. As much of the inorganic turbidity sediments out of the euphotic zone in the progression downstream along the length of the reservoir towards the dam and outlet, the depth and productivity of the trophogenic zone increase. However, if the retention time of the reservoir is short, i.e., the flow-through rate is high, there may be insufficient time for populations to develop fully before being flushed out of the reservoir.

Some reservoirs are so constructed that outflow is drawn from the hypolimnion and lower strata; then the increased retention time of surface waters can lead to a more typically lake-type of phytoplanktonic community development and primary productivity. Removal of hypolimnetic water, laden with nutrients from sedimenting organic matter and decomposition, can reduce the nutrient loading to the basin.

Annual Rates of Primary Production

The number of measurements of the primary productivity of phytoplankton in lake systems is prodigious. Only a few examples have been selected for general comparative purposes in Table 14–10, based on a rather subjective division into oligotrophic, mesotrophic, and eutrophic categories of

(*Text continued on p. 352*)

TABLE 14–10 Comparison of Rates of Primary Production of Phytoplankton in Selected Fresh Waters of Varying Fertility[a]

Lake	Remarks	Mean Daily Productivity for Entire Year (mg C m⁻² day⁻¹)	Range Observed (mg C m⁻² day⁻¹)	Annual Productivity (g C m⁻² year⁻¹)
OLIGOTROPHIC				
Char, N.W.T., Canadian arctic (lat. 74°) (Kalff and Welch, 1974)	80% of total production by benthic flora	1.1	0–35	4.1
Meretta, N.W.T., Canada (Kalff and Welch, 1974)	Polluted by sewage	3.1	0–170	11
Castle, Calif. (Goldman, pers. comm.)	Deep, alpine	98	6–317	36
Lunzer Untersee, Austria (Jónasson, 1972)	Small, alpine	(123)	–	45
Lawrence, Mich. (Wetzel; cf. Table 14–7)	Small, hardwater; 7-year average	112.6	5–497	41.1
Ransaren, Sweden (Rodhe, 1958)	Small lapplandic	–	23–66	–
Lake Superior, USA–Canada (Putnam and Olson, 1961)	Most unproductive of Laurentian Great Lakes	–	50–260	–
Lake Huron, USA–Canada (Vollenweider, et al., 1974)	Offshore stations	–	150–700	ca 100
Borax, Calif. (Wetzel, 1964)	Large, shallow saline lake (25% of total productivity)	249	10–524	91
Lake Michigan, USA (Vollenweider, et al., 1974)	Offshore stations	–	70–1030	ca 130
Lake Ontario, USA–Canada (Vollenweider, et al., 1974)	Offshore stations	–	60–1400	ca 180
MESOTROPHIC:				
Erken, Sweden (Rodhe, 1958)	Large, deep, naturally productive	285	40–2205	104
Clear, Calif. (Goldman and Wetzel, 1963)	Very large, shallow	438	2–2440	160
Esrom, Denmark (Jónasson and Mathiesen, 1959)	Large, moderately deep	370	23–422	260

Table 14-10 Continued on following page.

TABLE 14–10 Comparison of Rates of Primary Production of Phytoplankton in Selected Fresh Waters of Varying Fertility[a] — (Continued)

Lake	Remarks	Mean Daily Productivity for Entire Year (mg C m⁻² day⁻¹)	Range Observed (mg C m⁻² day⁻¹)	Annual Productivity (g C m⁻² year⁻¹)
Fureso, Denmark (Jónasson and Mathiesen, 1959)	Large, deep, many macrophytes	462	0–1380	168
Walter, Ind. (Wetzel, 1973)	Series of 4 interconnected marl lakes			
Basin A		418	102–1395	153
Basin B		210	12–535	77
Basin C		276	30–1048	101
Basin D		437	98–1458	160
Oliver, Ind. (Wetzel, 1973)	Large, deep, marl lake	336	32–775	123
Olin, Ind. (Wetzel, 1973)	Large, deep, marl lake	374	89–996	137
Martin, Ind. (Wetzel, 1973)	Deep, stained marl lake	561	27–1708	205
Pretty, Ind. (Wetzel, 1966b)	Moderate-sized, deep, marl lake			
1963		440	68–1850	161
1964		305	6–895	111
Crooked, Ind. (Wetzel, 1966b)	Large, deep, hardwater lake			
1963		469	142–1364	171
1964		359	23–870	131
Little Crooked, Ind. (Wetzel, 1966b)	Small, deep, kettle lake			
1963		618	263–1903	226
1964		598	9–2451	218

	Description			
Goose, Ind. (Wetzel, 1966a)	Small kettle lake	729	166–1753	266
Lake Erie, USA–Canada (Vollenweider, et al., 1974)				
Western stations		—	30–4760	(310)
Central stations		—	120–1690	(210)
Eastern stations		—	140–1440	(160)
EUTROPHIC:				
Wintergreen, Mich. (Wetzel, et al., unpublished)	Shallow, extensive nutrient loading	1012	60–2240	369
Frederiksborg Slotssø Denmark (Nygaard, 1955)	Shallow, enriched	1030	12–4160	376
Minnetonka, Minn. (Megard, 1972)	Extremely complex basin, large, deep	(820)	–	(300)
Sollerød Sø, Denmark (Steemann Nielsen, 1955)		1430	0–3800	522
Sylvan, Ind. (Wetzel, 1966a; cf. Table 14-9)	Complex basin, large, shallow	1564	9–4959	570
Lanao, Philippines (Lewis, 1974b)	Large, deep tropical lake	1700	400–5000	620
Victoria, Africa (Talling, 1965)	Large, deep, equatorial lake	1750	1700–3800	640
DYSTROPHIC:				
Kattehale Mose, Denmark (Nygaard, 1955)	Very shallow, acidic, peat bog	80	0–400	29
Smith Hole, Ind. (Wetzel, 1973)	Shallow, humic stained	194	24–5960	71
Store Gribsø, Denmark (Nygaard, 1955)	Deep, acidic, humic stained	230	4–680	84
Grane Langsø, Denmark (Nygaard, 1955)	Deep, acidic, humic stained	248	20–880	91

[a]Data approximated in some cases; estimates given in parentheses.

increasing fertility and dystrophic waters of high dissolved organic matter, especially of humic compounds.

The lowest annual productivity values for phytoplankton of natural fresh-water systems appear to be those from Char Lake at latitude 74° in the Canadian Arctic. The annual primary productivity of phytoplankton is about 4 g $C\ m^{-2}\ year^{-1}$ but productivity of mosses and epilithic algae increases this figure to about 20 g $C\ m^{-2}\ year^{-1}$ for the entire lake (Kalff and Welch, 1974; Kalff and Wetzel, in preparation). The low productivity results from low light and temperatures but also from low nutrient availability. For example, Meretta Lake, nearby to Char Lake, receives domestic pollution from a small village which is sufficient to increase primary productivity of phytoplankton by 5 times (Table 14–10). To be sure, lower values exist. To date all waters analyzed, even amictic, permanently ice-covered lakes, contain a limited algal flora and exhibit some net primary productivity.

A general categorization of the characteristics of lakes between the general extremes of very oligotrophic and very eutrophic systems has been attempted many times. The criteria employed, largely resulting from man's innate desire for orderliness in the natural world, have led to classification of lakes on the basis of nearly every parameter, including geomorphology, chemical constituents, nearly every type of organism from bacteria to fish, and even the waterfowl associated with different types of lakes. The most realistic parameter for such categorization, at least in a general way, is one that can be quantitatively determined directly as a rate of growth, and one that integrates the host of environmental parameters controlling the synthesis of organic matter that enters the system. Rates of autochthonous primary production are basic to such an evaluation, and those of the phytoplankton have been proposed many times as the best existing criterion (see, for example, Rodhe, 1969).

The general ranges of primary productivity of phytoplankton commonly associated with oligotrophy and eutrophy are given in Table 14–11, along with several related characteristics of the phytoplankton. Such groupings can only be used in a general way, since there are many exceptions. The general relationships, however, predominate in a majority of inland waters. The dystrophic lake category, discussed at length in Chapter 18, is very deviant and variable. Productivity of phytoplankton in these lakes generally is reduced greatly by complex interactions between light, high concentrations of dissolved humic compounds, and inorganic nutrient availability.

The greatest objection to the use of primary productivity of phytoplankton as a criterion of trophic state is the assumption that organic matter inputs from the littoral flora and allochthonous sources are small and insignificant in comparison to those of the phytoplankton. One of the major objectives of the 2 chapters that follow is to place these components of aquatic systems in proper perspective with the real world. This does not decrease the importance of phytoplankton; rather, the primary productivity of phytoplankton must be viewed as a variably important contributor to overall lake and stream metabolism.

Earlier discussion pointed out the relative preponderance of small algal species of the phytoplankton in oligotrophic lakes, and that in general the

TABLE 14–11 General Ranges of Primary Productivity of Phytoplankton and Related Characteristics of Lakes of Different Trophic Categories[a]

Trophic Type	Mean Primary Productivity[b] (mg C m⁻² day⁻¹)	Phytoplankton Density (cm³ m⁻³)	Phytoplankton Biomass (mg C m⁻³)	Chlorophyll a (mg m⁻³)	Dominant Phytoplankton	Light Extinction Coefficients (η m⁻¹)	Total Organic Carbon (mg l⁻¹)	Total P (µg l⁻¹)	Total N (µg l⁻¹)	Total Inorganic Solids (mg l⁻¹)
Ultra-oligotrophic	<50	<1	<50	0.01–0.5		0.03–0.8		<1–5	<1–250	2–15
Oligotrophic	50–300		20–100	0.3–3	Chrysophyceae, Cryptophyceae, Dinophyceae, Bacillariophyceae	0.05–1.0	<1–3			
Oligo-mesotrophic		1–3						5–10	250–600	10–200
Mesotrophic	250–1000		100–300	2–15		0.1–2.0	<1–5			
Meso-eutrophic		3–5						10–30	500–1100	100–500
Eutrophic	>1000		>300	10–500	Bacillariophyceae, Cyanophyceae, Chlorophyceae, Euglenophyceae	0.5–4.0	5–30			
Hypereutrophic		>10						30–>5000	500–>15000	400–60000
Dystrophic	<50–500		<50–200	0.1–10		1.0–4.0	3–30	<1–10	<1–500	5–200

[a]Modified from Likens (1975), after many authors and sources.
[b]Referring to approximately net primary productivity, such as measured by the ^{14}C method.

size of algae increases among more fertile lakes. A number of analyses of primary productivity in which the relative contribution of differently sized algae were size-fractionated show that a greater proportion of the total primary productivity is by smaller forms in less productive waters (Rodhe, et al., 1958; Goldman and Wetzel, 1963; Wetzel, 1964, 1965a, and others). These contributions of the smaller algae also shift seasonally to a point at which the nannoplankton and ultraplankton (<10 μm) contribute to nearly all of the planktonic primary productivity.

An inverse relationship generally exists between total algal biomass and productivity per unit biomass in the phytoplankton. Accordingly, it is observed that communities dominated by populations of species with small cells have a greater primary productivity per unit algal biomass than communities dominated by large-cell algal populations (Findenegg, 1965). As demonstrated by microautoradiography, the renewal time (generation time) is related inversely to cell size; those cells that are small with a low carbon content tend to have shorter renewal times and be more metabolically active than larger algae (Stull, et al., 1973). Therefore, some small species of relatively minor contribution to the algal community biomass are turning over rapidly and contributing more to the total primary productivity than larger species.

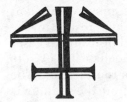

the littoral zone

CHARACTERISTICS OF THE LITTORAL ZONE

The littoral region consists of an interface zone between the land of the drainage basin and the open water of lakes. Its magnitude in relation to the size of the open water or pelagial region varies greatly among lakes, and depends on the geomorphology of the basin and the rates of sedimentation that have occurred since its inception. Most lakes of the world are relatively small in area and shallow. As such the littoral flora constitute a major source of synthesis of organic matter that contributes significantly to the productivity of lakes and the regulation of metabolism of the whole lake ecosystem.

The aquatic macrophytes are frequently demarcated into 3 distinct zones within the littoral. The shallower region, to a depth of about one meter, consists of highly productive emergent macrophytes that utilize the resources of both the terrestrial and aquatic habitats. High water and nutrient availability from the aqueous sediments, combined with much greater availability of oxygen and carbon dioxide from the atmosphere than is available to plants under water, result in high growth. The emergent angiosperms form some of the most productive plants of the biosphere. Lakeward, often between water depths of 1 to 3 meters, there is a zone in the lower littoral of floating-leaved vegetation. These plants, exemplified by the water lilies, usually are perennials that are rooted firmly with extensive rhizome systems; they extend their floating leaves to the surface by long, flexible petioles. Below this area, extending from about half a meter of depth to the limit of the photic zone, is the region of submersed rooted or adnate macrophytes. Submersed angiosperms rarely exceed water depths of 10 meters, in part as a result of the effects of hydrostatic pressure on internal gas transport. Submersed mosses and macroalgae, although they grow at various depths, are capable of growing at low light intensities and often penetrate to the lower limits of the photic zone. Nonrooted floating macrophytes consist of angiosperms, ferns, and some liverworts that range in size from minute duckweeds to massive floating populations which occasionally dominate protected waters.

Of terrestrial origin, the aquatic angiosperms exhibit a number of major adaptations in the transition to complete submergence. Submersed angio-

355

sperms tend towards structural reduction as the need for supportive tissue is reduced. Conversely, limitations of aqueous gas exchange result in the massive development of internal lacunae that permit rapid diffusion of gases within the plant tissue. Within the plants, physiological mechanisms exist to retain much of the metabolic oxygen and carbon dioxide for reutilization. The plants obtain nutrients from the water by their foliage, and from the sediments by their root and rhizoid systems. Translocation occurs in both directions. A portion of the nutrients and organic products of photosynthesis is released from the plants into the water and epiphytic complex of attached algae and bacteria.

Submersed macrophytes and aqueous portions of emergent and floating macrophytes provide an enormous surface area that is colonized by microflora. In addition, all nonliving substrates within the photic zone are colonized by microflora that are more or less attached. An extremely diverse spectrum of microhabitats occurs in the littoral zone among substrates of sand, rock, organic sediments, and macrophytes. The complex interactions of the associated microflora and their substrates are poorly understood. It is clearly evident, however, that metabolic advantages of resource acquisition exist in the proximity of microflora attached to or living in close juxtaposition to living or nonliving substrates that are not as prevalent among planktonic microflora living in a dilute medium. Sessile populations tend to be more stable than their planktonic counterparts.

Although about 90 per cent of all algal species are represented among the littoral algae, diatoms dominate on all substrates among moderately productive ecosystems. Blue-green algae and flagellates increase proportionately in organically enriched lakes. Algae living on loose sediments that are subject to disturbance and displacement by water movements exhibit an ability to migrate vertically in response to changes in light. This vertical migration is phased with diurnal rhythms in cell division and photosynthetic capacity. Within the littoral zone, photosynthesis of attached algae is influenced largely by light availability, although compensatory mechanisms exist that permit active growth at very low levels of irradiance.

When comparisons of growth among algal populations associated with different substrates are possible, they show that photosynthetic carbon fixation is commonly higher among those attached to sand grains than those associated with organic sediments. Algae epiphytic on macrophytes are often more productive than algae associated with nonliving substrates by an order of magnitude or more. Evidence indicates that a complex and highly dynamic metabolic relationship exists among the epiphytic algae, bacteria, and the supporting macrophyte.

Productivity rates of littoral algae are not much greater than those of phytoplankton, and turnover rates tend to be somewhat slower. However, the massive surface area available for colonization, especially among submersed macrophytes, can result in very high contributions of attached littoral algae to the total primary productivity of many freshwater systems. When this productivity is coupled with the very high rates prevalent among the emergent macrophytes, the littoral primary productivity can form a major input of organic matter to lake and certain stream systems.

AQUATIC MACROPHYTES OF THE LITTORAL ZONE

Classification of Aquatic Macrophytes

Any inquiry into the botanical aspects of lakes, separate from the phyto-plankton, immediately finds an array of rather arbitrary definitions of the sessile flora of aquatic systems, a subject that has been reviewed at some length by Arber (1920), Gessner (1955, 1959), and Sculthorpe (1967). Many definitions emanate from looking at the reproductive characteristics in rela-tion to the aquatic stages of the life history of groups of angiosperms. From the strictly botanical viewpoint, such definitions are quite satisfactory. From an ecological standpoint, however, such species-orientated categorizations are unrealistic and ignore major system interrelationships. Alternatively, words such as hydrophytes are broadly ambiguous, even though the term hydro-phyte generally refers to vascular aquatic plants, and could include any form of aquatic plant. The term *aquatic macrophyte*, as it is commonly used, in-cluding in this work, refers to the macroscopic forms of aquatic vegetation, and encompasses macroalgae (e.g., the alga *Cladophora*, the stoneworts such as *Chara*), the few species of pteridophytes (mosses, ferns) adapted to the aquatic habitat, and the true angiosperms. Division on the basis of size ad-mittedly also is arbitrary but, as discussed further on, when combined with the definition of the attached microflora, it permits a meaningful separation of the producer components of the littoral zone.

Numerous lines of evidence clearly indicate that the origin of truly aqua-tic angiosperms is from the land. A number of adaptations have occurred in the transition from terrestrial to aquatic conditions. Adaptation and specializa-tion have been achieved by only a few representatives of angiosperms (<1 per cent) and pteridophytes (<2 per cent). Many possess relics of their terres-trial heritage, such as a thin cuticle, functionless stomata, and a poorly ligni-fied xylem tracheary structure. Most are rooted, but a few species float freely in the water. Their dissimilar origins from many diverse groups, extreme plasticity in structure and morphology in relation to changing environmental conditions, and the very heterogeneous conditions of their littoral habitat frustrate attempts to classify this group more precisely.

Numerous biological classification systems have been proposed and vari-ously used. The primary groups of aquatic angiosperms, rooted and nonrooted, have been subdivided according to types of foliage and inflorescence, and whether these organs are emergent, floating on the water surface, or sub-mersed (Arber, 1920). Definitions of the extent of emergence or submergence and the manner of attachment or rooting to the substratum have led to the creation of numerous complex terms (cf. Hejný, 1960; Luther, 1949; den Hartog and Segal, 1964). Although useful for certain phytosociological analyses, these terms are cumbersome and lines of demarcation are rarely distinct.

On the basis of morphological and physiological grounds, rationale exists to retain a simple organization of aquatic macrophytes on the basis of attach-ment, after Arber (1920) and Sculthorpe (1967):

(A) *Aquatic macrophytes attached to the substratum:*

1. *Emergent macrophytes:* These occur on aerial or submersed soils,

from the point at which the water table is about 0.5 m below the soil surface to where the sediment is covered with ca. 1.5 m of water; they are primarily rhizomatous or cormous perennials (e.g., *Glyceria, Eleocharis, Phragmites, Scirpus, Typha, Zizania*); in heterophyllous species submersed and/or floating leaves precede mature aerial leaves; many species may exist as (usually sterile) submersed forms; all produce aerial reproductive organs.

2. *Floating-leaved macrophytes:* These are primarily angiosperms that occur on submersed sediments at water depths from about 0.5 to 3 m; in heterophyllous species submersed leaves precede or accompany the floating leaves; reproductive organs are floating or aerial; floating leaves are on long flexible petioles (e.g., the water lilies *Nuphar* and *Nymphaea),* or on short petioles from long ascending stems (e.g., *Brasenia, Potamogeton natans).*

3. *Submersed macrophytes:* These comprise a few pteridophytes (e.g., the fern *Isoetes*), numerous mosses and charophytes (stonewort algae *Chara, Nitella*), and many angiosperms. They occur at all depths within the photic zone, but vascular angiosperms occur only to about 10 m (1 atm pressure); leaves are highly variable, from finely divided to broad; reproductive organs are aerial, floating, or submersed.

(B) *Freely floating macrophytes:* A highly varied group that is typically not rooted to the substratum, but lives unattached in the water; diverse in form and habit, ranging from large plants with rosettes of aerial and/or floating leaves and well-developed submersed roots (e.g., *Eichhornia, Trapa, Hydrocharis*), to minute surface-floating or submersed plants with few or no roots (e.g., Lemnaceae, *Azolla, Salvinia*); reproductive organs are floating or aerial (e.g., aquatic *Utricularia*) but rarely submersed (e.g., *Ceratophyllum*).

Lacustrine Zonation

The bottom of a lake basin is separable from the free open water, the *pelagial zone,* and is further divisible into a number of rather distinct transitional zones from the shore to the deepest point (Fig. 15–1). The *epilittoral zone* lies entirely above the water level and is uninfluenced by spray; the *supralittoral zone* also lies entirely above the water level, but is subject to spraying by waves. The *eulittoral zone* encompasses that shoreline region between the highest and lowest seasonal water levels, and is often influenced by the disturbances of breaking waves. The eulittoral zone and the *infralittoral zone* collectively constitute the *littoral zone.* The infralittoral zone is subdivided into three zones in relation to the commonly observed distribution of macrophytic vegetation: *upper infralittoral* or zone of emergent rooted vegetation; *middle infralittoral* or zone of floating-leaved rooted vegetation; and *lower infralittoral* or zone of submersed rooted or adnate macrophytes.

Below the littoral is a transitional zone, the *littoriprofundal,* occupied by scattered photosynthetic forms, which is often adjacent to the metalimnion of stratified lakes. The upper boundary of the littoriprofundal zone at the lower

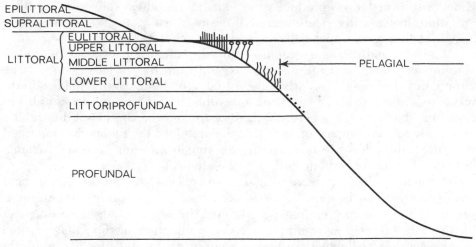

Figure 15–1   Lacustrine zonation (see text for discussion.) (After Hutchinson, 1967.)

edge of macrovegetation of the lower infralittoral is usually quite distinct. The lower boundary of the littoriprofundal includes a gradient of benthic algae, especially blue-green algae, and photosynthetic bacteria that is less sharply demarcated. The remainder of the sediments, consisting of exposed fine sediment free of vegetation, is referred to as the *profundal zone.*

Limitations of space do not permit treatment of the littoral flora in the detail these plants deserve. Because of the major impact of the littoral vegetation on the metabolism of a majority of lake systems, however, a brief résumé of the main aspects of their morphological and physiological adaptations is called for in order to appreciate their role in the productivity of fresh waters. Excellent reference works include Arber (1920), Gessner (1955, 1959), and Sculthorpe (1967), from which much of the ensuing discussion is drawn.

Emergent Flora

Aerial stems and leaves of emergent macrophytes, largely monocotyledons in the upper infralittoral zone, possess many similarities in both morphology and physiology to related terrestrial plants. The emergent monocotyledons such as *Phragmites* and the cattail *Typha* produce erect, approximately linear leaves from an extensive rhizome system of anchorage. Epidermal cells are elongated parallel to the long axis of the leaf, which lends flexibility to bending, and the cell walls are heavily thickened with cellulose, which provides the necessary rigidity. The mesophyll is generally undifferentiated, and contains large air spaces traversed at intervals by diaphragms. These lacunae are separated from each other by thin walls of parenchyma cells. Vascular bundles show typical anatomy: the xylem consists of scattered tracheids and parenchyma cells, and the phloem of sieve tubes, companion, and parenchyma cells is ensheathed with supporting sclerenchyma fibers.

Emergent dicotyledons produce erect leafy stems which show greater differentiation anatomically. The mesophyll tissue of leaves is divided into typically dicotyledonous upper palisade and lower spongy layers.

The root and rhizome systems of these plants exist in permanently anaerobic sediments, and must obtain oxygen from the aerial organs for sustained development. Similarly, the initial growth of young foliage under water must be capable of respiring anaerobically for a brief period until the aerial habitat and supply is reached, since the oxygen content of the water is extremely low in comparison to that of the air. Once the foliage has emerged, the intercellular lacunae increase in size, thus facilitating gaseous exchange between the photosynthetic cells and the atmosphere.

A linear gradient of decreasing oxygen concentration occurs in the gases of the intercellular lacunae between the aerial foliage and the underground organs. The rate of oxygen transport through the lacunal system is influenced by the rate of net photosynthetic oxygen production in the foliage, which undergoes diurnal fluctuation, by the respiratory demands for oxygen of the root system, and by the resistance to diffusion afforded by diaphragms of the lacunae. Experimental analyses have demonstrated that roots and rhizomes can tolerate appreciable periods, of as long as a month, of very reduced oxygen supply or anaerobiosis without any apparent harmful effects.

Transpiration

The epidermal cuticle of the foliage of emergent and floating leaves is generally well-developed. Cuticular transpiration is much lower in these plants than in meso- and xerophytic land plants (Table 15–1). Cuticular transpiration is generally greater from floating-leaved macrophytes than from aerial foliage. As a result, transpiratory water loss to the atmosphere by emergent macrophytes is restricted largely to the stomata.

Rates of transpiration by emergent macrophytes are extremely high, and result in a quantitative efflux of water vapor from the leaves that is much greater than an equivalent area of water (cf. Chapter 4, Table 4–3). Evaporation from the water surface is lower when the water surface is densely covered by macrophytes, such as water lilies or duckweeds (*Lemna*), that restrict contact of moving air with the water surface. However, this effect is counterbalanced by transpiration rates that increase markedly when the

TABLE 15–1 Comparison of the Ratio of Maximal to Minimal Cuticular Transpiration among Aquatic Macrophytes and Terrestrial Flora[a]

Aquatic Flora		Meso- and Xerophytes	
Potamogeton natans	1.4	*Quercus ilex*	4.6
Eichhornia crassipes	1.5	*Syringa vulgaris*	4.9
Alisma plantago	1.7	*Viola tricolor*	7.6
Nymphaea marliacea	3.2	*Laurus nobilis*	8.4

[a]Modified from Gessner, 1959.

vegetation is exposed to moderate wind velocities. As a result, the relative humidity of the atmosphere in and above an emergent or floating-leaved stand of vegetation is greatly increased in comparison to that of air overlying open water. The differential increase in relative humidity of air over a macrophyte stand in relation to that over open water changes diurnally as well, being maximal at the midday peak of net photosynthesis. Stomata of aquatic macrophytes generally do not close to nearly the extent observed in terrestrial flora during the light period, and result in higher transpiration rates than occur in land plants. Stomata perform a similar function in both aquatic plants and terrestrial flora: they prevent desiccation in the event of water fluctuations.

Many aquatic plants with emergent and floating leaves, which necessitate transpiration of water vapor, have developed means of increasing surface area and evapotranspiration. Such structures include papillae, either single-celled or multicellular extensions of the epidermis, or perforations into the leaf blade, termed stomatodes. Many emergents extrude water (guttation) from enlarged ends of veins (hydathodes) around the margins of leaves.

The very high transpiration rates among emergent vegetation, much greater than rates among floating-leaved forms, vary greatly diurnally and seasonally with the changing morphological development of the plant species. For example, among the reeds *Phragmites* and *Glyceria,* daily transpiration patterns are directly correlated with increasing temperature, decreasing relative humidity, and increasing evaporation rates. Transpiration is usually at a lower rate in the basal leaves than in apical foliage, resulting in a general tendency towards xerophytic properties with increasing age of the leaf.

Among terrestrial plants, water availability is a distinct factor influencing growth. Among aquatic macrophytes, population density is important in the regulation of transpiration rates by self-imposed control of environmental conditions of the littoral habitat. Evapotranspirational water losses by emernent macrophytes in the littoral zone of lakes and marshes can be so effective that their metabolism can significantly reduce the water levels of the surrounding terrestrial area and result in diminished terrestrial plant growth.

Increases in water level, to the point when large portions of or the entire emergent plant are submerged, result in a varied number of morphological and physiological adaptations. Although many species can survive submergence for long periods of time, growth is greatly reduced under these conditions of diminished light and oxygen. Leaves are generally reduced in size, and tend to lose rigidity and elongate. Mesophyll tissue often is reduced, with a corresponding increase in spongy tissue and intercellular air spaces. The xylem and lignified tissues of phloem fibers also are generally decreased, concomitant with an increase in stem diameter and lacunal air spaces.

Floating-Leaved Macrophytes

The macrophytes that are attached to the substratum and possess leaves that float on the water surface are nearly all angiosperms, most conspicuously represented by the ubiquitous water lilies. The surface of the water is a region subject to severe mechanical stresses from wind and water movements

that demand effective morphological resistance. Adaptations in these macrophytes, exhibiting parallel evolution from many unrelated groups, tend
towards strong, leathery, and peltate leaves, circular in shape with an entire
margin, hydrophobic surfaces, and with long pliable petioles. In spite of these
adaptations, severe winds and water movements restrict these macrophytes
to sheltered habitats in which there is little water movement.

Floating leaves exhibit well-developed dorsiventral organization, in
which the mesophyll usually is differentiated into an upper photosynthetic
palisade tissue and an extensive lacunate tissue. Localized masses of spongy
tissue aid buoyancy and offer resistance to tearing, in combination with vascular tissues. The venous network of the leaves of aquatic macrophytes is
much less extensive than among terrestrial plants. Reduction is greatest
among submersed angiosperms (Table 15–2).

Positioning of leaf surfaces horizontally on the water surface creates a
vigorous competition for space to expose maximum leaf area to incident light.
Leaf growth some distance from the root or rhizome system is accommodated
by long, very pliable petioles. Between 10 cm and 4 m water depth, a complete proportionality is found between water depth and the length of the leaf
petioles of water lilies. As root growth progresses to deeper water, petioles
compensate proportionally with about 20 cm excess length, slack that permits
movement on the surface among undulating waves. Stems of flowers, separate
from the leaves in many species, grow to the surface in the same way. If
gaseous exchange to the leaf surface is restricted experimentally, elongation
of the petioles continues, and the leaves become aerial instead of floating.
Although the control mechanisms of elongation are poorly known, results indicate that neither oxygen nor carbon dioxide availability is involved, but
rather that petiole or stem growth is arrested at the surface when the leaves
begin to lose ethylene to the atmosphere (see page 364). Under crowded conditions, elongation frequently continues to the point at which leaves become
somewhat aerial above the water surface.

TABLE 15–2   Vein Length Densities (cm cm^{-2}) among Different Plant Types[a]

Aquatic Plants		Terrestrial Plants	
FLOATING LEAVES:		*Rosa canina*	108
Nelumbo:		*Acer campestris*	102
Aerial leaf:		*Fraxinus excelsior*	88
Edge	104	*Quercus cerris*	86
Middle	91	*Salix rubrum*	52
Floating leaf:			
Edge	68		
Middle	71		
Nymphaea mexicana	44		
Salvinia auriculata	27		
SUBMERSED LEAVES:			
Potamogeton praelongus	14		
Nuphar luteum	7		

[a]After Gessner, 1959.

Distribution of stomata varies among different species. The most common pattern, however, is a restriction of stomata to the upper leaf epidermis. A few stomata are found on the underside of leaves; these are considered relict features and have no function.

The epidermis of the leaf underside of many floating-leaved species and submersed angiosperms contains groups of many smaller cells nested around a pore about 0.05 μm in diameter and overlain with an extremely thin cutin lamella. These "organs," termed hydropoten, assume a three-celled lenticular form in some water lilies or shield-like hairs in certain submersed angiosperms. The hydropoten function in ion absorption in both submersed and floating leaves in contact with water by an active mechanism analogous to that of root absorption (Lüttge, 1964). Absorbed ions are translocated to veins of the mesophyll.

Submersed Macrophytes

The submersed macrophytes are a heterogeneous group of plants that include (a) filamentous algae (e.g., *Cladophora*) that under certain conditions develop profusely and become pseudo-attached to the substrata of the littoral zone in massive mats, (b) certain macroalgae (e.g., the calcareous stonewort Charales) that can constitute dominating littoral populations in hardwater lakes, (c) mosses that are occasionally the major macroflora of softwater lakes and streams, (d) a few totally submersed ferns (e.g., *Isoetes*) that rarely occur in large populations and are restricted to clear, softwater lakes, and (e) vascular submersed macrophytes of slightly less than 20 diverse families, mostly monocotyledons.

Among the vascular submersed macrophytes, numerous morphological and physiological modifications are found in adaptation to a totally aqueous environment. Except for some highly modified species, stems, petioles, and leaves contain little or no lignin even in vascular tissues. Sclerenchyma and collenchyma are usually absent. No secondary growth occurs and no cambium can be recognized. Conditions of reduced illumination under the water are reflected in numerous characteristics: an extremely thin cuticle, leaves only a few cells in thickness, and concentration of pigments in chloroplasts in epidermal tissue. Leaves tend to be much more divided and reticulated than those of terrestrial or other aquatic plants. The vascular system is greatly reduced, and all major conducting vessels have been lost from stems. As a rule, conducting bundles have coalesced to axial vascular bundles in both mono- and dicotyledon submergents. Phloem and xylem cannot be differentiated, since phloem and woody parenchyma are nearly absent.

Leaves of submersed macrophytes occur in three main types: entire, fenestrated, and dissected. Entire leaf form is found most commonly throughout all groups and habitats. Entire leaves exhibit a strong tendency towards elongation to ribbonlike and filiform morphology in which length greatly exceeds width, even among more lanceolate-type leaves. Elongation to considerable length with pliability affords mechanical advantages in water movements, maximally utilizes the reduced available light, and increases the ratio of surface area to volume, and presumably also the efficiency of gaseous exchange and nutrient absorption.

Fenestrated leaves among submergents are rare and occur only among a few tropical monocotyledons. The adaptive advantages of such perforated, lace-like foliage are unclear. Dissected leaves, often to extremely reticulated positioning of leaves, are common among submersed dicotyledons. The most common form is extreme dissection with segments in whorls radiating from the petiole. Both the fenestrated and dissected leaf forms offer reduced resistance to water movements and greatly increase the surface area to volume ratios.

Heterophylly

Although vegetative polymorphism or heterophylly is not unique to submersed macrophytes, extremes in foliar plasticity are particularly conspicuous among aquatic macrophytes that normally grow in the submersed habitat in shallow water but, with fluctuations in water level, can adapt readily with floating or aerial foliage. A marked polymorphism of leaves often may be found on the same stem or petiole, which tends to shift from finely divided submersed leaves in the submersed habitat to more coalesced entire leaves in the transition to surface-floating or aerial leaves. Transitions are sometimes abrupt between forms; but among other species, or under differing conditions on the same plant, the transition may be a gradual elaboration of size and structure. Leaf form and anatomy can vary widely with age, water depth, current velocities, nutrient supplies, light intensity and day length, temperature, and other factors. Heterophylly is particularly conspicuous among such common genera as *Ranunculus*, *Callitriche*, and *Potamogeton*, and creates taxonomic chaos when this is based on leaf morphology. For example, leaf morphogenesis of the amphibious buttercup *Ranunculus* responds conspicuously to submergence and to temperature changes (Fig. 15–2). The lobate tripartite aerial leaf of the semi-aquatic phase is approximated in the submersed phase at high temperatures. Both lobe numbers and blade lengths increase conspicuously with growth at decreasing temperatures. Analogous morphological differentiation has been shown to be directly related to carbon dioxide concentrations in at least three taxonomically distinct amphibious species (Bristow, 1969). Under conditions of high aqueous CO_2 concentrations, the plants developed highly dissected submersed leaves, whereas when the CO_2 content was reduced, lobate aerial foliage developed rapidly. These plants were unable to utilize bicarbonate. The results are similar to the response among water lilies of continued elongation of petiole length under conditions of low CO_2 of the water until the relatively high levels of atmospheric CO_2 are reached (Gessner, 1959).

Recent evidence has demonstrated that when submersed, ethylene concentrations reach high levels in the internal intercellular lacunae in several aquatic angiosperms (Musgrave, et al., 1972). These high levels of ethylene are necessary to enhance the sensitivity of the tissue to the hormone gibberellic acid and to promote elongation. When the tissues reach the water surface,

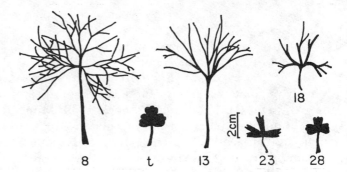

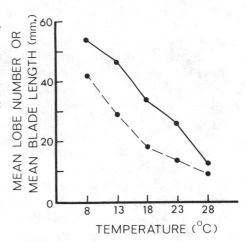

Figure 15-2 *Upper:* Silhouettes of leaves of a single clone of *Ranunculus flabellaris* grown in the semi-aquatic phase (*t*) and submersed at temperatures of 8, 13, 18, 23, and 28°C. *Lower:* Mean blade length (——) and mean lobe number (----) in relation to temperature. (Modified from Johnson, M. P.: Temperature dependent leaf morphogenesis in Ranunculus flabellaris. Nature (London), 214:1354–1355, 1967.)

gaseous contact is made between the intercellular air and the atmosphere, and the accumulated ethylene rapidly dissipates into the atmosphere. Gibberellic acid activity in promoting cell elongation is then reduced.

Numerous other factors influence the physiological manifestation of leaf form among aquatic macrophytes (cf. Sculthorpe, 1967; Gaudet, 1968; Hutchinson, 1975). Generalizations are difficult to make, however, because among some plants polymorphism occurs irrespective of external factors, is distinguishable at the earliest primordial stages, and is apparently determined genotypically and by the nutritional status of the shoot meristem.

Further morphological variations are found among aquatic macrophytes of the temperate zone that overwinter by the formation of winter buds (turions). In the fall many common macrophytes (e.g., *Utricularia, Myriophyllum, Potamogeton*) form amorphous, semispherical masses of aborted leaves with very short internodes in the axils of lower leaves. These turions separate from the mother plant and sink or float some distance away, and serve as a means of vegetative propagation. Many submergents persist for many years by this means without a sexual cycle. Turion formation is absent among the same species in the tropics, but can be easily induced when plants near maturity are exposed to water temperatures less than 10°C.

Freely Floating Macrophytes

The freely floating macrophytes, submersed or on the surface, also exhibit great diversity in morphology and habit. Several of these plants, notably *Lemna, Pistia, Salvinia, Eichhornia,* and *Trapa,* develop profusely and so densely populate waterways and lakes that they inhibit the commercial use of these systems (Hillman, 1961; Moore, 1969; Mitchell and Thomas, 1972).

The most elaborate of the free-floating life forms consists of a rosette of aerial and surface-floating leaves, a greatly condensed stem, and pendulous submerged roots. Most rosette species are perennials and are free-floating throughout growth, except for initial stages of seedling development. Among other free-floating groups, there is a strong tendency for structural reduction from the rosette habit. In the duckweeds (Lemnaceae), extreme reduction is seen in the trend towards elimination of roots and lack of separation between stem and leaf. Similarly, among the submersed free-floating "carnivorous" *Utricularia* stem and leaf are not differentiated and vegetative organs are greatly modified for entrapment of microfauna. Some species of *Lemna* and *Utricularia* are, like the ubiquitous *Ceratophyllum,* submersed during vegetative growth and rootless. Free-floating macrophytes generally are restricted to sheltered habitats and slow-flowing rivers. Their nutrient absorption is completely from the water; most of these macrophytes are found in waters rich in dissolved salts.

Most floating plants possess little lignified tissue. Rigidity and buoyancy of the leaves are maintained by turgor of living cells and extensively developed lacunate mesophyll tissue (often > 70 per cent air by volume). Vascular tissues of the leaves are very poorly differentiated; in most, the protoxylem is represented by a lacuna. Vegetative propagation by the production of lateral stolons that develop into new rosettes is a common mode of rapid expansion in this group. All free-floating rosette plants form well-developed adventitious roots, lateral roots, and epidermal hairs. The root system of *Eichhornia,* for example, represents 20 to 50 per cent of the plant biomass.

Gaseous Metabolism by Aquatic Macrophytes

Carbon Assimilation

Atmospheric carbon dioxide is clearly the dominant source of carbon among emergent, floating-leaved (rooted), and freely floating macrophytes. A few species of freely floating angiosperms, such as the duckweed Lemnaceae, utilize both atmospheric and aqueous carbon sources (Wohler, 1966; Wetzel and Manny, 1972). Heterophyllous aquatic macrophytes presumably are similarly adapted to use both atmospheric CO_2 and aqueous CO_2–HCO_3.

Rates of diffusion of gases in water are several orders of magnitude slower in water than in air (cf. Chapter 10) and present particular stresses on availability from aqueous solution. Morphological adaptations among submersed leaves, stems, and some petioles towards thinness of leaves to only 1 to 3 cell layers, reduced cuticle development, and extreme reduction or elimination of mesophyll with dense chloroplasts distributed in epidermal cells, all increase utilization and exchange of gases. Massive intercellular spaces in leaves, stems, and petioles facilitate rapid internal diffusion.

The question of limitation of photosynthesis by the slow diffusion of CO_2 in water has received much attention, especially in relation to the alternate utilization of bicarbonate ions. In large submerged angiosperms and algae, the diffusion path through the cells of the thallus or leaf is long (up to 50 μm), to which must be added an effective gradient zone of about 100 μm of unstirred water at the cellular surfaces; further, diffusion rates are much slower than among microalgae (Steemann Nielsen, 1947; Raven, 1970). Low velocity currents increase photosynthetic rates among submersed angiosperms and are related, in part, to disruption of the stagnant surrounding zone and increased diffusion rates (e.g., Barth, 1957; Westlake, 1967).

The ability to assimilate bicarbonate ions as a carbon source supplementary to carbon dioxide has been demonstrated clearly in a number of investigations. Space does not permit detailed discussion of the conflicting results of some of these studies, a subject most comprehensively reviewed by Raven (1970). Physiological mechanisms of bicarbonate uptake are not fully understood. However, among certain submersed angiosperms the process is as follows: (a) Carbon dioxide and bicarbonate ions are taken up through both leaf surfaces. (b) Carbon dioxide is fixed, and hydroxyl ions equivalent to the amount of bicarbonate used pass out through the adaxial leaf surface. (c) A quantity of cations, primarily calcium, equivalent to the amount of bicarbonate taken in through the abaxial leaf surface, is transported from the abaxial to the adaxial leaf surface, thereby achieving stoichiometry and charge balance in all compartments. (d) The passage of bicarbonate, calcium, and hydroxyl ions to the adaxial leaf surface usually results in precipitation of $CaCO_3$ on that surface. The carbonate deposits encrusting the submersed parts of macrophytes in calcareous hard waters often exceed the weight of the plant material (Wetzel, 1960). Among emergent and floating macrophytes that utilize atmospheric carbon dioxide, encrustation on submersed portions was found to be highly variable and proportional to the extent of development of epiphytic algae. Deposits on submersed species were much heavier and were correlated with morphological variations and the ability to utilize bicarbonate ions.

The ability to assimilate bicarbonate ions as well as carbon dioxide can be regarded as an adaptation by those plants possessing a long diffusion path from the medium to biochemical sites of utilization, when they are living in waters containing more bicarbonate ions than free carbon dioxide, or when the free CO_2 compensation point is above the CO_2 concentration in equilibrium with the air. In algae and especially larger macroalgae, the ability to utilize bicarbonate ions in photosynthesis is widespread, but not universal. The extensive work on this subject among mosses and angiosperms by Steemann Nielsen (1944, 1947) and Ruttner (1947, 1948, 1960) has shown that the freshwater red alga *Batrachospermum* and all assayed genera of freshwater submersed mosses utilize only free CO_2 and cannot assimilate bicarbonate. These latter plants are restricted almost universally to soft waters of relatively low pH and streams in which CO_2 concentrations are relatively high.

Aeration and Respiration in Submersed Organs

It has been assumed that as photosynthesis proceeds in submersed macrophytes, oxygen is released from the plants into the surrounding water in a general proportion to rates of photosynthesis. While such a relationship is approximately accurate for micro- and macroalgae and apparently also for bryophytes, such is not the case for angiosperms. Among angiosperms during the photoperiod oxygen accumulates rapidly in the lacunal intercellular atmosphere on a diurnal basis and only slowly diffuses out into the surrounding water (Hartman and Brown, 1966). Relative diffusion rates of oxygen into the water are not directly correlated with the intensity of photosynthesis, and accumulations in the intercellular lacunae can be utilized for respiration during both the photoperiod and in darkness, without effect on the oxygen concentrations of the medium. Although release of bubbles, largely oxygen, from submerged plants can be significant during periods of intensive photosynthesis in warm water, this release should not be used as a measure of photosynthesis (see Odum, 1957, among others) because it represents only a very small proportion of total photosynthetic oxygen output. Most of the oxygen diffuses into the water or is retained within the plant.

The magnitude of the internal lacunal system is highly variable among species, but is extensive almost universally and constitutes a major portion of the total plant volume, often exceeding 70 per cent. The fragility of the lacunal tissue and possibility of it becoming filled with water if the plant is damaged at any point are protected against by a number of types of lateral plates and watertight perforated diaphragms that interrupt the lacunae at intervals (Arber, 1920; Sculthorpe, 1967). Although experimental evidence is very meager, it is known that much of the oxygen produced in photosynthesis and retained in the lacunal system diffuses from the leaves through the petioles and stems to underground root and rhizome systems, whose respiratory demands are high.

Photosynthetic carbon fixation of aquatic macrophytes is balanced by losses of respiratory CO_2 and secretion of soluble organic compounds, and photosynthetic efficiency is influenced directly by rates of these processes. In calculations of primary productivity, the respiration rate is usually assumed to be the same in light as in the dark. Physiological evidence, however, indicates that such an assumption is erroneous in many cases. The normal process of mitochondrial (dark) respiration may be inhibited in the light in some plants, perhaps by suppression of glycolysis (Jackson and Volk, 1970). Furthermore, efficient refixation of respired CO_2 in the light is probably a universally common event in submersed aquatic plants. These processes would restrict the loss of respiratory CO_2.

Alternately, the phenomenon of photorespiration, well-known in terrestrial plants (Goldsworthy, 1970; Jackson and Volk, 1970; Hatch, et al., 1971), enhances loss of CO_2 and may be a significant factor in reduction of photosynthetic efficiency of aquatic macrophytes. In this process CO_2 is generated in the light as a result of glycolic acid, a direct product of C_3 Calvin cycle photosynthesis. The rate of this reaction is highly influenced by, and is proportional to, oxygen concentration, light intensity, and temperature. Glycolate metabolism is also enhanced when low CO_2 limits photosynthesis. The rate at which photorespired CO_2 is lost from the plant depends on the efficiency of CO_2 refixation.

Terrestrial plants in which all cells photosynthesize by the C_3 Calvin cycle can lose up to 50 per cent of fixed carbon immediately in photorespiration, depending on environmental conditions. Those plants in which photosynthesis proceeds through the C_4 β-carboxylation pathway in mesophyll cells efficiently refix CO_2 both from dark respiration and from photorespiration in the C_3 bundle sheath cells, and little or no CO_2 is lost from these plants in the light.

C_3 and C_4 plants can be distinguished by the following combination of characteristics, any one of which is often reasonable evidence for making the distinction (Black, 1971): C_4 plants have highly developed bundle sheath cells in leaf cross-sections, with unusually high concentrations of organelles and starch accumulation (elucidated by use of iodine-potassium iodide stain); C_3 leaves lack this differentiation. Photosynthesis is difficult to light saturate in C_4 plants, while in C_3 plants saturation illuminance is in the range of 10,000 to 45,000 lux. Photosynthetic temperature optima are 30 to 40°C in C_4 plants, 10 to 25°C in C_3 plants. Photosynthetic CO_2 compensation points are low for C_4 plants (0 to 10 ppm CO_2), and high for C_3 plants (30 to 70 ppm CO_2). Response of apparent photosynthesis to O_2 concentration is not detectable in C_4 plants, and shows increasing inhibition above 1 per cent O_2 in C_3 plants. Response of photorespiration (CO_2 release in light) to O_2 concentration is not detectable in C_4 plants, but shows increasing enhancement with increasing O_2 in C_3 plants. Glycolate synthesis and glycolate oxidase activity are low in C_4 plants as compared to C_3 plants. An extensive and growing list of C_3 and C_4 species exists (Hatch et al., 1971), which has obvious implications for determining presence or absence of photorespiration, but the list consists entirely of terrestrial plants; little is known of the role of photorespiration in aquatic plants.

Emergent hydrophytes are exposed partially to environmental conditions similar to those of terrestrial plants, and photorespiration undoubtedly can be extensive. The C_4 photosynthetic system thus is very likely of adaptive value in many emergent hydrophytes, particularly in regions of high temperature and high light intensity, and also in situations in which the salt content of the environment adversely affects internal CO_2 and water balance (Hatch et al., 1971). In the latter context, *Spartina*, the dominant plant of the salt marshes of the east coast of the United States, appears to be a C_4 plant (Black, 1971). There is some evidence that *Typha latifolia* is a C_3 plant with relatively low rates of photorespiration (McNaughton, 1966a, 1969; McNaughton and Fullem, 1969); limiting mechanisms are not known in this case.

Submersed hydrophytes are exposed to lower maximum O_2, light, and temperature relative to air, and photorespiration correspondingly may be of lesser magnitude than in terrestrial plants. Also, the greater resistance of water to diffusion of CO_2 relative to air and the presence of massive internal gas lacunae most likely retard loss of CO_2 from submersed hydrophytes and facilitate refixation of CO_2 regardless of the presence or absence of the C_4 photosynthetic system. The C_4 system may not be of adaptive value in submersed hydrophytes generally. There is no question that the evolution of an extensive internal lacunal system in submersed angiosperms constitutes an array of interrelated morphological and physiological adaptations related to efficiency of gas utilization, buoyancy-light availability interactions, plasticity in relation to water movements, etc.

Using a ^{14}C-assay developed to evaluate photorespiration in submersed macrophytes, Hough and Wetzel (1972; Hough, 1974) found that *Najas flexilis* refixes respired CO_2 efficiently by photosynthesis in the light. Similar results were suggested for the free-floating submergent *Ceratophyllum* (Carr, 1969). While extensive refixation of CO_2 is characteristic of plants with the C_4 photosynthetic pathway, analyses of leaf anatomy and early ^{14}C fixation products of photosynthesis indicated that *Najas* is a C_3 plant with Calvin cycle photosynthesis and glycolate metabolism. Respiration in the light increased with increasing dissolved oxygen concentration, which indicated the presence and enhancement of photorespiration, and that net photosynthesis would decrease with increasing oxygen concentration.

Rates of photosynthetic carbon fixation in submersed macrophytes generally are correlated with the intensity of solar radiation on a daily basis. However, a midday and afternoon depression of carbon fixation often is observed in which highest rates of photosynthesis are skewed to the morning (Wetzel,

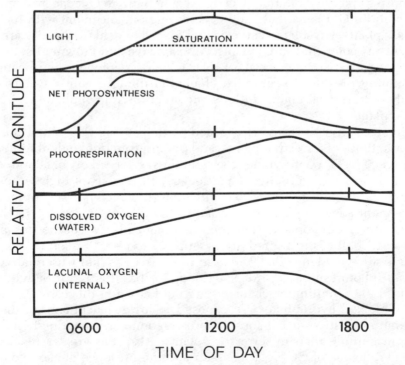

Figure 15–3 Commonly observed relative diurnal variations in light intensity, dissolved oxygen concentrations of the water and the internal lacunal spaces of submergents, net photosynthesis of submersed macrophyte (and some phytoplankton) populations, and proposed concurrent variations in photorespiration. (Modified from Hough, 1974.)

1965b; Sculthorpe, 1967; Goulder, 1970; Hough, 1974). In situ experiments with the submergent *Najas* showed that the afternoon decrease in net photosynthesis was correlated with an increase in photorespiration. The results of numerous studies suggest that photorespiration may increase through the day with increasing light intensity, oxygen tension of photosynthetic origin, increasing temperature, and possibly decreasing CO_2 availability (Fig. 15–3). Conditions conducive to accelerated photorespiration develop in this manner in the littoral zone, particularly among dense macrophytic populations and in calm weather with little turbulence. An increase in photorespiration would reduce net photosynthesis.

With cessation of photosynthesis during darkness, internal oxygen content decreases rapidly and carbon dioxide increases, and respiration is limited by the rate at which oxygen diffuses from the water to the cells. Dark respiration rates increase directly with rising temperatures and are generally, but not in all cases, enhanced by high oxygen in the surrounding water, especially if agitated slightly by low velocity currents (Pannier, 1957, 1958; Owens and Maris, 1964; McIntire, 1966; Westlake, 1967; McDonnell, 1971; Hough and Wetzel, 1972; Stanley, 1972; Prins and Wolff, 1974). In the daytime, the effect of oxygen on photorespiration and dark respiration would be additive on the release of CO_2.

Photorespiration and dark respiration were found to be 10-fold greater in the fall than in the summer as the annual submergent *Najas* was entering senescence (Hough, 1974). Photorespiration in the submersed perennial *Scirpus subterminalis* was relatively low in summer and increased under the ice in late winter.

Secretion of Dissolved Organic Matter

The loss of major fractions of recently synthesized organic carbon and nitrogen as secreted dissolved organic compounds has been demonstrated for a few submersed and freely floating angiosperms (Wetzel, 1969; Wetzel and Manny, 1972; Hough and Wetzel, 1972, 1974). Well-known among the planktonic algae (see Chapter 14), this loss of organic carbon during active photosynthesis of submersed macrophytes represents a significant reduction in photosynthetic efficiency. Secretion rates of dissolved organic carbon vary greatly (0.05 to over 100 per cent of carbon photosynthetically fixed) under an array of experimental conditions of light and ionic composition of the medium. Most values, however, and those from in situ analyses were between 1 to 10 per cent of photosynthetically fixed carbon, increasing somewhat during the daylight period and about doubling in darkness.

The secretion may well represent an incomplete adaptation of submersed macrophytes to a totally aqueous medium. Functionally, however, the secretion of labile organic compounds by submersed macrophytes enhances the development of a highly productive epiphytic community of microflora (Wetzel and Allen, 1970). Nutrient interactions, both inorganic and organic, take place between the submersed macrophyte and epiphytic algal and bacterial populations (cf. Allen, 1971, and Allanson, 1973); these interactions are lacking or greatly reduced in the diluted planktonic regime. This apparent decrease in photosynthetic efficiency by loss of dissolved organic carbon could be viewed as a symbiotic interaction between community components and, moreover, as the evolution of an ecosystem community adaptation to environmental conditions that impose physiological stress upon the macrophytes.

Ionic Absorption and Nutrition

The question of the role of the rooting systems in aquatic macrophytes, primarily angiosperms, has been a long-standing question in limnology (reviewed in Gessner, 1959; Wetzel, 1964; Sculthorpe, 1967; Bristow, 1974b). The two views held, each with supporting evidence, are first that the rooting systems function in absorption of nutrients from the substratum, and second that the roots function merely as organs of attachment. Among emergent and floating-leaved angiosperms with active transpiratory root pressure systems, nutrient absorption and translocation from the roots to the foliage are clearly operational. The anaerobic sediments, while necessitating a well-developed active aeration system from aerial organs, are highly reducing with high solubility and availability of ions. However, macroalgae, liverworts, mosses, ferns, and other lower macrophytes that float on the surface or in the water or are variously attached to the substratum presumably obtain nutrients by absorption by foliage rather than by rhizoidal structures. An exception is the macroalga *Chara*, which absorbs phosphorus equally well in all parts, and the phosphorus absorbed by the rhizoids is translocated to other parts of the plant (Littlefield and Forsberg, 1965).

In rooted submergents a positive absorptive capacity is present, with guttation always occurring in small quantities, increasing from root tip to the basal portions of the plant and increasing from the apical parts to the base (reviewed by Stocking, 1956). Water exits via hydathodes on the upper portions of many submergents, and nutrients can enter apically through the numerous porous hydropoten. The extensive development of roots and root hairs among aquatic plants (>95 per cent of 200 species of 105 genera and 54 families) emphasizes the probable dependency of these plants on roots for absorption of solutes as well as for anchorage (Shannon, 1953).

The absorption of phosphorus by roots of submergents both from the water and from the sediments is well-known (cf. Chapter 12, and reviews of Gessner, 1959, Schwoerbel and Tillmanns, 1964a). Among most submergents, phosphate absorption rates by foliage are proportional to and dependent upon the concentrations in the water. Very large quantities of several mg 1^{-1} are rapidly assimilated in excess of requirements until water levels are reduced to about 10 μg 1^{-1}. These relationships of luxury consumption, known especially for phosphorus but also for nitrogen, have led to the application of elemental analysis of tissues evaluating nutrient supplies for algae and aquatic angiosperms, a technique long used in agriculture (Gerloff, 1969; Fitzgerald, 1972). Tissue analysis is based on the assumption that concentration of elements in an organism varies over a wide range in response to the

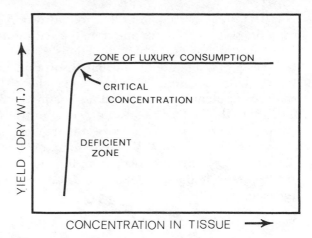

Figure 15–4 General relationship between aquatic plant biomass and the concentration of an essential element in plant tissue. (Modified from Gerloff, 1969.)

concentration in the environment, and that over part of this range yield (biomass) of the organisms is related to tissue content of the element. Tissue analysis depends on establishing for each species the critical concentration for each element of interest, that concentration which is just adequate for maximum growth (Fig. 15–4). Below this point in the "deficient zone" plant yield is increasing markedly, but the elemental concentration in the organism is changing little. Above this point in the "zone of luxury consumption" the content of the element of the organism is increasing but plant yield is not. The method assumes that plant content of an element below a critical concentration indicates growth is being limited by the supply of that element. Such an assumption is of course true in a general way, and readily demonstrable in crop plants, and some data suggest a similar relationship among algae and aquatic plants grown hydroponically in completely aqueous media. The correlation between nutrient concentration of the water and tissue content among submergents rooted in sediments, however, is less clear. Other investigations have shown, for example, in two species of *Myriophyllum* and in *Elodea* that from 60 to over 90 per cent of the phosphate in the shoots was not absorbed from the ambient medium but was derived from the roots (Bristow and Whitcombe, 1971). Similarly, iron and calcium were absorbed by roots and translocated to shoot tissues (DeMarte and Hartman, 1974). Absorption of phosphate and sulfate by the leaves of *Elodea* is limited at low concentrations of the water by the rate of diffusion through the gradient in the microlayer surrounding the leaves, by the process of active uptake at intermediate concentrations, and by the rate of diffusive influx into the leaves at high external concentrations (Jeschke and Simonis, 1965). Rates of nitrate assimilation are considerably less than those of ammonium assimilation by the leaves of several aquatic macrophytes, especially at high pH values, and ammonium ions are absorbed readily by roots and translocated to apical tissues (Schwoerbel and Tillmanns, 1964b, 1964c, 1972; Toetz, 1973b, 1974). Further, it is clear that nitrogen from N_2-fixing bacteria of the sediments and plant rhizosphere can serve as a major nitrogen source for aquatic plants (Bristow, 1974a).

It is therefore apparent that ion absorption in submergents occurs both from the water by foliage and from the sediments by root and rhizoid systems. Translocation occurs in both directions. Certainly in a majority of cases, roots serve as a primary site of nutrient absorption from the sediments (cf. Denny, 1972). This fact significantly negates the value of presently used tissue analyses of element concentrations in aquatic macrophytes as an index of the fertility of the lake or river water, since nutrient content of the water can be quite unrelated to plant growth of those species having ready access to the abundant nutrient supply of the sediments. Also, as was indicated earlier (Chapter 12), the rooted plants can function as an excellent "pump" of nutrients from the sediments which are lost to the water both during active growth and decomposition.

Rates of Photosynthesis and Depth Distribution of Macrophytes

Light

The photoinhibitory effects of high light intensities in the surface waters, commonly observed among planktonic algae, are not generally found among submersed macrophytes. Results from the few studies on in situ rates of photosynthesis in relation to light indicate a tendency towards increased photosynthesis at 1 to 2 meters below the surface on bright clear days where surface light intensities are reduced from 25 to 50 per cent. On the other hand, exposure of some submergents to extremely high light intensities (greater than 80,000 lux) did not significantly alter rates of photosynthesis; others are clearly adapted to growth at low light levels. Many submergents exhibit distinct morphological variations in relation to light intensity. Shade-adapted leaves are finely divided, whereas on the same plant growing at higher light intensities (shallower depths) leaves can be larger and much more lobate. Photosynthetic rates can be significantly higher near the surface, where overlying foliage of dense populations shades the lower leaves. Shade-adapted plants also have been shown to be sensitive to ultraviolet light, which is rapidly removed below the first meter of water, whereas photosynthetic rates of plants adapted to high light intensities were unaffected by ultraviolet light that is found in surface layers of water.

Gessner (1955) attempted to clarify the many physiological reactions of submersed macrophytes to light by grouping the plants into adaptation types either genetically fixed or phenotypically plastic (Fig. 15–5). Type (A) is a strictly shade-adapted form, such as the fenestrated tropical *Aponogeton*, which is physiologically active only at low light intensities. In type (B) numerous submergents are strictly light-adapted and require high light intensities for optimal metabolism. Type (C) is a group of submersed macrophytes of broad adaptation to light intensities, including some (D) which adapt to light or shade conditions depending on the area of development, others which are generally tolerant of high light but develop best in weak light (E), and still others which are shade-adapted but photosynthesize optimally at

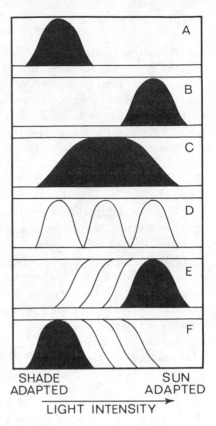

Figure 15–5 Diagrammatic scheme of shade and light adaptation among submersed aquatic plants. *Black*: genetically established forms; *White*: plastic growth forms. (Modified from Gessner, 1955.)

intermediate littoral light intensities (F). While such a scheme does not explain the physiological and morphological bases underlying such adaptations, it does demonstrate the extreme plasticity and adaptability to the highly variable underwater light conditions found among submersed macrophytes.

Some insight into the mechanisms of photosynthetic adaptation of submersed macrophytes has been gained by the investigations of Spence and Chrystal (1970a, 1970b) on two species of *Potamogeton*. The rates of net oxygen production at various irradiances, dark respiratory oxygen uptake, chlorophyll content, specific leaf area (cm² leaf area per mg leaf dry weight), and leaf thickness were measured for a shallow-water, "sun-adapted" species (*P. polygonifolius*) and a deepwater, "shade-adapted" species (*P. obtusifolius*). From estimates of relative rates of photosynthesis and respiration per leaf area and per pigment content of the species, and sun and shade leaves of the vertical length of each species, it was concluded that (a) surface leaf area increases and respiration and leaf thickness decrease with depth and light reduction, and (b) higher net photosynthetic capacity per unit area of shade-adapted leaves and shade species of *Potamogeton* at low irradiances (ca 1 per cent of summer daylight) is achieved by lowered respiration per area. The latter may result from a reduction in leaf weight per unit area. The recently observed high rates of photorespiration in aquatic plants, discussed earlier, were not considered in these studies.

Pressure

An abrupt inhibitory growth response occurs when submersed angiosperms are exposed to even very moderate increases in hydrostatic pressure (Fig. 15–6). Leaves become shorter, stems are much thinner and possess greatly reduced lacunal intercellular spaces, and growth of adventitious roots is inhibited (Ferling, 1957). Increased pressure of as little as 0.5 atm, equivalent to about 5 m depth, also induces increased growth of internodes, similar to the effects of low light intensities. Increased pressure also inhibits flower formation, which correlates well with general growth to the surface before the development of inflorescence. Most submersed angiosperms are wind pollenated.

Growth inhibition is not usually proportional to pressure, and many species exhibit an irreversible "all or none" effect at about 1 atm pressure (ca 10 m depth equivalency). Little can be said about the physiological effects of increased pressure. One effect is the inhibition of movement of intercellular gas exchange to the root system. Short-term effects include increased secretion of soluble organic matter by excessive pressure. Such increased cellular permeability may lead to losses of nutrients and hormonal growth substances. Golubić (1963) demonstrated in a very clear, transparent lake that a pressure of 0.8 atm limited the photosynthesis of submersed angiosperms. The macroalgae *Chara* and *Nitella*, lacking intercellular gas systems, are unaffected by pressure differences.

Primary Productivity of Macrophytes

Changes in Biomass

Evaluations of the productivity of macrophytes are most commonly based upon changes in biomass (Westlake, 1965b). The initial biomass of seeds of annual macrophytes is negligible. Biomass changes follow a typical sigmoid

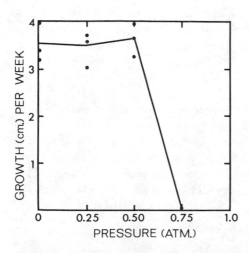

Figure 15–6   Average growth (cm) per week of *Hippurus* in relation to hydrostatic pressure. (Modified from Gessner, 1955.)

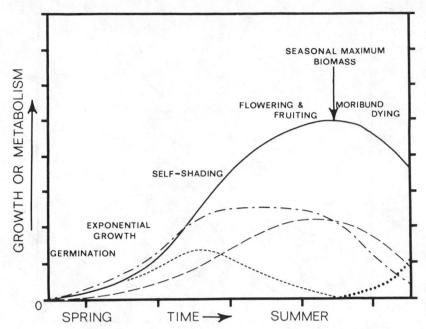

Figure 15–7 Generalized growth and metabolic patterns for a typical annual aquatic macrophyte. ——— = biomass; — · — · — = current gross productivity; ----- = current net productivity; — — — = current respiration rate; · · · · · = death losses. (After Westlake, 1965b.)

growth curve and decrease at later stages (Fig. 15–7). In this idealized example, gross productivity reaches a plateau and later declines in older tissues; net productivity decreases and becomes negative as respiration continues to increase with biomass. Maximum biomass, the maximum cumulative net production, is reached when the current daily net productivity becomes zero. Subsequently, both the biomass and cumulative net production decrease, and terminal net production is the material that has not been respired by the time the whole plant is dead. Although numerous variations are found among natural populations, this general pattern of growth and metabolism prevails among annuals of temperate and subtropical macrophyte communities.

If the initial biomass is negligible, or if the plant dies before the seasonal maximum is reached, and losses other than respiration are negligible, as is commonly the case, the seasonal maximum biomass is equal to the maximum cumulative net production (Fig. 15–8, curves A and B) (Westlake, 1965b). This annual regrowth is characteristic of many plants that die down to a small quantity of perennating organs, but becomes less obvious when much of the plant survives until the next spring growing season. If the initial biomass persists without appreciable losses until after the seasonal maximum, the annual net production may be obtained from the difference between the final and initial biomass (Fig. 15–8, curve C), but more commonly a variable proportion of the initial biomass is lost (curve D). The annual net production in the latter case, as seen for example in populations of the macroalga *Chara*, is commonly 50 to 80 per cent of the maximum seasonal biomass. When losses of the current seasonal production are appreciable, the estimation of production from bio-

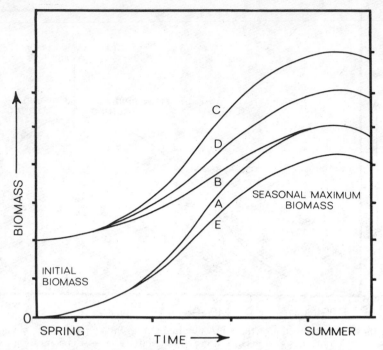

Figure 15–8   Types of growth curves among aquatic macrophytes. *A*, True annual or a plant with manifest annual regrowth; *B*, Plant with obscured annual regrowth; *C*, Plant with spring biomass persisting until seasonal maximum; *D*, Plant with only part of spring biomass persisting until seasonal maximum; *D*, Plant with only part of spring biomass persisting until seasonal maximum; *E*, Plant with annual regrowth and losses from current year's biomass before seasonal maximum. (After Westlake, 1965b.)

mass changes is complicated (Fig. 15–8, curve *E*). The biomass remains more or less constant in extreme cases, such as in tropical communities, despite continuous growth. In populations with a high mortality, e.g., in some reed-swamp emergent macrophytes (Westlake, 1966), much of the production does not survive to be measured as terminal maximum biomass. Similar situations have been found among submersed macrophytes (Borutskii, 1950; Rich, et al., 1971). Demographic methods developed for analyses of fish populations subject to high mortality have been applied successfully to dense macrophyte populations, in which, for example among *Glyceria* populations, net production was some 3.4 times greater than seasonal maximum biomass (Mathews and Westlake, 1969).

Underground Biomass

The roots or rhizoids of submersed freshwater macrophytes generally constitute a small but significant proportion of the total plant biomass (Table 15–3). Among floating-leaved and emergent macrophytes, however, the extensive system of roots, rootstocks, and rhizomes constitutes a major portion of total biomass. Clearly, accurate analyses of productivity must include an eval-

TABLE 15–3   Percentage of Total Biomass Found in Underground Organs of Rooting Systems of Mature Aquatic Macrophytes[a]

Type and Species	% of Total Biomass
SUBMERSED:	
Chara	<10
Ceratophyllum	< 5
Elodea canadensis	2.6
Littorella uniflora	46–55
Thalassia testudinum (marine)	75–85
FLOATING:	
Eichhornia crassipes (submersed roots)	10–48
FLOATING-LEAVED:	
Nuphar lutea and N. pumilum	50–80
Nymphaea candida	48–80
EMERGENT:	
Zizania aquatica	7–8
Cyperus fuscus	7–8
Alisma plantago-aquatica	40
Equisetum fluviatile	40–83
Glyceria maxima	>30–67
Phragmites communis	>36–96
Schoenoplectus lacustris	>46–90
Sparganium spp.	>25–66
Typha angustifolia	>32–59
Typha hybrid	64
Typha latifolia	43–50

[a]After Westlake, 1965b, 1968, from numerous authors.

uation of growth of the root system; measurements of only the above-ground foliage (= standing crop, see definitions in Chapter 14) are inadequate and can be quite misleading.

Determination of the biomass of underground organs is difficult but can be analyzed effectively by randomized coring and quadrat excavation (Westlake, 1968; Fiala, et al., 1968; Fiala, 1971, 1973). Growth patterns and discrimination of annual growth among perennial root systems can be determined by a combination of tagging growth experiments in which biomass increments of age classes are analyzed. The ratio of biomass of underground parts to above-sediment parts is relatively constant, at least among certain species at maturity in a particular habitat (e.g., Szczepański, 1969).

Losses During Growth

Losses of plant biomass by depth, damage (e.g., from water movement), and grazing before the seasonal maximum biomass are generally considered small among temperate annual macrophytes (Westlake, 1965b; Sculthorpe, 1967). Senescence and death during an annual period of active growth estimated for several macrophytes ranged between 2 to 10 per cent of the maximum biomass. Pathogenic decimation of entire populations of macrophytes

is rare; usually, at worst, natural diseases only cause gaps in existing populations, with an increase in heterogeneous distribution (see, for example,
Klötzli, 1971). While many animals, especially immature insects, eat live
aquatic macrophytes, few cause extensive damage (see especially the extensive review of Gaevskaya, 1966). Fish, birds, and a few mammals can, under
exceptional circumstances, cause extensive damage or consume significant
portions of the annual production. Values of grazing loss, however, usually
range from about 0.5 to 8 per cent of the total. Under most circumstances,
most of macrophyte production is consumed by bacterial and fungal degradation as the populations decline.

Chemical Composition of Aquatic Macrophytes

An extensive collation of existing data on the general chemical composition of aquatic macrophytes (Straškraba, 1968) provides useful information on
the relative potential contribution of these plants to the production and decomposition in aquatic systems. Morphological and physiological differences
among the different plant types are reflected in higher water, ash, and protein
content and reduced fiber content in the submergents as compared to emergent macrophytes (Table 15–4). Content of carbohydrate and fats is similar
among the two groups. Floating-leaved macrophytes exhibit approximately
intermediate composition values but contain significantly high lipids, which
is reflected in somewhat higher average caloric values. The caloric values of
aquatic macrophytes are approximately 20 per cent lower than the mean for
phytoplankton (ca 6000 cal g^{-1} organic matter).

Comparative Macrophyte Productivity Among Habitats

Comparison of productivity among freshwater ecosystems is difficult because of the variations in analytical and sample techniques employed. This

TABLE 15–4   Average and Range of Chemical Composition and Caloric Values of
the Three Primary Types of Aquatic Plants[a]

Ecological Type	Water (%)	Ash (% of dry weight)	% of Organic Matter				Caloric Value (cal g^{-1})
			Protein	Ether Extract	Crude Fiber	Carbo-hydrate	
Emergent	79	12	13	2.1	32	50	4480
	(70–85)	(5–25)	(3.5–26.7)	(0.4–6.1)	(17.6–45.4)	(23–69)	(4207–4987)
Plants with floating leaves	82	16	26.5	4.0	27	42	4770
	(80–85)	(10–25)	(18–44)	(2.8–5.7)	(14–38)	(31–64)	(4560–5140)
Submersed	88	21	22	2.2	27	51	4580
	(85–92)	(9–25)	(7.5–34.7)	(4.4–5.1)	(14.0–61.6)	(20–70)	(4165–5201)
Average all groups	83	18	19	2.4	29	50	4570

[a]From Straškraba, 1968, from numerous sources.

variability is especially great among evaluations of macrophyte productivity. Sampling of total biomass, including the rooting systems, has only been done seriously in recent years. Species distribution and production are varied in response to an array of physical and chemical characteristics of both the water and the sediment. The result is extremes in heterogeneity in both distribution and productivity, spatially and temporally. Further, results are commonly biased towards the growing season, which is often much shorter than a year.

TABLE 15–5 Representative Estimations of Annual Above-Ground Biomass and Productivity of Aquatic Macrophytes

Type and Lake	Seasonal Maximum Biomass or Above-Ground Biomass (g dry m^{-2})	Productivity (g m^{-2} year^{-1})	Source
SUBMERGENTS DOMINATING:			
Trout L., Wisc. (softwater)	0.07	—	Wilson, 1941
Sweeney L., Wisc. (softwater)	1.73	—	Wilson, 1937
Weber L., Wisc. (softwater)	16.8	—	Potzger and Engel, 1942
Lowes L., Scotland (dystrophic)	32	—	Spence, et al., 1971
Spiggie L., Scotland (dystrophic)	100	—	
L. Mendota, Wisc. (hardwater)	202	—	Rickett, 1921
Lawrence L., Mich. (hardwater)			
Submersed *Scirpus subterminalis*	338	565	Rich, Wetzel, and Thuy, 1971
Chara	110	155	
Annuals	130	199	
Croispol, Scotland	400	—	Spence, et al., 1971
Borax L., Calif. (saline lake, *Ruppia*)	60	64	Wetzel, 1964
River Ivel, England (*Berula*; very fertile)	500	—	Edwards and Owens, 1960
River Test, England (*Ranunculus*)	100–400	—	Owens and Edwards, 1961
River Yare, England (*Potamogeton*)	380	—	Owens and Edwards, 1962
Saline channels, Puerto Rico (*Thalassia*)	700–7300	—	Burkholder, et al., 1959
FLOATING:			
New Orleans, La. (*Eichhorina*)	630–1472	1500–4400	Penfound and Earle, 1948
EMERGENTS DOMINATING:			
Ladoga L., USSR	0.4–10.7	—	Raspopov, 1971
Onega L., USSR	0.1–33	—	
Blanket bog, England (*Sphagnum*)			
on hummocks	—	180	Clymo and Reddaway, 1971, 1974
in pools	—	290	
on "lawns"	—	340	
Polish lakes (emergent species)	440–830	—	Szczepańska, 1973
Minnesota wetlands (*Carex*)	850	738	Bernard, 1973
Surlingham Broad, England (*Glyceria, Typha*, and *Phragmites*)	800–1100	—	Buttery and Lambert, 1965
Opatovický Pond, Czechoslovakia (*Phragmites*)			
Aerial	1100–2200	—	Dykyjová and Hradecká, 1973
Below ground	6000–8560	—	
Cedar Creek, Minn. (*Typha*)	4640	2500	Bray, et al., 1959

A year is a well-defined unit in which the variability of annual productivity is much less than for shorter periods, since most communities have definite natural periodicities related to the year.

Despite these difficulties, approximate comparisons of macrophyte productivity are possible when accompanied by critical evaluation of necessary corrective adjustments and criteria of data transformation (see especially Westlake, 1963; Sculthorpe, 1967; Květ, 1971). The few exemplary figures given in Table 15–5 indicate the general range of above-ground maximum annual biomass and productivity among macrophyte communities. Submersed macrophytes generally exhibit conspicuously lower productivity than plants of floating or emergent communities. Annual net primary productivity of submersed macrophytes ranges from 1 to 7 metric tons per hectare (100 to 700 g m^{-2} year^{-1}), the lower values being more common in oligotrophic and moderately fertile waters, and ranging from 4 to 7 mt ha^{-1} year^{-1} in fertile lakes. Values of greater productivity are found among submersed communities of subtropical regions, from 10 to as high as 30 mt ha^{-1} year^{-1} in the marine *Thalassia* community of Puerto Rico. Few accurate data are available for productivities of floating or floating-leaved macrophytes, although it is apparent that this community reaches levels of 11 to 33 mt ha^{-1} year^{-1} in warmer climates, and easily can achieve greater productivity under optimal tropical conditions (Westlake, 1963).

Numerous investigations of the productivity of emergent macrophytes of fertile reedswamps and marshes indicate very high values, even though the range is quite large. Temperate zone emergents can attain productivity levels of 20 to 45 mt ha^{-1} year^{-1} and in subtropical and tropical regions reach 40 to at least 75 mt ha^{-1} year^{-1}.

Inspection of the annual net productivity of the plant communities (Table 15–6) shows that the emergent reed communities are as productive as terrestrial communities, or more productive under fertile conditions with a continuous growing season. In the temperate zones, the organic productivity of emergent communities is comparable to that of perennial agricultural crops and coniferous forests, and appreciably greater than that of deciduous forests. In the tropical zones emergent communities are among the most productive. By comparison, the submersed macrophytes are relatively unproductive, with the possible exception of tropical marine angiosperms, and similar to the productivity of phytoplankton. In spite of the number of adaptations to aqueous conditions found among submersed macrophytes, the reductions in rates of diffusion and spectrally selective attenuation of light by the water preclude intensive productivity. In certain respects, the submersed habitat is less severe than the terrestrial, e.g., in terms of water availability and thermal stability. These ameliorated parameters permit growth of perennial populations of submersed macrophytes to continue all year, but they are not sufficient advantages to offset the strictures of being totally submersed.

It should be emphasized in comparative analyses of productivity of macrophyte communities, as done briefly above and by others (e.g., Keefe, 1972), that geographically diverse ecosystems are not fully distinguished. The importance of geographic and genetic differences of the same species

TABLE 15–6   Annual Net Primary Productivity of Aquatic Communities on Fertile Sites in Comparison to That of Other Communities[a]

Type of Ecosystem	Approximate Organic (Dry) Productivity (mt ha^{-1} year^{-1})	Range (mt ha^{-1} year^{-1})
Marine phytoplankton	2	1–4.5
Lake phytoplankton	2	1–9
Freshwater submersed macrophytes		
Temperate	6	1–7
Tropical	17	12–20
Marine submersed macrophytes		
Temperate	29	25–35
Tropical	35	30–40
Marine emergent macrophytes		
(salt marsh)	30	25–85 .
Freshwater emergent macrophytes		
Temperate	38	30–45
Tropical	75	65–85+
Desert, arid	1	0–2
Temperate forest		
Deciduous	12	9–15
Coniferous	28	21–35
Temperate herbs	20	15–25
Temperate annuals	22	19–25
Tropical annuals	30	24–36
Rain forest	50	40–60

[a]After Westlake, 1963, 1965b.

is reflected in their productivity, which varies in a given environment from population to population as well as among populations from one environment to another. These multidimensional variations among ecotypes of the same species are well-characterized in the detailed studies of *Typha* populations of McNaughton (1966b, 1970) in which phenotypic adaptations to numerous environmental parameters, especially temperature and photoperiod, are seen in marked differences in productivity.

Productivity of Aquatic Macrophytes within Habitats

The literature on qualitative phytosociological composition among aquatic macrophytes is prodigious (e.g., Hejný, 1960). While these analyses of changes in dominance yield much information on the structure of the plant communities, the criteria used, e.g., percentage areal cover, do not give much insight into the causal mechanisms underlying competitive interactions among species.

Seasonal changes in compositional dominance commonly are observed among submersed, floating-leaved, and emergent species, but there is also a clear tendency to form extensive monospecific populations. Vegetative reproduction is prevalent among aquatic macrophytes, and rapid, relatively complete expansion into favorable habitats usually occurs. An example of this competitive growth is seen in the interactions between two emergent reeds, *Glyceria* and *Phragmites* (Buttery, et al., 1965a, 1965b). *Glyceria* grows rapidly in the spring, and under optimal growing conditions completely suppresses *Phragmites* by the production of dense foliage that effectively shades its slower-growing competitor. Under more anaerobic water saturated conditions of the sediments, however, *Glyceria* growth is reduced and *Phragmites*, with taller and more erect foliage that absorbs more incident light, can increase more effectively and eventually exclude *Glyceria*. Experimental manipulation of nutrient supply had no effect on the competitive interaction. Among submersed macrophytes of calcareous hardwater lakes, the ratio of organic matter to calcium carbonate of the sediments was implicated in the competition between the macroalga *Chara* and rooted angiosperms (Wohlschlag, 1950). Higher organic matter content of manipulated lake plots favored the production of rooted angiosperms.

The cycle of seasonal distribution among aquatic macrophytes can be exemplified by the common emergents *Phragmites* and *Typha*, a submersed annual *Myriophyllum*, and submersed perennial macrophytes of a hardwater lake. Many other examples exist with numerous variations from these typical temperate zone patterns.

Growth analyses of the production of emergent vegetation commonly are made in reference to changes in total plant biomass as well as to changes in their components (species, plant parts), and changes in size of assimilatory parts of the stand, usually expressed as leaf area index (LAI), or the ratio of leaf area per unit ground area (Kvĕt, 1971). The reedswamp plants exemplified in Figure 15–9 demonstrate the marked differences in growth rates found among perennial emergent plant stands. In *Phragmites*, maximum biomass (W) occurred in midsummer, whereas that of *Typha* increased over the entire growing season with little difference between the maximum values (1810 and 1620 g m^{-2}, respectively). The average rate of increase in shoot biomass was 9.4 g m^{-2} day^{-1} in the cattail population, and nearly 20 g m^{-2} day^{-1} in *Phragmites*. In both species, stem biomass (W_s) formed an increasingly greater proportion of the total (W), with decreasing proportion of leaf biomass (W_1). In *Phragmites*, however, both stem and leaf biomass decreased in the latter half of the growing season, whereas they continued to increase in *Typha*. The leaf area index followed the changes in leaf biomass, and, coupled with changes in shoot density, it is clear that later emerging shoots of *Typha* had a better chance to survive in the loose stands than in this dense *Phragmites* stand, in which shorter shoots were handicapped by competition for light.

Monodominant stands of emergent macrophytes in which invasion by other species is resisted for long periods of time are common. Density decreases with increasing accumulations of litter, increases when the area is flooded by rising water levels and increased inorganic nutrients, and decreases with grazing of buds and moderate increases in salinity (Rudescu, et al., 1965; Björk, 1967; Haslam, 1971a; Rodewald-Rudescu, 1974). High light, temperatures, and water levels promote optimal growth in dense, monospecific stands. Among mixed stands of emergents the final balance of domi-

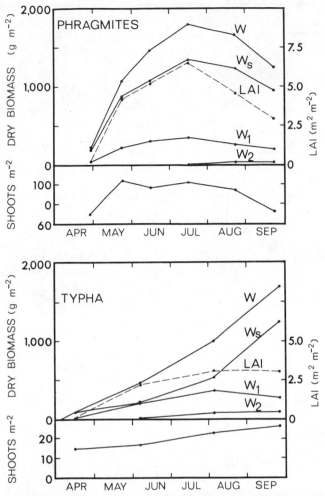

Figure 15–9 Changes in total plant biomass (dry, W); stem, sheaths, and stubble (W_s); leaf laminae (W_1); inflorescence (W_2); leaf area index (LAI); and the stand densities (*shoot* m^{-2}) of *Phragmites communis* and *Typha latifolia*, southern Czechoslovakia. (Modified from Květ, Svoboda, and Fiala, 1969.)

nance is determined by a combination of (a) physical factors and inherent properties of the plants, such as soil and water regime, growth form, size, and presence of reproductive organs, (b) physiological characteristics of the species, such as shade tolerance, requirements for germination and establishment, temperature in relation to water and litter cover, and (c) physical events such as opening of the litter mats and diseases (Haslam, 1971b, 1973).

The littoral community of eutrophic Lake Wingra, Wisconsin, is dominated by the submersed annual macrophyte *Myriophyllum* (Adams and Mc-Cracken, 1974). In situ measurements of photosynthetic rates demonstrated a rapidly formed spring maximum, followed by a marked summer reduction and subsequent maximum in late summer (Fig. 15–10). The spring maximum

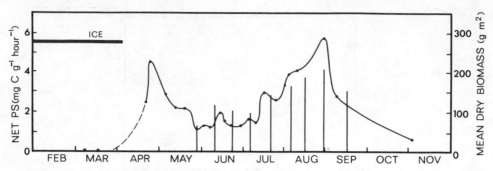

Figure 15–10   Rates of net photosynthesis (mg C gram dry weight^{-1} hour^{-1}) of 15 cm tip sections of *Myriophyllum spicatum* (curved line) and mean ash-free organic weight (g m^{-2}) (vertical bars) in eutrophic Lake Wingra, Wisconsin, 1971. (From data of Adams and McCracken, 1974.)

of photosynthetic rates was not reflected completely in the cumulative rates of biomass per time, a phenomenon similarly observed for the single submergent of a saline lake (Wetzel, 1964). The biomass maxima occurred at the time of flowering in late summer. Approximately 1.6 times the maximum biomass was lost by night-time respiration and sloughing of leaves of the submersed macrophytes.

Within the *Myriophyllum* community, the upper portion of the water column was the most important in regard to total biomass, leaf surface area, and photosynthetic activity. Light reduction with depth is definitely a major factor in the marked decrease in photosynthetic rates observed by many investigators. Only rarely is surface photoinhibition, so frequently found in phytoplankton photosynthesis, seen among submersed macrophytes. Other factors, such as temperature reduction, greater age of tissue, and encrustations on older leaf surfaces contribute to reduced photosynthetic rates at depth.

The importance of light to submersed macrophytes is further emphasized by the depth distribution of the annual biomass of the dominating perennial macrophyte *Scirpus subterminalis* (76 per cent of total biomass) of oligotrophic Lawrence Lake, Michigan (Fig. 15–11). Two annual maxima are evident from two dissonant peaks at different depths. The larger fall maximum consisted of a major peak at 2 m in October and at 4 m in the terminal midsummer maximum during highest insolation at depth. Under ice-cover and markedly reduced light availability at depth, maximum growth occurred at 2 m. A fall maximum of the macroalgae of the Characeae (16 per cent of total biomass) was evident within a relatively constant seasonal biomass distribution, and is similarly best described as a perennial population (Figs. 15–11 and 15–12). The appearance of *Nitella* at 7 m in late summer in this hardwater lake was interpreted as a response to an interaction between light, and thermal and carbon stratification. *Nitella* is a macroalga known to be limited by dissolved CO_2 availability above pH 7.3 (Smith, 1967). During July, the pH at 7 m in the deep water column shifted from pH 8.3 to 8.1 as summer stratification progressed. The shift may have been even greater at the 7 m contour, at which lake sediments were in contact with the 7 m water stratum.

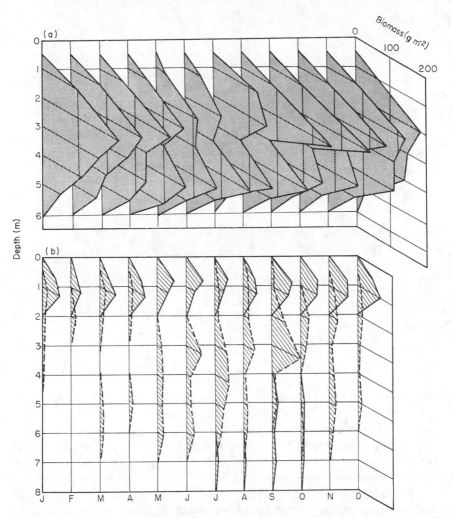

Figure 15–11 Annual biomass (g m^{-2} ash-free dry organic weight) of a mean transect with depth distribution of (*a*) *Scirpus subterminalis*, and (*b*) Characeae (———) and annual macrophytes (-----), Lawrence Lake, Michigan. (After Rich, P. H., Wetzel, R. G., and Thuy, N. V.: Distribution, production and role of aquatic macrophytes in a southern Michigan marl lake. Freshwater Biol., *1*:3-21, 1971.)

The disappearance of the *Nitella* population at the end of October coincided with autumnal overturn and the reestablishment of higher pH values. The utilization of CO_2 of the sediments was demonstrated by the submersed angiosperm *Lobelia,* a common plant of softwater lakes of extremely low inorganic carbon content of the water (Wium-Andersen, 1971).

The submersed annual macrophytes (8 per cent of total macrophytic biomass) of Lawrence Lake, largely resulting from two species of *Potamogeton,* exhibited two maxima and died back in the fall (Figs. 15–11 and 15–12).

The perennial biomass levels of macrophytes throughout the year in this example are probably much more common characteristics of lakes than is

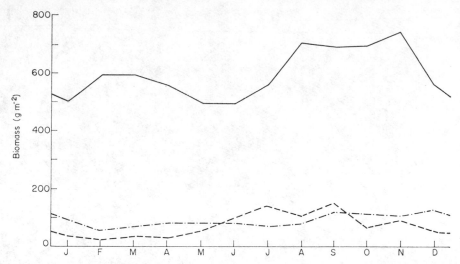

Figure 15–12 Annual biomass (g m⁻² ash-free dry organic weight) of the important macro-phytic groups as a mean transect summed with respect to depth, Lawrence Lake, Michigan. ——— = *Scirpus subterminalis*, - ·· -- ·· · = Characeae, - - - - - = annuals. (After Rich, P. H., Wetzel, R. G., and Thuy, N. V.: Distribution, production and role of aquatic macrophytes in a southern Michigan marl lake. Freshwater Biol., *1*:3-21, 1971.)

generally recognized. Turnover rates of emergent and submergent annuals are relatively low (2 to 20 per cent of maximum biomass) on an annual basis. Among perennial populations, turnover rates are higher (range 0.5 to 5.0 times maximum biomass). As a result, even among oligotrophic lakes the macrophytic productivity levels of perennial populations of submersed plants can approach those of much more productive annual macrophytes (Table 15–7). Comparisons of the relative contribution of the macrophytes to other components of the total primary productivity of lakes will be deferred until the conclusion of the following discussion of littoral microalgae.

ALGAE OF THE LITTORAL ZONE

In contrast to the immense amount of study on the systematics, physi-ology, and ecology of the phytoplanktonic algae of aquatic systems, a dearth of information exists on the algae attached to substrata or loosely aggregated in the littoral regions of lakes or shallow zones of streams. Although the taxa of these largely sessile algae are well-known, little information exists on the geographical distribution of species, seasonal population dynamics, utilization of microhabitats, responses to parameters of water movement or water and substratum chemistry, or interactions among species and utilization by grazing organisms.

Much of the problem centers on the extreme heterogeneity in distri-bution of the algae among a highly variegated spectrum of microhabitats,

TABLE 15–7   Comparison of Annual Primary Productivity of Submersed Macrophytes Among Several Lake Systems

Lake	g C m^{-2} (of Lake) Year^{-1}	Remarks
Wingra, Wisconsin	117	*Myriophyllum;* eutrophic lake (Adams and McCracken, 1974)
Lawrence, Michigan	87.9	Perennials; oligotrophic lake (Rich, Wetzel, and Thuy, 1971)
Narock, USSR	59.4	Largely charophyte perennials; mesotrophic lake (Vinberg, et al., 1972)
Marion, British Columbia	50.8	Annuals; mesotrophic lake (Davies, 1970)
Borax, California	1.2	Single annual *Ruppia;* saline lake (Wetzel, 1964)

which in turn are subjected to much more variable environmental physico-chemical and biotic parameters than occur in the open water. The problems of spatial and temporal heterogeneity of phytoplankton of the pelagial zone are minor in comparison to those of the attached algae. Nonetheless, despite all of the numerous problems associated with reasonable quantitative analyses, emerging information indicates high sustained rates of productivity by the algal populations specialized to these habitats. Our lack of knowledge of the complex interactions between the sessile flora and their substrata, and their contributions to the total system productivity, represents a major void in contemporary limnology that warrants intensified study.

Terminology and Zonation

The philology associated with bacteria and algae variously attached to substrata in aquatic systems is very involved (cf. reviews by Naumann, 1931; Cooke, 1956; Sládečková, 1962; Wetzel, 1964; Round, 1964a, 1965a; and Hutchinson, 1967). The number of terms is large and many are unnecessarily complex and quite confusing, especially when removed from their original definitions. The subject is rather unrewarding because of the great variation in microhabitats and association of organisms with substrata. The extensive usage of a few terms, however, demands brief discussion.

Perhaps the initial juncture should focus on the term *benthos,* originating from the Greek word for "bottom," which initially was defined broadly to include the assemblage of organisms associated with the bottom or, better, the solid-liquid interface of aquatic systems. Benthos is now nearly uniformly applied to animals associated with substrata.[1] The terms *Aufwuchs* (German: growth upon), and to a lesser extent haptobenthos, generally connote all organisms broadly associated adnate to, but not penetrating, a solid surface.

Periphyton, although variously used, usually refers to microfloral growth upon substrata. Reference to more specific subdivisions by complex phrases (e.g., epiphytic periphyton) results in involved and often redundant nomina. A much more explicit manner of expression, and one championed by Round, is to refer to the organisms with appropriate modifiers descriptive of the substrata upon which they grow in natural habitats. Hence, among the algal communities, one can readily differentiate the following (Fig. 15–13): (a) *epipelic* algae as the flora growing on sediments (fine, organic); (b) *epilithic* algae growing on rock or stone surfaces; (c) *epiphytic* algae growing on macrophytic surfaces; (d) *epizooic* algae growing on surfaces of animals; and (e) *epipsammic* algae as the rather specific organisms growing on or moving through sand. The general word *psammon* refers to all organisms growing or moving through sand (cf. Cummins, 1962, for a detailed discussion of particle size differentiation among sediments).

A group of algae found aggregated in the littoral zone is the *metaphyton,*[2] which is neither strictly attached to substrata nor truly planktonic. The metaphyton commonly originates from true phytoplankton populations that aggregate among macrophytes and debris of the littoral zone as a result of wind-induced water movements. In other situations, the metaphytonic algae derive from fragmentation of dense epipelic and epiphytic algal populations. A surprisingly large number of descriptions exist of clustering of metaphy-

---

[1]Herpobenthos as used by Hutchinson (1967) is synonymous with benthos when the term benthos is unqualified and means growing on or moving through sediment.

[2]Metaphyton (Behre, 1956) is essentially synonymous with the terms tychoplankton and pseudoplankton used much earlier by Naumann (1931), and the term pseudoperiphyton used by Sládečková (1960).

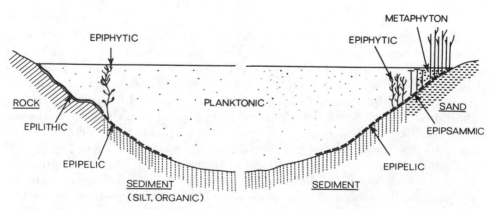

Figure 15–13 Major terms of algal communities associated with different substrata in fresh waters.

tonic algae and macrophytes into "lake balls," densely packed aggregations of algae or plant parts, or both. These balls are formed by the alternating rolling movements of wave action in the littoral zone (Nakazawa, 1973).

General Distribution of Littoral Algae

A survey of floristic distribution of algae among subcommunities of the littoral zone results in long lists of species in various associations that reflect differences in parameters of light, temperature, nutrient availability, water movement, substratum, and other factors. Detailed floristic analyses indicate specific characteristics of, and interactions among, species that provide a foundation upon which experimental approaches can be designed (Round, 1964a). The initial evaluations should lead to recognition of the various habitats and their associations, and enable quantitative sampling and the separation of casual species from the associations. All of these requisites are difficult to carry out, especially among algae of the sediments, but have been done in numerous cases in detailed analyses of littoral communities.

The attached benthic algal flora often forms the dominant algal biomass in small streams and shallow lakes, since the plankton are poorly developed in these waters. Probably over 90 per cent of all algal species grow in the attached habitat, and include practically all the species of pennate diatoms, and a majority of the many species of the Conjugales, Cyanophyta, Euglenophyta, Xanthophyceae, and Chrysophaceae (Round, 1964a).

Epilithic and epiphytic algae have many species in common, but there is relatively little interchange of species between those of the mud-living epipelic subcommunity and the others. The epipelic algae are largely motile forms, since motility is essential to enable species to move to the surface after any disturbance of the sediments. The algal species attached to sediments of streams and rivers with moderate to fast flow are almost exclusively motile.

Epipsammic algae, consisting largely of small diatoms and blue-green algae attached more or less firmly to the surface of sand grains in crevices, exhibit highly variegated, heterogeneous distribution (Fig. 15–14). These algae tend to be less motile than epipelic algae of finer, organic sediments (Round, 1965a; Meadows and Anderson, 1966, 1968). In the shallows of a temperate English lake, Moss and Round (1967) found that both epipelic and epipsammic algae showed marked spring population peaks. The biomass of the epipsammic algae of the surface 5 cm of sediments was consistently greater than that of the epipelic algae. Algal populations of the interstitial water of a sandy beach of a southern Michigan lake were also found to be large and exhibited a spring maximum (Davies, 1971). Percentage water content was a dominant factor determining algal composition at the zone of the water's edge.

Within the sand, light is attenuated very rapidly and essentially extinguished within half a centimeter. Yet viable epipsammic algae have been found attached to sand grains as deep as 20 cm in sandy beach zones exposed to heavy wave action. It is believed that wave action is adequate to mix the

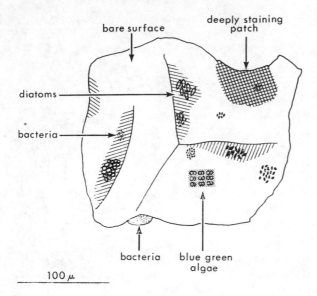

bare surface

deeply staining
patch

diatoms

bacteria

bacteria

blue green
algae

100 μ

Figure 15–14 Diagrammatic representation of typical localized distribution of microorganisms and staining (carbol fuchsin) patches of organic materials on the surface of a sand grain. (From Meadows, P. S., and Anderson, J. G.: Micro-organisms attached to marine and freshwater sand grains. Nature (London), 212:1059-1060, 1966.)

sand grains and return attached algae to a lighted zone sufficiently often to permit active, although low, photosynthesis (Steele and Baird, 1968).

Vertical Migration and Photosynthetic Rhythms

Algae growing adnate to more or less unconsolidated sediments are constantly at the mercy of shifting sediments and in danger of being removed from light, which is attenuated to zero in a few millimeters.[3] It is not known whether these epipelic algae supplement photosynthetic growth by heterotrophic utilization of organic substrates. Adaptation to the rigors of this habitat apparently is largely facilitated by an ability to move vertically within the sediments in response to light availability.

Persistent vertical migration rhythms that are exclusively diurnal have been shown in groups of epipelic diatoms, flagellates, and blue-green algae from flowing waters and shallow lakes (Round and Happey, 1965; Round and Eaton, 1966). Cell numbers on the sediment surface start to increase before dawn and reach a maximum about midmorning. Thereafter, surface cell numbers decrease and reach a minimum before the onset of darkness. This rhythm persists for a limited period in continuous darkness and continuous light, although the synchrony is lost more rapidly under the latter condition.

Studies of the diurnal rhythms of freshwater epipelic algae have shown that the maxima of cell emergence were coincident with photosynthetic capacity, but that the increase in photosynthetic capacity preceded cell emergence (Brown, et al., 1972). The photosynthetic rhythm occurred at light intensities well below saturation; minimum values were found in mid-

---

[3]Burial of cells under about 3 mm of sediment reduces incident radiation to about 1 per cent (Palmer and Round, 1965).

afternoon, after which the rate increased again, a rise that preceded the re-emergence of cells and continued throughout the night period. The maximum photosynthetic rate occurred at the time of maximum cell numbers present on the sediment surface.

Seasonal Population Dynamics

Quantitative evaluations of attached benthic algal populations are much more difficult than qualitative floristic analyses. Recognition of the various habitats and their associations is a problem shared by both endeavors, but is accentuated by ineffective sampling techniques that are difficult at best. Epipelic algae are difficult to separate from sediments for enumeration. The use of phototactic responses can assist in this separation; alternatively, other biomass methods such as evaluation of pigment content (correcting for pigment degradation products common to sediments) have been used. Quantitative sampling of epilithic and epiphytic algal populations is even more difficult, but a number of devices have been invented (see the review of Sládečková, 1962) that permit some degree of quantification. The problems of quantitative sampling of attached algae are further compounded by the settling of casual species originating from the plankton and metaphyton into the benthic populations. Only detailed sampling of the different communities combined with metabolic assays of growth can resolve this problem.

Use of Artificial Substrata.   Because of the extreme variability of algal habitat and community heterogeneity, there has been a strong tendency to place artificial substrata of uniform composition and colonizable area into aquatic systems for an estimation of attached algal growth. The number of substrate types that have been used includes glass slides, concrete blocks, rocks and rock plates, sand, various plastic and metal plates, and wood.

There are several restrictions inherent in the use of artificial substrata. Marked differences in colonization rates and biomass accrual have been found in relation to the position of the supported substrata and the location of the substrata within the aquatic system. For example, vertically positioned plates on which sedimenting organisms from the plankton or metaphyton could not accumulate contained less but more realistically simulated natural population characteristics than those held horizontally (e.g., Castenholz, 1961; Pieczyńska and Spodniewska, 1963). The influence of water movement or simply the process of retrieving the slides following incubation can lead to significant losses of attached microflora. Some organisms will not adhere to artificial substrata such as glass, or at least not until early successional forms have colonized the substratum, which may or may not be within the period of incubation. Apparently glass is not seriously selective for the attachment of diatoms (Patrick, et al., 1954). Moreover, grazing can lead to alterations of attached communities that may influence the results, depending on whether the questions being addressed concern specific attached flora or the entire community.

Some studies employing artificial substrata have positioned the materials in the pelagic zone of lakes or main surficial flow of streams quite removed from sites of natural substrates. While these analyses may have some value in a relative sense, such as in evaluating a response to toxicants in different water types, they are of little value in evaluating natural attached communities.

Perhaps the most serious criticism that can be directed against studies using artificial substrata centers on the implicit assumption they make that any metabolite, inorganic or organic, of the living or nonliving substrate has no appreciable effects on the attached community. And similarly, it is assumed that the metabolism of the attached microfloral community has no reciprocal effects upon the substratum. Although recent evidence for such synergistic effects is meager, it is sufficient to indicate that more than a passive relationship exists. The difficulties of quantitatively evaluating natural population growth on natural substrata then present themselves. Even though the synergistic metabolic relationships are real, especially among epiphytic algae and the supporting macrophytes, measuring the effects of the heterogeneous attached populations on the highly variable substrata often exceeds the capacities of existing methods. The problem is a technical one that must be approached without the use of artificial substrata.

On the other hand, when used critically artificial substrata can be a reasonably meaningful tool for the approximate estimation of biomass accrual of many attached microorganisms in analyses of system productivity. Much of our existing information on these communities is based on artificial substratum analyses within quite a range of critical application. Only a few recent investigations have addressed the in situ rates of algal productivity by means of metabolic techniques.

Factors Affecting Growth.   The detailed investigations of the population dynamics of epipelic algae of several lakes of the English Lake District (Round, 1957a, 1957b, 1957c, 1960, 1961a, 1961b) serve to illustrate several characteristics of these populations and potential spatial, temporal, chemical, and biotic factors affecting growth. Diatoms completely dominated the epipelic populations, and patterns of distribution could be discerned on the basis of general chemical features of the sediments and overlying water. Other algal groups were less well-represented. The Volvocales were very sparse and the Chlorococcales were represented only by the genera *Pediastrum* and *Scenedesmus*. Although not well-developed, the desmid flora was richer in species than other groups. Euglenoids were much better developed on more organic-rich sediments. Little correlation was found between the growth cycles of the epipelic and planktonic blue-green algae, and in some lakes a conspicuous development occurred on the sediments but not in the plankton. Epipelic blue-green algae developed equally well in alkaline, neutral, and slightly acidic waters.

In general, high diatom biomass corresponded to high organic matter content of the sediments, and growth was better on sediments with high leachable silica. Moderate organic matter content of the sediments favored blue-green algal population development, while low populations were associated with sediments of very high and very low organic matter content. Populations of the poorly represented Chlorophyceae and the flagellated forms were distinctly greater on organic-rich sediments. Although little cor-

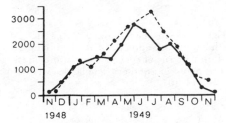 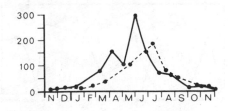

Figure 15–15 Seasonal cell counts of epipelic diatoms (*left*) and blue-green algae (*right*), in which all the depth stations are added together for each sampling date in the photosynthetic zone (1–6 m) of Lake Windermere (———) and Blelham Tarn (-----), England. (Modified from Round, 1961b.)

relation was found between population numbers and sediment content of sodium, potassium, iron, or manganese, diatom, blue-green, and flagellated algal populations were directly correlated to the calcium content. Algal growth was also greater on sediments with high phosphate content.

Although population numbers exhibited large variations among sediments of different lakes, sites within lakes, and depth, certain consistent patterns emerged. The main growth period of epipelic diatoms and blue-green algae of lakes of this moderate temperate region[4] was in the spring, commencing about February (Fig. 15–15). Except for the wave-active shallow zone (0 to 1 m depth), population numbers were relatively constant on sediments between 1 to 6 m depths; below this depth, numbers decreased precipitously to practically nil below a depth of 8 m. Growth of the dominant benthic diatom *Tabellaria flocculosa* was shown experimentally to be largely related to light intensity, attenuating with depth (Cannon, et al., 1961). Strictly phototrophic diatoms are more sensitive to light than blue-green algae, whose depth distribution usually extends considerably deeper than that of the diatom. Thus, the general seasonal biomass of epipelic algal populations in the photosynthetically illuminated zone followed the general curve of incident light and temperature.

Succession of species within the overall general annual cycle of epipelic algae is a much more elusive and complex matter. Among planktonic populations of algae, discussed in some detail in the foregoing chapter, the changes in physical and chemical parameters, particularly associated with periods of circulation in stratified lakes, are very large and related to dominance and competition of shifting species. The sudden collapse of populations seasonally is related, in part, to these "shock" events.[5] Epipelic populations tend to occur largely above the metalimnion, so that the effects of stratification may be less on these populations than among the plankton. The seasonal sequence of species is, however, just as pronounced as that found among the plankton (Round, 1972).

No species extends over the entire year in significant population numbers. Epipelic algal populations tend to increase quite rapidly and then to

---

[4] Ice-cover is rare and usually short-lived, a few days in duration.
[5] A concept discussed by Round (1971).

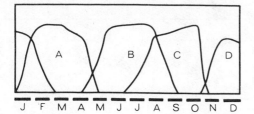

Figure 15–16 Generalized succession of growth curves of epipelic diatom populations based on patterns of species associations in two shallow pools in southern England. (Modified from Round, 1972.)

decline equally rapidly. Often 4 periods of species associations tend to dominate, each phasing into the next when live cells remain in the habitat for some time after cell division has ceased (Fig. 15–16). The substrata of attached algae provide innumerable microhabitats, certainly many more than exist under planktonic conditions. Therefore, one would expect a large number of coexisting species at almost any time of the year. This condition is aggravated by sampling techniques that tend to aggregate algae from microhabitats.

Annual and seasonal cycles of epipelic algae are marked and complex (cf. Round, 1964a). The generally observed diatom growth commences in early spring and reaches maximum cell numbers in April to May in the temperate zone, after which a decline occurs in midsummer prior to a smaller autumnal peak that is over by November. While this pattern is similar to that of planktonic species, the epipelic flora are slower to react. On shallow sediments, the epipelic algal populations tend to exhibit early winter and spring peaks, followed by midsummer maxima. Spring maxima are common among attached algal populations of streams, although these populations are subject to irregular decimation and disturbance by molar action of periodic flooding during spates (see, for example, Douglas, 1958). In woodland streams the seasonally changing light availability from the terrestrial canopy can shift the seasonal productivity from one of significant autotrophic productivity in the spring to one of essentially heterotrophic dominance later in the season[6] (see Kobayasi, 1961, and also discussion in Chapter 17).

The seasonal population dynamics of epiphytic algae growing on submersed portions of macrophytes are much more involved because the surface area for colonization and growth is continually changing as the plants grow, senesce, and enter detrital phases. Computations of the population changes of the total epiphytic algal community are rare, but changes obviously are quite different among different supporting macrophyte communities. For example, changes in the biomass of attached algae associated with a zone of the emergent bulrush *Scirpus acutus* were over an order of magnitude less and quite out of phase with those of algae attached to substrates in a submergent shallow zone of submersed plants of *Najas flexilis* and *Chara* (Fig. 15–17). Similar results were found in the seasonal dynamics of plant carotenoids (Allen, 1971).

---

[6]The author acknowledges the fruitful discussions on this subject of much unpublished work of Donna King of Michigan State University.

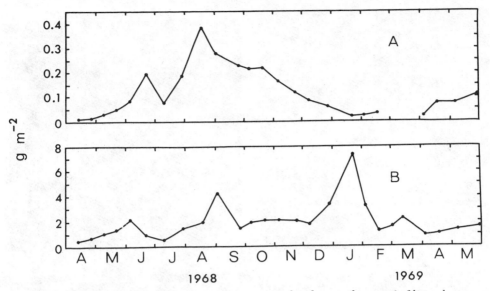

Figure 15–17 Chlorophyll *a* concentrations, extrapolated upward to g m⁻² of littoral zone on the basis of estimated surface areas of the macrophytes, of algae attached to artificial substrates in dense stands of the littoral zone dominated by A, the bulrush *Scirpus acutus*, and B, the submergents *Najas flexilis* and *Chara*, Lawrence Lake, Michigan. (After Allen, 1971.)

Only correlative data exist for the benthic algae regarding potential factors affecting population structure, and the control of seasonal variations of the population. Unlike the planktonic habitat, which is relatively isotropic, physical selective factors are involved in the benthic habitat. Within the overriding control factors of light and temperature, the complex and dynamic interactions of inorganic and organic chemical parameters of the substrata and surrounding water surely play major roles. Up to the present time, chemical correlations have centered on relatively extreme chemical habitats of sediments that permit only crude generalizations. Some correlations drawn between changing water chemistry and benthic algae may be helpful, as in the case of epilithic algae that can obtain relatively meager nutrients from their substratum, or that of certain epiphytic flora. An example of the latter are the previously discussed (see Chapter 13) results of Jørgenson (1957), who found that diatoms epiphytic on the reed *Phragmites* competed with plankton for silica of the water early in the major growing season. Later in summer and fall, apparent utilization by the diatoms of silica from the *Phragmites* stems gave them a competitive advantage over planktonic forms (see Fig. 13–19).

Assumptions that the overlying water has appreciable direct controlling chemical effects on epipelic algae have not been verified. For those species adapted to epipelic existence, the proximity to the totally different chemical milieu of the interstitial water within and immediately adjacent to the sediment-water interfacial zone can give them a distinct advantage that is quite unrelated to chemical events occurring in the overlying water. Yet the epi-

Figure 15–18 *See opposite page for legend.*

pelic forms are not exempt from characteristics and events of the overlying water. Examples are many. A heavy dissolved organic matter content or a seasonally variable development of planktonic algae, or of populations of macrophytic vegetation would affect markedly the quality and quantity of light reaching the sediments, the thermal regime of the sediments, water movements near and along the sediments, etc. While it is intuitively apparent that these changing events influence the species succession and productivity of epipelic algae, studies of these interactions are essentially nonexistent. The whole area of investigation is uncharted and difficult, but is one that is approachable by systematic application of modern research techniques.

Insight into the community structure of epiphytic microflora of submersed macrophytes can be gained from electron photomicrographs[7] of a typical association in Figure 15–18. The surface structure of the epiphytic community (Fig. 15–18a) indicates the manner in which the loosely woven diatomaceous component is held in place by a stranded matrix formed largely from gelatinous stalks of mucoid substances (Fig. 15–18b), which in places extends as a fragmented membrane above the deposits of calcium carbonate covering the cell wall (Allanson, 1973). While some algae (diatoms, blue-greens) and bacteria are loosely associated within and on the calcite and mucoid complex of the organic compounds of bacterial and algal origin, others penetrate through the complex to the macrophyte, attaching themselves directly along its entire surface or by means of simple or branched mucilagenous stalks. Much of the matrix is associated with the mucopeptide cell walls of the epiphytic bacteria and blue-green algae, an association lost in the techniques used to prepare these scanning electron photographs but seen in regular electron microscopy.

Most of this relatively rich community of microflora epiphytic upon macrophytes probably enters detrital stages. Massive accumulation rates on the supporting plants, *Chara* and *Potamogeton* (Allanson, 1973), and differences seen in the fine structure of the epiphytic-carbonate-mucoid complex between parts of the plant as it grows and provides new surfaces for colonization, indicate that little of the community is utilized by grazing animals. In places, however, evidence shows that the association can be utilized (Fig. 15–18d, e). Little is known of the in situ feeding processes, nutrition obtained from the epiphytic association, or effects of the grazing organisms on

---

[7]The scanning electron photomicrographs were kindly provided by Dr. B. R. Allanson, Rhodes University, South Africa; some are from his investigations reported in Allanson, 1973.

Figure 15–18   The structure of a microfloral community epiphytic upon the macroalga *Chara* from Wytham Pond, Oxford, England. A, Surface view of microflora on a stem (× 75); B, The epiphytic community after exposure to pH 4.5 to remove calcite deposits showing the dominant diatom *Achnanthes minutissima* and mucoid membranes attached to the wall of the host plant (× 500); C, Intimate association between the diatom component and supporting mucoid matrix, calcite deposits removed (× 1,000); D, Transverse feeding tracks left by nymphs of the baetid mayfly *Cloeon dipterum* (× 500); E, A chironomid larva, found enclosed in fronds of *Chara*, which may have been feeding in this position (×90). (Photographs courtesy of B. R. Allanson, 1973. From Allanson, B. R.: Fine structure of the periphyton of *Chara* sp. and *Potamogeton natans* from Wytham Pond, Oxford, and its significance to the macrophyte periphyton metabolic model of R. G. Wetzel and H. L. Allen. Freshwater Biol., 3:535–541, 1973.

the productivity of the epiphytes or macrophytes by the feeding process (e.g., the solubilization of certain nutrients that in turn are used by the epiphytes or supporting macrophyte).

Metabolic Interactions

Relationships Between Algae and Bacteria

The metabolic relationships between the attached algae and bacteria and the substrate are complex. The detailed investigation of Allanson (1973) on the fine structure of the epiphytic microflora of macrophytes demonstrated the structural proximity required for the metabolic interactions hypothesized earlier for these associations (Fig. 15–19) by Wetzel and Allen (1970) and Allen (1971). The macrophytes release large quantities of both inorganic compounds (e.g., oxygen, carbon dioxide, phosphates, silica, etc.) and organic compounds (secreted during active photosynthesis and autolysis during senescence of plant parts) into the immediate proximity of adnate epiphytic microflora and littoral waters. This proximity alone is sufficient for the transfer and exchange of nutrients (PO_4, CO_2) between the epiphyte and host before they completely move away to and are diluted by the surrounding water (see, for example, Harlin, 1973). In moderately hard waters precipitation of calcium carbonate is induced by photosynthetic activity (Chapter 10) to form a matrix intermixed with the microflora and mucoid substrates of polysaccharides and peptides. Organic substrates released by the macrophytes and algae in part are actively utilized by the epiphytic bacteria. Dissolved organic compounds not utilized by this association, or adsorbed within or to monocarbonate surfaces, enter the pool of littoral dissolved organic matter for further bacterial processing. In turn, epiphytic bacteria produce CO_2 and certain organic micronutrients, e.g., vitamin B_{12} (cf. Wetzel, 1969, 1972) that can serve as growth factors for the macrophytes or algae, or both. Photoassimilative active uptake of labile organic substrates has been demonstrated by a submersed angiosperm in axenic culture and by a species of the macroalga *Nitella* (Wetzel, unpublished; Smith, 1967).

There is evidence to support the conclusion that most phytoplanktonic algae do not actively augment photosynthesis by heterotrophic utilization of organic substrates when forced to compete with the active uptake of the substrates by bacteria. The immediate juxtaposition of attached algae to the metabolically active area of a living plant and bacterial concentrations can give them a distinct competitive advantage over their planktonic counterparts which live in a habitat where nutrients are much more dispersed (see, for example, the studies of Allen, 1971).

Analogous interactions certainly exist within the epipelic associations of algae and bacteria. Unfortunately, even less is known about the qualitative and quantitative metabolic interrelationships of these attached forms and their substrate than is known about the epiphytic microflora. Potential sources of inorganic nutrients and labile dissolved organic substrates are likely to be

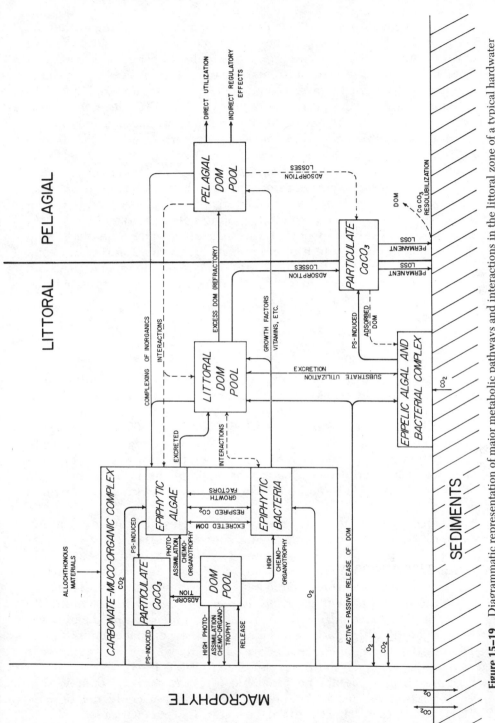

Figure 15–19 Diagrammatic representation of major metabolic pathways and interactions in the littoral zone of a typical hardwater lake; see text for discussion. DOM = dissolved organic matter; PS = photosynthesis. (Modified from Wetzel and Allen, 1970.)

greater under reducing conditions of the sediment-water interface than at the surfaces of macrophytes.

The movement of dissolved organic matter, continually being further degraded, flocculated, and adsorbed en route, from the littoral surfaces of active metabolism into the water of the littoral and pelagial zones is of particular importance. Dissolved organic substrates emanating from littoral sources or allochthonously from the drainage basin are utilized or sedimented in various ways as they travel to the pelagic zone (cf. Chapter 17). The rates of this "processing" are highly variable within the dynamics of seasonal and individual differences in physical and chemical parameters of lakes. The littoral region can be viewed as functioning as an effective but selective metabolic sieve or sink for certain dissolved organic compounds and inorganic nutrients from littoral and external sources before they reach the pelagic zone. During this transport, the more easily degradable organic substrates tend to be decomposed in the littoral zone, and compounds of greater resistance (i.e., slower turnover rates) tend to move towards the open water. Therefore, even though the amounts, rates of utilization, and fate of inorganic and organic compounds released from the littoral zone are highly variable among lakes and seasonally, their direct and indirect effects on the metabolism of the pelagial microflora can be great in a majority of lakes.

Inhibition of Phytoplankton by Submersed Macrophytes

A number of investigations have demonstrated a distinct inhibition of phytoplanktonic development in the littoral zones containing heavy stands of emergent and submersed macrophytes (e.g., Schreiter, 1928; Hasler and Jones, 1949; Hogetsu, et al., 1960; Stangenberg, 1968; Goulder, 1969; Dokulil, 1973). Circumstantially, these investigations implied that excreted organic compounds function in an inhibitory antibiotic way on the growth of phytoplankton. However, competition on the part of the macrophytes for nutrients, light, and other substances as a further inhibiting factor was not satisfactorily ruled out in these studies.

A detailed study of the influence of four species of submersed macrophytes on phytoplankton within the stands showed that the decrease in rates of phytoplankton photosynthesis was caused by shading (Brandl, et al., 1970). Average reductions over three years in the rates of phytoplankton production among the plant stands as compared to rates in the open water were: *Potamogeton pectinatus* 60 per cent, *P. lucens* 46 per cent, *Batrachium aquatile* 40 per cent, and *Elodea canadensis* 12 per cent. Phytoplankton production was markedly reduced when they were taken from the pelagial zone and incubated among the macrophytes, and increased greatly when littoral algae were incubated at the same depth in the open water (Table 15–8).

The presence of membrane filtratable (1.2 μm porosity) organic substances inhibiting the photosynthesis of phytoplankton could not be confirmed. However, increasing the pH of the water, as occurs with a reduction in available CO_2 in dense stands of actively photosynthesizing submersed macrophytes (Figure 15–20), reduced the rates of phytoplankton productivity. This decrease could be removed completely by small additions of CO_2.

TABLE 15–8   Changes in Rates of Primary Production of Phytoplankton from the Pelagial Incubated among Submerged Macrophyte Stands of the Littoral and Littoral Phytoplankton Incubated in the Pelagial, Czechoslovakian Shallow Lakes[a]

Submersed Macrophyte Species Dominating Littoral Area	Percentage Reduction in Littoral Zone	Percentage Increase in Pelagial Zone
Potamogeton pectinatus	56	250
Potamogeton lucens	46	270
Batrachium aquatile	39	230
Elodea canadensis	27	500

[a]From data of Brandl, Brandlová, and Poštolková, 1970.

While much of the apparent inhibition of phytoplankton growth among dense stands of macrophytes most likely is related to competition for light and nutrients, the possible production of biotically inhibitory or stimulatory organic substances cannot be excluded totally. For example, the presence of macrophytes favors the biomass and dominance of some blue-green algae (Guseva and Goncharova, 1965) while it is inhibitory to other groups (Kogan and Chinnova, 1972), and perhaps is influential in the successful competition of blue-green algae with other algal groups by heterotrophic utilization of certain excreted organic compounds. Zooplankton of many groups are inhibited strongly or repelled by substances secreted into the water by submersed angiosperms and macroalgae (much of the scattered literature is reviewed by Pennak, 1973). For example, *Daphnia* is repelled strongly by water containing *Elodea, Myriophyllum,* or *Nitella.* The magnitude of the negative responses varies with the plant species and is more pronounced at higher temperatures. Sexually mature *Daphnia* respond more rapidly than immature instars. The inhibitory or insecticidal effects of organic compounds secreted by macrophytes, especially the macroalgae Characeae, on mosquito larvae have been reported widely (and critically reviewed by Hutchinson, 1975). The active compounds are probably of the general allyl sulfide group, analogous to those derived enzymatically from garlic oil.

Figure 15–20 Changes in the pH (– – –·) and alkalinity (·———·) in the surface water along a transect across Sangwin Pond, in the afternoon (June) through intermittent areas of dense cover of the submersed macrophyte *Ceratophyllum demersum* (indicated by hatching). (Modified from Goulder, 1969.)

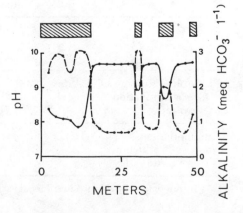

Effects of Light Availability

Among dense stands of emergent macrophytes, the rapid attenuation of light reaching the water plays a major role in the reduction of photosynthesis of littoral algae. For example, in their extensive studies of the subject, Straš-kraba and Pieczyńska (1970) found in reed stands of *Phragmites* a rectilinear relationship between the percentage transmission of light through the emergent parts of the reed and its density, expressed as shoots per m². As much as 96 per cent of the incoming radiation was removed above the water, whereas underwater the density of the stand had little influence on the light availability. The rather low light intensities occurring in reed stands, especially those of an average or high density, cause a substantial reduction in photosynthesis and biomass of phytoplankton entering the stands from the pelagial region. Photosynthetic activity of attached algae is also reduced in these dense stands. Cutting experiments showed a large increase in photosynthetic activity of each group that was directly proportional to the percentage improvement of light conditions (Table 15–9). In less dense stands of emergent plants, as for example in low-growing sedges, light reduction is not as serious, but a strong limitation of photosynthesis of phytoplankton still has been observed (Straškraba, 1963). Some decompositional products or secreted compounds of the plants are believed to play a causal role in this reduction.

Attached Algae as a Source of Phytoplankton

The concept that the presence of littoral benthic algae gives rise to plankton algae emerged many years ago. It was assumed that plankton algae formed resting spores that overwinter in the sediments, and that under favorable conditions germination and population development on the sediments served as the inoculum for the phytoplankton. In moderate-size to large lakes, there is no evidence for any conspicuous buildup of plankton on the sedi-

TABLE 15–9   Effect of Increased Light Availability on Photosynthetic Activity of Phytoplankton and Attached Algae in a Dense Stand of Phragmites communis Before and After Removal of Plant Shoots, Smyslov Pond, Czechoslovakia, in September[a]

	With *Phragmites*		*Phragmites* Cut	
	0 cm	*30 cm*	*0 cm*	*30 cm*
Phytoplankton (mg O_2 1^{-1} day^{-1})				
Gross production	0.14	0.06	7.9	3.0
Respiration	0.17	0.78	2.6	1.5
Attached algae				
Gross production		3.1		8.7
(mg O_2 dm^{-2} day^{-1})				

[a]From data of Straškraba and Pieczyńska, 1970.

ments before a plankton bloom, but only afterward when the dead plankton settle out onto the sediment (Round, 1964).

If such conditions are the case, the plankton algae always must be present in the water, although in extremely low concentrations, so that these few remnants of past populations can serve as the inoculum for a rapidly developing population increase under favorable growing conditions. Such seems to be the case for most species.[8] In one of the better studies of algae, resting spores of the diatom *Asterionella formosa* were not observed, and no evidence exists that appreciable numbers originate from shallow areas (Lund, 1949). Observed massive population developments originate from exponential growth of live cells present in the pelagial waters.

Some phytoplankton do settle out during periods of reduced turbulence, and remain in a resting state on the sediments until water movements are again sufficient to reintroduce the resting cells back into the plankton.[9] *Melosira italica* is a large diatom that clearly does enter a resting stage on the sediments, particularly under ice-cover and during summer stratification (Lund, 1954, 1955). There is no evidence, however, for active growth on the sediments, which are often below the photic zone. Resuspension of filaments during autumnal and spring circulation serves as a large inoculum for the rapidly developing pelagic population.

In shallow lakes, the separation between epipelic algae and littoral metaphyton is less distinct. It is common to find strictly benthic algae or predominately planktonic forms intermixed, especially following irregular storm turbulence (see, for example, Lehn, 1968; Moss, 1969a; Moss and Abdel Karim, 1969; Brown and Austin, 1973a). In shallow lakes in which thermal stratification is irregular and intermittent, shifts in the percentage composition of largely planktonic or of epipelic algae can be rapid. For example, in a pond of southern England, nonmotile species formed the predominant phytoplankton populations only when the water column was relatively turbulent. Motile species persisted in the water column, with the largest biomass of all species found when the water column was stratified. In the shallow Obersee of the Lake of Constance, southern Germany, littoral diatoms and algae of the metaphyton are transported to the open water, especially during spring storms (Müller, 1967).

Productivity of Littoral Algae

Quantitative Evaluation

Quantitative evaluation of attached algal material is basic to analyses of population changes and productivity, but is confounded seriously by the simultaneous collection of much debris and retention of interred cells. Count-

---

[8]Termed *holoplankton* in older literature, in contrast to *meroplankton* that have resting stages in sediments and are not observed in the open water for long periods of time, i.e., they are planktonic only at certain times in their life history.

[9]See page 323.

ing and identification, with all proper precautions and statistical evaluations, have a number of advantages for detailed analyses (Lund and Talling, 1957; Lund, et al., 1958; Vollenweider, 1969b). Developed for lake plankton, these methods of analysis usually are equally applicable to studies of littoral algae. Enumeration can be extended for productivity evaluations by estimations of more meaningful biomass parameters such as cell volumes or carbon at the species level (cf. Chapter 14). Use of cellular constituents such as pigment concentrations of attached algal populations can be used effectively as an estimate of biomass (e.g., Szczepański and Szczepańska, 1966). It is particularly imperative, however, to correct for pigment degradation products (Wetzel and Westlake, 1969), and not to overextend these estimates, which vary significantly in response to an array of physiological variables.

Biomass Measurements

To a limited extent, a temporal series of biomass measurements permit estimates of rates of primary production. These estimates are minimal, and often underestimate by representing net production after losses for excretion of organic compounds, respiration, mortality, decomposition, grazing, etc. The observed rates also are influenced by the rate of inoculation and colonization of new propagula, which is generally faster when conditions for growth are better. Hence, the observed accumulation of biomass is a composite of colonization and production. The saturated accumulations of biomass on substrata yield little insight into the production dynamics that have occurred before the cumulative point.

The rate of population turnover is an average of several variable species rates, and is difficult to determine from biomass analyses. Subjective errors also occur in determinations of attached algae of lakes and streams that are dependent on the frequency of sampling and estimation of what degree of colonization represents a climax stage of the community (Sládeček and Sládečková, 1964). Higher turnover rates result from increased frequency of sampling, especially when the intervals are less than a month.

Attempts to compare the productivity of littoral algae among lakes, within aquatic systems, and seasonally are therefore complicated by the number of techniques that have been employed. Very few in situ metabolic measurements have been made on these communities. As mentioned earlier, many analyses have employed artificial substrata to reduce the natural heterogeneity of the algae. The degree of simulation of benthic algae on these substrates, as compared to those on adjacent natural substances, is highly variable and deviates significantly both qualitatively and quantitatively, with seemingly minor differences in substrata, environmental differences in microhabitats, and particularly rates and duration of colonization. Although the number of studies, many of which have been conducted most critically, has been large, the discrepancies found between populations on natural and artificial substrata are sufficient to necessitate a thorough evaluation of each study in which artificial substrata are used (see, for example, Tippett, 1970; Warren and Davis, 1971). The extent of compromise must be determined in any ecological method, and this problem becomes especially acute among littoral communities.

Therefore, the admixture of estimates of productivity values in the ensuing discussion should be viewed as exemplary of the range encountered under natural conditions. Although no attempt has been made to summarize all of the literature, which is highly diverse and of variable precision, the examples given point out major characteristics and emphasize the major contributions that the attached flora can make to freshwater systems.

Spatial and Temporal Variations

The rates of primary production of attached algae are obviously dependent upon the substrate area available for colonization within the zone of adequate light. In some lakes the littoral substratum may be relatively uniform for much of the perimeter, but in most cases the variation is great. Moreover, the living macrophytic substrata are constantly changing, especially in the more seasonal temperate regions. An example of this spatial heterogeneity in productivity can be seen in a two-year analysis of littoral components — macrophytes, littoral phytoplankton, and attached, largely epiphytic, algae — in Mikolajskie Lake, central Poland (Pieczyńska, 1965, 1968; Pieczyńska and Szczepańska, 1966; Kowalczewski, 1965). Simultaneous measurements in midsummer along numerous points of the littoral zone showed much variation, which was accentuated during variable mass appearances of sessile blue-green *Gleotrichia* and the filamentous green alga *Cladophora* (Fig. 15–21). Similar spatial variations in numbers and species assemblages have been demonstrated in the littoral of a eutrophic lake of British Columbia (Brown and Austin, 1973b).

Seasonal fluctuations in primary productivity of attached algae are also quite variable, analogous to changes in biomass discussed earlier. In general

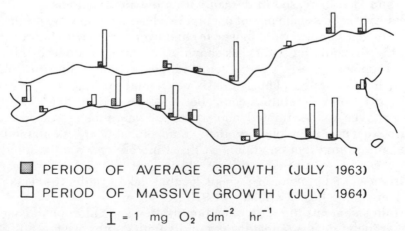

☒ PERIOD OF AVERAGE GROWTH (JULY 1963)

☐ PERIOD OF MASSIVE GROWTH (JULY 1964)

$I = 1$ mg O_2 dm^{-2} hr^{-1}

Figure 15–21   Spatial variations in the simultaneously measured rates of primary production of attached algae in different areas of the littoral zone of Mikolajskie Lake, central Poland. (After data of Pieczyńska and Szczepańska, 1966.)

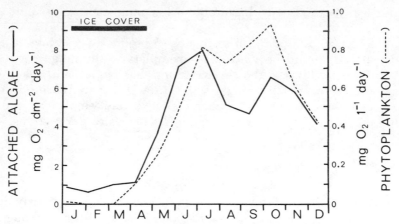

Figure 15–22 Rates of primary production of attached algae and phytoplankton in the littoral zone dominated by the emergent macrophyte *Phragmites*. (Modified from Pieczyńska and Szczepańska, 1966.)

terms, estimates of biomass accrual such as chlorophyll and in situ measurements of production rates by carbon uptake or oxygen production are fairly similar, particularly during periods of active growth (see Pieczyńska and Szczepańska, 1966; Allen, 1971; Hunding and Hargrave, 1973; Hickman, 1971a). Reductions in temperature and light under winter conditions clearly are the dominant causal mechanisms limiting photosynthesis both seasonally and vertically with depth (see Figs. 15–22 and 15–23). The productivity of attached algae collected under ice and exposed to increased light was found to increase greatly and approached values obtained under summer conditions. In the example from Mikolajskie Lake, productivity in the spring increased rapidly in both the attached algae and littoral phytoplankton, decreased in midsummer during maximum development of the littoral macrophytes, and then increased in the fall with a bloom of diatoms.

Photosynthetic enhancement of epipelic algae was shown experimentally when sediment samples were moved to shallower depths with improved light (Table 15–10). Hunding (1971) has shown that benthic diatoms exhibit light-saturated photosynthesis over a wide range of light intensities. During summer, light-saturated photosynthesis was found even at very high irradiances, and no photoinhibition could be demonstrated up to 38 klux (=30.4 μW cm^{-2}). The light-saturated photosynthetic value decreased in autumn, indicating that maximum photosynthetic rates probably are maintained under reduced prevailing light conditions at that time of year. Similar depth relationships can be seen in comparisons of pigment biomass and primary productivity of epipelic algae of a small English lake (Table 15–11), and in a clear, deep alpine lake (Capblancq, 1973).

Within sediments the rates of carbon fixation and chlorophyll concentrations of epipelic algae attenuate very rapidly with depths of a few millimeters (Table 15–12), whereas those of epipsammic algae penetrate somewhat deeper. As discussed earlier, light-mediated metabolism of these algae is

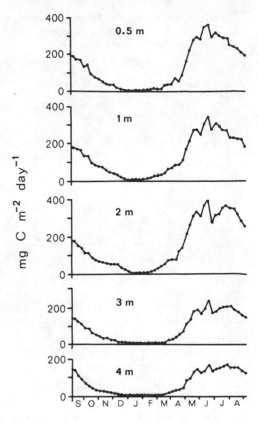

Figure 15–23 Weekly mean values of gross primary production rates of epipelic algae in shallow Marion Lake, a composite from 5 littoral stations, southern British Columbia. (Redrawn from Gruendling, 1971.)

probably intermittent as wave action circulates sand grains up to the sediment surface. Biomass and rates of production of epipsammic algae were consistently greater throughout the year than those of epipelic algae in the same lake system.

Seasonal cycles of primary productivity of epiphytic algae differ considerably in relation to the host plant. In the Lawrence Lake example of Figure 15–

TABLE 15–10   Enhancement of Photosynthesis of the Epipelic Algal Community of Marion Lake, British Columbia, in June by Moving Intact Sediments from Different Depths to 0.5 m[a]

Sample Depth (m)	Solar Radiation (g cal cm^{-2} hr^{-1})	Temperature (°C)	Incubated in situ at Depth (ml O$_2$ m^{-2} hr^{-1})	Incubated at 0.5 m Depth (ml O$_2$ m^{-2} hr^{-1})
0.5	32.6	23.0	19.9	19.9
1.0	23.3	22.0	22.7	26.4
2.0	14.4	17.0	19.4	28.8
3.0	10.6	14.0	14.5	19.4
4.0	5.5	13.0	2.4	5.7

[a]From data of Gruendling, 1971.

TABLE 15–11   Annual Mean Biomass and Rates of Primary Production of Epipelic Algae of Abbot's Pond, Southern England[a]

Year	Littoral Depth Category (m)	Mean Biomass (mg chlorophyll a m^{-2})	Mean Rate of Primary Production (mg C m^{-2} hr^{-1})
1966	0–1.0	11.7	5.50
	1.0–2.5	7.3	2.89
	2.5–4.0	4.0	0.56
1967	0–1.0	9.09	4.13
	1.0–2.5	1.38	0.75
	2.5–4.0	1.47	0.17
1968	0–1.0	13.6	6.23
	1.0–2.5	2.14	0.70
	2.5–4.0	1.47	0.23

[a]From data of Moss, 1969b, and Hickman, 1971a.

24, the annual maxima occurred during the midsummer period of maximum macrophyte development. At this time conservative values reached over 8 g C m^{-2} littoral zone day^{-1} of carbon fixed by epiphytic algae in this relatively oligotrophic lake. Even though the rates of production per substrate area were similar among the emergent bulrush *Scirpus acutus* site and that of the much more dissected submersed macrophytes *Najas* and *Chara,* the much greater area for colonization in the submersed macrophyte area increased the rates greatly per littoral zone area. In both cases annual maxima were found in August near the sediments at the basal portions of the macrophytes. Although the winter rates of carbon fixation by attached algae among the *Najas-Chara* were 100 to 200 times greater than those among the *Scirpus,* the values per substrate area among the more open *Scirpus* littoral were consistently much higher, ranging from double to an order of magnitude higher per area of plant. Causes for this difference are unknown, but the more open *Scirpus* area has greater light and water movement than is found among the dense Chara beds.

TABLE 15–12   Relative Rates of Carbon Uptake by Epipelic and Epipsammic Algae at Different Depths within Sediment, Expressed as a Percentage of the Maximum Value, Shear Water, England[a]

Depth (cm)	% of Maximum Values			
	Epipelic Algae	*Epipsammic Algae*		
	17 Mar. 69	*23 Nov. 67*	*17 Mar. 68*	*9 Sept. 68*
0–1.0	100	100	100	100
1.0–2.0	11.6	57.4	42.1	6.6
2.0–3.0	0	33.4	5.7	0.2
3.0–4.0	0	30.5	3.1	1.2
4.0–5.0	0	14.3	2.5	0.3

[a]All samples incubated at the same light conditions. From data of Hickman and Round, 1970.

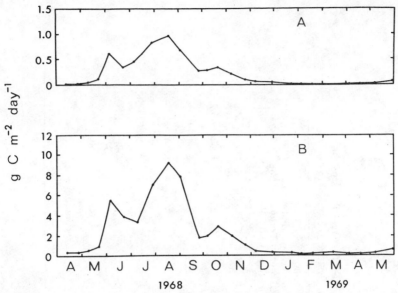

Figure 15–24   In situ primary production of attached algae in g C m⁻² of the littoral zone day⁻¹ among *A*, emergent *Scirpus acutus*, and *B*, submersed macrophytes *Najas flexilis* and *Chara*, Lawrence Lake, Michigan. (Modified from Allen, 1971.)

Rates of epiphytic bacterial utilization of simple organic substrates (glucose and acetate) followed first-order active transport kinetics (Allen, 1971). These rates were nearly an order of magnitude greater in the submersed macrophyte littoral area than among the emergent plants. This difference is potentially related to the greater bacterial exposure and adaptation to greater amounts of organic substrates released by the submersed plants.

Comparison of production rates of attached algae among lakes and climatic regimes is difficult because of the variety of methods for estimation employed in the analyses. However, there is a clear tendency for increasing rates of primary productivity of attached algae with decreasing latitude and extended growing season (Tables 15–13 and 15–14). The rates of turnover time of the attached algal populations also tend, not unexpectedly, to decrease under the better growing conditions. The highest values recorded are from subtropical springs in which constant temperatures and high flow and nutrient sources are conducive to high rates of production. A similar relationship is seen among streams (cf. Wetzel, 1975b), although available data are few and very difficult to compare in a satisfactory manner.

Productivity Rates of Attached Algae vs.
Phytoplankton and Macrophytes

Only a handful of detailed studies exists in which in situ measurements of productivity of attached algae and of phytoplankton and macrophytes have been made simultaneously (Table 15–15). However, these studies are im-
(*Text continued on page 415.*)

TABLE 15–13 Rates of Production of Attached Algae and Estimated Turnover Rates of the Algal Populations of Several Exemplary Aquatic Systems

Lake	Type of Algae	Average Biomass (g dry m^{-2})	Average Annual Estimated Net Productivity (mg dry m^{-2} day^{-1})	Estimated Turnover Time (days)	Source
Sodon Lake, Mich.	Glass slides, horizontal, open water	1.23	37.5	32.8	Newcombe, 1950
Falls Lake, Wash.	Glass slides, horizontal, at sediment interface (0.4 m)	3.24	148	21.8	Castenholz, 1960
Alkali Lake, Wash.	Glass slides, horizontal, at sediment interface (0.4 m)	2.94	131	22.4	Castenholz, 1960
Sedlice Reservoir, Czechoslovakia	Glass slides, vertical, open water, several depths	12.7	213	59.6	Sládeček and Sládečková, 1963
Lake Glubokoye, USSR	Epiphytic on *Equisetum*, biomass, and O$_2$ production	18.7	128.8	21.7	Assman, 1953
Silver Springs, Fla.	Epiphytic on *Sagittaria*, biomass, and O$_2$ production	177.	12,300.	14.4	Odum, 1957
Laboratory analyses, diatoms of Lake Balaton, Hungary	Epilithic diatoms, biomass, and O$_2$ production	—	—	ca 1 (optimal conditions)	Felföldy, 1961

TABLE 15–14 Comparisons of Net Rates of Production of Attached Algae, Expressed as Organic Matter Accrual Over a 10-Day Period, of Several Different Fresh Waters[a]

Lake	Mean Net Production Rate (mg org. mat. dm^{-2} day^{-1})	Depth of Maximal Net Production Accrual	Remarks
Sodon Lake, Mich.	0.5	0.2 m	Horizontal glass slides, 3 summer months; Newcombe, 1950
Falls Lake, Wash.	1.28	0.4 m	Horizontal glass slides, 22 months; Castenholz, 1960
Lenore Lake, Wash.	1.00	0.4 m	17 months
Alkali Lake, Wash.	1.76	0.4 m	12 months
Soap Lake, Wash.	1.67	0.4 m	17 months
Walnut Lake, Mich.	1.69	0.6 m	Horizontal glass slides, 2 summer months; Newcombe, 1950
Sedlice Reservoir, Czechoslovakia	2.13	3.0 m	Vertical slides, 10 months; Sládeček and Sládečková, 1964
Lake Tiberias, Israel	2.20	1.2 m	Vertical glass slides, 14 months; Dor, 1970
Shallow ponds, central Mich.	3.63	–	Glass slides, 2 summer months; Knight, et al., 1962
Borax Lake, Calif.	14.63	0.2 m	^{14}C methods, 12 months, in situ on epilithic algae; Wetzel, 1964
Red Cedar River, central Mich.	21.2	–	Plexiglas plates, summer months; King and Ball, 1966
Tiberias Hot Springs, Israel	73.0	0.1 m	Horizontal glass slides, 2 months; Dor, 1970
Silver Springs, Fla.	96.7	–	O_2 methods, epiphytic on *Sagittaria*, 12 months; Odum, 1957

[a]Techniques employed vary widely and only an approximate comparison is possible.

TABLE 15–15   Examples of Annual Net Productivity of Phytoplankton, Littoral Algae, and Macrophytes of Several Lakes in Which Productivity Estimates of Attached Algae Were Made on Natural Substrata

Lake	Area (ha)	Mean Depth (m)	Annual Mean (mg C m⁻² day⁻¹)	Annual Mean (kg C lake⁻¹ day⁻¹)	kg C ha⁻¹ of Lake Surface year⁻¹	(%)	Remarks
Borax, Calif.	39.8	<0.5					Saline lake; benthic algae, primarily epilithic, some epiphytic and metaphyton; single macrophyte species *Ruppia maritima*; ¹⁴C methods for all components (Wetzel, 1964)
Phytoplankton			249.3	101.0	926	(56.8)	
Littoral algae			731.5	75.5	692	(42.5)	
Macrophytes			76.5	1.36	12	(0.7)	
					1630		
Marion, British Columbia	13.3	2.2					Softwater, oligotrophic lake; benthic algae, primary epipelic; O₂ techniques, from which net production was estimated (Efford, 1967; Hargrave, 1969; Gruendling, 1971)
Phytoplankton			21.9	0.29	8	(1.6)	
Littoral algae			109.6	11.3	310	(62.2)	
Macrophytes			49.3	6.5	180	(36.1)	
					498		
Lake 239, Ontario	56.1	10.5					Softwater, oligotrophic lake; probably underestimates since winter production is not included; benthic algae, primarily epilithic; macrophytes probably insignificant; CO₂ utilization methods (Schindler, et al., 1973)
Phytoplankton					823	(99.0)	
Littoral algae					8.1	(1.0)	
Macrophytes					N.D.		
					ca 831		
Lake 240, Ontario	44.1	6.1					(Same as above for Lake 239)
Phytoplankton					501	(98.2)	
Littoral algae					9.0	(1.8)	
Macrophytes					N.D.		
					ca 510		
Lawrence, Mich.	5.0	5.9					Hardwater, oligotrophic marl lake; benthic algae, primarily epiphytic on sparse submersed macrophytes; ¹⁴C methods (Wetzel, et al., 1972)
Phytoplankton			118.9	2153	434	(25.4)	
Littoral algae			2003.	1977	399	(23.3)	
Macrophytes			240.8	4360	879	(51.3)	
					1712		
Wingra, Wisc.	139.6	ca 2					Large, shallow hardwater eutrophic lake; large littoral zone with dominant submersed macrophyte *Myriophyllum* and metaphytic mats of macroalga *Oedogonium*; ¹⁴C methods for all components; mostly only summer values (McCracken, et al., 1974; Adams and McCracken, 1974; J. F. Koonce, personal communication)
Phytoplankton			1200.	1675.	4380.	(78.6)	
Metaphyton (*Oedogonium*) (Summer, 1971)			3.0	4.2	11.1	(0.4)	
(Summer, 1972)			5.5	7.6	19.9		
Macrophytes			320.5	447	1170	(21.0)	
					5581.		

portant in that they demonstrate the magnitude of effect that the littoral productivity can have on lake systems. The range is great. In the large, shallow Borax Lake of northern California, epilithic algae are a major source of organic matter that totally dominates on an areal basis. Because the littoral area is small, when it is expanded for the whole lake surface the attached algal contribution is reduced but still constitutes nearly one-half of total primary productivity. In Marion Lake the primarily epipelic algae dominate the lake system. In both Lawrence Lake, Michigan, and Lake Wingra, Wisconsin, the submersed macrophytes are major components of the total primary productivity. Submersed macrophytes and largely epiphytic algae of Lawrence Lake constitute nearly three-fourths of the lake productivity, in spite of the limited littoral area of this small lake. In contrast, the littoral contribution to the total productivity of the two Ontario lakes constitutes only a few per cent because the geomorphology of these lakes does not lend itself to extensive littoral development.

A few other less complete studies which do not permit full comparisons should be mentioned to relate the relative rates of photosynthesis of attached algae to those of phytoplankton. Slow-growing submerged mosses, almost exclusively *Marsupella aquatica*, cover some 40 per cent of the bottom from between 2 to 35 m of clear, low mountain Lake Latnajaure in Swedish Lappland (latitude 68°). This large, deep ($\bar{z} = 16.5$ m, $z_m = 43.5$ m) lake is covered with over a meter of ice for about 10 months of the year, and has a mean annual water temperature of 2°C. During the period of investigation in August and September, the average primary productivity for the whole lake was between 3.5 and 4.0 kg C per day (Bodin and Nauwerck, 1969). The average turnover time of the biomass of the perennial moss was about 30 years, between extremes of about 15 years at a depth of 5 m, and more than 125 years at depths greater than 30 m. Of the total primary productivity of the lake, the phytoplankton were responsible for 60 per cent, the moss for 20 per cent, the epipelic diatoms for 15 per cent, and the epiphytic algae on the moss for 5 per cent. Similar results have been found in Char Lake of the Canadian Arctic at latitude 75° N (Kalff and Wetzel, in preparation).

Evaluation of the productivity of phytoplankton and attached algae of a large deep alpine lake of the central Pyrenees Range also indicated the major contribution of the macroalga *Nitella* and of epilithic algae to the total primary productivity (Capblancq, 1973). The biomass of attached algae of the littoral zone (0 to 6 m) and *Nitella* (6 to 19 m) formed a biomass some 140 times greater than the mean biomass of the phytoplankton during summer, and *Nitella* constituted about 80 per cent of the total benthic algae. Based on in situ productivity measurements, the contribution of the benthic algae to the total primary productivity of the lake was estimated to be 30 per cent.

Among shallower lakes, the work of Straškraba (1963) on two large fishponds of southern Czechoslovakia deserves particular mention. Based on extensive annual cycles of the nitrogen biomass and dynamics of all plant and animal components of these shallow lakes, it was determined that about 73 per cent of the primary productivity resulted from largely emergent macrophytes, 20 per cent from attached, mainly epiphytic algae, and 7 per cent from

TABLE 15–16   Mean Annual Rates of Primary Production of Phytoplankton in Relation to Those of Epipelic Algae and Algae Epiphytic on the Horsetail Equisetum fluviatile, Priddy Pool, England[a]

Phytoplankton (mg C m^{-2} hr^{-1})	Epipelic Algae (mg C m^{-2} hr^{-1})	Epiphytic Algae (mg C m^{-2} of substratum hr^{-1})
1.55	1.71	63.9

[a]Based on data of Hickman, 1971b.

phytoplankton. It was shown that less than 10 per cent of the primary productivity was utilized by higher trophic levels.

The number of studies evaluating the relative photosynthetic rates of algae of the littoral in comparison to the rates of phytoplankton are few. But the suggestion that the most productive site is epiphytic on macrophytes repeatedly appears, as shown for example in Table 15–16 for a shallow English lake, Lawrence Lake, discussed earlier, and in the littoral of Mikolajskie Lake (Table 15–17). These results support earlier discussion on the importance of factors of light, wave action-sediment movement, and inorganic-organic nutrient exchanges between macrophytes and attached algae. It is apparent that growth conditions are often better in the littoral attached habitat than in the planktonic mode. How significant littoral productivity is to a lake system depends to a great extent on the physical conditions of the lake morphometry and substrate characteristics.

Changing Littoral Productivity and Eutrophication

As the water of lakes receives increasing nutrient loading, there is a strong tendency for phytoplankton growth to increase to the maximum

TABLE 15–17   Percentage Contribution of Various Producers to the Annual Net Primary Productivity per m^2 in the Littoral Zone, Mikolajskie Lake, Poland[a]

Zone/Component	Percentage Contribution
Eulittoral	
Macrophytes	28
Planktonic	10
Metaphyton	21
Attached algae	41
Littoral overgrown	
with emergent vegetation	
Macrophytes	57.2
Planktonic	19.6
Metaphyton	0.1
Attached algae	23.1

[a]After Pieczyńska, 1970.

capacity within existing limitations of temperature and available light (cf. Chapter 14). However, it is imperative that eutrophication of aquatic systems is not viewed in the restricted sense of phytoplanktonic productivity. Within obvious geomorphological restrictions on littoral development, the common situation is for littoral productivity to play a major role in the early and final stages of increasing fertility *of the lake system as a whole.* Exceptions certainly exist, but these conditions are certainly widespread in a large percentage of lakes of the world.

Submersed macrophytes often constitute elements of increasingly greater importance to the total primary productivity of lakes, until a point is reached at which the fertility of the whole system is subject to a state of severe light attenuation (Fig. 15–25). This light limitation is usually associated with intense phytoplankton productivity. Within a given latitude and climate, maximum growth of phytoplankton is reached rapidly when population densities become self-limiting as a result of self-shading effects; growth can be increased beyond these limits only by greater turbulence and light availability than occurs under natural conditions (Wetzel, 1966b).

As the emergent vegetation assumes greater dominance in a lake ecosystem, and eventually encompasses a majority of the lake basin (see Chapter 18), an exceedingly productive combination of littoral macrophytes and attendant microflora develops (Fig. 15–25). Attached, largely epiphytic algae, eulittoral algae, and metaphyton develop in strong association with the emergent flora.

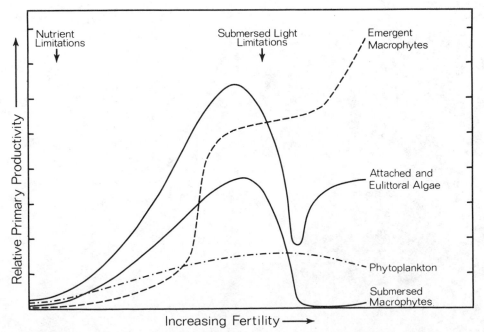

Figure 15–25 Generalized relationship of primary productivity of submersed and emergent macrophytic flora, attached algae and metaphyton, and phytoplankton of lakes of increasing fertility of the whole lake ecosystem. (After Wetzel and Hough, 1973.)

Natural changes in this general sequence are usually slow and extend over centuries and millennia, depending on the basin morphometry. The process can be accelerated greatly by increased nutrient loading, either artificially (see Smith, 1969) or more gradually (see Mattern, 1970).

A large majority of the lakes of the world are small, and their morphometry is such that the ratio of colonizable littoral zone to pelagic zone of production is large. In these cases, there is little question that macrophytic vegetation, and importantly, its attendant microfloral community, have a major impact on the lake ecosystem. The distinction between what constitutes allochthonous production (e.g., eulittoral, terrestrial, etc.) and what is part of true autochthonous lake production is a highly artificial one and has, indeed, led to an artificial treatment of mechanisms of ecosystem metabolism. It has long been recognized that all components of the drainage basin are influential in regulation of lake metabolism. Integrated data on the *functional* impact of these components on the entire system are essentially nonexistent. The concept of a lake as a microcosm must finally be laid to rest. In most lakes, the littoral complex of macrophytes and associated microflora is foremost in regulation of rates of eutrophication and in functional impact on the system as a whole.

zooplankton, benthic fauna, and fish community interactions

The animal components of fresh waters constitute an extremely diverse assemblage of organisms represented by nearly all phyla. Analyses of their functional roles within aquatic systems necessitate a reasoned balance between the general modes and timing of growth and reproduction in relation to food availability and utilization. Evaluation of the population dynamics, and certain important adaptive behavioral characteristics that influence these dynamics, is fundamental to formulations of productivity of individual species populations and of entire communities. Underlying all evaluations of productivity of the animals are their food or trophic relations with plants and other animals, and competitive and predatory interactions that lead to greater success of one species over another.

ZOOPLANKTON

The animal components of the freshwater plankton are dominated by 3 major groups: the rotifers, and 2 subclasses of the Crustacea, the Cladocera and Copepoda. These groups have been studied in some detail and form the bulk of the ensuing discussion. It is probable that under a majority of circumstances, the rotifers and especially the limnetic crustaceans overwhelmingly constitute the dominants of zooplanktonic productivity.

Protozoa

Among the protistian fauna, the population dynamics and productivity of the protozoa are poorly understood. Although generally a minor part of

419

the zooplankton both numerically and in biomass, it is apparent that at times the protozoa can constitute a significant component of the zooplanktonic productivity. For example, in midsummer in large Lake Dalnee, Kamchatkan, USSR, the flagellate and ciliate protozoans made up a substantial part of the pelagial zooplanktonic population (Sorokin and Paveljeva, 1972). Their maximal biomass was observed during the period of the decline of early summer algal populations and the simultaneous intensive development of bacterial populations. Vertically, protozoan maxima occurred in distinct layers of the water column, generally deep between 10 and 20 meters, at which levels their biomass reached about 3 g m^{-3}, approaching the total biomass of other zooplankton at their maximum development. Protozoan utilization of planktonic bacteria in this lake in July constituted a major pathway of energy flux among the fauna in the pelagial (discussed further in Chapter 17).

The lobosan rhizopod *Difflugia* is a rather common planktonic protozoan of both oligotrophic and eutrophic lakes. An example of a common life cycle is seen in the dynamics of *D. limnetica* in several German lakes (Schönborn, 1962). The planktonic populations increased in early summer and attained a maximum in later summer (Fig. 16–1). Much of the population then sinks as the fat globules are metabolized and density increases. Many die and decompose, some encyst, and others remain active throughout the winter in the littoral sediments. The sources from the sediments increase in numbers as benthic organisms, and then in late spring become planktonic by reducing their density by means of the formation of fat inclusions and gas bubbles. The tests of this species and some others are covered with minute sand grains from the sediments, and in the planktonic phase with diatom frustules or quartz grains that are circulated into the upper strata (Fig. 16–1). Many other protozoans exhibit this type of meroplanktonic[1] characteristic, in that only a portion of their life cycle is planktonic.

A number of ciliates are common to the zooplankton, although it is rare for them to dominate except in certain situations, e.g., in very shallow lakes or in the deeper strata of nearly or completely anaerobic hypolimnia. Ciliates can move much more rapidly (200 to 1000 μm sec^{-1})[2] than other Protozoa (0.5 to 3 μm sec^{-1} among those with pseudopodia; 15 to 300 μm sec^{-1} among those with flagellae), which contributes significantly to their greater dispersal and higher feeding rates. Although a few ciliates are mixotrophic and supplement nutrition by photosynthesis, most are holozoic and feed on bacteria, algae, particulate detritus, and other protozoans. A few are carnivorous, feeding on small metazoans.

Several protozoans have been shown to feed actively on algae. The importance of planktonic amoeboid forms, known to consume diatoms (e.g., Canter, 1973), is not fully understood. Two species of benthic ciliates of the genus *Loxodes* were studied in a shallow eutrophic lake in relation to feeding and digestion of algae (Goulder, 1972). Feeding rates were low (0.4 to 13

---

[1]See footnote, p. 405.
[2]The swimming rates are temperature dependent, e.g., 2 species of *Loxodes* averaged 270 μm sec^{-1} at 10°C and 430 μm sec^{-1} at 20°C (Jones and Goulder, 1973).

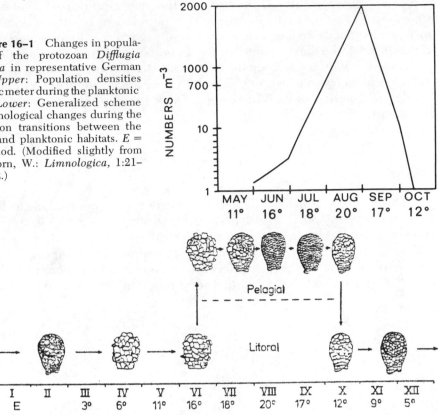

Figure 16–1 Changes in populations of the protozoan *Difflugia limnetica* in representative German lakes. *Upper:* Population densities per cubic meter during the planktonic phase. *Lower:* Generalized scheme of morphological changes during the population transitions between the littoral and planktonic habitats. E = Ice period. (Modified slightly from Schönborn, W.: *Limnologica*, 1:21–34, 1962.)

diatom cells per hour for the larger species), but there was an indication that the protozoan was able to distinguish among 3 species of the dominating alga *Scenedesmus* and feed selectively. From estimates of the feeding rates and the biomass of the algal and protozoan populations, it was concluded that maximum grazing of the ciliates consumed less than 1 per cent of the algal population per day, and had no appreciable effects.

The nature and quantity of available food have been implicated as major controlling factors in the population dynamics of ciliates. For example, in a shallow pond in Pennsylvania, 3 population surges of ciliates occurred and correlated with different nutritional habits (Bamforth, 1958). The small holophyid *Urotricha* developed rapidly under ice and fed actively on a dense population of chrysomonads. Oxytrichid ciliates appeared as the *Chlamydomonas* algal populations were declining rapidly in spring. Ciliate populations, mainly of the genus *Holophyra,* appeared and persisted during the summer, coincident with euglenoid algal blooms, following spring maxima of phytoplankton.

Although nearly all Protozoa are aerobic, a majority can grow very well in greatly reduced oxygen or anaerobic conditions (see, for example, Bragg, 1960). This microaerophillic ability is conspicuous among the planktonic and benthic ciliates, and is attested to by their major development in organic-rich and polluted waters (cf., for example, the monographic review of saprobic organisms by Sládeček, 1973). Appreciable populations of ciliates often develop in strata greatly reduced in or devoid of oxygen in which bacterial populations tend to be dense, such as the monimolimnion of meromictic lakes.

A few coelenterates, larval trematode flatworms, gastrotrichs, mites, and larval insects and fish occasionally occur among the true zooplankton, if only for a portion of their life cycles. These groups, however, rarely if ever are quantitatively significant.

Rotifers

The Rotifera (Rotatoria) is a large class of the pseudocoelomate Phylum Aschelminthes, clearly arisen from fresh water; only 2 significant genera and a few species are marine. About three-quarters of the rotifers are sessile and associated with littoral substrates. Approximately 100 species are completely planktonic, but these rotifers form a significant component of the zooplankton and are the most important soft-bodied invertebrates of the plankton

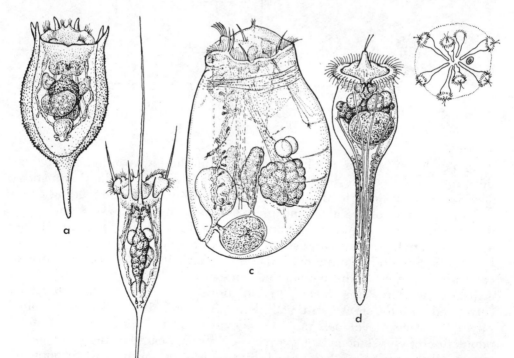

Figure 16–2   Exemplary planktonic rotifers: *a, Keratella cochlearis; b, Kellikottia longispina; c, Asplanchna girodi; d, Conochilus unicornis,* singly and in a colony. (From Ruttner-Kolisko, A.:III. Rotatoria. Das Zooplankton der Binnengewässer. I. Teil. Die Binnengewässer, 26:99–234, 1972, from various sources.)

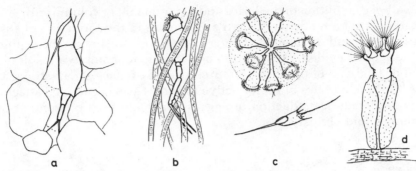

Figure 16–3 Exemplary types of rotifers of different habitats. *a*, Psammic rotifer (*Bryceella*) among sand grains; *b*, A littoral form (*Scaridium*) among algal filaments; *c*, Planktonic forms (*Conochilus* and *Kellikottia*); *d*, *Collotheca* epiphytic on the stem of a macrophyte. (From Ruttner-Kolisko, A.: III. Rotatoria. Das Zooplankton der Binnengewässer. I. Teil. Die Binnengewässer, 26:99–234, 1972.)

of lakes and streams. The general characteristics of the group have been treated recently in some detail by Pennak (1953), Hyman (1951), Hutchinson (1967), and especially Ruttner-Kolisko (1972).

The rotifers exhibit a very wide range of morphological variations and adaptations. In a majority, the body shape tends towards elongation, and regions of the head, trunk and foot usually are distinguishable (Fig. 16–2). The cuticle generally is thin and flexible, but in some rotifers it is thickened and more rigid and is termed a lorica; the lorica is of taxonomic importance in some groups. The anterior end or corona of rotifers is ciliated; in some species the periphery is ciliated as well. The movement of the cilia functions both in locomotion, especially among planktonic forms, and in directional movement of food particles towards the mouth. The mouth, although variously located, generally is anterior. The digestive system contains a complex mastax, a set of sclerotized jaws or trophi unique to the rotifers that functions to seize and disrupt food particles.

Most rotifers, both sessile and planktonic, are nonpredatory. Omnivorous feeding occurs by means of ciliary direction of living and detrital particulate organic matter into the mouth cavity. Predatory species, such as the common *Asplanchna*, are usually large and prey upon protozoa, other rotifers, and all micrometazoa of appropriate size.

Being mostly littoral inhabitants, the rotifers are largely sessile and associated with substrata. Population numbers are highest in association with submersed macrophytes, especially finely divided forms; densities commonly reach 25,000 per liter (Edmondson, 1944, 1945, 1946). Similarly high or greater densities are found in the interstitial water of sand of beaches at or slightly above the waterline (Pennak, 1940). With reduced sites for attachment and presumably less protection from predation, planktonic rotifer populations are much less dense. Densities of 200 to 300 ℓ^{-1} are common, in the range of 1000 ℓ^{-1} occasionally, exceeding 5000 ℓ^{-1} under natural conditions only very infrequently.

Several changes characterize the transition from the predominantly sessile to the planktonic life forms (Fig. 16–3). Weight reduction is common as a result of diminution of the lorica and enlargement of volume with gelatinous materials. The tendency toward an increase in the formation of suspension processes and swimming organs in the form of immovable spines or movable setae is evident. A reduction of attachment organs as a result of diminution or total loss of the foot structures also takes place. Adaptations which reduce the sinking rates of reproductive products also occur, e.g., attachment of eggs to the adult, production of eggs that contain high oil content or are ornamented, and vivipary.

Reproduction

The reproductive life history of typical planktonic rotifers, all of the subclass Monogononta, is characterized by a large number of generations in which reproduction is parthenogenetic by females. These amictic females are completely diploid in number of chromosomes, and produce amictic eggs that develop into further amictic females (Fig. 16–4). There may be up to 20 to 40 of such amictic generations. With an egg development time of about one day under warm, optimal conditions, and without the need to encounter males for fertilization, the populations of amictic females can develop rapidly in 2 to 5 days under good growing conditions.

This nonreductional division occasionally is broken, often only once or twice per year, by the development of a morphologically indistinguishable mictic female. The eggs of the mictic females undergo normal double meiotic division (Fig. 16–4). If these mictic females are fertilized by males, the eggs develop into thick-walled resting eggs that undergo a prolonged diapause and are highly resistant to adverse environmental conditions. If the mictic female is not fertilized, however, the much smaller eggs rapidly develop into males. The males are greatly reduced in size and complexity, and for all practical purposes are short-lived sperm-producing individuals of the populations. Extremely active, they are capable of copulation within an hour of hatching. Hatching of resting eggs is induced by biochemical factors, but the factors involved under natural conditions are poorly understood. The resting mictic eggs always result in parthenogenetic amictic females. The diapause extends over a period of several weeks or months, and hatching is related to changes in temperature, osmotic pressure, water chemistry, and oxygen content.

The mechanisms underlying the production of mictic females are unclear, although a number of environmental conditions have been proposed (reviewed by Hutchinson, 1967, and Ruttner-Kolisko, 1972). Although the stimuli are quite species-specific, significant factors include crowding of amictic females in relation to food and accumulation of substances such as pheromones, produced by the females. In clones of *Brachionus* reproducing amictically at 20°C, brief exposure to a shock temperature of 6°C resulted in nearly 50 per cent of the clones becoming mictic in less than 2 days (Ruttner-Kolisko, 1964). *Asplanchna,* the large predatory rotifer, is certainly among the best example studied in this regard. Gilbert, in a series of superb papers (1967a, 1968, 1972, 1973; Gilbert and Birky, 1971; Gilbert and Thompson,

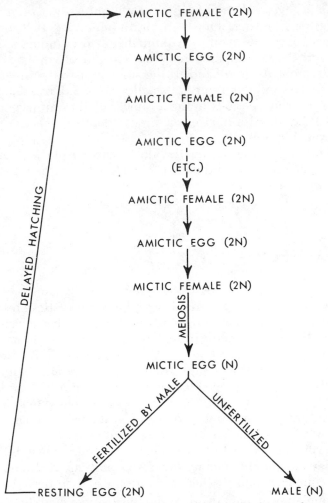

Figure 16–4 Diagrammatic sequence of the reproductive life history characteristic of most planktonic and many benthic rotifers. (Redrawn from Pennak, 1953.)

1968; Riggs and Gilbert, 1972), has demonstrated the subtle reproductive controls in this species that have far-reaching implications among the zooplankton. When fed the protozoan *Paramecium*, *Asplanchna* produced only amictic females, and increasing the population densities of the rotifer to extraordinary levels of crowding produced no mictic females. Adding low populations of the algae *Chlamydomonas* or *Euglena* to the dense paramecia produced significant proportions of mictic females. It was also shown that the algal cells had to be eaten to induce the reproductive change in *Asplanchna*; extracellular algal products would not induce the change. It was shown subsequently that the important dietary component of the plants was *d*-α-tocopherol (vitamin E), which induced the transition from parthenogenetic to sexual reproduction and is essential for spermatogenesis or male fertility. The female offspring were also larger and changed in morphology, exhibiting 4 outgrowths, partially retractile, of the body wall.

These "humps" were elicited by d-α-tocopherol within a range of 5×10^{-13} mol (0.2 ng) in a linear fashion to a maximum of 5×10^{-11} mol (20 ng) per female. Tocopherol, as well as cannibalism, also induced a third polymorph, a very large campanulate type. The adaptive significance of the larger forms probably is related to the ability of the population to utilize large prey species in the absence of smaller ones and, by means of cannibalism, to exist for short periods in the absence of other prey species. It is also possible that the increase in size is a direct, adaptive response to an increase in size of available prey.

Other rotifers, such as the sessile bdelloids, none of which are planktonic, are anabiotic, that is, they reproduce exclusively by parthenogenesis. Resting eggs are not produced.

Population Dynamics

To start with, the analysis of the population dynamics of zooplankton necessitates sound sampling methods with known statistics on sampling variance as well as population heterogeneity within the aquatic system. This subject is outside the purview of the present work and, moreover, has been treated excellently in methods manuals edited by Edmondson and Winberg (1971) and Winberg (1971). Once animal samples of the population are obtained, a number of methods have been employed to evaluate population size and fluctuations. Enumeration can be adequate to address certain questions, on population dynamics and interactions, but in order to evaluate secondary productivity, precise knowledge of changes in biomass throughout all stages of the organisms' life cycles is mandatory. Development times must be known and, although they are variable among organisms and environmental conditions, sampling frequencies must be taken within those time periods.

The population dynamics of rotifers are somewhat simplified because of the very short time that intervenes between hatching and attainment of reproductive capacity. Analyses can be made in terms of birth and death rates by estimations of the reproductive rate and rates of development. The reproductive rate can be determined indirectly from the ratio of eggs per female (Edmondson, 1960, 1965, 1968). The ratio of eggs per female (E) observable in a sample at any given moment can be converted to a finite measure of population birth rate of eggs (B), per female per day, by:

$$B = \frac{E}{D} \tag{1}$$

when D is the duration of development of the eggs, or the mean time an individual spends in the egg stage. This relationship assumes that eggs present at a given time will be added to the population during the next period of time, D. If the age distribution of the eggs is uniform, the fraction of eggs hatching during a day is $1/D$. If the population is changing size, the value of

B is exact only if $D = 1$; the bias introduced by longer durations can be estimated (Edmondson, 1968).

From the reproductive rate B, the instantaneous coefficient of reproduction, b, and the instantaneous growth rate of the population, r, can be calculated based on the conventional exponential growth model in which positive or negative growth over a short time interval is:

$$N_t = N_0 e^{rt} \text{ and } r = b - d \tag{2}$$

when $N_0 =$ the initial population at time zero,

   $N_t =$ the population size at a later time t,

   $r \quad =$ the growth rate coefficient or intrinsic rate of growth, and

   $d \quad =$ the instantaneous rate of mortality.

The effective rate of population increase is the difference between the natural logarithms divided by the time increment $(t - 0)$:

$$r = \frac{\ell n \, N_t - \ell n \, N_0}{t} \tag{3}$$

which is in essence the difference between natality (b) and mortality (d) over the time interval t.

Alternately, birth rate, estimated by the egg ratio method, gives the daily increment to the population. If the population grew by the amount E/D in one day with no deaths, its growth would be:

$$b = \ell n \, (B + 1) \tag{4}$$

Here B is estimated at the beginning of the time interval on the basis that for each female in the population there will be $(E/D) + 1 = B + 1$ females one day later. Assuming no mortality, the population would grow in one day at a rate of:

$$\ell n \, (B + 1) - \ell n \, (1) = \ell n \, (B + 1) \tag{5}$$

Thus by Equation 4, a finite per capita birth rate, B, from the egg ratio can be obtained by dividing the latter by the development time, from which the instantaneous rate, b, can be estimated. It assumes that B is small; under such circumstances both B and the E/D are approximately equal to the instantaneous birth rate b.

As pointed out by Edmondson (1968) and rederived by Paloheimo (1974), for moderately large values of r and bD, the finite per capita birth rate and the egg ratio/D can diverge considerably. Further, the relationships are most applicable to planktonic animals that carry their eggs until hatching, at which time free swimming progeny are liberated, i.e., the eggs are subjected to the same mortality as the adults. Paloheimo (1974) has shown that:

$$E/D = (e^{bD} - 1)/D \qquad (6)$$

when D is the development time of the eggs, which is algebraically equivalent to:

$$b = \ell n \; [(C_t/N_t) + 1]/D \qquad (7)$$

when C_t is the total number of eggs counted at time t, which estimates the instantaneous birth rate b from egg counts C_t or egg ratios C_t/N_t. This formula assumes that steady state conditions prevail, which is not usually the case. However, development time is temperature-dependent, and in analyses of natural populations this primary factor can be compensated for by good experimentally determined changes with temperature. Examples of estimates of instantaneous birth rates are given in Table 16–1, in which the C_t/N_t ratio has been held at 1. The egg ratio method has been used in several detailed population analyses of planktonic rotifers (especially by Edmondson, 1960, 1968), from which the following examples are drawn, as well as for cladoceran and copepod zooplanktonic populations to be discussed later.

Development Time

The rate of development of eggs, required to determine the reproductive rate B, has been shown to be an inverse function of temperature among a large number of rotifers and planktonic crustaceans. Examples of this relationship for a number of rotifer species are seen in Figure 16–5. It has been shown that the development time is quite independent of the type and quantitative nutrition of the adult female (see, for example, King, 1967) carrying the eggs.

The instantaneous growth rate, r, is the effective rate of population increase per unit time and is estimated from the parameters of the birthrate

TABLE 16–1   Estimates of Instantaneous Birth Rates at Differing Development Times in Days by Equation 7 when the Egg Ratio $C_t/N_t = 1$[a]

Development Time, D (Days)	Instantaneous Birth Rate, b
0.5	1.386
1	0.693
2	0.346
3	0.231
4	0.173
5	0.139
10	0.069

[a]After data of Paloheimo, 1974.

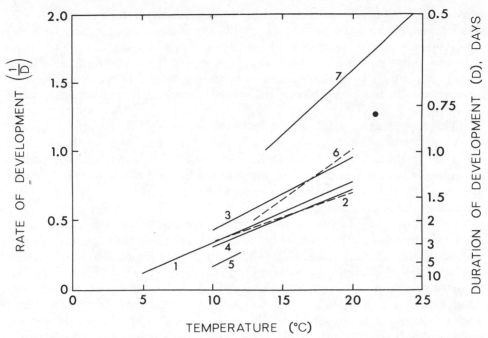

Figure 16–5 Average rates of development per day of rotifer eggs in relation to temperature. Line *1*, Least squares regression line for *Keratella cochlearis* and *Kellicottia longispina; 2,
Polyarthra vulgaris; 3, Keratella quadrata; 4, Kellicottia bostoniensis; 5, Ploesoma truncatum; 6,
Ploesoma hudsoni; 7, Hexarthra fennica,* •, *Euchlanis dilatata.* (From data of Edmondson, 1960,
1965, and King, 1967.)

coefficient *b*. A measure of death rate, *d*, is calculated by difference. The procedure assumes continuity of birth and death rates during the sampling interval. The method demands that the sampling interval is short in relation to the
life cycle and immaturity time of the animals.

Food and Reproductive Rate

The planktonic rotifers feed largely by sedimenting seston particles
into their mouth orifice by means of the pulsating action of their coronal
cilia (reviewed at length by Pourriot, 1965, and Hutchinson, 1967). Size of
food consumed is quite variable. Most food particles taken are small, less
than about 12 μm in diameter, although larger cells, up to approximately
50 μm, are sometimes seized, ruptured, and particulate parts are ingested.
Feeding is in part related to food sizes and the shape of algal cells, and some
evidence points to the possibility that certain algae, such as some species of
Chlorella, are eaten less actively and may be inhibitory.

Some rotifers are raptorial, seizing and ingesting whole prey organisms
or drawing in the cell contents after the cell or body wall is punctured. The
largest rotifer *Asplanchna,* predatory on algae, rotifers, and small planktonic
crustaceans, has the ability to shift size in response to changes in food size,
as was discussed earlier. Hence, there is a large range in size of food particles

consumed by rotifers, with a reasonable separation of rotifer species into size classes of available food niches. This separation is congruous with the observed co-occurrence of several species within the pelagial zone of lakes. It is probable that the fairly discrete niches within the particulate planktonic community are adequate to permit co-occurrence without severe competitive interactions, to the point of elimination of components of the rotifer community.

The reproductive rate of rotifers, however, is related strongly to quality and abundance of food as well as to temperature (Edmondson, 1946, 1965; King, 1967; Halbach and Halbach-Keup, 1974). It is clear that these factors are of major importance in analyses of seasonal fluctuations in the populations in terms of the balance between increase by reproduction and losses from predation and other causes of death.

In addition to effects on the rate of development of eggs, temperature obviously influences the rates of biochemical reactions, feeding, movement, longevity, and rates of reproduction. The composite effects can be seen in the relative rates of changes in concentrations of rotifers when plotted against temperature, both in natural and culture conditions (Fig. 16–6). A number of other factors are involved among the natural populations, but temperature is clearly a major factor affecting birth rate. It is also apparent that species are variable in their responses, some being very restricted stenotherms, while others are fairly eurythermal.

A number of studies under both natural and laboratory conditions have shown that the abundance and species of food exert a major influence on the population dynamics of rotifers. For example, in investigations of the dynamics and reproduction of two clones of *Euchlanis dilatata* fed different algal species at various food concentrations, King (1967) found several relationships that also have been shown to varying degrees by others. Two genetically

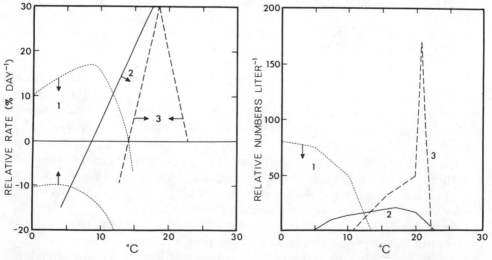

Figure 16–6   Relative rate of change of concentration of rotifers *(left)*, and concentration per liter *(right)*, in relation to temperature. *1, Synchaeta lakowitziana; 2, Euchlanis dilatata; 3, Pompholyx sulcata.* (Drawn from data given in Edmonson, 1946, after Carlin.)

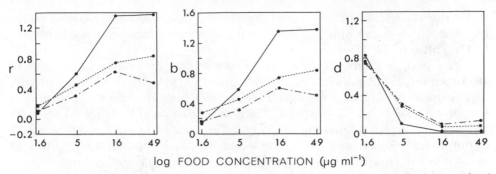

Figure 16–7 The instantaneous rate of increase *(r)*, birth rate *(b)*, and death rate *(d)* of young orthoclones of the rotifer *Euchlanis dilatata* as a function of food level and food species. ────── = *Chlamydomonas reinhardi;* ------- = *Euglena gracilis;* ─ · ─ · ─ · ─ = *Euglena geniculata.* (Drawn from data of King, 1967.)

identical clones of *Euchlanis* were derived from a single parent by parthogenesis, so that in succeeding generations the individuals of the young orthoclone were always derived from young parents and those of the old orthoclone from old parents. The clones were fed different concentrations of 3 algal species, *Chlamydomonas reinhardi, Euglena gracilis,* and *E. geniculata.* Growth rates of individual animals were found to depend on food concentration and not on age or food species of these algae. The rate of population increase was related directly to the density of food for concentrations up to 16.4 μg ml^{-1}, beyond which no further changes were observed. In *Brachionus,* filtration rate decreases at high algal densities; this is related to clogging of the filtration apparatus, reduced digestion, or inactivation of the corona by algal toxicity (Halbach and Halbach-Keup, 1974). Within the population, however, growth rates were related to both food species and food concentration (Fig. 17–7). *Euchlanis* fed on *Chlamydomonas* had higher rates of population increase than when feeding upon *Euglena gracilis. E. gracilis,* on the other hand, led to higher rates of population increase than *E. geniculata.* Individuals of the young orthoclone exhibited higher rates of population increase than those of the old orthoclone. The different rates of increase resulted more from corresponding differences in net reproduction (the number of eggs laid by the average female in her lifetime) than from the frequency of egg laying or survivorship. The results, confirmed by others, implicate differences in assimilation or in the chemical quality of the food, or both.

Seasonal Population Changes

Changes in the seasonal distribution of planktonic rotifer populations are complex and generalizations are difficult to make. A number of perennial species occur that commonly exhibit maximal densities in early summer in the temperate region. Other species are distinctly seasonal, of 2 general types: (a) cold stenotherms that develop greatest populations in winter and early spring, and (b) species that develop maxima in summer, often with 2 or more maxima, especially in late summer in conjunction with the development of certain blue-green algal populations. Some of the variations in seasonal succession can be seen in certain well-studied examples from several lakes. Additional examples from several older works are discussed critically in Hutchinson (1967).

The zooplanktonic community of Vorderer Finstertaler See, a deep, alpine drainage lake above the timber line in the Central Alps of Austria, was[3] extremely simple. Two rotifers *(Keratella hiemalis* and *Polyarthra dolichoptera)* and a single copepod *(Cyclops abyssorum)* constituted over 99 per cent of the zooplankton. The *Keratella* species of rotifer, known as a hypolimnetic cold stenotherm, developed maximal population densities in midwinter, decreasing conspicuously during spring and early summer (Fig. 16–8). The species of *Polyarthra* is also a coldwater form that tolerates reduced oxygen concentrations, commonly found in hypolimnia, to the exclusion of other species of this genus in some lakes. Its dominant maximum was seen in midsummer in this lake.

Studies of the rotifer populations in 2 small lakes in southern Ontario showed that the seasonal distribution varies considerably among lakes, and is synchronous among lakes only in a general way (George and Fernando, 1969). Paradise Lake, for example, is a small (7 ha), shallow (6 m), dimictic lake that exhibits reduced oxygen concentrations in the hypolimnion in late summer, but does not become anaerobic (Fig. 16–9). Summer stratification extended from early June through September in the year of the investigation. Of the 4 species[4] of *Keratella, K. canadensis* is a cold stenotherm found in the upper strata in early winter. *K. quadrata*, also a cold form, occurred from

---

[3]The lake has since been modified drastically by a large hydroelectric project.
[4]*Keratella canadensis*, which may not be a valid species, is very similar to *K. quadrata*.

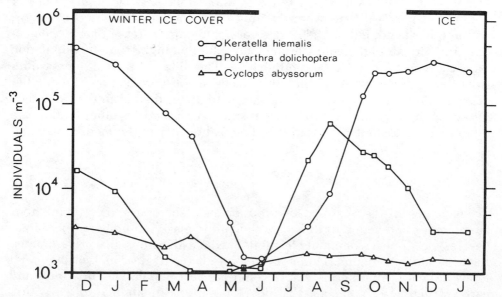

Figure 16–8   Seasonal variations in the population densities of three dominating zooplankton species (> 99% of the total) of Vorderer Finstertaler See, Central Alps of Austria, 1969. (Modified from Pechlaner, et al., 1970.)

early winter in the upper layers and moved to the cooler hypolimnion in midsummer before declining. *K. cochlearis* is also a cold stenotherm, and exhibited 2 maxima in this lake, one in winter and another in late spring, but the seasonal distribution of this species varies greatly among different lakes. *K. hiemalis,* known as a winter and early spring species, was common during colder periods but was reduced abruptly as temperatures reached approximately 15°C.

   Polyarthra vulgaris, P. euryptera, Conochilus unicornia, Gastropus stylifer, and *Keratella crassa* are largely summer and autumn species, although a considerable displacement and spatial separation of the populations occurred temporally with depth (Fig. 16–9). The predatory species of *Asplanchna* generally was a minor component of the rotifer community, in relatively uniform distribution in winter and early summer. The 2 species of *Kellicottia,* both perennial, were distributed uniformly with depth throughout much of the year, except that *K. bostoniensis* was restricted to cooler areas during much of the summer. Both species exhibited summer population maxima.

   These examples serve to illustrate some of the variations found in seasonal succession. Variations are very common, and in warm temperate regions very little is known about seasonal succession among rotifers. Interactions among species of rotifers are poorly understood; a notable exception is the predation of *Brachionus* by *Asplanchna,* discussed earlier. Clearly, temperature and food quality and quantity are dominant factors in the regulation of rotifer reproductive rates and population succession. Analyses of interactions in situ under natural conditions are rare. A detailed regression analysis of the seasonal dynamics of reproductive rates of planktonic rotifers of 4 lakes of the English Lake District in relation to those of food organisms verified the results of culture analyses (Edmondson, 1965). The abundance of food at a given time is an imperfect measure of food supply to the feeding organisms; the food consumed during a period of time depends upon the ability and rate of grazing, as well as the predator's reproductive rate, which is under the influence of still other environmental variables. Among the 3 dominant rotifer populations analyzed by Edmondson, reproductive rates were strongly related to temperature, most strongly of all in *Keratella.* A small flagellate, *Chrysochromulina,* was of major importance in the success of both *Keratella* and *Kellicottia.* Limited data on the bacterial densities showed no significant correlations with reproductive rates of the rotifers. The reproductive rate of *Polyarthra* was directly correlated with the abundance of the small alga *Cryptomonas.*

   The significance of food organisms to the rotifers was found to be partly but not wholly related to size. Some food organisms within an acceptable size range were of little significance. The green alga *Chlorella* was consistently negatively correlated as a food organism. Similarly, *Eudorina* was not eaten and may exert an inhibitory effect.

Diurnal Migration

   Although the diurnal migration is known in some detail for the cladocerans and copepods, and will be discussed further on, relatively little in-

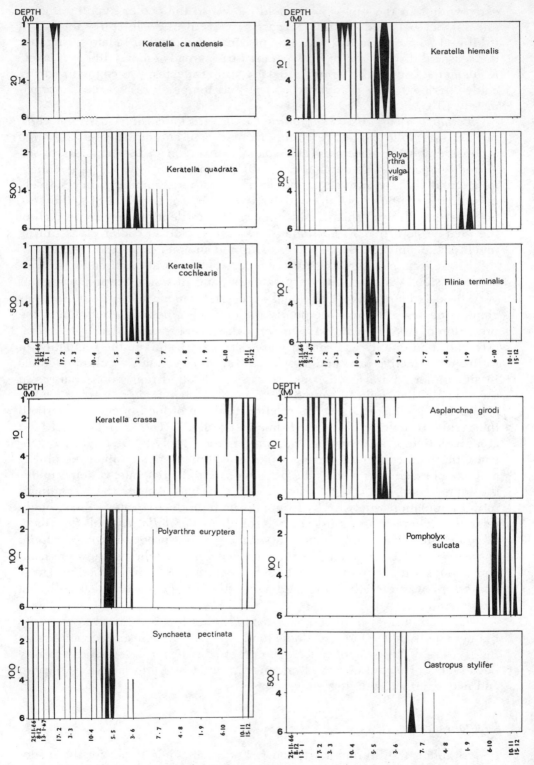

Figure 16–9 *See legend on opposite page.*

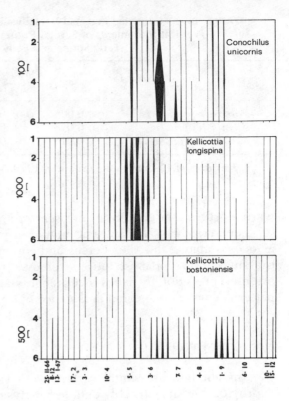

Figure 16–9 Seasonal distribution of 15 dominant species of planktonic rotifers of Paradise Lake, southern Ontario. The vertically written numbers and dimension bars, to the left of and different in quantity for each species, refer to the number of individuals per liter of each of vertical profiles within the annual figures. (From George, M. G., and Fernando, C. H.: Seasonal distribution and vertical migration of planktonic rotifers in two lakes in eastern Canada. Verhandlungen Int. Ver. Limnol., 17:817–829, 1969.)

formation exists on daily movements of planktonic rotifers. Commensurate with their relatively small size in comparison with the crustacean zooplankton, the movements of rotifers are relatively slow (Table 16–2). The maximum observed velocity of ascent in *Filinia* of 0.018 cm sec^{-1} is nearly equivalent to that of some of the slowest movements among the crustacean zooplankton (George and Fernando, 1970). The maximum velocities of vertical movement, as well as the amplitudes, vary among species with descent or ascent and seasonally (Table 16–3). The general tendency is toward upward movement during the daylight period, although the limited evidence indicates an up-

TABLE 16–2 Amplitude of Vertical Migration in Rotifers of Sunfish Lake, Ontario, Expressed in Meters Moved Vertically per 24 Hours[a]

Date	Polyarthra vulgaris	Filinia terminalis	Keratella quadrata
7–8 Jun. 1967	3.0 m	1.7 m	1.0 m
19–20 Jul. 1967	5.6	0.6	2.5
30–31 Aug. 1967	3.6	2.8	[b]
7–8 Feb. 1968	3.0	[b]	[b]
6–7 Mar. 1968	1.9	[b]	3.0
3–4 Apr. 1968	1.1	4.0	0.2

[a]After data of George and Fernando, 1970.
[b]Population negligible.

TABLE 16–3   Estimated Maximal Velocities in cm sec^{-1} ($\times 0.6 =$ m min^{-1}) of
Vertical Movements of Several Rotifers, Sunfish Lake, Ontario, in
Early Spring and Early Summer[a]

| | Day | | Night | |
Rotifers	Ascent	Descent	Ascent	Descent
Polyarthra vulgaris	0.006	0.005	0.007	0.01
Filinia terminalis	0.018	0.009	0.005	0.003
Keratella quadrata	0.01	0.014	0.004	0.003

[a]After data of George and Fernando, 1970.

ward nocturnal migration at one season and the reverse at another season
(cf. Pennak, 1944). No clear-cut migration pattern is seen among the rotifers,
as occurs among the cladocerans and copepods (cf. further on). Presumably,
the smaller size of most rotifers removes much of the pressure of visually
oriented predation by fishes that probably is associated with nocturnal ascent
patterns among cladoceran and some copepod zooplankton.

Horizontal Variations in Distribution

Insight into horizontal variations in the spatial distribution of rotifers
is provided by the superbly detailed analyses of zooplankton in a single cross-
section across the northern arm of Skärshultsjon, southern Sweden (Berzins,
1958). The eastern shore of the basin along this 145-meter transect is steep-
sided, dropping precipitously to the maximum depth of about 13 m. The western, more gently sloping shore supported a well-developed littoral zone of
submergents *Myriophyllum* and *Potamogeton*, grading through *Nuphar* lilies
to the emergent macrophytes of *Phragmites, Equisetum,* and *Juncus.* The
numerous samples were collected within a 2-hour period in midafternoon
on 7 August 1950. At this time the lake was stratified, with a freely circulating
epilimnion (19°C) to 3 m, below which the metalimnetic thermal discontinuity layer extended to 7 m, at which point the hypolimnion of 8 to 10°C extended to the bottom. The oxygen content of the hypolimnion was reduced
(< 2 mg ℓ^{-1}, < 20 per cent saturation) but not anoxic.

Of the 41 species of rotifers found, several conspicuous patterns of distribution were noted and are summarized in Figure 16–10, although only a
few of the dominant species are considered. The epilimnetic rotifers tended
to be more concentrated near the littoral regions on both sides or either side
of the basin (*Kellicottia longispina, Collotheca mutabilis, Polyarthra vulgaris,*
and *P. eurptera*). The dominant epilimnetic *Collotheca pelagica* was most
concentrated in the open water. Two major species, *Polyarthra major* and
Trichocerca similis, were restricted to and developed almost completely in
the metalimnion, and several others had distinctly hypolimnetic populations (*Polyarthra longiremis, Keratella hiemalis, Filinia longiseta*).

Within the population of *Keratella longispina,* Berzins found a distinct
separation according to size. Larger individuals of the population occurred

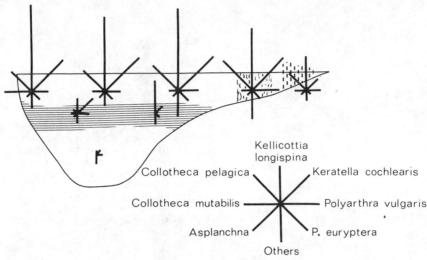

Kellicottia
longispina

Collotheca pelagica   Keratella cochlearis

Collotheca mutabilis ———   ——— Polyarthra vulgaris

Asplanchna   P. euryptera

Others

Figure 16–10 The relative quantitative distribution of dominant rotifers in per cent in different strata along a transect across Skärshultsjön, southern Sweden, 7 August 1950. (Modified from Berzins, 1958.)

at greater depth, which suggests a greater sinking rate, although this species has some power of vertical movement which helps it avoid sedimentation to deeper waters. A similar separation according to size was found with vertical distribution of several species of *Polyarthra*.

Cyclomorphosis and Predation

Cyclomorphosis or seasonal polymorphism is a phenomenon that is particularly conspicuous among planktonic Cladocera, but also is fairly common among the protozoa, the dinoflagellates, and the rotifers. The only comprehensive treatment of the subject of cyclomorphosis among plankton is the detailed résumé of Hutchinson (1967). Although it is more difficult to study the genetic continuity of seasonal forms of the same species of rotifers than of the planktonic crustaceans, cyclomorphosis in several rotifers has been analyzed in some depth.

Under conditions leading to seasonal polymorphism, the shape of some part of the organism changes markedly in relation to some standard dimension of its size. These morphological changes can be expressed in several ways in rotifers. The common changes in growth form among some rotifers include the following. (a) Elongation in relation to body width. In some species of *Asplanchna* midsummer populations can be some 5 times as long as wide, markedly changed from nearly spherical late spring morphology. These elongated forms are nearly always sterile and die back, not to reappear until the next spring. (b) Enlargement, with the formation of body wall outgrowths or humps. As discussed earlier, at least in *Asplanchna sieboldi*, this seasonal change in growth appears to be caused by tocopherol content of plant-derived food and most likely is an adaptive response to cope with larger sized food in summer. (c) Reduction in size, usually at higher temperatures in summer,

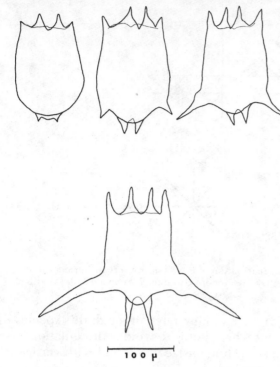

Figure 16–11 Spine variation in a clone of the rotifer *Brachionus calyciflorus* induced by increasing concentrations of a proteinaceous substance produced by its predator *Asplanchna*. (From Gilbert, J. J.: Asplancha and postero-lateral spine production in Brachionus calyciflorus. Arch. Hydrobiol., 64:1–62, 1967b.)

100 µ

with a disproportionate reduction in length of lorical spines. This form change is common among several species of *Keratella*. Its adaptive significance is unclear, since reduction in spine length increases sinking speed at higher summer temperatures, and is compensated for only partly by the reduction in body size. The causal mechanisms for the reduction in spine length are unclear. (d) Production of lateral spines. The best studied case of this type of cyclomorphosis is in *Brachionus calyciflorus*, in which the posterior spines elongate in the presence of its major predator, the large rotifer *Asplanchna*. *B. calyciflorus* always has 2 pairs of anterior spines and 1 pair of postero-median spines. A pair of postero-lateral spines may or may not be present (Fig. 16–11).

Gilbert (1967b; cf. also Pourriot, 1974) has demonstrated that extension of pre-existing anterior and postero-median spines and the de novo induction of large postero-lateral spines are caused directly by a substance released into the medium by the predatory *Asplanchna*. These form changes cannot be explained by temperature induction or by an allometric growth response in which the substance works indirectly on the spines by influencing body size. Changes in body size were found to be independent of postero-lateral spine production. The substance was shown to affect the spine-form change only at the egg stage in precleavage condition. The substance released by *Asplanchna* is a relatively thermolabile proteinaceous compound of unknown composition.

The shape and movements of long postero-lateral spines in *B. calyciflorus* significantly decrease predation by *Asplanchna*. Adult *A. girodi*, capturing about 25 per cent of newly hatched, spineless *Brachionus* with which they make direct contact, were completely unable to capture newly hatched, long-spined individuals. Adult *A. sieboldi* can capture nearly 100 per cent of adult, spineless *B. calyciflorus* contacted, but only about 78 per cent of the adult, long-spined forms. This capture rate drops to less than 15 per cent of the spined form by young *A. sieboldi*.

A number of analyses of natural populations substantiate the fact that the *Asplanchna*-released substance can induce postero-lateral spine production in *B. calyciflorus* populations (Gilbert and Waage, 1967; Green and Lan, 1974). Postero-lateral spine lengths were found to be unrelated to water temperature, and varied directly both with the density of *Asplanchna* and the presence of threshold quantities of *Asplanchna*-released substance.

CRUSTACEAN ZOOPLANKTON

Members of the crustacean arthropods are almost entirely aquatic; most are marine. Respiration is accomplished through the body surface or gills. The body generally is separated into 3 distinct regions, but the tendency is towards fusion of abdominal and thoracic segments until, in the Cladocera and Ostracoda, apparent body segmentation is lost. In many crustaceans the body, bearing paired, usually biramous, jointed appendages, is covered wholly or in part by a carapace.

In fresh water, the truly planktonic Crustacea are dominated almost completely by the cladocerans and Copepoda, to which most of the ensuing discussion will be devoted. A few species of this predominantly benthic group have become planktonic. Similarly, only a few insects are planktonic in immature stages; the larvae of *Chaoborus* (Diptera) is a notable exception that demands special consideration.

The freshwater Branchiopoda (fairy and clam shrimps) are common inhabitants of shallow lakes, particularly of temporary, saline inland waters. All members of this primitive group are distinctly segmented and bear many pairs of swimming and respiratory appendages. The tadpole shrimps (Notostraca) are essentially benthic and most often restricted to shallow, temporary lakes of arid regions. The fairy shrimps (Anostraca), lacking a carapace, and the clam shrimps (Conchostraca), compressed laterally with a bivalved flexible carapace, commonly are planktonic in shallow playa lakes of semiarid regions. The members of the latter groups are bisexual, although parthenogenesis is known to occur in brine shrimp *Artemia* under some conditions. Resistant eggs can be subjected to long periods of desiccation and hatch when wetted again, probably the result of a reduction in osmotic pressure at the egg surface (Hutchinson, 1967). In semipermanent lakes of more humid regions, hatching and reproductive rates are related strongly to temperature.

Cladocera

The fourth order of the Branchiopoda, the Cladocera, includes mainly microzooplankton. With the exception of 2 species, nearly all the cladocerans range in size from 0.2 to 3.0 mm. All have a distinct head, and the body is covered by a bivalve cuticular carapace. Light-sensitive organs usually consist of a large compound eye and a smaller ocellus. The second antennae are large swimming appendages and constitute the primary organs of locomotion. The mouth-parts consist of: (1) large, chitinized mandibles that grind food particles, (2) a pair of small maxillules, used to push food between the mandibles, and (3) a median labrum that covers the other mouthparts.

Feeding

Usually 5 pairs of legs are attached to the ventral part of the thorax. The legs are flattened and bear numerous hairs and long setae. Complex movements of these setose legs create a constant current of water through the valves, oxygenating the body surface, and force a stream of food particles anteriorly. The food particles, filtered by the setae, collect in a ventral food groove between the bases of the legs and are impelled forward towards the mouth to be mixed there with oral secretions. The importance of size of food particles in relation to morphological limitations of the filtering apparatus and food selectivity will be discussed later.

The primarily littoral chydorid cladocerans have modified legs that are somewhat prehensile in picking up larger pieces of detrital material. Feeding by filtration occurs as well. Two common genera, *Polyphemus* and *Leptodora*, contain members that are predaceous and feed mainly by seizing relatively large particles, such as protozoa, rotifers, and small crustaceans, with their prehensile legs.

Reproduction and Development

Somewhat analogous to that of the rotifers, reproduction in the cladocerans is parthenogenetic during a greater part of the year. Until interrupted by sexual reproduction, females produce eggs which develop into more parthenogenetic females without fertilization. The number of eggs produced per clutch varies from 2 in the Chydoridae to as many as 40 in the larger Daphnidae. The eggs are deposited into a brood chamber, or pouch, a cavity dorsal to the body bounded by the valves of the carapace. The eggs develop in the brood pouch and hatch in a small form similar to that of the parent. As a result, in contrast to the copepods, there are, with one exception, no free-living larval forms among the Cladocera. One clutch of eggs normally is released into the brood pouch during each adult instar.

The extent of molting is variable. Four preadult instars are common within a range of from 2 to about 8 in some *Daphnia*. Longevity, from the time of egg release into the brood pouch until death of the adult, varies widely with species and environmental conditions. Longevity and time between molts are approximately inversely related to temperature. For example, longevity in *Daphnia magna* averaged 108, 88, 42, and 26 days at 8, 10, 18, and 28°C, respectively (MacArthur and Baille, 1929). Longevity also is affected by food

TABLE 16–4   Instantaneous Rates of Population Increase per Day (r) $\pm$ S.E. of Daphnia galeata mandotae at Different Temperatures and Levels of Food Supply[a]

Relative Food Levels	Temperature		
	$11°C$	$20°C$	$25°C$
Low (1/4)	0.07 ± 0.005	0.23 ± 0.002	0.36 ± 0.015
Medium (1)	0.10 ± 0.003	0.30 ± 0.002	0.46 ± 0.013
High (16)	0.12 ± 0.006	0.33 ± 0.016	0.51 ± 0.006

[a]After data of Hall, 1964.

availability, and commonly increases with a decrease in food consumption, short of starvation which results in rapid death. But food supply is of secondary importance in comparison to the effects of temperature, as can be seen in studies of *Daphnia galeata mendotae* (Table 16–4) that were fed differing concentrations of mixed green algae at different temperatures (see Hall, 1964).

In the adult cladocerans, the number of molts is more variable than in juveniles, ranging from a few to well over 20 instars. Each instar ends with the release of young from the brood pouch, molting, rapid increase in size, and deposition of a new clutch of eggs into the pouch, all occurring in a matter of minutes. Therefore, an individual can produce several hundred progeny in a life span under favorable growing conditions of temperature and food supply.

Increases in temperature result in an immediate effect by increasing the rate of molting and brood production, whereas increases in food supply, within limits, affect the rate of development of the population by increasing fecundity, i.e., the number of eggs per brood. The latter process is not immediate and therefore this lag can be contributory to population oscillations. When the production of the population exceeds available food supply, reproduction decreases, with reduced replacement of adults. Reduction of the population below the carrying capacity of food availability results in an increase in the population after a lag period. The extent of population oscillations depends greatly on the duration of the time lag. These relationships have been demonstrated in detailed studies under culture conditions (Frank, 1952, 1957; Slobodkin, 1954), and will be shown further in discussion of seasonal populations later on. Under conditions of competition of multiple species for the same food resources, the decline of species population is responsive to the interactions of food quality and supply, feeding capabilities, and the effect of temperature on reproductive responses. Decline of a population does not eliminate that population, but is compensated for by a shift to an inactive state by the formation of resting stages.

Parthenogenesis continues until unfavorable conditions arise, whether they are physical, such as temperature reductions or drying, or biotic, i.e., induced through crowding by competition for food supply, or decrease in quality-size of food organisms. As the production of parthenogenetic eggs declines, some of the eggs develop into males, and other eggs that are haploid require fertilization. Females producing sexual eggs, as in the rotifers, are morphologically similar to parthenogenetic females, but usually produce only one or a few of the resting eggs. Following copulation and fertilization, the carapace surrounding the brood pouch thickens and encloses the egg(s), forming a semielliptical, saddle-shaped *ephippium.* During the subsequent molt the ephippium either is shed, as in the daphnids, or remains attached, as in the chydorids. If females with ephippia are not fertilized, the resting eggs are resorbed. One species, the arctic *Daphnia middendorffiana,* is known to be able to produce ephippial eggs parthenogenetically.

The ephippia either sink or float, and can withstand severe conditions such as freezing or drying. It is not unusual to find large accumulations of ephippia along windward shorelines. It is easy to envision entanglement and transport of ephippia by birds to other water bodies. A large percentage of the ephippia hatch under favorable but poorly understood conditions, always into parthenogenetic females. In *Daphnia pulex,* light was required for termination of diapause in a laboratory-cultured strain regardless of temperature or duration of ephippial storage (Stross, 1966). In naturally occurring ephippia of a lake population of the same species, diapause also was broken by light, but prolonged storage in darkness eliminated the need for light. The time required was 5 to 6 months at 3.5°C and only 3 to 6 weeks at 22°C.

Males are smaller and only slightly modified in morphology from females (larger antennules and clasping modifications of the first leg for copulation). Mechanisms underlying the induction of periodic production of males and ephippial females are unclear, although evidence indicates that the stimuli are different (reviewed by Hutchinson, 1967). Male production is correlated with crowding and a rapid reduction of food supply, as opposed to

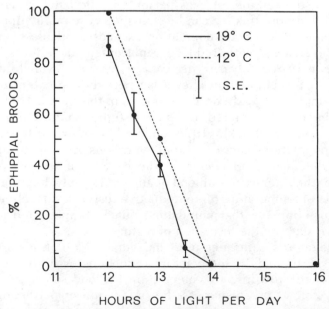

Figure 16–12 Influence of photoperiod on the production of ephippial egg broods in *Daphnia pulex* at high population densities. (Modified from Stross and Hill, 1965.)

a constant low food supply which simply inhibits reproduction (Slobodkin, 1954). Short-day photoperiods (12 hrs light, 12 hrs dark) increased the production of ephippia markedly in *Daphnia pulex,* in contrast to longer light photoperiods, which would be characteristic of midsummer in temperate latitudes (Fig. 16–12). If universal, this phenomenon must vary among species. Production of ephippial eggs has no genetic relationship to production of males.

Filtration and Food Selectivity

The entire subject of feeding and filtration and ingestion rates in Cladocera and, to a lesser extent, in copepods, is an area of current intensive investigation. Recent results indicate not only a number of physical and biotic factors that affect rates of ingestion, but that some organisms possess at least minimal abilities to select food for consumption. The implications of these studies are major not only from the standpoint of effectiveness of food ingestion and assimilation by the animals, but also because they point up the effects of the consumption on the food populations. This latter subject is most complex, but it is now apparent that at times zooplankton, as a result of direct cropping, can have appreciable effects on phytoplanktonic populations, and, through selective grazing, can influence the seasonal succession of the phytoplankton. This area of research has great potential for demonstrating subtle but major community interactions of significant impact on the overall productivity of lake systems.

Filtration of water to remove particulate organic matter is the dominant mode of collection and ingestion of food among most cladocerans and copepods. The *filtration rate* of a zooplankter refers to volume per unit time, and is defined as the volume of ambient medium containing the number of cells eaten by the animal in a given time (cf. Rigler, 1971). This term does not imply that the volume of water passed over the filtering appendages is known, that all particles of any given type have been removed from the water, or that all particles retained by the filtration apparatus have been consumed. The terms filtration rate, grazing rate, and filtration capacity have been used synonymously with filtering rate. In contrast, *feeding rate* is a measure of the quantity of food ingested by an animal in a given time (Rigler, 1971), whether it is measured in terms of number of cells ingested, volume, dry weight, carbon, nitrogen, and so forth, of the food.

The size of particles that can be cleared from the water is a function of the morphology of the setae of the moving appendages or of entrapment as locomotion of the animal brings particle-laden water to the setae. Measurements of particle removal are made by observing changes in the number of particles by grazing over time, or more recently by the rate of removal of radioactively labelled food particles (see the excellent review of Rigler, 1971). The latter approach has been extended most effectively to natural populations by use of an in situ grazing chamber (Haney, 1971) that permits estimates of grazing rates within short periods of time (< 10 minutes) before ingested food passes through the gut. These methods measure food actually taken into the gut after losses, either through active rejection of food or losses of food particles during the maceration process. Ingestion rates give little insight into assimilation rates of incorporated food, which can be highly variable with food type and concentrations.

A number of studies indicate that the rate of feeding stabilizes or decreases as the concentration of food particles increases. For example, in *Daphnia magna* feeding rate is proportional to concentration of food particles below a critical level (McMahon and Rigler, 1963, 1965). Feeding rate is constant above the incipient limiting concentration of food particles. At concentrations of food above this limiting level, rates of movement of the thoracic appendages that collect the food decrease, movements of mandibles and swallowing of food remain about the same or decrease slightly, and the rate of rejection of food increases (Burns, 1968b, and personal communication). For adult *D. rosea* feeding on a yeast, the critical concentration for feeding was between 0.75×10^5 and 1.0×10^5 cells ml^{-1} (Burns and Rigler, 1967). Critical concentrations (cells ml^{-1}) for feeding rate appear to increase with decreasing particle size, although other factors may be involved as well.

Within a given range of food concentrations, filtering efficiency was found to be independent of size of food particles (0.9 μm^3 to 1.8×10^4 μm^3 for *D. magna*) (McMahon and Rigler, 1963, 1965; see also Kersting and Holterman, 1973). At concentrations of yeast cells above 0.25×10^5 cells ml^{-1}, filtering rates of *D. rosea* decreased.

Similarly, the filtering rate of *Ceriodaphnia reticulata* decreased directly with increasing concentrations of phytoplankton within a size range of 3 to 24 μm from low to high densities (10^4 to 10^6 ml^{-1}) (O'Brien and deNoyelles, 1974). This relationship was not evident in cultures of *Ceriodaphnia* at low densities (10^4 to 10^5 ml^{-1}) of algae (Czeczuga and Bobiatyńska-Ksok, 1970), although the latter methods (very high zooplanktonic densities in a small volume with long incubation periods) do not permit comparison.

Filtering rates have been found to increase with increasing body length and with increasing temperatures (Fig. 16–13). Temperatures above a given point result in a decrease in the filtration rate. The temperature of maximum filtration rates apparently differs among species; for *D. rosea* the maximum was found to be about 20°C (Figure 16–13), for *D. magna* 28°C, and for *D. galeata mendotae* at or above 25°C (Burns, 1969b; Burns and Rigler, 1967; McMahon, 1965).

Brooks and Dodson (1965) have emphasized that the filtering rate should be proportional to the square of the body length in filter-feeding Cladocera. The increase in filtration rate relative to body length, however, varies among species, for example, about to the square in *D. magna* and to the cube of the length in smaller *D. rosea* (Burns, 1969b). Although the area of the filtering setae of these 2 *Daphnia* species increases approximately as the square of the body length, the filtering area and the filtering rates of *D. rosea* become increasingly greater proportionately than those of *D. magna* as the body length increases (Egloff and Palmer, 1971). The difference in the calculated rate of flow of water through the filtering setae, however, remains nearly constant at all body lengths.

The relationship between body size and the maximum size of particles

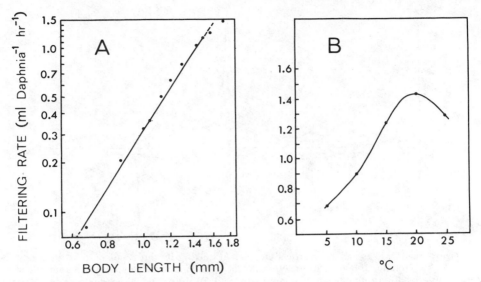

Figure 16–13 Relationship between filtering rate and A, body length at a constant food supply at 20° C, and B, water temperature (body length 1.65–1.85 mm) of *Daphnia rosea.* (Modified from Burns and Rigler, 1967.)

that could be ingested was studied in 6 species of *Daphnia* and the smaller *Bosmina longirostris* with spherical beads within a size range from < 1 to 80 μm in diameter (Burns, 1968a, 1969a). A strong positive correlation between increasing body size and increasing size of the largest particle ingested was evident for the species assayed (Fig. 16–14). Experiments in which particles were kept in suspension or were allowed to sediment to the bottom of flasks showed that under the latter conditions some species, forced to forage for particles, were able to select particles of different sizes better than others.

Figure 16–14 Relationship between body size and the diameter of the largest particle ingested by seven species of Cladocera. Broken lines equal 95 per cent confidence limits. (Modified from Burns, 1968a.)

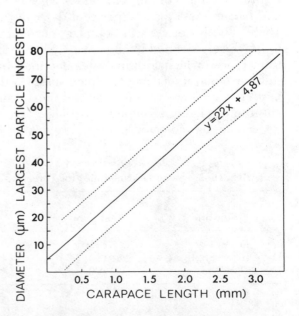

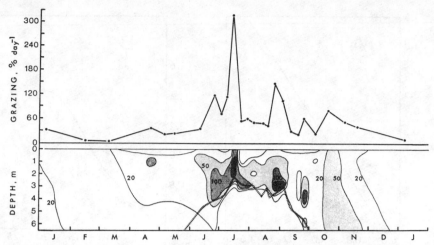

Figure 16–15 Seasonal and vertical changes in grazing by the zooplankton community in Heart Lake, Ontario, 1969. Mean grazing rates of upper figure calculated for the aerobic stratum. Isopleths of grazing rates (lower figure) are at 20, 50, 100, and 200 per cent per day. Broken line is the 1 mg l^{-1} dissolved oxygen isopleth. (After Haney, J. F.: An in situ examination of the grazing activities of natural zooplankton communities. Arch. Hydrobiol., 72:87–132, 1973.)

Leverage provided by the substratum aids in the ingestion of somewhat larger sized particles.

Extension of these and other results from laboratory feeding studies to natural populations by observations of in situ grazing rates has demonstrated some striking relationships. In analyses from small, eutrophic Heart Lake of southern Ontario, seasonal grazing patterns (Fig. 16–15) were found to be similar over a 2-year period (Haney, 1973). At certain depths during periods of the summer, grazing rates exceeded 100 per cent of the water volume per day, but during much of the year values were less than complete filtration per day and became less than 10 per cent day^{-1} during the winter (Fig. 16–15; Table 16–5). The lower vertical boundary of zooplanktonic filter feeding was found to be closely defined by the 1 mg l^{-1} isopleth of dissolved oxygen concentration, below which filtering rates declined precipitously. Grazing rates with different food items showed some seasonal differences. Smaller bacteria-size particles apparently were ingested most rapidly in summer, whereas in the autumn the larger sized algal particles were eaten most rapidly.

Since, as indicated earlier, the grazing rates of cladoceran zooplankton

TABLE 16–5 Mean Grazing Rates for Three Seasons in Heart Lake, Ontario, 1969[a]

Season	Total Days	Mean Grazing Rate (% day^{-1})	% of Total Grazing
Jan.–May	151	19.2	17.4
June–Sept.	122	80.1	61.4
Oct.–Jan.	92	35.2	20.9

[a]After data of Haney, 1973.

TABLE 16–6 Comparison of Filtering Rates of Various Cladoceran Zooplankters

Species	Type of Food	Animal Size Range (Length in mm)	Average Filtering Rate (ml animal^{-1} day^{-1})	Source
Daphnia				
D. rosea	In situ phytoplankton	1.3–1.6	5.5	Haney, 1973
			3.6	Burns and Rigler, 1967
D. galeata		1.5–1.7	6.4	Haney, 1973
			3.7	Burns and Rigler, 1967
D. parvula		0.7–1.2	3.8	Haney, 1973
D. longispina			2.3	Nauwerck, 1963
Ceriodaphnia				
C. quadrangula		0.7–0.9	4.6	Haney, 1973
Diaphanosoma				
D. brachyurum		0.9–1.4	1.6	Haney, 1973
Bosmina				
B. longirostris		0.4–0.6	0.44	Haney, 1973
Chydorus				
C. sphaericus		0.1–0.2	0.18	Haney, 1973

vary greatly with body size, food, and temperature, the in situ rates of filtration vary considerably seasonally. The average rates given in Table 16–6, however, show the general relationship to body size. Returning to the example from Heart Lake, the grazing contribution of each species can be calculated by multiplying filtering rates of each species at a given time by its population density at that time. In Heart Lake in 1969, *Daphnia rosea* and *D. galeata* were the most important grazers, and together they accounted for about 80 per cent of the total annual grazing activity (Haney, 1973).

Filtering rates of the major zooplankton were also studied, although less intensively, in 2 much less productive lakes of southern Ontario: Halls Lake, a deep oligotrophic lake, and Drowned Bog Lake, a typical *Sphagnum* acidic bog lake (cf. Chapter 18) with high concentrations of humic dissolved organic matter. The filtration rates of the 2 dominant zooplankters, *Bosmina* and *Holopedium,* were very different, but the small *Bosmina* formed a dominant percentage of the total grazing in the autumn because of its very large population densities (Table 16–7). In contrast to the bog lake and to eutrophic Heart Lake, in which intense grazing generally was limited vertically to the upper 3 meters, the grazing rates in oligotrophic Halls Lake were extremely low and distributed uniformly throughout the water column. The dominant zooplankter of this oligotrophic lake was the copepod *Diaptomus,* and in general nonpredatory copepods have much lower filtration rates than the Cladocera (Tables 16–6 and 16–8).

In studies in which species of widely different algae were fed to cladoceran and copepod zooplankton at concentrations above critical limitations to

TABLE 16–7 Filtering Rates and Contribution to Total Grazing of Species
Dominant Zooplankton of Acidic Drowned Bog Lake, Ontario, in Early September[a]

Species	Filtering Rates (ml animal^{-1} day^{-1})		Species Contribution to Total Grazing (%)	
	Sept. 1968	Sept. 1969	Sept. 1968	Sept. 1969
Bosmina longirostris	0.46	0.45	85	44.8
Holopedium gibberum	—	9.4	12	46.2
Daphnia parvula	—	1.6	0.1	7.5
Diaptomus oregonensis	—	2.1	2.0	0.9
Diaphanosoma brachyurum	—	1.2	0.1	0.4

[a]Extracted from data of Haney, 1973.

feeding, filtering rates decreased markedly with increases in the food concentration (Fig. 16–16). Ingestion rates were calculated as the product of the filtration rate (ml animal^{-1} day^{-1}) and the food concentration (cells ml^{-1}) (Infante, 1973). These results, coupled with measurements of radioactively

TABLE 16–8 Comparison of Filtering Rates of Various Copepods

Species	Type of Food	Particle Concentration (cells $\times 10^{-3}$ ml^{-1})	Filtering Rate (ml animal^{-1} day^{-1})	Source
Diaptomus				
D. graciloides	Natural phytoplankton	—	0.3–2.8	Nauwerck, 1959
	Scenedesmus	13.6	4.1	Malovitskaia and Sorokin, 1961
D. siciloides	Pandorina and Chlamydomonas	—	2.0	Comita, 1964
D. oregonensis	Chlamydomonas	1.5–25.0	2.5	Richman, 1966
		25.0–52.0	2.5–1.4	
	Chlorella	52.0–198.0	1.4–0.3	
	In situ phytoplankton	—	1.4–0.00	Haney, 1973
D. gracilis	Melosira and Asterionella	24.2–52.0	1.92–1.96	
		198.0	0.68	Malovitskaia and Sorokin, 1961
	Chlorella	<30.0	0.61 (5°C)	Kibby, 1971
			1.51 (12°C)	
			2.40 (20°C)	
	Scenedesmus	<30.0	0.94 (12°C)	
			1.32 (20°C)	
	Diplosphaeria	<30.0	1.76 (12°C)	
			2.54 (20°C)	
	Ankistrodesmus	<30.0	1.61 (12°C)	
			2.45 (20°C)	
	Carteria	<30.0	0.87 (20°C)	
	Nitzschia	<30.0	1.96 (20°C)	
	Pediastrum	<30.0	0.02 (20°C)	
	Haematococcus	<30.0	2.16 (20°C)	
	Bacteria	<30.0	0.19 (20°C)	
Limnocalanus				
L. macrurus	Scenedesmus	—	2.45 (<5°C)	Kibby and Rigler, 1973
	Chlamydomonas	—	1.24 (<5°C)	
	Rhodoturula (yeast)	—	0.1 (<5°C)	

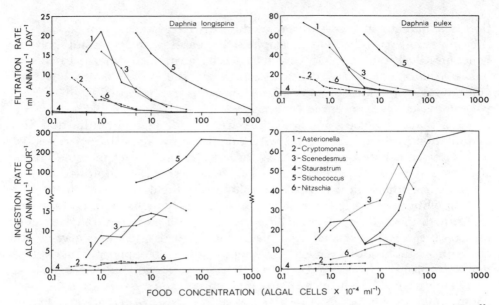

Figure 16–16 *Upper*: filtration rates (in ml animal⁻¹ day⁻¹); *Lower*: ingestion rates (in cells animal⁻¹ hour⁻¹ as the filtration rate times food concentration in cells ml⁻¹), in *Daphnia longispina* and *D. pulex* exposed to different algal species. Ingestion rate of *Stichococcus* (5) for *D. pulex* is drawn × 0.1 that found. (Drawn from data of Infante, 1973.)

labelled carbon from the algae that were incorporated into the animals as an estimate of assimilation, showed marked differences with algal species. The diatom *Asterionella* was ingested more readily and assimilated more easily than *Nitzschia*. Algae with heavy cell walls, such as *Scenedesmus* and *Stichococcus*, were ingested rapidly (Fig. 16–16) but were not utilized well. The Cladocera digested some of the protoplasm of such cells without visible changes in the cell walls, whereas the copepods assayed (*Eudiaptomus* and *Cyclops*) were unable to digest them and the algae which left the gut were essentially intact. *Staurastrum* and *Cryptomonas* were poorly utilized. Differences also occurred among zooplanktonic species. *Daphnia longispina*, for example, incorporated nearly twice the number of *Nitzschia* as *D. pulex* did, but only one-half the amount of *Scenedesmus*.

Similarly, when the numbers of grazing Cladocera and copepods were increased experimentally among natural phytoplanktonic populations, small algae such as cryptomonads, certain diatoms, and other nannoplankton decreased, gelatinous green algae such as *Sphaerocystis* increased, and large or unpalatable species such as the blue-green *Anabaena* remained unchanged (Porter, 1973). While the size of particles grazed is limited by the morphological structure of the setae (Monokov and Sorokin, 1961), large or unpalatable algae result in reduced filtering rates. For example, the ingestion, assimilation, survivorship, and reproduction rates of *Daphnia* that were fed blue-green algae were lower than those fed green algae (Arnold, 1971). Therefore, even though food species of a usable size may be present, they may be too rare or selectively rejected by the grazer at high algal densities. This selection is apparently at least partly related to relative abundance of particles of different biomass. Selection in *Daphnia* increases with increasing particle concentrations until particle consumption is maximized when particles of one size dominate in mixed food (Berman and Richman, 1974). When concentrations of all particles were approximately equal, selection favored small particles that were present in high numbers.

Ingestion of algae does not necessarily result in their utilization and assimilation by the grazers. Algae with durable cell walls, gelatinous sheaths, or masses of colonial cells can pass through the gut intact in some grazer species and remain viable (Porter, 1975). Other species can utilize broken colonies or partially decayed cells of colonies by breaking them into smaller pieces, while intact colonies pass through relatively unaffected. Increases in the growth rates of certain algae after ingestion and passage through grazers may be the result of several interrelated mechanisms. Nutrients may be obtained from the degradation and digestion of other algae; phosphorus, in particular, is probably a major nutrient that is made more available in this process.

Selective grazing and utilization can remove or reduce species in the algal community competing for the given resource base. Alternatively, grazer utilization of an algal community species can result in enhancement of primary productivity of that species, in that reduction of that population biomass can lead to an increase in turnover rate. This effect, as pointed out by Cooper (1973), can be quite independent of increased nutrient regeneration rate, if any, brought about by grazing.

Assimilation

In evaluating the population dynamics of any organism, estimating food incorporation or ingestion and the utilization of ingested food is imperative. As used when referring to aquatic animals, assimilation means the absorption of food from the digestive system and the efficiency of assimilation refers to the percentage fraction of ingested food that is digested and absorbed into the body. Assimilation efficiency is not constant, and varies greatly with the food quality and rates of food ingestion.

Measurements of assimilation are based on the simple relation:

$$\text{Assimilation} = \text{ingestion} - \text{egestion, or}$$

$$\text{Assimilation} = \text{growth} + \text{respiration.}$$

While these relationships are simple, accurate measurements among natural populations of zooplankton and larger organisms are extremely problematic. Methods employed are discussed critically in Edmondson and Winberg (1971), and Winberg (1971). Although ingestion can be measured in situ with reasonable sophistication, egestion rates are often beyond the capacity of contemporary technology. Growth is used in a general sense in that it includes the production of eggs or young, in much the same way that respiration includes excretion and other losses as well as normal biochemical respiration.

TABLE 16–9 Energy Budget of Adult Daphnia pulex After 18 Instars (40 days) of Growth at Four Food Concentrations of Chlamydomonas[a]

Food Concentration ($cells \times 10^{-4}$ ml^{-1} day^{-1})	mg ml^{-1}	Energy Consumed (cal)	Energy of Growth and Young (cal)	Energy of Respiration (cal)	Energy of Egestion (cal; by difference)	% Assimilation	% Energy Consumed as Growth and Young	% Energy Assimilated as Growth and Young	% Energy of Growth and Young as Respiration
25	6.2	6.14	1.07	0.84	4.23	31.1	17.4	56.0	78.6
50	12.4	13.59	1.77	0.94	10.87	20.0	13.1	65.3	53.0
75	18.6	20.74	2.35	1.02	17.37	16.3	11.4	69.8	43.3
100	24.8	29.24	2.93	1.08	25.23	13.7	10.0	73.0	37.0

[a] After data of Richman, 1958.

Assimilation and its efficiency can be approached using the first equation given by measurement of changes in radioactivity of the animal after feeding first on radioactive food, and then after feeding on nonradioactive food (cf. Sorokin, 1968; Saunders, 1969; Edmondson and Winberg, 1971). Growth and respiration can be estimated by the second equation by knowing the feeding rate (e.g., in calories per time), the growth rate, and respiration by oxygen consumption, which can be converted to energy units if the respiratory quotient (CO_2 produced/O_2 consumed) is known.

A number of studies have demonstrated the importance of food quality and concentrations in the assimilation efficiencies. Among the better studies is that of *Daphnia pulex* (Richman, 1958), among which assimilation efficiency was shown to decrease markedly with increasing food concentrations (Table 16–9). Growth as average length, number of young per brood, and total young increased with increasing food concentrations at the levels used. The energy used for growth was nearly constant at all food levels in preadults, because the energy of egestion increased and assimilation efficiency declined as food consumption increased. In adults, however, the major portion of stored energy went into the production of young. Similar results have been found for a number of other cladocerans.

In addition to increasing assimilation rates with increasing temperature, several studies show that the assimilation efficiency increases with higher energy content of the food (Schindler, 1968; Pechen'-Finenko, 1971). Foods of low caloric value, such as detritus of plant origin in the range of 2 to 4 cal mg^{-1} dry weight, are assimilated slower than algae and bacteria in a caloric range of 5 to 6 cal mg^{-1}. Although the assimilation efficiency of detritus is low, that of algae and bacteria by zooplankton is variable. The few studies on the subject show that in general algae are assimilated more efficiently (15 to 90 per cent) than bacteria (3 to 50 per cent) (Monakov and Sorokin, 1961; Saunders, 1969; Winberg, et al., 1973). The percentage of assimilation varies seasonally with shifts in algal composition. In contrast, when given mixed diets of green algae and a photosynthetic bacterium, *Ceriodaphnia* assimilated less of the algae than when fed on algae alone (Gophen, et al., 1974). Utilization of bacteria was relatively unchanged when the algae were also included in the diet. There are several indications that algae alone do not meet the nutritional requirements of cladocerans (see, for example, Taub and Dollar, 1968). In spite of the relatively high assimilation rates of algae and bacteria and low assimilation rates of detritus, as will be pointed out later it is unusual for bacteria to dominate energy income for zooplankton. Moreover, in the few cases in which it has been analyzed in some detail (cf. Chapter 17), detritus forms a major source over much of the year.

Seasonal Population Dynamics

The seasonal succession of the Cladocera is quite variable among species and within species among different lake conditions. Some are perennial species that overwinter in low population densities as adults rather than as resting eggs. These species may exhibit 1, 2, or more irregular maxima. Some

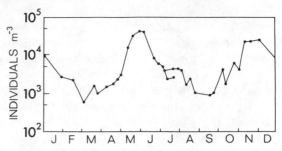

Figure 16–17 Population density of *Daphnia galeata mendotae* in Base Line Lake, Michigan, 1960–61. (Redrawn from Hall, 1964.)

perennial species exhibit maxima in surface layers only during colder periods in the spring, and in the cooler hypolimnetic and metalimnetic strata during summer stratification. The aestival species that have a distinct diapause in a resting egg stage commonly develop population maxima in the spring and summer at relatively high water temperatures. Although one population maximum is general, a second population peak often occurs in the autumn.

An example of a typical case of population dynamics can be taken from the well-studied zooplankton of Base Line Lake, Michigan (Hall, 1964), consisting mainly of *Daphnia*. Two population maxima occurred in the population of *Daphnia galeata mendotae*, one in the spring and a second in late autumn (Fig. 16–17). Although a few ephippia were produced, this species, in contrast to *D. pulex* in the same lake, overwintered in the free-swimming stage with little to no reproduction. The initial spring maximum in May-June and that in the fall were preceded by a maximum in brood size, the ratio of eggs to mature *Daphnia* (Fig. 16–18). Degenerate eggs were observed frequently in the brood pouches of *Daphnia* and had no apparent relationship to normal brood size.

Employing the instantaneous rate of increase of the population discussed earlier, the population dynamics of *D. galeata mendotae* were evaluated in

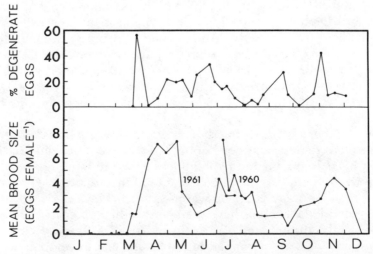

Figure 16–18 *Upper,* percentage of degenerate eggs in the population, and *lower,* mean brood size (mean number of eggs per adult) of *Daphnia galeata mendotae,* Base Line Lake, Michigan, 1960–61. (Redrawn from Hall, 1964.)

Base Line Lake. The instantaneous birth rate (b) was estimated from the finite birth rate of newborn occurring over a time interval in relation to the existing population size. The development time of eggs was determined experimentally in relation to temperature. The instantaneous rate of population increase, r (see Equation 2), and birth rate were both positive and nearly equal during the spring and fall months (Fig. 16–19). Maximum values of b occurred in the summer months, whereas r was negative during this period of population decrease and increased death rate, d. During the winter, b was zero as reproduction ceased and r was negative as a slow death rate, d, gradually reduced the population until March.

The summer minimum, with a loss rate of 28 per cent per day, apparently was associated with predation on the juvenile stages of *Daphnia galeata* by the large cladoceran *Leptodora* which reached its maximum in summer (Fig. 16–19). *D. retrocurva* reached its maximum density, greatly exceeding *D. galeata*, in late summer, well after water temperatures exceeded 20°C. Its successful competition for available food resources in fall most likely was contributory to the reduced birth rate of *D. galeata* in the early fall. Analyses of fish consumption showed that while fish fed extensively on the large *D. pulex* in spring, the smaller *D. galeata* was not taken significantly. *Daphnia catawba*, which like *D. retrocurva* exhibits one late summer maximum and overwinters with resting eggs, is also known to coexist with *D. galeata* in late summer (Tappa, 1965). Its presence and competition for the same food base appeared to be instrumental in delaying the autumnal maximum of *D. galeata*.

In a detailed study of the zooplanktonic populations of shallow ($\bar{z}$ = ca 1 m), extremely productive Sanctuary Lake in northwestern Pennsylvania, 3 species of *Daphnia* occurred and exhibited a large spring maximum and a small autumn peak (Cummins, et al., 1969), similar to that discussed in the Base Line Lake example. *D. galeata mendotae* dominated these populations and *D. retrocurva* became somewhat of a codominant in late summer. With the decline of the spring maximum, there occurred a massive development of the predacious *Leptodora kindtii*, reaching biomass maxima in June in one year and in June, in July, and in August the following year. Only the larger *Leptodora* of from 6 to 12 mm are predacious, ingesting the fluids of their prey. The population dynamics of *Leptodora* were correlated with loss rates of prey, and this group apparently shifted seasonally to preying on *Daphnia*, *Ceriodaphnia*, *Bosmina*, and *Chydorus* and the copepods *Cyclops* and *Diaptomus*.

A few cladocerans are distinctly cold-water species, inhabiting shallow lakes of northern regions and developing large populations only in the early spring and in cooler meta- and hypolimnetic strata in more temperate lakes. These species generally are perennial, and tend to tolerate lower oxygen concentrations than the more common warm-water species.

The chydorid cladocerans have not been studied extensively. The group is predominantly littoral and distinctly aestival. A few species, notably *Chydorus sphaericus*, are perennial in the littoral and develop in the pelagic zone during midsummer, usually coinciding with the late summer phytoplanktonic maximum.

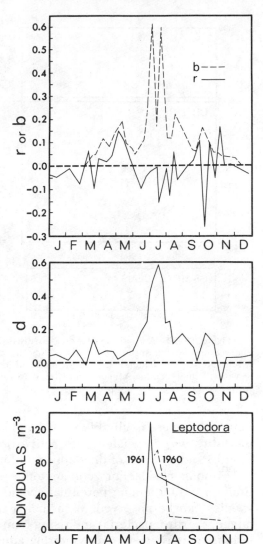

Figure 16–19 *Upper,* instantaneous birth rate (natality), *b,* and rate of population increase or decrease, *r, Daphnia galeata mendotae; middle,* instantaneous death rate, *d (b* minus *r),* of *Daphnia galeata mendotae; Lower,* population density of the predatory *Leptodora,* Base Line Lake, Michigan, 1960–61. (Redrawn from Hall, 1964.)

Among the best analyses of littoral chydorids is the detailed study of the population dynamics of 4 species of a hardwater lake of southern Michigan (Keen, 1973). *Chydorus sphaericus* reached its maximum population density in the spring, declined to a low level in the summer, and then increased to a smaller fall peak preceding a winter plateau (Fig. 16–20). It is common for *Chydorus* to become abundant in the open water in summer, coincident with the minimal littoral populations. As indicated in the upper portion of the figure, production of males and ephippial females was very small in this littorally perennial species. In contrast, *Graptoleberis, Acroperus,* and *Camptocerus* are aestival and after a winter absence appeared in spring, from ephippial eggs, and attained maximal densities in late summer and autumn (Fig. 16–20). The latter peaks were terminated with the massive development of ephippial females and males prior to winter. The factors of natural death and

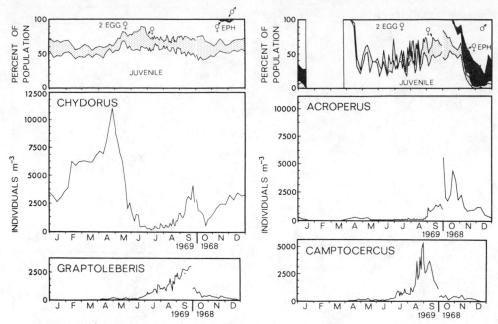

Figure 16–20 Annual population densities of the four major species of littoral chydorid cladocerans among submersed beds of *Scirpus subterminalis,* west littoral, Lawrence Lake, Michigan. Upper panels show percentage of the population of *Chydorus* and *Acropterus* seasonally in juveniles, females, females with eggs, males, and ephippial females. (Redrawn from Keen, 1973.)

emigration were negligible in this and several similar studies. High summer mortality was associated with active predation by small fishes (< 4 cm in length) and nymphs of dragonflies (Odonata).

Among cladoceran zooplanktonic populations in general, low populations in winter among perennial species and a near absence among aestival species can be observed. With increasing food supply from photosynthesis (algae, detritus, bacteria) and rising temperatures in spring, cladoceran populations increase from overwintering adults or resting eggs. Temperature increases the rate of molting and brood production, while rising food supply increases the egg number per brood. Population cycles in the summer are more variable, being influenced by reduced food supply, shifts in the quality of food to less palatable species, and predation by other zooplankton, fishes, and other organisms. Mortality of cladocerans, especially juveniles, normally is highest in summer, as is competition for available food resources among coinhabiting species.

Cyclomorphosis

The phenomenon of seasonal polymorphism occurs in the Cladocera perhaps more conspicuously than in any other group. The subject has been studied extensively, especially among the genus *Daphnia* (reviewed in detail by Hutchinson, 1967).

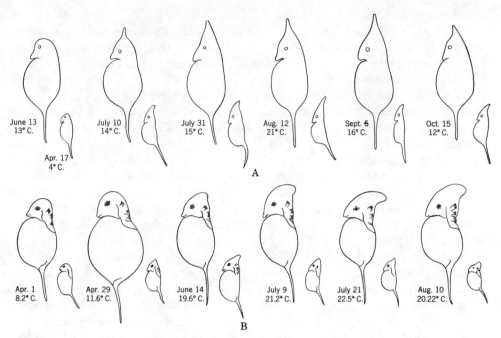

Figure 16–21 Cyclomorphosis in A, *Daphnia cucullata* of Esrom Sö, Denmark, and B *Daphnia retrocurva* of Bantam Lake, Connecticut. The small individuals at the right are first instar juveniles drawn to the same scale as the adults. (A: From Hutchinson, G. E.: A Treatise on Limnology. Vol. 2, New York, John Wiley & Sons, Inc., 1967; B: From Brooks, John Langdon: Cyclomorphosis in Daphnia: Ecological monographs, 16:1946. Copyright 1946 by Duke University Press.)

With increasing water temperatures in the spring, the common seasonal pattern of variation among successive generations in temperate lakes is a gradual extension of the anterior part of the head to form a crest or helmet (Fig. 16–21). Carapace length, on the other hand, changes little or decreases slightly between the spring and summer, and then increases again somewhat in autumn. The number of parthenogenetic eggs per individual generally decreases markedly in the strongly helmeted summer forms, although the number of instars increases with higher temperatures. The extent of head development and change of head shape are extremely variable among species and within the same species under different environmental conditions. In subtropical and tropical waters, cyclomorphosis is weak or does not occur, except when there is a large temperature change between winter and summer. An increase in the length of the tail spine also is observed in some species during summer. Cyclomorphosis in the genera *Bosmina, Ceriodaphnia,* and *Chydorus* is much less distinct than in *Daphnia,* and consists of slight reductions in body length and of the antennule in summer. In *Bosmina* enlargement often occurs by means of the formation of transparent dorsal humps, with no increase in length.

The causal mechanisms stimulating cyclomorphosis in *Daphnia* have received much study. It is clear that temperature is the primary stimulus affecting the height of the head helmet, and that it is effective during the middle of embryogenesis. In a number of species, helmet extension does not occur at water temperatures below 10 to 15°C. Carapace length is maximal in winter, and the ratio of head length to carapace length increases with increasing temperatures. Transfers of animals reared at one temperature to a lower or higher temperature result in a reduction or increase in head length in relation to that of the carapace in later instars. Food supply also affects the specific growth rate and has a relatively greater effect on growth of the head.

Water turbulence also has been shown to be a significant factor in cladoceran cyclomorphosis (see, for example, Brooks, 1947; Jacobs, 1962). In either turbulent cultures or naturally turbulent environments such as the epilimnion of larger lakes, helmet length relative to that of the carapace is increased significantly in the light in comparison to in quiescent situations. Antennal beating rate also increases significantly in turbulent conditions, probably in order to maintain a vertical orientation to light. Helmet development is distinctly more conspicuous in species typically inhabiting epilemnetic turbulent strata (Brooks, 1964). Populations genetically capable of considerable phenotypic variation do not exhibit their extreme phenotype when they live in lower water strata in which growth is slower.

Such a conspicuous seasonal polymorphism of daphnids has been assumed to have a major adaptive significance for these organisms. However, that significance is unclear. Early investigators assumed that cyclomorphic forms would have greater resistance to sinking, since the viscosity of the water decreases in summer at higher epilimnetic temperatures. Cyclomorphosis is almost totally confined to epilimnetic species, and turbulence there decreases the small advantage gained from form resistance. Furthermore, such resistance is costly in terms of energy in vertical diurnal migration, discussed further on.

Recent evidence indicates that significant selective value and the primary adaptive significance of the daphnid cyclomorphic growth focus on keeping the central portion of the body (thorax, abdomen, and basal portion of the head) that is visible to predators small. Growth continues by expansion of transparent peripheral structures (Brooks, 1965, 1966; Jacobs, 1966). Thus, when size-dependent planktivory is most intense in summer, the central visible portion of the daphnid body is proportionately smallest and less susceptible to being eaten by fish. Cyclomorphomic organisms are, through natural selection for high rates of growth in less visible peripheral structures, able to maintain high assimilation rates during midsummer without the disadvantage of greater visible size when predation is intense. In the winter, when the rate of predation is negligible and turbulence essentially is eliminated under ice-cover, evidence indicates that cessation of molting in juveniles and adults results from the lack of a turbulence cue. With the loss of ice-over, even at low temperatures, molting, growth, and egg production immediately begin from the overwintering population of large females.

Copepoda

The free-living copepods of this order of the class Crustacea are separable into 3 distinct groups: the suborders Calanoida, Cyclopoida, and Harpacti-

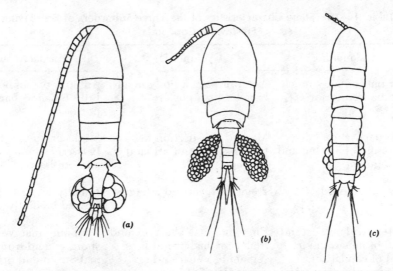

(a)

(b)

(c)

Figure 16–22 Diagrams of the three major types of free-living freshwater copepods (females). *a*, Calanoid. *b*, Cyclopoid. *c*, Harpacticoid. (From Wilson, M. S., and Yeatman, R. C.: Free-living copepoda. *In* Edmonson, W. T., ed., Freshwater Biology. 2nd ed., copyright © 1959 by John Wiley & Sons, Inc. New York, 1959. Reprinted by permission of John Wiley & Sons, Inc.)

coida. Although accurate identification is based largely on morphological details of appendages, several general characteristics delineate the major groups. The body consists of the anterior metasome (cephalothorax), divided into the head region, bearing 5 pairs of appendages of antennae and mouthparts, and the thorax, with 6 pairs of mainly swimming legs. The posterior urosome consists of abdominal segments, the first of which is modified in females as the genital segment, and terminal caudal rami bearing setae.

The 3 suborders of free-living copepods can be distinguished by structure of the first antennae, urosome, and fifth leg (Fig. 16–22 and Table 16–10). The harpacticoid copepods are almost exclusively littoral, habitating macrovegetation, mosses in particular, and the littoral sediments and particulate organic matter. Certain species have life histories with a diapause similar to that discussed further on for the cyclopoid copepods. Although the cyclopoid copepods are primarily littoral benthic species, those few members that are predominantly planktonic form major components of the copepod zooplankton, especially in small shallow lakes. The calanoid copepods are almost exclusively planktonic in the pelagial zone.

Feeding

The mouthparts of harpacticoids are adapted for seizing and scraping particles from the sediments and macrovegetation. The food and feeding behavior of the cyclopoid copepods have been studied in detail in 2 masterful works by Fryer (1957a, 1957b). No filtration mechanisms occur in the free-living Cyclopoida. Feeding is raptorial; plant or animal food particles are seized by mouthparts and brought to the mouth. The maxillules hold and

**Table 16–10   Some Characteristics of the Three Suborders of Free-Living
Freshwater Copepoda[a]**

Calanoida	Cyclopoida	Harpacticoida
Anterior part of body much broader than posterior	Anterior part of body much broader than posterior	Anterior part of body little broader than posterior
Marked constriction between somite of 5th leg and genital segment	Marked constriction between somites of 4th and 5th legs	Slight or no constriction between somites of 4th and 5th legs
One egg sac, carried medially	2 egg sacs, carried laterally	Usually 1 egg sac, carried medially
First antennae long, extend from end of metasome to near end of caudal setae, 23–25 segments in female	First antennae short, extend from proximal 3rd of head segment to near end of metasome, 6–17 segments in female	First antennae very short, extend from proximal 5th to end of head segment, 5–9 segments in female
5th leg similar to other legs	5th leg vestigial	5th leg vestigial
Planktonic, rarely littoral	Littoral, a few species planktonic	Exclusively littoral, on macrovegetation and sediments

[a]Modified from the summary of Wilson and Yeatman, 1959.

pierce the prey and force particles between the mandibles; intermittent oscillating movements mascerate some of the food. Some particles are swallowed intact and are differentially digested. Diatoms tend to be digested, while some green algae, if they are not ruptured, pass through the gut undigested.

Carnivorous cyclopoids include many species of the major genera *Macrocyclops, Acanthocyclops, Cyclops,* and *Mesocyclops.* The food of these carnivores includes microcrustaceans, dipteran larvae, and oligochaetes, many of which are larger than the copepod which preys on them. Herbivorous cyclopoids include many species of *Eucyclops,* some *Acanthocyclops,* and *Microcyclops,* which feed on a variety of algae ranging from unicellular diatoms to long strands of filamentous species. Carnivorous species tend to be larger than herbivorous species. Random chance encounter appears to be the dominant mode of finding food in both carnivorous and herbivorous species by discontinuous irregular movements in the water or over the substratum. Herbivorous species apparently employ gustatory chemoreceptor organs, which may help in food seeking if only to facilitate discrimination between inorganic and organic particles encountered by chance.

While locomotion in most copepods is in the form of short jerky swimming movements, propelled by rapid movement of most appendages simultaneously, swimming is more continuous in the Calanoida copepods. Their gliding movement results from rotary motion of the antennae and mouth appendages. The movements set up small vortical currents that carry particles to the maxillae, which are modified to filter the water. Selective feeding also exists among the calanoid copepods. For example, 2 closely related calanoids, *Diaptomus laticeps* and *D. gracilis*,[5] were found to coexist in the plankton of Lake Windermere, England (Fryer, 1954). The species are separated by size differences which were correlated with differences in food consumed. The larger *D. laticeps* fed chiefly on *Melosira*, and the smaller *D. gracilis* consumed mainly minute spherical green algae and particulate detritus. Neither fed upon the then abundant diatom *Asterionella*. Competition for food by these species was virtually nonexistent. Such small differences in feeding habits, based on morphological and behaviorial variations, are common in the calanoids and sufficient to separate species into different food niches, even though they occupy the same volume of water.

Similar results were found for *Diaptomus* feeding on natural phytoplanktonic populations (McQueen, 1970). Filtration rates were low for cells smaller than 100 μm^3, increased to a maximum of 12.9 ml per animal per day for cells ranging from 102 to 333 μm^3, and remained constant with increased cell volume and decreased cell concentrations.

Reproduction and Development

Reproduction in free-living copepods is similar in spite of widely varying species differences in sexual behavior and periodicity of breeding. Some species reproduce throughout the year, others only briefly at specific times of the year. Copepods are bisexual and copulation occurs by brief male clasping of the female and transfer of spermatophores to the ventral side of the female genital segment. Fertilization may take place immediately or several months after copulation. Fertilized eggs are carried by the female in 1 or 2 egg sacs (Fig. 16–22). The number of eggs in cyclopoid copepods, carried in 2 lateral egg sacs, is very variable among populations of different lakes and seasonally, and has not been associated clearly with changing environmental conditions. Egg number, up to 72, is generally lower in summer and increases in the autumn.

In the calanoid copepods, eggs, carried in a single egg sac, are variable in number from about 1 to 30. Eggs are differentiated into subitaneous and resting eggs, both being fertilized in many species. Resting eggs usually are dropped to the sediments, where they undergo a period of diapause. Clutch size varies greatly in calanoids, usually being maximal in the spring and the fall, separated by a summer minimum. Higher clutch size is correlated with lakes of greater primary productivity and spring conditions of higher seasonal food supply. As in the Cladocera, food availability directly influences the clutch size and temperature determines the rate of egg production (Elster, 1954; Comita and Anderson, 1959; Burgis, 1970; Geiling and Campbell,

---

[5]Hutchinson (1967) differentiates these species into *Arctodiaptomus laticeps* and *Eudiaptomus gracilis*, respectively.

1972). In this way a small number of females with a few eggs for a short period at high temperatures can produce more young than many females with large clutches for a long period at low temperatures.

Copepod eggs hatch into small, free-swimming larvae, termed nauplii, and then develop by molting through a number of subsequent larval stages. The initial nauplius has 3 pairs of reduced appendages (first and second antennae, mandibles). The successive naupuliar stages, 6 in all, result in feeding, growth, and molting, with the acquisition of further appendages. After 6 naupliar molts, the next molt results in an enlarged and more elongated form, the first copepodite instar. There are 5 copepodite stages during which additional appendages and body segments develop. The sixth and final cope-podite stage results in the adult. The time required to complete the larval stages is highly variable among species and under seasonal conditions.

Seasonal Dynamics

A number of cyclopoid copepods are known to enter various periods of diapause, either at the egg stage or in the copepodite stages with or without encystment, at the sediments (see, for example, Wierzbicka, 1966). In this way, for reasons that are not very clear, the annual cycle of the copepod populations is interrupted by a diapause that persists from one to several months. The resting stage may occur in midwinter, as in *Cyclops bicolor,* or in summer, as is seen in *Cyclops strenuus strenuus* (Fig. 16–23). Nauplii develop from egg-bearing females in this species in a bimodal population in winter and in spring. Then, in summer, the larger population of copepodite stage IV aesti-

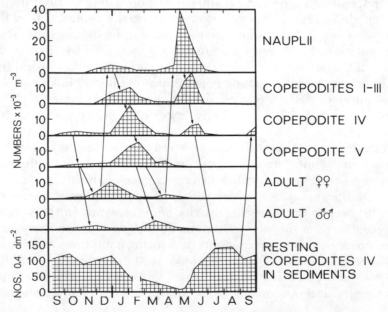

Figure 16–23 Development and resting stages of *Cyclops strenuus* in Bergstjern, Norway. Midsummer diapause is indicated by aestivation of copepodites IV in the sediments. (Drawn from data of Elgmork, 1959.)

vates in the sediments for about 2 months in midsummer before becoming pelagic again in the autumn and developing adults. In contrast, *Cyclops strenuus abyssorum* exhibits only one effective generation in each year in several lakes of the English Lake District (Smyly, 1973). In the deepest of the 4 lakes studied in great detail, individuals of this generation hatch from eggs laid in the spring and reach the adult stage in early winter, pass the winter in the planktonic zone, and start laying eggs early in the following year. In the 3 other lakes, most individuals of the spring generation reach the fifth copepodite stage by midsummer, and spend the next 8 months aestivating in the profundal zone. They leave this zone in February or March to return to the planktonic zone, become adults, and start breeding. Temperature, oxygen concentrations, light intensity, and day length are believed to be associated with the initiation and termination of this diapause (Smyly, 1973). Co-occurring species of cyclopoid copepods in the same water are known to have their maxima and diapause periods at different times. This alternation presumably minimizes competitive interactions for the same food resources.

The seasonal life cycles of the calanoid copepods are generally somewhat longer than those of the cyclopoids. In the temperate region, most species exhibit prolonged reproductive periods with several generations per year that are indistinguishable from one another. An example of these seasonal dynamics is seen in the populations of *Diaptomus* in a small, shallow beaver pond of southern Ontario (Fig. 16–24). The *Diaptomus* hatched from resting eggs in early May and produced 4 complete generations, the last maturing in late autumn (Carter, 1974). Development of spring and summer generations required about 4 weeks, while the time was about 6 weeks in autumn generations. Population characteristics were similar over a 3 year period. In one year, when water levels were reduced, resulting in increased mean heat content per unit volume, the development time of the generation was accelerated and population production increased significantly.

A number of studies have demonstrated that most adult cyclopoid copepods are carnivorous, and that their predatory activities can play a significant role in the population dynamics of other copepod species. For example, *Mesocyclops* was found to be selectively predacious on copepodites of *Diaptomus* rather than on cladocerans (Confer, 1971). Similarly, adult *Cyclops* preys heavily on nauplii of *Diaptomus* and its own species (McQueen, 1969). Some 30 per cent of the naupliar recruitment has been observed to be lost by this copepod predation and by cannibalism. Juvenile mortality often is very high and is density-dependent, increasing conspicuously at high population levels (Elster, 1954).

The coexistence of several different congeneric species of copepods in the same volume of waters has been a subject of much interest. It is assumed that when coexistence does occur, the species have slightly differing tolerances and optima to existing environmental conditions. Mechanisms for coexistence include: (a) seasonal separation, discussed earlier; (b) vertical separation in relation to stratification and distribution of food resources; and (c) size differences in relation to available food particles utilized. All of these mechanisms were implicated as factors permitting the coexistence of 3 species of *Diaptomus* in an Ontario lake (Sandercock, 1967; see also Hammer

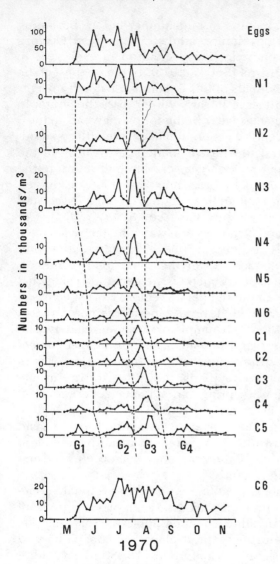

Figure 16–24 Seasonal cycles of *Diaptomus reighardi* in Black Pond, Ontario. Broken lines separate the estimated limits of successive generations, *G*. (From Carter, J.C.H.: Life cycles of three limnetic copepods in a beaver pond. J. Fish. Res. Bd. Canada, 31: 421–434, 1974. Reprinted by permission of the Journal of the Fisheries Research Board of Canada.)

and Sawchyn, 1968). *D. minutus* was separated vertically from *D. sanguineus*, and the latter is distinctly smaller than the former species. *D. minutus* and *D. oregonensis* also were separated to some degree by size differences and exhibited different seasonal maxima. *D. sanguineus* and *D. oregonensis* were separated vertically and had different seasonal maxima. Therefore, in this lake, it appears that a combination of 2 of the 3 factors permitted coexistence in each of the 3 congeneric species, but different combinations were in effect in each case.

Cyclomorphosis

Seasonal polymorphism among adult copepods is minor in comparison to that found among the parthenogenetic Cladocera and rotifers. The most conspicuous feature is a slight inverse relationship of size to increasing temperature, so that, in the few cases studied, animals of summer populations tend to be somewhat smaller than animals of colder seasons.

Vertical Migration and Spatial Distribution
of Cladocera and Copepods

One of the most conspicuous features of the cladoceran zooplankton, and to a lesser extent the copepods, is the marked vertical migration of these small animals over large distances on a daily basis. There is no question that light is the stimulus for the vertical migration (Siebeck, 1960; Ringelberg, 1964), but the adaptive significance of this movement is less clear. Movements vary considerably with lake conditions in relation to underwater light character-istics, season, and the age and sex of a species. However, certain generaliza-tions do emerge.

Most species migrate upward from deeper waters to more surficial strata as darkness approaches. Maximum numbers are found in the surface layers in this *nocturnal* migration in a single maximum some time between sunset and sunrise. In other cases, in *twilight* migration 2 maxima occur in surface layers, at dawn and at dusk. In a few species, *reverse* migration occurs with a single surface maximum during the daylight period.

The pattern of vertical migration commonly found may be seen in an example from relatively transparent Lake Michigan (Fig. 16–25). At this time the *Daphnia* population migrated from 24 m through the thermal discon-tinuity of the metalimnion at dusk at a rate of over 10 m hr^{-1}. Some sinking occurred after the initial ascent, a characteristic common to the twilight type of migration that could be associated with reduced nighttime activity and passive sinking. For example, narcotized *Daphnia galeata mendotae* sink passively between 3.6 and 14.4 m hr^{-1}, in an inverse relationship to body length (Brooks and Hutchinson, 1950). The in situ populations generally do not sink that rapidly during the night period (usually less than 0.5 m hr^{-1}), which indicates that they are sufficiently active so that sinking exceeds up-ward movements only slightly. After another surface maximum before sun-rise, the daphnids left the upper waters abruptly about sunrise at velocities of about 5 m hr^{-1}. Both the evening ascent and morning descent occurred when light intensity at the mean depth of the population was changing rapidly (Fig. 16–25).

Daphnia has trichromatic vision with spectral peaks of sensitivity in the ultraviolet (370 nm), deep blue-violet (435 nm), yellow (570 nm), and possibly in the red (685 nm) (Fig. 16–25). At the midday depth of habitation, the photo-environment at depth is predominantly blue-green. The photosystem is most sensitive in the blue, and may be best adapted for detecting changes in the in-tensity at low energy levels (McNaught, 1966). All photosystems may function in the detection of changes in wavelength and intensity at higher levels during daytime vision. The responses to light changes are conspicuous, but there is no evidence that the zooplankters are attempting to remain in a constant photoenvironment.

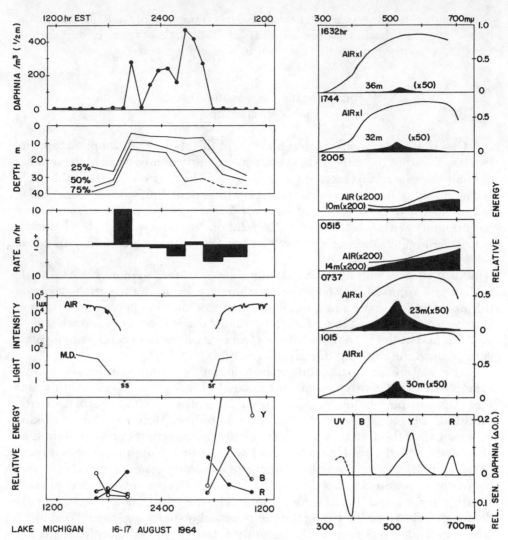

Figure 16–25 The vertical distribution and rate of movement of *Daphnia retrocurva* in relation to light quality at depth. *Left, top to bottom:* number of *Daphnia* m⁻³ at 0.5 m depth; quartile distribution; rate of movement (m hr⁻¹); light intensity (lux) in air and at mean population depth *(M.D.)*; and relative energy received by blue *(B)*, yellow *(Y)*, and red *(R)* photosystems of *Daphnia. Right, top to bottom:* relative energy versus wavelength in air and at mean population depths at different times of the diurnal period; spectral sensitivity of *Daphnia*, expressed as change in optical density *(O.D.)* of visual pigments extracted with aqueous digitonin upon bleaching. *sr* = sunrise; *ss* = sunset. (From McNaught, D.C., and Hasler, A.D.: Photoenvironments of planktonic Crustacea in Lake Michigan. Verhandlungen Int. Ver. Limnol., 16:194–203, 1966.)

The vertical migrations also occur under much less transparent conditions of more productive lakes. Under these situations, even though the amplitude of movement is much less and often only a meter or 2, the pattern is conspicuously similar (Table 16–11; Fig. 16–26). The mean morning maximum at the surface occurred 60 minutes (SD ± 37 minutes) after sunrise and the mean evening maximum at the surface occurred 19 ± 61 minutes before sunset. Deviations from the mean pattern increased under less distinct light conditions of moderate to heavy cloud cover. But, the cue is light intensity. Re-

TABLE 16–11 Examples of the Estimated Velocity and Amplitude of Vertical Migration of Several Zooplankters[a]

Group/Species	Lake	Maximum Observed Rate (m hr^{-1}) Ascent	Descent	Maximum Observed Amplitude (m)	Source
Cladocera					
Daphnia galeata mendotae	Michigan	—	—	10	Wells, 1960
	Mendota, Wisc.	1.4	1.4	1.5	McNaught and Hasler, 1964
D. longispina	Grand, Colo.	—	—	5.8	Pennak, 1944
	Lucerne	—	—	31	Worthington, 1931
	Victoria	—	—	50	
D. retrocurva	Michigan	10.6	5.0	24.3	McNaught and Hasler, 1966
D. pulex	Summit, Colo.	—	—	2.7	Pennak, 1944
D. schoedleri	Mendota	1.4	1.4	1.5	McNaught and Hasler, 1964
Bosmina longirostris	Michigan	19	12	20	McNaught and Hasler, 1966
	Silver, Colo.	—	—	1.8	Pennak, 1944
Polyphemus pediculus	Michigan	3.6	2.0	13	McNaught and Hasler, 1966
Holopedium gibberum		6.9	—	13.7	
Leptodora kindtii		4.2	2.0	8.5	
Copepoda					
Limnocalanus macrurus	Michigan	18.0	9.8	24	McNaught and Hasler, 1966
Diaptomus shoshone	Summit	—	—	8.8	Pennak, 1944
Cyclops bicuspidatus	Silver, Colo.	—	—	3.2	
Amphipoda/Mysids					
Pontoporeia affinis	Michigan	11.7	13.9	40	McNaught and Hasler, 1966
Mysis relicta		18	13	40	
		30–48	30–48	76	Beeton, 1960

[a] See also Tables 16–2 and 16–3 for rotifer movements.

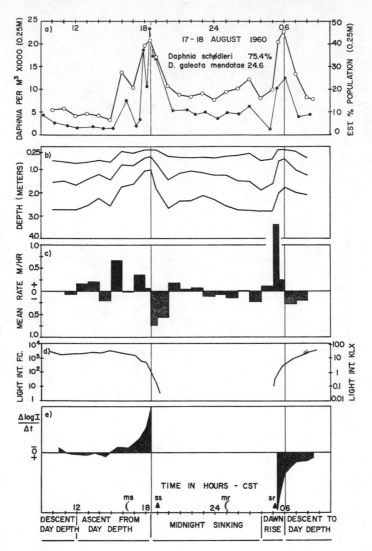

Figure 16–26 Diurnal vertical migration of two *Daphnia* populations in Lake Mendota in August in relation to light conditions. a, Number (•) and estimated population (o) at 0.25 m depth; b, quartile distribution; c, net rate of vertical movement (m hr⁻¹); d, light intensity (lux and foot-candles) at 0.25 m depth; and e, rate of change in the logarithm of the light intensity. ss = sunset; sr = sunrise; ms = moonset; mr = moonrise. (From McNaught, D.C., and Hasler, A. D.: Rate of movement of populations of *Daphnia* in relation to changes in light intensity. J. Fish. Res. Bd., Canada, 21:291–318, 1964. Reprinted by permission of the Journal of the Fisheries Research Board of Canada.)

sponse can be shown experimentally and under abnormal natural light conditions, e.g., the response is identical under twilight conditions during a solar eclipse (see, for example, Bright, et al., 1972).

Variations in vertical migration are great. Generally, young stages have been found to migrate greater amplitudes than adults both in cladocerans and copepods. Differences also occur in the migratory patterns of females and males, especially among copepods. Males tend to be less migratory than females, although the pattern is not consistent. The amplitude and rates of migration vary considerably seasonally during the ice-free season. Some of this change may be related to progressive decreasing oxygen content of lower layers as well as changing light conditions as the planktonic turbidity shifts. Although temperature influences rates of locomotion, metabolism, and regeneration of photochemical visual pigments, it is minor in effecting vertical migrations in comparison to light. Migrations continue under ice-cover and are influenced strongly by light reductions from snow-cover. Reverse migration is more common under ice-cover, especially among copepods (see, for example, Cunningham, 1972). Zooplanktonic concentrations are high in surface layers of low light below heavy ice and snow.

Another conspicuous feature of the movements of pelagic cladocerans and copepods is a distinct "avoidance of shore." At the shore the littoral and near-littoral waters of lakes are nearly free of pelagic crustaceans. When released within the littoral zone, these animals migrate in a horizontal plane away from the shore area if illumination is adequate. This aspect of spatial orientation has been the subject of intensive experimentation (especially Ringelberg, 1964; Siebeck, 1968; Siebeck and Ringelberg, 1969; Daan and Ringelberg, 1969).

The behavioral analyses showed that these crustaceans orient themselves towards the elevation of the horizon and the position of the sun. The direction of migration coincides with the plane of symmetrical light stimuli in the underwater angular light distribution. Near the shore this plane of symmetry of angular light depends primarily on the position of the elevation of the horizon. With increasing distance from the shore, the effect of the elevation of the horizon on the angular light distribution decreases and finally becomes insignificant, at which point the animals no longer select a directional movement away from shore. The higher the elevation of the horizon, as in mountain lakes, the wider the zone near shore with a reduced concentration of pelagic crustacean zooplankton.

Light from the sun and sky incident on water is refracted in the direction of the vertical axis as it penetrates downward into the water. The maximum angle of penetration is about 49 degrees from the vertical. The angular light distribution operates in effect in an inverted conical fashion, in which light impinging at greater angles is absorbed selectively more rapidly and forms a light-dark boundary or contrast at an angle of about 49 degrees to the vertical axis. The contrast or light gradient is used in orientation of the body axis in swimming. For example, *Daphnia magna* is unable to swim normally or maintain its normal body position when light from every direction is of equal intensity. Normal swimming reappears when a light gradient is introduced. The body and swimming positions are orientated by distinct pairs of ommatidia of the compound eye that detect contrasts. The eye is turned in a dorsal body direction so that the light contrast can be projected on the same dorsal part of the eye. The body is then turned in order to orient the animal so that a physiologically fixed angle between the eye axis and the body axis is maintained.

In the littoral zone, the angular light distribution consists of a dark area along the landward side within the cone of illumination. This distribution results in 1 of the 2 contrasts present at 49 degrees in the pelagial being shifted to the vertical, and swimming would be away from the shore (Fig. 16–27). In a plane parallel to the shoreline, this contrast difference is absent and the contrasts are the same as in the open water.

Evidence for changes in the rates of feeding during the diurnal migration of crustacean zooplankton is meager. At certain times of the year in the seasonal succession of zooplanktonic species, the grazing rates are several times greater during the dark period (Haney, 1973). At other times, the zooplankton undergo diurnal migration without altering their grazing rates. However, simply because of the massive diurnal movements of major components of the zooplanktonic populations to surface strata at night, a marked increase in grazing pressure on the algal, detrital, and bacterial populations occurs at night. Certain algae, such as motile unicellular flagellates, can move downward at night, perhaps a survival adaptation to increased predation pressure.

Study of the adaptive significance of nocturnal vertical migration has led to a number of hypotheses, all of which probably interact to varying degrees. First, predation by fish and other predators is largely a visual process requiring light. Movement upward into the trophogenic zone in darkness or periods of low light intensities would avoid much of this predation pressure. Second, the food quality of algae is variable diurnally, as is discussed further in the following chapter (see Fig. 17–19). Carbohydrate synthesis of photosynthesis occurs during the photoperiod, while protein synthesis reaches its maximum during the night period, coinciding with maximum grazing pressure in that area. Third, growth efficiency is somewhat greater at low tempera-

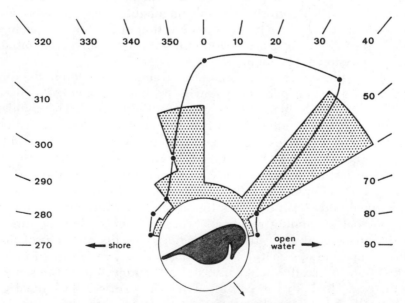

Figure 16–27 Orientation of a daphnid zooplankter along a plane perpendicular to the shore. Angular light distribution (solid line), magnitude of contrast expressed as the difference between two succeeding measurements (shaded bars), and the most probable orientation and movement of the daphnid. (From Siebeck, O., and Ringelberg, J.: Spatial orientation of planktonic crustaceans. Verhandlungen Int. Ver. Limnol., *17*:831–847, 1969.)

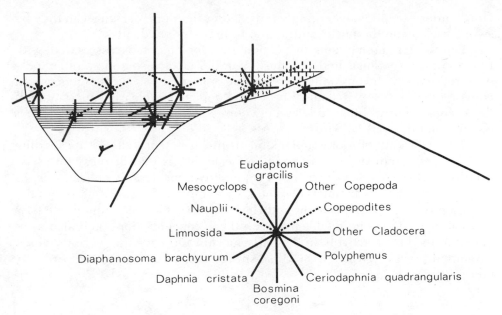

Eudiaptomus
gracilis

Mesocyclops   Other Copepoda

Nauplii   Copepodites

Limnosida   Other Cladocera

Diaphanosoma brachyurum   Polyphemus

Daphnia cristata   Ceriodaphnia quadrangularis
Bosmina
coregoni

Figure 16–28 The relative quantitative distribution of crustacean zooplankton in per cent in different strata along a transect across Skärshultsjön, southern Sweden, midafternoon, 7 August 1950. (Modified from Berzins, 1958.)

tures, such as are found in the lower strata as soon as summer thermal stratification occurs (McLaren, 1963). Similarly, feeding rates and efficiency are greater at higher temperatures. According to this differential growth-feeding hypothesis, a significant advantage would be obtained with only a degree of temperature difference between the water strata. Since the photoperiod would have little influence on advantages conferred by these mechanisms, the adaptations most likely are coupled with the advantages of reduced predation and higher food quality that are related to the photoperiod. The energy expenditure in such migrations is very small (Vlymen, 1970), and probably represents less than one per cent of the total over a life cycle.

In addition to the avoidance of the littoral regions by cladocera and copepods discussed earlier, a number of detailed analyses have shown a nonrandom horizontal distribution of zooplankton in the pelagic zone of lakes. An example is seen in the intensive 1-day analysis of Berzins (1958) across Skärshultsjön in Sweden (Fig. 16–28), the same study discussed earlier in regard to spatial variations of rotifers (Fig. 16–10). While *Polyphemus* is distinctly littoral, many of the cladocerans and copepods avoid the littoral and are not uniformly distributed in the pelagic zone horizontally.

Nonrandom dispersion of phytoplankton and zooplankton within the pelagial zone in many cases is caused by hydrographic water movements. Microcrustacea tend to accumulate in the epilimnion at the leeward end of the basin and to depauperate at the windward end whenever a fairly strong wind persists for an appreciable period of time (see, for example, Langford and Jermolajev, 1966). Once at the lee side of the basin, weak locomotion presumably is adequate for most of the animals to maintain depth distribution there.

Only under exceptionally strong winds and epilimnetic currents can they be swept back along the lower epilimnion to the windward side.

Part of the nonrandom dispersion of zooplankton is associated with Langmuir convectional helices whose parallel clockwise and counterclockwise rotations in the epilimnion create linear alternations of divergences and convergences (see pp. 97ff and Fig. 7–6). Zooplankton clearly concentrate in these areas of upwelling midway between foamline streaks at convergences (George and Edwards, 1973).

This concentration of zooplankton in linear patches affects the feeding behavior of certain fishes. An excellent example is that of the white bass (*Roccus chrysops*), a pelagic fish that is planktivorous, largely on *Daphnia*, in its younger stages in summer. *Daphnia* were found to be concentrated in the linear convergences (McNaught and Hasler, 1961). The schooling white bass were able to locate and feed actively on these concentrations, and also were keyed to feed most actively in the early morning and evening periods which coincided precisely with the maximal surface concentrations of the vertically migrating *Daphnia*.

Predation by Fishes and Size Selectivity

Thus far in this discussion, competitive interactions among zooplanktonic populations have centered on differences in spatial and temporal distributions and on food-size requirements that permit coexistence in the same body of water. The succession and competitive success of zooplanktonic populations can be materially influenced by predation of other zooplankters, a point which was demonstrated earlier among the rotifers and the large cladoceran *Leptodora*. A number of recent studies have demonstrated the importance of planktivorous fish in regulating zooplanktonic populations and in causing a distinct shift favoring survival of species that are smaller in size. As noted briefly earlier, Hrbáček and collaborators were the first to demonstrate unequivocally the importance of fish in the regulation of the size and species composition of zooplankters (Hrbáček, 1958, 1961, 1962, 1965; Novotná and Kořínek, 1966). When large, shallow ponds were stocked heavily with cyprinid fish, the zooplankton consisted of small cladocerans, *Bosmina*, *Ceriodaphnia*, and rotifers. Transparency increased with the predominant development of small nannoplanktonic algae. Upon selective removal of the fish, the zooplankton changed suddenly to larger cladocerans, particularly *Daphnia longispina*, and rotifer abundance decreased. Transparency decreased as the phytoplankton shifted to smaller densities of larger species.

Another series of remarkably simple but eloquent studies conducted by these Czechoslovakian workers involved experimental manipulation of portions of the same small lakes by fencing off portions of the littoral to fish and ducks (Straškraba, 1963, 1965, 1967; Poštolková, 1967). Within the protected littoral enclosures, submersed vegetation increased and afforded zooplankton greater protection when exposed to fish predation than in the open water. Zooplanktonic biomass increased in the littoral as the small *Chydorus* and other cladocerans were replaced by larger *Daphnia*, *Simocephalus*, and many copepods. Differences could not be related to changes in food supply but rather to changes in fish predation.

All planktivorous fish have closely spaced gill rakers. The example of the alewife of Figure 16–29 shows the closely spaced gill rakers of the first branchial arch of the planktivorous *Alosa pseudoharengus* in comparison to those of the closely related *Alosa mediocris*, which feeds principally on small fish. Studies of the pharyngeal sieves of the planktivorous yellow perch *(Perca flavescens)* and the rainbow trout *(Salmo gairdneri)* demonstrated that the trout and perch could remove few zooplankters of a size smaller than 1.3 mm (Galbraith, 1967; Wong and Wood, 1972). It must be emphasized, however, that all freshwater planktivorous fish examined so far actively search for and visually select each plankter that they ingest (Brooks, 1968). However, in the normal respiratory behavior of passing water through the mouth and across the gills, some zooplankters are collected by gill rakers if they are sufficiently close together to be retained. Fish such as the alewife are more or less obligately planktivorous, while others such as the trout and perch eat large zooplankters, and if these are unavailable they readily shift to alternate food sources in the lake. Size selection of prey appears to be characteristic

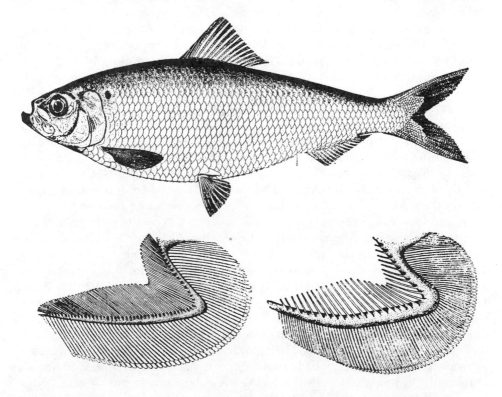

Figure 16–29 The alewife, *Alosa pseudoharengus*. *Upper*: A mature specimen, 300 mm in length. *Lower left*: First branchial arch with closely spaced gill rakers that act as a plankton sieve. *Lower right*: First branchial arch with widely spaced gill rakers of *A. mediocris*, a species that feeds primarily on small fish. (From Brooks, J. L., and Dodson, S. I.: Predation, body size, and composition of plankton. Science, 150(3692):28–35, 1965. Copyright 1965 by the American Association for the Advancement of Science.)

of both the obligate planktivores such as some of the alewifes, and those species which feed facultatively on the plankton.

The relation between size selection of prey and foraging efficiency was shown in the bluegill sunfish *(Lepomis macrochirus)*, a common centrarchid of many temperate waters that is known to select prey on the basis of size (Werner, 1974; Werner and Hall, 1974). Growth rates are known to increase significantly in direct relation to food size (see, for example, LeCren, 1958; Hall, et al., 1970), and much of this differential growth has been attributed to the efficiency of foraging, that is, the expenditure of energy or metabolic cost in obtaining food relative to the return. In the bluegill, size selection of prey is related to the optimal allocation of time spent searching for and handling the prey. In other words, the size range of prey that maximizes the energy return per unit of energy expended depends on the abundance and handling times of different prey in the environment. At times of low prey abundance, when search time is large, prey of different size are eaten as encountered. As prey abundance is increased and search time decreases, smaller sized classes are eaten less frequently or ignored, so that the predator may forage in an optimal fashion, and overall return per time increases (see also Brooks, 1968). The handling time per prey increases markedly in an exponential fashion as the ratio of prey size to mouth size increases.

The earlier Czechoslovakian observations on the effects of planktivorous fish on the zooplanktonic species composition have been demonstrated excellently by a study of the changes in the zooplanktonic populations of Crystal Lake, Connecticut (Brooks and Dodson, 1965). Prior to the introduction of the alewife into the lake, zooplankton included the large calanoid copepod *Epischura, Daphnia,* and the cyclopoid *Mesocyclops,* as well as numerous smaller *Diaptomus* and *Cyclops* copepods (Fig. 16–30). Some 10 years after the alewife invaded the lake, the larger forms of zooplankters had been replaced and dominants included very much smaller forms, the cladoceran *Bosmina* and 2 cyclopoid copepods *Tropocyclops* and *Cyclops* (Fig. 16–30). The modal length of the numerically dominant forms had shifted from 0.8 to 0.3 mm in the zooplanktonic assemblages. Larger forms were found only in the littoral zone and near the sediments, areas that are avoided by the pelagic *Alosa.*

Similar size-specific predatory elimination of larger zooplankters, particularly *Daphnia,* from lakes by other planktivorous fish has been shown. The planktivorous smelt *(Osmerus mordax)* is particularly effective in this regard (see, for example, Reif and Tappa, 1966, and especially Galbraith, 1967). However, the relationship of size and planktivore predation is not always that simple. If prey are large but relatively transparent, such as *Leptodora* or planktonic larvae of the dipteran *Chaoborus,* they often are overlooked and predation is reduced (see, for example, Costa and Cummins, 1972). Furthermore, the slower, steady movements of cladocerans render them more vulnerable to predation than the jerky, irregular movements of copepods.

The importance of visibility in planktivorous predation by fish is demonstrated by the tropical *Ceriodaphnia cornuta,* which shows 2 distinct polymorphic forms of the same body length within the same lake (Zaret, 1972a, 1972b). One form has pointed, horn-like extensions of the exoskeleton on the head, body, and tail regions with a small area of black pigmentation in

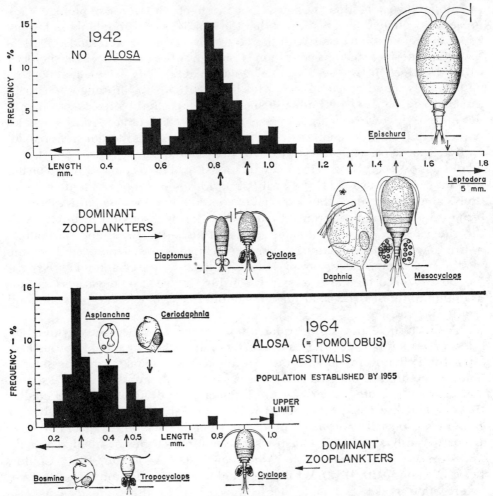

Figure 16–30 The composition of the crustacean zooplankton of Crystal Lake, Connecticut, before (1942) and after (1964) a population of *Alosa aestivalis* became well-established. Each square of the histograms indicates that one per cent of the total sample counted was within that size range. Larger zooplankters are not represented because they were relatively rare. The specimens depicted represent the mean size (length from posterior base lines to the anterior end) of the smallest mature instar. The arrows indicate the position of the smallest mature instar of each dominant species in relation to the histograms. The predaceous rotifer *Asplanchna* is the only noncrustacean included in this study. (From Brooks, J. L., and Dodson, S. I.: Predation, body size, and composition of plankton. Science, 150(3692):28–35, 1965. Copyright 1965 by the American Association for the Advancement of Science.)

the compound eye. The other phenotype is unhorned but possesses a large, pigmented eye. Predation by the dominant planktivore fish, a silverside minnow (Atherinidae; *Melaniris chagresi*) was strongly in favor of the form with the large, pigmented eye. This form of this species of *Ceriodaphnia* has a superior reproductive potential and a more rapid population growth with greater longevity than that of the horned, small-eyed form. Without fish predation, the large-eyed form can rapidly outcompete the less conspicuous form, but under predation pressure the form with the small-eye, although growing more slowly, can coexist because of reduced visibility to the predator.

When size selection by fish is not in effect and large zooplankters are present, a common observation is that the smaller sized zooplankton are not generally found to co-occur with the larger forms. Brooks and Dodson (1965) have hypothesized that this condition is caused by the differences in efficiency of food collection by "herbivorous" zooplankters. This "size-efficiency hypothesis" is based on competition for food particles in the range of 1 to 15 μm that can be used by both the small and large herbivores. According to this proposal, however, larger species are more efficient at feeding on small particles and exclude smaller-sized species by reducing the density of small particles to below the starvation level for small species. Yet, in effect, one would not expect this size-efficiency relationship of food competition to be the general case because of (1) the rarity of periods when zooplanktonic populations can even approach the potential of completely removing particulate material from the water of certain strata by filtration as discussed earlier, and (2) the presence of generally greater excess of algae, bacteria, and particulate detrital organic matter than can ever be consumed (most synthesized organic matter is decomposed in nonpredatory pathways; see Chapter 17). Experimental evidence on these food relationships is weak or negative (see, for example, Sprules, 1972; Dodson, 1974a), but does not exclude possible food quality interactions.

A much more plausible explanation is again a size-selective predation, but a predation by the larger zooplankters rather than by planktivorous fish. Such predation occurs among the size range of zooplankton that is smaller than can be utilized effectively by fish. A large number of studies have shown that the invertebrate predation can be sufficient to eliminate certain smaller species, and that this predation, as with some fish, is size-selective. Larger copepods especially and other predators such as *Chaoborus* and *Leptodora* are particularly effective in causing significant mortalities of small zooplanktonic species (see, for example, McQueen, 1969; Anderson, 1970; Confer, 1971; Dodson, 1970, 1972). While fish planktivory is visual and size-selective, invertebrate predation is more tactile since eyes of rotifers, crustaceans, and most insects detect light intensity and movements but do not form images. This relationship of size-selectivity and tactile responses in consumption suggests to Dodson (1974b) that shape of prey within the correct size range can influence whether the prey is taken or rejected. Thus cyclomorphic polymorphism would have an adaptive advantage for prey in both invertebrate predation by morphology and vertebrate predation by reduction of visibility while still permitting growth.

Zooplanktonic Productivity

The production rates of specific populations of zooplankton refer to the net productivity or the sum of the growth increments of all specimens of the population. This net productivity excludes maintenance losses (respiration, excretion), and it includes the growth increments of the animal itself as well as the biomass produced as exuviae of molting and gametes. In some larger invertebrates and some vertebrates, productivity values of specific animal populations are influenced markedly by emigration and immigration. The productivity measurements are complicated by predation in which a significant portion of the population is removed and this loss to the population is difficult to evaluate accurately.

The methods of estimating the rates of production of a specific population demand an accurate evaluation of the distribution of the organisms, the different stages of development and age, and the generation times. All of these vary from species to species, seasonally, and under changing environmental conditions which greatly influence the methods of sampling, analyses of biomass within each age group, and frequency of sampling. Sampling intervals must be kept several times shorter than generation time of the animals.

The manner in which production rates of a specific population are estimated depends on the particular life cycle, reproductive characteristics, and generation times (discussed at length in Edmondson and Winberg, 1971; Winberg, 1971). The productivity of a species with a long life cycle and short period of reproduction is determined relatively easily if individuals are of the same age, or separate cohorts can be recognized and their biomass determined. In this situation, after the brief period of reproduction no recruitment occurs to confound estimates of growth and mortality. When reproduction and recruitment to the population are continuous, estimation of production is more involved. In this latter case, which applies to most zooplanktonic populations, cohorts overlap, which makes it difficult to separate changes in their temporal abundance and biomass. Thus, it is necessary to estimate finite birth and growth rates of individuals and evaluate changes in biomass over the life cycle from birth to death. Several simple examples were cited for growth rates of specific cohorts in earlier discussion when these values could be expanded to multiplication of the different weight groups (or other measures of biomass, such as calories or nitrogen content) by abundance.

An index of productivity that has become widespread in European and Russian literature in recent years is the P/B coefficient, the ratio of production to biomass. Among populations with a constant age structure and biomass, which are a rarity in nature, the P/B ratio is relatively constant. The ratio varies in most situations when age structure, biomass, and growth are discontinuous and what is obtained is an average ratio over an interval of time. Multiplication of the P/B ratio by biomass yields a rough estimation of production during that interval of time. This ratio is analogous to the turnover rate, and also its reciprocal the turnover time, the average duration of life of a species under a given set of conditions.

The P/B ratio is an expression of some value in a comparative way on an annual basis among water bodies. However, it masks the large variations that occur within the population dynamics over a year. When used per se, then, it relates few of the internal population characteristics or, more importantly, of the causal interrelationships leading to these dynamics and productivity. Used appropriately, especially over brief periods of time under relatively steady environmental conditions, the turnover rates reflect population characteristics of relative growth.

TABLE 16–12 Examples of the Production of Herbivorous and Predatory Zooplankton Communities

Lake Type	Period of Investigation	Biomass (B)[a] g m^{-3}	Biomass (B)[a] kcal m^{-3}	Production (P)[a] g m^{-3}	Production (P)[a] kcal m^{-3}	Biomass Turnover Time (days) Average	Biomass Turnover Time (days) Range	Remarks and Source
OLIGOTROPHIC								
Lake Baikal, USSR	June–July Sept.	0.136 0.43						Primarily *Epischura*; Moskalenko and Votinsev, 1970
Clear Lake, Ontario	Annual (0–50 m) Annual	– 0.20		3.44 3.02	– 16.435	25	12–333	Herbivorous zooplankton (rotifers, Holopedium, Daphnia, Bosmina, Diaptomus); Schindler, 1970
MESOTROPHIC								
Lake 239, Ontario	May–Nov.			0.61	3.331	22		Arctic lake; Winberg, 1970
Lake Krugloe, USSR	Annual		0.405	0.94	5.116	29	10–91	
Taltowisko Lake, Poland[a]	May–Oct.	0.12		3.04		14.3		29% of total lake area in littoral zone; Kajak, et al, 1970; Kajak, 1970
Herbivores				4.68	25.43	10.2		
Predators				0.46	2.50	9.1		
Lake Naroch, USSR	Annual	0.07	0.38	1.12	6.11	22.4		Winberg, 1970
Lake Krasnoe, USSR	Annual	0.14	0.76	3.09	16.82	16.5		
EUTROPHIC								
Lake Mikolajskie, Poland	May–Oct.							Kajak, et al., 1970; Hillbricht-Ilkowska, et al., 1970
Herbivores				6.45	35.09	9.2	4.0–12.5	
Predators				1.32	7.18	25.0		
Lake Śniardwy, Poland	May–Oct.							
Herbivores				3.08	16.78	14.9	8.3–33	
Predators				0.50	2.71	14.3		
Kiev Reservoir, USSR								
Herbivores		0.35	1.9	9.15	49.8	13.9		Winberg, 1970
Predators		0.022	0.12	1.16	6.3			
Severson Lake, Minnesota	May–Oct.							Comita, 1972
Herbivores				2.51	13.66			
Predators				0.11	0.60			
DYSTROPHIC (high dissolved organic matter)[b]								
Lake Flosek, Poland	May–Oct.							*Sphagnum* bog; high littoral and allochthonous organic inputs; Kajak, et al., 1970; Hillbricht-Ilkowska, et al., 1970
Herbivores				25.68	139.7	6.3		
Predators				1.16	6.3	25.0		

[a]Estimated using the mean caloric value for microconsumers (Cummins and Wuycheck, 1971).
[b]See discussion in Chapter 18.

A large number of studies have appeared in which zooplanktonic productivity estimates have been made. The number of estimates will increase greatly as the results of the International Biological Program, in preparation at the time of this writing, are synthesized and summarized. A few examples will illustrate the ranges encountered in lakes of differing primary productivity (Table 16–12).

In general, a direct correlation exists between the rates of primary production of phytoplankton and of the nonpredatory zooplankton (see, for example, Winberg, et al., 1970). However, this statement does not necessarily imply that the herbivorous zooplankters are consuming the algae directly in proportion to their biomass and growth. As was discussed earlier, size, quality, and other factors influence both the ingestion and assimilation of algae. Moreover, much of the algal production enters the detrital pathways of nonpredatory particulate and dissolved organic matter (cf. Chapter 17). Although highly variable from lake to lake, dead particulate organic matter constitutes between 60 to 85 per cent of the total seston during most seasons of the year.

In an extremely detailed and perceptive study of the relationships between phytoplankton and zooplankton of mesotrophic Lake Erken, southern Sweden, Nauwerck[6] (1963) demonstrated that algal productivity was quite inadequate to support the herbivorous zooplanktonic productivity. The filter feeding zooplankton, amounting to a total productivity of 0.22 g m^{-2} day^{-1} on an annual mean basis, consisted of primarily *Eudiaptomus* (60 per cent), with a turnover time of about 36 days (annual mean; 24 days during the ice-free period). Utilization of bacteria and particulate detrital organic matter was demonstrated in numerous subsequent studies, even though the caloric content of the detrital material is considerably reduced. The result, then,

---

[6]Certain computational errors have been found in the original paper, but these do not alter the general conclusions.

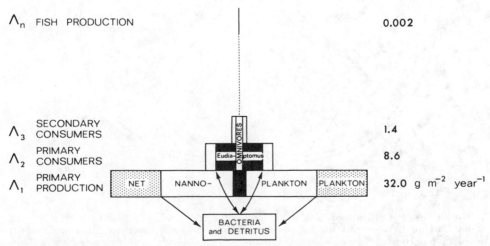

Figure 16–31 Diagrammatic representation of the trophic relationships in Lake Erken, Sweden. (Modified from Nauwerck, 1963.) Productivity estimates in g dry weight per m^2 per year of pelagic zone only.

is a considerably modified trophic pyramid in which the transfer to higher food levels is short-circuited, to use the words of Nauwerck, via the bacteria and particulate detrital organic matter (Fig. 16–31). In similar fashion, the greater part of the herbivorous zooplanktonic productivity is not utilized by predatory zooplankton and fish, but enters the nonpredatory detrital pathways of decomposition and is reutilized partially by filter feeders as bacteria and detritus. Therefore, Lindeman's classical conceptualization of the generalized lacustrine food-cycle relationships must be modified to emphasize the importance of nonpredatory detrital and bacterial pathways (Fig. 16–32); these pathways were underestimated because of inadequate data availability in the period when the cycle was formulated. This change does not alter

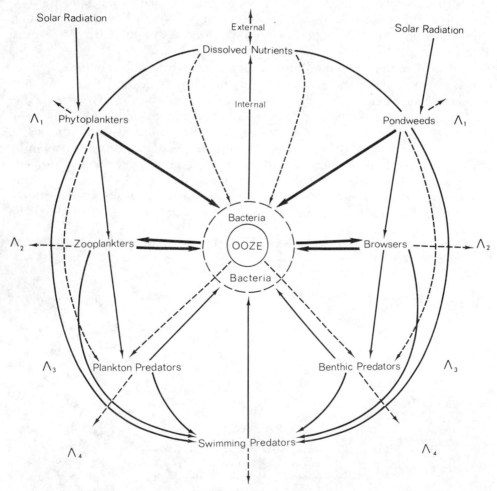

Figure 16–32 Generalized lacustrine food-cycle relationships after Lindeman (1942). Added darkened arrows indicate pathways ignored or underestimated in Lindeman's actual calculations, a consequence of accepting existing data on chemical assay of crude fiber in place of direct assimilation or egestion measurements. (From Wetzel, R. G., Rich, P. H., Miller, M. C., and Allen, H. L.: Metabolism of dissolved and particulate detrital carbon in a temperate hardwater lake. Mem. Ist. Ital. Idrobiol., 29: Suppl., 185–243, 1972.)

the fundamental contribution of Lindeman, but rather places the proper emphasis on the major contributions of bacterial and detrital feeding in the trophic relationships of primary consuming animals, and on the major losses of organic matter through decomposition that occur without animal consumption (cf. Chapter 17).

Other comparisons of the proportions of herbivorous zooplankton utilized by predators demonstrate similar results. In detailed studies of 9 very different lakes of the Soviet Union, the biomass of predators varied from 6.5 to 52 per cent of the total biomass of zooplankton, averaging about 30 per cent (Winberg, 1970). The productivity of the filter feeding zooplankters is distinctly higher than that of the predacious zooplankters (Table 16–13). Variation is great, however, particularly in relation to temperature, food quantity and quality, and other factors. Comparisons over 2 or more years with good sampling show the extent of variation that can occur with the same species in the same lake (Table 16–13; Figure 16–33). These differences are much greater than can be attributed to sampling and measurement errors. Perhaps the lowest production rates of zooplanktonic populations of a permanent lake are those of polar Char Lake at latitude 74° in the Canadian Arctic (Rigler, 1974). The dominant zooplankters *Limnocalanus macrurus* and *Keratella* produced 261 mg ash-free dry weight per m^2 per year and 3 mg m^{-2} yr^{-1} respectively. This is equivalent to an annual average of about 0.00006 g m^{-3} day^{-1} an exceedingly low value.

In the particular case of *Daphnia hyalina* in an eutrophic reservoir (Fig. 16–33), maximum biomass was found nearly 2 months before highest numerical density in 1970, but numerical and biomass maxima were coincident the following year (George and Edwards, 1974). The periods of maximum spring biomass occurred during blooms of green algae in late April and early May, when many *Daphnia* survived and grew for some time after reaching maturity. Increases in production coincided with, or lagged somewhat behind, increases in biomass. In the second year the production was nearly one-half that of the first year, and the average biomass turnover time also decreased (Table 16–13).

The efficiency of energy transformations between components of animal populations is an area of intensive investigation. At present few cases have been studied in detail in natural populations; much of the necessary information of ingestion, respiration, and egestion is determined under controlled laboratory conditions that have been applied variously to natural populations.

A certain percentage of the food ingested is not utilized. As a percentage, this egestion rate varies greatly with age, food quantity and quality, and other factors within an observed range of about 30 to 90 per cent loss (Table 16–14). How much of this egested detrital material is reingested is unclear, at least among the pelagic organisms. Consumption of feces after varying periods of colonization and decomposition by bacteria and fungi is a fairly common phenomenon among benthic fauna of lakes and streams.

The rates of assimilation and respiratory loss in biochemical maintenance also vary widely with environmental conditions (Table 16–14). Assimilation efficiency is generally greater among young animals and decreases with

TABLE 16-13 Examples of Productivity of Herbivorous and Predatory Forms of Zooplankton

Type/Species	Lake/General Productivity	Period of Investigation	Production Estimates[a] $g\,m^{-3}\,day^{-1}$	$kcal\,m^{-2}\,day^{-1}$	Biomass Turnover Time (Days) Average	Range	Source
FILTER FEEDERS:							
Cladocera:							
Daphnia hyalina	Eglwys Nynydd Reservoir, Wales; eutrophic	Annual, 1970	0.57		21.3 }	3.8–333	George and Edwards, 1974
		Annual, 1971	0.32		15.9 }		
D. parvula	Severson Lake, Minn.; eutrophic	Annual	0.010	0.102	—		Comita, 1972
D. galeata mendota	Sanctuary Lake, Pa.; eutrophic reservoir	May–Nov., 1966	0.407	3.026			Cummins, et al., 1969
		May–Nov., 1967	0.030	0.223			
	Canyon Ferry Reservoir, Mont.: eutrophic	April–Sept.	0.114		10.0		Wright, 1965
D. schodleri	Canyon Ferry Reservoir, Mont.; eutrophic	April–Sept.	0.227		6.7		Wright, 1965
D. longispina	Lake Sevan, southern USSR	Annual	0.006		58.9		Meshkova, 1952, in Winberg, 1971
Bosmina longirostris	Severson Lake, Minn.	Annual	0.007	0.071	—		Comita, 1972
B. longirostris and *B. coregoni*	Sanctuary Lake, Pa.	May–Nov., 1966	0.183	1.361			Cummins, et al., 1969
		May–Nov., 1967	0.067	0.498			
Ceriodaphnia reticulata		July–Nov.	0.031	0.154			
Chydorus sphaericus		July–Aug., 1966	0.004	0.020			
		July–Aug., 1967	0.047	0.233			
Cladocera	Naroch Lake, USSR	May–Oct.	0.0026	0.117	13.7		Winberg, et al., 1970
	Myastro Lake, USSR	May–Oct.	0.015	0.403	10.9		
	Batorin Lake, USSR	May–Oct.	0.033	0.484	10.5		
Copepods:							
Cyclops strenuus	Buttermere, England; oligotrophic	Annual	0.0004				Smyly, 1973
	Rydal Water, England; eutrophic	Annual	0.0005				
	Grasmere, England; eutrophic	Annual	0.0006				
	Esthwaite Water, England; eutrophic	Annual	0.0017				
	Lake Sevan, southern USSR	Annual	0.0007		79.3		Meshkova, 1952, in Winberg, 1971

Eudiaptomus graciloides	Naroch Lake, USSR	May–Oct.	0.0010	0.044	24.7	Winberg, et al., 1970
	Myastro Lake, USSR	May–Oct.	0.0065	0.174	20.2	
	Batorin Lake, USSR	May–Oct.	0.0070	0.104	15.4	
	Severson Lake, Minn.	Annual	0.0046	0.045	—	Comita, 1972
Mesocyclops edax			0.0067	0.067	—	
Diaphanosoma leuchten-bergianum						
Diaptomus siciloides	Lake Sevan, southern USSR	Annual	0.0342	0.341	—	Meshkova, 1952, in Winberg, 1971
Arcthodiaptomus (2 species)			0.0014	—	162	
Rotifers:						
Keratella quadrata	Severson Lake, Minn.	Annual	0.0021	0.021	—	Comita, 1972
K. cochlearis			0.0007	0.0074	—	
Filinia longiseta			0.0011	0.0112	—	
Brachionus sp.			0.0075	0.0752	—	
Polyarthra sp.			0.0010	0.0103	—	
Rotifers	Naroch Lake, USSR	May–Oct.	0.0027	0.120	1.7	Winberg, et al., 1970
	Myastro Lake, USSR	May–Oct.	0.0024	0.065	3.9	
	Batorin Lake, USSR	May–Oct.	0.0105	0.156	2.6	
PREDATORY FEEDERS:						
Cladocera:						
Leptodora kindtii	Sanctuary Lake, Pa.	May–Nov., 1966	0.003	0.022	—	Cummins, et al., 1969
		May–Nov., 1967	0.013	0.097	—	
Cladocera	Naroch Lake, USSR	May–Oct.	0.0003	0.013	11.3	Winberg, et al., 1972
	Myastro Lake, USSR		0.0009	0.023	3.9	
	Batorin Lake, USSR		0.0002	0.003	10.8	
Copepods:						
Cyclops sp.	Naroch Lake, USSR	May–Oct.	0.0008	0.034	9.7	
	Myastro Lake, USSR		0.0023	0.062	19.4	
	Batorin Lake, USSR		0.0094	0.140	14.2	
Rotifers:						
Asplanchna priodonta	Naroch Lake, USSR	May–Oct.	0.0014	0.061	2.9	
	Myastro Lake, USSR		0.0061	0.163	2.5	
	Batorin Lake, USSR		0.0105	0.156	3.2	
Asplanchna sp.	Severson Lake, Minn.	Annual	0.0031	0.031	—	Comita, 1972
Synchaeta sp.			0.00009	0.0009	—	
Insect larvae:						
Chaborus punctipennis			0.0001	0.001	—	

[a]Conversions estimated using the caloric mean of microconsumers (Cummins and Wuycheck, 1971), when mean depths available.

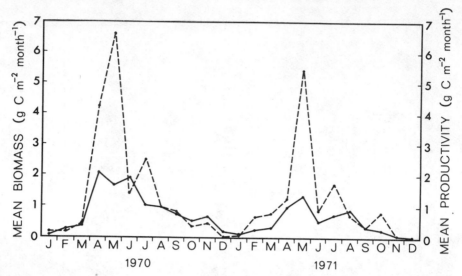

Figure 16–33   Seasonal monthly mean biomass (———) and productivity (-----) of *Daphnia hyalina* in g C m⁻² month⁻¹, calculated from daily measurements, of Eglwys Nyndd, Great Britain. (Drawn from data of George and Edwards, 1974.)

age; average values, mostly from experimental conditions better than those in situ, are in the range of 10 to 50 per cent. Respiratory costs are high, and as would be anticipated increase greatly under adverse conditions such as low temperatures. For example, the population respiration of the dominant zooplankter of the Canadian arctic Char Lake, with a maximum of about 4°C for one month of the year, far exceeded net production (Table 16–15). The ratio of production to assimilation was about 13 per cent.

Of the production that is utilized in growth and reproduction, usually a small portion is utilized in the diapause stages such as in overwintering eggs. Most of the production enters the nonpredatory pool of detrital organic matter to be decomposed by microorganisms. A variable proportion is utilized by predators. The percentage of the primary consumers to appear as net production of secondary consumers also varies, but is generally around 10 per cent or less (e.g., Comita, 1972) on an annual average. As was discussed earlier, however, at certain periods of the year the impact of herbivores and predators can be quite significant and literally can decimate component populations of the communities.

In the original conceptualization of productivity within the trophic level system developed by Lindeman (1942) in a greatly oversimplified fashion (Fig. 16–32), organisms were separated into trophic categories. The productivity of each trophic level $\wedge_n$ was considered to be the rate of change of energy content entering (λ_n) and leaving (λ'_n) that trophic level

$$\frac{d \wedge_n}{dt} = \lambda_n + \lambda'_n.$$

λ_n is the rate of energy contribution at trophic level $\wedge_n$ from the previous

TABLE 16-14 Exemplary Estimates of Efficiencies of Food Utilization by Various Animals[a]

Organisms	% of Ingested Food Utilized in:				% of Assimilated Energy Expended in:		Source
	Egestion	Assimilation	Respiration	Growth and Reproduction	Respiration	Growth and Reproduction	
ZOOPLANKTON							
Daphnia pulex	69–86	14–31	4–14	10–17	27–44	56–73	Richman, 1958
Ceriodaphnia reticulata	–	10.6	1.8	–	–	–	Czeczuga and Bobiatyńska-Ksok, 1970
Simocephalus vetulus							Klekowski, 1970
Juveniles (♀)	27.6	72.4	19.5	52.9	26.9	73.1	
Reproducing (♀)	68.3	31.7	11.2	20.5	35.3	64.7	
Leptodora kindtii	60.0	40.0	–	–	92.7	7.3	Cummins, et al., 1969; Moshiri, et al., 1969
Mesocyclops albidus	–	20–75	ca 20	ca 25	ca 50	ca 50	Klekowski and Shushkina, 1966
10 Herbivores	52.4	47.6	40.1	7.5	–	–	Comita, 1972
BENTHIC ANIMALS							
Asellus aquaticus	69.7	30.3	24.7	5.6	81.8	18.2	Klekowski, 1970
Lestes sponsa	63.4	36.6	13.2	23.5	36.0	64.0	Klekowski, et al., 1970
FISH							
Phytophagous carp, Ctenopharyngodon	86.0	14.0	12.2	1.9	86.0	14.0	Fischer, 1970
Predatory perch, Perca fluviatilis	64.2	35.8	16.2	19.5	45.5	54.5	Klekowski, et al., 1970

[a]Methods of analysis and experimental conditions vary greatly and are comparable only approximately.

TABLE 16–15 Mean Production and Respiration of the *Limnocalanus macrurus* Population in Polar Char Lake, Canadian Arctic, Over a 4-Year Period[a]

Production and Respiration	Mean mg m^{-2} year^{-1}	% of Total
Growth	220	11.1
Eggs	5.6	0.3
Exuviae	34.9	1.8
Total production	261	13.1
Respiration	1660	83.4
Assimilation	1990	

[a]After data of Rigler, et al., 1974.

trophic level ($\wedge_{n-1}$), gross primary productivity among producers, and assimilation at animal levels. Efficiency is simply the ratio of energy or material leaving a system to the energy entering the system. In this case efficiency would equal the ratio of these productivities, $\wedge_n / \wedge_{n-1}$ (discussed critically by Kozlovsky, 1968).

A great number of different interpretations and definitions of efficiency have evolved, some of which are relatively meaningless and confusing (cf. Kozlovsky, 1968). Determinations of ecological efficiencies demand independent measurement of energy and material input and output. The criteria can rarely be achieved among animal populations in natural conditions, forcing the worker to rely heavily on laboratory evaluations. While these experimental results are valuable and suggestive of in situ potential efficiencies, they must be used with extreme caution.

In discussion of the trophic level concept, it must be emphasized that it is a theoretical construct that has helped in the evaluation of interrelationships among animal organisms. The concept has been justly criticized for being greatly oversimplified, with critics maintaining that real analyses call for measurements of all rates of species production and utilization among trophic levels (Ivlev, 1945). Trophic levels consist of population groups of different species of organisms utilizing the same mode of nutrition, many of which change their mode of feeding to another trophic level in the course of their life cycle. As was emphasized earlier in this chapter, animal nutrition is strongly dependent upon food density, quality, and many other factors. In turn, the efficiency of food utilization for growth and maintenance varies with the rate of food consumption and population densities. These efficiencies are only reasonably accurate when the populations studied are in a fairly fixed state of equilibrium (Slobodkin, 1960); however, this condition rarely exists in natural populations, since they undergo large fluctuations even though the tendency in nature is towards equilibrium of population densities.

While under controlled conditions the metabolic efficiencies of food utilization of particular species have rigorous meanings from a physiological point of view; extrapolation to the complex dynamics of natural systems can only be done in a general way. The evaluation of ecological efficiencies among trophic levels is really a theoretical way of viewing the interrelationships among organisms. Such an evaluation loses precision as soon as it is

TABLE 16–16 Utilization and Transfer of Energy from Trophic Levels of Several Freshwater Systems, Expressed in g cal cm^{-2} Year^{-1}[a]

Parameter[b]	Lake Mendota, Wisc.				Temperate Cold Spring			Silver Springs, Fla.				Cedar Bog Lake, Minn.			Severson Lake, Minn.		
	Λ_1	Λ_2	Λ_3	Λ_4	Λ_1	Λ_2	Λ_3	Λ_1	Λ_2	Λ_3	Λ_4	Λ_1	Λ_2	Λ_3	Λ_1	Λ_2	Λ_3
Ingestion (I)	118,872	55.4	3.4	0.3	1,095,000	—	—	1,700,000	—	—	—	118,872	19.6	3.4	89,586	57.7	—
Assimilation (A)	501.7	44.3	3.1	0.3	710	2318	242	20,810	3368	383	21	120	16.8	3.1	581.4	27.4	1.2
Egestion (nonassimilation)	118,370	11.1	0.3	0.0	1,094,290	—	—	1,679,190	—	—	—	118,752	2.8	0.3	89,005	30.2	—
Respiration (R)	125.3	16.9	1.8	0.2	55	1746	89	11,977	1890	316	13	30	6.4	1.8	309.1	23.1	0.8
Net productivity (A-R)	376.4	27.4	1.3	0.1	655	576	155	8833	1478	67	6	90	10.4	1.3	272.4	4.3	0.4
Production	55.4	3.4	0.3	0.0	(655)	208	—	—	—	—	—	19.6	3.4	0.0	57.7	—	—
Decomposition losses	321.0	24.0	1.0	0.1	—	—	—	—	—	—	—	70.4	7.0	1.3	—	—	—

[a]From data of Juday (1940), Teal (1957), Odum (1957), and Lindeman (1942), as interpreted by Kozlovsky (1968), and from Comita (1972). The lake systems are planktonic only and do not consider littoral and allochthonous inputs.

[b]I = insolation for producers or food ingestion for organisms.
Egestion includes nonutilized light and egestion.
R includes all energy losses; also urine in higher organisms.
Production refers here to that portion of productivity passed on to next trophic level, some of net productivity being dissipated in decomposition, tissue accumulation, and loss from the system.

extended beyond the solar radiation level of primary productivity. Such analyses can be instructive, however, if one can keep in mind the limitations imposed by their simplicity and by their gross aggregation of differently operating functional units of the composite populations of each level. By way of examples, 5 freshwater systems are compared in Table 16–16, based on mostly early, rather limited data. In a detailed analysis of the numerous types of efficiency relationships developed by many workers, Kozlovsky (1968) used most of these limited data to illustrate changes taking place in the passage from one trophic level to the next. Some of his conclusions follow, the most important of which is that when the net productivity and assimilation of one level are compared to the net productivity and assimilation of the previous level, the efficiency of transfer is practically constant at about 10 per cent. In more physiologically based comparisons of food utilization in animals, in relation to food ingested, assimilation and respiration increase at higher trophic levels. It then follows that in relation to ingestion and assimilation in animals, production decreases at higher trophic levels.

BENTHIC AND FISH COMMUNITIES

The fauna of lakes and streams are rather poorly studied. Investigations have not gone a great deal beyond necessary descriptive analyses of their types and distribution within fresh waters. When applied to natural populations, the strengths of physiologically oriented experimental analyses have not been utilized to nearly the extent that they have been in studies of planktonic communities. The population dynamics and trophic interrelationships of the benthic fauna are poorly understood, with the exception of benthic communities found in running waters (Hynes, 1970).

Much of the difficulty is the result of the heterogeneous distribution of the diverse fauna within lakes in relation to their requirements for feeding, growth, and reproduction. These requirements interact with and are altered by changes in the substratum and overlying water on a seasonal basis, e.g., changes in oxygen content, and in the inputs of living and dead organic matter for food. The benthic organisms either possess adaptive mechanisms to cope with these changes, enter relatively dormant stages until more physiologically amenable conditions return, move, or die. Within their physiological limits, the adaptive capabilities of the benthic animals to the dynamics of environmental parameters and food are basic to their distribution, growth and productivity, and reproductive potential.

Another major problem encountered in effective analyses of benthic animal communities is the difficulty of sampling quantitatively. Substrate heterogeneity leads to a patchy, nonrandom distribution. Sampling methods are discussed in some detail in the methods of the International Biological Program (Edmondson and Winberg, 1971; Holme and McIntyre, 1971). Once samples are obtained, the problems of quantitative separation of the organisms from the sediment and evaluation of cohorts among continuously reproducing populations are severe. Emigration and immigration of the populations of certain groups, especially among the insects, force more elaborate sampling methods. Furthermore, the taxonomy of many groups is confusing and in some cases incomplete. In spite of these problems, careful and detailed analyses have generated quantitative evaluations of certain populations which provide insight into controlling environmental and biotic interactions.

The following brief discussion hardly does justice to the diversity of the benthic animal groups. In the brief space available, we can only point out some of the major groups, giving examples of their population dynamics in relation to lake systems. Important characteristics of the sediments have been discussed elsewhere in relation to oxygen (Chapter 8), redox (Chapter 13), microbial metabolism of organic matter (Chapter 17), and organic composition of different sediments (Chapter 17).

Benthic Animal Communities

Perhaps the least understood groups of benthic animals that occur in massive numbers on and in surficial sediments are the Protozoa and the ostracod crustaceans. General characteristics of the protozoans were mentioned earlier in relation to planktonic populations. In spite of the abundant information that we have on protozoan morphology, physiology, genetics, and behavior, very little is known concerning their population dynamics and contributions to lake productivity in the sediments. A majority of the Protozoa are attached to substrata and they are particularly abundant in habitats of active oxidative decomposition. The importance of ciliate protozoans in sewage treatment facilities and organically polluted streams as a major metabolizing organism of dissolved and particulate organic matter is well-known (Cairns, 1974). The very large diversity of species of Protozoa, which possess wide ranges of tolerance to environmental extremes and varied feeding capabilities (including algae, bacteria, particulate detritus, and other protozoa), coupled with their large population densities on aerobic organic-rich sediments, all point to their significant metabolic role in freshwater systems. Because of their low population biomass, it generally has been assumed that their contributions are small. However, their short generation times and rapid turnover under the more optimal trophic conditions of the sediments suggest that the Protozoa contribute appreciably to the degradation processes of lakes.

Although many factors are involved in the distribution and growth of free-living Protozoa (cf. review of Noland and Gojdics, 1967), oxygen is of paramount importance. Few species can tolerate the anaerobic conditions of the sediments for any appreciable period. Distribution and abundance of chlorophyll-bearing forms are regulated to a large extent in ways analogous to the regulation of algae, by inorganic and organic nutrient availability and light. Among heterotrophic protozoa, the distribution and quality of dissolved and particulate organic matter for food are important. The nutritional requirements of only a few species are known. The significance of vitamins and other growth substances in the population dynamics is just beginning to be appreciated as a component of competition.

Porifera

The freshwater sponges are very uncommon among the predominantly marine Porifera. Although some species in Lake Baikal have been known to grow to magnificent lobed structures of about half a meter in size, most are small, inconspicuous, and morphologically variable. The flagellated chambers create weak currents that pass through the internal chambers and entrap bacteria and particulate detrital material for intracellular digestion. Zoochlorellae, ingested algae that persist inter- and intracellularly in the tissue, can grow and metabolize appreciably within certain sponges (Gilbert and Allen, 1973). Although some symbiotic relationships between the organisms exist, the zoochlorellae are not essential to the survivial of the sponge, at least in *Spongilla.*

Growth occurs by proliferation of the tissue over the substratum, with the addition of new chambers. Under unfavorable conditions, such as in the autumn in the temperate zone, tissue is reduced by partial or total deterioration. Highly resistant resting structures termed gemmules, usually less than a millimeter in diameter, are formed. Gemmules contain undifferentiated mesenchymal cells, and have an outer layer of compacted spicules within a tough sclerotized membrane. They can withstand freezing and desiccation. Germination usually is associated with warmer temperatures above 15°C in the early spring (Harrison, 1974). Syngamic sexual reproduction also occurs in warmer periods of the year.

Among the few animals requiring large amounts of silica for spicule development, sponges are restricted largely to waters of reasonable silica content (>0.5 mg 1^{-1}) (Jewell 1935). Species distribution is also correlated with the calcium content of the water (Jewell, 1939; Strekal and McDiffett, 1974), although causal mechanisms remain obscure and probably are related to food abundance rather than to a direct demand for calcium. Freshwater sponges, which usually occur only in relatively clear, unproductive waters, rarely, if ever, abound as a major component of benthic communities. Their significance in benthic productivity is minor.

Coelenterata

The hydroids and "jellyfish" of the Class Hydrozoa belong to a predominately marine group that is poorly represented in fresh waters. The common *Hydra* occurs only as the tentacled polyp stage, while the rare *Craspedacusta* or freshwater jellyfish undergoes metagenesis with alternation between medusoid and polyp stages common to the marine Hydrozoa. Epidermal cells contain nematocysts or stinging cells that function in capture of prey; it is then moved to the mouth orifice by the tentacles. The hydroids are sessile on substrates and almost completely carnivorous on zooplankters. One species of green hydra (*Chlorohydra*) contains unicellular green algae in a symbiotic relationship, intracellularly in gastrodermal cells. About 10 per cent of photosynthetically fixed carbon is released and assimilated by the hydra (Muscatine and Lenhoff, 1963) and is clearly of nutritional significance.

Development in quiet waters is rapid in spring, and greatest densities usually are found in early summer. Only rarely do hydroid densities become appreciable, however, and their contribution to total benthic productivity is considered negligible.

Flatworms

The only important free-living members of the Platyhelminthes are the turbellarian flatworms. Most members of the group, the tapeworms and the flukes, are entirely parasitic. The free-living flatworms generally possess abundant cilia that assist them in movement over the substrata of shallow lakes and streams. Upon the animal's encountering food organisms or detrital animal matter, the ventral pharynx is extruded through the mouth to incorporate small invertebrates (Young, 1973). Some of the rhabdocoels contain symbiotic zoochlorellae algae in the parenchyma and gastrodermis cells.

Reproduction is either asexual by budding or fission, and/or sexually by sex organs that usually develop over winter with the release of egg capsules in the later winter or spring in permanent bodies of water. Nearly all flatworms are hermaphroditic, with both sex organs occurring in the same individual. Laid eggs are encapsulated or sometimes enclosed in a stalked shell.

Winter eggs are encapsulated more heavily and able to withstand low temperatures but not desiccation. Development is strongly temperature dependent in both the egg and immature stages, although no larval stages are found in most species.

Most flatworms, especially the triclads, are negatively phototactic and occur in shaded areas among debris, rocky substrata, and macrophytes. Among 4 species of lake dwelling triclads studied extensively by Reynoldson (1966), all occupied the same general substratum in the shallow littoral zone and fed on the prey organisms there. Feeding mechanisms of all the species were similar, and wide overlap existed in the type and size of prey eaten by young and adults of the species. The species populations exhibited restricted fluctuations in numbers and lived under conditions of food shortage for long periods of the year. This resistance to death by starvation is accomplished by their ability to resorb and then regenerate tissues, and because of low predation pressure by other organisms.

The distribution and abundance of the individual triclad species are determined primarily by interspecific competion for food (Reynoldson and Bellamy, 1971). Despite the high degree of overlap of food prey species, each triclad was found to feed on a particular kind of prey to a greater extent than others. Thus the competition depended largely on the pattern of distribution and abundance of important types of prey, such as gastropod mollusks, amphipods, and oligochaetes, each of which dominated the ingestion of triclad species.

The abundance of these triclad species was directly correlated with the general productivity of the waters and amount of food available in the littoral zone, which in turn was directly related to chemical parameters of concentrations of calcium and total dissolved matter (Reynoldson, 1966). As a result, minerally poor lakes of low productivity support smaller populations and fewer species through food competition than lakes of higher productivity, in which both the diversity and abundance of food species increase.

Similar relationships were found among stream-dwelling flatworms (Chandler, 1966). The shallow streams also exhibited wide fluctuations in temperature that were shown to influence feeding activity, reproductive capacity, and acclimation capacity differently, and to provide further separation of species distribution. Additionally, triclad abundance was related inversely to velocity of water flow. Mechanical abrasion of shifting substrata by water movement, both in streams of high flow and in the littoral zone of breaking waves, is a generally unfavorable habitat for flatworms and many other benthic animals. Although adults may be able to tolerate shifting substrata, egg or immature stages often cannot.

Nematodes

Of the classes of the pseudocoelomate Phylum Aschelminthes other than the rotifers, the Nematoda or roundworms constitute a significant component of the benthic fauna. Although most of the large number of nematode species are parasitic, the many free-living species are distributed widely in all types of freshwater habitats. The taxonomy of this group is difficult, incomplete, and has not improved the relative paucity of knowledge on the ecology and productivity of nematodes.

The nematode's external proteinaceous cuticle, noncellular and layered, is molted 4 times during its life cycle. This ridged cuticle is interfaced with longitudinal muscle cells that permit active snakelike movements along and through sediments.

Great diversity in feeding habits occurs among the nematodes. Some members are strictly detritivores, feeding solely on dead plant or animal particulate matter, or both. Others are herbivorous, and have specialized mouthparts for chewing living plant material or for piercing and sucking cytoplasm from plants. Mouthparts are more specialized among carnivorous nematodes, enabling them to seize, rasp, and macerate small prey animals such as other nematodes, protozoans, gastrotrichs, tardigrades, and oligochaetes.

Reproduction is variable, with some species being parthenogenetic or hermaphroditic. Most species, however, are syngamic with separate females and males. Fertilized eggs, usually laid outside of the body, develop rapidly (1 to 10 days) and hatch into young which are nearly fully developed except for the reproductive organs. Reproduction and egg laying by adults is more or less continuous. Little is known of the fecundity of free-living nematodes under natural conditions, but it is apparent that a large proportion of net productivity is diverted into reproduction in a normal life cycle. Several hundred eggs with a biomass several times greater than that of the producing female are apparently common. The duration of the life cycle is also variable, from about 2 to 40 days; most species, however, require 20 to 30 days for development and reproduction.

In a detailed study of the nematode populations of the sediments of a typical dimictic, relatively oligotrophic alpine lake, Bretschko (1973) found nematodes to be concentrated between 3 and 4 cm of depth in the sediments. Few penetrated below 6 cm into the sediments. Although the 4 molting instars could not be separated, analyses of the seasonal population dynamics of the many species demonstrated 3 generations per year: (a) The first hatched in September-October and matured in November-December just after ice-cover formed. (b) The second generation was the most productive, and reached maximum abundance and biomass in January (Table 16–17). The peak of relative abundance of adults of this generation occurred in April. (c) Their offspring formed the third generation, with a second lesser maximum in biomass and abundance in May (about 50 per cent of that peak in January). As the ice-cover was lost in May and the lake circulated, for reasons not known, the entire community of nematode populations collapsed. Over a 2 year period, the average population density was 235,000 m^{-2} under ice-cover but only 60,000 m^{-2} during the ice-free period.

Nematode species distribution and biomass production were strongly correlated with the type of substratum (Table 16–17). Over 80 per cent of production occurred in the deeper zone of deep, fine sediments in which gravel was scarce and boulders were absent. Within this zone, 90 per cent of the biomass production was by one species of *Tobrilus*. Causal relations for the observed high productivity in early winter are unclear, but presumably were related to good food conditions after the ice-free period of higher primary productivity and to reduced predation by fishes and possibly by certain predacious cyclopoid copepods.

TABLE 16–17 The Total Production of Biomass (kg dry wt) Seasonally of Nematodes in Vorderer Finstertalersee, Austria, in Different Zones of Substratum, October 1968 to September 1969[a]

Substratum Zone and Its Area	Seasonal Period				Per Cent of Total Biomass
	Oct.–Jan.	Feb.–May	June–Sept.	Oct.–Dec.	
Littoral-slope zone, fine silt; 58,868 m²	1.24	0.16	0.34	1.74	13
Current-angle zone, reduced silt covering gravel and boulders; 47,213 m²	0.24	0.26	0.32	0.82	6
Flat-depth zone, below 20 m, deep fine mud, gravel rare, boulders absent; 51,520 m²	8.64	1.56	0.44	10.64	81
Entire lake, 157,600 m²	10.16	1.96	1.08	13.20	100

[a] From data of Bretschko (1973); dry weight estimated from fresh weight values given by a factor of 20%; totals do not sum parts exactly because of rounding.

Nematodes of the littoral zone can reach very high densities among the attached algae of dense emergent and submersed macrophytes. In a detailed study of the seasonal dynamics of nematode populations in the littoral zones of several lakes, Pieczyńska (1959, 1964) found maxima occurred in May and June, when the populations numbered over one million individuals per square meter of substrate surface, a level brought about by mass reproduction of the dominant species. The same species reproduce throughout the year, although not as intensely. During the other seasons of the year, the nematodes occurred in smaller, more or less constant densities. Variations in the development of the spring maximum from year to year were correlated with warming temperatures of the water and the duration of ice-cover in the littoral.

Species of *Prochromadorella* and *Punctodora* dominated the colonization of littoral substrates, especially shoots of higher aquatic plants, attaching readily by the secretion of an adhesive substance from their caudal glands and by their ability to become entangled in colonies of littoral algae. Littoral ice-cover was found to be especially destructive to the substrata and nematode populations. Wave action caused little mortality and was found to be instrumental to the rapid colonization of newly growing algal and macrophytic surfaces by nematodes of nearby populations. The densities of nematodes were correlated directly with the development and density of attached algae.

Bryozoans

The colonial bryozoan members of the primarily marine Entoprocta are rarely animals of quantitative importance in fresh waters. These sessile forms, however, occasionally form massive colonies that can become conspicuous members of shallow eutrophic lakes and open areas of swamps for brief periods. The microscopic zooids contain ciliated tentacular crowns that project into the water and create water currents. Particles as large as the size of microcrustaceans are directed into the lophophore and mouth. Although hermaphroditic sexual reproduction occurs in summer periods, asexual reproduction permits rapid proliferation on the outer edge of the colonies, which occasionally form amorphous ball-like masses one-quarter of a meter in diameter. These colonies are common to small farm ponds in water less than a meter in depth, and their growth is influenced by fish predation (Dendy, 1963). When protected from fish predation, colonies were branched and more productive; when exposed to fish predation, unbranched, cropped colonies persisted. A large number of invertebrate animals, from protozoans to large insect larvae, particularly chironomids, live within the colonies of bryozoans. Predators of the bryozoans consume these coinhabitants as well as the bryozoan parts.

Extensive studies of 12 species of bryozoans in 122 lakes, streams, and shallow waters of Michigan showed correlations of species distribution with several general limnological parameters such as pH, current and substratum type (Bushnell, 1966; cf. also the detailed review of Bushnell, 1974). Some bryozoan species were associated with particular macrophyte substrata, and others were distributed more ubiquitously. Live colonies were found under ice at water temperatures of <2°C. Growth was often erratic, but luxuriant colonies of *Plumatella* were found to grow rapidly, doubling in 3.4 to 4.9 days in summer and from 4.0 to 7.4 days in the spring. Growth ceased when temperatures decreased below 9°C. A majority of the mature colonies broke up and died after a period of growth, for reasons not completely understood. Predation on actively growing colonies was often very high, especially by caddisfly larvae and snails. *Plumatella* lived from 4 to 53 days; polypide death rate was highest (80 per cent) between 21 and 36 days after statoblast germination, and few lived longer than 6 weeks.

Aquatic Worms

Two major groups of the annelids, or segmented worms, are represented in fresh waters. The first of these, the Oligochaeta or aquatic "earthworms," often form a major component of the benthic fauna, particularly of lakes. The other group, Hirudinea or leeches, is a diverse class of annelids of much biological interest. Although the significance of leeches as a component of total benthic animal productivity is unclear at this time, their predatory habits can materially influence the population dynamics of other benthic organisms, especially oligochaetes. By far the most comprehensive study of the aquatic oligochaetes is the work of Brinkhurst and Jamieson (1971). While largely a global taxonomic work, much of the presently rather limited knowledge on the anatomy, embryology, distribution, and ecology of this group is reviewed in their important book.

Oligochaetes. Oligochaetes are typically segmented, bilaterally symmetrical, hermaphroditic annelids with an anterior ventral mouth and a posterior anus. Size ranges from < 1 mm to about 40 cm (maximum > 200 cm), but most freshwater forms are less than 5 cm in length. Each segment, except for the first and a few terminal segments, bears 4 bundles of setae. Sexually mature Oligochaeta possess an anteriorly located thickened region in the body wall which is glandular and produces a cocoon at oviposition.

Much of the information on oligochaete populations focuses on the geographical distribution, habitat selection, and effects of organic pollution. Several species usually are found in each habitat and differ little in streams and in lakes of the same region. The number of species is often greater in larger lakes, perhaps because of the greater number of different microhabitats common in larger lakes. Correlations between species distribution and the general level of lake productivity are variable, some being restricted to relatively oligotrophic waters while others are distributed widely in lakes of greatly differing productivity, from oligotrophic to extremely eutrophic (see, for example, Milbrink, 1973a, 1973b). Obviously a number of interacting parameters are influential within the physiological tolerances of individual species. Some of these factors are discussed further on.

The vertical distribution of oligochaetes on sediments within the same lake also indicates that a large number of species coexist in the same general substrata. Different species, however, do reach maximal abundance at different depths (Fig. 16–34). Although differences in the composition of the

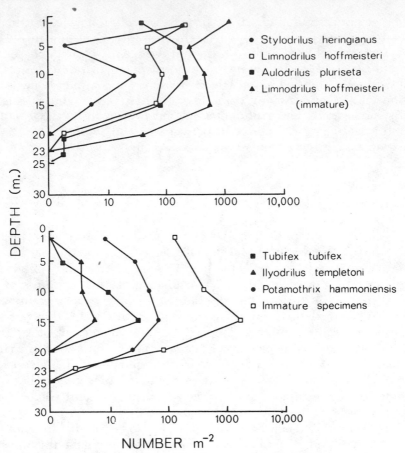

Figure 16–34 Distribution of various species of oligochaetes in relation to depth in hypereutrophic Rostherne Mere, southern England. The lake was devoid of oxygen below a depth of 25m. (Drawn from data of Brinkhurst and Walsh, 1967.)

sediments were not critically analyzed in this study of this eutrophic lake, a number of studies have suggested that the particulate composition and organic matter content of the sediments are correlated with the distribution and abundance of oligochaetes (Timm, 1962a, 1962b). In a large, oligotrophic Italian lake, the number of oligochaetes was generally less in sediments of relatively large particle size (0.11 to 0.12 mm) than in more organic-rich sediments of finer particle size (0.07 to 0.08 mm) (Della Croce, 1955). A similar relationship has been found to exist in relation to the total organic content of the sediments (Brinkhurst, 1967; Wachs, 1967). Experimental studies on the ubiquitously distributed *Tubifex tubifex* verified these general relationships. Therefore nutrition and the availability of food are factors influencing distribution and abundance of oligochaetes.

Oligochaetes ingest surficial sediments containing organic matter of autochthonous and allochthonous origins colonized with bacteria and other microorganisms. Some species actively graze attached microorganisms growing epiphytically on macrophytic vegetation. In a majority of studies on sediment feeding of this group (reviewed in Brinkhurst and Jamieson, 1971), it has been found that the organic carbon, caloric, and total nitrogen contents of the ingested sediments are reduced slightly. However, critical experiments with mixed species cultures failed to demonstrate any reduction in the percentage of organic matter, nitrogen, or caloric content in feces as opposed to that in sediment provided as food (Brinkhurst, et al., 1972). These and other experimental studies on the utilization of bacterial species in sediments indicated that this species of worms is able to selectively utilize different microbial components of the sediment (Brinkhurst and Chua, 1969; Wavre and Brinkhurst, 1971).

As lakes and streams become organically polluted, it is common to find an abundance of tubificid oligochaetes (Brinkhurst and Cook, 1974). Associated with such organic enrichment is acute reduction or elimination of oxygen, which is lethal to a majority of benthic animals. The number of species of tubificid worms also decreases with higher productivity and increased organic pollution. However, as long as some oxygen is available periodically and toxic products of anaerobic metabolism do not accumulate, the combined conditions of a rich food supply and freedom from other competing benthic animals permit rapid growth.

Many oligochaetes and especially the tubificids burrow headfirst into the sediments but leave their tails, containing a majority of gills or respiratory appendages, projecting and undulating in the water above the sediment-water interface. The caudal respiratory movements lessen in frequency and vigor as the oxygen content of the water decreases. In general tubificids are able to adapt respiration in relation to oxygen concentration down to a critical level of the range of 10 to 15 per cent saturation. Hemoglobin of tubificids can be loaded at oxygen concentrations to about a 15 per cent saturation level. Feeding and defecation are then reduced greatly or cease. Many tubificids can tolerate anaerobic conditions for at least a month or longer, but only if they are exposed intermittently to some oxygen since they cannot respire anaerobically for any appreciable length of time. Water movement is apparently an important factor in removal of toxic metabolic wastes, since anaerobic conditions can be tolerated longer in moving situations than when the water is stagnant.

Reproduction in oligochaetes is variable, but even when more or less continuous some periodicity in breeding intensity is common. Oligochaetes lack discrete age classes and production is often approximately continuous. The limited data from natural populations suggest that maturation takes as long as a year in some species, and up to 2 to 4 years in others. Variation is great, however, in the same species, and even from site to site in the same body of water. Breeding intensifies in late winter and spring as temperatures increase above about 10°C. A majority of adult oligochaetes die after sexual reproduction. It is therefore difficult to impose the idea of alternations in periodicity of species abundance as an interacting mechanism permitting coexistence of species. Selective predation by other animals as a factor in coexistence has been studied very little in natural populations. However, it is clear that selective breeding and metabolic adaptations by oligochaetes permit cohabitation. For example, when 3 species were combined, population respiration of mixed cultures decreased about 35 per cent, while lowering the population density of a single species had no such effect (Brinkhurst, et al., 1972). More energy was used for growth when the 3 species were mixed than when they were maintained separately in single-species culture, and the rate of respiration lessened in the mixed populations. Isolated single species spend more time burrowing and searching for their preferred food, which is feces of other oligochaete species, rather than feeding (Chua and Brinkhurst, 1973; Brinkhurst, 1974). Such interspecific interactions probably are causal factors in the frequently observed clumped distribution of several species of oligochaetes.

In the deep profundal sediments of the Danish eutrophic Lake Esrom, some tubificid *Potamothrix hammoniensis* can first reproduce at 3 years of age, but most must wait until they are 4 years old (Jónasson and Thorhauge, 1972). The life span of this species can reach 5 or more years. Mature specimens in this lake laid eggs in May-July, and this process was related to moderate increases in sediment temperatures. From 2 to 13 embryos hatched from the cocoons after about 2 months in August. Growth was most rapid in the autumn and spring, and no weight increase occurred during the summer period of stratification. Numbers of this species varied between 10,000 and 25,000 m^{-2} over a 6-year period. In spite of these large numbers, biomass varied between 0.75 and 1.80 g dry weight m^{-2}, usually greater than 1.0 g m^{-2}. These values are similar to those observed in other lakes, for example as shown in Table 16–18.

Mortality in the Lake Esrom populations was clearly related to death after reproduction. However, predation by the dipteran larvae of *Chironomus anthracinus*, studied in great detail in this lake, caused significant reduction of the oligochaete populations. The reproduction of the oligochaete populations was depressed severely every second year as a result of the 2-year life cycle of the *Chironomus* larvae (see also Loden, 1974).

TABLE 16–18   Average Population Densities and Biomass of 29 Species of Oligochaetes of Chud-Pskov Lake, Estonia, USSR[a]

	Average Population Density (No. m^{-2})	Per Cent of Total Density of Benthic Animals	Biomass[b] (g m^{-2})	Per Cent of Total Biomass of Benthic Animals
Total lake	370–1400	—	—	—
June	859	47	1.307	23
Fine grained silt of profundal	287	58	0.433	44
Silt with fibrous particulate detritus	1452	62	2.776	16
Sandy with little silt; no vegetation	1015	22	1.409	36
Upper littoral	1185	20	0.133	6

[a]From data of Timms, 1962a.
[b]Assumed to be wet weight; methods not stated (see Jónasson and Thorhauge, 1972, for discussion of errors involved with preserved vs. fresh wet weights).

Oligochaete species are separated within the sediments. Naidid oligochaetes are concentrated at the sediment-water interface, rarely more than 2 to 4 cm below the surface (Milbrink, 1973c). Tubificid oligochaetes are most dense between 2 to 4 cm of sediment depth, occasionally as deep as 15 cm. The movements of benthic organisms within the sediments are important because of the possibility that their activity can disrupt the oxidized microzone of the sediment-water interface and thereby alter rates of chemical exchange between the sediments and overlying water. Experiments with tubificids have shown that the activity can alter the stratigraphy of surface sediments (Davis, 1974), which is of some interest in micropaleontological work (Chapter 18). As will be discussed in the following chapter, it appears that the intensive microbial activity at the sediment-water interface is sufficient to reestablish redox conditions quickly or almost immediately after such disturbance.

Hirudinea. An excellent general summary on the structure, physiology, and ecology of leeches is given by Mann (1962) and especially Sawyer (1974). In general leeches intermittently consume blood and body fluids of vertebrates greatly in excess of their body weight as ectoparasites. They then enter a period of fasting, during which the consumed food is utilized over a period that can be as long as 200 days, with progressive loss of body weight. Other leeches are predatory on invertebrates such as oligochaetes, and entirely consume their prey (Elliott, 1973a).

Reproduction is usually initiated in the spring or early summer, and apparently is governed by temperature, the density of the populations, and age. Many leeches breed at 1 year of age and die after breeding, although some pass through 2 generations in 1 year (Tillman and Barnes, 1973). If they do not breed in the first year, other species live 2 years, overwinter and breed in the second, and die soon thereafter (Elliott, 1973b). In the first 3 months after hatching, mortality of the young from predators (carnivorous invertebrates, amphibia, many fish, and birds) is heavy (ca 95 per cent; see Table 16–19). Thereafter mortality lessens, but is still severe. When population densities become high, size can be regulated by high mortality of the eggs, which are consumed by adult leeches (Elliott, 1973a).

The abundance of leeches is highly variable among different habitats of lakes and streams. A general direct correlation exists between leech abundance and lake productivity. This relationship probably is associated with the increasing diversity of substrate types among the macrophytes and sediments, with correspondingly greater amounts of invertebrate food sources for the predacious leeches and birds, and vertebrates for the blood-consuming leeches (Sander and Wilkialis, 1972).

Estimates of the production rates of leeches are extremely rare. The figures cited in Table 16–20 are for a moderately productive stream, and point out some of the population characteristics. Production of this leech species varied considerably between years and year-classes. Production was always greatest in the first year of the life cycle, even though the mean biomass was very similar in both years of the life cycle. The rate of production was most rapid during March to July, after which production declined sharply.

Overwintering leeches of the species *Helobdella stagnalis* were found

TABLE 16–19   Life History and Mortality of Two Common Freshwater Leeches in England[a]

	Glossiphonia complanata	Erpobdella octoculata
Breeding period	Mar.–May	June–Aug.
Average number of young per parent	26	23.5
Mortality in first 6–9 months (%)	97	91
Proportion of year-group breeding at 1 year old (%)	70 (or less)	87
Mortality from 6 to 18 months (%)	33	49
Proportion of year-group breeding at 2 years old (%)	100	100
Mortality from 18 to 30 months (%)	84–88	69
Proportion of total population surviving to 3 years (%)	5–6	4

[a]After Mann, K. H.: Leeches (Hirudinea). Their Structure, Physiology, Ecology and Embryology. Oxford, Pergamon Press, 1962, from several sources.

to reproduce in May in a shallow, eutrophic reservoir of South Wales (Learner and Potter, 1974). These postreproductive adults then died. The spring generation grew rapidly, and about one-third of this generation reproduced in July. The nonbreeding portion of the spring generation matured later and reproduced at the same rate as the summer generation in the following spring. The mean biomass was 96 mg (dry) m^{-2} in 1970, and 393 mg (dry) m^{-2} in 1971. Production was estimated to be 310 mg (dry) m^{-2} yr^{-1} in 1970 and

TABLE 16–20   Production, Mean Biomass and the Ratio of Production: Biomass (P/B) for Populations of the Leech Erpobdella octoculata in an English Stream[a]

Population	Production g Wet Weight m^{-2}	Biomass g Wet Weight m^{-2}	P/B Ratio
1964 year-class			
Second year	8.25	10.35	0.80
1965 year-class			
First year	23.42	10.02	2.34
Second year	8.52	11.99	0.71
Both years	31.94	11.01	2.90
1966 year-class			
First year	12.34	4.61	2.68
Second year	4.60	5.29	0.87
Both years	16.94	4.95	3.42
1967 year-class			
First year	6.54	2.47	2.65
Annual estimates			
1966	33.46	20.96	1.60
1967	19.77	14.36	1.38
1968	11.63	7.82	1.49

[a]After data of Elliott, 1973b.

1190 mg (dry) m⁻² yr⁻¹ in 1971, with production to biomass ratios of 3.2 and 3.0 in the respective years.

Ostracods

The ostracods, small bivalved crustaceans usually less than one millimeter in size, are widespread in nearly all aquatic habitats. Their small size, distribution in the surface sediments with few planktonic species, and difficult taxonomy have all contributed to a very poor understanding of their ecology, population dynamics, and productivity in fresh waters. Their valves, which superficially resemble clam shells, are held apart when undisturbed. Between the open valves protrude the cladoceran-like crustacean appendages used for locomotion. Most ostracods move about on the sediments by beating movements of the antennae and the caudal ramus. Ostracods are omnivorous and, like the cladoceran microcrustacea, feed on bacteria, algae, detritus, and other microorganisms by means of filtration. Very little is known of the effects of ostracods on the turnover and metabolism of benthic microflora. The large population numbers of ostracods suggest that their role in the metabolism of surficial sediments is underestimated.

Most ostracod reproduction is parthenogenetic for much of their life cycle. In some species, males have not been found; in others sexual reproduction with both females and males is common. Egg development is variable from days to months, and is strongly temperature dependent. The larva hatches as a nauplius with a reduced number of appendages, and undergoes a series of growth and molting stages, usually 8 molts to reach the ninth, adult stage, during which morphology becomes more complex and appendages develop.

Little is known of the population dynamics of ostracod populations, although many exhibit distinct seasonal periodicity. Some species exhibit a single generation per year, others 2 or 3 per year. The detailed study of the reproductive potential and life history of a species of *Darwinula* in a deep, dimictic temperate lake is illustrative of some of the characteristics of this group (McGregor, 1969). *D. stevensoni* was abundant throughout the littoral sediments of this mesotrophic lake, and reached a maximum density at a depth of 6 m throughout the year. The majority occurred between depths of 3 and 6 m and densities decreased markedly between 6 and 9 m. Only a few individuals were found below a depth of 9 m, and none below 12 m. More than 95 per cent of the adults and juveniles present in the sediments occurred in the upper 5 cm of sediment.

The reproductive period of this *Darwinula* species in this Michigan lake began in May and was effectively completed by October (Fig. 16–35). The number of young per individual increased from a maximum of 3 in May to 15 in August. Although the reproductive potential of a given individual was correlated strongly with temperature and varied somewhat with depth, most adults produced only one brood per year of about 11 young per individual at 3 to 6 m of depth (13 at 9 m).

Eggs of *Darwinula* are released and develop within the carapace of the

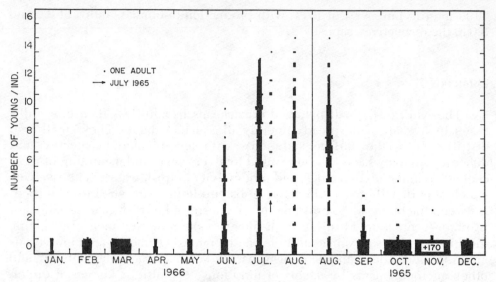

Figure 16–35 Annual reproductive cycle and reproductive potential of the benthic ostracod *Darwinula stevensoni* in Gull Lake, southwestern Michigan. (From McGregor, D. L.: The reproductive potential, life history and parasitism of the freshwater ostracod Darwinula stevensoni (Brady and Robertson). *In* Neale, J. W., ed., The Taxonomy, Morphology and Ecology of Recent Ostracoda. Edinburgh, Oliver & Boyd, 1969.)

parent (McGregor, 1969). Juveniles produced by first and second reproductive season adults are released during the summer and early autumn, and overwinter as juveniles. These young mature and reproduce the following summer. Surviving members of this age class overwinter as adults during the second winter and reproduce again the following summer. Most adults of the second reproductive year die following release of young, and are replaced by overwintering juveniles entering their first reproductive period. Thus, a nearly complete turnover of the adult portion of the population occurs each year.

Although much of the *Darwinula* population was parasitized by ectoparasitic rotifers, this did not appear to affect the reproductive potential. Effects on mortality by predation of other benthic organisms and bottom-feeding fish are unclear in this and other species. Evaluations of densities of ostracod populations are few, and range from an average of 57 m^{-2} in oligotrophic Great Slave Lake in the Canadian subarctic (Tressler, 1957) to about 10,000 m^{-2} of the *Darwinula* of mesotrophic Gull Lake. Much higher population densities of >50,000 m^{-2} are common in more productive lakes.

Other Crustacea

Of the many members of the malacostracean crustaceans, 4 groups have distinct freshwater representatives of some interest and importance. All have a definite and fixed number of body segments.

Mysids. The mysids or opossum shrimps, reaching a maximum length of about 3 cm, resemble crayfish. Their appendages, however, are elongated, contain abundant setae, and generally are greatly modified for active swimming. Their mode of feeding is by filtration of small zooplankton, phytoplankton, and particulate organic matter via setose appendages, particularly 2 pairs of maxillipeds. Water currents caused by these movements carry food aggregations anteriorly to the mouth. Gut analyses of *Mysis relicta* showed that at night adults were voracious predators, feeding on *Daphnia*, other cladocerans, and the rotifer *Kellicottia* during their vertical migration (Lasenby and Langford, 1973). *Mysis* had an assimilation efficiency of about 85 per cent when feeding on *Daphnia*, but this efficiency dropped to zero when feeding on debris from mosses growing in the lake.

The freshwater mysids are distinctly stenothermal, coldwater forms, and in stratified lakes are restricted to hypolimnetic strata of less than 15°C. Reproduction is restricted to colder periods of autumn, winter, and early spring. Eggs number to 40 per clutch, and are kept within the female brood pouch (marsupium). After hatching, juveniles are retained within the female for considerable periods of time (up to 3 months); hence the name opossum shrimps.

A comparison of the life history, growth, and respiration of *Mysis relicta* in an arctic lake (Char Lake, Northwest Territories, Canada) and a temperate lake (Stony Lake) of southern Ontario is most instructive (Lasenby and Langford, 1972). *Mysis* of the arctic lake take 2 years to reach maturity, while those of the temperate lake mature in one year. *Mysis* from both lakes exhibited a trimodal distribution of size classes. In the arctic lake, the population consisted of 8 to 9 month old immature mysids, 20 month old adult males and females, and 32 month old females. In contrast, in the temperate lake the cohorts consisted of 2 month old immatures, 1 year old females, and a few 2 year old females. Respiration measurements showed no differences between *Mysis* from the 2 lakes over its environmental temperature range. Over a 2 year period from hatching to maturity, the energy used by the arctic population in growth and reproduction was approximately 68 per cent of that used by the temperate mysids in growth and reproduction over a 1 year period from hatching to maturity (Table 16–21). Of much interest is the finding that the amount of energy (200 cal) used by the mysids for growth and respiration in order to become reproducing adults was about the same for the arctic population in 2 years as for the temperate population in 1 year. Similarly, size-frequency distributions of *Mysis relicta* of the Superior, Huron, and Ontario basins of the Great Lakes showed differences in life cycles in the 3 lakes (Carpenter, et al., 1974). In the colder Lake Superior the generation time appeared to be 2 years, while in lakes Huron and Ontario the generations matured in 18 months.

Lacking gills, mysids respire by exchange through a thin carapace. The high respiratory demands for oxygen and cold thermal requirements restrict their distribution to oligotrophic lakes in which the oxygen content of the hypolimnion is not reduced appreciably.

Members of the 2 major genera, *Mysis* and the brackish water *Neomysis*, all exhibit distinct, rapid diurnal migrations. During the day the mysids are on the sediments or in the strata immediately overlying the sediments (Beeton,

TABLE 16–21 Energy Budgets in Calories per Individual Female Mysis relicta from Two Canadian Lakes, Assuming an Assimilation of Efficiency of 85 Per Cent[a]

Energy Budget	Arctic Char Lake	Temperate Stony Lake
Ingestion[b]		
First Year	62	245
Second Year	180	510
Total	242	755
Growth		
First Year	10	39
Second Year	18	39
Total	28	78
Reproduction		
First Year	0	14
Second Year	8	14
Total	8	28
Respiration		
First Year	43	156
Second Year	127	376
Total	170	532
Egestion		
First Year	9	36
Second Year	27	81
Total	36	117

[a] Based on data of Lasenby and Langford (1972).

[b] Based on ingestion = growth + reproduction + respiration + egestion, when $0.85 \, I = G + R$.

1960; Herman, 1963). *Mysis relicta* ascends when surface light intensities decrease to about 160 to 10 lux, and migrates into or below the metalimnion, in which temperatures are well below 15°C. This species rarely migrates into warmer surface waters. Descent occurred in populations of the Great Lakes when surface light intensities were increasing from 0.01 to 0.1 lux. Rates of the upward and downward migrations were similar at 0.5 to 0.8 meters per minute over distances of at least 75 meters, the fastest rates known among small invertebrates (Table 16–11). *Mysis relicta* is very well-adapted to the photic environment of these relatively clear, oligotrophic lakes in which light in the range of 490 to 540 nm penetrates the deepest. The peak photosensitivity of *Mysis* is to light at 515 nm and a threshold response of at least 0.00009 lux (Beeton, 1959), well within the light intensities at 100 m in lakes of low extinction coefficients even late in the daylight period. Langsby and Langford (1972) indicate that the respiratory expenditure of energy for the daily migration is very small, about 3 per cent of the total over a life span of 2 years.

Isopods. The isopods or sow-bugs are largely marine or terrestrial, but occasionally constitute significant members of the benthic organisms of lakes and streams. These small organisms (< 2 cm) are flattened dorsoventrally and well-adapted to substrata exposed to water movements of streams or the littoral of lakes. Their 7 pairs of walking legs are well-developed. The first is modified for grasping particles eaten in their omnivorous diet of plant and animal matter. Isopods are generally not active predators on other benthic fauna.

Reproduction is more active in periods of warmer temperatures, but otherwise is similar to that mentioned for the mysids. The number of eggs per female is fairly high (to several hundred), and both eggs and young are retained in the brood pouch for about a month. Relatively little is known about their life cycle and population dynamics; generation time is about 8 to 12 months but quite variable. Reynoldson (1961) has shown a strong direct correlation between the distribution of *Asellus* and the concentrations of calcium and total dissolved matter. This isopod was distinctly absent in soft, unproductive waters (< 5 mg Ca l^{-1}; < 70 mg l^{-1} of total dissolved matter). Isopods are common among the benthic invertebrates of saline inland waters and can tolerate high concentrations of salinity (see, for example, Eriksen, 1968; Ellis and Williams, 1970).

Decapods.   The decapod Crustacea, of which the lobster is a familiar example, is again largely a marine order. The few freshwater crayfish and shrimps are characterized by their approximately cylindrical body, heavily sclerotized with a translucent shell, and laterally compressed rostrum. Their 19 pairs of appendages include well-developed antennae and 5 pairs of large walking legs, the first 3 of which are clawed and the first greatly enlarged with a strong pincer chela used for crushing food. While the few shrimps are more or less continuous swimmers, the crayfish move slowly over the sediments and can move backward quickly by flexing their abdomen. Crayfish are omnivorous but primarily herbivorous on algae and larger aquatic plants, and occasionally are scavengers (Momot, 1967a).

The population dynamics and productivity of the common crayfish of temperate lakes have been studied in some detail (Momot, 1967a) and serve as an example for this group. In the crayfish population of a hardwater lake, exhibiting a year-class fluctuation in age structure, males had a higher growth rate than females. After the second year, mortality rates for females were greater than for males. Both sexes matured after the midsummer molting in the second year, then mated until early autumn, and eggs were laid the following spring. The average number of ovarian eggs carried by females in the spring, a measure of the reproductive capacity, was 58 per cent of the eggs carried in the preceding fall. Within the maximum life span of 3 years of this species, *Orconectes virilis*, 2 year old females produced most (92.5 per cent) of the eggs. Newly hatched young left females in the spring and early summer and remained in shallow water in this lake. Adult females then molted and migrated to deeper waters (3 to 7 m) and most remained there all summer. Migration of males, somewhat later, then followed to deeper water. Population size appears to be regulated by natural mortality at molting and cannibalism of both males and females on each other at high population densities. Predation by fish was not an important population control mechanism in this lake (Momot, 1967b). A number of pollutants of lakes and streams, particularly those that reduce oxygen content or alter the substratum, cause changes in the population distribution or seasonal migrations (Hobbs and Hall, 1974). The successive year-class estimates of the crayfish biomass of this hardwater lake are given in Table 16–22. Seasonal changes in biomass are small because of compensatory changes among age groups. The estimated net production of the crayfish population was 127.1 kg ha^{-1} between the summer of 1962 and that of 1963.[7] These values are about intermediate within the range of biomass values found among the few crayfish populations studied in other waters.

Amphipods. The amphipods or scuds, also chiefly marine, are represented in fresh waters by a few important species. With the exception of *Pontoporeia,* discussed further on, the amphipods are benthic on the sediments. Most species are small (5 to 20 mm), with a laterally compressed, many-segmented body. At the base of their 7 pairs of thoracic legs many have gills that, like the lateral gills on some species, are exposed to currents of water created by beating of pleopod appendages on the abdomen. Amphipods are generally omnivorous substrate feeders that consume bacteria, algae, and particulate detritus of animal and plant remains; only rarely are they predacious on living animals (Minckley and Cole, 1963; Marzolf, 1965a; Hargrave, 1970a).

In a detailed study of the population dynamics of the amphipod *Hyalella azteca* in a eutrophic lake of southeastern Michigan, as well as related effects such as natality, growth, and mortality of temperature, light, and animal densities, Cooper (1965) was able to estimate production rates and biomass turnover of the natural populations (Table 16–23). The amphipod popula-

---

[7]A corrected value; see footnote of Table 16–22.

TABLE 16–22   Successive Estimates of the Annual Biomass (in kg) of the Crayfish Populations of West Lost Lake, Michigan[a]

Age Group	Sex	Summer 1962	Total ♂♂ and ♀♀	Spring 1963	Total ♂♂ and ♀♀	Summer 1963	Total ♂♂ and ♀♀
0	♂♂	8.7	–	6.2	–	4.2	–
0	♀♀	9.7	18.4	6.2	12.4	5.2	9.4
I (2nd year)	♂♂	36.2	–	23.7	–	41.9	–
I	♀♀	24.5	60.7	18.9	42.6	27.8	69.7
II (3rd year)	♂♂	49.7	–	25.9	–	39.0	–
II	♀♀	8.0	57.7	16.5	42.4	10.0	49.0
III (4th year)	♂♂	9.8	–	25.7	–	7.7	–
III	♀♀	3.7	13.5	1.6	27.3	1.8	9.5
Total, kg		150.3		124.7		137.6	
kg ha^{-1}		100.3		88.1		91.8	

[a]Data of Momot (1967a) with corrections for computational errors that appeared in the original publication (Momot, personal communication). West Lost Lake has an area of 1.5 ha and a maximum depth of 13.4 m (mean depth, 8.5 m).

TABLE 16–23   Estimated Mean Production Rates and Numerical Turnover of the
Amphipod Hyalella azteca Population from 4 Stations of Sugarloaf Lake,
Southeastern Michigan, May to October, 1962[a]

Age Group	Number of Animals m^{-2}	Mean Dry Weight (μg)	Numerical Yield m^{-2} day^{-1}	Mean Population Density (mg m^{-2})	Per Cent Numerical Turnover day^{-1}
11–13 antennal segments	1259.4	15.3	39.8	19.3	3.2
14–16 antennal segments	1383.0	35.6	24.8	49.2	1.8
New adults	2044.6	133.3	71.8	272.5	3.5
Old adults	183.8	200.0	4.8	36.8	2.6
Total	4870.8	–	141.2	–	2.9

[a]Compiled from data of Cooper (1965).

tions showed positive intrinsic rates of increase at temperatures above 10°C
with maxima between 20 to 25°C. Reproduction began when water temperatures reached 20°C in the spring. Estimates of mortality rates from size and
age structures of the population indicated that a size-specific mortality factor
was operating on the large, reproductively mature amphipods during the
summer months. It was established that the highly size-specific feeding behavior of the yellow perch was taking a significant proportion of the adult
amphipods. A similar situation was found in the benthic invertebrate populations of shallow experimental ponds (Hall, et al., 1970), in which the fish
(*Lepomis*) preyed upon larger particles, selectively removing the larger
adult stages of *Hyalella* and the terminal aquatic stages of many benthic insects. There is some evidence that heavy predation pressure by fish can lead
to inheritable selection for smaller sized individuals of adult amphipods
(Strong, 1972).

In spite of the relatively low biomass levels of the *Hyalella* populations,
the rapid growth rates and short generation times (33 days to maturation at
25°C; 98 days at 15°C) result in a mean production rate of about 0.13 kg ha^{-1}
day^{-1}, which is significant in comparison to other components of the benthic
fauna. The observed percentage turnover rate of about 3 per cent day^{-1} is
close to the maximum of 4 per cent day^{-1}, estimated from experimental studies, that the amphipod population could maintain over the course of the
summer at temperatures above 20°C with any degree of population stability.

The extensive studies of *Hyalella azteca* in shallow (z_m = 5 m) Marion
Lake, British Columbia, by Hargrave (1970a, 1970b, 1970c) demonstrated that
growth, density, and body size of this deposit-feeder depended on the quantity of epipelic algae and sediment microflora. The highest concentrations of
sedimentary chlorophyll and microflora, as well as the lowest concentrations
of nondigestable lignin-like organic material, occurred in the upper 2 cm of
sediment, which was also the limit of the vertical distribution of the *Hyalella*
in the sediments. Increased growth of *Hyalella* during June was independent
of temperature in Marion Lake and correlated closely with the increased
rates of epipelic primary productivity. Egg production began in May as
growth rates increased, and the maximum density of *Hyalella* was reached
in August.

TABLE 16–24   The Estimated Energy Budget of Adult Amphipod Hyalella azteca of 700 μg Dry Weight at 15°C Feeding on the Surface Sediment of Marion Lake, British Columbia[a]

Energy	Calories Amphipod^{-1} Hour^{-1}
Ingestion	0.0525
Production (growth, molts, and egg production)	0.0012
Respiration	0.0039
Egestion	0.0430
Excretion (soluble organic matter)	0.0029

[a]From data of Hargrave (1971). The mean caloric value of adult *Hyalella* was 3850 cal g^{-1} ash-free dry weight; therefore a 700-μg amphipod contains 2.7 calories.

Bacteria and algae, except for blue-green algae, were assimilated with about a 50 per cent efficiency; cellulose and lignin-like compounds were not assimilated at all by *Hyalella* (Hargrave, 1970a). Overall, sediment and associated microflora were assimilated with a 6 to 15 per cent efficiency. In the estimated energy budget for adult *Hyalella* at 15°C in Marion Lake, 49 per cent of calories assimilated were respired, 36 per cent were lost as soluble excretory products, and 15 per cent were accumulated as growth, egg production, and molts (Table 16–24). The fecal pellets produced were colonized rapidly by heterotrophic microorganisms and the dissolved organic compounds excreted by *Hyalella* significantly increased the rate of recolonization of the fecal material (Hargrave, 1970c). Epipelic algal production was stimulated somewhat by the feeding activities of the amphipod at natural densities, even though less than 10 per cent of the daily microfloral production was required to supply the energy necessary for observed rates of amphipod growth, respiration, and egg production. The mechanism of stimulation of algal productivity on the sediments is unclear. It may emanate directly from inorganic-organic nutrient sources from the amphipod excretions, indirectly from increased metabolism of the heterotrophic microorganisms of the sediments, or both. Alternately, the mechanical disruption of the oxidized microzone of the sediment-water interface may be instrumental in encouraging release of certain reduced, and more soluble, nutrients into the proximity of the epipelic algae. While these compounds could not migrate far into overlying aerobic water before being oxidized and decreasing in solubility, the algae directly at the interface could benefit from such disturbance.

Pontoporeia affinis is exceptional among the amphipods in that it migrates extensively into the pelagic zone at night. Having some similarities with the behavior and requirements of the mysid *Mysis* discussed earlier, *Pontoporeia* is stenothermally restricted to relatively oligotrophic cold waters and migrates only into the upper hypolimnion and metalimnion, in which temperatures are usually less than 15°C (Marzolf, 1965b; Wells, 1968). Only a portion of the predominately benthic population migrates into the water. During daytime, *Pontoporeia* is largely on and in the sediments or in the immediately overlying water.

Pontoporeia is a burrowing amphipod and a deposit-feeder on the surficial sediments. The distribution of this amphipod, which is the predominant macrobenthic invertebrate of the upper Great Lakes, in Lake Michigan showed no significant correlation with depth, particle size, or organic matter content of the sediments (Marzolf, 1965a). Distribution of *Pontoporeia* was, however, directly correlated with the number of bacteria in the sediments. In experiments with different sized sediment particles, *Pontoporeia* selected sediments with a particle size smaller than 0.05 mm that contained higher bacterial densities and organic matter.

With the bacteria of the sediments as the predominant food source of *Pontoporeia* and no evidence for planktonic feeding, the adaptive significance of the nocturnal vertical migration of a small portion of the population is unclear. Marzolf (1965b) suggests that the migrations have adaptive value in maintaining genetic continuity in otherwise isolated benthic populations by dispersal and possibly copulation with populations from other areas within the community.

Mollusks

The freshwater Mollusca are separable into 2 distinct groups, the univalve snails (Gastropoda) and the bivalve clams and mussels (Pelecypoda). Among the univalve mollusks, the shells of snails generally are spirally coiled, while those of the few freshwater limpets are conically shaped. The head is distinct, with a pair of contractile tentacles and a ventral mouth with a chitinous, sclerotized jaw and a chitinous internal radula containing numerous transverse teeth. The radula is extended from the mouth and moves back and forth rapidly, scraping and macerating food particles. Respiration in the snails occurs by gills (Prosobranchia) in many aquatic forms, and by pulmonary cavities or "lungs" in the pulmonate snails. The latter group can stay submersed for long periods of time. Some of the most common pulmonate snails are able to stay submersed for a very long time or for their entire life cycle without filling their pulmonary cavities with air. Cutaneous respiration through the body membranes is common to all freshwater snails (Ghiretti, 1966). Even though the lung is retained in pulmonate freshwater snails, it is of secondary function and is used only when the oxygen concentration drops to low values. Then the snails come to the surface to breathe by means of movement or flotation. Other pulmonates have their lung filled with water through which gaseous exchange occurs, much the same as it does with a gill. Locomotion in snails is by muscular movements of the ventral surface of the body, the "foot."

The body of clams or mussels is enclosed in 2 symmetrical, opposing shell valves and lacks a head, tentacles, eyes, jaws, and a radula. The body is enclosed by membraneous tissue, the mantle, which secretes the shell valves. At the posterior end of the clam, the mantle has 2 openings, the siphons. The lower ciliated siphon draws water into the body cavity which aerates the gills and carries in food particles. Water then exits from the upper siphon. The muscular foot can be extended from the valves in front of the clam, implanted in the sediment, and then contracted to draw the animal forward. The food of the Pelecypoda clams consists primarily of particulate detritus and microzooplankters of the sediments (see review of Fuller, 1974). There is no evidence to support the view that clams utilize living phytoplankton; living and active diatoms apparently pass through the alimentary canal unaffected.

The life cycle of freshwater snails in temperate regions, particularly the smaller species, tends to be annual (Harman, 1974). One reproductive period may occur in the spring or fall, or 2 or more reproductive periods may occur throughout the summer, during which the original cohort is replaced or supplemented. Some species overwinter as juveniles or adults and reach maturity the following spring or summer. Larger species tend to have life cycles that extend over 2, 3, or even 4 years. A few examples of the growth and population characteristics of some freshwater snails illustrate these variations.

In the common pond snail *Physa gyrina* of the mid-temperate region, copulation occurred in April (DeWitt, 1954). Egg masses adhere to substrata and the number of eggs laid, about 100, varies with the size of the snail. After about a week of embryological development at 20 to 30°C, the young snails hatch and grow very rapidly, attaining most of their adult size of ca 12 mm in about 8 weeks. Beyond this period, growth is very slow for the remainder of the 12 to 13 month life span. Growth in *Physa* is not determinate but continues throughout the life of the individual.

Analyses of the population size classes of the pulmonate snail *Lymnaea* that lives along the shoreline littoral regions of fresh waters showed a similar pattern, in which growth of overwintering individuals was linear and very rapid in the spring and early summer (McCraw, 1961, 1970). Little growth occurred before mid-April or after early July, when most of the production is shunted into reproductive development. Small individuals overwinter and almost all of the spring and summer population is derived from snails ovipositing in late summer or autumn of the previous year. Although changes in population densities of *Lymnaea stagnalis* did not affect rates of food consumption, egg production under experimental conditions was significantly lower at high population densities (Mooij-Vogelaar, et al., 1973). Eisenberg (1966) found in both natural and experimental populations of *Lymnaea elodes* that both food quantity and quality limited the population densities. Adult fecundity increased some 25-fold and numbers of young increased 4 to 9-fold when abundant food of high quality was available. Mortality among natural populations of young snails was very high (93 to 98 per cent), especially by dipteran larvae, but food limitations affecting fecundity were considered to be the primary mechanism affecting the final population structure.

The efficiency of assimilation (expressed as the percentage of ingested organic carbon that is assimilated) of the stream snail *Potamopyrgus* varied from 3.7 to 9.0 per cent, depending on the type of sediment used as food (Heywood and Edwards, 1962). The average was 4 per cent for adult snails, in which about 7 μg organic carbon was assimilated per snail per day. In contrast, the tropical snail *Pila*, feeding on the macrophyte *Ceratophyllum*, had much higher absorption rates, possibly[8] greater than 50 per cent (Vivekanandan, et al., 1974). Utilization efficiency of the plant food decreased as the supply of food increased.

The sperm of male clams is shed into the water and drawn into the female mantle cavity by the normal water current of the siphon system. Fertilized ova are incubated in portions of the gills, in which mortality is very high from bacteria and protozoans. The mature larvae hatch after periods of a few months to a year as glochidia, with 2 chitinous valves bearing ventral hooks in some species. The glochidia are discharged in large numbers, fall to the sediments, and die within a few days if they do not come into contact with a suitable fish host. On contact the glochidia attach to surface membranes or to the gills of fish; some species of clam are specific to a particular species of fish. During this parasitic stage, lasting from about 10 to 30 days, internal development occurs although there is little change in size while the glochidia is encysted on the host fish. Juveniles break encystment and fall to the sediments. Development ensues if substrate and food conditions are within tolerable limits. Large clams live as long as 15 years, while the smaller "fingernail clams" (Sphaeriidae) have a longevity of about a year or slightly longer. Mortality is exceedingly high in the egg, larval-glochidial stages, and juvenile stages. The adult populations and their distribution can be modified greatly by fish predation, certain birds, and a few mammals such as the muskrat (cf. Fuller, 1974).

Study of the distribution of 5 species of sphaeriid clams of the common genus *Pisidium* showed that 4 coexisted among the more heterogeneous substrata of Lunzer Untersee, Austria (Hadl, 1972). The profundal sediments were dominated by a single species. The littoral species are exposed to large seasonal fluctuations in temperature and are gravid in spring, release young in the summer during which rapid growth occurs, and then the new generation overwinters to reproduce the following spring. In contrast the deepwater species, exposed to relatively constant temperatures throughout the year, exhibited no definite reproductive cycle. Embryos were found throughout the year in mature females. The larger clams, for example the *Anodonta*, move seasonally on the sediments of different water depths (Burla, 1971). Species of this genus were found to move uphill to the upper littoral in spring and early summer until late in the year. In autumn and early winter most returned to deeper water. These movements were not correlated with temperature but rather with photoperiod, which may be related more to food availability.

In a very detailed study of the benthic fauna of a eutrophic, lowland lake of southern Norway, Ökland (1964) studied the mollusk fauna in particular (13 species of gastropod snails and one dominant bivalve clam *Anodonta piscinalis*). The average quantitative data of Table 16–25 are included to point out a typical distribution with depth of the sediments. The greatest quantity of gastropods occurred in the zone of submersed plants at a depth of 1.5 m.

---

[8]The methods employed in measurements of changes in plant matter are questionable, and the reported values of loss most likely are overestimates.

TABLE 16–25   Average Numerical Density and Wet Weight Biomass of Mollusks of Eutrophic Lake Borrevann, Norway, in Relation to Other Benthic Macroinvertebrates at Different Depths[a]

Zone and Depth	Anodonta piscinalis $Ind.\ m^{-2}(\%)$	$g\ m^{-2}(\%)$	Other Mollusks $Ind.\ m^{-2}(\%)$	$g\ m^{-2}(\%)$	Other Benthic Fauna $Ind.\ m^{-2}(\%)$	$g\ m^{-2}(\%)$
Littoral						
0.2 m	– (–)	– (–)	121 (5)	3.3 (18)	2422 (95)	14.6 (82)
1.5 m	11 (–)	186 (80)	2041 (57)	28.7 (12)	2705 (43)	17.0 (7)
2 m	20 (–)	361 (94)	1182 (26)	6.6 (2)	3398 (74)	14.6 (4)
3 m	31 (2)	1185 (99)	93 (5)	0.3 (–)	1648 (93)	6.1 (1)
Sublittoral						
5 m	24 (2)	500 (99)	12 (1)	0.2 (–)	1242 (97)	5.4 (1)
6 m	11 (1)	73 (96)	5 (1)	– (–)	785 (98)	2.8 (4)
7 m	– (–)	– (–)	– (–)	– (–)	810 (100)	3.3 (100)
Profundal						
10 m	– (–)	– (–)	4 (–)	– (–)	1143 (100)	7.0 (100)
15 m	– (–)	– (–)	– (–)	– (–)	1598 (100)	5.9 (100)

[a]Extracted from data of Ökland (1964).

The bivalve *Anodonta* was restricted to the littoral, and all mollusks were intolerant of the reduced oxygen concentrations of the deep strata during both summer and winter periods of stratification. The *Anodonta* constituted a trivial amount of the numerical population density of the total benthic fauna (Table 16–25), but because of their large size and heavy shell, they completely bias the quantitative biomass comparisons. Such data, while instructive in a comparative way, are meaningless for productivity estimates unless the rates of turnover of each of the components constituting the categories are known. The much smaller and more numerous nonmolluscan fauna have a small instantaneous biomass but in general a higher turnover rate (shorter turnover time). The gastropod snails reached maximum numbers and biomass in the fall (October). The sphaeriid clams attained maximum biomass per area in early summer (June) prior to maximum densities in late summer (mid-August).

Aquatic Insects

By far the most abundant and diverse group of animals of the earth is the insects, most of which are terrestrial. Of those that are aquatic, nearly all are fresh water. Some orders are entirely aquatic. The characteristics of the insects are well-known, and no attempt will be made here to summarize the group differences except to point out salient features of the life cycles that are important to feeding and reproduction as related to benthic productivity and distribution.

The chitinous exoskeleton of arthropods necessitates molting for continued growth. The true bugs (Hemiptera), dragonflies (Odonata), stoneflies (Plecoptera), and mayflies (Ephemeroptera) are orders of winged insects that undergo gradual metamorphosis. Wings develop as external pads in the young, termed *nymphs*, and increase in size with each molt. The other orders, including the flies (Diptera), caddisflies (Trichoptera), the alderflies and dobsonflies (Megaloptera), beetles (Coleoptera), a few species of moths (Lepidoptera), and spongeflies (Neuroptera) undergo complete metamorphosis. In this development the wing pads develop internally in early *larval* instars and then evert to the outside in the preadult instar *pupal* stage. In nearly all of the important aquatic insects only the immature stages live in the water; the adults and in some groups the pupae are terrestrial. Only some of the beetles and hemipterans have adapted to the point at which both the adults and immature stages can live in the water.

A brief summary of the major life history characteristics follows. In generalizing interrelationships among so diverse a group of insects it is important to divert attention from taxonomic categorization to more functional relations of feeding, reproduction, and productivity within aquatic systems. Since much of Hynes' (1970) compendium on running waters is devoted to the invertebrates of stream and river systems, remarks here are directed more to lakes and reservoirs.

Dragonflies and Damselflies.   After fertilization, the eggs of dragonflies and damselflies (Odonata) are deposited into the water, onto substrata in or near the water, or into submersed parts of macrovegetation. Egg development, as in most invertebrates, is temperature dependent; nymphs hatch after about 2 to 5 weeks of egg development. Growth in the nymphal stages is very variable, especially in relation to temperature and food supply. Within a range of about 5 weeks to 5 years, 10 to about 20 instar moltings occur. The mature nymphs then leave the water on some emergent substratum as aerial adults. The odonate nymphs are almost entirely littoral in habitat, living among macrovegetation and littoral sediments, and burrowing into surficial sediments. The nymphs have fairly high respiratory demands and oxygen requirements.

Mayflies.   The mayflies (Ephemeroptera) are almost totally aquatic. The adult longevity is very brief (1 to 3 days), during which no feeding occurs. After mating in flight, fertilized eggs are laid in the water or on submersed objects; the time to hatching varies from a few days to many weeks. Growth is relatively rapid, from less than 1 mm at hatching to about 2 cm in nymphal stages of many species. From 20 to over 40 instars of molting occur over an average life cycle of about 1 year. A few species live as nymphs for 2 years or longer. All mayflies are restricted to waters of relatively high oxygen content but are widely distributed even in waters with moderate organic loading (Roback, 1974). Some species can regulate respiratory movements of gills in response to changing oxygen concentrations (Eriksen, 1963a). Mayfly nymphs are characteristic of shallow streams and littoral areas of lakes, and are distributed widely. However, many species are restricted to specific substrata of macrophytes, sediments of waveswept or moving stream areas, or sediments of specific sized particles (Eriksen, 1963b).

Stoneflies.   The stoneflies (Plecoptera) are terrestrial insects but in the nymphal stages they are strictly aquatic, and most are restricted to flowing waters of relatively high oxygen concentrations. Fertile eggs, laid into the water, require 2 to 3 weeks for hatching in many species, and several months among some larger forms. The nymphal instars, from 10 to over 30 moltings, occur in 1 to 3 years.

Hemiptera. The Hemiptera or true bugs are essentially terrestrial; a few are semiaquatic and very few species have adapted to submersed conditions. Most hemipterans overwinter as adults in moist sediments or vegetation. Eggs are laid on semiaquatic substrata or in aquatic macrophytes and develop rapidly in 1 to 4 weeks. Nymphs also develop rapidly in 1 to 2 months, commonly with 5 instar stages, and generally have a 1-year life cycle. Few hemipterans are truly aquatic, with cutaneous respiration; most are dependent on atmospheric oxygen for respiration. They inhabit substrata either near or above the water, possess adaptations for obtaining air while underwater, or carry entrapped pockets of air with them on underwater excursions. Several large families of these hemipterans are adapted to stride over the water surface without disruption of the surface tension (cf. page 12).

Diptera. The most important group of the aquatic insects with complete morphogenesis includes the flies, midges, and mosquitoes (Diptera). The dipterans form major constituents of the benthic invertebrates of many standing and running waters, and the chironomid larvae are particularly ubiquitous, as will be seen in some of the examples that follow. Much variability is found in the morphology, reproductive biology, and respiration of dipteran larvae. Adults essentially are never aquatic, but most of their life cycle is as immature forms in fresh waters. The larval stage, with about 3 or 4 growth molts, extends from several weeks to at least 2 years in different species, and many overwinter in the larval stage. Most species have 1 generation per year, some have 2 per year, and fewer of those known species have a 2-year life cycle. Most larvae respire cutaneously or by means of "blood gills," which also function in ionic regulation. Fewer larvae rely on air, and possess various adaptive structures to obtain air above the water or from the internal lacunae of aquatic angiosperms. Some chironomid larvae possess a type of hemoglobin in their blood that functions efficiently at low oxygen concentrations.

Caddisflies and Moths. The caddisflies or Trichoptera generally have a 1-year cycle. Adults emerge in the warmer periods of the year, often from overlapping cohorts, from May to October. Eggs are laid under water on submersed substrata and develop in about 1 to 3 weeks. Many caddisfly larvae build beautifully intricate cases from substrate particles of sand, small stones, leaf fragments, and the like, and are highly specific to types of substratum (cf. Cummins, 1964; Cummins and Lauff, 1969). Some construct a net that traps microorganisms and detrital particles in flowing water. After 6 or 7 larval instars, pupating occurs underwater within a cocoon. The pupal stage generally lasts only a few weeks, after which the pupa leaves the cocoon, moves to an aerial substratum, and emerges as an adult.

Few species of the moths (Lepidoptera) have aquatic larval stages and most are of the *Pyralididae.* Many characteristics of the life history of the "aquatic caterpillars" are similar to those of the closely related caddisflies.

Spongeflies, Alderflies, Dobsonflies. The spongeflies (Neuroptera) and alderflies and dobsonflies (Megaloptera) are closely related and often are grouped together under the first Order. Both groups are largely terrestrial but, because of the large size of some megalopteran larvae, they may represent a significant portion of the biomass of some benthic communities. Eggs generally are laid on aerial substrates overhanging the water. After a rapid development period (1 to 2 weeks), the larvae drop into the water and feed actively. The aquatic neuropterans (Sisyridae) are restricted to and parasitic on freshwater sponges and can have multiple generations annually. The megalopterans have numerous molting instars and most of their life cycle of 1 to 3 years is spent in the larval stage. The pupal stage lasts up to one month. The megalopteran larvae are vicious predators on other insects, and some can reach a size of several centimeters.

Aquatic Beetles.   There exists such great diversity among the aquatic beetles (Coleoptera) that even broad generalizations are difficult to make. Generally, the beetles adapted for larval and adult existence in water occur among the more phylogenetically primitive coleopteran groups. The life cycles of many (Gyrinidae, Haliplidae, Dytiscidae, and Hydrophilidae) are annual, with 3 larval instars. Eggs laid on or in macrophytes or sediments hatch in 1 to 3 weeks. Larval development is relatively rapid, about a month, and pupation takes place on some nearby terrestrial or aerial substratum. Overwintering generally occurs in the adult stage. Nearly all adult beetles are dependent on atmospheric oxygen and carry an air supply with them on the ventral side of their bodies, or are variously adapted to obtain air from the surface or from aquatic angiosperms. Some larvae have gills to obtain oxygen directly from the water. Although they generally are omnivorous, the feeding habits of beetles often change markedly from the larval to the adult stages.

Trophic Mechanisms and Food Types

The great diversity of food ingested by the aquatic insects and their various feeding mechanisms can be organized within a functional framework of combined properties of the feeding mechanisms and particle size of the dominant food utilized (Table 16–26). Distinctions can be made between herbivory, defined as the ingestion of living vascular plant tissue or algae, carnivory, the ingestion of living animal tissue, and detritivory, the intake of nonliving particulate organic matter and the heterotrophic microorganisms associated with it (Cummins, 1973). The organic matter in any of these categories can be ingested either in particulate form (by swallowing, biting, or chewing) or in dissolved form (by piercing or sucking). While association of aquatic insects with a particular substrate usually is related directly to feeding on that substrate or on the associated microflora, this is not always the case. For example, the insect may be associated with aquatic angiosperms as a source of air or as a site for reproduction or protection from predation. However, the relationship between insects and certain substrates, e.g., angiosperms, is often highly specialized and even species specific (cf. the compilation of Gaevskaia, 1966). Other insect groups, especially large ones such as the dipteran Chironomidae, show great diversity in feeding mech-

TABLE 16–26 A General Categorization of the Trophic Mechanisms and Food Types of Aquatic Insects[a]

General Category Based on Feeding Mechanism	General Particle Size Range of Food (μm)	Subdivision Based on Feeding Mechanisms	Subdivision Based on Dominant Food	Aquatic Insect Taxa Containing Predominant Examples
Shredders	> 10^3	Chewers and miners	Herbivores: living vascular plant tissue	Trichoptera (Phryganeidae, Leptoceridae) Lepidoptera Coleoptera (Chrysomelidae) Diptera (Chironomidae, Ephydridae)
		Chewers and miners	Detritivores (large particle detritivores): decomposing vascular plant tissue	Plecoptera (Filipalpia) Trichoptera (Limnephilidae, Lepidostomatidae) Diptera (Tipulidae, Chironomidae)
Collectors	< 10^3	Filter or suspension feeders	Herbivore-detritivores: living algal cells, decomposing particulate organic matter	Ephemeroptera (Siphlonuridae) Trichoptera (Philopotamidae, Psychomyiidae, Hydropsychidae, Brachycentridae) Lepidoptera Diptera (Simuliidae, Chironomidae, Culicidae)
		Sediment or deposit (surface) feeders	Detritivores (fine particle detritivores): decomposing organic particulate matter	Ephemeroptera (Caenidae, Ephemeridae, Leptophlebiidae, Baetidae, Ephemerellidae, Heptageniidae) Hemiptera (Gerridae) Coleoptera (Hydrophilidae) Diptera (Chironomidae, Ceratopogonidae)
Scrapers	< 10^3	Mineral scrapers	Herbivores: algae and associated microflora attached to living and nonliving substrates	Ephemeroptera (Heptageniidae, Baetidae, Ephemerellidae) Trichoptera (Glossosomatidae) Helicopsychidae, Molannidae, Odontoceridae, Goreridae) Lepidoptera Coleoptera (Elmidae, Psephenidae) Diptera (Chironomidae, Tabanidae)
		Organic scrapers	Herbivores: algae and associated attached microflora	Ephemeroptera (Caenidae, Leptophlebiidae, Heptageniidae, Baetidae) Hemiptera (Corixidae) Trichoptera (Leptoceridae) Diptera (Chironomidae)
Predators	> 10^3	Swallowers	Carnivores: whole animals (or parts)	Odonata Plecoptera (Setipalpia) Megaloptera Trichoptera (Rhyacophilidae, Polycentropidae, Hydropsychidae) Coleoptera (Dytiscidae, Gyrinnidae) Diptera (Chironomidae)
		Piercers	Carnivores: cell and tissue fluids	Hemiptera (Belastomatidae, Nepidae, Notonectidae, Naucoridae) Diptera (Rhagionidae)

[a]Modified slightly from Cummins, K. W.: Trophic relations of aquatic insects. Ann. Rev. Entomol., 18: 183–206, 1973.

anisms and substrates ingested (Table 16–26). Most aquatic insects tend to be nonselective in their food habits.

The ingestion of food bears little relation to rates of assimilation. The relationship may not even differ with variations in the caloric content of food material. For example, fortuitous ingestion of inorganic sediment by fine-particle detritivores obviously has little nutritive value, but coatings of organic compounds and attached microflora may be of considerable value or aid in digestion by means of mechanical disruption of cells within the gut. Little is understood about the digestive capabilities or efficiencies of food utilization of aquatic insects (see review of Cummins, 1973). Much of the material ingested by litter and deposit-feeding invertebrates probably is indigestible by the animals, and must be degraded by the enzymatic activity of enteric symbiotic microflora and fauna of the gut. A number of carbohydrases have been found in the digestive tracts of several species of Trichoptera, the carnivorous amphipod *Gammarus,* and 2 species of sediment-feeding dipteran *Chironomus* (Bjarnov, 1972). All species degraded mono- and disaccharides and species differences were found only in polysaccharide degradation. Cellulose was found to be degraded only by the amphipod, and chitinase activity was found only in carnivorous species and in 1 detritivorous caddisfly. Klug and Meitz (personal communication) have demonstrated recently that certain aquatic larvae possess anaerobic bacteria in intestinal diverticulae that are able to degrade cellulose.

If digestion of particulate organic detritus is slow or impossible, as appears to be the case, an alternate mode of degradation of tough plant materials, such as of terrestrial windblown leaves, has been observed. These materials are colonized by bacteria and especially by fungi with exoenzymes of proteolytic and cellulolytic capabilities that degrade the material prior to animal ingestion. While most of this degraded material is not eaten by animals and eventually is further degraded to CO_2, in feeding on the colonizing microflora the shredding activities of benthic insects can accelerate the reduction of particle size and microbial degradation during the ingestion process.

Biomass and Feeding Rate

In viewing the relationships between biomass and feeding, it is found that the biomass of aquatic insects is relatively constant. They are supplied with consistent and abundant food supplies of similar caloric and protein content throughout the year (Cummins, 1973). The rate of insect biomass turnover is controlled primarily by temperature, and is mediated mainly by the positive relation between temperature, feeding rate, and respiration. The ratio of feeding and respiration to growth are relatively constant. In general, streams experience lower seasonal fluctuations in temperatures than do the littoral regions of lakes, and therefore much of the feeding and growth of aquatic insects of streams occurs in the fall and winter, even in the temperate zones. Assimilation efficiencies seem to be fairly constant over the broad range of foods consumed by insects. There is a suggestion (Welch, 1968), based on limited and variable data, that this efficiency[9] is somewhat higher in carnivores than in herbivores and detritivores. However, the net or tissue growth efficiencies[10] of carnivores tend to be lower than those of herbivores and detritivores.

Littoral and Profundal Benthic Fauna

Some of the important relationships among the benthic macroinvertebrates can be seen by a comparison of the littoral fauna with that of the profundal. As lakes become more productive and the hypolimnetic water strata undergo periods of oxygen reduction and increases in the metabolic products of microbial decomposition, the number of animals adapted to these conditions decreases precipitously. Those adapted to these conditions are exposed to relatively homogeneous conditions of temperature and substratum throughout the year, and competitive pressures for available resources and predation are reduced considerably. Therefore, a commonly observed community structure consists of a rich fauna with high oxygen demands in the littoral zone above the metalimnion. Substratum heterogeneity increases greatly in the littoral, and species diversity and competitive interactions are more complex. By contrast, the profundal zone is more homogeneous and becomes more so as lakes become more productive and species diversity decreases (cf. Jónasson, 1969).

The production of phytoplankton and macrophytes is a major determinant of the subsequent conditions of food supply, oxygen, ionic composition, pH, and numerous other factors that delimit the range and competitive abilities (cf. general review of Macan, 1961) of benthic fauna. The effects are seen not only in the qualitative distribution, but also in the quantitative aspects. Two maxima of abundance are observed with depth, one in the littoral zone and another in the profundal. Examples of this distribution are very numerous. The data presented in Table 16–27 show these 2 maxima both in numbers and biomass, as well as the reduction in species diversity in the progression from the littoral to the profundal zones. The same type of distribution is illustrated by the example from Lake Washington (Fig. 16–36). If the lakes become extremely enriched, to the point that the population densities of the phytoplankton become so great that they shade out the submersed macrovegetation, then the habitat diversity of the littoral decreases. Correspondingly, the diversity and quantity of littoral macroinvertebrates decrease, and often a peak in animal biomass can be observed only in the lower profundal zone. With further increases in eutrophication and lengthening of the period of hypolimnetic oxygen reduction and depletion and associated chemical changes, the rates of respiratory activity of the adapted benthic animals are reduced. Rates of growth and survival are also reduced,

---

[9]Assimilation efficiency is defined here as assimilation/ingestion, or growth plus respiration/ingestion.
[10]Defined as growth/assimilation.

TABLE 16-27 Average Numerical Density and Wet Weight Biomass of the Principal Groups of Benthic Invertebrates at Different Depths of Eutrophic Lake Borrevann, Southern Norway[a]

Benthic Invertebrates	Littoral																		Sublittoral						Profundal	
	0.2 m		1.5 m		2 m		3 m		5 m		6 m		7 m		10 m		15 m									
	No. m⁻²	g m⁻²	No. m⁻²	g m⁻²	No. m⁻²	g m⁻²	No. m⁻²	g m⁻²	No. m⁻²	g m⁻²	No. m⁻²	g m⁻²	No. m⁻²	g m⁻²	No. m⁻²	g m⁻²	No. m⁻²	g m⁻²
Oligochaeta	114	0.6	1370	8.9	1652	8.4	897	4.5	252	1.6	280	1.4	245	0.8	378	1.4	406	0.7
Hirudinea	126	2.9	163	2.1	368	2.2	63	0.5	2	0.02	—	—	—	—	—	—	—	—
Ephemeroptera	1416	5.0	150	1.1	18	0.04	9	0.01	—	—	—	—	—	—	—	—	—	—
Trichoptera	391	5.2	203	2.0	112	0.6	193	0.3	80	0.2	10	0.02	—	—	—	—	—	—
Chaoborus	—	—	—	—	2	0.01	3	0.01	—	—	5	0.02	60	0.1	172	0.5	958	3.0
Chironomidae	279	0.5	643	1.2	1138	2.5	463	0.7	900	3.5	490	1.4	505	2.5	575	5.0	299	2.2
Gastropoda	105	3.2	1964	27.9	902	5.2	60	0.2	6	0.2	—	—	—	—	2	—	—	—
Other Groups	112	0.5	253	2.5	388	2.3	53	0.2	14	0.1	5	0.02	—	—	20	0.1	5	0.01
Total	2543	17.9	4746	45.8	4580	21.3	1741	6.4	1254	5.6	790	2.9	810	3.4	1147	7.0	1668	5.9

[a]From data of Ökland (1964); bivalve mollusk *Anodonta* excluded.

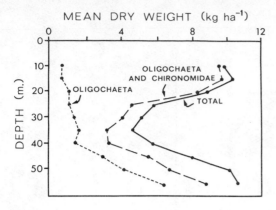

Figure 16–36   Vertical distribution of the mean biomass of oligochaetes, of oligochaetes and chironomids, and of the total benthic macroinvertebrates of Lake Washington, Washington. (Redrawn after Thut, 1969.)

and some insect larvae increase their life cycles from 1 to 2 years. As hypolimnetic strata of hypereutrophic waters undergo extreme eutrophicational or pollutional loading of organic matter and decomposition, essentially all of the aquatic insects may be eliminated. Practically the only group of benthic fauna adapted to these conditions is the oligochaete annelids. Variations in the vertical distribution of biomass of benthic invertebrates are discussed in some detail by Deevey (1941).

The changes in the species composition, especially among the dipteran larvae, and the quantity of benthic fauna in the profundal zone of lakes as they become more eutrophic have been subjects of intensive study since shortly after the turn of the century. Prompted by the perceptive work of Thienemann and other German workers, especially Lenz and Lundbeck, an array of typological schemes of lake classifications were developed on the basis of indicator species of dominant benthic fauna. Particular emphasis was given to the distribution of species within the ubiquitous dipteran family Chironomidae of midges and the oxygen content of the hypolimnion. Much of the early work on the classification of lakes in relation to the Chironomidae is reviewed in the book by Thienemann (1954; see also the critical discussion by Deevey, 1941). While these analyses did much to stimulate study of the benthic fauna, such classification systems are of limited value since they do not fully consider the complex interactions of morphometry, chemical differences among sediments and water, biotic interactions such as predation, and an array of other factors.

As lakes become more eutrophic, shifts occur in percentage composition of the 2 dominant groups of benthic animals of the profundal zone of lakes, the Chironomidae and the oligochaetes. The few examples given in Table 16–28, which could be greatly expanded, simply show the general reduction in the number of chironomids and other benthic animals and a concurrent increase in oligochaete worms among more productive lakes. The transition can occur in the same lake and can be accelerated greatly by the eutrophicational activities of man, as was shown dramatically in the case of Lake Erie (Table 16–28) in a period of less than 30 years.

TABLE 16–28   Comparison of the Relative Composition of the Dominant Benthic Macroinvertebrates of Several Lakes of Differing Productivity Based on Other Criteria[a]

Lake	% Chironomidae	% Oligochaeta	% Sphaeridae	% Others	Source
OLIGOTROPHIC					
Convict, Calif.	65.3	30.8	0.4	3.5	Reimers, et al. (1955)
Bright Dot, Calif.	77.5	3.1	19.1	0.3	
Dorothy, Calif.	69.5	23.3	3.5	3.7	
Constance, Calif.	56.9	20.5	20.5	2.1	
Cultus, British Columbia	65.0	24.0	–	–	Ricker (1952)
Lake Ontario	1.8	6.4	3.4	88.4[c]	Johnson and Brinkhurst (1971)
Lake Erie (1929–1930)	10	1	2	87[b]	Wright (1955)
EUTROPHIC					
Washington, Wash.	43	51	3	3	Thut (1969)
Lake Erie (1958)	27	60	5	8	Beeton (1961)
Glenora Bay, Lake Ontario	42.3	29.4	6.2	22.1	Johnson and Brinkhurst (1971)

[a] Data are only approximately comparable because of different methods employed.
[b] Mostly the mayfly *Hexagenia*.
[c] Mostly amphipods.

Benthic Fauna of Lake Esrom

Certainly one of the foremost analyses of the profundal benthic fauna can be found in the work of Jónasson (1972 and numerous earlier works reviewed therein). The environmental factors, and especially the phytoplanktonic productivity of the moderately large (17.3 km²) and deep (22 m), eutrophic Lake Esrom, a dimictic lake of Denmark, were studied in relation to the food, growth, life cycles, and population dynamics of 3 profundal detritivores and 2 carnivores. Because of the completeness of these studies, it is instructive to discuss them in some detail.

The dipteran detritivore *Chironomus anthracinus* was found to feed at the sediment surface. Its growth was limited to 2 very short periods, one in spring during the phytoplanktonic maximum when the hypolimnion was oxygen-rich, and the other after the autumnal circulation when oxygen was available again but food production was declining. Growth continued during winter as long as the lake was ice-free, but essentially ceased under ice-cover. During the summer, growth stopped when the oxygen concentrations of the hypolimnion were reduced slightly below 1 mg O_2 l^{-1}. Weight biomass, protein content, and fat content of larvae of this species were directly correlated seasonally with the influx of oxygen and phytoplanktonic detritus to the hypolimnion.

Continuous quantitative data extending over 17 years showed that *C. anthracinus* exhibited a 2-year life cycle in the deep (20 m) populations. The 2-year cycle is maintained by alternate years of high initial population densities in which a large percentage of the population does not emerge as adults in the first year. These larvae prevent a new generation from becoming established by their feeding activity, which inflicts severe mortality on new eggs laid on the sediments. As a result at any given time the population de-

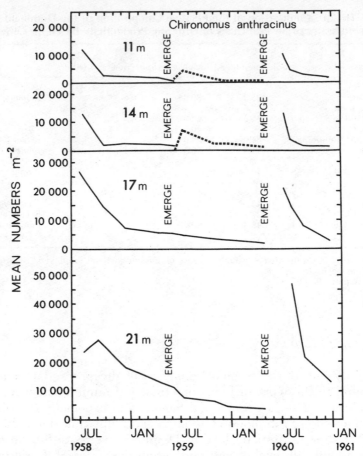

Figure 16–37   Seasonal fluctuations in populations of the midge larvae *Chironomus anthra-cinus* in sediments of different depths of Lake Esrom, Denmark. One-year life cycles were found at 11 and 14 m, two-year cycles at 17 and 21 m depths. (Modified from Jónasson, 1972.)

rives from a single generation of eggs. In alternate years, when the density of larvae declines below 2000 m⁻² after the first year's emergence of adults, a new generation is able to establish itself, because now the larvae are not able to feed over the whole of the sediment surface and most eggs survive. During these years of recruitment of a new generation, the larval population is a mixture of 2 generations (Fig. 16–37).

The larvae of this species in the metalimnetic sediments (11 and 14 m) grew longer because of greater oxygen concentrations, with greater weight per individual. There was more exposure to predation, however, and population numbers were lower than in the profundal zone. Those at 11 m depth had a 1-year life cycle; the transitional zone between 1 and 2-year life cycles was between 14 and 17 m sediments. In sediments of the latter depth, recruitment of larvae every year was not successful (Fig. 16–37). At 14 m the larvae succeeded in having a yearly recruitment in some years. Population size increased with depth, probably primarily the result of reduced predation.

TABLE 16–29   Estimates of the Average Annual Production, Mortality, Respiration, and Emergence of the Dominant Profundal Benthic Fauna of Lake Esrom, Denmark[a]

Species	Respiration (minimal)	Mortality	Emergence	Net Production
	kcal m^{-2} year^{-1}			
Dipterans:				
Chironomus anthracinus	361.8	35.9	39.4	75.3
Chaoborus flavicans	51.2	2.6	11.6	14.4
Procladius pectinatus	7.3	1.3	1.7	2.9
Oligochaete:				
Ilyodrilus hammoniensis	7.3	–	–	6.0
Mollusks:				
Pisidium casertanum	4.8	–	–	1.5

[a]Composite from data of Jónasson (1972).

The carnivorous phantom midge larvae, *Chaoborus flavicans* and *Procladius pectinatus*, were similar to each other in growth patterns and consistently exhibited a 1-year life cycle. The detritivorous tubificid oligochaete *Ilyodrilus hammoniensis* lives at least 5 years and does not begin to reproduce until after 3 years; the detritivorous small clam *Pisidium* also exhibits a long life cycle.

Detailed analyses of the population dynamics, survivorship, emergence, and respiration permitted an evaluation of the net production of the profundal macroinvertebrates of eutrophic Lake Esrom (Table 16–29). About 83 per cent of the production was by the 3 benthic detritivores, mostly by the *Chironomus* species. Jónasson was able to extend these data, making a number of reasonable assumptions, to a partial energy flow budget for this moderately eutrophic lake to show that an appreciable portion (perhaps 20 per cent) of the phytoplanktonic primary production reaches the profundal sediments and, while being decomposed, is utilized by profundal invertebrate fauna along with the microflora (Table 16–30). In preparing this table from his data, the unknown category of littoral and allochthonous organic matter has been included intentionally. As discussed at length in the ensuing chapter, these inputs can be very significant and often exceed that of the phytoplankton. Jónasson himself emphasizes that these figures represent only part of the whole. In spite of the excellence of his studies, further study is needed to evaluate how significant these littoral and allochthonous inputs are to profundal invertebrate productivity in a lake of moderately large size such as Esrom.

Food supply is of major importance in the regulation of the population dynamics of the detritivore species (cf. also Iyengar, et al., 1963), especially in the profundal zone in which predation shifts from other animals to cannibalism of eggs. Among carnivorous chironomid larvae, the young (first instar) are largely detritivores, but beyond the second instar they shift to feeding on small chironomid larvae, small crustacea, and some plant food. Although *Procladius* are carnivorous, their diet is very variable depending on accessibility (Kajak and Dusoge, 1970). Mature larvae consume about 10 per cent of their body weight per day and have a relatively high degree of utilization

TABLE 16–30 Estimates of Energy of Biomass and Flow from Components
of Profundal of Lake Esrom, Denmark[a]

Components	kcal m^{-2} year^{-1}	Per Cent
Littoral and Allochthonous Organic Matter	?	–
Phytoplankton		
Net production	1630	73
Respiration	610	27
Zooplankton		
Net production	120	40
Respiration	180	60
Detritivorous Benthic Fauna		
Respiration		
Micro benthic fauna	210	28
Macro benthic fauna	374	50
Mortality-predation	43	6
Emergence	40	5
Net production (macro)	83	11
Carnivorous Benthic Fauna		
Respiration	58	63
Mortality-predation	4	4
Emergence	13	14
Net production	17	18

[a]Generated from data of Jónasson (1972). Respiratory values of phytoplankton and zooplankton estimated.

of animal food. Abundant food availability increases growth and development greatly, in contrast to conditions of food limitation.

The Chironomidae form the dominant macroinvertebrate fauna of all arctic lakes studied (Hobbie, 1973). It is noteworthy that those few studied have a 2-year cycle and are detritivores.

Phantom Midges

In addition to the chironomid larvae, oligochaetes, and the small clam *Pisidium*, another major component of the profundal zone of lakes is the phantom midge *Chaoborus*. The extensive studies of *C. flavicans* demonstrate some of the major characteristics of this group (Parma, 1971). This midge dominated the benthic fauna of the profundal zone of Lake Vechten, a small (5 ha), moderately deep (12 m) lake of the Netherlands that is thermally stratified from April to November and mixes all winter (very brief ice-cover). The lake is moderately eutrophic and the hypolimnion becomes anaerobic during stratification.

The life span of the adult midge is very brief (< 6 days). Eggs (about 500 in number) are laid on the water in rafts and most (97 per cent) hatch in 2 to 4 days. The larvae can develop to the fourth instar in 6 to 8 weeks. The first and second instars are always limnetic and positively phototactic, and they develop rapidly in a few weeks. The third instar, mostly limnetic but also occurring in the sediments, is much longer in duration and can overwinter in this stage (Stahl, 1966, Roth, 1968). After a variable period of up to several months, ecdysis to the fourth instar occurs; this instar is limnetic much of the time.

The fourth instar of many species of *Chaoborus* undergoes strong vertical migrations diurnally (see, for example, Northcote, 1964; Malueg and Hasler, 1966; LaRow, 1968; Goldspink and Scott, 1971). A typical rhythm is movement from the sediments or their daytime depth to the surface strata at sunset; they stay there until about sunrise before descending. Rates of ascent and descent are about the same at 4 to 6 m hr^{-1}. Migration does not occur as actively at temperatures below 5°C. When the larvae are burrowed in the sediments, they push up to the surface of the mud as if to sample the light conditions. If they are below critical light intensities, they will emerge from the sediments and begin the migration (LaRow, 1969). In some interesting experiments, LaRow (1970) obtained evidence that the migration was influenced in part by oxygen tension of the water near the sediments. When the oxygen concentrations were high, most of the populations stayed in the sediments; when there was less than 1 mg O_2 l^{-1}, about one-third of the population migrated into the water above the sediments and about one-third migrated to the surface during the night. Migration varies with season, being highest in midsummer and lowest in winter.

Since the early instars are small and very transparent, predation pressure probably is low. With increasing size, predation pressure becomes greater and results in a more benthic habit. Oxygen stresses in the hypolimnion impose a demand for oxygen which is obtained in the upper strata, but this is best accomplished at night when predation pressure from fish is reduced.

Pupation takes from 1 to 2 weeks, and occurs from May to October in *C. flavicans*. The pupae migrate daily. Pupation and emergence as adults result in a reduction in the total benthic population during summer. Highest densities of larvae, 1400 to 1800 m^{-2} in Lake Vechten, occurred in November. Mortality in the winter was high (65 per cent) from reductions in food supply, and a total mortality of about 97 per cent occurred before the fourth larval stage was reached. Net production of the benthic part of the population was estimated to be 8 to 12.5 kg (dry) ha^{-1} [70 to 90 kg wet weight ha^{-1}] for *C. flavicans* in this lake per year.

Chaoborus larvae are predatory on pelagic zooplankton and certain small benthic animals. Their daily ration of food, amounting to about one crustacean zooplankter per hour, is variable but amounts to approximately 10 per cent of their body weight (Kajak and Ranke-Rybicka, 1970). It was estimated that *Chaoborus* could consume about 12 per cent of the macrozooplanktonic biomass on the average, and a much greater percentage at certain times of the year, such as late in the period of summer stratification.

It has been demonstrated further that *Chaoborus flavicans* is conspicuously selective in the food it consumes (Swuste, et al., 1973). Among zooplankton the copepod was actively selected over the cladoceran *Daphnia*; the pelagic ostracod *Cypria* was accepted but then immediately rejected. When given the choice between benthic oligochaetes and zooplankton, the oligochaetes were taken and predation on *Daphnia* ceased.

The coexistence of 2 or more species of *Chaoborus* within the same niche, without any apparent great differences in their characteristics or requirements that would prevent their competition, has been noted often (Stahl, 1966a, 1966b). Previously, it was assumed that the abundance of food precluded serious competition. Roth (1968), however, observed severe competitive interactions among 3 species of *Chaoborus* in the same lake, and presented evidence to show that slight differences in morphology of the larvae and their distribution, especially during nocturnal migration, were adequate to permit the 2 quantitatively minor species to coexist with the dominant species. Similarly, 2 coexisting species of *Chaoborus*, the dominant profundal fauna of Lake George of tropical Africa, were found to have interspecific differences in the mean number of eggs per adult female, the nature of egg batches, the morphology of mouthparts, as well as differences in the vertical distribution of the larvae and seasonal occurrence of the adults (McGowan 1974).

Productivity of Benthic Fauna

The extreme heterogeneity of the vertical depth of spatial distribution and the seasonal population dynamics of the benthic invertebrate fauna of lakes and streams have impeded detailed analyses of their productivity. Moreover, mortality by natural causes and predation is complex and related to changes in the types and intensity of predation, since the macroinvertebrates change their size and distribution during their life cycles. Several aspects of these predation-related changes have been discussed already, especially among the oligochaetes, chironomids, and the phantom midge *Chaoborus*.

Much of the existing quantitative data on the benthic fauna is expressed on the basis of biomass per unit area. A comparative set of examples of the average biomass among many lakes is set out in Table 16–31; the number of examples of this nature could be expanded manifold. Such data, however, are of quite limited value owing to great differences in the effectiveness of sampling times and in frequency and duration of sampling. As has been pointed out several times in the foregoing discussion, the summer period of low population densities and high predation pressure is the poorest time for analyses. Many studies are conducted only during the ice-free period. In both streams and especially the lower strata of lake sediments, the maximum period of growth usually occurs during the winter months in the temperate zone. More importantly, a large majority of these studies yield insufficient data on size-frequency distribution and other population characteristics to permit an evaluation of the most important parameter of turnover rate. Animals with small biomass but high turnover rates (shorter generation times) can obviously be more productive and contribute more to the system than animals with high biomass but low turnover rates.

Thus, average data such as are summarized in Table 16–31 must be viewed with caution. One not too surprising trend is observed, namely, that in general the biomass of benthic fauna increases as the general productivity

TABLE 16–31   Comparisons of the Biomass of Benthic Macroinvertebrates of Exemplary Lakes[a]

Lake	Wet Weight (kg ha^{-1})	Dry Weight (kg ha^{-1})	Notes and Source
Great Slave, N.W.T.	20.0	2.5	Rawson (1955)
Athabaska, Alberta-Saskatchewan	(32.8)	4.1	
Minnewanka, Alberta	(36.0)	4.5	
Simcoe, Ontario	(99.2)	12.4	
Waskesiu, Saskatchewan	(198.0)	24.7	
Wyland, Ind.			
Mean 1955	–	8.05	Gerking (1962)
Mean 1956	–	8.60	
July	56.0	10.3	
August	31.6	5.8	
75 Finnish Lakes	23.6	(3.0)	Deevey (1941) and Hayes (1957) after numerous sources
5 Swedish lakes	31.1	(3.9)	
3 New Brunswick lakes	25.4	(3.2)	
10 Lakes of USSR	41.3	(5.2)	
38 Lakes of USA, mostly of	87.2	(10.9)	
Connecticut and New York	(10–348)	(1–44)	
13 Lakes of northern Canada	88.9	(11.1)	
43 European alpine lakes	76.1	(9.5)	
64 Lakes of northern Germany	115.0	(14.4)	(includes some mollusks)
19 Eutrophic Polish lakes	28	3.5	Pieczyńska, et al. (1963)
	(2–56)	(0.3–7.0)	
15 Pond-type Polish lakes	101	12.6	
	(20–370)	(2.5–46.3)	
Dystrophic Store Gribsø, Denmark	94	(11.8)	Berg and Peterson (1956)
Weber, Wisc. 1940	553	(69)	Juday (1942)
1941	147	(18)	
Nebish, Wisc. 1940	122	(15)	
1941	590	(74)	
Shallow fish culture ponds, Czechoslovakia	222–272	(28–34)	Lellák (1961)
Parvin Reservoir, Colo.	582	99.8	Buscemi (1961)
Green, Wisc.			
0–1 m	36.6	8.7	Juday (1924)
0–10 m	82.2	15.9	
10–20 m	166.0	29.9	
20–40 m	171.1	33.5	
40–66 m	149.6	30.9	
	605.5	118.9	
Mendota, Wisc.			
0–7 m	433.9	53.8	Juday (1942)
>20 m	696.8	76.6	
Average for lake	414		Juday (1921)

[a]Values are comparable only approximately because of the differences in methods and frequency and duration of sampling; all exclude mollusks unless stated to the contrary. Values in parentheses were estimated using the relationship of dry weight equals 12.5 per cent of wet weight (see discussion of Thut, 1969); Winberg, et al. (1971) recommend 10 per cent and Gerking (1962) and others found about 20 per cent more realistic.

of the system increases. The fauna of oligotrophic lakes of northern latitudes and alpine regions are quantitatively less than those of lowland lakes of temperate and tropical regions. Within some rather obvious limits of increasing food supply by autochthonous productivity and increasing the diversity of littoral habitats, productivity of benthic animals increases with greater eutrophication. If the organic loading becomes so great that the conditions of hypolimnion are intolerable to even the most adapted fauna, as occurs in extremely hypereutrophic lakes, productivity of the animals decreases. Generalizations from only biomass data weaken rapidly beyond this level.

Changes in the total amount of biomass of benthic fauna with enrichment have been demonstrated dramatically many times in fertilization experiments. For example, high phosphate and nitrogen fertilization of fish culture ponds resulted in a 42 per cent greater yield of benthic invertebrates, and 3.3 times greater yield of zooplankton, than yields of unfertilized ponds (Ball, 1949). In a similar study on ponds in Michigan, fertilization increased biomass of benthic invertebrates by about 75 per cent and the food organisms utilized by the common bluegill sunfish (*Lepomis*) by about 70 per cent (Patriarche and Ball, 1949). However, in a more detailed analysis of fertilization relationships in these shallow waters, Hall, et al. (1970) demonstrated the complex

TABLE 16–32 Estimates of the Biomass and Productivity of Benthic Invertebrate Fauna of Several Lakes

Lake		Biomass (B)	Production (P)	P/B Ratio	Source
			$g\ m^{-2}\ year^{-1}$		
Śniardwy, Poland	6 m	5	51	12.9	Kajak, et al. (1970)
	8 m	3	45	14.0	
Mikolajski, Poland	4 m	18	143	10.7	
	8 m	6	73	13.0	
Taltowisko, Poland	4 m	35	367	13.5	
	8 m	8	110	13.2	
			$kcal\ m^{-2}\ per\ Vegetative\ Season$[a]		
Naroch, USSR					Winberg, et al. (1970)
Nonpredatory		5.2	12.3	2.4	
Predatory		0.3	0.7	2.3	
Myastro, USSR					
Nonpredatory		0.7	3.3	4.7	
Predatory		0.2	0.7	3.5	
Batorin, USSR					
Nonpredatory		2.4	11.5	4.8	
Predatory		0.3	1.4	4.7	
Red, USSR					Andronikova, et al.
Nonpredatory		4.56	19.4	4.2	(1970)
Predatory		0.2	1.0	5.0	
Krugloe, USSR					Alimov, et al. (1970)
Herbivores-detritivores		2.15	4.50	2.1	
Carnivores		0.236	0.72	3.0	
Krivoe, USSR					
Herbivores-detritivores		1.51	1.85	1.23	
Carnivores		0.15	0.39	2.6	

[a]The "vegetative season" generally refers, as far as can be ascertained, to the ice-free period from May to October, a period of 180 days.

interactions between shifting development of macrophytic vegetation and increasing densities of phytoplankton, and the effects of these substrata on species composition and predation pressure from fish. Although increases in productivity of benthic invertebrates were found in more nutrient-rich waters, in large part the production was compensated for by shifts in species composition with changing substrata and mortality by greater fish production.

In the foregoing discussion examples of estimates of the production rates of benthic macrofauna have been given for individual groups or for specific regions of lakes, e.g., the profundal zone. Studies of the rates of production of the total invertebrate benthic fauna of whole lake systems, for which sufficient detail is available to calculate turnover rates over the entire year, are rare. A few estimates have emerged from the data of the International Biological Program, and more will appear when these data are synthesized more completely. The data exemplified in Table 16–32 can be treated only as approximate estimates, but they do show the general relationships commonly observed between nonpredatory detritivore-grazing benthic fauna and the less productive predatory fauna. The P/B ratios were obtained by summation of the production rate estimates of individual dominant species and then by dividing by the total biomass. The productivity estimates of the nonpredatory benthic fauna were found to be 5 to 35 times less than those of nonpredatory zooplankton (Winberg, 1970). As would be anticipated, the productivity of predatory benthic fauna was consistently found to be much lower than that of the nonpredatory benthic fauna. In general the productivity of the nonpredatory benthic fauna accounts for less than 10 per cent of the primary productivity, and decreases appreciably as the productivity of macrophytes increases and the loading of allochthonous organic matter to the lake increases in proportion to phytoplanktonic productivity.

Fish and Predation on Benthic Fauna

The subject of fish as animals and the fish fauna of lakes and streams is an enormous field to which much attention has been devoted. The diversity in the morphology, physiology, behavior, reproductive capacity, growth, and feeding of freshwater fishes is a field of study that has been treated in detail in many recent works many times the size of this book. Even though this separation of the treatment of fish biology, ecology, and population dynamics from limnology is justifiable from the standpoint of giving due attention to these organisms of major economic importance, and because the subject matter is so extensive, the schism is unfortunate in many ways.

Fish are an integral component of freshwater systems. Although the amount of energy and carbon flux to the fish trophic level is exceedingly small in most aquatic systems, this does not justify their neglect as an operational component within the system. Their impact on the operation of the system in terms of carbon flux and nutrient regeneration at times can be quite significant. For example, as has already been discussed, the shift of fish species feeding on larger particle-sized food organisms to planktivorous species can have marked effects on zooplanktonic composition and productivity. This change can in turn influence the species composition of phytoplankton and their productivity at the primary level.

Another conspicuous example is the effect of the general eutrophication of lake systems, in which the fish composition shifts dramatically from salmonid and coregonid species of quite stringent low thermal and high oxygen requirements to more warmwater species that are more tolerant of eutrophic conditions. The warmwater fish are restricted to the epilimnion during summer stratification. Certain fish, such as the carp, have omnivorous feeding habits and can be very effective in modifying the littoral substrata to the point at which many submersed macrophytes are eliminated. These activities often can so disturb the sediments that abiogenic turbidity is increased, and transparency and phytoplanktonic productivity are reduced.

A large number of studies have indicated the major effects of fish predation on the productivity of benthic macroinvertebrates. In 4-ha Third Sister Lake, Michigan, the littoral benthic fauna increased to a maximum concentration in early winter (Ball and Hayne, 1952). The invertebrate population declined steadily after the formation of ice-cover until spring loss of ice, when a further reduction occurred immediately as a result of emergence of the adult forms of the aquatic insects. From the start of summer stratification, the populations and their biomass progressively increased from this annual low to the autumnal and winter maxima. Experimental removal of all of the fish from the lake not only resulted in a large increase in the biomass and numbers of benthic fauna, but the low population characteristics of early summer were not nearly as severe as under conditions of normal pressure of fish predation. Experimental manipulation of fish populations in half-hectare ponds showed that the presence of fish predation resulted in a very significant increase in the rate of production of benthic invertebrate fauna and a reduction in their biomass (Hayne and Ball, 1956). In the absence of fish, the apparent production rate of benthic fauna decreased and stabilized at a higher level of biomass than under conditions with fish predation.

While there are great differences in the qualitative composition of invertebrate fauna and seasonal succession of fauna among littoral stands of emergent vegetation, little difference is found between the quantity of macrofauna and the production of higher emergent vegetation in productive lakes or ponds (e.g., Dvořák, 1970). The presence or absence of submersed vegetation, however, results in large differences and these plants can be influenced markedly by fish and waterfowl. A number of studies have experimentally excluded the feeding and disturbance activities of these vertebrate animals from the littoral vegetation (see, for example, Kořínková, 1967; Kajak, 1970). As a result of both the effects of predation and reduction of submersed vegetation, when exposed to active fish, such as the carp, and birds, especially ducks, the biomass and production rate of benthic macrofauna were severely reduced, usually by at least 50 per cent. Although mortality of the natural production of benthic invertebrates can be very high from fish predation (at least half), food supply for the invertebrates generally is the dominant controlling factor influencing the population dynamics and productivity of benthic invertebrate fauna (see Lellák, 1965; Hall, et al., 1970).

Studies of the profundal macroinvertebrate fauna, for example in the

TABLE 16–33   Estimated Rates of Production of Profundal Macroinvertebrate Fauna of Lake Beloie, USSR, Over an Annual Period, 1935–36, and the Fate of the Fauna Within the Lake[a]

	kg (dry weight) lake^{-1}	Per Cent
Net production	4010	100
Emerging insects	256	6
Predation mortality (mainly by fish)	542	14
Natural mortality and decomposition	2213	55
Surviving population at end of year	999	25

[a]Generated from data of Borutskii (1939).

early very detailed analyses by Borutskii (1939), showed that predation mortality, chiefly by fish, was relatively low (Table 16–33). Only a small percentage lived to the adult stage and emerged. His studies indicated that the turnover rate was about 2 times the biomass annually, which is in general agreement with other studies (production rates equal about 2 to 4 times the mean annual biomass). There is evidence that the values of turnover are considerably higher in the littoral zone (4 to 8 times the annual biomass) (see, for example, Anderson and Hooper, 1956).

In an excellent study by Gerking (1962), the production rates of zooplankton and macroinvertebrate fauna were determined both directly and by measurements of utilization by the dominant predator, the bluegill sunfish (*Lepomis macrochirus*). Population dynamics and mortality analyses showed that the dense fish population grew slowly in this Indiana eutrophic lake, and became overcrowded and undernourished. The diet of the fish was varied, but the dominant prey were midges (45 per cent) and *Daphnia* (26 per cent) among all size classes of fish (Table 16–34). Each year the diet shifted from a high proportion of *Daphnia* in July to a high proportion of midge larvae in August as the cladoceran populations decreased drastically during the summer. The estimated fraction of midge production that was consumed by fishes during the summer was 0.50 (Table 16–34), which indicates a high cropping efficiency of the benthic fauna, much higher than "losses" through emergence as adults.

The production rate of the bluegill fish was estimated by multiplying the instantaneous rate of growth by the average summer biomass, and adding the recruitment to this value (Table 16–34). The minimum summer production of benthic fauna was computed as the necessary amount to replace the losses by predation. Midges and other dipteran larvae accounted for more than one-half of the production of food that was derived from the bottom fauna. Therefore it was estimated that minimum food production was 4 to 5 times the production of bluegills during the summer, and probably near 10 times on an annual basis.

As is common among fish of the temperate zone, both reproduction and growth of the bluegill sunfish are limited to the spring, summer, and early autumn. Spawning occurs for about a month in the spring, depending on the temperature. Growth occurs for about 5 summer months, during which time nearly all of the annual production occurs, 91 kg ha^{-1} in this example (Table

Table 16–34 Relationships Between the Bluegill Sunfish (Lepomis macrochirus) and Its Food Supply of Benthic Fauna and Zooplankton in Wyland Lake, Indiana[a]

	Bluegill Sunfish Weight (kg ha⁻¹)			Food Supply Weight (kg ha⁻¹)	
	Protein	Live Weight		Protein	Live Weight
Biomass:			July biomass:		
June	17.0	98	Benthic fauna	6.1	56
Average	12.2	72	Zooplankton	9.2	191
			Total	15.3	247
Percent diet constituents:			August biomass:		
Benthic fauna	–	74	Benthic fauna	3.4	33
Zooplankton	–	26	Zooplankton	1.1	23
Midges	–	45	Total	4.5	56
Monthly summer food turnover:			July-August decline:		
Benthic fauna	6.2	64	Benthic fauna	2.7	24
Zooplankton	2.2	23	Zooplankton	8.1	168
Total	8.4	87	Total	10.8	192
Annual production:			Minimum summer		
Growth	6.5	37	production rate		
Recruitment	9.1	54	Benthic fauna	17.8	196
Total	15.6	91	Midges	9.2	71

	Production Turnover Ratio	
	Protein	Live Weight
Monthly summer food turnover compared with:		
July Biomass:		
Benthic fauna	0.99	1.14
Zooplankton	0.24	0.12
July-August decline:		
Benthic fauna	2.30	2.67
Zooplankton	0.27	0.14
Summer efficiencies based on total production:		
Production biomass/intake	0.15	0.09
Production fish/production food	0.23	0.19
Intake by fish/production food:		
Benthic animals	0.48	0.47
Midges	0.50	0.55

[a]Modified from Gerking (1962).

16–34). Growth and feeding are attenuated abruptly in the autumn for the remainder of the year. In contrast, the production of benthic fauna continues for much of the year and increases greatly in the fall, under reduced predation pressure, until about the time of ice-cover formation. On an annual basis, the benthic faunal production is much larger than that utilized, at least 196 kg ha⁻¹ in this example (Table 16–34), and probably less than 10 to 20 per cent is utilized by carnivorous fish.

Fish Communities of the Great Lakes: A Case History of Responses to Man

The fish communities of the Laurentian Great Lakes have experienced a remarkably rapid succession of species in response to a series of eutrophicational changes and manipulations brought about by man's activities. Since many freshwater fish utilize a diverse diet and their growth is quite elastic in response to changes in food abundance, one of the first apparent changes in the early stages of eutrophication is the enhancement of growth of fish (cf. review of Larkin and Northcote, 1969). As eutrophication continues, however, with the accompanying progressive acceleration of decomposition and oxygen reduction in the tropholytic zone, stenothermal fish such as the trout and whitefishes (coregonids) are restricted to smaller and smaller volumes of water in the oxygenated metalimnion (see, for example, Frey, 1955b). Eventually, these species are eliminated and replaced by warmwater tolerant fish species, often coarse fish that are less desirable for food by man. Many of the changes associated with eutrophication (e.g., changing littoral zone, increased organic deposition) also affect the reproductive success of fish populations by reducing the survival of the young.

The effects of accelerated eutrophication of lakes and streams resulting from the activities of man, often somewhat incorrectly termed cultural eutrophication, have been documented in a great number of cases (see, for example, Hasler, 1947, and numerous more recent reviews). The effects also can be seen in the massive Laurentian Great Lakes, which have undergone rapid changes in less than a century.

The major changes in the fish communities of the Great Lakes system began in the early 1800s with the access of marine species to the interior 4 lakes (Erie, Huron, Michigan, and Superior), brought about by engineering projects planned to bypass the great Niagara Falls (Aron and Smith, 1971). The Erie Canal was opened to Lake Ontario in 1819 and to Lake Erie in 1825, and the Welland Canal connected Lake Ontario to Lake Erie much more directly in 1829.

The drainage basin of Lake Ontario, the lower and most eastern of the 5 Great Lakes, was the first to be settled and modified by immigrants to the region in the early 1800s. After the construction of the Erie Canal, the anadromous marine alewife (*Alosa pseudoharengus*) became widely established in the drainage of Lake Ontario by the late 1800s. Its direct access to the Ontario lake system via the St. Lawrence River apparently had been prevented earlier by the abundance of large piscivorous fish, particularly the Atlantic salmon (*Salmo salar*) and the lake trout (*Salvelinus namaycush*). Intensive fishery exploitation reduced the salmon and the trout of Lake Ontario by the 1860s, and the salmon became extinct in the lake by 1900 (Fig. 16–38). The salmon failed to move further upstream in spite of the route of the Welland Canal (Smith, 1972a). The alewife continued its rapid expansion, and there occurred a simultaneous precipitous reduction in the once very abundant lake herring (*Leucichthyes artedi*) and other coregonid species, related to intensive fishery exploitation. Similarly, exploitation reduced certain of the percid fishes, notably the walleye and blue pike.

Simultaneously with these events, the parasitic sea lamprey (*Petromyzon marinus*), probably entering through the Erie Canal, became established in Lake Ontario by the 1890s (Fig. 16–38) but caused little concern because its primary prey, the salmon and the trout, had already declined. The sea lamprey entered the other Great Lakes via the Welland Canal and became established in Lake Huron in the 1930s and in lakes Michigan and Superior by the 1940s. Prior to establishment of the sea lamprey, the lake trout was relatively

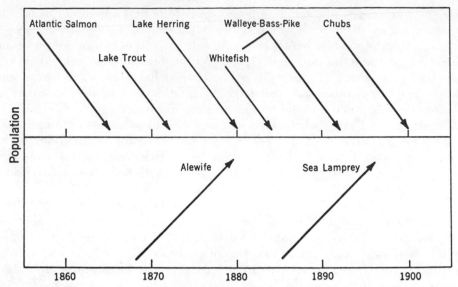

Figure 16–38 Trends in the fish populations of Lake Ontario for the period preceding and following the establishment and increase of the alewife and sea lamprey. (From Aron, W. I., and Smith, S. H.: Ship canals and aquatic ecosystems. Science, 174(4004): 13–20, 1971. Copyright 1971 by the American Association for the Advancement of Science.)

abundant in the upper Great Lakes and stable to intensive fishing exploitation. It was the only abundant large species that inhabited the colder metalimnetic regions of the lakes preferred by the sea lamprey. Increasing parasitic predation by the lamprey, as well as intensified fishery exploitation, led to the rapid decline and near extinction of the lake trout in the mid 1950s.

With the decline of the lake trout, fishery exploitation and predation by the lamprey on the lake herring and chubs (*Leucichthys*) were intensified. Two of the largest species now are apparently extinct in Lake Michigan, and several others have been reduced severely (Fig. 16–39). In the meantime, the alewife populations increased markedly in the absence of heavy predation and exploitation, to a point at which they totally dominated the fish biomass of lakes Michigan and Huron. Massive international efforts to control the sea lamprey by selective chemical treatment of the immature stages in spawning and breeding streams have been effective in reducing lamprey populations, but require continued treatment. These selective control measures probably owe some of their success to the sharp decline in abundant prey of large species.

As the lamprey populations declined, in the 1960s efforts began to restore the large piscivore populations by large-scale introduction of the lake trout and salmonids, such as the chinook and coho salmon from the Pacific Coast of North America. When the alewife is abundant, piscivore growth is exceptionally good; when it is not, such as in Lake Superior, only fair growth and survival have been reported. Rehabilitation will require extreme reduction of the alewife, and restoration of an interacting complex of deep- and shallow-water

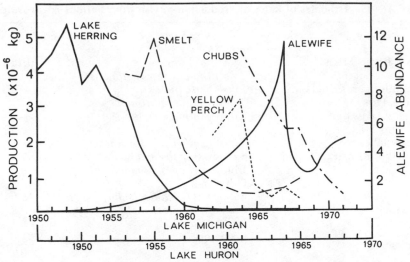

Figure 16–39 Successive declines in the production of lake herring, smelt, and yellow perch in Lake Michigan, and of chubs in Lake Huron, superimposed on the general population increase of the alewife from lakes Michigan and Huron. (Modified from Aron and Smith, 1971.)

forage species and minor and major piscivore species (Smith, 1970, 1972b)

Lake Erie, being much more shallow than the other Great Lakes, has received the greatest concentration of nutrient loading per unit volume and is extremely eutrophic, especially in the major shallower western end of the basin. The species populations of the fish community of Lake Erie have undergone a succession similar to that discussed for the other lakes (Regier and Hartman, 1973). Exploitation now is centered largely on the smelt (*Osmerus mordax*) and the yellow perch (*Perca flavescens*). Although the exploited species have changed dramatically over the last century, the total amounts of fish taken have been relatively stable over the last 50 or more years.

Thus, several combined factors have led to marked changes in the fish community of the Laurentian Great Lakes (Smith, 1972a): (1) Initial intensive selective exploitation of certain species. (2) Extreme modifications of the drainage basins and tributaries that have made many areas of the littoral and of streams unsuitable for spawning and natural reproductive success. (3) Establishment of marine species by invasion or introduction. (4) Physical and chemical modification of the lakes as a result of urbanization and industrialization within the immediate adjacent drainage basins. Much of the decline of oligotrophic piscivore species can be related to deforestation of much of the Great Lakes basin which, coupled with pollution and impoundment, led to warming and elimination of spawning areas. While these species were affected negatively, warm, sediment-laden streams were conducive to rapid development of lamprey spawning and larval growth. The potential to rehabilitate these waters surely exists, but not without both proper treatment of the inflowing water and prudent exploitation measures.

The changes seen in the examples given for the Great Lakes are reason-

TABLE 16–35 Estimates of Rates of Production of Fish from Various Fresh Waters[a]

Water	Fish	g m^{-2} year^{-1}
Manistee River, Mich.	Chestnut lamprey (*Ichthyomyzon castaneus*)	0.15
Lake Windermere, England	Pike (*Esox lucius*)	0.14–0.51
Reservoir, Ore.	Trout (*Salmo gairdneri*)	5.3
Dystrophic lakes of Wisc. and Mich.	Trout (*Salmo gairdneri*)	1.9–8.4
New York lakes	Trout (*Salvelinus fontinalis*)	3.3–6.5
Cultus Lake, British Columbia	Salmon (*Oncorhynchus nerka*)	5.9
Wyland Lake, Ind.	Bluegill (*Lepomis macrochirus*)	9.1
	Other fish	13.6
Eutrophic Reservoir, Ore.	Salmon (*O. tshawytscha*)	15.6
Lake Naroch, USSR[b]	Plankton feeders	2.2
	Benthos feeders	2.1
	Predators	1.8
	Total	6.1
Lake Batorin, USSR[b]	Plankton feeders	0.8
	Benthos feeders	8.6
	Predators	2.4
	Total	11.8
River Thames, England	5 species	42.6
Horokiwi Stream, New Zealand	Trout (*Salmo trutta*)	54.7

[a]From data cited in Chapman (1967), and Winberg, et al. (1970), from numerous sources.

[b]Assumed to include period of 180 days (May-Oct.); therefore values have been halved for the annual figures given.

ably well-understood because of the magnitude of the economics involved. Analogous changes have and continue to occur among hundreds of fresh waters throughout the world. It is imperative that these changes be understood, so that effective control and management practices can be implemented in accordance with demophoric pressures. Rehabilitation of fresh waters for water supplies or for fish exploitation after serious degradation is much too costly to undertake on a global basis when preventive measures can be taken prior to total degradation at much less energy expense.

Fish Production Rates

Although an enormous literature exists on the yield of fish per given area of freshwater habitat, relatively few analyses exist of the growth and mortality over annual periods from which reasonable estimates of production rates can be made. Growth is highly variable among different species of fish. In general, the instantaneous growth coefficient is correlated inversely with biomass. At greater biomass, most of the food energy is utilized for maintenance, while in young of lower biomass more is diverted into growth. Hence, the ratio of production to biomass usually decreases with age and greater size.

The few estimates of net production rates of various fish set out in Table 16–35, although comparable only approximately, demonstrate the general range encountered in the temperate region among largely carnivorous fish. Production rates would be considerably higher in tropical waters in which growth is not restricted to about half of the year. In fertilized standing waters the production rates of herbivorous fish in the tropics would be expected to have rates of several hundred g m^{-2} year^{-1} (Chapman, 1967). In standing waters of temperate regions, in which a single species predominates, the range of rates is from 1 to about 20 g m^{-2} year^{-1}. The rates in streams usually are higher, to about 50 g m^{-2} year^{-1} in the temperate areas, but per volume of water involved they are considerably lower in productivity than those in lakes.

chapter 17

organic carbon cycle and detritus

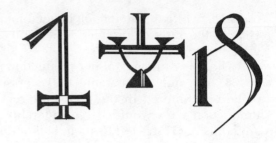

GENERAL CHARACTERISTICS

The general characteristics of the distribution and transformations of organic carbon in aquatic systems indicate that (a) the major pool of organic carbon is in soluble form; (b) 3 major sources of particulate organic carbon occur: allochthonous, and from 2 distinct zones of carbon flux, the littoral and the pelagic; (c) allochthonous inputs from the drainage basin to, and exports from, lakes occur largely as dissolved organic carbon and represent a major flux of carbon through the metabolism of the lake; and (d) major sites of metabolism of detrital organic carbon occur in the sediments of the benthic region, where in many lakes a majority of the particulate organic carbon is decomposed, and en route during sedimentation from the pelagic zone. Secondary metabolism of organic carbon is thereby displaced from sites of production or introduction by sedimentation in the case of particulate organic carbon, and by aggregation and coprecipitation with inorganic substances in the case of dissolved organic carbon.

Allochthonous inputs of organic matter to streams in the form of dissolved and particulate detrital organic carbon commonly form a major driving source of material and energy for stream metabolism. As in lakes, the dissolved organic carbon of streams is about 10 times that in particulate form. Labile components of the dissolved organic matter pool are decomposed rapidly in comparison to particulate organic matter. More refractory components of the soluble organic matter flocculate to the particulate phase, or are transported in the dissolved phase to downstream ecosystems. Large particulate organic matter is decomposed slowly, sometimes partly accelerated by faunal consumption, and has a longer retention time within a particular reach of the stream system.

Nonhumic organic substances, including carbohydrates, proteinaceous compounds, fatty acids, and other low molecular weight compounds, are labile and easily degraded by microorganisms. The resulting rapid rates

538

of utilization and turnover result in low instantaneous concentrations under aerobic conditions. Humic substances, aromatic polymeric compounds of partially degraded plant and animal material, form the bulk of organic matter of soils and water. These compounds are relatively resistant to microbial degradation and tend to persist with long residence times.

The biochemical origins of dissolved organic carbon in aquatic ecosystems are largely photosynthetic. Superimposed upon the variable inputs of allochthonous organic matter is that organic carbon produced autochthonously by plankton of the open water and photosynthetic production of the littoral zone. Dissolved organic compounds released extracellularly during active growth of algae and macrophytes and on autolysis during senescence and at death are decomposed rapidly. The vertical distribution of concentrations of dissolved organic carbon is relatively uniform with depth and seasonally. The dissolved organic carbon pool consists largely of refractory compounds that are resistant to microbial degradation, and inputs of these substrates are approximately equal to their slow microbial degradation.

The spatial and temporal distribution of pelagic particulate organic carbon correlates closely with the productivity and biomass distribution of phytoplankton and bacteria during stratified periods. In periods of water circulation, particulate organic carbon often increases markedly throughout the lake as surficial sediments are disturbed and resuspended. The littoral components of the production input to the particulate organic carbon pool of lakes increase in importance in relation to the whole in the transition from nutrient-limited conditions of oligotrophic lakes to more fertile conditions of eutrophic lakes.

Although a number of analyses of the spatial and temporal distribution of bacterial populations exist, they generally are sampled at intervals greatly exceeding the generation times of the bacteria and offer little in situ insight into control mechanisms of the population dynamics. The development of methodology for the direct measurements of growth of natural populations and of regulating variables is in its infancy. It is clear, however, that the microflora control the movement of carbon on a daily basis within the limits of constantly changing environmental conditions in a series of simultaneously oscillating reactions. The sources of organic matter are coupled with the qualitative composition of organic substrates, and losses by grazing and sedimentation.

Particulate organic matter formed in the pelagic zone is decomposed more readily and completely during sedimentation than the more refractory lignified tissue of littoral flora. As the littoral flora is decomposed and fragmented by successive populations of fungi and bacteria, redistribution often occurs in which the partly degraded material is displaced widely in the sediments. Bacterial populations and metabolic activity of surficial sediments increase several orders of magnitude over those of the overlying water, and decrease markedly within sediments at greater depth. Under anaerobic conditions within the sediments and lower hypolimnion of stratified lakes, rates of carbon mineralization are slower. Under these anoxic conditions, alternate electron acceptors are being reduced instead of molecular oxygen. Organic matter is converted anaerobically mainly to organic acids and to methane, hydrogen, and CO_2. These gases accumulate in the overlying water and are oxidized microbially under aerobic conditions. Under certain productive conditions bubble formation can be sufficient to generate currents effective in distributing nutrients and degradation products to overlying strata.

The metabolism of detrital organic matter results in a complex carbon cycle that dominates both the structure and function of aquatic ecosystems. Much of the organic matter entering the system from photosynthetic sources is not utilized in the grazer and detritus food chains. Lack of appropriate recognition of nonpredatory losses of organic carbon from autotrophs and nonpredatory losses from heterotrophs represents a major void in contemporary system analyses. Metabolism of detrital dissolved and particulate organic carbon from allochthonous, littoral, and pelagic components, often largely in the sediments, provides a fundamental stability to ecosystems. While the components of the biotic (living), trophic dynamic structure of the system generally undergo rapid oscillations in relation to an array of dynamic factors governing their growth, the detrital metabolism is slower and more evenly sustained on a much larger organic reserve. The trophic dynamic structure of the systems is almost totally dependent for its energy upon the detrital dynamic structure. In many lake systems and reservoirs, the littoral and allochthonous components are the dominant sources of syntheses and of inputs of detrital organic carbon, and are implicated as the major driving force.

ORGANIC CARBON

Nearly all of the organic carbon of natural waters consists of dissolved organic carbon (DOC) and dead particulate organic carbon (POC). The ratio of DOC to POC approximates 6:1 to 10:1 almost universally in both lacustrine and stream systems (Wetzel and Rich, 1973). DOC has been separated from POC by many techniques, such as by sedimentation, centrifugation, and filtration. However, DOC is defined rather arbitrarily in most studies by the practical necessity of fractionation of POC from DOC by filtration at the 0.5 μm size level; hence, DOC concentrations can include a significant colloidal fraction, in addition to truly dissolved organic carbon. Living POC of the biota constitutes a very small fraction of the total POC. The metabolism of these biota, however, mediates a series of reversible fluxes between the dissolved and particulate phases of detrital carbon.

SOURCES OF ALLOCHTHONOUS ORGANIC MATTER

The sources and composition of organic matter are diverse and poorly understood. Production of dissolved and particulate organic carbon is, in part, a result of autotrophic and heterotrophic metabolism. Instantaneous measurements of the chemical mass of DOC and POC, however, are highly biased towards refractory compounds (that is, compounds that are relatively chemically resistant, of low solubility, and resistant to rapid microbial degradation), components of detrital organic carbon that accumulate in fresh waters. The readily utilizable labile components cycle rapidly at low equilibrium concentrations, but represent major carbon pathways and energy fluxes. Moreover, in streams and most lakes, much of the detrital organic carbon, mostly dissolved, is of terrestrial origin. A majority of lakes are small, with a high proportion of their surface area as littoral zone. Allochthonous and littoral sources of dissolved and particulate detrital carbon form major inputs to the lacustrine system, and can markedly influence metabolism of the open-water pelagic zone. In streams, major metabolic carbon pathways for processing dissolved organic carbon are by the benthic microflora and especially the planktonic bacteria (Cummins, et al., 1972; Wetzel and Manny, 1972b). In the relatively static waters of lakes, gravity acts as an important selective agent by which greater rates of sedimentation displace a major portion of carbon metabolism to the sediments (Vallentyne, 1962; Wetzel, et al., 1972). To approach the complex carbon cycle that dominates both the structure and function of lakes and streams, one must have a complete representation of the productivity of all components of the ecosystem, as well as an understanding of the origins and metabolism of detrital organic carbon.

Detrital DOC and POC have long been known to exceed by many times the amount of organic carbon of the bacteria, plankton, flora, and fauna (Birge and Juday, 1926, 1934; Saunders, 1969; Wetzel, et al., 1972). An example from a lake in southeastern Michigan demonstrates the general relationships found among many lakes (Fig. 17–1). The amount of carbon in higher organisms is even smaller than the smallest amount indicated here, and trivial in comparison to the whole amount of carbon in the lake. These relationships are measures of chemical mass and, as indicated earlier, bear little relationship to the metabolic fluxes of carbon, discussed at length later.

In spite of the relatively crude techniques for separation of particulate from dissolved organic matter and approximate chemical analyses available to earlier workers on the subject, the massive data of Birge and Juday (1926, 1934) on a large number of Wisconsin lakes provide an initial orientation to the general relationships. Organic seston consisting of living and mostly dead particulate organic matter was separated by centrifugation from that in colloidal and true solution. The proximate composition was analyzed for

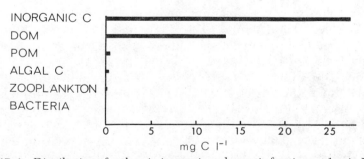

Figure 17–1   Distribution of carbon in inorganic and organic fractions at 1 m in Frains Lake, Michigan, 22 July 1968. (Modified from Saunders, 1972.)

total nitrogen of protein and other nitrogen-containing compounds, lipid compounds in ether extracts, and the residual carbohydrate. The range of total organic carbon content of natural waters is 1 to 30 mg l^{-1}; higher values usually are encountered only under polluted conditions. The average values from over 500 Wisconsin lakes were: dissolved organic matter, 15.2 mg l^{-1}, and particulate organic matter 1.4 mg l^{-1} or a ratio of 11:1. The average composition of the dissolved organic matter (detailed in Table 11–8 of the earlier discussion on dissolved organic nitrogen) was:

Crude protein	15.6 per cent
Lipid material	0.7 per cent
Carbohydrate	83.7 per cent

The importance of these and related studies to the present discussion lies in their relationship to the magnitude of allochthonous organic carbon entering most lakes from terrestrial and littoral marsh areas, and the qualitative changes in composition of this organic matter en route. As the total organic matter of lake water increases, the percentage in the dissolved fraction increases disproportionately to that of the particulate fraction. The nitrogen content progressively decreases with increasing DOC as the proportion of allochthonous to autochthonous organic matter increases. As a result, the organic C:N ratios increase and are indicative of an increased input of organic compounds low in nitrogen as well as decreased decomposition rates of the more refractory organic compounds, which permits them to accumulate. Dissolved organic matter of allochthonous origin contains a very low percentage of nitrogen (C:N about 50:1), while that produced autochthonously within the lake by algae and macrophytes has a much higher initial nitrogen content (C:N about 12:1). It is clear that much of the DOC of allochthonous terrestrial and marsh plant origin has humic acid characteristics, discussed further on, that impart a stained brown color to the water. Numerous workers have shown a nearly direct relationship between color units (Chapter 5) and dissolved organic matter or carbon.

The colloidal fraction of measured "dissolved" organic matter probably is small, although data are very few (Krogh and Lange, 1932). Nitrogen content decreases and carbon content increases in the progression from particulate to colloidal to truly dissolved organic matter fractions (Table 17–1). The small colloidal fraction is high in lakes rich in DOC such as bog waters, and in hard waters in which organic compounds are adsorbed onto carbonate colloids (cf. Ohle, 1934b; White, 1974; White and Wetzel, 1975).

Both allochthonous and autochthonous sources of detrital particulate (POM) and dissolved (DOM) organic matter, therefore, are instrumental as variable inputs to aquatic systems. Within the lake, autochthonous sources include: (1) littoral sources of POM and DOM by active secretion and autolysis of the macrophytes and attached microflora, and (2) primary producers of the pelagic zone, primarily the phytoplankton. Under some circumstances photosynthetic and chemosynthetic bacteria are also significant. Rapid transformations between POC and DOC by heterotrophic microflora result in qualitative transformations of organic matter with progressive losses as CO_2

TABLE 17–1   Annual Average Particulate, Colloidal, and Dissolved Organic Fractions of Lake Furesø, Denmark[a]

Organic Fraction[b]	Total Dry Weight[c] (mg l^{-1})	Organic Weight[c] (mg l^{-1})	Crude Protein (%)	Lipid Material (%)	Carbohydrate (%)
Particulate	2.1	1.56	53.2	9.9	36.9
Colloidal	3.6	0.67	41.5	11.7	46.8
Dissolved	ca 70.	8.8	37.5	–	62.5

[a]After Krogh and Lange, 1932.
[b]Colloidal and dissolved fractions separated by ultrafiltration.
[c]Based on a limited number of largely surface and near-sediment samples.

and heat. The amount of organic carbon utilized and transformed by animals is a quantitatively small portion of that of the whole system. Much of the heterotrophic metabolism, which is almost completely microbial, occurs in the open water and is displaced to the sediments, the latter being the major site of transformation in most lakes, especially as depth and volume decrease.

Terrestrial Sources

Allochthonous sources of organic matter to aquatic systems are primarily of terrestrial origin from photosynthetic production. The organic carbon of this plant origin is transformed variously by animal utilization and microbial degradation in transport to and within runoff water. Much of the input of terrestrial organic matter to streams is in the form of dissolved organic matter. This input can occur either directly in leachates from the flora, or in soluble compounds carried in runoff from the plant material in various stages of fungal and microbial decomposition. Particulate organic matter, again mostly of plant origin, can fall directly into the river water from overhanging canopies, be transported by runoff water, or be windblown into the stream. Foliage from trees and ground vegetation can provide very significant inputs of organic matter to streams, both as POM and as leached DOM from the dead POM. The large and fine particulate matter of terrestrial vegetation entering streams is leached of a significant portion of its organic matter as dissolved compounds, the amount varying with the plant species (Fig. 17–2). This DOM leachate is metabolized very rapidly by bacterial populations that increase markedly to the DOM loading. For example, in large experimental streams, leaf leachate was demonstrated to have a bacteriologically labile dissolved organic carbon (DOC) fraction that was rapidly decomposed ($T_{1/2}$ 2 days), and a refractory DOC fraction ($T_{1/2}$ 80 days) (Wetzel and Manny, 1972b). Most of the refractory dissolved organic nitrogen compounds persisted in a relatively unmodified state for at least 24 days. A quantity of the dissolved leachate precipitates (Lush and Hynes, 1973). The rate of precipitation and size of particles that result depend upon leaf species and water chemistry.

Aggregations of large particulate organic matter, such as leaf packs

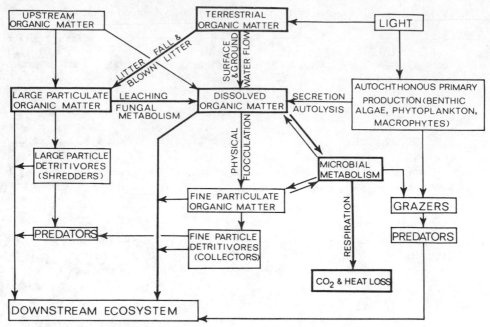

Figure 17–2 Simplified compartment model of the structure of an idealized stream ecosystem. Heavier lines indicate dominant transport and metabolism pathways of organic matter. (Composite of modified figures after Fisher and Likens, 1973, and Cummins, et al., 1973.)

trapped among stream substrata, undergo colonization by fungi in complex successional patterns within the detrital microhabitats (M. J. Klug and K. Suberkropp, personal communication). As the resistant plant material is degraded, solubilized products of decomposition are utilized by bacteria living in highly stratified populations in the steep redox gradients of the compacted plant material (Fig. 17–2). The detrital material and its associated microflora serve as a major nutritive source for numerous aquatic invertebrates, especially the immature insect fauna (Kaushik and Hynes, 1971; Cummins, 1973; Cummins, et al., 1973; Iversen, 1973). There is no question that the shredding, collecting, and grazing activities of aquatic insects in complex trophic relationships play a role in quickening the reduction of the size of POM and subsequent microbial degradation. It is also clear that a major portion of the faunal nutrition is obtained from the microflora attached to the resistant particulate detrital organic matter.

Detailed analyses of the metabolism of organic carbon in streams are few. However, there is sufficient evidence to indicate that (1) Allochthonous inputs of terrestrial organic matter, in the form of detrital dissolved and particulate organic matter, commonly form a major driving source of material and energy for stream ecosystems. (2) Approximately 10 times more organic matter occurs as dissolved organic matter than occurs in particulate form. (3) The metabolic flux rates of processing of the DOM are rapid (days) in comparison to those for particulate organic matter (leaves in weeks, woody material in years). (4) The labile components of DOM are processed rapidly, but more refractory components are exported to downstream ecosystems. Large particulate organic matter is decomposed slowly, and has a longer retention time within a particular reach of the stream system.

Estimations of the rates and importance of autochthonous primary production in streams by the attached benthic algae, lotic phytoplankton, and larger aquatic plants are very difficult (reviewed by Wetzel, 1975b). It generally is assumed that terrestrial photosynthesis and importation of this organic matter to stream ecosystems are the driving carbon and energy sources of these systems, that is, that streams are largely heterotrophic. Such is indeed the case in heavily canopied woodland and forested streams in which autochthonous primary production is very low or negligible. In other streams and

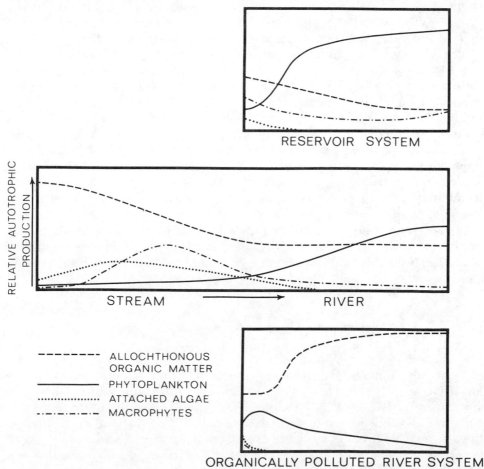

Figure 17–3 Generalized scheme of the relative contributions of allochthonous organic matter and autotrophic production by attached algae, phytoplankton, and aquatic macrophytes in the transition of a stream to a river system, if it is impounded where velocity of flow is reduced or practically eliminated, or it is organically polluted. (From Wetzel, R. G.: Primary production. *In* Owens, M., and Whitton, B., eds., River Ecology. Oxford, University of Cambridge Press, 1975.)

as rivers increase in size and decrease in velocity of flow, the significance of primary productivity of lotic phytoplankton and macrophytes increases (Fig. 17–3). Lotic waters clearly exhibit a spectrum of metabolic states in their proportion of heterotrophic and autotrophic metabolism. The relative significance of these types of metabolism is dynamic and varies considerably at local levels within a river system, and seasonally with shifts in many physico-chemical parameters and shifts in loading with allochthonous (natural or artificially by pollution) organic matter.

It is also becoming increasingly apparent, as will be emphasized further on, that the trophic dynamic structure of aquatic systems depends operationally on a dynamic detrital structure (Wetzel, et al., 1972; Wetzel and Rich, 1973). From the standpoint of carbon fluxes, the majority of energy and carbon of the systems is of autotrophic origin, dead, and undergoing microbial degradation at varying but serially decreasing rates with increasing organic refractility and resistance. In lakes in general, the major supportive metabolic base of carbon and energy is detrital in origin. Since a large majority of lakes of the world are small to very small, much of the autochthonous detrital production of lakes originates in the benthic, littoral flora. Similarly, in streams detrital heterotrophic metabolism dominates, and usually the major sources are of allochthonously derived, autotrophically produced terrestrial material. Common to both lake and stream systems is the total dominance of detrital metabolism, which gives the systems a metabolic stability. The trophic structure above the producer-decomposer level, with all of its complexities of population fluctuations, metabolism, and behavior, has a relatively minor impact on the total carbon flux of the system. The detrital system provides stability to streams: the slower, relatively consistent metabolism of detritus by decomposition underlies the more sporadic autochthonous metabolism that responds rapidly to and depends to a greater extent on environmental fluctuations. Autochthonous primary production generally is minor and variable, but combined with allochthonous, autotrophically produced detritus, they drive the lotic system. The functional operation of lentic and lotic systems converges at this point of similarity in detrital metabolism.

Allochthonous Organic Matter Received by Lakes

The dissolved organic matter of surface runoff is therefore composed of relatively refractory organic compounds resistant to rapid microbial degradation as it is processed in stream systems prior to reaching lakes. The amounts of DOC and POC reaching a lake and the quality of these organic compounds vary seasonally with the volume of flow in relation to time in the stream, the growth and decay cycles of the terrestrial and marsh vegetation through which runoff flows, and other factors, especially climatic ones. The influxes of allochthonous detrital organic carbon, in soluble and particulate form, to a small temperate lake were analyzed over an annual period in relation to their fate within and losses from the lake (Wetzel and Otsuki, 1974). Detrital organic carbon influxes were determined in water from an

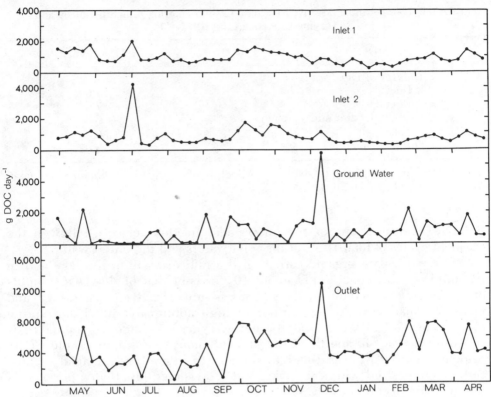

Figure 17–4 Dissolved organic carbon entering Lawrence Lake, Michigan, via the two stream inlets and groundwater and leaving the outlet, 1971–72. (From Wetzel, R. G., and Otsuki, A.: Allochthonous organic carbon of a marl lake. Arch. Hydrobiol., 73:31-56, 1974.)

inlet of spring origin and after it had traversed briefly through a marsh to the lake, in water from another inlet stream at the headwaters, and also after it had crossed the marsh to the lake. Samples were also taken from groundwater when it entered the lake, and from the outlet. Concentrations of organic carbon of inputs and losses were converted to budgetary figures by a detailed annual water budget (Fig. 17–4). The DOC fractions increased significantly (2 times) as the inlet water of the streams passed through the marshes prior to entering the lake. A similar result was found in a larger stream as it passed through an expansive marsh system (Manny and Wetzel, 1973). Lowest inputs of DOC occurred during the summer period of low precipitation and active marsh growth, and increased markedly in the autumn and early winter. DOC of ground water was very low, but because of the volume of this inflow it constituted one-third of the annual influx of allochthonous DOC (Table 17–2). Analogous to the stream exports, a net loss of 15 g DOC m^{-2} year^{-1} from the outflow represents a major carbon pathway of removal from the lake. Analyses of the heterogeneous DOC of inputs and outflow water by spectrophotometric absorption of ultraviolet light and fluorescence indicated a high percentage composition of humic and particularly of yellow organic acids.

TABLE 17–2 Annual Dissolved Organic Carbon Entering Allochthonously and Leaving Lawrence Lake, 1971–72[a]

DOC	g C m^{-2} year^{-1}
Influxes of DOC:	
Inlet 1	7.00
Inlet 2	7.94
Ground water and seepage	6.01
Total	20.95
Outflow DOC:	35.82

[a]From Wetzel, R. G., and Otsuki, A.: Allochthonous organic carbon of a marl lake. Arch. Hydrobiol., 73:31–56, 1974.

Relative concentrations of the humic acid fractions closely followed the seasonal cycles of total DOC of the inlet waters and the outlet water.

The particulate organic carbon of the influxes, within the lake and in the outflow, was consistently about 10 per cent that of the DOC. Inputs of POC were minimal during the summer months of active marsh growth and increased markedly during periods of high autumnal rainfall and spring runoff (Fig. 17–5). POC of ground water was very low, but on an annual basis constituted over one-fourth of the allochthonous POC input (Table 17–3). Allochthonous POC was small in relation to littoral and pelagial organic carbon synthesized within the lake, and represented less than 5 per cent of the gross production of POC of the lake system.

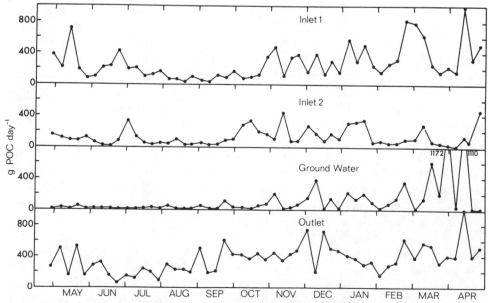

Figure 17–5 Particulate organic carbon of allochthonous origin and the outflow, Lawrence Lake, Michigan, 1971-72. (From Wetzel, R. G., and Otsuki, A.: Allochthonous organic carbon of a marl lake. Arch. Hydrobiol., 73:31-56, 1974.)

TABLE 17–3   **Annual Particulate Organic Carbon Entering Allochthonously and Leaving Lawrence Lake, Michigan, 1971–72**[a]

POC	$g\ C\ m^{-2}\ year^{-1}$
Influxes of POC:	
Inlet 1	1.99
Inlet 2	0.962
Groundwater and seepage	1.15
Windblown POC	0.00
Total	4.10
Outflow POC:	2.75

[a]From Wetzel, R. G., and Otsuki, A.: Allochthonous organic carbon of a marl lake. Arch. Hydrobiol., 73:31–56, 1974.

The quantity of windblown POM to this sample lake was negligible. Leaf litter fall and material blown can become significant in small lakes or ponds in heavily forested areas, but generally this input is small (Szczepański, 1965) in comparison to the major import of organic matter from many streams. Domestic and industrial pollution of lakes by organic wastes is an exceedingly individualistic parameter that lends itself poorly to generalization. Pollution of this nature normally is transported to lakes and reservoirs via stream systems. The types of organic pollution and its processing and effects are treated excellently by Hynes (1960, 1970). As pointed out by Pieczyńska (1972) organic metabolism of lakes that receive a high proportion of allochthonous organic matter generally is more variable and complicated than metabolism of predominantly autotrophic lakes. It should be stressed, however, that most lakes derive a highly significant amount of organic carbon and many nutrients from terrestrial sources of the surrounding drainage basin. The magnitude of allochthonous inputs and, more importantly, their metabolic role within the lake, is difficult to study and an area in great need of further investigation.

Composition of Organic Matter

Before discussing our limited knowledge of the utilization and degradation of detrital dissolved and particulate organic matter in lake systems, a brief consideration of the heterogeneous composition of such materials should be presented. The detailed reviews of Vallentyne (1957) on the distribution and composition of organic matter in lakes, of Breger, et al. (1963) on organic geochemistry of all major components, and of Konanova (1966) on soil organic matter are especially recommended. Numerous compendia have appeared on organic matter in natural waters, which are most relevant to this involved subject (e.g., the lengthy summations of marine and littoral organic compounds of Datsko, 1959, Duursma, 1961, Hood, 1970, and Khailov, 1971, and the freshwater symposia, e.g., Maistrenko, 1965, Shtegman, 1969, Trifonova, et al., 1969, and Faust and Hunter, 1971). Humic substances are reviewed comprehensively by Swain (1963), Haworth (1971), and especially Schnitzer and Khan (1972).

It is perhaps most direct to view organic matter of soils and waters in a general way as a mixture of plant and animal products in various stages of decomposition, of compounds synthesized biologically and chemically from the breakdown products, and of microorganisms and their decomposing remains. This inclusively complex system may be simplified by separation into 2 categories: (a) *nonhumic substances,* a class of compounds that includes carbohydrates, proteins, peptides, amino acids, fats, waxes, resins, pigments, and other low molecular weight organic substances. These substances generally are labile, that is, relatively easily utilized and degraded by microorganisms, and exhibit rapid flux rates within the ecosystems. Because of rapid rates of utilization and turnover, the instantaneous concentrations of nonhumic substances in water are usually very low. (b) *Humic substances,* which form most of the organic matter of soils and waters, and consist of amorphous, brown or black dark colored hydrophylic and acidic complexes of molecular weights ranging from hundreds to many thousands. Humic substances are formed largely as a result of microbial activity on plant and animal material. The compounds resulting from this decomposition are relatively resistant to microbial degradation and tend to persist in aquatic systems, with long residence times.

Humic Substances. Separation of fractions of humic substances usually is accomplished empirically by chemical fractionation into base-soluble *humic acid* which precipitates upon acidification of the alkaline extract, *fulvic acid,* which is soluble in both acid and base, and *humin,* a fraction insoluble in both dilute base and acid, similar in structure and properties to humic acid, and of highest molecular weight. These 3 fractions are structurally similar but differ in molecular weight and functional group content, with the fulvic acid fraction having a lower molecular weight than the humic acid and humin fractions, and a higher content of oxygen-containing functional groups. This fractionation is arbitrary and not very satisfactory, because the fractions are molecularly heterogeneous and often contain various substances such as sugars and phenols, depending on the source material and its state of degradation.

Decomposing humus materials exhibit a colloidal structure that is important in the physical behavior of humic solutions. The colloidal particles (ca 0.02 μm in size) can be fractionated by dialysis and electrophoresis, and exhibit Brownian movement. A basic characteristic of the humic colloids, as well as of dissolved humic and fulvic acid fractions, is their association with metal bases and acids by adsorption or peptization. The importance of this association of humates with Fe, P, Ca, and other ions has already been indicated in previous chapters. The pH of the solution governs both the extent of condensation and colloidal size, as well as adsorptive site availability. At low pH, such as in acidic bog waters, a concentration of humic solutions occurs.

Humic material of the series humic acid–hymatomelanic acid (an alcohol-soluble component of humic acid)–fulvic acid–humin consists of polymeric micelles whose basic structure is aromatic rings which are variously bridged into condensed form by —O—, —NH—, —CH$_2$—, or —S— linkages (Swain, 1963; Schnitzer and Khan, 1972). Attached hydroxy groups provide acidity, hydrophilic properties, base exchange capacity, and tannin-like character (for example, they react with proteins). Fulvic acid is the least polymerized of the fractions, while humin is the most condensed component of humic acids. Fulvic acid is highly oxidized, stable, and water-soluble (Schnitzer, 1971). About 60 per cent of the weight of the fulvic acid fraction is composed of functional groups such as carboxyls, hydroxyls, and carbonyls attached to a predominantly aromatic nucleus ring structure. Fulvic acid is a naturally occurring metal complexing agent of di- and trivalent metal ions that can bring into and maintain stable solution metal ions from practically insoluble hydroxides and oxides.

Humic substances contain a variable amount of nitrogen (1 to 6 per cent), which is largely hydrolyzable to amino compounds. Most of the combined amino acid nitrogen occurs in amino acid bound by peptide linkages. The remainder is in the form of amino sugars, ammonia, and amine linkages.

Formation of Humic Compounds.   The formation of humic substances is primarily microbiological, occurring in the degradation of plant material. Carbohydrates serve as the main microbial source of energy and carbon in the synthesis of protein and hemicellulose intracellularly. Polyphenolic lignin is modified in the substrates during this degradation, and humic-like substances of high molecular weight are synthesized. Further degradation of these compounds results in the generation of humic and fulvic acids and various products of decomposition (fats, amino compounds, and CO$_2$, H$_2$, CH$_4$, N$_2$, NH$_3$, and H$_2$S) in relation to the microflora and existing environmental conditions. Autolysis of the microorganisms themselves represents a major source of humic-like substances.

It is clear that the fungi play a major role in the degradation of plant material and the synthesis of humic substances. Fungi, as well as cellulolytic actinomycetes and myxobacteria, oxidize enzymatically lignin-derived polyphenols which condense with proteinaceous degradation products to form humic acids. Terrestrial plant and soil systems originate humic substances largely from cellulose and lignin. The needle and other products of coniferous origin are decomposed mainly by fungi which reduce celluloses and hemicelluloses, but not lignin. This coniferous litter is easily leached, especially of organic acids (Nykvist, 1963). Litter of deciduous vegetation, however, is decomposed more readily by fungi and bacteria to lignoprotein complexes. When allochthonous input of particulate organic matter from higher plants is high, as in the case of many river systems and many lakes with extensive littoral vegetation, fungi probably play a major role in the degradation of lignified and cellulolytic tissue for subsequent bacterial decomposition. In the open water of lakes, in which autotrophic production by algae results in plant constituents composed mainly of hemicelluloses, proteins, and carbohydrates, the demand for cellulolytic activity is reduced. As would be anticipated, fungal populations increase greatly in sediment accumulations, in littoral regions of higher plant densities, and in streams receiving high amounts of terrestrial plant debris. Sparrow (1968) provides an excellent review of the distribution and general ecology of the freshwater fungi.

The phenols, quinones, phenol carboxylic acids, and other similar phenolic compounds are common constituents of humic acid complexes, and can be quite inhibitory to bacterial fermentation and to fungi. These compounds, and associated acidity and low redox potentials, lead to frequent accumulations of humic substances undergoing very slow rates of degradation. The antiseptic qualities of humic substances from plant remains result in accumulations in aquatic depressions, for example in the peat of bogs and lake sediments, over long periods of time without significant degradation. The superb preservation of human bodies placed into peat bogs in Europe over 2000 years ago by the "bog people" (Glob, 1969) attests to the effectiveness of these combined properties and conditions.

DISTRIBUTION AND SOURCES OF ORGANIC MATTER IN LAKES

The implications of the preceding discussion are that a large portion of the particulate and dissolved detrital organic matter that enters lakes from allochthonous sources has undergone microbial stripping of more labile compounds. The residual and bulk of organic detrital carbon is in the form of humic substances that are relatively resistant to microbial degradation. It is then necessary to superimpose upon this variable allochthonous input of detrital organic matter the organic matter that is produced in the body of water itself from decomposition. The autochthonous organic matter falls into 2 broad categories: planktonic production of the open water, and photosynthetic production of the littoral zone.

Autolysis

As algae and macrophytes senesce and die, it would be expected that *autolysis* of cytoplasmic cellular contents would release significant amounts of dissolved organic matter rapidly, and that this release would be followed by rapid utilization of energy-, carbon-, and nitrogen-rich substrates. The loss of total substance is very high immediately after "death" in all organisms (Table 17–4).

Closer inquiry into the autolytic release of DOM initially by algae, aquatic plants, and zooplankton and progressive decomposition thereafter has been undertaken (Otsuki and Hanya, 1968, 1972a, 1972b; Golterman, 1971; Otsuki and Wetzel, 1974; Krause, 1964; Botan, et al., 1960). A few examples illustrate the general relationships believed to prevail. Organic carbon of dead green algal cells decomposing under aerobic conditions was reduced to about 45 per cent in 5 days (20°C), and decreased more slowly thereafter. Cell morphology changed little during the first 50 days, indicating that the cellular contents were decomposed initially and the cell wall was relatively resistant to microbial degradation. Organic nitrogen of the cells was reduced by about 70 per cent during the first 30 days; thereafter, decomposition of both organic C and N was slower. Dissolved organic C and N released by initial autolysis and from bacterial intermediate products of decomposition

TABLE 17–4   Average Percentage Loss of Total Substance of Different Aquatic Organisms Immediately and 24 Hours After Death, Aerobic, 20°C[a]

Organisms	Immediate Loss (%)	Loss After 24 Hours (%)
Planktonic green algae	11	33
Planktonic diatoms	11	15
Leaves of aquatic plants:		
Callitriche hamulata	9	16
Scirpus subterminalis[b]	35	–
Mixed zooplankton:	15	52
Cladocera (*Daphnia*)	29	–
Copepods (*Cyclops, Diaptomus*)	8	–
Sediment-living worms (*Tubifex*)	5	68
Small fish (*Lebistes*)	7	28

[a]After Krause, 1962.
[b]Otsuki and Wetzel (1974b).

are degraded rapidly, and then rates decrease with time (Table 17–5). Similar results were found with the decomposition of submersed vascular aquatic plants.

Kinetic evaluations of aerobic bacterial decomposition of algae approximated first-order reaction rates during the initial 30 days of degradation. Decomposition of cellular nitrogen and the production of dissolved organic nitrogen demonstrated that the dead algal organic nitrogen is separable into labile and refractory constituents according to their relative resistance to bacterial degradation. Further dissolved organic nitrogen is generated by bacteria through reassimilation of mineralized nitrogen. The general sequence of decomposition of dead algal material, the major organisms involved in this degradation, and the simultaneous regeneration of organic nitrogen are indicated in Figure 17–6. Following proteolysis by means of various enzyme systems, amino acids are deaminated by deaminases to ammonia, which then proceeds through a series of nitrification stages to hydroxylamine and nitrite to nitrate. Organic nitrogen is synthesized in the reduction of nitrate, denitrification, and nitrogen fixation processes (see Chapter 11) that

TABLE 17–5   Changes in Composition of Dissolved Organic Products During the Aerobic Decomposition of Scenedesmus Cells at 20°C[a]

Time (days)	Organic C (mg l^{-1})	Organic N (mg l^{-1})	C:N Ratio
0	213	22	11
35	184	21	10
185	87	17	6

[a]From Otsuki, A., and Hanya, T.: Production of dissolved organic matter from dead green algal cells. I. Aerobic microbial decomposition. Limnol. Oceanogr., 17:248–257, 1972.

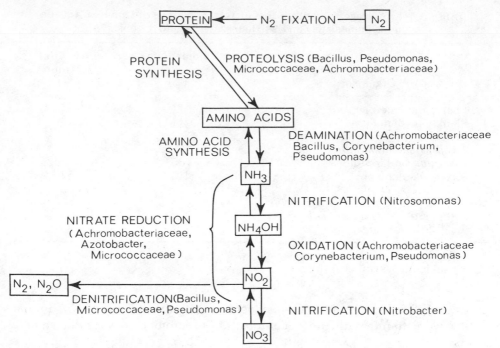

Figure 17–6 Decomposition and utilization stages of nitrogenous organic matter of aquatic organisms. (Modified from Botan, et al., 1960.)

vary with numerous environmental parameters. In turn, lysis of the microbial flora generates further dissolved organic nitrogenous compounds.

Decomposition rates of organic matter are quite different under anaerobic conditions. For example, after 60 days of anaerobic decomposition of green algal cells at 20°C, 30 per cent of algal cell carbon was transformed into dissolved organic carbon (DOC) and 20 per cent was mineralized; 50 per cent remained as particulate matter. Of the algal cell nitrogen, only 8 per cent was transformed to soluble form, 48 per cent was mineralized, and 44 per cent remained in particulate form after 60 days. The DOC in solution consisted largely of yellow organic acids very resistant to further degradation. The decomposition rates and production of dissolved organic compounds from dead algal cells were only one-fourth as great under anaerobic as compared to aerobic conditions, and the rate constant of cell carbon and nitrogen decomposition decreased to less than one-half under anoxic conditions.

The rates of decomposition under both aerobic and anaerobic conditions are influenced strongly by temperature (usually higher rates with increasing temperatures in the natural range), pH (commonly lower at acidic pH values), the major cation concentrations and cationic ratios, and other factors. The rates of cycling and utilization of the numerous dissolved organic intermediate products of bacterial and fungal degradation, such as amino acids, keto acids, various fatty and nonvolatile organic acids, phenolic derivatives, mercaptanes, and others, are highly variable (see, for example Krause, et al., 1961).

Secretion of Organic Compounds

In addition to autolysis and leaching of organic compounds that include an array of labile substrates (for example, sugars, amino acids, and so forth) as intermediate products of bacterial decomposition, and a proportion of resistant substances analogous to humic compounds formed in terrestrial situations, dissolved organic substrates also originate by *secretion* from actively growing algae (Chapter 14) and large aquatic plants (Chapter 15). Cellular organization is not so perfect as to prevent the release of extracellular soluble organic substances. This release varies widely in quantitative and qualitative composition among species and under various physiological conditions. Simple organic acids, sugars, and more complex carbohydrates, amino acids, peptides, pigments, and probably enzymes are among the compounds known to be liberated during active growth. Most of these secreted compounds are assimilated rapidly by the microbial flora so that their concentrations in the water would always be expected to be low. Some extracellular products, such as polypeptides, are more resistant to rapid degradation and could form a more significant portion of the total dissolved organic matter at any given time.

DISTRIBUTION OF ORGANIC CARBON IN LAKES

Dissolved Organic Carbon

As one views the depth-time distribution of total dissolved organic carbon in a lake (Fig. 17–7), a conspicuous feature is the relatively small amount of change with depth and seasonally, even though the lake is experiencing

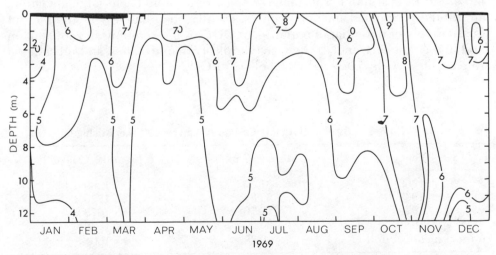

Figure 17–7 Depth-time diagram of isopleths of dissolved organic carbon (mg C l^{-1}), Lawrence Lake, Michigan, 1969. (From Wetzel, R. G., Rich, P. H., Miller, M. C., and Allen, H. L.: Metabolism of dissolved and particulate detrital carbon in a temperate hard-water lake. Mem. Ist. Ital. Idrobiol., 29:Suppl., 185-243, 1972.)

dramatic metabolic changes spatially and temporally throughout the period. The constancy of the DOC pool persists regardless of whether the sampling at each meter of depth is done at weekly or biweekly intervals. It is apparent that the DOC pool consists primarily of refractory organic carbon compounds that are relatively resistant to bacterial decomposition, and inputs of these organic substrates are approximately equal to their slow microbial degradation. Highest concentrations occur during summer stratification in the epilimnion, and consistently fluctuate to a greater extent in the upper strata, especially near the surface, than do concentrations in the hypolimnion. It is evident from earlier discussion (see Chapters 14 and 15), that secretion of DOC by phytoplanktonic algae and littoral flora is associated with a portion of the higher epilimnetic concentrations of DOC. Decomposition of largely labile secreted organic compounds often is very rapid (e.g., < 48 hours), and their dynamics would not be delineated by the sampling frequency employed for this generalized picture of the DOC pool. The constancy of the DOC pool is reflected further in annual average values per square meter and for the whole lake calculated by integration of each meter depth and compensation for changes in volume of water at those depths (Table 17–6). Horizontal variations of DOC within this lake were very small (< 10 per cent deviation from the mean of the central depression). Certainly in some lakes, however, and especially in reservoirs, point sources of DOC input, such as river influxes high in DOC, would cause greater horizontal variations.

It should be emphasized that the biochemical origins of DOC in aquatic ecosystems are largely photosynthetic, either autotrophically synthesized within the water or allochthonously generated in terrestrial systems of the drainage basin and transformed in transport to recipient basins. Superimposed upon the major sources to the DOC pool from (1) photosynthetic inputs of the littoral and pelagic flora and DOC subsequently lost to the pool through secretions and autolysis of cellular contents, and (2) allochthonous DOC, largely terrestrial humic substances refractory to rapid bacterial degradation, are (3) the very much smaller amounts, usually quantitatively negligible but qualitatively of potential importance, of DOC from excretions of zooplankton and higher animals, and (4) chemosynthesis of organic matter, and subsequent release of DOC by bacterial metabolism of CO_2.

TABLE 17–6 Dissolved Organic Carbon of Lawrence Lake, Michigan[a]

Years	Mean g C m^{-2}	Mean kg C lake^{-1}
1968	31.62	1569.5
1969	34.84	1729.3
1970	27.76	1378.2
1971	26.80	1330.1
Annual mean, period	30.28	1501.8

[a]From Wetzel, R. G., et al.: Metabolism of dissolved and particulate detrital carbon in a temperate hard-water lake. Mem. Ist. Ital. Idrobiol., 29, Suppl.: 185–243, 1972.

Chemosynthesis

Microbial production by chemosynthesis occurs to a marked extent in layers having contact with anaerobic zones of an aquatic system, especially in boundary layers between anaerobic and aerobic zones. As indicated in earlier discussions of cycling of several elements, the anaerobic processes of decomposition of organic matter provide reduced inorganic compounds that serve as energy substrates for the chemoautotrophic bacteria. Chemosynthetic production, a type of secondary production, becomes significant primarily in steep gradient layers of redox potential (Sorokin, 1964a, 1965, 1970). Outside of these layers, the significance of chemosynthesis in relation to total bacterial production is very low (Romanenko, 1966). Chemosynthesis by bacteria is normally very low in infertile waters and lakes of intermediate productivity, and becomes a significant contribution to the whole only in productive or meromictic lakes exhibiting steep redox gradients.

In the context of the increasing fertility of lake systems and the shifts in proportional contributions of phytoplanktonic and littoral flora, first of the submersed macrophytes and attached algae and in later stages of the emergent macrophytes and eulittoral algae (Chapter 15, Fig. 15–25), it is apparent that phytoplanktonic productivity, and allochthonous sources from the drainage basin are the primary sources of the DOC pool of oligotrophic waters (Fig. 17–8). In moderately large to very large bodies of water, phytoplanktonic photosynthetic metabolism can almost dominate as a source of the DOC pool.

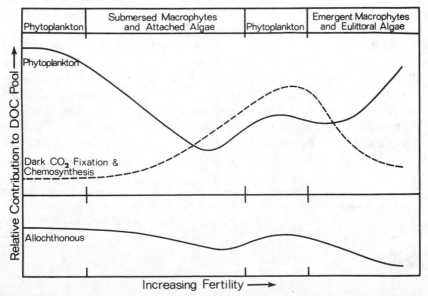

Figure 17–8 Generalized relative contributions of phytoplankton, dark and chemosynthetic CO_2 fixation, and allochthonous sources of dissolved organic carbon (*DOC*) to lakes of increasing fertility in the progression of dominating pelagic and littoral flora. Upper panel of generalized changes in productivity by autotrophic flora refers to that detailed in Figure 15–25. (Slightly modified from Wetzel, R. G., and Rich, P. H.: Carbon in freshwater systems. *In* Woodwell, G. M., and Pecan, E. V., eds., Carbon and the Biosphere. Springfield, Va., National Information Service, 1973.)

A majority of lakes, however, are small, and the ratio of littoral to pelagic photosynthetic productivity increases greatly as the mean depth of the basin decreases. If the drainage basin is not manipulated extensively by man, allochthonous inputs of DOC are relatively constant for a given lake; their relative contribution to the system decreases as autochthonous sources increase (Fig. 17–8). In reservoir systems the allochthonous inputs of DOC can exceed by several times that of autochthonous organic matter and DOC production within the reservoir (Romanenko, 1967).

The ratio of photosynthetic fixation of CO_2 to dark CO_2 fixation (almost all by heterotrophic metabolism of bacteria; see Gerletti, 1968; Romanenko, 1973) and bacterial chemosynthesis generally is large in oligotrophic waters. For example, the 4-year average of dark CO_2 fixation was 16.3 per cent of phytoplanktonic fixation in the light in a small Michigan lake (Table 17–7). This percentage is much less for the lake, about halved, if the littoral photosynthesis of this lake is considered in addition to that of the phytoplankton. The ratio of photosynthetic to dark CO_2 fixation decreases in the transition to planktonic eutrophy and hypereutrophy. With further transition of the system to, or in lakes with a predominance of productivity by emergent macroflora and associated attached and eulittoral microflora, the relative contribution of dark CO_2 fixation to the DOC pool apparently decreases.

Particulate Organic Carbon

Examination of the general spatial and temporal distribution of pelagic particulate organic carbon (POC) of lakes of varying degrees of productivity permits a few generalizations. In oligotrophic to moderately productive lakes, the observed depth-time distribution of pelagic POC follows the productivity and biomass distribution of phytoplankton rather closely during stratified periods. In the example from Lawrence Lake (Fig. 17–9), when the lake is stratified, the POC maxima follow peaks in productivity from several days to about 2 weeks, especially when sedimentation of plankton is slowed in thermal density gradients of the metalimnion. Under ice-cover, a period of reduced primary production and allochthonous inputs of POC, a gradual reduc-

TABLE 17–7   Dark CO_2 Fixation in the Pelagic Zone of Lawrence Lake, Michigan, Integrated for the Water Column at the Central Depression, 1968–71[a]

Years	g C m^{-2} year^{-1}	Mean mg C m^{-2} day^{-1}	% of Light CO_2 Fixation
1968	5.0	13.7	10.7
1969	10.2	28.1	23.6
1970	8.0	22.0	17.6
1971	5.1	14.0	13.4
Average:	7.1	19.4	16.3

[a]From Wetzel, R. G., et al.: Metabolism of dissolved and particulate detrital carbon in a temperate hard-water lake. Mem. Ist. Ital. Idrobiol., 29, Suppl.: 185–243, 1972.

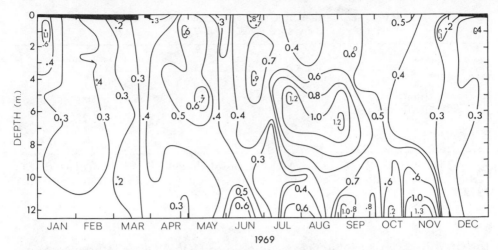

Figure 17–9 Depth-time distribution of pelagic particulate organic carbon (mg C l^{-1}), Lawrence Lake, Michigan, 1969. (From Wetzel, R. G., Rich, P. H., Miller, M. C., and Allen, H. L.: Metabolism of dissolved and particulate detrital carbon in a temperate hard-water lake. Mem. Ist. Ital. Idrobiol., 29: Suppl., 185-243, 1972.)

tion of POC often is observed. A period of increase of POC generally is observed during periods of circulation when surficial sediments are disturbed and being resuspended into the water column. Hypolimnetic POC of less productive lakes generally does not increase greatly unless the lower hypolimnion becomes anaerobic, for example, in the latter portions of summer stratification when specialized bacterial populations or algae, for example, euglenophytes, may develop in profusion. In hypereutrophic lakes that receive large inputs of planktonic and littoral POC and in which the hypolimnia are rapidly rendered anoxic (as exemplified by Wintergreen Lake, Fig. 17–10), bacterial productivity contributes to marked increases in POC.

The amounts of pelagic POC of a lake are relatively constant from year to year on a short-term basis (Table 17–8), as long as the lake system is not per-

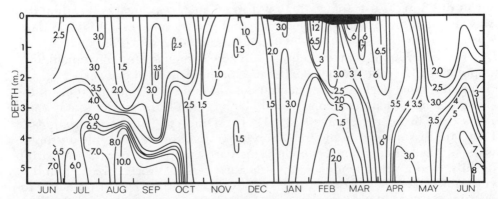

Figure 17–10 Depth-time distribution of pelagic particulate organic carbon (mg C l^{-1}) of hypereutrophic Wintergreen Lake, Michigan, 1971-72. (From Wetzel, et al., 1975.)

TABLE 17–8   Particulate Organic Carbon of the Pelagic Zone of
Lawrence Lake, 1968–71[a]

Years	Mean g C m^{-2}	Mean kg C lake^{-1}
1968	2.1	102.8
1969	2.6	130.8
1970	2.2	108.3
1971	2.1	104.3
Average	2.25	111.6

[a]From Wetzel, R. G., et al.: Metabolism of dissolved and particulate detrital carbon in a temperate hard-water lake. Mem. Ist. Ital. Idrobiol., 29, Suppl.: 185–243, 1972.

turbated by external influences. The ratio of DOC:POC is rather constant at about 10:1 in most unproductive to moderately productive aquatic systems. Deviations from this 10:1 ratio with depth and season are small in less productive waters (compare, for example, Figs. 17–7 and 17–9). As lakes become more eutrophic, the ratio of DOC:POC fluctuates greatly seasonally and spatially with depth. In the Wintergreen Lake example (Fig. 17–10), the annual average is about 5:1, with periods of intensive algal (epilimnetic) and bacterial (metalimnetic and hypolimnetic) profusion when the ratio decreases to 1:1 or less, and then increases to about 10:1 during the fall period of circulation (see also Weinmann, 1970). In another productive lake of Michigan, Saunders (1972) found that the particulate organic detritus constituted from 1.3 to 16.9 times the phytoplanktonic biomass and made up more than 50 per cent of the seston. Inorganic matter, for example, particulate $CaCO_3$, and silica, contributed a major portion of the remainder of the seston.

Estimates of the replacement of algal-cell carbon in the pelagic POC have been made from the biomass of algal-cell carbon and the rates of net primary production (Miller, 1972). Compensating for respiratory loss of carbon, the daily net accumulation of POC can be estimated from net primary production. Total suspended epilimnetic POC of Lawrence Lake had an average replacement time (turnover) of 40.7 days (range 8.1 to 544 days). Of this POC pool, a mean of 83 mg C m^{-2} (range 30 to 241 mg C m^{-2}) was algal-cell carbon that was replaced by primary productivity in 1.1 days (range 0.30 to 2.55 days) during the ice-free seasons and with an annual mean of 3.6 days.

Generalizations on the cycling of organic carbon among phytoplanktonic populations are difficult to make because of the paucity of data, especially on detailed annual cycles, under natural conditions. POC concentrations of the pelagic zones usually are considerably larger, double, or greater in eutrophic lakes than in infertile waters (Fig. 17–11). Differences in concentrations of DOC are less marked in the transition from oligotrophic to highly eutrophic waters. Algal-cell carbon commonly increases nonlinearly with increasing fertility, and a slight tendency towards algal cells of greater size is found in eutrophic lakes. Replacement times of algal-cell carbon by net primary production are usually greater (slower turnover times) in eutrophic than in oligotrophic waters, but are not proportional to increases in cell carbon. Hence the

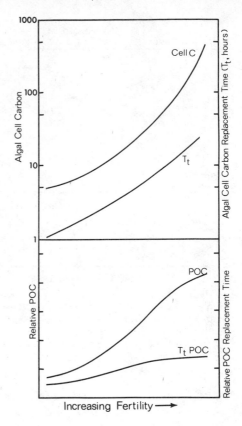

Figure 17–11 Generalized relationship of pelagic algal-cell carbon and particulate organic carbon (POC), and their relative replacement times (T_t) in lakes of increasing fertility. (From Wetzel, R. G., and Rich, P. H.: Carbon in freshwater systems. In Woodwell, G. M., and Pecan, E. V., eds., Carbon and the Biosphere. Springfield, Va., National Information Service, 1973.)

algal cells are photosynthesizing more per unit cell carbon, i.e., have a greater carbon flux, in less fertile waters.

All indications then point to the importance of autochthonous primary production by the phytoplanktonic and littoral flora as the major contributor to the POC of natural lake systems. Allochthonous POC is relatively small in contrast to major inputs of DOC, except for special cases, for example, in small lakes or ponds in heavily forested areas, or reservoirs that are small in volume in relation to inputs and flow-through. The sources of POC shift in their relative contributions to the POC pool as the freshwater systems frequently progress through stages of increasing fertility, or as seen among lakes at different stages of development (Fig. 17–12). Generalizations are again problematic because of the high degree of lake individuality. The importance of the littoral components to the production of the POC pool of the lake increases greatly and changes markedly in the transition from nutrient-limited conditions of oligotrophic lakes, to light limitations imposed by biogenic turbidity, to dominance by the emergent macrophytes and associated microflora.

UTILIZATION OF ORGANIC MATTER

The biochemical transformations of particulate and dissolved detrital organic matter by microbial metabolism are fundamental to the dynamics of nutrient cycling and energy flux within aquatic ecosystems. Although the entire subject of microbial ecology in aquatic systems, as well as of detrital feed-

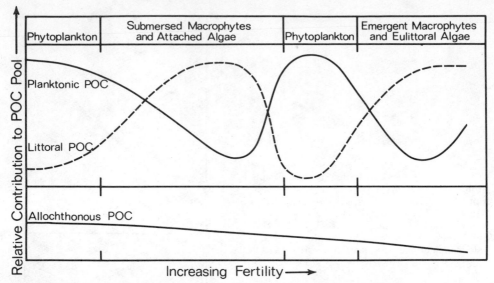

Figure 17–12   Generalized relative contributions of planktonic, littoral, and allochthonous sources of particulate organic carbon (*POC*) to lakes of increasing fertility as the dominance of pelagic and littoral primary productivity shifts (as detailed in Fig. 15–25). (From Wetzel, R. G., and Rich, P. H.: Carbon in freshwater systems. *In* Woodwell, G. M., and Pecan, E. V., eds., Carbon and the Biosphere. Springfield, Va., National Information Service, 1973.)

ing by higher organisms, is in a stage of comparative infancy, representative examples of the utilization and transformations of organic matter offer insight into the qualitative and quantitative structure and operation of the organic carbon cycle.

It is perhaps most appropriate to begin discussion with the generalized composite organic carbon cycle in a lake (Fig. 17–13), modelled after the general discussions of Kuznetsov (1959, 1970), and to attempt systematically to provide a quantitative range for the reactions and processes involved among lakes of varying productivity. The sources and general composition of allochthonous and autochthonous organic matter in dissolved and particulate form have been treated in some detail in this and in the previous chapters. Dead organic matter forming the detritus is separable into dissolved and particulate fractions, but it must be emphasized that this demarcation is only an operational division made by the investigators. From a functional standpoint of the system, the energetic transformations of organic carbon are similar whether they are operating on a large particle or a dissolved organic compound within a size spectrum from the large pieces of organic debris to the smallest of organic molecules. Only the rates of transformation differ.

The decomposition rate of organic substances is greatly dependent on solubility (Vallentyne, 1962). Preservation will be greatest for those organic compounds or complexes that occur in environmental concentrations exceeding the saturation level in the surrounding water. Degradation rates of soluble organic compounds vary, but for the compound to really be preserved and removed from the organic carbon cycle, sedimentation must occur. The prerequisite for sedimentation by gravity is insolubility. Sedimented, relatively

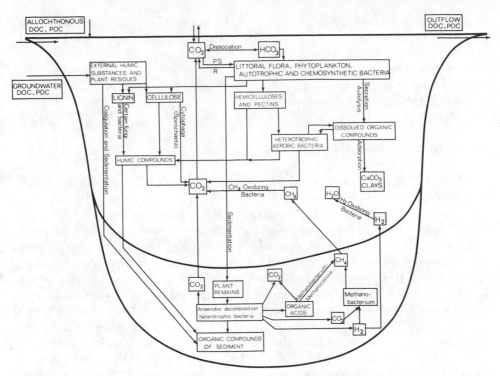

Figure 17–13   Simplified organic carbon cycle of a typical freshwater lake. *DOC* and *POC* = dissolved and particulate organic carbon; *PS* = photosynthesis; *R* = respiration. (Modified from Kuznetsov, 1959, 1970.)

insoluble compounds become buried in anaerobic sediments in which further degradation is slow. Although cellulose can be decomposed very rapidly by anaerobic bacteria under certain conditions, for example, sewage digesters and rumen, in anaerobic aquatic sediments acidic fermentation products rapidly accumulate and lower the pH sufficiently to inhibit bacterial metabolism (Brock, 1966). Additionally, the temperatures generally are low in many aquatic sediments. Lignin is even more resistant to degradation than cellulose, particularly under anaerobic conditions, and both form humic substances. The humic acids of the water coagulate to some extent, sediment, and form a portion of aquatic sediments.

The primary intermediate products of anaerobic decomposition of sedimented organic matter (especially cellulose) are fatty organic acids which are further degraded to CO_2 and hydrogen. Methane is biogenically generated from CO_2 and from hydrogen, and then is lost as bubbles or oxidized to CO_2 in less reduced strata of the sediments or water.

Much of the decomposition of organic compounds occurs in the aerobic waters prior to sedimentation to the bottom of the basin. The extent of degradation en route is governed by an array of physical (morphometry, stratification patterns, temperature) and chemical conditions in relation to the magnitude of allochthonous and autochthonous inputs of organic matter. The amounts of organic inputs to the water mass of oligotrophic lakes are generally small and coupled with oxic conditions for long distances (greater time); degradation of sedimenting organic matter is relatively complete and organic sediment accumulation slow. Massive inputs of organic matter in eutrophic lakes result in rapid sedimentation (shorter distances), less volume of aerobic water, and rapid accumulation of organic matter in anaerobic hypolimnia and sediments.

Amino Acids and Related Compounds

Insight into the complex differential utilization of more labile organic compounds, such as amino acids, is gained from several detailed analyses of the amino acid and carbohydrate cycling in lakes. Concentrations of dissolved "free" amino acids of lake water and interstitial water of sediment generally are very low (< 10 μg 1^{-1}) at any given time (Brehm, 1967), although they can increase to considerably higher levels in eutrophic pond waters (Zygmuntowa, 1972). All of the protein amino acids as well as several nonprotein amino acids have been found. Over an annual period, and in several different types of waters, concentrations of serine and glycine were relatively high, those of alanine, aspartic acid, and threonine occurred in less abundance, and those of all remaining amino acids and glutamic acid were very low. Maximum epilimnetic concentrations were found in the winter, especially just below the ice, and during the summer period of maximum algal production. Concentrations of the metalimnion were generally very low and increased somewhat in the hypolimnion, except just above the sediments. Much of the dissolved amino acids was found to be adsorbed to colloidal carbohydrates liberated by the microflora. These colloids were most abundant in the epilimnion during the summer and were reduced to extremely low concentrations in the hypolimnion, in which intensive bacterial metabolism occurred.

High molecular weight peptides of the free water and interstitial water of the sediments were associated with humic substances and increased in surface waters during autumnal influxes of allochthonous plant material (Brehm, 1967). These compounds and, additionally, mucopeptides of blue-green algal and bacterial origin, undergo relatively slow bacterial degradation. Amino acids are adsorbed selectively by these humic substances of dead particulate matter, and by the microcomplexes of the cell walls of bacteria and blue-green algae. These polypeptides remain relatively undegraded after autolysis of algae. In contrast, certain amino acids, especially aspartic and glutamic acids, are released very rapidly by autolysis and assimilated immediately by heterotrophic bacteria. Other amino acids are released in autolysis more slowly, resulting in a higher relative concentration in the dead plankton. The bacterial decomposition of plankton is effected first on the autolytically soluble products. Most of the cell decomposition occurs in the epilimnion; more resistant cell components are degraded more slowly after the soluble autolytic substances are utilized. The concentrations of amino acids found are nearly always < 50 μg 1^{-1} in a variety of aquatic situations. Measurements with individual amino acids of the uptake kinetics demonstrated accordingly that those in least abundance were assimilated rapidly (< 20 hours), and conversely that those

TABLE 17–9   Dissolved Organic Nitrogen Compounds in the Epilimnion of Two Lakes of Northern Germany, 27 June 1967[a]

Dissolved Organic Nitrogen Compounds	Schuënsee (Mesotrophic)	Pluss-See (Eutrophic)
Extracellular amino compounds:		
Free amino acids	$39 \ \mu g \ 1^{-1}$	$71 \ \mu g \ 1^{-1}$
Peptides	226	250
Colloids	118	414
Glucosamines	31	126
Total	$414 \ \mu g \ 1^{-1}$	$861 \ \mu g \ 1^{-1}$
Amino-N as a percentage of the total extracellular organic nitrogen	32.2%	23.7%
Amino acids and amino sugars as a percentage of the total dissolved organic matter	10.3%	5.5%
C:N of dissolved organic matter	7.5:1	10.7:1

[a]From data of Gocke, 1970.

compounds found in higher concentrations exhibited greater turnover times (the time required for the plankton to remove all the substrate), about 30 to 90 hours (Hobbie, et al., 1968). These results were reinforced by Gocke's (1970) demonstration of the rapid bacterial decomposition of free amino acids (down to 2 per cent within 6 days) and peptides (to 13 per cent in 6 days) by natural bacterial populations, while amino-colloidal complexes are resistant to degradation and are accumulated. This relative degradation of amino components is reflected in the absolute concentrations found (Table 17–9).

It is also apparent that the simple nitrogen organic compounds are utilized more effectively in more productive lakes (faster turnover times; lower instantaneous concentrations as a percentage of the total organic nitrogen). This observation suggests with high probability that the enzymatic systems of bacteria of eutrophic lakes are adapted or keyed to respond to a much greater variety of substrates than those of oligotrophic waters exposed to conditions of chronic substrate limitation. The oligotrophic species, upon receipt of an influx of a specific substrate for which it does not possess an enzymatic uptake system, must adapt to that substrate. Hence, the observed assimilation rates initially are slow but can increase markedly as adaptation occurs. Those species of eutrophic waters which are exposed more frequently or continuously to a variety of substrates respond to and assimilate the substrates immediately.

Carbohydrates and Simple Organic Acids

The distribution of carbohydrates and simple organic acids has been investigated in several lake systems (see, for example, Saunders, 1963; Walsh 1965a, 1965b, 1966; and especially Weinmann, 1970). Consistently, as with simple amino compounds, the instantaneous concentrations of individual organic acids and carbohydrate compounds are very low (usually $< 30\ \mu g\ l^{-1}$) and represent a small portion of the total dissolved organic matter. Free monosaccharides and oligosaccharides are assimilated readily by bacteria and occur at very low concentrations ($< 10\ \mu g\ l^{-1}$). Dissolved polysaccharides, consisting of numerous hexoses, pentoses, and methylpentoses, and especially phenolic and cresolic compounds, occur in greater abundance. Various organic acids (lactic, citric, oxalic, malic, glycolic, and short-chain fatty acids) have been found but also at low levels.

It is clear that major sources of these compounds are secretion products of phytoplankton (Chapter 14) and littoral flora (Chapter 15), autolysis of the plants and microflora, and intermediary products of microbial degradation. In bacterial-free algal cultures, carbohydrates, especially mono- and disaccharides, and total dissolved organic matter increase in proportion to growth. In the presence of a natural bacterial flora, the extracellular products are degraded rapidly (hours). This reduction is selective when the labile carbohydrates, organic acids, and amino acids decrease much more rapidly than the total dissolved organic matter (Weinmann, 1970; Gocke, 1970; Poltz, 1972).

The relative rates of microbial degradation discussed earlier, while following from the wealth of physiological information on bacterial utilization of organic substrates, are exceedingly difficult to quantify in natural populations of aquatic bacteria. The assimilation measurements on heterogeneous bacterial populations, consisting of many species with varying affinities and specificities of their transport systems for substrates, are fraught with technical methodological problems (cf. Jannasch, 1969, 1974). Nevertheless, the kinetics of organic substrate utilization by heterogeneous planktonic populations, determined by the uptake rates of radioactivity labelled organic compounds *at substrate concentrations that occur naturally*, are instructive in a relative if not absolute way.

The uptake of inorganic ^{14}C-labelled carbon by natural populations over short intervals of time has been used widely as a measure of rates of carbon fixation in photosynthesis. Parsons and Strickland (1962) employed labelled organic substrates to measure uptake by natural heterotrophic populations in the dark by the relationship:

$$v = \frac{c \cdot f \cdot (S_n + A)}{C \mu t}$$

when:

v = rate of uptake (mg C m^{-3} hr^{-1}),
c = radioactivity of the filtered organisms (cpm, counts per minute),
f = a correction for isotopic discrimination (1.05 or 5 per cent slower uptake of ^{14}C of greater mass than ^{12}C),
S_n = in situ concentration (mg l^{-1}) of the organic substrate,
A = concentration (mg l^{-1}) of added substrate (labelled and unlabelled),
C = cpm from 1 μCi of ^{14}C-labelled substrate on the radioassay instrumentation used,
μ = quantity of ^{14}C added to the sample,
t = incubation time (hours).

This equation assumes that natural substrate concentrations (S_n) are much less than A. When the uptake of a solute is mediated by a transport system located on or in the cell membrane, the rate of uptake can be described by Michaelis-Menten kinetics:

$$v = \frac{V\,(S)}{K_m + S}$$

when:

  $v =$ velocity at a given substrate concentration S,
  $V =$ maximum velocity, attained when uptake sites are continually saturated with substrate,
  $K_m =$ Michaelis constant, by definition the substrate concentration when the velocity is exactly one-half the maximum velocity V. Also called K_t, the transport constant, a measure of the affinity of the uptake system for the substrate.

As developed by Wright and Hobbie (1966; cf. also Allen, 1969), this nonlinear uptake relation over substrate concentration can be transformed by the Lineweaver-Burk equation to yield:

$$\frac{C\mu t}{c} = \frac{(K_t + S_n)}{V} + \frac{A}{V}$$

With this equation, data from uptake measurements from algae and bacteria at low substrate concentrations can be plotted as $C\mu t/c$ versus A, giving values for $(K_t + S_n)$ and V (Fig. 17–14). The negative intercept on the abscissa is equal to $(K_t + S_n)$, and the reciprocal of the slope is V, the maximum rate of uptake. The ordinate intercept is equivalent to the turnover time (T_t), the time required for complete removal of the natural substrate by the microflora

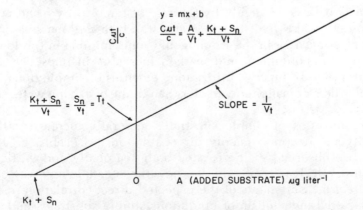

Figure 17–14 Graphical analysis of bacterial uptake at low organic substrate concentrations following Michaelis-Menten enzyme kinetics. A plot of $C\mu t/c$ against increasing substrate concentrations, S, illustrating derivation of (1) maximum natural substrate concentrations, $K_t + S_n$, as $\mu g\ l^{-1}$, (2) maximum velocity, V, as $\mu g\ l^{-1}\ hr^{-1}$, and (3) turnover time for substrate regeneration (T_t) in hours. (After Allen, H. L.: Chemo-organotrophic utilization of dissolved organic compounds by planktic algae and bacteria in a pond. Int. R. ges. Hydrobiol., 54:1-33, 1969.)

(see Hobbie, 1967). The $(K_t + S_n)$ approximates the maximum natural substrate concentration (S_n) if K_t is very small, as is often, but not always, the case. These relationships assume that a constant rate of regeneration of the organic solute is occurring in situ, or that steady state conditions exist. An appreciable portion of ^{14}C organic substrate is respired rapidly by the microflora, and corrections for this loss must be made (Hobbie and Crawford, 1969).

In contrast to the already discussed nonlinear active uptake velocities at low substrate concentrations, uptake of organic compounds by natural planktonic populations at high substrate concentrations ($>$ ca 0.5 mg 1^{-1}) does not exhibit rate limitation kinetics or saturation of uptake sites. Passive uptake velocity continually increases with rising substrate concentrations. The slope of the response line is constant (K_d), derived from diffusion kinetics, and has been used to estimate diffusion uptake by natural populations of algae.

Analyses of utilization rates of organic substrates under in situ conditions with natural populations have been applied in only a few cases, and employing only a few simple substrates such as glucose, other sugars and isomers of glucose, acetate, glycolate, and amino acids. These analyses rarely are extended over an annual period (Wetzel, 1967; Allen, 1969; Hobbie, 1967, 1971). However, despite inherent physiological limitations, the analyses do yield insight into utilization of dissolved organic substrates. Foremost, they demonstrate that dark algal uptake of simple organic substrates for use in heterotrophic growth at low natural substrate concentrations by diffusion is always less than 10 per cent of active permease-mediated uptake by bacteria. Stated in another way, the algae are generally ineffective in competing for the available organic substrates at the substrate concentrations maintained by active bacterial heterotrophic uptake.

Maximum velocities of uptake are quite variable among lakes and substrates, but generally occur after spring and late summer algal maxima. Rates decrease by about an order of magnitude during winter in temperate lakes. Concentrations of substrates such as glucose, acetate, and amino acids remain low throughout the year as the rate of inputs to the labile dissolved organic pool is balanced roughly by the rate of removal by bacteria. Vertical distribution of uptake velocities indicates trends toward maxima during and following algal maxima in the trophogenic zone, minima in the lower metalimnion-upper hypolimnion, and marked increases in rates near the sediments. Within the sediments, utilization velocities of simple organic solutes are several orders of magnitude greater than in the overlying water (Harrison, et al., 1971).

Utilization rates of simple substrates generally increase with greater productivity in an approximately direct relationship (Table 17–10). Great variations are observed in these rates both seasonally and vertically with changes in depth. It is impossible to view these data in any way other than general *direct* measurements of the expected direct correlation between increased bacterial metabolism of ubiquitous simple substrates and waters of increasing inputs of organic matter. Methodological problems, such as competitive inhibition of one substrate on the uptake of another (Burnison and Morita, 1973), and corrections for respiratory losses of CO_2, preclude detailed explanation of interactions.

TABLE 17-10 Comparison of Approximate Rates of Turnover of Organic Substrates by Natural Bacterioplankton in Lakes of Increasing Productivity Based on the Range of Maximum Phytoplanktonic Photosynthetic Rates of Carbon Fixation (P_{max})

Lake	Substrates	P_{max} (mg C m^{-3} day^{-1})	T_t (hours)	Source
OLIGOTROPHIC				
Lapplandic lakes, Sweden (summer)	Glucose, acetate	1–30	< 10,000	Rodhe, et al., 1966
Lawrence, Mich. (annual)	Glucose	1–80	40–300	Wetzel, et al., 1972
	Acetate		10–120	
MESOTROPHIC				
Crooked, Ind. (annual)	Glucose	63–110	80–470	Wetzel, 1967, 1968
	Acetate		20–350	
Klamath, Ore. (summer)	Water { Glucose	–	220	Harrison, et al., 1971
	{ Acetate		250	
	Sediments { Glucose	–	2.25	
	{ Acetate		0.75	
EUTROPHIC				
Erken, Sweden				
Summer	Glucose	40–130	10–100	Hobbie, 1967
Winter	Glucose	2–20	100–1000	
Wintergreen, Mich. (summer)	Glucose	80–120	< 1–20	Saunders, pers. comm., Wetzel, unpublished
Little Crooked, Ind. (annual)	Glucose	190–205	36–232	Wetzel, 1967, 1968
	Acetate		24–190	
Duck, Mich. (annual)	Glucose	10–320	8–50	Miller, 1972
	Acetate		4–40	
Pamlico River (estuary) N.C.	Amino acids	–	1.5–26	Hobbie, 1971
Lötsjön, Sweden				
Summer	Glucose	< 100	0.4–5	Allen, 1969
Winter	Glucose	< 20	20–300	

Fatty Acids and Related Compounds

Although relatively little is known about the distribution and metabolism of lipid compounds in fresh waters, the detailed investigations of Poltz (1972) on lakes of northern Germany demonstrate relationships that probably occur frequently. It is not known at this time how generally applicable these results are.

The distribution of lipid fractions among the largely living net plankton, sedimenting particulate matter, and sediments shows a distinct reduction towards the sediments (Table 17–11). The range of variations about these means, as well as among fatty acid patterns, was large on a short-term basis in response to the rapid changes in species composition of phytoplankton and zooplankton and changes in their distribution by both normal population oscillations and vertical migrations. Long-term changes in the lipid content and fatty acid compositional patterns of the plankton over an annual period were not as great, however, because of opposing changes in lipid content of the phyto- and the zooplankton. The relative amount of short-chained fatty acids, especially of the 16:1 isomer, increased from summer to winter so that the mean length of the total fatty acid composition decreased. The same fatty acids were found in the triglycerides of the plankton as in the total fatty acid

TABLE 17–11 Average Percentage by Weight (Dry) of Lipid Fractions in the Net Plankton, Sedimenting Particulate Organic Matter, and Sediments of Several Lakes of Northern Germany[a]

Fractions	Net Plankton (>20 μm)	Sedimenting Particulate Matter	Sediment[b]
Total organic matter per dry weight	83.1%	36.7%	16.0%
Total lipids per organic matter	22.3	9.0	6.8
Total fatty acids per total lipids	61.2	32.5	12.1
Triglycerides of total lipids	30.8	8.0	1.8

[a]After Poltz, 1972.

[b]Grosser Plöner See only; mean values; values greatest near surficial interface and decreased progressively within first meter.

fractions, but the relative concentrations of highly unsaturated fatty acids were less in the triglycerides.

Decomposition of organic matter and fatty acids was shown to be most rapid under the warmer, aerobic conditions of the epilimnion. While the mineralization of the total particulate organic matter of the pelagial zone was rapid and usually exceeded 85 per cent, the intensity of decomposition of triglycerides was greatest (> 98 per cent) in the epilimnion and followed the general sequence of degradation: triglycerides > total fatty acids > lipids > total organic substances (Poltz, 1972). Rates of decomposition decreased markedly in the hypolimnion. Only small amounts of the substances produced in the plankton were found settling to the sediments: 4 to 10 per cent of particulate organic matter, 1 to 2.5 per cent of lipids, 0.1 to 0.5 per cent of total fatty acids, and less than 0.1 per cent of triglycerides. After deposition, further degradation rates were most rapid at the surface interface zone of the sediments, although the rates were slower than those occurring in epilimnetic waters during sedimentation.

These analyses were limited to the pelagic zone of lakes, and consequently do not reflect the massive amounts of particulate organic matter that can reach the sediments of many lakes when the littoral flora dominate production. The sediment interface is clearly the site of most intensive respiratory metabolism in many lakes (see discussion further on), and much of the organic matter undergoing degradation there originates from littoral and allochthonous sources in addition to residual sedimenting seston. Part of the lipid derivatives originates from the bacteria and fungi and intermediate products of their metabolism during the degradation of this accumulated organic matter (cf. Schulz and Quinn, 1973).

Humic Compounds

The structural composition of humic compounds is complex, and generally bacterial degradation of phenolic linkages is difficult. Particulate and especially dissolved humic compounds, derived from both autochthonous and largely allochthonous sources, tend to have long residence times in lakes and are metabolized only very slowly. Haan (1972) demonstrated that a species of *Arthrobacter* was able to grow slowly on humic fractions and that it was able to hydrolyze the phenolic ether linkages. Studies of the degradation of dissolved humic compounds of heavily stained Finnish lake waters also provide evidence that these compounds, although relatively resistant, are mineralized slowly (Ryhänen, 1968; Sederholm, et al., 1973). Increased concentrations of inorganic nitrogen and phosphorus accelerated this degradation.

Studies of the effects of a fulvic acid fraction, isolated from a lake in the Netherlands, on the growth of a bacterial species of *Pseudomonas* showed that the presence of fulvic acid caused an increase in the cell yield and in the cell yield per respiratory consumption of oxygen as compared to growth in organic media lacking the fulvic fraction (Haan, 1974). These results suggest that the stimulating effect of fulvic acid on bacterial metabolism operates as a compound coupled to the metabolism of less refractory organic substrates. Cometabolism of many very refractory compounds, e.g., herbicides, is well-known (Horvath, 1972), and is probably an important mechanism for the degradation of the more refractory compounds of fresh waters. The mechanisms of such cometabolism deserve greatly increased investigation.

Distribution of Bacteria in Lakes

Based on the preceding discussion of direct measurements of utilization rates of selected simple organic substrates, a general increase in bacterial biomass with increased loading of the lake system with organic substrates would be expected, provided that the organic matter was of a quality subject to reasonable degradation under the typical range of limnological conditions. The most direct approach to evaluation of microbial biomass is enumeration, either by isolation with cultures or by direct microscopic observation (cf. detailed discussions of techniques by Kuznetsov and Romanenko, 1963; Rodina, 1965, 1971; Sorokin and Kadota, 1972; Rosswall, 1973; and Romanenko and Kuznetsov, 1974).

Without elaborating on the limitations of techniques to any extent, the culturing of bacteria on media, while mandatory for physiological identification purposes, is extremely selective for specific bacteria under most conditions. These methods underestimate heterogeneous indigenous bacterial populations by factors from 1000 to 100,000 times (Fig. 17–15). Direct enumeration by microscopy has the inherent difficulty of discriminating viable cells from detritus and cells that are dead, but recent techniques, such as epifluorescence microscopy (Francisco, et al., 1973), provide great improvements.

Measurements of bacterial biomass separate from nonliving detrital organic matter can also be approached by measurement of a uniformly distributed cellular constituent. Adenosine triphosphate (ATP) fulfills these criteria in that it decomposes readily on death, and can be measured by bioluminescent reactions at extremely low concentrations. The ATP concentra-

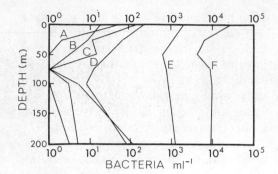

Figure 17–15 Bacterial populations of California coastal waters obtained by different methods: *A*, Agar pour plate method, *B*, Macrocolonies on membrane filters, *C*, Extinction dilution method, *D*, Microcolonies on membrane filters, *E*, Direct microscopic enumeration on membrane filters, and *F*, Cholodny method of direct enumeration of microorganisms on glass slides after filtration. (After Jannasch and Jones, 1959.)

tions permit an estimation of total biomass in terms of organic carbon or dry weight, because the ATP levels of microbial cells are fairly uniform (Holm-Hansen and Paerl, 1972).

A general relationship has been observed in which the numbers and biomass of bacteria increase with increasing concentrations of inorganic and organic compounds among lakes (Table 17–12). In spite of great seasonal variations in response to individualistic characteristics, bacterial numbers and biomass increase from oligotrophic to eutrophic lakes. Highest concentrations have been observed in eutrophic reservoirs. Numbers decrease

TABLE 17–12   Range of Bacterial Numbers in the Surface Waters in Summer Among Lakes of Differing Productivity[a]

Lake Type	Lake	Bacterial Numbers $(\times 10^{-3}\ ml^{-1})$
Oligotrophic	Baikal	50–200
	Onezhskoe	240–340
	Ladozhskoe	100–300
	Pert	130
	Konch	170
	Dal'nee	120
Mesotrophic	Glubokoe	1000–1400
	Imolozh'e	450
	Kolomenskoe	1100
Eutrophic	Beloe	2200
	Chernoe	4000
	Baturin	3500–8000
	B. Medvezh'e	3700
	B. Krivoe (saline)	12,300
	M. Umreshchevo (saline)	12,300
Eutrophic Reservoirs	Rybovodnye (pond)	1000–40,000
	Dubasarskoe	3000–16,000
	Kakhovskoe	to 40,000
	Kashkorenskoe	7800–57,900
Dystrophic	Torfianoi (quarry)	2300
	Piiavochnoe	1070
	Chernoe	430

[a]After Kuznetsov, 1970.

TABLE 17–13 Relationship Between Average Rates of Phytoplanktonic Primary Production and Bacterial Numbers at 1 m Depth in Four Lakes of Northern Germany[a]

Lake	Primary Productivity (mg C m^{-3} day^{-1})	Saprophytes (no. ml^{-1})	Total Bacteria ($\times 10^{-6}$ ml^{-1})
Schöhsee	16	300	0.5
Schluënsee	17	350	0.5
Grosser Plöner See	35	650	1.0
Pluss-See	57	1000	1.1

[a]Modified from Overbeck, 1965.

markedly in acidic dystrophic lakes of high concentrations of humic organic matter. In a similar manner, production of bacteria as calculated from changes in biomass over short time periods increases with increasing productivity of lakes and the average generation time decreases among more eutrophic lakes. A directly proportional relationship has been demonstrated between the average rates of phytoplanktonic productivity and bacterial numbers and production (Table 17–13).

Seasonal and Vertical Distribution

The seasonal distribution of bacterial populations is highly variable from lake to lake and from year to year within a lake. A few generalizations, however, can be advanced even from population data of sampling frequencies which exceed by far the generation time of the bacteria. The most conspicuous is the often-found rather close correlation between the numbers of phytoplanktonic algae and heterotrophic bacteria both seasonally (Fig. 17–16) and vertically (Fig. 17–17). But while there is no doubt that the

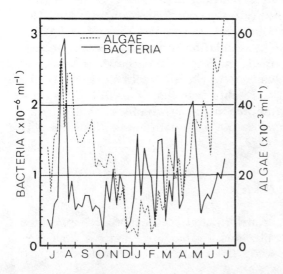

Figure 17–16 Annual cycle of heterotrophic bacteria and plankton algae in the surface water (0.5 m) of a pond in Potsdam, Germany, 1958-59. (After data of Overbeck and Babenzien, 1964.)

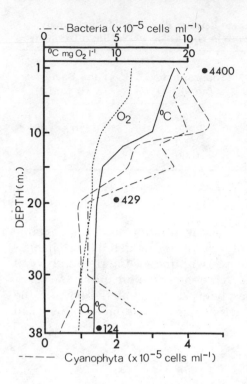

Figure 17-17 Vertical distribution of heterotrophic bacteria in relation to plankton algae in Schluënsee, northern Germany, 18 August 1965. Circles = saprophytes ml^{-1}. (After data of Overbeck, 1968.)

heterotrophic metabolism of the bacteria depends to a significant extent on organic substrates produced by the phytoplankton, such correlations are often weak. More frequent sampling has demonstrated that the vertical distribution of bacteria in dimictic lakes can change very rapidly (Rasumov, 1962; Saunders, 1971). For example, the rapid changes observed in Figure 17-18 became particularly acute in midJune following a phytoplanktonic maximum when bacterial numbers increased sharply and decreased again in a few days. The rapid inversions in vertical distribution of the waxing and waning bacterial populations indicate that much shorter term variables and processes controlling distribution are operational (Saunders, 1971). Correlations of a general nature suggest relationships between production sources of organic matter and, at times, grazing by zooplankton, but yield little insight into the control mechanisms of the interactions. Further, the frequency of sampling must be considerably less than the life of the organism, a few hours in this case, to approach the mechanisms that regulate microbial metabolism. Studies of this nature on in situ populations are very few.

Decomposition of Particulate Organic Detritus

Experimental studies on the transformation of artificially generated radioactive algal detritus under conditions approaching natural lake conditions are instructive in understanding the relationships involved in lake

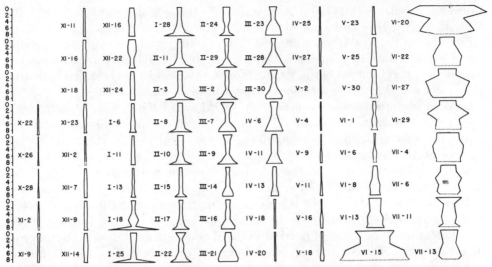

Figure 17–18   Relative vertical distribution of bacteria by direct enumeration of different dates in Frains Lake, southeastern Michigan, 1959-60. Ordinate: depth in meters; abscissa: relative numbers. (From Saunders, G. W.: Carbon flow in the aquatic system. *In* Cairns, J., Jr., ed., The Structure and Function of Fresh-Water Microbial Communities. Blacksburg, Va., Virginia Polytechnic Institute, 1971.)

systems (Saunders, 1972). As was indicated earlier (Fig. 17–1), the living components of the water constitute a very small portion of the total organic matter; most organic matter occurs as dissolved organic matter (DOM) and in the particulate phase as detrital particulate organic matter (POM). For example, in the surface water of a productive lake, the summer POM averaged 57 per cent of the total seston (inorganic and organic; Table 17–14). The phytoplanktonic dry weight ranged from between 5 and 25 per cent of the organic particulate detritus most of the time, and only occasionally exceeded 40 to 50

TABLE 17–14   Ratio of Dry Weight of Organic Particulate Detritus to Phytoplankton, and Organic Particulate Detritus as a Percentage of Seston, Frains Lake, Michigan[a]

Time Period	Ratio of Detrital POM to Phytoplankton	Organic Particulate Detritus as a % of Seston
21 Jan. 1967	5.7	—
27 Apr. 1967	4.6	45
3 May 1967	6.6	72
9 May 1967	16.9	65
16 May 1967	15.1	67
23 May 1967	2.3	21
30 May 1967	12.0	58
6 June 1967	4.5	57
14 June 1967	4.9	61
20 June 1967	8.4	68
27 June 1967	4.8	62
22 June 1968	1.3	44

[a]After Saunders, 1972.

per cent. Similar results have been shown for the seston composition of Lake Erie (Leach, 1975) and several Russian lakes (Winberg, et al., 1970). Considering that the POM constitutes only about 10 per cent of the total detrital organic matter (ca 90 per cent as DOM), and that living organic matter makes up only a small fraction of the particulate component, it is clear that quantitatively organic detritus is a major component of aquatic systems.

Most of the soluble components of DOM are utilized rapidly by bacteria in a period of usually less than 5 days. Decomposition of the particulate organic detritus, generated from planktonic communities of surface lake water, occurs more slowly at rates on the order of 10 per cent per day (Saunders, 1972). Soluble radioactive carbon derived from the particulate organic detritus amounted to about 1 per cent of the initial radioactivity after one day, values that were about 2 to 6 times greater than that of dissolved organic matter released extracellularly by phytoplankton in the trophogenic zone.

Decomposition rates of detrital POM of course are influenced by an array of compositional and environmental factors. Certainly differences in the organic composition of algal cell walls lead to differences in degradation rates. For example, POM of dead blue-green algae decomposes much more rapidly than that of green algae and diatoms; desmid algal POM is especially resistant (Gunnison and Alexander, 1975). The concentrations and species diversity of bacteria also are involved, being functions in part of the amount and diversity of organic substrates available in the aquatic system. Among environmental parameters, temperature is paramount, and availability of inorganic nutrients, especially nitrogen, can be an important variable.

The control of decomposition rates of organic detritus (soluble or particulate) by bacteria can be considered simply as the rate of change between bacteria and detrital substrates, compounds, or particles, and can be expressed as a bimolecular second-order reaction (Saunders, 1972b):

$$(m_1, m_2 \ldots) \qquad (T, [N], n_1, n_2 \ldots)$$
$$\downarrow \qquad \qquad \downarrow$$
$$\frac{-d\,[S]}{dt} = C_1\,[S] \qquad C_2\,[B]$$

when:
  [S] = concentration of detrital substrate,
  [B] = concentration of bacteria,
   T = temperature,
  [N] = concentration of inorganic nutrients,
  $m_1, m_2 \ldots$; $n_1, n_2 \ldots$ = other variables,
  C_1, C_2 = coefficients.

The rate of decomposition of detrital substrates is a function of their concentration and of the enzymatic activity concentration of the bacteria dispersed in the water or attached to surfaces of detrital particles (Saunders, 1972b). The coefficients C_1 and C_2 are obviously highly dynamic and variable operators on the detrital-bacterial interactions. Control variables that operate on the substrate coefficient C_1 include biodegradability factors (Alexander, 1965b). Examples are many. A structural characteristic of the molecule of the substrate or of the particle surface parameters (that is, area, particle size, adsorption sites) can prevent the enzymes from acting. Certain requisite enzymes can be inactivated by adsorption to surfaces of minerals or colloids, or inhibited by phenolic and polyaromatic substrates or their derivatives. Inaccessibility of the substrates includes not only surface characteristics, but also transport of the substrates to microenvironments in which conditions preclude or greatly reduce bacterial activity or enzymes from reaching the substrates.

Control variables that operate on the bacterial component C_2 include obvious parameters such as temperature T, inorganic nutrient concentrations [N], as well as other essential growth factors, for example, a biologically utilizable terminal electron acceptor. The environment can be variably toxic to bacterial growth in the presence of biologically generated organic inhibitors, microbially formed inorganic inhibitors, high salt concentrations, temperature extremes, acidity, or other environmental conditions outside the range suitable for microbial proliferation (Alexander, 1965b). Moreover, a microbial community, because it lacks the appropriate enzyme system, may not be able to metabolize certain compounds, or the substrate may not be able to penetrate into cells that possess the appropriate enzymes. Thus the decomposition rates of organic substrates in natural systems are a function of the physical and biological structure at any instant in time, as well as of the substrate characteristics and physiological state of the microflora and their enzymatic systems.

Within this general decompositional framework, it is then desirable to know the flux rates of organic carbon among the different components of the system. As we have seen earlier (Fig. 17–1), the system structure is dominated by 2 very large pools of nonliving carbon, the inorganic and dissolved organic carbon. Algae and particulate organic detritus components contain much less carbon, zooplankton and other fauna contain even less, and bacteria contain about 10 per cent of that of the zooplankton (Saunders, 1971). This approximate general distribution of carbon mass of the trophogenic zone of lakes bears little relation to carbon flux rates, however.

Taking the carbon structure of the lake's pelagic system on the same day, metabolic flux rates indicate that the microflora control the movement of carbon (Fig. 17–19). The biochemical processes of the biota utilize the carbon of the pools within the limits of varying environmental conditions. These

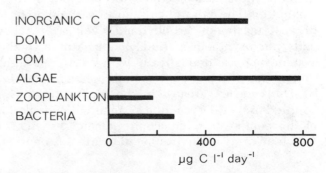

Figure 17–19 Input and assimilation rates of carbon in μg C l^{-1} day^{-1} by different microbial components at 1 m in Frains Lake, Michigan, on 22 July 1968. (Modified from Saunders, G. W.: Carbon flow in the aquatic system. *In* Cairns, J., Jr., ed., The Structure and Function of Fresh-Water Microbial Communities. Blacksburg, Va., Virginia Polytechnic Institute, 1971.)

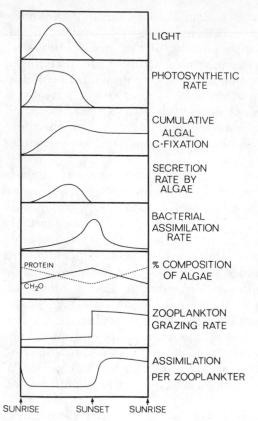

LIGHT

PHOTOSYNTHETIC
RATE

CUMULATIVE
ALGAL
C-FIXATION

SECRETION
RATE BY
ALGAE

BACTERIAL
ASSIMILATION
RATE

% COMPOSITION
OF ALGAE

ZOOPLANKTON
GRAZING RATE

ASSIMILATION
PER ZOOPLANKTER

SUNRISE   SUNSET   SUNRISE

Figure 17–20 Generalized diagrams of the distribution of concentration and rates of processes from sunrise to sunrise in the upper trophogenic zone of a lake system in summer. (Modified from Saunders, G. W.: Carbon flow in the aquatic system. *In* Cairns, J., Jr., ed., The Structure and Function of Fresh-Water Microbial Communities. Blacksburg, Va., Virginia Polytechnic Institute, 1971.)

environmental variables are changing constantly on a daily basis, resulting in continually changing rates of transfer among the carbon pools. The changes in concentrations and metabolic rates in the surface waters in summer over a 24-hour period can be generalized in an exemplary way in Figure 17–20. Photosynthetic rates of algae are coupled with light incidence, and usually exhibit rates skewed toward the morning period of high light intensity, decreasing precipitously with decreasing light in evening (Saunders and Storch, 1971). Coupled with this secretion cycle, bacterial assimilation rates increase and reach maximal levels about 3 hours following the secretion maximum. The ratio of protein to carbohydrate of the algae also shifts diurnally; maximum carbohydrate levels are reached at the conclusion of the daily photosynthetic period. Zooplanktonic grazing rates on the epilimnetic system increase dramatically with the loss of light and rapid diurnal migration of many zooplankters to the trophogenic zone. Assimilation rates of algae by *Daphnia* are greater in the night than during the day. This series of coupled oscillating reaction systems collectively forms the major components of the composite microscopic pelagic system in the trophogenic zone, on which an array of environmental variables operate.

Sedimentation and Decomposition of Pelagic
Particulate Organic Matter

The vertical distribution of bacteria in stratified lakes, with greatest numbers and activity found in the trophogenic zone underlain by a meta-limnetic minimum and an increase in the lower hypolimnion, also is expressed in the decomposition of sedimenting organic matter produced in the epilimnion. A number of devices have been used to measure the amount of tripton, the term for all nonliving suspended matter (cf. methods discussed in Edmondson and Winberg, 1971). Containers of various types are suspended at several depths to entrap sedimenting materials over a period of 1 to several weeks, depending on the sedimentation rates. After a period of collection, kept as short as possible to reduce complicating interferences, chemical analyses of the collected seston permit an approximate indirect evaluation of changes in the organic matter that have occurred during sedimentation by decomposition.

The soluble organic matter synthesized by planktonic primary production and released by secretion and autolysis decomposes rapidly to a major extent (> 75 per cent) at the trophogenic site of generation and release, as discussed previously. The particulate organic detritus undergoes slower degradation. During stratification periods in lakes of moderate depth, 75 to > 99 per cent of the particulate organic matter synthesized in the trophogenic zone is decomposed in the water column by the time it reaches the sediment interface of the lower tropholytic zone (Table 17–15). Similarly, in Lawrence Lake, Michigan, an annual mean of 88 per cent of the sedimenting particulate organic carbon was decomposed by the time it reached the lower epilimnion (5 m; Wetzel, et al., 1972). Little further decomposition occurred on further hypolimnetic sedimentation; an annual mean of 10 per cent of planktonic particulate organic carbon of the trophogenic zone was found at 10.5 m. Similar results were obtained by Lawacz (1969) using different methods.

Even though the degradation of particulate organic matter is fairly complete with increasing depth, seasonal variations are great (Fig. 17–21). The increasing concentrations of sedimenting organic matter with depth must be considered in relation to the decreasing area of each stratum as the basin morphometry constricts with depth. This relationship is illustrated diagrammatically in Figure 17–22 for Schluënsee, in which the funneling effects of the basin morphology increased the concentration of organic carbon per area, even though the total amount of synthesized particulate organic matter was greatly reduced (> 97 per cent) by decomposition at a cumulative depth of 40 m (Ohle, 1962). As the size of the basin decreases, the significance of the funneling effects of concentration increases.

As lakes become more shallow and the importance of littoral production in relation to that of the phytoplankton increases, a greater proportion of the synthesized organic matter is shifted to the sediments for benthic metabolism by decomposition. Similarly, related to the time required for decomposition of more refractory components of algal cells (see, for example, Jewell and McCarty, 1971) and the distance (and time) available during sedimenta-

TABLE 17–15 Percentage of Organic Carbon Produced by Phytoplanktonic Primary Production That Was Mineralized by Bacterial Decomposition per Area at Depth and at the Bottom of the Water Column in 3 Lakes of Northern Germany[a]

Schluënsee Dates	Depth (m)	% Mineralization Per Area	Total	Schöhsee Dates	Depth (m)	% Mineralization Per Area	Total	Pluss-See Dates	Depth (m)	% Mineralization Per Area	Total
9 June–6 July	20	58.8		11 April–18 May	10	52.8		11 May–30 May	5	93.1	
	30	33.5			20	30.7	83.5		15	5.0	
	40	5.0	97.3						25	0.9	98.9
6 July–16 Aug.	20	73.7		18 May–4 July	10	75.1		30 May–23 June	5	90.6	
	30	12.2			20	16.9	92.0		15	7.6	
	40	11.7	97.6						25	1.3	99.5
16 Aug.–7 Sept.	20	77.6		4 July–10 Aug.	10	26.8		23 June–8 Aug.	5	89.8	
	30	9.8			20	54.0	80.8		15	9.1	
	40	1.5	98.5						25	0.9	99.7
7 Sept.–5 Oct.	20	87.8		10 Aug.–5 Sept.	10	75.3		8 Aug.–16 Sept.	5	83.7	
	30	4.2			20	18.8	94.1		15	13.5	
	40	6.8	98.8						25	2.6	99.7
		Mean = 98.1%		5 Sept.–10 Oct.	10	17.2		16 Sept.–4 Nov.	5	70.0	
					20	77.2	94.4		15	26.5	
									25	2.5	99.1
				10 Oct.–17 Nov.	10	12.9				Mean = 99.4%	
					20	80.7	93.6				
				17 Nov.–5 Dec.	10	8.8					
					20	74.3	83.1				
						Mean = 88.8%					

[a]After Ohle, 1962.

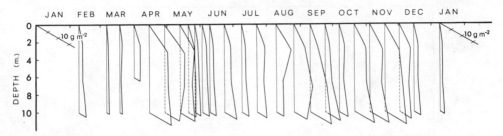

Figure 17–21 Seasonal changes in the organic content of sedimenting seston (g m^{-2}), Lawrence Lake, Michigan, 1972. (From White and Wetzel, unpublished.)

tion for aerobic decomposition in deeper lakes, we would expect decomposition to be more complete in the water column of larger, oligotrophic lakes. Empirical evidence supports this conclusion.

Resuspension and Redeposition

A conspicuous feature of the annual cycle of sedimentation rates of seston among dimictic lakes is the marked increase during periods of spring and autumnal circulation (Fig. 17–23). The extent of resuspension of surficial sediments into the water column by the circulating water is of course variable among lakes of differing morphology and according to variations in meteorological conditions prior to initiation of stratification (Pennington, 1974).

Figure 17–22 Schematic diagrams of (*upper*) the relationships between planktonic primary production and sedimentation of organic particulate carbon in Schluënsee, Germany, 16 July–16 August 1960, and (*lower*) the morphometric funnelling effect on sedimentation in lakes. (Modified from Ohle, 1962.)

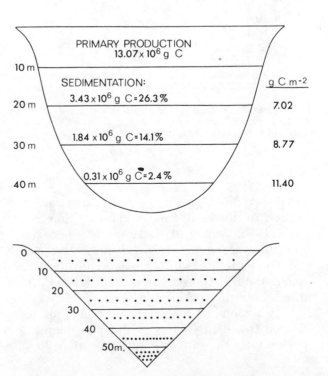

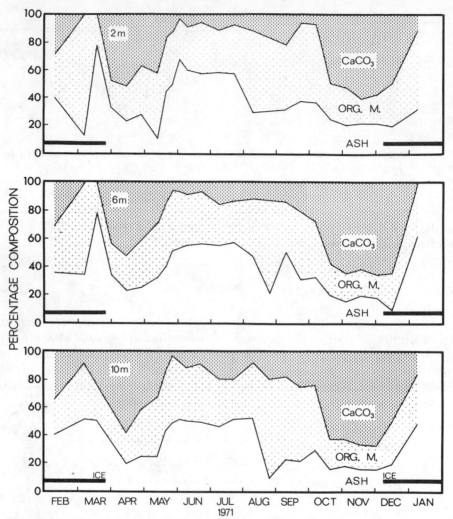

Figure 17–23   Seasonal changes in the percentage composition of calcium carbonate, particulate organic matter, and ash sedimenting to 2, 6, and 10 meter depths, Lawrence Lake, Michigan, 1971. (From White and Wetzel, unpublished.)

However, it is clear that even the deepwater sediments of many lakes are disturbed to depths of from several millimeters to several centimeters (Davis, 1968, 1973; Wetzel, et al., 1972). This fine-grained particulate matter is swept into the water column. In calcareous lakes, as exemplified in Figure 17–23, calcium carbonate forms a major constituent of the seston during the periods of circulation. Resuspension of surface sediments occurs at irregular intervals throughout the year in nonstratified lakes.

Horizontal differential sedimentation of particulate matter has been demonstrated in studies of the movement, deposition, and redeposition of pollen grains of different sizes and densities by Davis and Brubaker (1973).

Larger particles with rapid sinking rates are deposited evenly onto the sediments throughout the lake basin. Smaller-sized particles with slower sinking rates in water are kept in suspension in the turbulent waters of the epilimnion, are carried across the lake in wind-induced water currents, and are deposited preferentially onto littoral sediments. Later, particularly during fall circulation, these fine particles are resuspended in the littoral deposits, mixed in the lake water, and redeposited over the entire basin. Larger particles of organic matter or seston are not extensively resuspended and transported from the littoral to the center of the lake, although they are frequently shifted about in the littoral region (Wetzel, et al., 1972). The result is a size-selective sorting and a transport of the finest-sized particles from the littoral, which sediment into the quiescent waters below the metalimnion (cf. also Sebestyén, 1949).

Littoral Decomposition and Microbial Metabolism

Evaluations of the horizontal distribution of bacteria in lakes demonstrate a conspicuous increase in densities in the littoral zone as compared to the open water. Although comprehensive studies of these spatial distributions are few, examples from both small and large lakes are consistent. The quantity of total bacterioplankton of the littoral zone of Lake Balaton, Hungary, determined by direct enumeration, was found to be 2 to 3 times greater than that of the open water (Oláh, 1969a, 1969b). Saprophytic bacteria[1] of the littoral zone were 40 to 120 times more abundant than in the open water. The observed differentiation was greatest during the spring period of active growth of the emergent macrophyte reed zone (*Scirpus* and *Phragmites*) and in the autumn as the littoral foliage senesced. High numbers of bacteria of the littoral zone extended out into the open water for considerable distances (1500 meters) in this large lake during the spring, but they were restricted largely to the macrophyte belt during the summer.

Similarly, the microbial activity of large, shallow Lake Mekhtev, USSR, was influenced strongly by the aquatic macrophytes and associated sessile algal flora (Aliverdieva-Gamidova, 1969). While temperature plays an overriding role in bacterial growth on a seasonal basis, and maximal populations and shortest generation times occur in midsummer (Table 17–16), bacteria clearly are more abundant and on the average grow faster in water among macrophyte stands than in open water free of littoral flora. While these studies indicate that massive amounts of organic substrates are being released by the macrophytes and their epiphytic microflora during growth and senescence, measurements of actual transfer rates are not available.

---

[1]The term saprophytic bacteria, as used in these and similar studies, refers to those bacterial colonies that grow on organic-rich agar such as Difco nutrient agar or casein-glucose agar. These techniques are highly selective but are a sensitive reflection of differences in concentration of organic substances easily assimilable by bacteria.

TABLE 17-16   Distribution and Activity of Bacterioplankton in Water Among Emergent Bulrush Vegetation and in Open Water Without Macrovegetation in Lake Mekhteb, Dagestan, USSR[a]

Period	Water Temperature (°C)	Water Among Macrophytes			Open Water		
		Total Bacteria ($\times 10^{-6}\ ml^{-1}$)	Generation Time (hours)	Total Saprophytes ($\times 10^{-3}\ ml^{-1}$)	Total Bacteria ($\times 10^{-6}\ ml^{-1}$)	Generation Time (hours)	Total Saprophytes ($\times 10^{-3}\ ml^{-1}$)
Dec.–Jan.: no bulrushes	0–2	0.55	19–17	5–12	0.55	19–27	5–12
Feb.–Mar.: young bulrushes	3–7	1.4	11–16	28–39	1.0	17	22–103
Apr.–May: developing stands	17–23	1.3–1.8	10–13	11–31	1.4	10–12	14–21
June: mature bulrushes	26	3.3	5.2	40	3.0	7.1	31
July	28	5.3	4.4	232	2.5	6.5	28
Aug: declining bulrushes	21	4.3	7.6	178	3.1	6.3	81
Oct.	11	1.5	11	21	1.1	23	17
Nov.	3	0.85	21	0.2	0.27	18	0.4

[a]After data of Aliverdieva-Gamidova, 1969.

TABLE 17–17   In Situ Percentage Loss of Dry Organic Matter of Aquatic Macrophytes Over Long Periods of Time

| Plant Type | 7–14 Days | % Loss of Dry Weight | | | | | | | Source |
		4 Weeks	6 Weeks	9 Weeks	12 Weeks	15–16 Weeks	24 Weeks	43 Weeks	
EMERGENT:									
Juncus effusus	5	—	—	—	30	60	65	—	Boyd, 1971
Typha latifolia	5	—	—	—	25	30	45	—	Boyd, 1970
Spartina alterniflora	2	—	—	—	10	30	50	85	Burkholder and Bornside, 1957
Carex gracilis	—	42	42	55	—	—	—	—	Koreliakova, 1959
Sparganium simplex	22	—	48	—	58	100	—	—	Koreliakova, 1958
Salix leaves (overwintered)	17–56	—	—	—	—	—	—	—	Pieczyńska, 1972
Phragmites communis									
Overwintered plants	40	—	42	—	50	53	—	—	Koreliakova, 1958
Fresh plants	—	43	48	63	—	—	—	—	Koreliakova, 1959
FLOATING-LEAVED:									
Lemnaceae	20	—	—	—	90	—	—	—	Laube and Wohler, 1973
Brasenia schreiberi									
Leaves in surface water	(2)	—	—	—	45	—	50	95	Kormondy, 1968
Leaves at sediment	20	—	—	—	20	—	45	90	
Nymphaea odorata									
Leaves in surface water	(15)	—	—	—	45	—	70	98	
Leaves at sediment	(20)	—	—	—	58	—	75	95	
Polygonum amphibium									
Overwintered plants	25	—	35	—	35	37	—	—	Koreliakova, 1958, 1959
Fresh plants	—	42	46	55	—	—	—	—	
SUBMERSED:									
Potamogeton lucens	6–92	—	—	—	—	—	—	—	Pieczyńska, 1972
P. perfoliatus	6–95	—	—	—	—	—	—	—	

Losses of Macrophytic Particulate Organic Matter

Earlier discussion indicated that the microbial utilization and degradation of dissolved organic matter released extracellularly during active plant growth and by autolysis was rapid and relatively complete. In contrast, the degradation of the particulate organic matter of the littoral zone is much slower. Studies are few and generally employ approximate techniques of following the rates of weight loss over periods of time in experimental or in situ situations.

Emergent macrophytic vegetation generally exhibits very slow rates of degradation in relation to those of floating-leaved and submersed aquatic plants (Table 17–17). Part of this greater resistance to decomposition is related to the greater extent of lignified tissue found in emergent flora (Sculthorpe, 1967). With adaptation to a largely or totally submersed existence, a conspicuous reduction in lignification of supportive and conducting tissues is found in floating and submersed vegetation. Lignin, consisting of aromatic rings with side chains and $-OH$ and $-OCH_3$ groups, exhibits a high degree of chemical stability as a group of polymerized amorphous substances of similar composition. Catabolism of lignin by oxidative depolymerization is restricted largely to fungi, especially the Hymenomycetales (Basidiomycetes), which possess the inductive exoenzymes for this cleavage (Trojanowski, 1969). The role of bacteria in the decomposition of lignin macromolecules is limited primarily to side chains or to monomers detached from the lignin by fungi.

The sources of accumulated organic matter in the littoral zone are variable among lakes in relation to morphometry and production by the littoral algal and macrophyte components. In most situations, a majority of the particulate organic matter originates from the macrophytes (ca. 90 per cent), most conspicuously from the emergent forms. Lesser quantities are derived from attached algae, loosely attached littoral algae, and windrowed planktonic algae. Rates of decomposition of the littoral particulate organic matter are quite variable in relation to conditions of accumulation (Table 17–18). In the example from Mikolajskie Lake, decompositional rates were lowest among plant material in the emergent part of the eulittoral zone on the surface of the sediments. Senescence and death of submersed species in the submersed littoral site delayed decomposition; when dead and decaying in pools and among reed heaps, decomposition was very rapid. Filamentous algae and accumulations of planktonic blue-green algae in the littoral decayed much more rapidly (> 95 per cent in 3 to 10 days) under comparable conditions (Pieczyńska, 1970).

Degradation Patterns of Littoral Flora

Lignified tissue of emergent hydrophytes is degraded slowly by a succession of fungal flora. For example, it has been observed that the mycoflora of the cattail *Typha latifolia* passes through early stages which are similar to those of terrestrial plants, with primary colonization by topical leaf surface

TABLE 17–18   Decomposition Rates of Macrophyte Particulate Organic Matter in Various Habitats of the Littoral Zone of Mikolajske Lake, Poland, 1967[a]

Plant Material	Period	Emergent, 1 m from Shore Line	Reed Heaps on the Shore Line	Small Pools on Shore Partly Isolated from Lake	In Water, 0.5 m Depth, 2 m from Shore Line
EMERGENT:					
Phragmites communis leaves	Mid June	4	25	43	21
	Late July	4	34	57	14
	Early Sept.	6	27	38	24
Salix leaves	Mid June	5	35	37	18
	Late July	7	28	56	25
	Early Sept.	4	30	39	17
SUBMERSED:					
Potamogeton lucens	Mid June	3	35	89	—
	Late July	5	41	92	5
	Early Sept.	6	—	77	4
Potamogeton perfoliatus	Mid June	1	38	83	7
	Late July	6	40	95	6
	Early Sept.	7	50	70	8

[a]After Pieczyńska, 1970. Percentage losses of dry weight after 10 days of in situ exposure.

fungi, followed by a secondary phase in which many species of *Leptosphaeria* dominate moribund leaves (Pugh and Mulder, 1971). The succession is associated with aging of plant material and substrate changes. Final stages of decomposition are accompanied by a dominance of fungi predaceous on nematodes. Similar successional patterns have been found on analogous substrates (Sparrow, 1968). The roots and rhizomes are very poorly colonized by fungi.

The movement of fine particles of particulate organic detritus of decomposing macrophytes from the littoral zone to the open water can occur by water movements and currents returning from the littoral to the pelagic zones, both along the surface and near the sediments (cf. Schröder, 1973). In addition to these fine particles, nutrients, particularly phosphorus, are transported in major quantities from the littoral flora to the open water. The maximum phosphorus inputs from the littoral occur in the summer period when the phosphate concentrations in the open lake water are minimal.

Very fine-sized (< 100 μm) particles of the emergent reed *Phragmites* were studied as they underwent decomposition in lake water (Oláh, 1972). The successional patterns of bacterial populations in the water containing leached dissolved organic matter from the particles and on the particles correlate with progressive utilization of organic substances (Fig. 17–24). Within a few hours large rods dominated the liquid phase, presumably utilizing the leached dissolved organic substrates of the detritus. A large population of small cocci succeeded these bacteria, and a few days later were displaced by large cocci; both were associated with the liquid phase and populated the surfaces of the small *Phragmites* fragments. Myxobacteria, well-known decomposers of living and dead particulate organic matter (Ruschke, 1968), colonized both the large cocci and the residual, stabilized detrital particles.

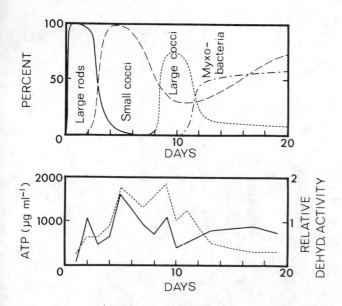

Figure 17–24 *Upper*, bacterial populations, and *lower*, activity associated with decomposition of fine particles of *Phragmites* detritus. ——— = ATP biomass carbon; ---- = relative dehydrogenase activity. (Drawn from data of Oláh, 1972.)

Bacterial biomass, measured as ATP biomass, coincided with the maximal development of the bacterial population succession and was maximal at the time of the mass development of small cocci (Fig. 17–24). Bacterial metabolism, determined by dehydrogenase activity, behaved similarly and decreased after about 10 days, which was related to a stabilized condition of decreasing quantities of easily assimilable organic substances in the residual particulate detritus.

Analogous sequences of microbial succession were found during the decomposition of emergent reeds (*Typha, Phragmites, Sparganium*), floating-leaved plants, and submersed macroflora (*Ceratophyllum, Myriophyllum, Najas*) (Gorbunov, 1953; Krasheninnikova, 1958). Degradation rates were greater among the submersed flora under similar conditions.

SEDIMENTS

Sediments of fresh water systems have been analyzed from a number of standpoints. Sediments of running waters are sorted in a gradient of substrate particle sizes in relation to velocity of water movement, and among many benthic invertebrates the inorganic and organic particle size is of major importance in the distribution and growth of these organisms (cf. Cummins, 1962). Because of the overriding importance of microbial metabolism in direct mineralization of organic matter, of inorganic biogeochemical cycling of nutrients, and of the amount of matter deposited without appreciable further degradation in permanent sediments, the organic composition of sediments in lake systems has received much attention.

General Composition

Sediments consist of 3 primary components: (a) organic matter in various stages of decomposition; (b) particulate mineral matter, especially quartz; and (c) an inorganic component of biogenic origin, e.g., diatom frustules and calcium carbonate. In many respects, the organic sediments of lakes are similar to the uppermost A_0 horizon in terrestrial soils (Hansen, 1959a). Heterogeneous humus in particulate form can be divided into 2 main types, acid humus and neutral humus (sapropel[2]). From a colloid-chemical viewpoint, acid humus is largely an unsaturated sol with negatively charged particles, while neutral humus is largely a gel in which anions are adsorbed to the surfaces of the humus particles.

The acid humus prevails in aquatic peat bog systems. Decomposition of peat plant materials, such as the moss *Sphagnum*, occurs forming unsaturated humus colloids (dopplerite). Aggregations of black, gelatinous dopplerite are highly soluble in water; they dry irreversibly. Nitrogen content of this humic material is very low (0 to 2 per cent).

---

[2]True sapropel is bluish-black, contains much hydrogen sulfide (H_2S) and methane (CH_4), and is deposited under strong anaerobic reducing conditions.

Dy and Gyttja

The terms *dy* (pronounced di) and *gyttja* (pronounced yit'-ja) were introduced in the middle of the 19th century by von Post (Hansen, 1959a, 1959b). Gyttja is a coprogenous sediment mixture of remains of all particulate organic matter, inorganic precipitations, and minerogenic matter. In a fresh state gyttja is very soft and hydrous, with a dark greenish-grey to black color, but never brown. In a dry state some gyttjas are hard and black, while others are more friable and lighter in color, depending upon the main constituent. The organic carbon content of gyttja is less than 50 per cent.

Dy[3] is a gyttja mixed with unsaturated humus colloids of littoral bog and allochthonous origin. Fresh dy is soft, hydrous, and brown in color. In a dry condition dy is very hard and dark brown. The organic carbon content of dy and peat is greater than 50 per cent.

Numerous comparative studies, related to the state of organic matter, have demonstrated a much lower proteinaceous and nitrogen content in acidic humus. The C:N ratio of *Sphagnum* peat is about 35; that of pure dopplerite varies from 46 to 52. Hansen concluded that if the C:N is less than 10, the humus is neutral humus and the sediment is gyttja. If the C:N exceeds 10, the gyttja is mixed with acid humus and the sediment is a dy.

The sources of particulate organic matter and their decomposition rates during sedimentation in the pelagic zone were discussed earlier. It is clear that in large oligotrophic lakes and also large eutrophic waters in which non-lignified algal phytoplanktonic productivity dominates in the system, humic inputs to the sediments will be relatively low in comparison to influxes from littoral and allochthonous sources. Gyttja sediments would be anticipated in these lakes, and this is generally borne out by empirical observations (cf., for example, Hansen, 1961). Associated with increasing littoral contributions to the total lake productivity and especially acidity increases derived from specific macroflora, e.g., *Sphagnum* mosses, an increase in the humic content often is found with a concomitant shift to dy sediments.

Humic Compounds Originating from the Littoral Flora

Bases formed by microbial degradation of lignin to quinones and amino acids can polymerize and form humus compounds (Flaig, 1964). As a result, it would be expected that the littoral macrophytes, and particularly the emergent lignified flora, would be a greater source of humic compounds than submersed hydrophytes or planktonic microflora. Analyses of the successive degradation of macroflora cast some light on the rates and products formed.

The carbon and nitrogen contents of fractions of (1) fresh, living plant material of *Stratiotes aloides* were compared to those in (2) semi-decomposed, dead plant material that was morphologically distinct but undergoing intermediate phases of decomposition, (3) young, thin sapropel formed by partial

---

[3]Also termed tyrfopel in older literature, meaning a fine-grained peat.

TABLE 17–19   Carbon, Nitrogen, and C:N Ratios Among Fractions of Decomposing Stratiotes Organic Matter in Lake Venematen, Netherlands (see text)[a]

Carbon, Nitrogen, and C:N Ratio	Total Organic Matter	Humic Organic Matter	Residual Organic Matter	Free Humic Substances	Free Humic Acids	Free Fulvic Acids	Sorbed Humic Acids	Bound Humic Substances	Bound Humic Acids	Bound Fulvic Acids
CARBON										
Fresh *Stratiotes*	57.0	40.4	16.7	20.6	6.5	14.1	8.9	10.8	4.3	6.5
Semi-decomposed plant material	60.6	38.6	22.0	20.6	10.2	10.4	3.1	14.9	9.3	5.6
Young, thin sapropel	34.0	34.1	—	14.4	10.7	3.8	3.7	16.0	16.0	0.00
Old, thick sapropel	34.6	34.6	—	14.7	10.3	4.4	3.9	16.0	15.2	0.7
NITROGEN										
Fresh *Stratiotes*	2.6	2.3	0.2	1.1	0.1	0.9	0.8	0.5	0.3	0.2
Semi-decomposed	3.0	1.4	1.6	0.8	0.2	0.6	0.2	0.4	0.3	0.1
Thin sapropel	1.3	0.6	0.7	0.2	0.1	0.06	0.3	0.2	0.2	0.00
Thick sapropel	1.7	0.6	1.1	0.1	0.1	0.04	0.2	0.2	0.2	0.03
C:N										
Fresh *Stratiotes*	22.3	17.5	69.4	19.7	46.7	15.5	11.7	21.7	14.3	32.7
Semi-decomposed	20.5	28.0	14.0	27.2	56.4	18.0	17.2	33.9	29.0	46.9
Thin sapropel	25.5	55.8	—	72.3	76.1	63.2	13.9	106.6	106.6	—
Thick sapropel	20.5	58.6	—	105.1	103.2	109.8	16.1	76.1	84.5	24.3

[a]After data of Úlehlová, B.: Decomposition and humification of plant material in the vegetation of *Stratiotes aloides* in NW Overijssel—Holland. Hidrobiologia, 12:279–285, 1971.

mineralization, and (4), old, thick, stratified sapropel resulting from nearly complete mineralization and humification (Úlehlová, 1970, 1971). Results (Table 17–19) show that the carbon and nitrogen content decreases markedly with progressive decomposition to sapropels with little change in the C:N ratio in the total decomposing community. Within the humic substances, the free humic substances exhibited a reduction in carbon in the sapropel phases of decomposition which occurred primarily in the fulvic acid fraction and was transferred to the free humic acid fraction. Among the chemically bound humic substances, carbon content increased with progressive degradation, again from the fulvic acid to the humic acid fraction. Nitrogen content of the total increased in the semi-decomposed stage and old sapropel decreased in the bound humic fractions. Most of the nitrogen occurred in humic substances from fulvic acid fractions. These results suggest that carbon of free humic substances of early phases is transformed, mediated by microbial metabolism, to chemically bound humic substances in later stages of decomposition. As the nitrogen of humic compounds is removed selectively and C:N ratios increase markedly, rates of further degradation are reduced permitting some net accumulation in the sediments.

Microflora of Sediments and Rates of Decomposition

A conspicuous feature of microbial populations in lakes is their great increase in numbers in the transition from the overlying water to the surficial

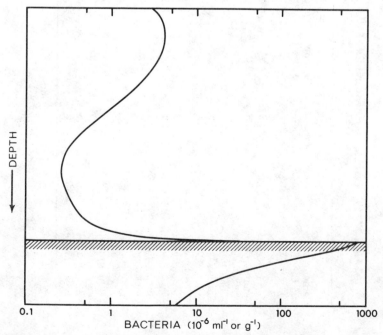

Figure 17–25 Generalized relative vertical distribution of bacterial numbers in water and sediments of a moderately productive lake. Depth scale in the water relative in meters and in the sediments in centimeters.

TABLE 17–20 Bacterial Numbers (Direct Enumeration) in the Surficial Layer of Sediments of Several Lakes of the Moscow Region of the USSR[a]

Lake	Ash Content of Sediments (%)	Bacterial Numbers $(10^{-6}\ g^{-1})$		Organic Substances of Bacteria in % of Organic Matter of Sediments
		Fresh Sediment	Dry Sediment	
Beloe (Kosino)	54.1	2326	54,219	7.8
B. Medvezh'e	58.3	1905	38,253	6.0
Chernoe (Kosino)	49.3	1285	35,109	4.5
M. Medvezh'e	47.3	1624	32,032	4.0
Sviatoe (Kosino)	18.6	922	29,790	2.3
Krugloe	79.2	1110	7991	2.5
Gab (Karelian region)	81.9	1883	15,680	5.8

[a]Modified from Kuznetsov, 1970, after Khartulari.

sediments (Fig. 17–25). The numbers of bacteria increase about 3 to 5 orders of magnitude from the water to the surface sediments (wet, fresh weight) and decrease rapidly within sediments at greater depth. Expressed in numbers per gram dry weight of sediment, bacterial populations can reach 6 to 7 orders of magnitude greater than in the overlying water (Table 17–20). Saprophytic bacteria decrease much more rapidly than the total bacterial numbers with increasing depth below the sediment-water interface, which suggests the rapid reduction of readily assimilable organic substrates below the interface zone.

Bacterial distribution varies horizontally within the surface sediments over the lake basin as well (Henrici and McCoy, 1938). When the littoral zone is covered with a well-developed macrophyte community, bacterial numbers of the sediments are much higher by several orders of magnitude than those found in profundal sediments of deepwater areas. Numbers in sandy, wave-swept littoral areas are lower than those in profundal sediments. A more detailed analysis of bacterial numbers of sediments within the macrovegetative littoral zone as compared to those of sediments free of vegetation shows similar relationships (Table 17–21), but with marked shifts seasonally in relation to development of organic matter by the macrovegetation. Nitrogen-fixing bacteria *Azotobacter* and the obligate anaerobe *Clostridium* were greater in numbers in sediments with macrovegetation than in sediments without vegetation or in the overlying water. Similar relationships were found among populations of hydrocarbon- and hydrogen-oxidizing bacteria, cellulose decomposing bacteria and actinomycetes, and fungi. Large numbers of methane-forming and sulfate-reducing bacteria developed under anaerobic conditions in sediments containing large quantities of decomposition products of higher aquatic plants.

Bacterial numbers of the deepwater profundal sediments show a fair correspondence with the productivity of the lakes (Table 17–22), and are much

TABLE 17–21 Comparison of Microbial Populations (Thousands per Gram of Sediment) of the Sediments Among Emergent Bulrush Vegetation (A) and of Open Water Free of Bulrushes (B) in Lake Mekhteb, USSR[a]

Period	Total Saprophytes		Actinomycetes		Fungi		Clostridium pasteurianum		Hydrogen Oxidizing		Methane Forming		Sulfate Reducing	
	A	B	A	B	A	B	A	B	A	B	A	B	A	B
Dec.–Jan.: no bulrushes	590	790	0	80	0	30	10	10	1	10	0	1	0.1	2
Feb.: young bulrushes	1800	2700	200	0	400	400	1	10	1	10	40	40	8	1
Mar.: young bulrushes	4400	4800	400	300	0	0	1	10	10	10	5	240	40	10
Apr.: developing stands	5300	3100	800	0	0	300	100	100	10	1	1500	140	50	1300
May:	1800	1800	1000	100	200	300	1000	10	1000	10	2000	15,000	100	100
June: mature bulrushes	7900	4800	2900	300	100	0	10	10	1000	10	2000	15,000	400	1200
July:	60,200	3800	0	0	1000	200	100	100	10	10	1200	2000	240,000	5000
Aug.: declining bulrushes	16,400	5400	1500	0	2800	0	10	10	100	100	1800	3000	1400	1500
Sept.:	3800	2700	300	0	0	0	10	10	10	10	170	13	1	1
Nov.:	230	80	0	0	0	9	1	1	10	1	70	10	0	0

[a] From data of Aliverdieva-Gamidova, 1969.

TABLE 17–22 Comparison of Relative Bacterial Numbers (Plate Culture Techniques) of the Profundal Sediments Among Several Wisconsin Lakes of Differing Productivity[a]

Lake	Type	Type of Sediment	Average Bacteria per cm³ of Surface Sediments	Total Bacteria per cm³ of First 18 cm of Sediment
Crystal	Oligotrophic	Gyttja	2160	38,880
Weber	Oligotrophic	Gyttja	2350	42,300
Big Muskellunge	Mesotrophic	Gyttja	10,930	196,740
Trout	Oligotrophic	Gyttja	29,790	536,220
Little John	Mesotrophic	Gyttja	39,050	642,900
Mary	Dystrophic	Dy	39,450	710,100
Helmet	Dystrophic	Dy	120,300	2,165,400
Alexander	Eutrophic	Marl gyttja	144,240	2,599,320
Mendota	Eutrophic	Marl gyttja	609,300	>10,000,000

[a]After data of Henrici and McCoy, 1938.

better correlated than bacterial numbers of the water, which undergo large, rapid fluctuations. In extremely oligotrophic lakes, total numbers of bacteria of the sediments can be less than the cumulative total of the entire overlying water, which indicates that in these lakes a majority of the readily decomposable organic matter has been processed before the organic residues reach the sediments.

Bacterial Activity of Sediments

While the relationship of increased bacterial populations and metabolic activity to greater organic matter in sediments may appear obvious, in situ measurements are very few. A significant correlation ($r = 0.93$; $p = 0.01$) between dehydrogenase activity and organic content of surface sediments has been found in a eutrophic reservoir (Lenhard, et al., 1962). These correlations, however, would not be expected to hold in highly organic sediments in which other conditions, for example, high acidity of bogs, influence microbial activity.

Data on oxygen consumption by the benthic community of bacteria, algae, micro- and macrofauna, in which oxygen utilization by these components can be partitioned approximately and corrected for nonbiotic chemical utilization, offer some insight into bacterial respiration rates of sediment communities. Under aerobic conditions of shallow Marion Lake, Canada, respiratory oxygen consumption by bacteria of sediments over an annual period was found to be related primarily to temperature (Fig. 17–26). The proportion of community respiration resulting from bacterial respiration varied with season (Fig. 17–26), and was lowest during maximal community respiration during the summer period of higher temperatures. This percentage certainly increases under other conditions among lakes, such as in deepwater sediments in which nearly all metabolism is microbial.

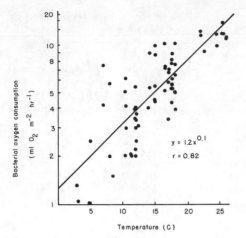

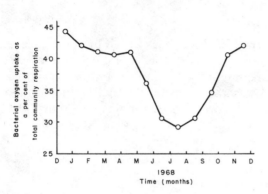

Figure 17–26 *Upper,* relationship of temperature and sediment bacterial respiration, and *lower,* bacterial respiration as a proportion of total community oxygen consumption in shallow water Marion Lake sediments during 1968. (After Hargrave, B. T.: Epibenthic algae production and community respiration in the sediments of Marion Lake. J. Fish. Res. Bd. Canada, 26: 2003-2026, 1969.)

Particulate organic detritus consumed up to 3 orders of magnitude more oxygen per unit dry weight than sand (Hargrave, 1972). Uptake rates were related inversely to surface area of particles (particle size) and related directly to particle organic content of carbon and nitrogen.

Relative assays of heterotrophic activity of benthic microorganisms of shallow water sediments by the uptake of labelled glucose, acetate, and glycine demonstrated that the greatest activity occurred in the summer months when the water temperatures exceeded 10°C (K. J. Hall, et al., 1972). Availability of substrates and slow diffusion rates exerted some control over the uptake and utilization of dissolved organic compounds of the interstitial water. The fraction of the substrate respired as CO_2 varied with the compound and was highest (average 63 per cent) with glycine in comparison to glucose (22 per cent) and acetate (13 per cent). Most of the heterotrophic activity occurred in the upper sediment layers, the sites of highest bacterial numbers in the sediments.

Anaerobic Decomposition in Sediments

Oxygen serves as the universal hydrogen acceptor for biochemical reactions of microbes under aerobic conditions. Under anaerobic conditions of decomposition of organic compounds, however, the relationships are much more complicated when the hydrogen acceptor is transferred to various substances and intermediate metabolic organic compounds. Often the same compound can serve as a hydrogen acceptor or donor, depending on the circumstances.

Methane Fermentation

Degradation of large quantities of organic matter occurs under anaerobic conditions by methane fermentation in the sedimentary accumulations of lakes, ponds, and streams, as well as in waste treatment facilities (oxidation ponds, anaerobic lagoons, and septic tanks). Organic matter is converted quantitatively to methane and carbon dioxide in 2 stages (Fig. 17–27). In the first stage, a heterogeneous group of facultative and obligate anaerobic bacteria, termed "acid formers," converts proteins, carbohydrates, and fats primarily into fatty acids by hydrolysis and fermentation (Deyl, 1961; McCarty, 1964). The methane-producing bacteria then utilize the organic acids, converting them to carbon dioxide and methane. Certain alcohols from carbohydrate fermentation can also be converted by methane-producing bacteria.

Utilization of organic compounds in suspension or solution by acid-forming bacteria in the first stage results only in the synthesis of bacterial cells and the generation of endproduct organic compounds such as organic acids. Removal of oxidizable organic compounds occurs in the second stage, and is directly proportional to the quantity of methane produced. Under some anaerobic conditions, the formation of hydrogen occurs (see further on) or results in the reduction of inorganic hydrogen acceptors, such as sulfates, nitrates, and nitrites, as discussed previously.

The first stage of acid formation results from hydrolysis and fermentation (McCarty, 1964). Proteins first are hydrolyzed with the aid of enzymes to polypeptides, and then to simple amino acids. Complex carbohydrates such as starch and cellulose are hydrolyzed to simple sugars, and fats and oils are hydrolyzed to glycerol and fatty acids. The amino acids, simple sugars, and glycerol formed by hydrolysis are soluble and fermented by acid-forming bacteria by oxidation and reduction of the organic matter. In the absence of

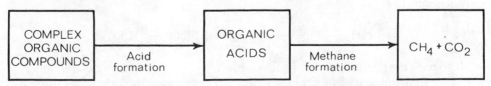

Figure 17–27   Two stages of methane fermentation of complex organic compounds. (Modified from McCarty, 1964.)

oxygen, a portion of the organic molecule is oxidized, while another portion of the same molecule or sometimes of another compound is reduced. These energy-yielding oxidation-reduction reactions result in reduced saturated fatty acid products and oxidized carbon as CO_2. The endproduct ammonia is also produced from amino acid fermentation.

The methane-producing bacteria are strictly anaerobic, and consist of 4 major genera: rod-shaped nonsporulating *Methanobacterium*, rod-shaped sporulating *Methanobacillus*, and the spherical *Methanococcus* and *Methanosarcina*. Species are differentiated on the basis of the substrates they are capable of using.

Methane is formed by 2 major processes. In the first, carbon dioxide serves as a hydrogen acceptor and is reduced to methane by enzymatic removal of hydrogen from the organic acids:

$$CO_2 + 8H \longrightarrow CH_4 + 2H_2O$$

In the second, acetic acid, a major intermediate produced by the fermentation of complex organic compounds, is converted to CO_2 and methane from the methyl carbon:

$$^*CH_3COOH \longrightarrow {}^*CH_4 + CO_2.$$

Fermentation of mixed organic compounds to methane results largely from the intermediate acetic acid (ca 70 per cent), and much of the remainder of the methane results from CO_2 reduction. Propionic acid, an intermediate produced in significant quantities from protein, carbohydrate, glycerol of fats, and other compounds, also can serve as a substrate for methane production. Longer-chain acids are of less importance.

Methane Oxidation

While the production of methane in sediments can be very intense and reach as much as 85 per cent of the total gas by volume formed in the deposits, little of the methane escapes to the atmosphere, owing to the presence and activity of methane-oxidizing bacteria in the overlying water (ZoBell, 1964). Methane oxidizers are distributed widely in most natural waters. Among the best-known of many species of methane-oxidizing bacteria is *Methanomonas methanica,* which, in addition to methane, can oxidize numerous other hydrocarbon compounds. All of this group appear to be strict aerobes, although certainly microaerophyllic. The overall empirical equation for methane oxidation is:

$$5CH_4 + 8O_2 \longrightarrow 2(CH_2O) + 3CO_2 + 8H_2O,$$

with a free energy yield of 195 kcal per mole.

The distribution of methane-oxidizing bacteria in dimictic lakes indicates that these forms are present throughout the year in significant numbers. The methane oxidizers increase during summer stratification, the highest numbers ($>5 \times 10^5$ liter^{-1}) occurring in metalimnetic and hypolimnetic waters of reduced oxygen concentrations and maximum concentrations of methane (Cappenberg, 1972). As methane oxidation proceeds, residual oxygen is reduced further. Numbers are reduced greatly as strata of the hypolimnion become totally anoxic. During periods of circulation, numbers of methane-oxidizing bacteria are evenly distributed at reduced levels throughout the water.

Hydrogen Metabolism

Molecular hydrogen is formed largely in the sediments under anaerobic conditions by fermentative degradation of carbohydrates, cellulose, and hemicelluloses to fatty acids, especially acetic acid. Hydrogen and CO_2 are generated by *Clostridium* from simple carbohydrates, and by *Achromobacter* and *Bacillus* from cellulose. Formate serves as the carbon source of some species of *Bacterium* in the formation of hydrogen.

Hydrogen is oxidized in several ways, so that, as is the case with methane, little reaches the surface and escapes to the atmosphere as bubbles. Certain strictly anaerobic sulfate-reducing bacteria of the genus *Desulfovibrio* oxidize hydrogen while utilizing CO_2 to synthesize at least part of their cell substance (ZoBell, 1973):

$$H_2SO_4 + 4\ H_2 \longrightarrow H_2S + 4\ H_2O\ (\Delta F = -73{,}130\ \text{cal mole}^{-1})$$

Such sulfate reducers were found to fix an average of 0.24 g CO_2–C per gram of H_2 oxidized, and nearly all of the CO_2–C was accounted for as bacterial biomass. Oxygen for the oxidation is derived from the sulfate.

Certain species of all of the major genera of the methane-forming bacteria can oxidize molecular hydrogen.

Carbon Monoxide Metabolism

Carbon monoxide results from certain microbial fermentations but generally, even under exceptionally favorable conditions, does not exceed 3 per cent of the gases formed by bacterial fermentation of organic wastes (ZoBell, 1964, 1973). The chemical reactivity of CO and the relative ease with which various bacteria reduce it to methane or oxidize it to CO_2 probably account for its absence or very low concentrations in most environments.

Aerobic CO-oxidizing bacteria are distributed widely, especially in organic-rich sediments. Species of *Carboxydomonas, Hydrogenomonas, Bacillus,* and certain methane-oxidizing bacteria oxidize CO (Hubley, et al., 1974):

$$CO + \tfrac{1}{2}\ O_2 \longrightarrow CO_2\ (\Delta F = -66\ \text{kcal mole}^{-1})$$

Additionally, several aerobic bacteria in mixed populations quantitatively convert carbon monoxide into methane in the presence of hydrogen:

$$CO + 3\ H_2 \longrightarrow CH_4 + H_2O\ (\Delta F = -46\ kcal\ mole^{-1}).$$

In the absence of H_2, aerobic bacteria (*Methanosarcina*) produce CO_2 and methane:

$$4\ CO + 2\ H_2O \longrightarrow 3\ CO_2 + CH_4.$$

Gaseous Metabolism in Sediments

Analyses of the distribution of the gases contained in lake sediments and water show that the total quantity of nitrogen, methane, and hydrogen in lake sediments increases with water depth (Table 17–23; cf. also Fig. 11–1). The total pressure of the gases was usually slightly higher than the corresponding hydrostatic pressure (Koyama, 1964).

Experimental investigation of the metabolism of fresh lake sediments under normal pressure demonstrated the following sequence (Koyama, 1955): (a) CO_2 production began immediately at high rates; (b) as the sediments became anaerobic, hydrogen was evolved, followed by increasing rates of methane production. The primary reaction of methane formation was the reduction of CO_2 by hydrogen and hydrogen donor compounds. Hydrostatic pressures that would be encountered in the sediments of a majority of lakes do not greatly alter bacterial production of gases. Methane production decreased slightly as pressures were increased from 0 to 50 atmospheres and then increased up to 300 atm, after which gradual decreases were observed.

Production of Gases in Sediments, Ebullition, and Their Relation to Lake Productivity

The distribution of fermentation gases within the water column of most productive lakes is one of very low concentrations in the surface layers and

TABLE 17–23   Average Content of Gases in Sediments of Several Japanese Lakes in Relation to Gas and Hydrostatic Pressures[a]

Lake	Depth (m)	ml Gas l^{-1} of Interstitial Water of Sediments					Average Total Gas Pressure (atm)	Hydro-static Pressure of the Bottom (atm)
		CO_2	O_2	N_2	CH_4	H_2		
Nakatsuna-ko	12	234	0	12	72	7	2.68	2.08
Kizaki-ko	29	137	0	17	110	17	4.13	3.71
Aoki-ko	56	89	0	18	161	24	5.14	6.32

[a]After Koyama, 1964.

markedly increasing levels in the hypolimnion and near the sediments. If production rates and pressure conditions are such (cf. p. 124ff) as to permit bubble formation, the gases can rise rapidly to overlying strata. Usually gases or bubbles of these reactive gases are redissolved and metabolized by microflora en route. Bubble release to the surface and atmosphere generally occurs in shallow aquatic situations such as marshes and littoral areas of intensive anaerobic fermentation in the surficial sediments.

As lakes become very eutrophic, the anaerobic fermentation of sediments can reach levels sufficient to result in the steady ebullition of gases. For example, Strayer (1973) found that an average of 21 m moles m^{-2} day^{-1} of methane left the anaerobic sediments by ebullition in hypereutrophic Wintergreen Lake, Michigan, during the period of late May through August. Ohle (1958) demonstrated by both chemical analyses and sonar echo-sounding that under stratified conditions of very productive lakes the ebullition is sufficient to induce "methane convection" currents by which sediment particles and dissolved substances of the hypolimnion can be transported to the trophogenic zone.

The return of nutrients from the sediment zone to the zone of primary production represents an *internal fertilization* of the system. The ebullition of gases from the sediments is closely related to lake productivity. When the oxygen content of the deeper water is exhausted, ebullition of fermentation gases increases. These gases reduce the oxygen content further through bacterial oxidation and circularly mediate conditions for further intensification of gas generation and ebullition. The result can be observed in a jump-like increase in eutrophication (*rasante Seen-Eutrophierung*) (Fig. 17-28). The accelerated biogenic response intensifies syntheses and inputs of organic matter to the sedimentary fermentation system and generation of gases.

Ohle (1958) further proposes a symbiotic relationship among the bacterial populations which utilize and generate the gases of fermentation. Sulfate-reducing bacteria of the sediments can utilize methane as an excellent energy

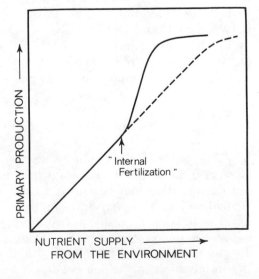

Figure 17-28 Accelerated primary production of lakes resulting from "internal fertilization" during the progression of increasing nutrient inputs from the surrounding environment. (After Ohle, 1958.)

and carbon source, but they are strict anaerobes. Their generation of hydrogen sulfide, even though inhibitory to many bacteria, serves as an efficient means to maintain anoxia against intrusions of aerated water.

CONSEQUENCES OF DISPLACED METABOLISM TO THE SEDIMENTS

Sealing a lake off from its drainage basin would alter it drastically. Although lakes are frequently referred to as ecosystems, the important influences of the drainage basin, that is, imported and exported detrital organic matter, inorganic inputs and losses, morphometry, and so forth, intimately link a lake with its surrounding terrestrial environment. Obviously a lake is only one component of a larger landscape unit which must include the entire drainage basin impinging on the lake per se. Use of the term "ecosystem" in reference to the lake alone is arbitrary although useful rhetoric. In discussions of the fluxes of organic detrital matter from its origins to its various fates, the term ecosystem is used to describe a level of organization within which structure and function operate only in an aqueous environment.

The Central Role of Detritus

Stoke's law states that particles of a given density sediment through water at a rate directly proportional to the square of their radii. Particulate organic carbon (POC) and dissolved organic carbon (DOC) therefore move with water, and POC ultimately will be deposited at the bottom of static water. DOC will also sediment if associated with inorganic particulate matter, such as, for example, DOC adsorbed to clay or $CaCO_3$ particles. These factors cause a displacement of some terrestrial production to the littoral zone of lakes with runoff, and the displacement of more littoral and pelagic production to the benthic sediments of lakes. In terms of energetics of the ecosystem, the medium of exchange is detritus, and the functional difference between POC and DOC is a smooth continuum from large particles to small molecules (Rich and Wetzel, 1975). Although the specific nature of the detritus modifies its behavior and its effects upon the environment, detritus inevitably carries energy from its point of origin to its place of destruction.

Chemosynthesis and Nutrient Regeneration

The generally low light intensities that reach the pelagial sediments of lakes and the sedimentation of primarily dead particulate organic carbon imply that benthic metabolism is mostly heterotrophic and detrital (Rich and Wetzel, 1975). The poor diffusion of oxygen into saturated sediments and its rapid utilization there mean that much metabolism is anaerobic, even when hypolimnetic oxygen depletion does not occur. Hutchinson (1941) terms this metabolism "pelometabolic," and describes its importance relative to "hydrometabolism" which occurs in the free water of the lake. Carbon di-

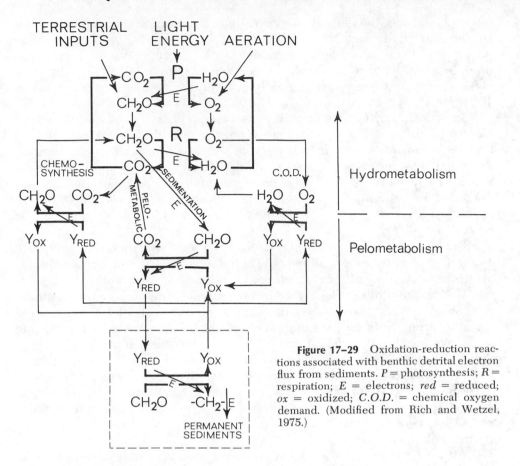

Figure 17–29 Oxidation-reduction reactions associated with benthic detrital electron flux from sediments. $P =$ photosynthesis; $R =$ respiration; $E =$ electrons; $red =$ reduced; $ox =$ oxidized; $C.O.D. =$ chemical oxygen demand. (Modified from Rich and Wetzel, 1975.)

oxide produced by anaerobic detrital pelometabolism represents the escape of the oxidized product of an oxidation-reduction reaction (Fig. 17–29). The continued production of CO_2 during anaerobiosis indicates that the respiratory quotient (CO_2 release/O_2 uptake) of benthic metabolism is greater than one, and that alternate electron acceptors are being reduced instead of molecular oxygen. Thus, alternate electron acceptors are receiving energy originally transferred during photosynthesis in which water is oxidized to oxygen and CO_2 is reduced to carbohydrate.

This accumulation of electrons in the sediments has several effects, which must be interpreted as functional with respect to the ecosystem (Wetzel, et al., 1972; Rich and Wetzel, 1975). The *detrital electron flux* into and ultimately out of the sediments derives from the same photosynthetic energy source as a predator's food and is a much larger proportion of the total source. Presumably, evolution, or at least integration of the ecosystem has incorporated this energy flow through predator-prey systems. Reduced ions, radicals, and molecules are generally more soluble, and net diffusion increases out of the sediments. Upon entry of these products into the oxidized layers of the lake, further oxidation may proceed either chemically or biologically. If the original detrital reduction involved a nutrient (sulfate, nitrate, or ferric phosphate), the detrital energy has effected nutrient regeneration. If the subsequent "hydrometabolism" is biological (chemosynthesis), the detrital energy has been reintroduced into the biota. The energy thus transferred must be considered heterotrophic rather than synthetic in that it is not pure wave energy, but is derived originally from primary production (cf. Sorokin, 1964; 1965).

In order to further understand the consequences of the benthic detrital electron flux, some basic considerations of the possible alternatives of benthic metabolism are worth considering. Material enters the sediments by sedimentation because it is particulate, that is, its solubility rates are low. If it remains insoluble it will enter the permanent sediments of the lake and decrease the volume of the lake, that is, reduce the life span of the lake. In order to escape the sediments it must become highly soluble, gaseous, or both.

Presumably, cellulose and similar material represent much of the organic material that enters the sediments. Their intermediary metabolic products are increasingly soluble, culminating in carbon dioxide which is both gaseous and highly soluble. Very reduced carbon compounds, for example, methane, are also gaseous, volatile, and/or soluble. The loss of dissolved organic carbon from the lake sediments is not well understood, but organic acids, acetic acid in particular, appear to be significant factors. As a carbohydrate, cellulose, for example, requires a molecule of oxygen for each carbon atom in order to be oxidized to CO_2 and water. The removal of a molecule of oxygen will release 2 carbon atoms when the product is methane. However, oxygen represents the heaviest atom in cellulose, and its simple removal of CO_2 reduces the mass of the molecule appreciably and results in less potential sediment.

As we have seen, CO_2 and methane are the major gases escaping from benthic metabolism. Highly reduced carbon compounds are also well-known products of long-term sedimentation, that is, coal and petroleum, and increasing concentrations of carbon, in excess of that found in carbohydrates, are measurable in lake sediments over much shorter time periods (Shiegl, 1972). Given this format, at least 2 additional ecosystem-level consequences of the benthic detrital electron flux may be hypothesized.

The redox potential in lake sediments brought about by anaerobic metabolism represents electrons released from biochemical (enzymatic) "restraints" or reaction specificity. Consequently, these electrons may diffuse out of the sediments in association with a number of inorganic compounds, as already described. The exhaustion of a particular class of electron acceptors causes the benthic redox potential to become more negative relative to molecular oxygen until new acceptors can quickly precipitate back into the sediments following hydrometabolic oxidation as highly insoluble compounds, for example, FeS, CuS, and so forth. On the other hand, the carbon and particularly the oxygen represented by CO_2 are the relatively massive products of the oxidized component of the anaerobic oxidation-reduction reaction. Thus, the net diffusion of reduced compounds out of the sediments appears to be controlled by redox potential which increases benthic concentrations, and the net diffusion appears to control the export of mass, that is, carbon and oxygen, from the lake sediments. In this manner, the life span of the lake is affected directly by reduction of mass through losses of large quantities of gases of carbon and oxygen. The rate of the detrital electron flux is fundamental to eutrophication rates of lake systems.

The importance of detrital electron flux in benthic metabolism is reflected further by the changes in ratios of CO_2 evolution to consumption of molecular oxygen of hypolimnia over an annual period. The oxygen uptake by the benthic community of sediments in lakes generally is considered to exceed the concomitant evolution of CO_2. Oxidations of proteins and fats have respiratory quotients ($RQ = CO_2/O_2$) of less than unity, as has been demonstrated often by caloric combustions of benthic organisms. An RQ value of 0.85 generally is accepted as an average value (Ohle, 1952; Hutchinson, 1957) that has been used in estimates of lake productivity by hypolimnetic CO_2 accumulation (cf. Chapter 10). Anaerobic bacterial fermentative metabolism produces excess CO_2, a positive CO_2 anomaly, and volatile organic compounds such as methane that diffuse out of the sediments.

As Rich (1975) points out, the sediments of lakes have an "oxygen debt" that is indicated by low and negative redox potentials and nonbiological chemical uptake of molecular oxygen by reduced substrates formed under anaerobic conditions. This evidence suggests that the oxidative respiratory portion leading to CO_2 may be proceeding while reduction to water may be impeded by limited O_2 diffusion (Wetzel, et al., 1972). Viewing the RQ concept at the benthic community level, which would include the anaerobic bacteria that utilize electron acceptors other than molecular O_2, the benthic respiratory quotients would be greater than those of organisms undergoing aerobic metabolism. Additionally, the in situ RQ values would be expected to vary seasonally with circulation periods and changes in redox gradients in response to varying oxygen diffusion rates. Such was the case in Dunham Pond, Connecticut, in which Rich (1975) found that during stratification the respiratory quotients of the hypolimnion varied inversely with the availability of oxygen (Fig. 17–30), from less than 1 after spring circulation and oxygen renewal to nearly 3 under anoxic conditions of the latter portion of summer stratification. Under anaerobic conditions, the high respiratory quotients probably are the result of the oxidation of organic carbon to CO_2 during the reduction of alternate electron acceptors which then appear as oxidizable substrates.

The existence of highly reduced carbon compounds escaping from and existing in lake sediments also implicates the redox potentials generated by benthic detrital electron flux. However, this appears to be an endometabolic bacterial process, since reduction potentials in excess of carbohydrate result. Reduced sediment mass also results from this process, in that soluble and gaseous products are formed which increase diffusion of reduced matter out of the sediments. Further, the removal of oxygen (reduction) from permanently sedimented carbon compounds represents the removal of more than half their mass.

To conclude, this hypothesis concerning the role of detrital electron flux suggests that existing lakes may sediment energy, although temporarily, rather than mass as an alternative which permits their continued existence.

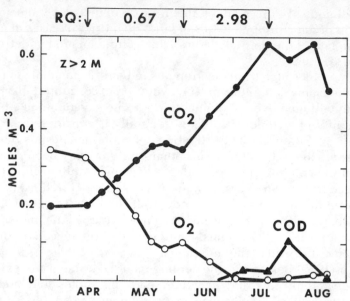

Figure 17–30   Changes in the hypolimnetic concentrations of CO_2, O_2, and chemical oxygen demand (COD) and the benthic respiratory quotients, Dunham Pond, Connecticut. (From data of Rich, 1975.)

Simultaneously, it appears that benthic detrital electron flux tends to close the carbon and particularly the oxygen cycles of lakes.

Detritus: Organic Matter as a Component of the Ecosystem

Studies that address the problems of detrital origins and metabolism directly, including the nonplanktonic and terrestrial components of the lacustrine ecosystem, are exceedingly rare. The heterogeneity and diversity of lacustrine detritus is reflected in the compound nature of most lake ecosystems. The role of allochthonous inputs in the metabolism and trophy of streams has been emphasized by numerous recent investigations (Ross, 1963; Darnell, 1964; Hynes, 1963; Cummins, 1973). Similarly, pelagic metabolism can be strongly influenced by edaphic factors and terrestrial metabolism in small lakes with high drainage area to volume ratios. The relatively recent invasion of fresh waters by angiosperms disputes Forbes' (1887) statement that a lake "... is an islet of older, lower life in the midst of the higher, more recent life of the surrounding region." Further, it is simply untrue, as Shelford (1918) has proposed, that "one could probably remove all the larger plants from a lake and substitute glass structure of the same form and surface texture without greatly affecting the immediate food relations." A great majority of lakes are small with high shoreline to surface area ratios, and pelagic metabolism is modified by littoral metabolism and inputs. Moreover, increased sedimentation in the relatively static waters of lakes results in a displacement of much lake metabolism to the sediments. Prerequisite to a study of lacustrine ecosystem structure is a complete representation of the productivity inputs of *all* components of the lake. The metabolism of detrital organic matter results in a complex carbon cycle which dominates both the structure and function of lake systems.

Definitions of Detritus and Its Functions

Organic detritus or "biodetritus" originally was described by Odum and de la Cruz (1963) as dead particulate organic matter inhabited by decomposer microorganisms. The existence and importance of detritus in many habitats have since become the subject of a large and widespread literature. However, the position of detritus in relation to the trophic dynamic concept of Lindeman (cf. Chapter 16), was not clarified until recently.

Balogh (1958) has demonstrated that detritus, as egested material, can be an important fraction of metabolism in several terrestrial soil and litter communities. On the other hand, Odum (1962, 1963) and others have emphasized that detritus originating as ungrazed primary production supports a "detritus food chain" which is essentially parallel to the conventional "grazer food chain" at succeeding trophic levels.

The importance of the divergence between the concepts of the grazer food chain and the detritus food chain has been emphasized and clarified by Wetzel and coworkers (1972). Part of the difficulty stems from the concept of ecological efficiencies as it grew out of the original statement of the trophic dynamic model by Lindeman (1942, and Kozlovsky, 1968). Lindeman defined trophic efficiency as the ratio between assimilation by one trophic level and the assimilation of the preceding trophic level, that is, $\wedge_n / \wedge_{n-1}$. This formulation does not recognize the existence of material egested or otherwise lost by trophic level n or postassimilatory, nonpredatory losses at $\wedge_{n-1}$ which may be lost to those trophic levels and to the grazer food chain, but which *are not lost to the ecosystem as a whole*. Although later studies on ecological efficiencies have used different formulae, for example, $\text{Ingestion}_n / \text{Ingestion}_{n-1}$ (Slobodkin, 1960, 1962), a failure to distinguish between egestion and other nonpredatory losses from respiration tends to persist. Therefore, most empirically derived aquatic ecological efficiencies that do not specify respiration in reality are agricultural or grazer food chain efficiencies, and not applicable to complete ecosystems in which detrital organic matter is a significant trophic pathway.

The more realistic system operation is integrated diagrammatically in Figure 17–31. Here autochthonous organic matter represents all carbon fixed by autotrophs living within the lake ecosystem, minus their respiratory losses. The actual immediate form of productivity may be living cellular material (primary particulate organic matter), which may become detrital secondarily, or dead or dissolved organic carbon, or both.

A 3-phase system of organic carbon is operational: dead organic matter, both particulate and dissolved (Fig. 17–31, *left*); living particulate organic material (*right*), which has the potential of entering either the POC or DOC pool upon death, and whose metabolism may either create or destroy all 3 phases; and CO_2. Nonliving equilibrium forces are also present, for example,

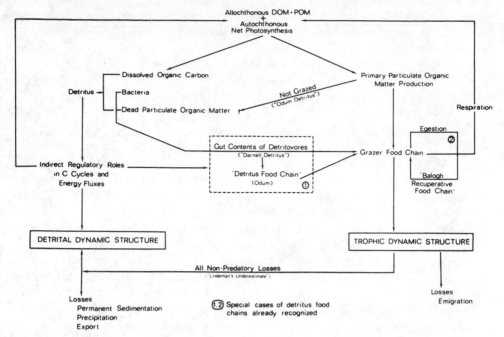

Figure 17–31 Generalized integration of the trophic and detrital dynamics of aquatic ecosystems. (From Wetzel, et al., 1972.)

photooxidation and hydrolysis of organic compounds, but they probably represent turnover rates several orders of magnitude smaller than biological metabolism under most conditions. Therefore, metabolism represents the organizing force in the structure of the organic system, and between the oxidized and reduced forms of carbon. The top of the diagram represents the zone of reduced carbon inputs, the middle portion the zone of oxidation to CO_2, and the lower portion represents losses of organic matter from boundaries of the system. The 2 parallel dynamic pools of organic carbon (dead, *left;* living, *right*) originate photosynthetically and are depleted by oxidation to CO_2 and exports. Although bacteria are shown as associated with detritus, they function as biota. Odum's detritus food chain is really a link between the 2 pools in the direction of the biota.

A major point to be emphasized is that much, and often at least 90 per cent of the organic matter entering the system from photosynthetic sources is not utilized in the grazer and detrital food chains. Among studies of the trophic dynamic food-cycle relationships, lack of appropriate recognition of nonpredatory losses from autotrophs ("Odum detritus") and nonpredatory losses from heterotrophs ("Lindeman's underestimate," see Fig. 16–31) represents a major void in contemporary system analyses.

Detritus and associated terms have been redefined (Wetzel, et al., 1972; Rich and Wetzel, 1975) in an attempt to make current terminology consistent with the ecosystem concept and to avoid perpetuating the oversight of nonpredatory pathways. The new definitions do not contradict the old, but are defined more broadly and avoid some of the ambiguity inherent in the ecosystem analyses since their origins. The following definitions complement Lindeman's original statement on trophic dynamics, as well as subsequent discussion of detritus by Odum and others.

Detritus consists of organic carbon lost by nonpredatory means from any trophic level (includes egestion, excretion, secretion, and so forth) or inputs from sources external to the ecosystem that enter and cycle in the system (allochthonous organic carbon). This definition removes the highly arbitrary "particulate" restriction from existing definitions of detritus. In terms of the ecosystem, there is no energetic difference between DOC lost from a phytoplankter or feces or other exudates from an animal. Detritus is all dead organic carbon, distinguishable from living organic and inorganic carbon. Secondly, the bacterial component is not combined with detritus ("biodetritus" = POM plus bacteria of Odum and de la Cruz, 1963) because this interferes with the general applicability of the term to situations in which detritus is not simply ingested. Such cases include the use of detrital energy in the regeneration of nutrients such as CO_2, N, and P, algal heterotrophy, losses by adsorption, flocculation, precipitation with $CaCO_3$ or both, chelation, and so forth.

The detritus food chain is any route by which chemical energy contained within detrital organic carbon becomes available to the biota. Detrital food chains must include the cycling of detrital organic carbon, both dissolved and particulate, to the biota by heterotrophy of DOC, chemo-organotrophy, absorption, and ingestion. The definition of "detritus food chain," emphasizing the actual trophic linkage between the nonliving detritus and living organisms, recognizes bacterial action on detrital substrates as a trophic transfer.

The special case of a detrital food chain, in which detrital energy is subsequently transferred by noncarbon substrates in an anaerobic environment, has been termed detrital electron flux. As discussed in the previous section, such energy may reenter the biota (chemosynthesis), or mediate chemical or physical phenomena, or both, such as increasing the availability of inorganic nutrients in an anaerobic hypolimnion or in the sediments. The term flux specifies the flow of electrons, not carbon, to alternate electron acceptors in the absence of molecular oxygen.

Detritus, as a component of the environment, can also affect facets of the chemical and physical milieu without clearly defined energetic transformations. In this situation, reference is made to effects of detritus that do not involve oxidation-reduction reactions. Examples include several indirect effects by which detrital organic carbon can influence and regulate the total energy and carbon flux of an aquatic system. Adsorption and coprecipitation of dissolved organic compounds with inorganic particulate matter (clays, $CaCO_3$, and so forth), and complexing of inorganic nutrients by dissolved organic substances, such as humic compounds, are examples of this interaction that have been discussed at some length earlier.

Detrital Dynamic Structure of Carbon in Lakes

There have been very few attempts to examine the dynamics of detrital organic carbon on a functional basis, including all nonpredatory losses of organic carbon from any trophic level and inputs from sources to the ecosystem that enter and cycle in the system. Obviously, a number of simultaneously interacting components are operational. It is clear, nonetheless, that most of the organic metabolism of lake systems is operational within the detrital dynamic structure, and that an understanding of this functional system is a prerequisite for meaningful evaluation of control mechanisms of the whole ecosystem. Stated another way, we must understand carbon flux rates among the components at the same time that we seek an understanding of analyses of regulatory parameters of metabolism.

Quantitative analyses of the major organic carbon transformations in lake systems have shown the following general characteristics: (1) The central pool of organic carbon is in the dissolved form. (2) Three major sources of particulate organic carbon occur, allochthonous, and 2 distinct major zones of autochthonous carbon flux, the littoral and the pelagic. (3) Allochthonous inputs from the drainage basin and exports occur largely as dissolved organic carbon, and represent a major flow of carbon through the metabolism of the lake. (4) The major areas of detrital metabolism are the benthic region, in which in many lakes a majority of particulate organic carbon is decomposed, and the pelagial area during sedimentation. Secondary metabolism thus is displaced from sites of production and introduction by sedimentation in the case of particulate organic carbon, and by aggregation and coprecipitation with inorganic matter in the case of dissolved organic carbon.

Supportive information for these generalizations can be obtained from analyses of an arctic tundra pond and of the pelagic zone of a eutrophic temperate lake for a single day, of a large reservoir over a 30-day period of the summer, and of a softwater and a hardwater temperate lake of low productivity over an annual period. Because of the complexity of these studies and the insights they offer into the central questions on metabolism, these analyses warrant somewhat closer scrutiny than has been given to other examples in this discussion.

The instantaneous carbon mass of the major components and transfer rates of carbon between them were evaluated for a typical midsummer day in a small arctic tundra pond by Hobbie, et al. (1972). These small ponds cover the northern tundra by the thousands, lying in depressions between low ridges formed above ice wedges and underlain by permafrost which prevents any outflow. Most of the ponds are very shallow (less than 40 cm) and are frozen solid for 9 months of the year. The tundra vegetation surrounding the ponds is dominated by 2 sedges (*Carex aquatilis, Eriophorum angustifolium*) and a grass (*Dupontia fischeri*). The dominant herbivore of the area is the brown lemming, which clips this vegetation close to the ground. Litter from these clippings and feces form a significant source of particulate and dissolved matter reaching the ponds.

The dissolved organic carbon pool contains most of the carbon in the aqueous portion of the system in that the inorganic carbon is quite low (Fig. 17–32). Although the primary productivity on an annual basis is very low, 50 per cent is contributed by the macroaquatic plants, 47 per cent by the benthic algae, and 3 per cent by the phytoplankton. The most active transfer path is

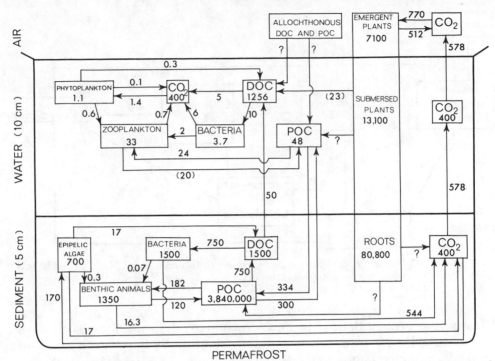

Figure 17–32 Carbon flow among components of a tundra pond ecosystem representative of approximately average midsummer conditions at coastal Barrow, Alaska (lat. 71°N). Data from Pond B on 12 July 1971 are used in the water portion of the diagram and those for Pond J on the same date for the sediment portion. Numbers in boxes refer to standing crops (mg C m⁻²) and along arrows the transfer rates (mg C m⁻² day⁻¹). *DOC* = dissolved organic carbon; *POC* = particulate organic carbon. (Modified from Hobbie, et al., 1972.)

the movement of detrital organic carbon from the aquatic plants to the sediments, in which decomposition results in major respiratory loss as CO_2 to the water and atmosphere. In the water the flux rates are much smaller than in the sediments, and among the organic carbon the transfer between particulate organic carbon and the zooplanktonic crustacea is significant. The photo-oxidation of dissolved organic compounds to CO_2, although small, is significant in this shallow habitat, as has been indicated for certain stream systems (cf. Gjessing, 1970; Gjessing and Gjerdahl, 1970). In most systems, however, such abiotic transformations would be very small in relation to other fluxes.

The carbon flux in the pelagic zone at midsummer in the surface water of eutrophic Frains Lake, southeastern Michigan (Saunders, 1972a), demonstrates an analogous relationship (Fig. 17–33). Dissolved organic carbon ranged within a 24-hour period from 2 to 6 times greater than that released extracellularly by the phytoplanktonic algae. Particulate organic detritus ordinarily was 5 to 10 times greater than that of the phytoplankton, constituted a much larger potential source of DOC than living phytoplankton, and was dispersed both in and outside of the photic zone. Similarly, detrital particulate carbon was, at this time and depth, much more important energetically than phytoplankton as a food source to the crustacean zooplankton.

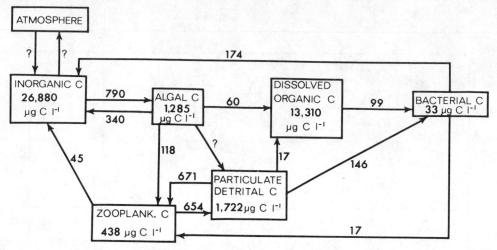

Figure 17-33 Pelagic carbon flow at one meter depth in Frains Lake, Michigan, on 22 July 1968. Numbers on arrows refer to transfer rates in μg C l^{-1} 24 hours^{-1}; numbers in blocks refer to carbon concentration in μg C l^{-1} at sunrise. (Modified from Saunders, 1972a.)

Clearly, the organic detrital matter is quantitatively a major constituent of this system, both in terms of mass and metabolism. Its importance increases markedly from the photic to the tropholytic zone.

Another example of the relationships among pools of organic matter, living components, and fluxes among components emerges from investigations of the pelagial zone of Dalnee Lake, eastern USSR (Sorokin and Paveljeva, 1972). This lake receives a major influx, about one-third, of organic matter from allochthonous sources (Fig. 17-34). During the month of July,

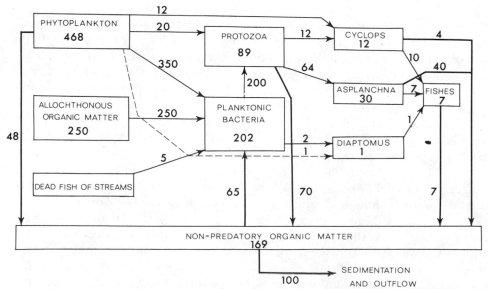

Figure 17-34 Generalized relationships of the average energy flow in the pelagic zone of Dalnee Lake, Kamchatka, USSR, during the month of July. Data expressed in calories m^{-2} per 30 days. (From data of Sorokin and Paveljeva, 1972.)

the average flux rates of organic matter, expressed as calories m^{-2} per 30 days, indicate that the planktonic bacterial populations are consumed largely by protozoa. Most of the protozoan organic matter enters the nonpredatory organic matter pool and undergoes mineralization. Only a very small portion of the total organic base of detritus and microflora is utilized by zooplankton and higher organisms. Of the zooplanktonic utilization of the microflora, about one-third was by the rotifer *Asplanchna*, predatory on protozoa at this time of year. It was estimated that well over 70 per cent of the organic matter entering the system from autochthonous photosynthetic and allochthonous sources becomes part of the nonpredatory organic matter pool and is decomposed subsequently or lost from the system by sedimentation and outflow without entering the animal components of the system.

A detailed evaluation of the carbon budgets of a whole lake system at meter depth intervals over several years was undertaken on a hardwater lake of southwestern Michigan (Wetzel, et al., 1972). In this system allochthonous inputs, exports, and dynamics in the littoral, pelagic, and benthic zones were evaluated over an annual period (Fig. 17–35). Even though the littoral flora of this lake is not well-developed in comparison to most lakes of similar morphometry, the single largest flow of organic carbon originates as primary particulate production of the littoral zone (Table 17–24). This material is distributed to the deeper sediments, and there it is supplemented by addition-

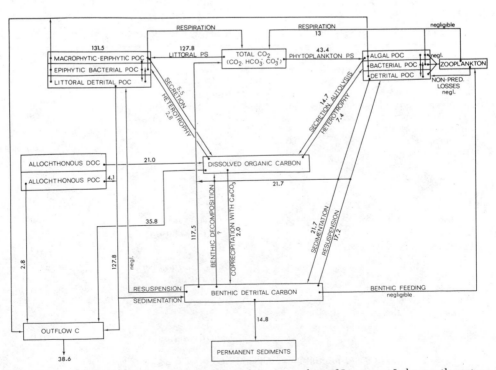

Figure 17–35   Detrital structure and flux of organic carbon of Lawrence Lake, southwestern Michigan. *DOC* = dissolved organic carbon; *POC* = particulate organic carbon; *PS* = photosynthesis. (After Wetzel, et al., 1972.)

TABLE 17–24 Littoral Carbon Budget, Lawrence Lake, Michigan[a]

Carbon	g C m^{-2} year^{-1}
Inputs:	
Macrophytes	87.9
Epiphytic algae	37.9
Epipelic algae	2.0
Heterotrophy	2.8
Resuspension	0.0
Total	130.6
Outputs:	
Secretion	5.5
Gross productivity:	125.1
Benthic respiration:	117.5
Net productivity:	+7.6

[a]From Wetzel, R. G., et al.: Metabolism of dissolved and particulate detrital carbon in a temperate hard-water lake. Mem. 1st. Ital. Idrobiol., 29, Suppl.: 185–243, 1972. Estimates are on a lakewide area basis rather than an arbitrarily defined littoral zone.

al particulate organic carbon sedimenting from the pelagic zone (Table 17–25). Allochthonous particulate organic carbon input to the sediments is a fraction of that sedimenting in the pelagic zone, and is included there.

Benthic metabolism of the sediments releases a major amount of the im-

TABLE 17–25 Pelagic Organic Carbon Budget, Lawrence Lake, Michigan[a]

Dissolved Organic Carbon	g C m^{-2} year^{-1}	Particulate Organic Carbon	g C m^{-2} year^{-1}
Inputs:		*Inputs:*	
Algal secretion ⎫	14.7	Phytoplankton	43.4
Algal autolysis ⎭		Resuspension	17.2
Littoral secretion:		Allochthonous	4.1
Macrophytes	3.5	Bacteria:	
Epiphytic algae	1.9	Chemosynthesis	7.1
Epipelic algae	0.1	Heterotrophy of	
Allochthonous DOC	21.0	DOC	7.4
Total	41.2	Total	79.2
Losses:		*Losses:*	
Outflow	35.8	Outflow POC	2.8
Coprecipitation with CaCO$_3$	2.0	Sedimentation	21.7
Total	37.8	Total	24.5
Gross production:	+3.4	*Gross production:*	+54.7
Respiration:		*Respiration:*	
Bacteria	−20.6	Algae	13.0
(Estimated as 50% of		Sedimenting POC	21.7
DOC production)			
		Total	−34.7
Net production:	−17.2	*Net production:*	+20.0

Total net production (DOC + POC) = +2.8 g C m^{-2} year^{-1}

[a]From Wetzel, R. G., et al.: Metabolism of dissolved and particulate detrital carbon in a temperate hard-water lake. Mem. Ist. Ital. Idrobiol., 29, Suppl.: 185–243, 1972.

TABLE 17–26 Annual Benthic Particulate Carbon Budget of Lawrence Lake, Michigan[a]

Particulate Carbon	$g\ C\ m^{-2}\ year^{-1}$
Inputs:	
Submersed macrophytes	+ 87.9
Epiphytic algae	+ 37.9
Epipelic algae	+ 0.0
Loess	+ 0.0
Precipitation of DOC with $CaCO_3$	+ 2.0
Sedimentation	+ 21.7
Outputs:	
Benthic respiration	−117.5
Permanent sedimentation	− 14.8
Balance (utilization):	+ 17.2

[a] From Wetzel, R. G., et al.: Metabolism of dissolved and particulate detrital carbon in a temperate hard-water lake. Mem. Ist. Ital. Idrobiol., 29, Suppl.: 185–243, 1972.

ported organic carbon to the dissolved CO_2–HCO_3^-–$CO_3^=$ pool in respiration (Table 17–26), which represents the primary flow of carbon to the photic zone. A much smaller amount of benthic organic carbon is utilized by the pelagic community, and approximately an equivalent amount was lost to permanent sedimentation. Utilization of benthic detrital carbon (resuspension

TABLE 17–27 Total Annual Budget of Carbon Fluxes in $g\ C\ m^{-2}\ year^{-1}$ for Lawrence Lake, Michigan[a]

Components	$g\ C\ m^{-2}\ year^{-1}$
Inputs:	
Macrophytes	87.9
Epiphytic algae	37.9
Epipelic algae	2.0
Heterotrophy	2.8
Algal secretion and autolysis	14.7
Phytoplankton	43.4
Allochthonous particulate	4.1
Allochthonous dissolved	21.0
Littoral plant secretion	5.5
Bacterial chemosynthesis	7.1
	226.4
Outputs:	
Benthic respiration	117.5
Outflow dissolved	35.8
Outflow particulate	2.8
Permanent particulate sedimentation	14.8
Coprecipitation with $CaCO_3$	2.0
Bacterial respiration dissolved organic matter	20.6
Bacterial respiration particulate organic matter	8.6
Algal respiration	13.0
	215.1

[a] Derived from Wetzel, et al., 1972.

of Fig. 17–35) was almost equivalent to sedimentation losses from the pelagic zone, and represents an important contribution to pelagic productivity of material originating from the littoral zone.

An integration of these budgets of inputs and outputs permits an evaluation of the annual carbon fluxes in a whole lake system in which the allochthonous, littoral, benthic, and pelagic organic carbon flows are considered (Table 17–27). The whole lake budget is within 5 per cent of being balanced, which is very good considering the magnitude of variances involved in the analyses under in situ conditions. Analyses of zooplanktonic grazing by experimentally increasing animal concentrations serially to 200-fold in situ (Wetzel, et al., 1972) and in situ measurements of grazing of phytoplankton (Haney, Wetzel, Manny, unpublished) yielded maximal values of ca 20 per cent (nighttime) and < 5 per cent (daytime) during the period of maximum development of zooplanktonic populations. At other times of the year the values were lower and therefore were considered negligible in this lake. The importance of the Lawrence Lake analyses of the dynamics of the annual carbon fluxes, of which only summary values are given in Tables 17–24 through 17–27, lies in their demonstration of the major inputs from the littoral flora and dominating respiration in the sediments, where this material is decomposed. When macrophytes are minor producers among the total autochthonous organic inputs, as is the case in Mirror Lake, New Hampshire (Table 17–28), the respiratory metabolism of the sediments is still a major site of decomposition. The outputs of Mirror Lake (Table 17–28), in which phytoplankton dominate the inputs of organic carbon, are nearly identical in percentages to those of Lawrence Lake (Table 17–27), in which the littoral inputs dominate. The importance of dissolved organic detritus in the inputs from allochthonous sources and in the outflow is in the same relationship to the whole in both lakes. This structure and metabolism of photosynthetically fixed carbon by the littoral and the sediments, which probably dominates in most lakes of the world since a vast majority of lakes are small to very small, relegates the role of animals as "decomposers" to a lesser category and shifts the burden of metabolism to the microflora. In large lakes in which the inputs from littoral sources are proportionately low, or in hypereutrophic lakes in which phytoplanktonic densities reduce light to the point of excluding much of the littoral flora, animals probably assume a somewhat greater importance in the degradation of particulate organic carbon. The bulk of organic carbon is dissolved, however, and decomposition of that organic matter is almost completely microbial.

The structure of biomass or carbon fluxes in flowing waters is much less well-documented. Cummins (1972) estimated from a detailed budget of the average daily biomass flow in a temperate woodland stream that 25 to 30 per cent of the particulate biomass input flows through the benthic animals, assuming an average assimilation rate of 40 per cent for nonpredators, and is decomposed in this manner. Allochthonous and autochthonous inputs of dissolved organic matter, which constitute at least 5 times and probably 10 times that of the particulate organic matter, were not considered in this budget. Experimental analyses in artificial streams have demonstrated that a large portion (>50 per cent) of this dissolved organic matter can be degraded

TABLE 17–28 Annual Organic Carbon Fluxes of Mirror Lake, New Hampshire[a]

Components	g C m^{-2} year^{-1}	Per Cent of Subtotal
Inputs:		
Autochthonous		
Phytoplankton	78.5	74.4
Attached algae	2.2	2.1
Macrophytes	2.8	2.7
Bacteria (chemosynthesis)	4.0	3.8
Allochthonous		
Particulate	6.6	6.3
Dissolved	11.3	10.7
Total Inputs:	105.4	100.0
Outputs:		
Respiration	87.5	83.0
Sedimentation	7.6	7.2
Outflow		
Dissolved	9.3	8.8
Particulate	1.0	1.0
Total Outputs:	105.4	100.0

[a]From data of Jordan and Likens (1975), and Jordan, personal communication.

within from 24 to 48 hours by planktonic and attached bacteria under optimal conditions. The amount of dissolved organic carbon that is flocculated to particulate forms, especially in hardwater streams as discussed earlier, or enters bacterial biomass that is then partially degraded by animals is unclear. This is an area of great interest in need of further research (cf. Willoughby, 1974). Less than 1 per cent of detrital organic inputs was estimated to be processed by macroinvertebrate fauna of a mountain stream of New England (Fisher and Likens, 1973). As in lakes, although probably even more in streams, the role of animals in comparison to microflora in decomposition of particulate organic matter varies rather widely.

It is of the utmost importance to emphasize that these examples of functional structure of lake and stream ecosystems are only a prelude to the important underlying questions of control mechanisms that operate to determine the observed structural relationships. The analyses of structure and dynamics of ecosystems are a prerequisite to analyses of operants regulating the observed and changing structure. But such analyses must not become an end in themselves. Sufficient underlying similarities between the structure and dynamics of ecosystems probably exist that will permit generalizations to emerge with less rigorous analyses. As a result, more effort could be devoted to experimental evaluation of control mechanisms. It is critical that the regulatory functional mechanisms of the system components among natural ecosystems be understood before manmade perturbations of the systems can be effectively evaluated and minimized. Meaningful applied research and wise management of aquatic resources cannot be undertaken effectively without a basic understanding of functional control mechanisms.

Net Ecosystem Production

The dissolved organic carbon component of lakes receives inputs from both pelagic and littoral photosynthesis. In the Lawrence Lake example, an equivalent amount was received allochthonously from the drainage basin. The major loss of dissolved organic carbon through outflow represented most of the net ecosystem production of Lawrence Lake relative to the surrounding environment (Table 17–29). This estimate of *net ecosystem production* yields essentially a measure of export from the system, for example by outflow, sedimentation, and emigration.

Woodwell and Whittaker (1968) originally defined net ecosystem production (NEP) as:

$$NEP = GP - [Rs_{(A)} + Rs_{(H)}],$$

when GP   = gross production,

     $Rs_{(A)}$ = respiration of autotrophs,

     $Rs_{(H)}$ = respiration of heterotrophs,

or simply NEP = (gross production + imports) − (respiration of the ecosystem). The essential feature of NEP is that it accounts for importation and losses via respiration.

The original expression was modified by Woodwell, et al. (1972), to account for import and export of organic material:

$$NEP = (GP + NP_{in}) - [Rs_{(A)} + Rs_{(H)} + NP_{out}],$$

when NP_{in} = net production from another ecosystem,

    NP_{out} = net production exported.

TABLE 17–29   Net Ecosystem Production of Organic Carbon, Lawrence Lake, Michigan[a]

	Dissolved Organic Carbon	Particulate Organic Carbon	Total (g C m^{-2} year^{-1})
Inputs:			
Inlet 1	7.0	2.0	9.0
Inlet 2	7.9	1.0	8.9
Groundwater	6.0	1.2	7.2
Total	20.9	4.2	25.1
Outputs:			
Outflow	35.8	2.8	38.6
Sedimentation	—	14.8	14.8
Total	35.8	17.6	53.4

Net ecosystem production (NEP): + 28.3 g C m^{-2} year^{-1}

[a]From Wetzel, R. G., et al.: Metabolism of dissolved and particulate detrital carbon in a temperate hard-water lake. Mem. Ist. Ital. Idrobiol., 29, Suppl.: 185–243, 1972.

Both formulations interpret NEP as the positive or negative increment of organic matter, either living biota or dead organic storage, after total respiration within the ecosystem. The latter definition of NEP, which specifically subtracts organic exports, limits application of the term to only that material which remains inside ecosystem boundaries.

The original definition of NEP (Woodwell and Whittaker, 1968), as used in Table 17–29, is more generally applicable to aquatic situations. Aquatic systems appear to have much more dynamic equilibria than forest ecosystems, the context in which the term was construed originally. In aquatic systems the biotic (living) structure is small compared to annual productivity, and turnover is very high. Thus the NEP equation is an awkward way to ascertain the status of the biota. Moreover, much aquatic productivity quickly enters the detritus pool as dissolved organic carbon, which represents the vast bulk of organic carbon in water. The turnover of this storage pool is also high, and unresolved estimates calculated by the NEP equation are relatively meaningless. As discussed earlier, a detrital dynamic structure exists in aquatic ecosystems which parallels the trophic dynamic structure (biota) originally described by Lindeman (1942). Much of the detrital material in flux through this system is respired to CO_2 by various components of the biota and falls into the respiratory categories $[Rs_{(A)} + Rs_{(H)}]$ of the NEP equation.

Biotic Stability and Succession of Productivity

The mechanisms and couplings of detrital metabolism, as elaborated in this 3-phase system (allochthonous, littoral, and pelagic) and exemplified in the dynamics of a whole lake system (Lawrence Lake), provide a fundamental stability to the entire biotic dynamics of lakes and streams (Wetzel, 1975b; Rich and Wetzel, 1975). Living components of the system generally undergo rapid oscillations of productivity in an opportunistic series of competitive responses to changes in availability of constraining nutritional and physical factors governing their growth. However, the overall metabolism of the aquatic systems, both lakes and streams, is operational to a very major extent on the detritus components of dead dissolved and particulate organic carbon which form the primary source of energy of utilization.

The detrital organic carbon is present in dominating quantities and forms an excess detrital reserve that is refractory to rapid metabolism. The relatively slow utilization rate of the detrital reservoir gives stability to aquatic systems, tiding the system over during periods of low detrital carbon inputs and recharging the reserve during excessive inputs.

The stability is afforded in part by the refractory chemical structure of the organic substrates and in part by the displacement of much of the organic matter to anoxic environments (Fig. 17–36). When oxygen is absent or its diffusion rate into a habitat is insufficient to fulfill the requirements of aerobic microflora, the rate of carbon mineralization is slower and the amount of microbial cells formed per unit of substrate is less than when oxygen is abundant (Alexander, 1971; J. B. Hall, 1971). Complex and often simple organic compounds such as fatty acids and amines can accumulate when the quantity

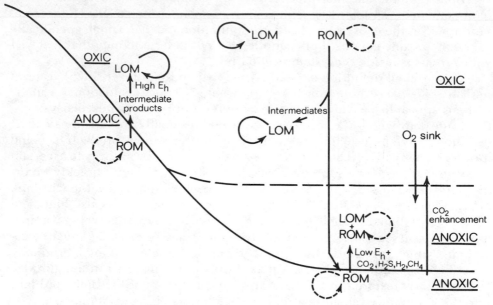

Figure 17–36   Relationships of the general degradation rates of organic compounds in relation to oxic and anoxic conditions. Solid circular arrows indicate relatively rapid rates of mineralization and turnover; dashed circular arrows relatively slow rates. *LOM* = labile, and *ROM* = refractory organic matter.

of readily fermentable carbon is large, as in the sediments. The low redox conditions lead to the generation of anaerobic fermentative products and enhancement of gases (CO_2, CH_4, H_2, H_2S) that simultaneously inhibit many
forms of microbial growth and, in biotic oxidation, serve as a major oxygen
sink. Degradation under these conditions is slowed and limited largely to
utilization of intermediate compounds of anaerobic microflora. When oxic
conditions are in close juxtaposition to anaerobic degradation, intermediate
products are more completely and efficiently degraded. Inhibitory conditions
of low E_h and toxic gaseous products are reduced greatly.

The overriding importance of detrital metabolism, within which the
trophic dynamic structure operates and utilizes a small portion, has been
alluded to in earlier works (see, for example, Lindeman, 1942; Odum and de
la Cruz, 1963; Wetzel, 1968, 1972; Wetzel and Allen, 1970; Rich and Wetzel,
1972; Wetzel, et al., 1972; Saunders, 1972a). But its demonstration for a whole
lake system (Fig. 17–35) has not been available previously. This detrital
metabolism and the stability it affords to the trophic dynamic structure must
be operational in all biotic systems, terrestrial and aquatic. The trophic
dynamic structure is almost totally dependent energetically upon the detrital
dynamic structure. The manner in which it is manifested differs in fine
detail among systems, but functionally, it must be the same in all ecosystems.

Within the operation of the detrital dynamic structure, the sources, quantities, and composition of detrital inputs are important and related to rates of
lake eutrophication. As has been discussed in detail earlier (Chapter 15),

the littoral plant complex of macrophytes and its associated microflora dominate autochthonous productivity of many lake systems. The contributions of phytoplankton, submersed and emergent macrophytes, and attached and eulittoral algae shift with increasing fertility. Dominance by the sessile flora is usually the case, except for certain extremely oligotrophic lakes, lakes of very large area and volume, and especially hypereutrophic lakes, in which biogenic turbidity severely attenuates submersed light conditions (see Fig. 15–25).

If the detrital dynamic structure is functionally controlling metabolism in a majority of lakes, and the littoral and allochthonous detrital inputs dominate the synthesis and inputs of detrital organic carbon, the littoral flora are implicated as a major driving metabolic force in lacustrine systems. Moreover, the detrital inputs and derived dissolved organic substrates vary with the flora adapted to individual lake conditions. For example, in hardwater marl lakes DOC interacts with inorganic components such as $CaCO_3$, which effectively suppress its subsequent pelo- and hydrometabolism. These suppressing mechanisms can be loaded to saturation, resulting in rapid transition to highly eutrophic conditions, geologically speaking. Similarly, but conversely, succession of certain macrophytes such as *Sphagnum* that function as effective ion exchangers with simultaneous release of refractory organic compounds (cf. following chapter) can create conditions in which detrital metabolism is slowed and permits accumulation of organic matter in particulate form. Underlying the entire metabolism and trophic structure are detrital metabolism and the rates and cycling of dissolved organic matter.

chapter 18

ontogeny and
evolution of
lake ecosystems

PERSPECTIVE

Analyses of the dynamic living aquatic systems necessitate a deep understanding of the physiology and biochemistry of organisms, integrated with
a knowledge of influencing environmental parameters and variables of physical and chemical disciplines. From these evaluations, the functional control
of contemporary metabolic states of organisms, populations, and aquatic
systems can emerge; this represents a major objective of limnologists today.
The ability to predict changes in the operation of the systems in response to
perturbations is essential to limnological efforts.

The effectiveness of predictive capability can be increased in many ways.
Certainly among the most informative approaches is to observe the responses
of aquatic systems from a previously known functional state to manipulation
of distinct parameters. This method is far from new, and has been practiced
in various forms for centuries, for example in the fertilization of ponds to
increase the growth rates and biomass of an organism of interest, such as
a particular species of fish. The early methods were largely trial and error,
without appreciable cognizance of controlling factors. Much of the same
approach permeates contemporary manipulation of water resources. Because of existing voids in the operational biology, engineering methods of
utilization often are implemented blindly. Some of these methods are margin-
ally operational, most are inefficient, and many are blatantly destructive of
water resources on a long-term basis. Insights into regulating parameters of
metabolism and population dynamics also have been obtained under much
more sophisticated, controlled conditions ranging from laboratory cultures,
isolation of small to very large portions of aquatic systems containing natural
biota, to the manipulation of whole lake systems. Much more of this integrated physiological and large manipulative system research is needed, and
certainly represents a major direction in contemporary limnology.

In the past, manipulations of aquatic systems have been brought about by

622

changes in climatic conditions, and more recently by man's activities, which have affected conditions of the drainage basins, water budgets, nutrient budgets, and of course the resultant productivity and rates of eutrophication. Records of these changes have been left in the sediments, which are relatively static derivatives of the dynamic systems. Paleolimnology focuses on the sedimentary changes and diagenetic processes that can alter that record, with the ultimate goal of gaining insights into past conditions, productivity, and changes in regulatory parameters that have caused the lake system to enter a different stage of productivity.

The mineralogy and structure of sediments, their organic and inorganic chemical constituents, and the morphological remains of organisms preserved in the sediments permit interpretations about past states and conditions of the system that have led to positive or negative alterations in productivity. As in all paleontological records, gaps occur and the record is incomplete. Furthermore, inputs from the drainage basin, redistribution of sediments within the lake basin, and differential preservation of interred remains demand critical interpretation of the record. The amount of information in the sediments, however, is large and has been appreciated fully only in recent years. Accurate interpretation depends strongly upon an understanding of biology and physico-chemical processes of contemporary systems.

General patterns of interacting causal mechanisms of autotrophic productivity demonstrate several ontogenetic pathways as lake basins evolve and are obliterated as aquatic systems. Primary factors focus on a transition from a dominance of allochthonous loading of nutrients and organic matter affecting planktonic autotrophy to accelerated loading of organic matter by autochthonous productivity. In the latter stages, autochthonous autotrophy first is dominated by phytoplanktonic productivity with some contributions by littoral production, but rapidly shifts in the terminal stages to a total dominance by littoral components. The transition from a lake system to wetlands is relatively slow in planktonic stages, in which organic sedimentation is balanced more closely by degradation. In shallow basins and in lake systems in which sedimentation has reduced the depth sufficiently, and tolerable humid climate prevails, massive littoral vegetative production shifts to a state in which organic matter accumulation increasingly exceeds decompositional capacity for oxidation and removal. Physiological characteristics of certain macrophytes, especially bryophytes of the genus *Sphagnum,* can greatly alter the decompositional capacity of the systems by increasing acidity and decreasing salinity. These conditions result in a further acceleration of organic matter accumulation.

Although inorganic nutrient loading, particularly of phosphorus and nitrogen, is fundamental to initial eutrophication and to maintaining high productivity of phytoplankton, the importance of dissolved organic compounds in recycling of nutrients and metabolism of aquatic systems is only beginning to be fully appreciated. The composition and quantities of dissolved organic matter are important to shifting states of trophy in that they regulate, in part, the cyclic regeneration of inorganic and organic nutrients by means of bacterial metabolism and by maintenance of increased inorganic nutrient availability for photosynthesis by complexing mechanisms.

PAST METABOLIC STATES OF LAKE SYSTEMS: PALEOLIMNOLOGY

Sedimentary Record

The fossil record of sediments includes both the biochemical substances produced by organisms or resulting from their degradation and morphological remnants of specific organisms, or both. The subject is large and the specific literature is voluminous. Recent reviews on paleolimnology from which much of the following résumé was drawn, are particularly instructive (Frey, 1964, 1969a, 1969b, 1974; Juse, 1966; Krode, 1966; Swain, 1965; Vallentyne, 1960).

Sediments that accumulate in lake basins consist of numerous source materials that are influenced to a great extent by the geomorphology of the lake basin and the drainage basin. Although many lakes were formed between the intervals of cyclic glaciation and deglaciation, most lakes of earlier glacial periods were obliterated by succeeding glaciations, filling with sedimentary materials or emptying by draining. It is only because of the very recent retreat of the last major glaciation phase that so many lakes exist at the present time. With the notable exceptions of ancient lakes, particularly in rift areas in Africa and Asia, a vast majority of lakes are very young ($<25,000$ years).

The primary sedimented materials are controlled by regional geology and climate, and modified by biological processes of both the drainage basin of the lake and the lake itself. Water movements by wave action and currents sort particulate matter in relation to size and density and the energy available for displacement. The result is a general gradient of coarser particles near shore regions to finer particles in sediments over deeper waters. This gradient can be disrupted by various mechanisms that often are related to the morphology of the lake basin. For example, accumulations of sediments in littoral regions or river deltas can exceed stability and slump in landslide fashion, or move in turbidity flows to deeper portions of the basin and overlay finer sediments.

Inorganic Chemistry

The external source materials in dissolved and particulate forms that leave the drainage basin and enter the recipient lake basin are influenced markedly by the vegetative cover. Numerous examples were discussed quantitatively in the preceding chapters, and need not be reiterated here. The long-term effects on the productivity of lakes are influenced by climatic variations as well as man-induced changes in the vegetation and weathering processes. An example of the effect of climate on weathering was demonstrated in some endorheic lakes of East Africa (Hecky and Kilham, 1972). Lakes in humid areas in which weathering rates of drainage are high receive greater inputs of silica and are largely bicarbonate-dominated. In more arid regions, sodium chloride originating from precipitation dominates the inorganic composition, and inputs of silica are low. The chemistry and climatic changes were reflected in the sedimentary record as well as the diatom populations of these different lake types.

Chemical constituents of the sediments have been used variously to interpret rates of both limnological activities of the lake as well as changes in climate and alterations of the drainage basin. The analyses assume that the chemical chronology represents, at least on a long-term basis, the composite changes in inputs to the lake and metabolic transformations that have occurred within the lake. For example, comparison of the chemical composition of numerous elements, mostly biologically nonessential, in lake sediments as well as in rock and soils of the drainage basins permits an effective evaluation of the chronology of erosion and leaching rates (Cowgill, Hutchinson, and collaborators, 1963, 1966a, 1966b, 1970, 1973, in a series of extraordinary studies). Inputs of extremely inert, conservative elements, such as titanium, permit an evaluation of erosion rates of the lake basins. Depending on the geology of the area, other elements may indicate disturbances of drainage basins. For example, in Laguna de Petenxil, a small lake in Guatemala, calcium, strontium, potassium, and to some extent sodium chronologies of the sediments were enriched during periods of active agricultural activity in the drainage basin. Manganese, iron, and phosphorus followed the fluctuations of agriculturally determined alkalies and alkaline earths in an analogous way.

Detailed investigations of the paleolimnology of a small closed lake of volcanic origin in Italy, Lago di Monterosi, demonstrated the dramatic effects of alterations to the catchment area by the construction of a road alongside the lake in early historical times (Hutchinson, et al., 1970). The lake was quite productive shortly after its formation for a period of about 2000 years, probably related to initial easy leaching of materials rich in potassium and of fairly high phosphorus content. By 24,000 years B.P., the sedimentation rates were very low. Low rates of deposition of inorganic constituents and organic matter, indicative of very low productivity, continued until at least about 5000 years B.P. and the Roman period. This interval was marked by a period of cold, dry conditions associated with the retreat of the last glacial activity in the area. Although a slight increase in productivity was discernible during the latter portion of this interval and a more recent period, most likely related to slight climatic amelioration and some human activity, a very large increase in nutrients and productivity occurred slightly over 2000 years B.P. (Fig. 18–1). The changes in the sedimentary records coincided rather precisely with the major disturbance of the drainage basin by the construction of the Roman road, the Via Cassia, in about 171 B.C. A great increase in inputs of alkaline earths, particularly calcium, and of phosphorus resulted and the lake rapidly became eutrophic with the deposition of much organic matter. Productivity of the lake since the period of major disturbance has declined with changing land uses in the area.

Records of such disturbances in the drainage basins are common and reflected in numerous studies of the sedimentary record. For example, the productivity of Grosser Plöner See in northern Germany was increased markedly in the early 13th century by the construction of a mill dam that raised

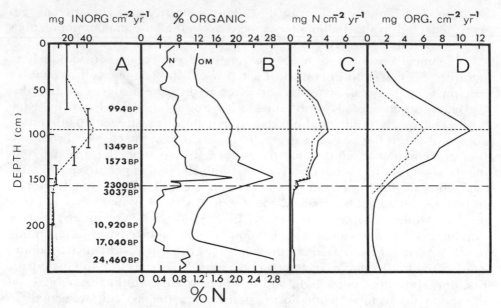

Figure 18–1 Stratigraphy of organic matter and nitrogen in the sediments of Lago di Monterosi, Italy. A, Estimation of sedimentation rates as a function of sediment depth and radiocarbon dates; B, Ignitable organic matter and nitrogen as percentages of dry sediment; C, Estimated rate of deposition of nitrogen, the broken line being corrected for organic nitrogen brought in by erosion; D, Estimated rate of deposition of organic matter, the broken line being corrected for organic matter brought in by erosion. (Redrawn from Hutchinson, et al., 1970.)

the water depth a few meters and flooded portions of the surrounding catchment basin (Ohle, 1972). In more recent times in the New World, the total clearance of forests over large portions of North America has lead to dramatic changes in productivity of lakes that are recorded in the chemical stratigraphy of sediments in the last 2 centuries. Often the increased leaching and nutrient loading leads to a marked acceleration of productivity. In contrast, when this increased leaching occurs in very calcareous regions, the extreme loading of calcium and carbonates can result in a negative effect on productivity by carbonate interactions and sedimentation of essential inorganic and organic nutrients (Wetzel, 1970).

The old concept that lakes progress from states of oligotrophy to eutrophy is not necessarily universal. As has been demonstrated repeatedly from chemical analyses of sediment stratigraphy, coupled with other indices of past productivity, lake productivity responds to changing nutrient incomes, climate, and morphometry. Accumulation rates of nitrogen, phosphorus, and organic matter commonly are high in early postglacial time, a period of high availability of nutrients, and then decline as the drainage basin gradually is impoverished by leaching. In recent times increases or decreases in certain geological regions usually are directly related to disturbances caused by the activities of man.

Organic Constituents

Chemical analyses of sedimentary organic constituents must address the problem of differentiating whether the organic compound originated within or outside of the lake basin. Further, many organic compounds undergo degradation during and after sedimentation; the extent of preservation is not always uniform with geologically changing conditions at and within the sediments. However, certain organic constituents of sediments, especially pigment degradation products, are very helpful in the interpretation of past episodes of productivity, particularly when coupled with other indices.

Total amino acid, carbohydrate, and chlorophyllous and flavinoid pigment residues in lake and bog sediments typically increase towards the surface (Swain, 1965). Carotenoid pigments also commonly increase upward in more recent lake sediments, but decrease markedly in the transition in the development of a lake to a bog (see discussions further on). Maxima of many organic compounds occur just beneath the surface of the sediment, probably associated both with increased microbial metabolism at the interface and with variable sorption of mineral matter and other organic substances.

Although much information is available on organic compounds and their distribution within sediments (Vallentyne, 1957b, 1960, 1969), except for pigment products this information provides little insight into paleolimnological events. Most biological polymers are not very stable and tend to be hydrolyzed enzymatically and chemically and then metabolized microbiologically. With many organic molecules it is very difficult to distinguish the time, origin, and mechanism of synthesis. Some significant progress has been made in the association of specific compounds with specific organisms, mostly with certain pigments. This area of study holds much potential and deserves intensified investigation. The problem of diagenetic changes in organic compounds after deposition, about which little is known, is a serious one, because these transformations can be confused with those resulting from succession or changes in the pattern of organisms that synthesize them.

Pigments

By far the most promising organic constituents of sedimentary stratigraphy that thus far have been investigated in detail are fossil pigments, a subject reviewed at length by Vallentyne (1957b, 1960) and Brown (1969). Upon senescence and death, the pigments synthesized by largely photosynthetic organisms undergo molecular transformations in which ions (such as the magnesium ion of chlorophylls) and side groups are lost progressively during physical or biological degradation. There is a tendency for the degradation products to increase in stability, that is, decrease in solubility and increase in relative resistance to further microbial and physical decomposition. Chlorophyll *a*, for example, degrades to pheophytin *a* with the loss of magnesium and then to pheophorbide *a* with the loss of the phytyl group. The alternate sequence in which the phytyl group is lost first leads to the formation of the intermediate product chlorophyllide *a*.

The number of pigments found in sediments is large. Nondegraded chlorophylls are rare but pheophytins, chlorophyllides, pheophorbides, and in some cases, bacteriochlorophyll degradation products have been identified and quantified. Additionally, a number of carotenoids have been found, some of which are highly specific to groups of organisms such as the blue-green algae or certain families of organisms. Their stratigraphy has been studied. The distribution of pigments, therefore, offers the possibility of identification and estimation of past plant populations as well as of environmental conditions of the water and sediment at the time of their deposition.

The analysis of fossil pigments in lake sediments has been applied to a large number of different lake systems, pioneered in the 1950s by Vallentyne and colleagues, and correlated with the stratigraphy of microfossils of organisms in interpreting the ecological record. The potential information that pigments can provide, with critical interpretation, is exemplified in analyses of Bethany Bog, Connecticut (Vallentyne, 1956). Myxoxanthin, a carotenoid specific to the blue-green algae, was present only in the sediments of the eutrophic phase of the lake and was not found in the more recent sediments of the lake as it entered its bog phase.

A number of analyses on several different types of lakes have demonstrated general correlations between sedimentary pigments and productivity. Most studies show maximum concentrations of pigments in early postglacial time after a period of arctic tundra conditions associated with retreating glaciation (Fig. 18–2). The peaks generally are attributed to early leaching of nutrients from the drainage basin during the ameliorated thermal period, which is characterized by transitions from tundra to coniferous and then deciduous vegetation. This initial surge in productivity usually is followed by an intermediate period of relative stability, with minor fluctuations associated with variations in climate, rainfall, and nutrient conditions. Subsequent fluctuations in pigment concentrations, and by inference in productivity, are variable but commonly have been lower until very recent times. This reduction can be associated with manipulations of the catchment area that increase the ratio of allochthonously derived organic matter to that sedimenting from production within the lake (Fogg and Belcher, 1961).

The rapid eutrophication of many lakes in recent times, often within the last century, generally is reflected in large increases in the pigment concentrations of recent sediments. This recent increase often coincides with renewed leaching of nutrients as the land was deforested for agriculture. In highly calcareous drainage basins, deforestation can accelerate inputs of calcium and bicarbonate, which interact with numerous inorganic and organic factors to suppress productivity. The inverse relationship between $CaCO_3$ and pigment stratigraphy has been demonstrated in several hardwater lakes (Wetzel, 1970). The abrupt decline in pigment concentrations within 200 years B.P. (Fig. 18–2, *upper*) coincided precisely with deforestation, massive increases in $CaCO_3$, and the initiation of *Ambrosia* ragweed pollen associated with agriculture.

Except for the diatoms, relatively few morphologically recognizable algal forms are well-preserved in freshwater sediments. Therefore, if pigments that are preserved are restricted to specific taxonomic groups that are

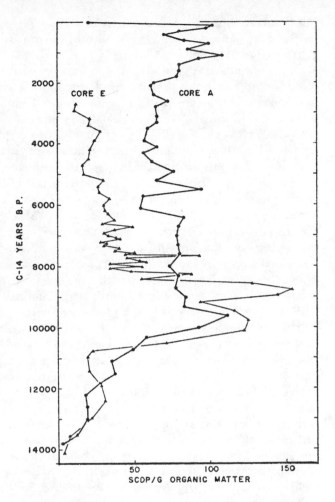

Figure 18–2 Sedimentary chlorophyll degradation products (*SCDP*) per gram organic matter in two cores from Pretty Lake, northeastern Indiana. *Core A* was from the deepwater central depression of the lake at a water depth of 25 m; *Core E* was from a shallow marl lakemount near shore that ceased accretion of sedimentation because of water movements at a water depth of about 1 m below the surface 2740 years B.P. (Wetzel, 1970, as modified by Frey, D. G.: Paleolimnology. Mitteilungen Int. Ver. Limnol., 20:95–123, 1974.)

characteristic of relatively distinct ecological events, their stratigraphy can be most helpful in interpretation of past conditions in the lake's ontogeny. The cosmopolitan distribution of chlorophylls among plant groups limits their specificity to protistan groups. Certain bacterial carotenoids and bacteri-ochlorophyllous degradation products of purple photosynthetic and green sulfur bacteria have been found in sediment stratigraphy and associated with events that led to eutrophication and meromixis (Brown, 1968; Czeczuga and Czerpak, 1968). Myxoxanthin, myxoxanthophyll, and oscillaxanthin have been used in this manner to infer the time of invasion and development of blue-green algae (Züllig, 1961; Brown and Coleman, 1963; Griffiths, et al., 1969). For example, quantitative measurements of myxoxanthophyll in sediment cores of 5 Swiss lakes provided a clear record of the first appearance of blue-green algae in the absence of morphological fossils during the marked recent eutrophicational changes in 3 of these lakes (Züllig, 1961). The appearance of oscillaxanthin that is specific to blue-green algae of the family Oscillatoriaceae in recent sediments of Lake Washington in Seattle was related directly to the development of dense populations of this algal group as the lake became excessively eutrophic from sewage effluent enrichment (Griffiths, et al., 1969).

Use of fossil pigments as a measure of qualitative and quantitative changes of former protistan populations must be done critically. Several assumptions often are made that are never completely valid and vary with specific conditions of individual lake systems (cf. reviews of Brown, 1969; Moss, 1968; Wetzel, 1970). The extent to which pigment products of terrestrial origin dilute those formed within the lake is not always clear. Chlorophyllous pigments decompose readily in soils and senescent leaves to pheopigments (Hoyt, 1966). The relative allochthonous input of inorganic and organic matter is largely a function of morphometry of the lake basin in relation to its drainage basin and the rate of erosion (cf. Mackereth, 1966). Although allochthonous inputs of organic matter can be quite significant in small, shallow lakes, the distribution and decomposition rates of chlorophyll and carotenoid derivatives in woodland soils, swamps, ponds, and lakes indicate that allochthonous contributions usually are small in relation to autochthonous sources (Gorham and Sanger, 1964, 1972; Sanger and Gorham, 1970, 1973). The latter workers introduced a pigment diversity index, based on the enumeration of pigment spots on two-dimensional thin-layer chromatograms of acetone extracts, as a means of assessing the origins of sedimentary organic matter. Particulate detritus of terrestrial origin contains 3 dominant pigments: pheophytin a, β-carotene, and lutein, while leaves of various deciduous trees at the time of abscission contain very low quantities (Sanger, 1971). The carotenoids in aerobic soils were degraded more rapidly than pheophytins. Therefore, the terrestrial vegetation has a low pigment diversity which decreases further on degradation in soils. Examinations of pigment diversities of upland vegetation, aquatic macrophytes, algae, and sediments showed a marked increase in diversity with the transition to lakes and autochthonous sources (Table 18–1). The pigment diversity in lake sediments increased over that of other source materials as a result of decomposition of the phytoplankton. Thus, the pigment diversity index is primarily a measure of pigment diagenesis when the initially high pigment diversity of algae is increased during

TABLE 18–1   Pigment Maxima and Diversity Indices Among Woodland Soil Detritus, Swamp Peats, and Sediments of a Series of Dimictic Lakes and a Meromictic Lake[a]

| System | Relative Units per Gram Organic Matter | | | |
	Chlorophyll Derivatives	Epiphasic (xanthophyll) Carotenoids	Hypophasic (carotenes) Carotenoids	Pigment Diversity
Woodland maximum	1.3	0.2	0.4	10
Swamp maximum	12.6	2.4	5.0	23
Dimictic lakes, minimum	1.1	0.4	0.8	24
Dimictic lakes, maximum	16.3	26.6	31.4	47
Meromictic lake, maximum	59.2	40.5	64.7	50

[a] Modified from Sanger and Gorham, 1973, and Gorham and Sanger, 1972.

decomposition (cf. Daley, 1973). The ranges commonly observed were 7 to 8 for upland vegetation, 12 to 15 for aquatic macrophytes, 10 to 21 for algae, and 24 to 47 for lake sediments. The highest diversity was found under ideal conditions for preservation of pigments in sediments of a meromictic lake. The results indicate that during oligotrophic conditions, much of the organic matter preserved in sediments is allochthonous in origin, while under eutrophic conditions pigments of autochthonous origins greatly exceed those sedimented from external sources.

Another assumption generally made is that the pigment concentrations of living algal populations represent a reasonable estimate of algal productivity, so that the sedimentary record reflects past conditions. Although chlorophylls have been used widely as a measure of contemporary productivity (Chapter 14), the pigment content of living populations also is influenced by numerous environmental factors. Over time the succession of species and changes in nutrient status may result in changes in the ratio of pigment to organic matter in sedimenting materials. In a general and relative way, however, pigments do reflect contemporary plant productivity.

Differential degradation of pigments during and after sedimentation is a more serious problem that is difficult to resolve. If pigment diagenesis is greatly variable, it is difficult to distinguish between high productivity and rapid rates of degradation or low productivity and low diagenesis. Detailed experimental investigations of the destruction of chlorophyll and derivative formation from lacustrine pigments demonstrated that several factors are involved (Daley, 1973; Daley and Brown, 1973). Photooxidative destruction of chlorophyll occurred in senescent phytoplanktonic cells and at an accelerated rate in cells lysed artificially, by bacteria, or by a virus. Chlorophyll *a* was degraded faster than chlorophyll *b*. In prolonged darkness, no destruction was observed. Enzyme-mediated chlorophyll degradation could not be detected, and from a comparison of available information it was concluded that photooxidation is one of the principal causes of chlorophyll destruction in situ.

Destruction of pigments and derivative formation were found to be unrelated processes. Pheophytins and pheophorbides accumulated only in lysed cells, and were accentuated by the presence of dilute acids and oxygen. Derivative formation is strictly chemical and governed by oxygen concentration and pH, decreasing precipitously at pH values above neutrality. Ingestion of algae by herbivores, such as *Daphnia* and the chrysophycean flagellate *Ochromonas*, resulted in the destruction of chlorophyll and the formation of pheophytins and pheophorbides.

Preservation of pigments also is favored by low temperatures such as would be found in hypolimnetic waters. Thus one would anticipate the best conditions for preservation under the anoxic, cold, and relatively dark conditions of the monimolimnia of meromictic lakes (cf. Table 18–1). In more typical lakes, however, the effects of circulation should be kept in mind. The pigments sedimented are exposed to resuspension of surficial layers and exposed to oxygenated, lighted conditions during circulation. Temperatures are relatively low in hypolimnia, during circulation periods in the temperate regions, and in winter. Hence, the conditions for chemically mediated derivative formation may be fairly similar over much of the history of the lake development. As the basin fills in and becomes too shallow to permit stratification or becomes meromictic, conditions for preservation and diagenesis can change markedly.

Therefore it must be clearly recognized that use of pigment concentrations and constituents as a paleolimnological index requires critical evaluation. The method has merit in interpreting past lacustrine events, but should complement morphological and other reconstructions rather than serve as a sole approach.

Morphological Remains

Lacustrine paleolimnology has a long history that began with analyses of morphological remains of aquatic and terrestrial organisms. Preservation is incomplete and variable among groups and with conditions at the time of deposition. Pollen and spores of terrestrial plants are most abundant, followed by diatoms and chrysomonad cysts (Frey, 1974). Some other groups of algae can occur, but many are not represented. The Cladocera and chironomid midges have the most abundant and diversified of animal remains, although all groups of animals are represented to some extent. Many remains can be matched to species, particularly among the diatoms, desmids, Cladocera, ostracods, and beetles. When only a stage in the life cycle is represented, for example, cysts of chrysomonads or resting eggs of rotifers, differentiation often is possible only to the point of the genus or family level.

Pollen

The stratigraphy of pollen and spores provides a relative chronology of the vegetation from which presumptive evidence for climatic changes can be obtained. These vegetational changes of the landscape and catchment area provide insight into temperature and rainfall changes, soil development, and changes in the drainage basin caused by the activities of man. Knowledge of the detailed vegetational history of the drainage basin is indispensable for interpretation of changes in the developmental stratigraphy of the lake through its sediments.

Nearly all detailed paleolimnological analyses augment their interpretative power by means of pollen diagrams. Hence, the number of studies on pollen stratigraphy is very large. A few examples will demonstrate their usefulness.

In a classical study of the now meromictic Längsee, Austria, Frey (1955) demonstrated that permanent meromixis began about 2000 years ago when man moved into the area and began clearing the drainage basin of the lake of forest vegetation for agriculture. Pollen stratigraphy showed marked regression in the dominant forest species and a sharp increase in pollen of smaller species, 2 cultivated plants, and a number of agricultural weeds. Evidence also was found for markedly increased erosion of fine clay particles that probably were sufficient to increase the density of lower water strata and initiate meromixis, which was maintained subsequently from biogenic sources.

Numerous other examples exist in which the pollen stratigraphy reflects how sensitively lake metabolism responds to changes in the drainage basin. The response of the Italian Lago di Monterosi to clearing operations for a Roman road, discussed earlier, also can be seen in the abrupt changes in pollen deposition from the terrestrial vegetation. More recent changes in lake drainage basins resulting from agricultural clearing of land are seen in many parts of North America. As the land was denuded, erosion and leaching was increased markedly, usually resulting in accelerated nutrient loading and productivity. This sedimentary horizon is demonstrated by many chemical and biotic changes, the most conspicuous being the sudden rise in pollen of the agricultural weed, *Ambrosia,* or common ragweed. The importance of the erosion rate in the drainage basin also was shown to be of major importance in the stratigraphy of sediments of several lakes that have been studied extensively in the English Lake District (Tutin, 1969). Pollen spectra throughout late and postglacial times correspond closely with the chemical history of the drainage basin. During the late glacial and early postglacial period, the erosion rate was controlled by low temperatures. In the mid-postglacial period of deciduous forest, erosion was at a minimum. Within the last 5000 years variations in or destruction of the deciduous forest cover among lake systems are evident from pollen stratigraphy and rates of erosional deposition.

Algal Remains

Diatoms, some green algae, Chrysophyceae, and dinoflagellates are well-represented in lake sediments, and the heterocysts and certain other parts of blue-green algae are sufficiently abundant to follow progressive changes in stratigraphy (Korde, 1966; Juse, 1966; Round, 1964b). By far the most attention has been directed to diatoms. Diatom profiles in sediments are complicated by differential rates of dissolution of their frustrules during and after sedimentation, which raises questions about how rigorously the remains found in sediments reflect the quantitative composition of producing populations. The few studies directed towards this question are conflicting, but it is apparent that dissolution rates vary among individual lake systems in relation to their chemical and morphometrically related history (cf. Chapter 13). For example, it has been found that frustrules sedimented in the littoral and shallower profundal zones dissolved at higher rates than those sedimenting in the deep profundal zone (Tessenow, 1966).

In spite of these interpretative problems, the stratigraphy of diatoms has been most useful in casting insight into past conditions and causes of change. Because of known ranges of ecological requirements for many species and knowledge of community composition from contemporary diatoms, inferences about past conditions are possible.

In very general terms, centric diatoms are associated with more oligotrophic waters, while pennate diatoms tend to be more characteristic of eutrophic waters. The ratio of these various indicator groups of diatoms in the stratigraphic record offers some help in interpreting the chronology of positive or negative changes in productivity. Returning to our example of Lake Washington in Seattle, early development of the population around the lake extended over the period from 1850 to the early 1900s. From about 1910 to the mid-1920s, raw sewage entered the lake in increasing amounts. Between 1926 and 1941, much of the primary effluent was diverted away from the lake, and then in 1941, with progressively increasing population expansion, large amounts of secondarily treated sewage discharge entered the lake. These effluents of treated sewage continued to enter the lake until the mid-1960s, when all was diverted to Puget Sound. Examination of the diatoms preserved in the recent sediments indicated detectable responses to the sequence of enrichment, reduction, and renewed enrichment (Stockner and Benson, 1967). Major changes in the diatom community occurred (Figs. 18–3 and 18–4), related largely to reciprocal fluctuations in the abundance of the centric *Melosira italica* and the pennate *Fragilaria crotonensis*. Relative composition of the diatoms deposited, as well as species diversity and redundancy values, was constant for the period prior to enrichment. Correlated with the pattern of sewage discharge into the lake, many of the species changed in proportion in accordance with their ascribed nutritional behavior as indicator species. It was concluded that use of indicator groups, here contrasting the reduction in centrales with the increase in araphidine diatoms in response to nutrient enrichment, was much more reasonable and reliable than indicator species.

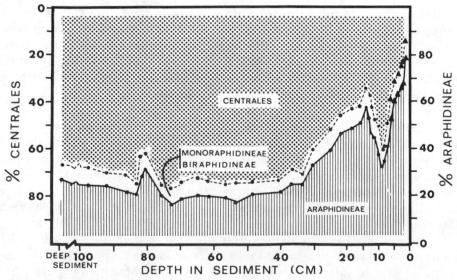

Figure 18–3 Changes in the percentage composition of the centric and araphidine diatoms in the surficial sediments of Lake Washington. The changes around 80 cm are associated with a clay turbidite layer peculiar to this core but not generally present throughout the lake. (Redrawn and modified from Stockner and Benson, 1967.)

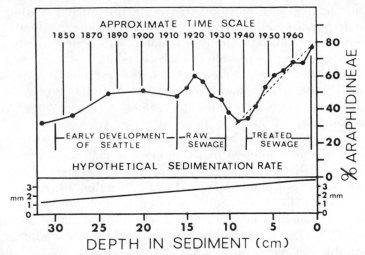

Figure 18-4 Changes in the relative abundance of araphidine diatoms in the recent sediments of Lake Washington in relation to the developmental history of the region surrounding the lake. The time scale is derived from an estimated average sedimentation rate of 2.5 mm per year over the upper 30 cm. (Redrawn and modified from Stockner and Benson, 1967.)

Animal Remains

Nearly all groups of animals leave at least some identifiable morphological remains in lake sediments. At the microscopic level, the most abundant animal remains in sediments are those of the Cladocera and midges, and under favorable circumstances those of rhizopods and ostracods can be found (Frey, 1964, 1969). At the macroscopic level, mollusk and beetle remains dominate. The remains of larger organisms generally are less abundant and require larger volumes of sediments for the recovery of significant numbers. Therefore, their stratigraphy must be based on coarser intervals than that of microscopic fossils.

Numbers of species or groups usually are expressed in relative percentages of the total population or, even better, related to the weight of organic matter or some other parameter. Varying rates of sedimentation can be determined from radiocarbon dating and certain other stratigraphic criteria that permit expression as an absolute rate of accumulation, and can result in quite different interpretations than those based on percentage composition (cf. Davis and Deevey, 1964).

Certain shifts in populations of organisms would be anticipated with changes in the characteristics of a lake in its ontogeny from a primitive to an advanced condition. Therefore, as a lake becomes more eutrophic with decreasing oxygen content of the hypolimnion, among profundal midges, for example, we might expect a succession of midge populations, because of their differing requirements for oxygen, from those that require high oxygen to those that tolerate decreasing oxygen content. In detailed studies of the stratigraphy of chironomid and *Chaoborus* midge faunas of several lakes of northern Indiana, Stahl (1959) demonstrated such a succession in these populations and offered strong evidence for the historical transition from aerobic to anoxic hypolimnetic conditions among the more productive lakes.

In contrast to some problems associated with selective dissolution of diatoms, the *Cladocera* exoskeletons preserve well. This group has been studied in the fossil record in great detail in the last decade. Analyses of past populations also have been made in which information theory at the community level has been applied, refining interpretations of responses to changing ecological conditions.

Species of chydorid Cladocera are largely littoral in distribution on substrata. Considerable evidence exists, however, that the littoral faunal remains are redistributed by currents within lake basins and become integrated with remains of planktonic Cladocera (Mueller, 1964; DeCosta, 1968; Goulden, 1969a). The populations recorded in profundal sediments therefore offer a reasonable integration of community dynamics over habitats and seasons. Changes with time or among lakes can then reflect changes in the communities in response to varying biotic conditions such as shifts in food quantity and quality, competition, or predators.

There are many examples of the usefulness of cladoceran remains in interpreting past conditions (Frey, 1969, 1974). Changes in species diversity of chydorid communities, as pioneered in analyses by Goulden (1969b), have been used to assess the stability and responses of the communities to disturbances. Equilibria of the population associations have been shown to be disrupted in response to the establishment of competing species, agricultural disturbances, climatic alterations, and shifts in the ratio of lake productivity contributed by littoral and phytoplanktonic sources (Whiteside, 1969). The ratio between planktonic and littoral cladoceran remains also is useful in interpreting past changes in water level and littoral development.

ONTOGENY OF AQUATIC SYSTEMS

Considerations of the dynamics of aquatic ecosystems, using the comparative method that has been employed throughout much of this treatment, have been strongly biogeochemical. Rationale for this input-output approach has focused on its power to provide insight into the operational mechanisms governing contemporary productivity. Clearly, the physiological and population characteristics of the heterogeneous biota of aquatic ecosystems must be integrated with the dynamics of the physical and chemical parameters influencing growth and behavior. However, it also should be readily emphasized that aquatic ecosystems commonly are not in steady state, except in a restricted short-term sense. The differences among aquatic ecosystems, as exemplified comparatively, obviously are related to an array of dynamic factors that change over time. Short-term changes, keyed to diel and seasonal fluctuations, are reasonably repetitious and form the basis for many steady state assumptions. The preceding few pages on paleolimnology, however, should make it clear that the long-term progression is not completely steady state.

Aquatic systems evolve. Erosion is a dominant geomorphological feature in running water systems in which the lotic metabolism shifts with the progressive physical and chemical changes towards the headwater regions. Geomorphology also is critical to the ontogeny of lake ecosystems. Geochemical inputs to lake basins, coupled with morphometric characteristics of the depression and changes in both the morphology and drainage patterns, influence their productivity.

A primary characteristic of this ontogeny is sedimentation that gradually fills the lake basin. Inorganic inputs to the sedimentary influx and deposition vary greatly with the parent materials of the surrounding catchment area. These inorganic inputs generally are slow, and a majority of organic inputs are mineralized prior to and after sedimentation. The ratio of organic to inorganic deposition can shift markedly in response to numerous changes in the nutrient incomes, drainage basin characteristics, and basin morphology. A point is reached at which organic deposition exceeds the capacity for its decomposition by increasing amounts. Sedimentation rates can increase very rapidly once a combination of characteristics is reached, and can result in an accelerated obliteration of the lake basin towards a terrestrial landscape. The mechanisms involved in this ontogeny are highly individualistic but certain similarities emerge.

The ensuing discussion of the evolutionary ontogeny of lake systems emphasizes the inputs of organic matter. The productivity of lake systems includes that produced in the littoral region. The boundary of the lake per se can often be delineated relatively easily by the shoreline and supralittoral area. The boundary in many lake systems, however, is much more diffuse. From a functional viewpoint, in both the metabolism of the pelagial region and of the entire lake, the saturated marsh areas surrounding many lakes constitute a major source of inorganic and organic inputs to the whole system. Separation of these massive organic sources from metabolism of the open water "lake" is functionally incorrect. Similarly, it is wrong to ignore the sensitivity of the lake per se to allochthonous inputs of nutrients and organic loading. To reiterate earlier statements, lakes cannot be treated as separate microcosms. While the importance of allochthonous inputs long has been known, especially from British work, and the importance of the littoral inputs has been advanced more recently, there is a persistent tendency, particularly among North American workers, to treat the pelagial planktonic community as an isolated, nonintegrated component of the lake system. This tendency has been accentuated by several recent modeling simulations that are extraordinarily simplistic, and detract from the powerful predictive capacity that these simulations potentially possess.

Planktonic Productivity

Viewing the development of natural lake systems in an evolutionary and successional manner predicates an understanding of the basic controlling factors of metabolism. The productivity of lakes is fundamentally autotrophic; hence the emphasis on inorganic and organic biogeochemical cycling. Attempts to generalize lake ontogeny in which photosynthetic productivity and degradation of planktonic and littoral sources are integrated are difficult but necessary. They must be treated separately initially and then integrated.

General Eutrophication

The word eutrophy, originating from the German adjective *eutrophe* and in general referring to nutrient-rich, is a greatly misused word in contemporary dialogue.[1] Naumann (1919) introduced the general concepts of oligotrophy and eutrophy, and distinguished them on the basis of phytoplanktonic populations. Oligotrophic lakes contained little planktonic algae, and were common in regions dominated by primary rocks. Eutrophic lakes contained much phytoplankton, and were common among more naturally fertile lowland regions in which human activity provided an increased supply of nutrients. Although at that time chemical methodology was crude or nonexistent for certain assays, Naumann emphasized that within a normal thermal range chemical factors, particularly phosphorus, combined nitrogen, and calcium, were primary determining factors.

Shortly after the turn of the present century, Thienemann (summarized in 1925) found in alpine and subalpine lakes that the midge larvae, largely of *Tanytarsus,* were characteristic of unproductive, deep lakes in which the hypolimnetic water lost little of its oxygen content during summer stratification. Eutrophic lakes (he later adopted Naumann's terms) were shallower, richer in plankton, and the oxygen-reduced hypolimnetic water was dominated by fauna, such as the midge *Chironomus,* that can tolerate very low oxygen concentrations. The extensive studies and conceptualizations of these earlier workers, coupled with those of numerous others, formed the fundamental basis on which much regional limnology of the 1930s was based.

A plethora of subsequent limnological studies was directed towards evaluation of the characteristics of the different lake types. The terminology for different lake types that developed in the ensuing two decades was indeed phenomenal. Many lake types were differentiated on the basis of indicator species. The basis for many of these descriptive classifications was founded in controlling physical and chemical parameters, while the basis for others was not. The terminology and dichotomies of classification that developed, while instructive, were excessive. Lakes were categorized in nearly every limnological aspect — geomorphological, physical, chemical — and by indicator species or aggregations of nearly every group of organisms from bacteria to fish. Even the extremes of lake types based on the waterfowl that commonly are associated with them were analyzed. As more lakes were studied, exceptions were found which led to further splitting and name generation. Early

---

[1]Hutchinson (1973) has written an excellent general, although classical, review of the subject and the development of the term in limnology. The superb semi-popular book *The Algal Bowl — Lakes and Man* by Vallentyne (1974) complements the recent collections of studies on eutrophication: *Eutrophication: Causes, Consequences, Correctives* (National Academy of Sciences, 1969), and *Nutrients and Eutrophication,* edited by Likens (1972).

integrative analyses, such as those of Thienemann (1925) and Naumann (1932), became less common as the terminology became more complex. If the details of classifying are carried to an extreme, lakes are so individualistic that one would soon approach the situation of requiring taxonomic keys for lake types (which has been proposed by Zafar, 1959).

The grouping of lakes must be left as a spectrum of states of lake metabolism between the extremes of oligotrophy and eutrophy. Historical studies show that in many small lakes of temperate, formerly glaciated areas, a succession is common from more inorganic sediment, containing oligotrophic-indicator fossils, to a more organic sediment, containing eutrophic-indicator fossils (Hutchinson, 1973). Evidence indicates that once organic sedimentation has become established after the initial oligotrophic phase, a type of trophic equilibrium or reasonably stable steady state occurs. This implies that nutrient inputs from the drainage basin are relatively constant over long periods of time, and undergo only minor changes with oscillations in climate and inputs from vegetative cover and erosion. The systems can, and commonly do, regress in productivity as the surficial soils are reduced in nutrient content by leaching. Conversely, productivity can be accelerated greatly by nutrient inputs, as has been discussed.

The importance of dissolved organic compounds in recycling of nutrients in natural systems is only beginning to be fully appreciated. Dissolved organic matter is important in shifting states of trophy by accelerating the cyclic regeneration of nutrients through bacterial metabolism and maintenance of increased inorganic nutrient availability for photosynthesis.

Major causal pathways governing the eutrophicational ontogeny of lakes are based on interacting mechanisms regulating autotrophic metabolism. Our discussion is directed primarily at the majority of temperate lakes of glacial, tectonic, and volcanic origin and of moderate size and depth. Exceptions certainly exist.

Oligotrophy to Eutrophy

Low rates of productivity of oligotrophic lakes are regulated to a large extent by low input of inorganic nutrients from external sources. Morphometric characteristics of relatively large size and depth that yield high hypolimnion to epilimnion volume ratios are common to oligotrophic lakes and influential in nutrient cycling. The low production of organic matter, resultant low rates of decomposition, and oxidizing hypolimnetic conditions result in relatively low nutrient release from the sediments in a cyclical causal system (Fig. 18–5). Low concentrations of relatively readily decomposable dissolved organic substrates from both planktonic and littoral sources result in generally low bacterial populations and slow rates of microbial metabolism. Synthesis of organic micronutrients, essential to most planktonic algae, would be correspondingly limited. Effective complexing of essential inorganic micronutrients by dissolved organic compounds, by which solubility and physiological availability could be partially maintained under oxidizing

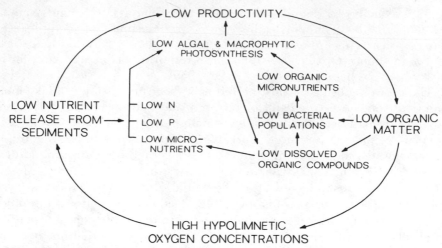

Figure 18–5 Major inorganic and organic interactions influencing the metabolism of phyto-plankton of oligotrophic lakes. (Modified from Wetzel and Allen, 1970.)

conditions, would be reduced under oligotrophic conditions of low synthesis of organic matter and high rates of degradation of available substrates.

Under a large majority of lake conditions, the most important nutrient factors causing the shift from oligotrophy to a more productive state are phosphorus and nitrogen. Typical plant tissue of aquatic algae and macrophytes contains phosphorus, nitrogen, and carbon in approximately the ratios (Vallentyne, 1974):

> 1 P : 7N : 40C per 100 dry weight or
> 1 P : 7N : 40C per 500 wet weight.

If one of the 3 elements is limiting and all other elements are present in excess of physical needs, phosphorus theoretically can generate 500 times its weight in living algae, nitrogen 71 (500:7) times, and carbon 12 (500:40) times.

Comparison of the relative amounts of different elements required for algal growth with supplies available in fresh waters illustrates the general importance of phosphorus and nitrogen (Table 18–2). Similar proportional ratios were demonstrated by comparison of the demand among terrestrial plants with the accessible supply of elements from the lithosphere (Hutchinson, 1973). Even though variations in conditions of solubility or availability at times may make very abundant elements such as silicon, iron, and certain micronutrients almost unobtainable, phosphorus, and secondarily nitrogen, both critical metabolic constituents of biota, are the first to generally impose limitation on the system. This relationship is emphasized further by consideration of average demand:supply ratios in late winter prior to the spring algal maximum common in temperate regions, and those in midsummer during maximum sustained algal productivity (Table 18–3).

Oligotrophic lakes often are limited by phosphorus and contain an excess of nitrogen. As the lakes become more productive, the primary effecting agent

TABLE 18–2 Proportions of Essential Elements for Growth in Living Tissues of Freshwater Plants (Requirements), in the Mean World River Water (Supply), and the Approximate Ratio of Concentrations Required to Those Available[a]

Element	Average Plant Content or Requirements (%)	Average Supply in Water (%)	Ratio of Plant Content: Supply Available
Oxygen	80.5	89	1
Hydrogen	9.7	11	1
Carbon	6.5	0.0012	5000
Silicon	1.3	0.00065	2000
Nitrogen	0.7	0.000023	30,000
Calcium	0.4	0.0015	<1000
Potassium	0.3	0.00023	1300
Phosphorus	0.08	0.000001	80,000
Magnesium	0.07	0.0004	<1000
Sulfur	0.06	0.0004	<1000
Chlorine	0.06	0.0008	<1000
Sodium	0.04	0.0006	<1000
Iron	0.02	0.00007	<1000
Boron	0.001	0.00001	<1000
Manganese	0.0007	0.0000015	<1000
Zinc	0.0003	0.000001	<1000
Copper	0.0001	0.000001	<1000
Molybdenum	0.00005	0.0000003	<1000
Cobalt	0.000002	0.000000005	<1000

[a]After Vallentyne, J. R.: The Algal Bowl—Lakes and Man. Miscellaneous Special Publication 22, Ottawa, Dept. of the Environment, 1974.

is increased loading of phosphorus. As discussed in Chapter 12, the instantaneous concentrations usually decrease and are quite variable, but the turnover rate increases markedly. Rates of losses increase, and high productivity requires sustained phosphorus loading of the system from allochthonous and littoral sources.

Nitrogen in combined form, although found in large quantities in the lithosphere, is largely unavailable to plants. Nitrogen supplies to freshwater

TABLE 18–3 Comparisons of the Ratios of Required Concentrations of Inorganic Nutrients to Average Supplies Available in Fresh Waters[a]

Element	Ratio of Demand:Available Supplies	
	Late Winter	*Midsummer*
Phosphorus	80,000	up to 800,000
Nitrogen	30,000	up to 300,000
Carbon	5000	up to 6000
Iron, silicon	Generally low, but variable	
All other elements	<1000	<1000

[a]After Vallentyne, J. R.: The Algal Bowl—Lakes and Man. Miscellaneous Special Publication 22, Ottawa, Dept. of the Environment, 1974.

systems are augmented more readily by inputs from external sources (Chapter 11) than are those of phosphorus. Under extremely eutrophic conditions, planktonic utilization of combined nitrogen can exceed inputs and literally deplete the trophogenic zone. At this time, molecular nitrogen fixation by heterocystous blue-green algae augments the income of nitrogen, but at a high metabolic cost of products of photosynthesis.

The potential for limitation of eutrophication by inorganic carbon availability has been a subject of much recent discussion (cf. Chapter 10). In a few exceedingly productive situations, such as sewage lagoons, in which phosphorus and nitrogen compounds are available in excess of any demands, carbon can become limiting to algal growth (Kerr, et al., 1972). Alternately, while there is some evidence for inorganic carbon limitation in exceedingly softwater lakes (see, for example, Allen, 1972), it is clear that diffusion of atmospheric CO_2 generally is adequate to sustain the carbon requirements of phytoplanktonic populations. It is extremely unlikely that carbon limitation is of significance in the limitation of algal populations in the majority of harder waters, in which dense algae develop under eutrophic conditions.

The importance of phosphorus in comparison to nitrogen and carbon has been particularly well-illustrated by large-scale fertilization experiments (Schindler, 1974). Lake 226 is located in northwestern Ontario in Precambrian Shield bedrock, and was chemically and biologically similar to more than 50 per cent of the waters draining to the Laurentian Great Lakes. The trophogenic zone of Lake 226 was partitioned into 2 lakes at a constriction in the basin (Fig. 18–6). One basin was fertilized with phosphorus, nitrogen, and carbon, and the other with equivalent concentrations of nitrogen and carbon.[2] The phosphate-enriched basin quickly became highly eutrophic, while the basin receiving only nitrogen and carbon remained at prefertilization conditions (Fig. 18–6). In this phosphate-enriched basin and in several other lakes receiving analogous treatments over a period of several years, algal biomass increased to 2 orders of magnitude over that of lakes receiving only nitrogen and carbon enrichment. Recovery to near prefertilization levels was very rapid when only phosphate additions were discontinued.

The circumstances of the Shield lakes are ideal for demonstrating the overwhelming importance of phosphorus limitation and rapid recovery from eutrophic conditions once the inputs were controlled. The Lake Washington and Swiss lake examples, discussed earlier, are analogous examples of the success in the abatement of eutrophication that is possible once phosphorus control measures are instituted. In spite of the obvious caution which must be exercised in making generalizations among so complex and varied systems as lakes, it is important to emphasize once more that phosphorus abatement will not return all lakes to preenrichment conditions. But the importance of the relation of phosphorus demand to supply for plant growth is such that its reduction in inputs to fresh waters is the first place to begin and most likely to succeed in a majority of cases. Further, if phosphorus inputs to surface

---

[2]Additions were equivalent to 3.16 g of NO_3-N and 6.05 g of sucrose C per m^2 per year in both basins, made in 20 equal weekly increments. The northeast basin additionally received 0.6 g m^{-2} yr^{-1} of PO_4-P. The N/P and C/P ratios were greater than in treated sewage, but the quantity of P added was not exceptionally high for lakes that commonly receive pollution from domestic sources.

Figure 18–6   Lake 226 of northwestern Ontario, which was partitioned at the constriction of the basin, 4 September 1973. The far, northeastern basin, fertilized with phosphorus, nitrogen, and carbon, was covered by a dense algal bloom within two months. No increases in algae or species changes were observed in the near basin, which received similar quantities of nitrogen and carbon but no phosphorus. (From Schindler, D. W.: Eutrophication and recovery in experimental lakes: Implications for lake management. Science, *184*:May 24, 897–899, 1974. Copyright 1974 by the American Association for the Advancement of Science. Photograph courtesy of D. W. Schindler.)

water are to be reduced most effectively, point sources should be eliminated as rapidly as possible. The millions of tons of phosphate currently introduced by synthetic detergents can be eliminated relatively easily in conjunction with application of technologically available methods of phosphate removal in wastewater treatment. The scientific basis for the importance of phosphate and nitrogen in eutrophication is so overwhelming that an international resolution was ratified at the 19th International Congress of Theoretical and Applied Limnology in 1974 (Wetzel, 1975a). This resolution emphasizes the critical role of phosphorus in the rapid eutrophication of inland waters and the need to control the addition of this element to any inland water by any means available. In addition to secondary treatment of sewage, methods of control include: (a) restrictions on the use of cleaning products that contain phosphates or other ecologically harmful substances, (b) removal of phosphate at sewage treatment facilities discharging effluents into such water, and (c) control of drainage from feedlots, agricultural areas, septic tanks, and other diffuse sources of phosphorus. Control measures for nitrogen also should be considered in basins in which there is evidence that such controls are appropriate. Implementation of such control measures is socially complex, but attainable with existing technology.

Under eutrophic conditions the loading rates of phosphorus and nitrogen, as well as other nutrients under less acute demand, are relatively high (Fig. 18–7). As the rates of photosynthetic energy fixation and productivity increase, the cyclic interactions of regeneration of inorganic nutrients and organic compounds increase. Planktonic productivity increases markedly and results in a compression of the trophogenic zone from light limitations. Reduction in depth of the trophogenic zone continues with intensification of eutrophication, until planktonic population densities impose self-shading light restrictions beyond which further increases are not possible under natural conditions (Fig. 18–8). The rate of planktonic production then reaches a plateau, with a gradual progression towards eventual extinction through sedimentation of organic matter in excess of decomposition. Under certain conditions of morphometry, variable meteorological conditions, and productivity, it is not rare for mesotrophic to eutrophic lakes of small surface area and moderate depth to undergo biogenically induced meromixis (Chapter 6). A reduction, usually temporary, in productivity can occur when redistribution of the hypolimnetic waters and their nutrient-rich content from reducing conditions is prevented.

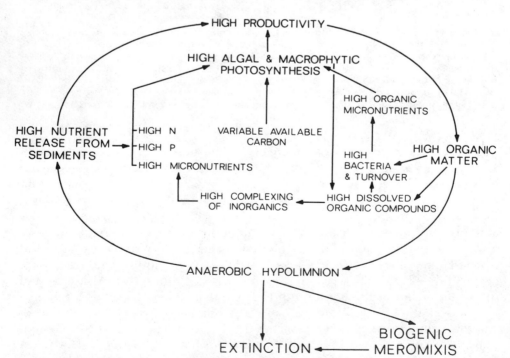

Figure 18–7   Major inorganic and organic interactions influencing the metabolism of phytoplankton of eutrophic lakes. (Modified from Wetzel and Allen, 1970.)

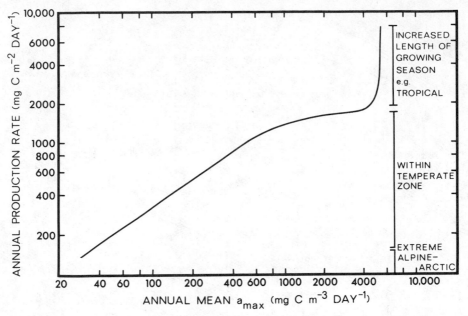

Figure 18–8 General relationship between the annual mean volumetric production rate at the depth of optimal growth and the annual mean areal production rates of phytoplankton. (After Wetzel, 1966a, and others.)

Hardwater Marl Lakes

The ontogeny of an oligotrophic lake towards eutrophic conditions can be altered markedly by the natural inputs of carbonates and associated cations from surrounding sedimentary rock formations and glacial till. Extremely calcareous hardwater lakes, common to large regions of the world, represent a type of maintained oligotrophy in which high calcareous inputs are sustained over long periods of time. In these waters, reduced productivity is maintained by decreased nutrient availability rather than "true" oligotrophy, in which nutrient inputs to the systems are deficient once the low temperature conditions of immediate postglaciation are passed.

Early postglacial eutrophication of marl lakes generally proceeded rapidly, as was characteristic of many lakes soon after their inception as the climate ameliorated. In these calcareous regions, however, high inputs of carbonates are maintained over long periods of time. Under excessively buffered bicarbonate conditions, the productivity of marl lakes is suppressed by an array of inorganic-organic interactions on the metabolism of the micro- and macroflora (Fig. 18–9). Major interactions have been discussed topically in earlier chapters and need be mentioned only briefly here.

Phosphate and essential inorganic micronutrients, particularly iron and manganese, form highly insoluble compounds and are effectively lost in entirety from the trophogenic zone of marl lakes (Wetzel, 1972). Combined nitrogen compounds tend to be in high concentrations resulting from relatively high inputs common in calcareous deposits and low rates of biotic utilization. It is not rare to find inorganic nitrogen concentrations in extreme

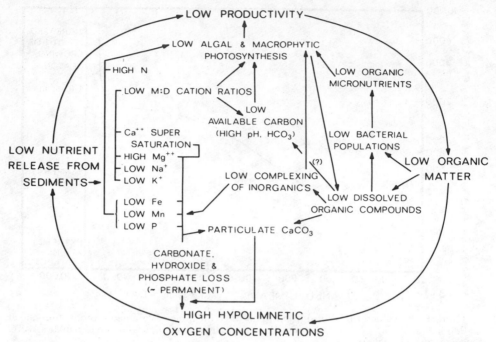

Figure 18–9 Major inorganic and organic interactions influencing the metabolism of producers in hardwater marl lakes. (From Wetzel, 1972.)

marl lakes that approach levels considered toxic for human consumption. Calcium and magnesium concentrations can be exceedingly high; calcium is often in apparent supersaturation and approaches or exceeds 100 mg l^{-1}. Intensive decalcification is characteristic of these hard waters, during which massive precipitation of particulate carbonate, largely biogenically induced, occurs in the trophogenic zone. Apart from inorganic adsorption of inorganic nutrients to colloidal and sedimenting particulate $CaCO_3$, certain labile and humic dissolved organic compounds are adsorbed and sedimented with the $CaCO_3$ from the photic zone. The major monovalent cations, sodium and potassium, commonly occur at low concentrations. Low levels of sodium, an essential element of certain blue-green algae, have been associated as a contributory interacting factor in reduced growth of heterocystous blue-green algae in marl lakes (Ward and Wetzel, 1975). Although inorganic carbon concentrations occur at very high levels, there is adequate evidence that the low availability of free CO_2 can be contributory to reduced photosynthesis of certain flora under stagnated conditions when the pH is excessively high (cf. Chapter 10). The capability for photosynthetic utilization of organic carbon sources, for example, carbamino carboxylic acids, under natural conditions is unclear. Recent evidence indicates that certain species of natural phytoplanktonic algae of a marl lake are capable of significantly augmenting photosynthesis by photoheterotrophy of simple organic substrates and compounds produced extracellularly by aquatic macrophytes (McKinley and Wetzel, in preparation).

Low photosynthetic productivity of marl lakes contributes to reduced inputs of dissolved organic compounds, such as extracellular loss from phytoplankton and macrovegetation, which is compounded by effective adsorptive losses to monocarbonates. Bacterial rates of metabolism are low because of the low concentrations and turnover of easily decomposable organic compounds, which can lead further to concomitant reduced synthesis of organic micronutrients. Certain organic micronutrients such as vitamin B_{12} have been shown to be contributory to reduced phytoplanktonic productivity of marl lakes; a portion can be metabolically inactivated by adsorption to $CaCO_3$. There is a tendency for available, more reactive organic substrates to be utilized rapidly by bacteria under deprived conditions and inactivated by adsorption to sedimenting $CaCO_3$. This reduction in less resistant organic compounds reduces the complexing capacity available to retain certain inorganic nutrients in a state available to photosynthesis, and also leads to an apparent accumulation of more resistant organic compounds. It is common to find very hardwater lakes with a stained coloration and high concentrations of humic compounds that are degraded slowly and persist in apparent accumulation.

Critical to increasing eutrophication of marl lakes are factors that reduce the buffering capacity and carbonate reservoir to a point at which direct inhibitory effects on inorganic nutrient availability are reduced, and stimulatory direct and indirect effects of dissolved organic matter are increased. Once the loading capacity of the carbonate reservoir system is exceeded, the cyclic interacting controls on productivity are reduced, and eutrophication can proceed relatively rapidly. Reductions in or depletion of bicarbonate and cation inputs from the drainage basin or increases in the loading of dissolved organic matter are 2 obvious means of altering the major causal controlling mechanisms. Morphometric changes accompanying gradual net increases in sedimentation rates also have significant effects on stratification patterns, hypolimnetic capacity for maintaining oxidizing conditions, and nutrient recycling.

In large areas of sedimentary rock that have been glaciated, it is common to find former marl lakes overlain by bog lakes and bogs, discussed in some detail further on. It is apparent that numerous marl lakes have shifted relatively rapidly, in the range of a millennium, from highly calcareous, alkaline conditions to a state of very acidic, organic rich conditions, which exhibit a marked paucity of divalent cations and bicarbonate. In nearly closed basins of moderate to small size, the ontogeny of marl lakes can be altered at a relatively rapid rate by the development of specialized littoral flora, particularly the mosses such as *Sphagnum*. These plants and associated encroaching vegetation function as particularly effective cationic sieves (Clymo, 1963, 1964). It is not uncommon to find pioneering small hummocks of *Sphagnum,* growing in the back-littoral reaches of marl lakes directly on sediments of nearly 50 per cent $CaCO_3$ content, overlain with a thin organic layer. The eulittoral of highly stained alkaline hardwater lakes, termed bog lakes in contrast to bogs (see further on), is often interspersed with stretches of *Sphagnum* in dense mats. These alkaline bog lakes represent a transitional stage between certain marl lakes and true *Sphagnum* bogs. The ionic exchange mechanisms of these mosses, accompanied by the simultaneous release of organic acids, for example, polyuronic acids, are effective in reducing the cation influxes from surface sources. As the littoral sieving flora circumscribe the basin, the buffering capacity of the marl lake system is reduced progressively. The subsequent vegetative development then can proceed relatively rapidly, in a rather classical pattern of bog succession from a lake to a terrestrial situation, by accelerated accumulation of organic matter under acidic, reducing conditions that are not particularly favorable to decomposition.

Dystrophy and Bog Systems

Early workers formulating relationships in studies of comparative regional limnology were acutely aware of the important differences among lakes in regard to the proportion of organic matter supplied from allochthonous sources and that from autochthonous production.[3] *Trophy* of a lake refers to the rate of organic matter supplied by or to the lake per unit time.[4] Trophy, then, is an expression of the combined effects of organic matter supplied to the lake.

As has already been indicated (Chapter 17), under natural conditions the relatively resistant humic substances of largely terrestrial plant origin represent the most common component of allochthonous organic matter. Lakes that receive large amounts of their organic matter supply from allochthonous sources commonly are heavily stained and have been referred to as "brownwater lakes." These lakes were termed *dystrophic*, in reference to their high content of humic organic matter. Productivity of most dystrophic lakes classically has been described as low. It must be emphasized, however, that this low to moderate productivity criterion *refers to the planktonic productivity* and, as it was developed, ignores the littoral plant components of the lake system. As we shall see, in dystrophic lakes that develop bog flora, the littoral plants completely dominate the metabolism of these lake systems as sources of dissolved and particulate organic matter.

As a consequence of uncritical use of the terms associated with dystrophy, bogs, bog lakes, and dystrophy have been loosely, and incorrectly, equated. Much of the unfortunate confusion emanates from workers who have described an apparent ontogeny of lake systems in their particular region of investigation. For example, the famous successional scheme put forth by Lindeman for Cedar Bog, Minnesota, has been more than once uncritically proposed in general ecology texts as the universal situation. In this way, the erroneous concept that all lakes become bogs and then land may become widely accepted. While some lakes do progress through this sequence, it is far from the rule. A large, heavily stained Finnish lake on a primary bedrock

---

[3]Birge and Juday (1927), for example, in their extensive studies on the origin and supplies of dissolved organic matter, differentiated between autotrophic lakes dominated by autotrophic inputs, and allotrophic lakes that receive a majority of their soluble organic matter from the drainage basin.

[4]A concise review of the development of the trophic concept is given by Rodhe (1969).

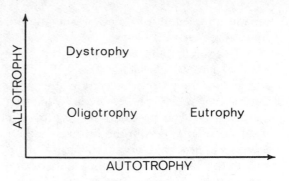

Figure 18–10 Classical trophic types of lakes based on the rate of supply of organic matter from autotrophic and allochthonous sources. In the original development of these concepts autotrophic productivity was viewed as only phytoplanktonic. (Modified from Rodhe, 1969.)

basin with no characteristic bog vegetation can correctly be termed dystrophic just as effectively as the *open pelagial water* of a quaking *Sphagnum* bog. Trophy refers to the rate of supply of organic matter.

These examples emphasize a most important point. As developed originally and as largely used today, the trophic concepts refer to the pelagial zone-planktonic *portion* of the lake ecosystem. The relative differentiation was between rate of organic matter input to the system from autotrophic phytoplanktonic sources and from allochthonous sources of the catchment basin (allotrophy) (Fig. 18–10). The littoral flora and its often dominating supply of autochthonous organic matter to the system was and usually still is ignored. Although conceptually it is easy to place the littoral productivity within such trophic schemes, quantitative evaluations of its contribution are

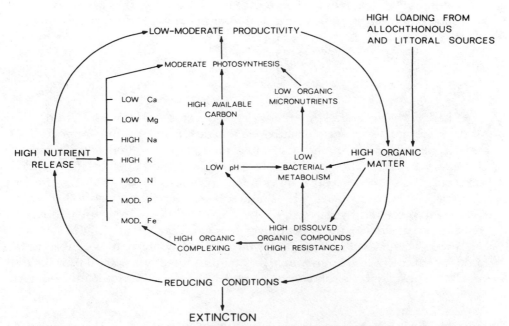

Figure 18–11 Dominant inorganic and organic interactions influencing the metabolism of phytoplankton in the pelagial waters of dystrophic and bog lakes. (Modified from Wetzel and Allen, 1970.)

meager. Sufficient information is available, however, to indicate that it cannot be ignored in most cases.

The phytoplanktonic productivity of dystrophic systems exhibits several general control characteristics (Fig. 18–11). These relationships are representative of algal productivity of the pelagial of certain highly stained open basins that receive much organic loading from allochthonous sources or of open water algal productivity of closed bog lakes in which organic matter inputs are largely from the littoral vegetation. Photosynthetic rates are generally rather low, but organic loading of relatively resistant dissolved organic substrates from allochthonous and littoral sources is very high. The dominance of humic substrates and relatively low pH values is not conducive to intensive bacterial metabolism. Because of the low inputs from the drainage basins or because of the effective cation exchange mechanism of littoral flora, monovalent to divalent cation ratios are high. The combined characteristics of acidity and reduced nutrient availability usually preclude high phytoplanktonic productivity in spite of high carbon availability as free CO_2.[5]

The relationships among the major causal mechanisms regulating the phytoplanktonic trophic states in the 4 main types of lakes may be connected further in an overall sequence of development (Fig. 18–12). It must be recalled, however, that this scheme of ontogeny refers primarily to changes in the functional interacting environmental parameters influencing algal, and to some extent macrophytic, productivity. The pathways vary greatly in relation to the geomorphology and macroclimate. Coupling these functional interactions within the pelagic zone with the geological ontogeny of extinction as the basin fills and is obliterated demands consideration of sedimentation. Consideration of accumulation rates of materials as a basin is filled requires an appreciation of the importance of littoral macrovegetation in culmination of the sequence.

Littoral Development

Shallow water bodies, correctly termed lakes but often referred to in older literature as ponds, usually are characterized by an abundance of aquatic macrovegetation and associated microflora attached to all surfaces. It is not unusual for the vegetation to extend over the entire basin, provided the depth does not exceed requirements, primarily light and pressure (cf. Chapter 15), for the plants. The shallow lakes represent basins that never have been preceded by a larger, deeper lake, and those basins which represent a terminal stage in the extinction of deeper lakes.

From a hydrological viewpoint, the shallow water bodies can be separated into those that are *permanent,* containing some water at all times of the year, and those that are *temporary,* in which the basin periodically becomes

---

[5]High productivity of phytoplankton in heavily stained dystrophic lakes is known, especially in lowland areas of Scandinavia. Järnefelt (1925) called these lakes mixotrophic, which would fit above eutrophy in Figure 18–10, but introduction of another term in this conceptual discussion is unnecessary.

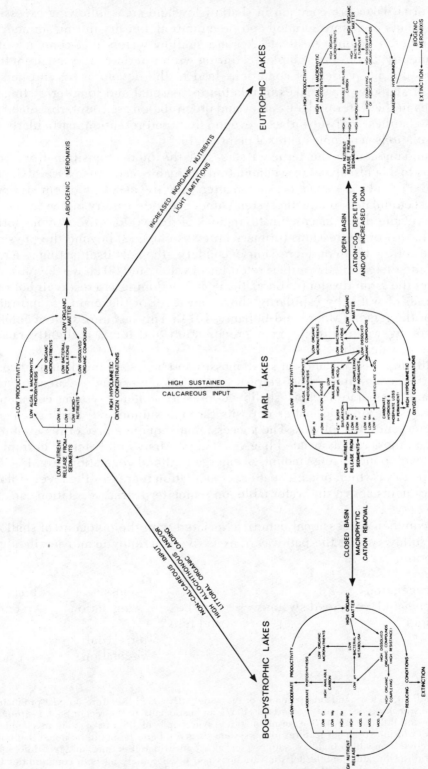

Figure 18–12 Common ontogeny of the four main types of lakes, each indicating major causal mechanisms regulating the phytoplanktonic trophic states. See text for discussion. (Extensively modified from Wetzel and Allen, 1970.)

dry. Vernal lakes are common in shallow lowland areas following excessive water inputs from spring runoff and precipitation, but dry during summer in the temperate region. Aestival lakes are shallow waters that contain water permanently, but freeze completely during winter periods. It is apparent that a nearly infinite variety of conditions lead to the development and persistence of shallow, small lakes in relation to seasonal and long-term climatic changes in the water budget. Inflow and precipitation arc counterbalanced by outflow (surface and subsurface), evaporation, and retention, particularly by accumulations of organic matter of plant origin.

Fundamental to the terminal stages of the biotic transition from lake systems to a landscape is an accumulation of organic matter in excess of degradation. Partially decayed organic matter, mainly of plant origin, termed *peat*[6], accumulates in aquatic systems under a wide variety of conditions in all except the driest macroclimatic regions of the world. Mire systems form: (1) in basins or depressions (primary mire systems); (2) beyond the physical confines of the basin or depression (secondary), the peat itself acting as a reservoir and increasing the surface retention of water; and (3) above the physical limits of the groundwater (tertiary), the peat functioning as a reservoir holding a volume of water by capillarity above the level of the main groundwater mass of the region (Moore and Bellamy, 1974). This last mire system obtains much of its water directly from precipitation and forms a slightly raised (perched) water table.

Although dense stands of small mosses and herbs are the dominant components of contemporary active peat formations over immense areas of subarctic and temperate regions, shallow depressions of lake systems commonly gain excessive organic matter deposits through submersed, floating-leaved, and emergent macrophytes. The successional sequence may or may not terminate in moss and associated bog vegetation. Many highly fertile eutrophic lakes develop increasing amounts of emergent littoral vegetation (Fig. 15–25), to the point at which organic matter accumulation increases the level of flora and sediments above the water table, on which terrestrial vegetation can encroach.

A common successional pattern associated with the ontogeny of shallow basins follows along the pathway of excessive macrophytic and sessile algal production:

macrovegetation
and associated $\longrightarrow$ reed swamps $\longrightarrow$ marshes $\nearrow$ moss $\longrightarrow$ bog
microflora   and fens $\searrow$ vegetation   systems

                                                          terrestrial
                                                          vegetation

---

[6]Excellent reviews and interpretation of the widely dispersed literature on peat-producing ecosystems or *mires* are given by Gorham (1957), and especially in the recent book *Peatlands*, by Moore and Bellamy (1974). The terminology and synonymity of this area of study are complex; only a brief résumé of major features of the systems is given here. It should be noted that mire is a collective term which includes both *bog* (= "moss" in British literature) and *fen*, differentiated according to rather subtle floristic variations and often containing both communities together. Mire is equivalent to the Swedish *myr* and the German *Moor*.

Reed swamps are characterized by closing vegetation of the littoral, often of polycorm, tall graminoid emergent plants, for example, *Phragmites, Cyperus papyrus, Scirpus,* and tall *Carex* species, with only occasional mixtures of submersed and floating-leaved macrophytes. Usually species diversity is low, and 2 herbaceous layers dominate within the main vegetative structure.

As organic accumulations increase to the displacement of the standing water, the water-logged peat habitat often is characterized by moderate-sized graminoid vegetation and small herbs in 2 or 3 layers. The distinction between the terms marsh and the European fen is largely made on the basis of phytosociological differences in floristic associations in which certain species or groups of species have a high fidelity. As marshes and fens dry, that is, depositions of organic matter exceed the mean water table, terrestrial very tall dicotyledonous vegetation increasingly colonizes and develops over the basin to the displacement of the emergent aquatic macrophytes. Some marshes and fens, under appropriate conditions of reduced available nutrients, possess a well-developed bryophyte layer, particularly of *Sphagnum* species. The moss vegetation eventually can dominate the system in the formation of bogs.

Early distinctions of various bog systems and types were made mostly on the basis of the primary, secondary, and tertiary stages of mire development discussed above in relation to water sources and floristic differences. The ontogeny of mire systems now is viewed generally within a combination of hydrological, phytosociological, and chemical processes. A sequence under ideal geomorphological and hydrological conditions illustrates common stages in the succession of mire systems in shallow surface depressions in which water accumulates (Moore and Bellamy, 1974 after Kulczynski). Water supply to the model lake system of Figure 18–13 consists of direct rainfall, runoff, and seepage from the immediate catchment of the lake basin, and a continuous flow of groundwater that enters by an inflow stream, of sufficient volume to affect the whole lake.

In the initial stages of these mire systems developing in a shallow drainage lake (Fig. 18–13), the system is under the influence of continuously or intermittently flowing groundwater. The ionic composition of these "rheophilous mires"[7] developing in mobile groundwaters is dominated by calcium and bicarbonate. Allochthonous organic matter inputs are high and result in accrual of peat within the central portion of the basin (Stage 2). In some cases, this peat accumulation is sufficiently buoyant to float as a mat. The main flow of water tends to be channelized to peripheral areas. In Stage 3, in a transitional stage of mire development,[8] continued accrual of peat deposits diverts inflow from the basin. Water supply is derived largely from the immediate catchment of the mire and direct precipitation on the surface. Ionic composition usually is dominated by calcium and sulphate. Further accrual of peat results in large areas of the mire surface that are not affected directly by moving water (Stage 4), but are inundated periodically when the water level rises during heavy rainfall.

---

[7]Synonymous with terms "low moor" and *Niedermoore.*
[8]*Ubergangsmoore* of older German literature.

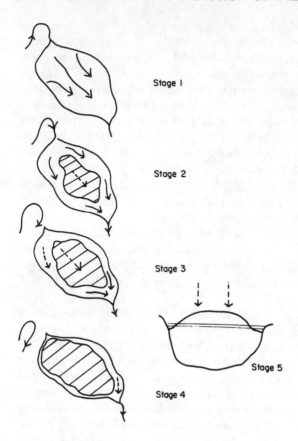

Stage 1

Stage 2

Stage 3

Stage 4

Stage 5

Figure 18–13 Idealized succes-sional stages in mire systems from a small, shallow drainage lake, through a seepage system, to a raised bog deriv-ing most of its water supply from pre-cipitation directly on its surface. Hatched area = peat accumulation. (From Moore, P. D., and Bellamy, D. J.: Peatlands. London, Paul Elek Scien-tific Books Ltd., 1974.)

Ombrophilous mires[9] (Stage 5) are no longer subject to the influence of flowing groundwater. Nearly all of the water, dominated by sulphate and hydrogen ions, is received from direct rainfall. The surface of the mire system is now above the vertical oscillations of the groundwater.

The change in ionic composition of water from dominance by calcium and bicarbonate to that of sulphate and hydrogen ions is a conspicuous feature characterizing these mire systems (Fig. 18–14). The active removal of ions from surface water supplies results in very low salinities with greatly re-duced buffering capacity. Under these conditions, relatively small additions of acids can result in a considerable reduction in pH.

The origin of acidity and control of cation exchanges of mire systems result from a combination of factors. The effectiveness of pH and cation re-duction is increased greatly in communities dominated by the mosses of genus *Sphagnum*. Water surrounding *Sphagnum,* a dominant group over immense areas of the temperate, boreal, and subarctic regions of the world (see, for example, Sjörs, 1961), has a pH usually below 4.5 and sometimes below 3.0. Much of the acidity of the aquatic environment of *Sphagnum* can be attributed to cation exchange (Anschutz and Gessner, 1954; Clymo, 1963,

---

[9]Termed "high moors" or *Hochmoore* in early works.

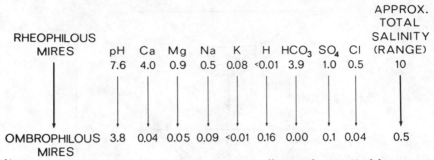

Figure 18–14 General shifts in ionic composition (milliequivalents 1^{-1}) of the water in the transition from rheophilous to ombrophilous mire systems of Western European mires. (From average values of Bellamy and of Sjörs given in Moore and Bellamy, 1974.)

1967; Brehm, 1970; and others). *Sphagnum* spp. behave as cation exchangers, even when dead, resulting from high concentrations (to 30 per cent of dry weight) of long chain polymers of unesterified uronic acids.[10] Although small amounts of free organic acids are released from *Sphagnum*, most evidence points to release of H^+ ions on $-COOH$ groups of freshly produced new growth of the plants as they are exchanged for cations in rain or groundwater that flows over the plants as the dominant source of acidity. It also has been suggested that some of the acidity of bog pools derives from H^+ concentrations of sulfur metabolizing bacteria in the anaerobic zones (Gorham, 1966; Clymo, 1965), but all evidence indicates that this contribution is a relatively small portion of the whole.

Bogs and Quaking Bogs

The importance of the littoral flora in the senescence of a lake therefore is clearly implicated as critical in production of organic matter in excess of decompositional removal. Bog (mire) systems additionally require climatic conditions of abundant precipitation and relatively high humidity over much of the annual period.

In summary, the stratigraphy of a common development of a small lake to a bog can be viewed as a sequence as depicted in Fig. 18–15. The depth of open water decreases continually with sedimentation. As the basin progressively fills, the littoral flora simultaneously advance toward the center and finally cover the entire area under what is loosely termed swamp conditions, with standing water among the macrovegetation. Excessive production of littoral vegetation generally accelerates in geological terms, and the sediments reach the surface of the original lake. The stage generally is referred to as a marsh, in which the sediments are saturated continually but little if any standing water is found among the vegetation. Decomposition is intense under these conditions, the mean pH decreases significantly, often below

---

[10]Similar to sugars, but the sixth carbon is part of a carboxyl group. These long chain molecules may be mixed polymers containing both sugars and uronic acids.

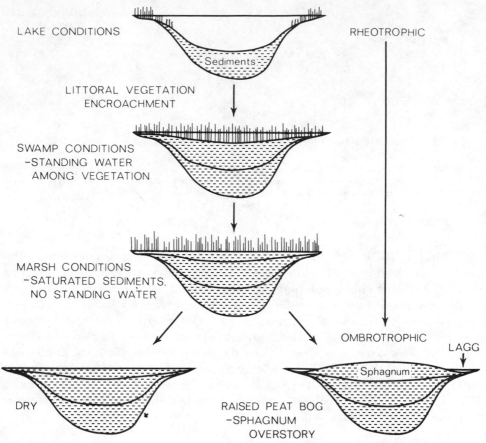

Figure 18–15 Frequently observed ontogeny of shallow lake systems through swamp and marsh stages to dry landscape or to raised peat bogs. See text for discussion.

neutrality, and the concentrations of relatively resistant humic compounds of the dissolved organic matter pool increase. Shifts in the dominance of vegetation also occur, with a tendency towards competitive advantage by acidophilic flora such as certain sedges and grasses.

The formation of a typical bog depends greatly on the prevailing climatic conditions of high humidity and precipitation. In moderately dry regions, the transition is from marsh conditions to terrestrial vegetation, without the characteristic development of mosses. Conditions of much of the temperate zone, however, are sufficiently moist to encourage the colonization and eventual dominance of *Sphagnum* and other mosses among the often predominantly sedge flora. Through a succession of mosses, especially *Sphagnum* species, cation exchange and acidity increases, which effectively decreases rates of decomposition (Chapter 17) and further accelerates net accumulation of organic matter. The accumulation of particulate organic matter in the understory, continued growth at the surface, and excellent characteristics of capillarity of the dead and living mass of mosses raise the general water level within the mat above that of the original lake level. In this way, the flora mat is capable of growing significantly (usually <50 cm) above the water table; hence, the term *raised bog*.

Lagg zones, or moat-like areas of shallow water, characteristically exist between the central peat mat and higher land as remnants of flowing groundwater diverted around the central peat mat. Flora of the lagg zone commonly consists of small graminoid species, bryophytes, and foliaceous liverworts.

Shifts in the relative importance of mineral nutrients from that of the soil of the drainage basin to that dominated by atmospheric precipitation and particulate fallout are a dominant feature of the succession to bogs. In mire systems in general, the terms rheotrophic for the former and ombrotrophic for the latter have been introduced[11] to emphasize the interplay of geomorphological, chemical, climatic, and biotic factors involved in this successional process. In regions under strong influence of oceanic contributions to ionic composition of precipitation (Chapter 9), such as western Britain and Ireland, ombrotrophic nutrition is only slightly inferior to weak rheotrophic nutrition, and little difference in vegetation results.

Quaking bogs represent a particular development of bogs within a relatively deep lake basin of small surface area. The development of *Sphagnum* mosses occurs early in the succession of littoral flora, and in geological context leads to a very rapid accumulation of organic matter in the littoral areas (Fig. 18–16). In more advanced conditions, the littoral development along the surface grows more rapidly than the deposition of peat deposits beneath it and floats in a thick mat (several meters) above the open water. Growth proceeds all along the periphery towards the lake center, and eventually encompasses the entire surface.

A marked feature of the floating mat is the pioneering development of the *Sphagnum* in the floating mat formation toward the lake center, with concen-

---

[11]Originally proposed as "minerotrophic" and "ombrotrophic" by DuRietz (Sjörs, 1961). A detailed discussion of the interrelationships and terminology is given by Moore and Bellamy (1974).

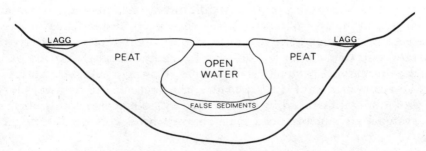

Figure 18–16   An idealized quaking *Sphagnum* bog in which the mat encroachment towards the center of the basin overlies littoral peat accumulations and much of the open water.

tric zonation of plant communities landward. Within the water there generally are certain submersed angiosperms (certain *Potamogeton* and *Utricularia* species) and floating-leaved macrophytes such as the water lilies *Nuphar* and *Nymphaea*. The lakeward *Sphagnum* mat contains sedges (especially *Carex* species), low shrubs such as the leatherleaf *Chamaedaphne,* followed by landward development of tall shrubs, and finally bog trees, for example, black sprucc and tamarack (Gates, 1942).

The older portions of the mat are grounded firmly in the basin by underlying deposits of peat. The floating mat, growing considerably above the water level often is attached only by the vegetative mass and can be displaced by the added weight of a person and "quake." The sensation is that of walking on an immense floating, saturated sponge. The sedimentation of much particulate organic matter in the open water areas from the littoral mat results in loosely aggregated, flocculant sediment that is not compacted sufficiently to support small weights (for example, an anchor), often for many meters; hence, the term "false bottom" or "false sediments" (Fig. 18–16).

Quantitative information on the productivity and population dynamics of biota of the pelagic and profundal regions of bog lakes and quaking bogs is extremely sparse. Much more knowledge exists on qualitative aspects because of the extremely diverse and interesting biota found adapted to the extreme conditions of bogs. The microflora of the plankton and of small pools within and among hummocks of *Sphagnum* and larger plants are characterized by a great diversity of species of algae, although few are ubiquitous and develop in dominant numbers. Most abundant is a great variety of desmids, often several hundred species. Species of blue-green algae, chrysomonads, dinophyceans, and diatoms are adapted to bog conditions but seldom are found in abundance. The microfauna are dominated by testaceous rhizopod protozoans and rotifers; very few cladoceran and copepod zooplankters are tolerant of the bog milieu.

Very few macrofauna have adapted well to the extreme acidity and low salinity of bog waters. Species diversity is very low and entire groups are lacking or poorly represented. For example, in acidic, quaking *Sphagnum* bogs, sponges, coelenterates, ostracods, hydracarnid mites, oligochaetes, Ephemeroptera, Malacostraca, mollusks, nematodes, flatworms, and fish are completely absent or very poorly represented.

The genesis and ontogeny of bog systems can be viewed as a transitional succession of lake basins of greatly differing geomorphology to wetlands in which organic matter of macrophytic vegetation accumulates in excess of complete degradation. The causal processes regulating productivity can be analyzed effectively from a short-term biogeochemical standpoint as operational in a reasonably steady state system. However, the very nature of excessive accumulation of organic matter, the successional pathways of which are profoundly influenced by climatic and hydrological factors, indicates that the systems are not balanced even in very brief geological time periods.

epilogue

The foregoing pages represent only a brief summary of contemporary knowledge of freshwater systems. In spite of concerted efforts for brevity, this book demonstrates that diversity in the biota is large and interactions among biotic populations and environmental parameters are numerous. Furthermore, it has been emphasized repeatedly how the metabolism of the biota often affects the total cycling and characteristics of the system. The entire thrust of this treatment, however, focused on generalized operational similarities among fresh waters, whether they are standing in basins or running in channels. Despite the heterogeneity and abundance of individuality and variation from site to site, one must appreciate that there is underlying unity of function.

An attempt has been made to state these generalized relationships in simple terms, without resorting to the advantages of mathematical rigor and the attendant disadvantages of oversimplification. General principles in limnology are, and will be, modified little by mathematical manipulations. Nonetheless, mathematical modeling and systems simulation possess intrinsic manipulative and predictive power if based soundly on adequate quantitative data. In no way, however, will such manipulation improve poor data or conceptualizations.

In the comparative approach taken, a concerted attempt was made to integrate the massive number of dynamic variables that govern and effect the resultant biotic productivity. Throughout this treatment, the dominant factors relating to the productivity of oligotrophic and eutrophic systems were contrasted. The underlying rationale was to emphasize the basic control mechanisms which operate on the progressive changes that lead to enrichment and increases in productivity per given quantity of water. A further intent was to emphasize what happens as aquatic systems are loaded excessively from man-induced changes.

As has been so succinctly described by Vallentyne (1974), a common result of misuse of the drainage basin and excessive nutrient loading of fresh waters is the acceleration of eutrophication, literally turning lakes into "algal bowls." It was emphasized in the present treatment that the metabolism of all aquatic systems, and indeed of a major portion of the biosphere, is detrital metabolism in operation. Accelerated eutrophication leads to accelerated pelagial and littoral primary productivity with progressive intensification of detrital metabolism, effectively relegating lakes to "detrital bowls" in an operational sense. Metabolically mediated changes in the environment leading to strata of prolonged anoxia and attendant reductions in catabolism of detrital organic matter result in decreased efficiencies of utilization and degradation of organic matter.

A conscientious individual must view these changes in his natural environment with concern. As the exploitative pressures of demophoric growth increase, man's concern must involve more than simply his aesthetic values and those of future generations of humans. The very survival of man centers on the wise utilization of finite freshwater resources; to think otherwise is naive and myopic.

Use of fresh water for purposes of aquaculture, to provide protein for burgeoning human demands, is not a very promising endeavor because of the relatively low efficiencies of conversion of solar to potential chemical energy. Terrestrial agricultural, in spite of its high freshwater requirements, is in general more efficient. The greatest demands for fresh water, however, are in the area of technology. Much of this water is returned to the environment in a seriously degraded form or removed from a source, such as groundwater, with slow renewal times and lost or reduced from the region of need.

The demands on water resources now are increasing exponentially, and necessitate a corresponding increase in management, as well as a degree of manipulation. Unless we comprehend the rudiments of their functional operation, any hope of using them effectively is lost. Contemporary management of freshwater resources operates much the same as a floundering infant; the rationale behind many practices is purely hydrological, without the slightest interest in or appreciation of the biotic interactions of the water bodies themselves, or the long-term related effects on the climate of large regional areas, and, in the end, on the biosphere. The metabolism of aquatic systems *does* matter and must be considered and understood in all manipulative measures taken.

Our knowledge of the operation of freshwater systems has come a great distance in the last 50 years. However, if we really scrutinize the existing information in an unbiased way, the only conclusion is that we remain quite ignorant of the complex operation of these interactions. Accrual of the needed operational information must be done in a greatly accelerated manner. There is no alternative to devoting a greater percentage of our intellectual and financial resources to fresh waters. Everything allocated to this critical resource thus far has been utter tokenism. We have passed the point at which we can continue to take unabatedly without casting back some comprehensive efforts in return.

APPENDIX I Units and Conversion Factors for Solar Radiation Data (See Chapter 5) [a]

1 watt = 1 joule s^{-1}
1 watt cm^{-2} = 1 joule s^{-1} cm^{-2}
1 watt = 14.3 cal min^{-1}
1 watt cm^{-2} = 14.3 langley min^{-1}
1 ly min^{-1} = 0.0698 watt cm^{-2}
1 ft-candle = 10.764 lux
1 lux = 0.0929 ft-candle
1 ly = 2.11 × 10^{15} × Å quanta cm^{-2} (where Å = wavelength of the quanta in Å units)
1 ly min^{-1} = 2.11 × 10^{15} × Å quanta cm^{-2} min^{-1}
1 ly min^{-1} = 3.50 × 10^{-9} × Å einsteins cm^{-2} min^{-1}

A: Energy [dimensions: mass length^2 time^{-2}]

	erg	cal	kcal	J(=Ws)	Wh	kWh	Btu*
erg	1	2.39 × 10^{-8}	2.39 × 10^{-11}	10^{-7}	2.78 × 10^{-11}	2.78 × 10^{-14}	9.48 × 10^{-11}
cal	4.19 × 10^{7}	1	10^{-3}	4.19	1.16 × 10^{-3}	1.16 × 10^{-6}	3.97 × 10^{-3}
kcal	4.19 × 10^{10}	10^{3}	1	4.19 × 10^{3}	1.16	1.16 × 10^{-3}	3.97
J(=Ws)	10^{7}	0.239	2.39 × 10^{-4}	1	2.78 × 10^{-4}	2.78 × 10^{-7}	9.48 × 10^{-4}
Wh	3.6 × 10^{10}	8.60 × 10^{2}	8.60 × 10^{-1}	3.6 × 10^{3}	1	10^{-3}	3.41
kWh	3.6 × 10^{13}	8.60 × 10^{5}	8.60 × 10^{2}	3.6 × 10^{6}	10^{3}	1	3.41 × 10^{3}
Btu*	1.06 × 10^{10}	2.52 × 10^{2}	2.52 × 10^{-1}	1.06 × 10^{3}	2.93 × 10^{-1}	2.93 × 10^{-4}	1

B: Power (energy per unit time) [dimensions: mass length^2 time^{-3}]

	erg s^{-1}	cal s^{-1}	cal min^{-1}	W(=Js^{-1})	Btu* min^{-1}	Btu* h^{-1}
erg s^{-1}	1	2.39 × 10^{-8}	1.43 × 10^{-6}	1.00 × 10^{-7}	5.69 × 10^{-9}	9.48 × 10^{-11}
cal s^{-1}	4.19 × 10^{7}	1	60	4.19	2.38 × 10^{-1}	3.97 × 10^{-3}
cal min^{-1}	6.97 × 10^{5}	1.67 × 10^{-2}	1	6.97 × 10^{-2}	3.97 × 10^{-3}	6.62 × 10^{-5}
W(=Js^{-1})	1.00 × 10^{7}	2.39 × 10^{-1}	14.3	1	5.69 × 10^{-2}	9.48 × 10^{-4}
Btu* min^{-1}	1.76 × 10^{8}	4.20	2.52 × 10^{2}	17.6	1	1.67 × 10^{-2}
Btu* h^{-1}	1.06 × 10^{10}	2.52 × 10^{2}	1.52 × 10^{4}	1.06 × 10^{3}	60	1

*British thermal unit

C: Radiant Flux Density and Irradiance Units ([dimensions: mass time^{-3}] [λ in cm])

	quantum cm^{-2} s^{-1}	einstein cm^{-2} s^{-1}	erg cm^{-2} s^{-1}	cal cm^{-2} s^{-1}	cal cm^{-2} min^{-1}	cal cm^{-2} h^{-1}
quantum cm^{-2} s^{-1}	1	1.66×10^{-24}	$1.99 \times 10^{-16}\,\lambda^{-1}$	$4.75 \times 10^{-24}\,\lambda^{-1}$	$2.85 \times 10^{-22}\,\lambda^{-1}$	$1.71 \times 10^{-20}\,\lambda^{-1}$
einstein cm^{-2} s^{-1}	6.02×10^{23}	1	$1.20 \times 10^{8}\,\lambda^{-1}$	$2.86\,\lambda^{-1}$	$1.72 \times 10^{2}\,\lambda^{-1}$	$1.03 \times 10^{4}\,\lambda^{-1}$
erg cm^{-2} s^{-1}	$5.03 \times 10^{15}\,\lambda$	$8.35 \times 10^{-9}\,\lambda$	1	2.39×10^{-8}	1.43×10^{-6}	8.6×10^{-5}
cal cm^{-2} s^{-1}	$2.11 \times 10^{23}\,\lambda$	$3.50 \times 10^{-1}\,\lambda$	4.19×10^{7}	1	60	3.6×10^{3}
cal cm^{-2} min^{-1}	$3.51 \times 10^{21}\,\lambda$	$5.83 \times 10^{-3}\,\lambda$	6.98×10^{5}	1.67×10^{-2}	1	60
cal cm^{-2} h^{-1}	$5.85 \times 10^{19}\,\lambda$	$9.71 \times 10^{-5}\,\lambda$	1.16×10^{4}	2.78×10^{-4}	1.67×10^{-4}	1
cal dm^{-2} h^{-1}	$5.85 \times 10^{17}\,\lambda$	$9.71 \times 10^{-7}\,\lambda$	1.16×10^{2}	2.78×10^{-6}	1.67×10^{-3}	10^{-2}
kcal m^{-2} h^{-1}	$5.85 \times 10^{18}\,\lambda$	$9.71 \times 10^{-6}\,\lambda$	1.16×10^{3}	2.78×10^{-5}	1.67×10^{-3}	10^{-1}
μW cm^{-2}	$5.03 \times 10^{16}\,\lambda$	$8.35 \times 10^{-8}\,\lambda$	10	2.39×10^{-7}	1.43×10^{-5}	8.6×10^{-4}
W cm^{-2}	$5.03 \times 10^{22}\,\lambda$	$8.35 \times 10^{-2}\,\lambda$	10^{7}	2.39×10^{-1}	14.33	8.6×10^{2}
W m^{-2}	$5.03 \times 10^{18}\,\lambda$	$8.35 \times 10^{-6}\,\lambda$	10^{3}	2.39×10^{-5}	1.43×10^{-3}	8.6×10^{-2}
Btu ft^{-2} min^{-1}	$0.95 \times 10^{21}\,\lambda$	$1.58 \times 10^{-3}\,\lambda$	1.89×10^{5}	4.53×10^{-3}	0.27	16.2

	cal dm^{-2} h^{-1}	kcal m^{-2} h^{-1}	μW cm^{-2}	W cm^{-2}	W m^{-2}	Btu ft^{-2} min^{-1}
quantum cm^{-2} s^{-1}	$1.71 \times 10^{-18}\,\lambda^{-1}$	$1.71 \times 10^{-19}\,\lambda^{-1}$	$1.99 \times 10^{-17}\,\lambda^{-1}$	$1.99 \times 10^{-23}\,\lambda^{-1}$	$1.99 \times 10^{-19}\,\lambda^{-1}$	$1.05 \times 10^{-21}\,\lambda^{-1}$
einstein cm^{-2} s^{-1}	$1.03 \times 10^{6}\,\lambda^{-1}$	$1.03 \times 10^{5}\,\lambda^{-1}$	$1.20 \times 10^{7}\,\lambda^{-1}$	$12.0\,\lambda^{-1}$	$1.20 \times 10^{5}\,\lambda^{-1}$	$6.34 \times 10^{2}\,\lambda^{-1}$
erg cm^{-2} s^{-1}	8.6×10^{-3}	8.6×10^{-4}	10^{-1}	10^{-7}	10^{-3}	5.28×10^{-6}
cal cm^{-2} s^{-1}	3.6×10^{5}	3.6×10^{4}	4.19×10^{6}	4.19	4.19×10^{4}	2.21×10^{2}
cal cm^{-2} min^{-1}	6×10^{3}	6×10^{2}	6.98×10^{4}	6.98×10^{-2}	6.98×10^{2}	3.69
cal cm^{-2} h^{-1}	10^{2}	10	1.16×10^{3}	1.16×10^{-3}	11.63	6.17×10^{-2}
cal dm^{-2} h^{-1}	1	10^{-1}	11.63	1.16×10^{-5}	1.16×10^{-1}	6.17×10^{-4}
kcal m^{-2} h^{-1}	10	1	1.16×10^{2}	1.16×10^{-4}	1.163	6.17×10^{-3}
μW cm^{-2}	8.6×10^{-2}	8.6×10^{-3}	1	10^{-6}	10^{-2}	5.28×10^{-5}
W cm^{-2}	8.6×10^{4}	8.6×10^{3}	10^{6}	1	10^{4}	5.28×10
W m^{-2}	8.6	8.6×10^{-1}	10^{2}	10^{-4}	1	5.28×10^{-3}
Btu ft^{-2} min^{-1}	1.62×10^{3}	1.62×10^{2}	1.89×10^{4}	1.89×10^{-2}	1.89×10^{2}	1

[a]Data modified from Strickland, 1958, and Šesták, Čatský, and Jarvis, 1971.

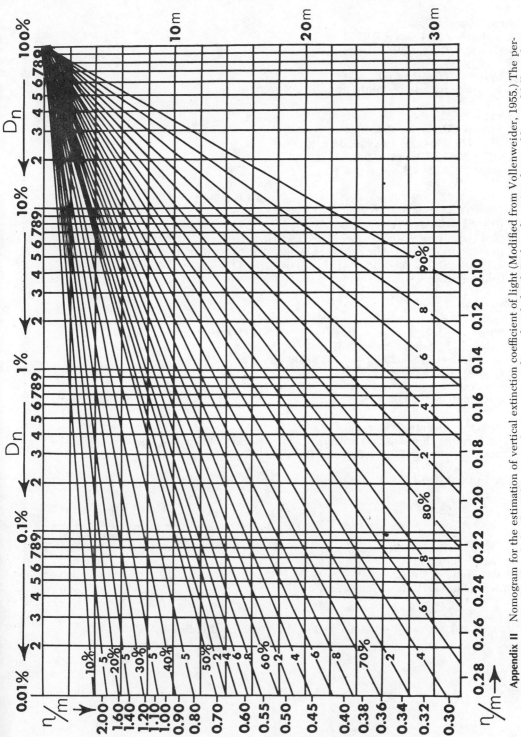

Appendix II Nomogram for the estimation of vertical extinction coefficient of light (Modified from Vollenweider, 1955.) The percentage transmission of surface light along the upper abscissa is coordinated with depth (right ordinate) to diagonal line and followed to η along the left ordinate and lower abscissa.

APPENDIX III Units and Conversion Factors of Length and Weight[a]

Length, meter, m
1 inch, in = 25.4 mm = 0.0254 m
1 foot, ft = 12 in = 304.8 mm = 0.3048 m
1 yard, yd = 36 in = 0.9144 m
1 mile = 1760 yd = 1.60934 km
1 international nautical mile (n mile) = 1.852 km

Area, m^2
1 in^2 = 6.4516 cm^2 = 6.4516 × 10^{-4} m^2
1 ft^2 = 929.0304 cm^2 = 9.290304 × 10^{-2} m^2
1 yd^2 = 8361.2736 cm^2 = 0.83612736 m^2
1 acre = 4046.86 m^2 = 0.40486 ha

Volume, m^3 (1 liter, 1, = 1 dm^3 = 10^{-3} m^3 and 1.0 ml = 1 cm^3)
1 in^3 = 16.3871 cm^3 = 1.63871 × 10^{-5} m^3
1 ft^3 = 28.3168 dm^3 (1) = 0.0283168 m^3
1 pint (UK), pt = 0.568261 dm^3 (1) = 5.68261 × 10^{-4} m^3 [× 1.201 = pints (US)]
1 quart (UK), qt = 1.13652 dm^3 (1) = 1.13652 × 10^{-3} m^3 [× 1.201 = quarts (US)]
1 gallon (US), US gal = 3.78541 dm^3 (1) = 3.78541 × 10^{-3} m^3
1 gallon (UK), gal = 4.54609 dm^3 (1) = 4.54609 × 10^{-3} m^3

Mass, kg
1 ounce (av), oz = 28.3495 g = 0.0283495 kg
1 ounce (UK), fluid; fl oz = 28.413 cm^3 = 2.84131 × 10^{-5} m^3
1 ounce (US), fluid, fl oz = 29.57 cm^3
1 pound, lb = 453.59239 g = 0.45359239 kg
1 quarter (UK, long) = 12.7006 kg
1 hundred weight (long), cwt = 50.8023 kg
1 ton (long), = 1016.05 kg

1 dz (Doppelzentner, German)
1 ctr (centner, Russian and Polish) } all = 100 kg
1 m.q. (metric quintal, French and Polish)

The quintal in the U.S.A. and Britain is a variable unit which may equal 100 or 112 lbs.

	Multiplying factors	
	to kg/m^2	from kg/m^2
mt/ha (metric tons or tonnes)	0.1000	10.00
mt/acre (metric tons)	0.2471	4.047
lt/acre (long or British tons)	0.2511	3.983
st/acre (short or United States tons)	0.2242	4.460
lb/yd^2	0.5429	1.842

Pounds per acre are best converted to short tons by dividing by 2000.

To Convert the Values in Units of the Top Line, Multiply by the Relevant Factors in Order to Obtain Corresponding Values of the Lefthand Column.

	Angstroms	Nanometers (millimicrons)	Micrometers (microns)	Millimeters	Centimeters	Kilometers	Meters	Inches	Feet	Miles
Angstroms	1	10	10^4	10^7	10^8	10^{13}	10^{10}	2.540×10^8	3.048×10^9	1.609×10^{13}
Nanometers (millimicrons)	10^{-1}	1	10^3	10^6	10^7	10^{12}	10^9	2.540×10^7	3.048×10^8	1.609×10^{12}
Micrometers (microns)	10^{-4}	10^{-3}	1	10^3	10^4	10^9	10^6	2.540×10^4	3.048×10^5	1.609×10^9
Millimeters	10^{-7}	10^{-6}	10^{-3}	1	10	10^6	10^3	2.540×10	3.048×10^2	1.609×10^6
Centimeters	10^{-8}	10^{-7}	10^{-4}	0.1	1	10^5	10^2	2.540	3.048×10	1.609×10^5
Kilometers	10^{-13}	10^{-12}	10^{-9}	10^{-6}	10^{-5}	1	10^3	2.540×10^{-5}	3.048×10^{-4}	1.609
Meters	10^{-10}	10^{-9}	10^{-6}	10^{-3}	10^{-2}	10^3	1	2.540×10^{-2}	3.048×10^{-1}	1.609×10^3
Inches	3.937×10^{-9}	3.937×10^{-8}	3.937×10^{-5}	3.937×10^{-2}	3.937×10^{-1}	3.937×10^4	3.937×10	1	12	6.336×10^4
Feet	3.281×10^{-10}	3.281×10^{-9}	3.281×10^{-6}	3.281×10^{-3}	3.281×10^{-2}	3.281×10^3	3.281	8.333×10^{-2}	1	5.280×10^3
Miles	6.214×10^{-14}	6.214×10^{-13}	6.214×10^{-10}	6.214×10^{-7}	6.214×10^{-6}	6.214×10^{-1}	6.214×10^{-4}	1.578×10^{-5}	1.894×10^{-4}	1

[a]Modified from Bickford and Dunn, 1972, Westlake, 1963, and Šesták, et al., 1971.

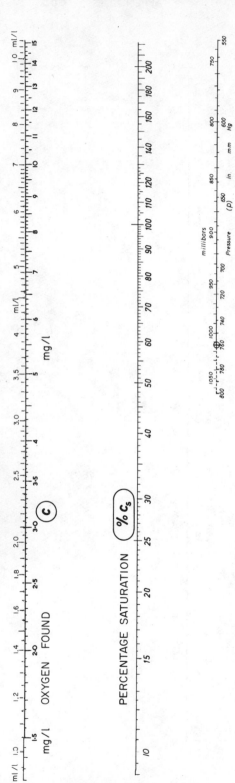

Oxygen Saturation Nomogram

For the estimation of the percentage saturation (%c_s) of dissolved oxygen in a lake water sample, for which the oxygen content (c), the temperature (t), and the altitude (h) of the lake surface (or an equivalent atmospheric pressure, p), are known.

Instructions for Zero Altitude
(or normal pressure, 760 mm Hg, 1013 mb)

With a straight line (taut thread or ruling on a transparent strip) join the measured oxygen content on scale c to the sample temperature on scale t. The corresponding percentage saturation value is where the line intersects the %c_s scale.

If the oxygen content is too high (or too low) for scale c, enter scale c with one-tenth (or ten times) the measured content and multiply the %c_s value by 10 (or 0.1).

Instructions for Nonzero Altitude
(or nonnormal atmospheric pressure)

Select the appropriate altitude scale[b] A or B (or the atmospheric pressure scale p). Take a pair of dividers, place the left leg (No. 1) on the zero of the altitude scale (or the normal pressure point 760 mm Hg) and set the right leg (No. 2) to the required altitude or pressure value. Now transfer the divider legs unchanged to the t scale, with leg No. 1 at the sample temperature. A straight line joining leg No. 2 on scale t and the sample oxygen content on scale c intersects the %c_s scale at the percentage saturation value appropriate to the selected altitude or atmospheric pressure.

[a]From Mortimer, C. H.: Addendum to Mitteilung Internationale Vereinigung für Limnologie, 6:120 pp., 1956 (revised 1975).

[b]Based on the simplifying assumptions of Schmassmann, H.: Schweiz. Zeitschrift für Hydrologie, 11:430, 1949, discussed in Mortimer, 1956.

references cited

Åberg, B., and W. Rodhe.
1942. Über die Milieufaktoren in einigen südschwedischen Seen.
Symbol. Bot. Upsalien., 5(3), 256 pp.

Adam, N. K.
1937. A rapid method for determining the lowering of tension of exposed water surfaces, with some observations on the surface tension of the sea and of inland waters.
Proc. Roy. Soc. London (Ser. B) 122:134–139.

Adams, M. S., and M. D. McCracken.
1974. Seasonal production of the Myriophyllum component of the littoral of Lake Wingra, Wisconsin.
J. Ecol., 62:457–465.

Albers, J.
1963. Interaction of Color.
New Haven, Yale Univ. Press, 75 pp.

Albrecht, M.–L.
1964. Die Lichtdurchlässigkeit von Eis und Schnee und ihre Bedeutung für die Sauerstoffproduktion im Wasser.
Deutsche Fischerei-Zeitung, 11:371–376.

Alexander, M.
1961. Introduction to Soil Microbiology.
New York, John Wiley & Sons, Inc., 472 pp.

Alexander, M.
1965a. Nitrification.
Agronomy, 10:307–343.

Alexander, M.
1965b. Biodegradation: Problems of molecular recalcitrance and microbial fallibility.
Adv. Appl. Microbiol., 7:35–80.

Alexander, M.
1971. Microbial Ecology.
New York, John Wiley & Sons, Inc., 511 pp.

Alimov, A. F., et al.
1970. Biological productivity of lakes Krivoe and Krugloe. In Z. Kajak and A. Hillbricht-Ilkowska, eds. Productivity Problems of Freshwaters.
Warsaw, PWN Polish Scientific Publishers, pp. 39–56.

Aliverdieva-Gamidova, L. A.
1969. Microbiological processes in Lake Mekhteb.
Mikrobiologiya, 38:1096–1100. (Translated into English, Consultants Bureau, New York, 1970.)

Allanson, B. R.
1973. The fine structure of the periphyton of Chara sp. and Potamogeton natans from Wytham Pond, Oxford, and its significance to the macrophyte-periphyton metabolic model of R. G. Wetzel and H. L. Allen.
Freshwat. Biol., 3:535–541.

Allen, H. L.
1969. Chemo-organotrophic utilization of dissolved organic compounds by planktic algae and bacteria in a pond.
Int. Rev. ges. Hydrobiol., 54:1–33.

Allen, H. L.
1971. Primary productivity, chemo-organotrophy, and nutritional interactions of epiphytic algae and bacteria on macrophytes in the littoral of a lake.
Ecol. Monogr., 41:97–127.

Allen, H. L.
1972. Phytoplankton photosynthesis, micronutrient interactions, and inorganic carbon availability in a soft-water Vermont lake. In G. E. Likens, ed. Nutrients and Eutrophication: The Limiting-Nutrient Controversy.
Special Symposium, Amer. Soc. Limnol. Oceanogr., 1:63–83.

Allen, M. B.
1952. The cultivation of Myxophyceae.
Arch. Mikrobiol., 17:34–53.

---

[1]A few discrepancies occur in the transliteration of Slavic languages. When the work of an author is translated into another language and I have cited only one of his or her works, that citation is recorded as printed even if the transliteration is incorrect. When more than one work of the same author is cited, but some are from the mother language, the name is cited consistently as it should be transliterated. An exception is Vinberg in the Russian; because of his European ancestry, the author uses Winberg in several important writings in the English. Conforming to the transliteration in this case could lead to confusion if the references were sought.

Allen, M. B., and D. I. Arnon.
1955. Studies on nitrogen-fixing blue-green algae. II. The sodium requirement of *Anabaena cylindrica*.
Physiol. Plant., 8:653–660.

Allgeier, R. J., B. C. Hafford, and C. Juday.
1941. Oxidation-reduction potentials and pH of lake waters and of lake sediments.
Trans. Wis. Acad. Sci. Arts Lett., 33:115–133.

Alsterberg, G.
1927. Die Sauerstoffschichtung der Seen.
Bot. Not., 1927:255–274.

American Public Health Association.
1971. Standard Methods for the Examination of Water and Wastewater. 13th ed.
Washington, D. C., American Public Health Association, 874 pp.

Anderson, D. V.
1961. A note on the morphology of the basins of the Great Lakes.
J. Fish. Res. Bd. Canada, 18:273–277.

Anderson, G. C.
1958. Some limnological features of a shallow saline meromictic lake.
Limnol. Oceanogr., 3:259–270.

Anderson, R. O., and F. F. Hooper.
1956. Seasonal abundance and production of littoral bottom fauna in a southern Michigan lake.
Trans. Amer. Microsc. Soc., 75:259–270.

Anderson, R. S.
1970. Predator-prey relationships and predation rates for crustacean zooplankters from some lakes in western Canada.
Can. J. Zool., 48:1229–1240.

Andronikova, I. N., V. G. Drabkova, K. N. Kuzmenko, N. F. Michailova, and E. A. Stravinskaya.
1970. Biological productivity of the main communities of the Red Lake. *In* Z. Kajak and A. Hillbricht-Ilkowska, eds. Productivity Problems of Freshwaters.
Warsaw, PWN Polish Scientific Publishers, pp. 57–71.

Anschütz, I., and F. Gessner.
1954. Der Ionenaustausch bei Torfmoosen (*Sphagnum*).
Flora, 141:178–236.

Anthony, E. H., and F. R. Hayes.
1964. Lake water and sediment. VII. Chemical and optical properties of water in relation to the bacterial counts in the sediments of twenty-five North American lakes.
Limnol. Oceanogr., 9:35–41.

Arber, A.
1920. Water Plants. A Study of Aquatic Angiosperms.
Cambridge, Cambridge University Press, 436 pp.

Arnold, D. E.
1971. Ingestion, assimilation, survival, and reproduction by *Daphnia pulex* fed seven species of blue-green algae.
Limnol. Oceanogr., 16:906–920.

Aron, W. I., and S. H. Smith.
1971. Ship canals and aquatic ecosystems.
Science, 174:13–20.

Aruga, Y.
1965. Ecological studies of photosynthesis and matter production of phytoplankton. II. Photosynthesis of algae in relation to light intensity and temperature.
Bot. Mag. Tokyo, 78:360–365.

Assman, A. V.
1953. Rol vodoroslevykh obrastanii v obrazovanii organicheskogo veshchestva v Glubokom Ozere.
Trudy Vsesoiuznogo Gidrobiol. Obshchestva, 5:138–157.

Ayers, J. C., D. C. Chandler, G. H. Lauff, C. F. Powers, and E. B. Henson.
1958. Currents and Water Masses of Lake Michigan.
Publ. Great Lakes Res. Div., Univ. Mich., 3, 169 pp.

Baas Becking, L. G. M., I. R. Kaplan, and D. Moore.
1960. Limits of the natural environment in terms of pH and oxidation-reduction potentials.
J. Geol., 68:243–284.

Bachmann, R. W.
1963. Zinc-65 in studies of the freshwater zinc cycle. *In* V. Schultz and A. W. Klement, Jr., eds. Radioecology. First National Symposium on Radioecology.
New York, Reinhold Publishing Corp., pp. 485–496.

Bachmann, R. W., and C. R. Goldman.
1965. Hypolimnetic heating in Castle Lake, California.
Limnol. Oceanogr., 10:233–239.

Ball, R. C.
1949. Experimental use of fertilizer in the production of fish-food organisms and fish.
Tech. Bull. Mich. State Univ. Agricult. Exp. Stat., Sec. Zool., 210, 28 pp.

Ball, R. C., and D. W. Hayne.
1952. Effects of the removal of the fish population on the fish-food organisms of a lake.
Ecology, 33:41–48.

Ball, R. C., T. A. Wojtalik, and F. F. Hooper.
1963a. Upstream dispersion of radiophosphorus in a Michigan trout stream.
Pap. Mich. Acad. Sci. Arts Lett., *48*:57–64.

Ball, R. C., and F. F. Hooper.
1963b. Translocation of phosphorus in a trout stream ecosystem. *In* V. Schultz and A. W. Klement, Jr., eds. Radioecology. First National Symposium on Radioecology.
New York, Reinhold Publishing Corp., pp. 217–228.

Balogh, J.
1958. On some problems of production biology.
Acta Zool. Acad. Sci. Hungaricae, 4:89–114.

Bamforth, S. S.
1958. Ecological studies on the planktonic protozoa of a small artificial pond.
Limnol. Oceanogr., 3:398–412.

Bandurski, R. S.
1965. Biological reduction of sulfate and nitrate. *In* J. Bonner and J. E. Varner, eds. Plant Biochemistry.
New York, Academic Press, pp. 467–490.

Barica, J.
1970. Untersuchungen über den Stickstoff-Kreislauf des Titisees und seiner Quellen.
Arch. Hydrobiol. Suppl., 38:212–235.

Barrett, P. H.
1957. Potassium concentrations in fertilized trout lakes.
Limnol. Oceanogr., 2:287–294.

Barth, H.
1957. Aufnahme und Abgabe von CO_2 and O_2 bei submersen Wasserpflanzen.
Gewässer Abwässer, 4(17/18):18–81.

Baumeister, W.
1943. Der Einfluss des Bors auf die Photosynthese und Atmung submerser Pflanzen.
Jb. wiss. Bot., *91*:242–277.

Baxter, R. M., R. B. Wood, and M. V. Prosser.
1973. The probable occurrence of hydroxylamine in the water of an Ethiopian lake.
Limnol. Oceanogr., *18*:470–472.

Bayly, I. A. E., and W. D. Williams.
1973. Inland Waters and Their Ecology.
Victoria, Longman Australia Pty. Ltd., 316 pp.

Bazin, M., and G. W. Saunders.
1971. The hypolimnetic oxygen deficit as an index of eutrophication in Douglas Lake, Michigan.
Mich. Academician, 3:91–106.

Beadle, L. C.
1943. Osmotic regulation and the faunas of inland waters.
Biol. Rev., *18*:172–183.

Beadle, L. C.
1957. Comparative physiology: osmotic and ionic regulation in aquatic animals.
Ann. Rev. Physiol., *19*:329–358.

Beadle, L. C.
1959. Osmotic and ionic regulation in relation to the classification of brackish and inland saline waters.
Arch. Oceanogr. Limnol. Suppl., *11*:143–151.

Beadle, L. C.
1969. Osmotic regulation and the adaptation of freshwater animals to inland saline waters.
Verh. Int. Ver. Limnol., *17*:421–429.

Beadle, L. C.
1974. The Inland Waters of Tropical Africa. An Introduction to Tropical Limnology.
London, Longman Group Ltd., 365 pp.

Beeton, A. M.
1958. Relationship between Secchi disc readings and light penetration in Lake Huron.
Trans. Amer. Fish. Soc., 87(1957):73–79.

Beeton, A. M.
1959. Photoreception in the opossum shrimp, *Mysis relicta* Lovén.
Biol. Bull., *116*:204–216.

Beeton, A. M.
1960. The vertical migration of *Mysis relicta* in lakes Huron and Michigan.
J. Fish. Res. Bd. Canada, *17*:517–539.

Beeton, A. M.
1961. Environmental changes in Lake Erie.
Trans. Amer. Fish. Soc., *90*:153–159.

Behre, K.
1956. Die Algenbesiedlung einiger Seen um Bremen und Bremerhaven.
Ver. Inst. Meeresforsch. Bremerhaven, *4*:221–283.

Benoit, R. J.
1957. Preliminary observations on cobalt and vitamin B_{12} in fresh water.
Limnol. Oceanogr., 2:233–240.

Benoit, R. J.
1969. Geochemistry of eutrophication. *In* Eutrophication: Causes, Consequences, Correctives.
Washington, D. C., National Academy of Sciences, pp. 614–630.

Bentley, J. A.
1958a. Role of plant hormones in algal metabolism and ecology.
Nature, *181*:1499–1502.

Bentley, J. A.
1958b. The naturally-occurring auxins and inhibitors.
Ann. Rev. Plant Physiol., 9:47–80.

Berg, K.
1938. Studies on the bottom animals of Esrom Lake.
K. Danske Vidensk. Selsk. Skr. Nat. Mat. Afd., 9, 8, 255 pp.

Berg, K., and I. C. Petersen.
1956. Studies on the humic, acid Lake Gribsø.
Folia Limnol. Scandinavica, 8, 273 pp.

Berger, F.
1955. Die Dichte natürlicher Wässer und die Konzentrations-Stabilität in Seen.
Arch. Hydrobiol. Suppl., 22:286–294.

Berman, M. S., and S. Richman.
1974. The feeding behavior of *Daphnia pulex* from Lake Winnebago, Wisconsin.
Limnol. Oceanogr., 19:105–109.

Bernard, J. M.
1973. Production ecology of wetland sedges: the genus *Carex*.
Pol. Arch. Hydrobiol., 20:207–214.

Berzins, B.
1958. Ein planktologisches Querprofil.
Rept. Inst. Freshwat. Res. Drottningholm, 39:5–22.

Bickford, E. D., and S. Dunn.
1972. Lighting for Plant Growth.
Kent, Ohio, Kent State University Press, 221 pp.

Biggar, J. W., and R. B. Corey.
1969. Agricultural drainage and eutrophication. *In* Eutrophication: Causes, Consequences, Correctives.
Washington, D. C., National Academy of Sciences, pp. 404–445.

Billaud, V. A.
1968. Nitrogen fixation and the utilization of other inorganic nitrogen sources in a subarctic lake.
J. Fish. Res. Bd. Canada, 25:2101–2110.

Birge, E. A.
1915. The heat budgets of American and European lakes.
Trans. Wis. Acad. Sci. Arts Lett., 18(Pt. 1):166–213.

Birge, E. A.
1916. The work of the wind in warming a lake.
Trans. Wis. Acad. Sci. Arts Lett., 18(Pt. 2):341–391.

Birge, E. A., and C. Juday.
1911. The inland lakes of Wisconsin. The dissolved gases of the water and their biological significance.
Bull. Wis. Geol. Nat. Hist. Survey, 22, Sci. Ser. 7, 259 pp.

Birge, E. A., and C. Juday.
1914. A limnological study of the Finger Lakes of New York.
Bull. U. S. Bur. Fish., 32:525–609.

Birge, E. A., and C. Juday.
1926. Organic content of lake water.
Bull. U. S. Bur. Fish., 42:185–205.

Birge, E. A., and C. Juday.
1927. The organic content of the water of small lakes.
Proc. Amer. Phil. Soc., 66:357–372.

Birge, E. A., and C. Juday.
1934. Particulate and dissolved organic matter in inland lakes.
Ecol. Monogr., 4:440–474.

Birge, E. A., C. Juday, and H. W. March.
1927. The temperature of the bottom deposits of Lake Mendota. A chapter in the heat exchanges in the lake.
Trans. Wis. Acad. Sci. Arts Lett., 23:187–231.

Bjarnov, N.
1972. Carbohydrases in *Chironomus, Gammarus* and some Trichoptera larvae.
Oikos, 23:261–263.

Björk, S.
1967. Ecologic investigations of *Phragmites communis*. Studies in theoretic and applied limnology.
Folia Limnol. Scand., 14, 248 pp.

Black, C. C.
1971. Ecological implications of dividing plants into groups with distinct photosynthetic production capacities.
Adv. Ecol. Res., 7:87–114.

Bland, R. D., and A. J. Brook.
1974. The spatial distribution of desmids in lakes in northern Minnesota, U.S.A.
Freshwat. Biol., 4:543–556.

Blanton, J. O.
1973. Vertical entrainment into the epilimnia of stratified lakes.
Limnol. Oceanogr., 18:697–704.

Bodin, K., and A. Nauwerck.
1969. Produktionsbiologische Studien über die Moosvegetation eines klaren Gebirgssees.
Schweiz. Z. Hydrol., 30:318–352.

Borutskii, E. V.
1939. Dynamics of the total benthic biomass in the profundal of Lake Beloie.
Proc. Kossino Limnol. Stat. Hydrometeorol. Serv. USSR, 22:196–218.
(Translated by M. Ovchynnyk, Michigan State University.)

Borutskii, E. V.
1950. Dinamika organicheskogo veshchestva v vodoeme.
Trudy Vses. Gidrobiol. Obshchestva, 2:43–68.

Botan, E. A., J. J. Miller, and H. Kleerekoper.
1960. A study of the microbiological decomposition of nitrogenous matter in fresh water.
Arch. Hydrobiol., 56:334–354.

Boyd, C. E.
1970. Losses of mineral nutrients during decomposition of *Typha latifolia*.
Arch. Hydrobiol., 66:511–517.

Boyd, C. E.
1971. The dynamics of dry matter and chemical substances in a *Juncus effusus* population.
Amer. Midland Nat., 86:28–45.

Bradford, G. R., F. I. Bair, and V. Hunsaker.
1968. Trace and major element content of 170 high Sierra lakes in California.
Limnol. Oceanogr., 13:526–530.

Bragg, A. N.
1960. An ecological study of the protozoa of Crystal Lake, Norman, Oklahoma.
Wasmann J. Biol., 18:37–85.

Brandl, Z., J. Brandlová, and M. Poštolková.
1970. The influence of submerged vegetation on the photosynthesis of phytoplankton in ponds.
Rozpravy Československ. Akad. Věd, Řada Matem. Přír. Věd, 80(6):33–62.

Bray, J. R., D. B. Lawrence, and L. C. Pearson.
1959. Primary production in some Minnesota terrestrial communities for 1957.
Oikos, 10:38–49.

Breger, I. A., ed.
1963. Organic Geochemistry.
New York, Macmillan Company, 658 pp.

Brehm, J.
1967. Untersuchungen über den Aminosäure-Haushalt holsteinischer Gewässer, insbesondere des Pluss-Sees.
Arch Hydrobiol. Suppl., 32:313–435.

Brehm, K.
1970. Kationenaustausch bei Hochmoorsphagnen: Die Wirkung von an den Austauscher gebundenen Kationen in Kulturversuchen.
Beitr. Biol. Pflanzen, 47:91–116.

Bretschko, G.
1973. Benthos production of a high-mountain lake: Nematoda.
Verh. Int. Ver. Limnol., 18:1421–1428.

Brezonik, P. L., J. J. Delfino, and G. F. Lee.
1969. Chemistry of N and Mn in Cox Hollow Lake, Wisc., following destratification.
J. Sanit. Eng. Div. Proc. Amer. Soc. Civil Eng., SA-5:929–940.

Brezonik, P. L., and C. L. Harper.
1969. Nitrogen fixation in some aquatic lacustrine environments.
Science, 164:1277–1279.

Brezonik, P. L., and G. F. Lee.
1968. Denitrification as a nitrogen sink in Lake Mendota, Wisconsin.
Environ. Sci. Technol., 2:120–125.

Bright, T., F. Ferrari, D. Martin, and G. A. Franceschini.
1972. Effects of a total soliar eclipse on the vertical distribution of certain oceanic zooplankters.
Limnol. Oceanogr., 17:296–301.

Brinkhurst, R. O.
1967. The distribution of aquatic oligochaetes in Saginaw Bay, Lake Huron.
Limnol. Oceanogr., 12:137–143.

Brinkhurst, R. O.
1974. Factors mediating interspecific aggregation of tubificid oligochaetes.
J. Fish. Res. Bd. Canada, 31:460–462.

Brinkhurst, R. O., and K. E. Chua.
1969. Preliminary investigation of the exploitation of some potential nutritional resources by three sympatric tubificid oligochaetes.
J. Fish. Res. Bd. Canada, 26:2659–2668.

Brinkhurst, R. O., K. E. Chua, and N. K. Kaushik.
1972. Interspecific interactions and selective feeding by tubificid oligochaetes.
Limnol. Oceanogr., 17:122–133.

Brinkhurst, R. O., and D. G. Cook.
1974. Aquatic earthworms (Annelida: Oligochaeta). *In* C. W. Hart, Jr., and S. L. H. Fuller, eds. Pollution Ecology of Freshwater Invertebrates.
New York, Academic Press, pp. 143–156.

Brinkhurst, R. O., and B. G. M. Jamieson.
1971. Aquatic Oligochaeta of the World.
Toronto, University of Toronto Press, 860 pp.

Brinkhurst, R. O., and B. Walsh.
1967. Rostherne Mere, England, a further instance of guanotrophy.
J. Fish. Res. Bd. Canada, 24:1299–1309.

Bristow, J. M.
1969. The effects of carbon dioxide on the growth and development of amphibious plants.
Can. J. Bot., 47:1803–1807.

Bristow, J. M.
1974a. Nitrogen fixation in the rhizosphere of freshwater angiosperms.
Can. J. Bot., 52:217–221.

Bristow, J. M.
1974b. The structure and function of roots in aquatic vascular plants. *In* The Development and Function of Roots.
New York, Academic Press (in press).

Bristow, J. M., and M. Whitcombe.
1971. The role of roots in the nutrition of aquatic vascular plants.
Amer. J. Bot., 58:8–13.

Broch, E. S., and W. Yake.
1969. A modification of Maucha's ionic diagram to include ionic concentrations.
Limnol. Oceanogr., 14:933–935.

Brock, T. D.
1966. Principles of Microbial Ecology.
Englewood Cliffs, N. J., Prentice-Hall, Inc., 306 pp.

Broecker, W. S.
1965. An application of natural radon to problems in ocean circulation. In Diffusion in Oceans and Freshwaters.
Symposium Proceedings, New York, Lamont Geological Observatory, Columbia University, pp. 116–145.

Broecker, W. S.
1973. Factors controlling CO_2 content in the oceans and atmosphere. In G. M. Woodwell and E. V. Pecan, eds. Carbon and the Biosphere.
Brookhaven, N.Y., Proc. Brookhaven Symp. in Biol. 24. Tech. Information Center, U. S. Atomic Energy Commission CONF-720510, pp. 32–50.

Broecker, W. S., J. Cromwell, and Y. H. Li.
1968. Rates of vertical eddy diffusion near the ocean floor based on measurement of the distribution of excess ^{222}Rn.
Earth Planet. Sci. Lett., 5:101–105.

Broecker, W. S., and T.-H. Peng.
1971. The vertical distribution of radon in the Bonex area.
Earth Planet. Sci. Lett., 11:99–108.

Brooks, J. L.
1947. Turbulence as an environmental determinant of relative growth in Daphnia.
Proc. Nat. Acad. Sci., 33:141–148.

Brooks, J. L.
1964. The relationship between the vertical distribution and seasonal variation of limnetic species of Daphnia.
Verh. Int. Ver. Limnol., 15:684–690.

Brooks, J. L.
1965. Predation and relative helmet size in cyclomorphic Daphnia.
Proc. Nat. Acad. Sci., 53:119–126.

Brooks, J. L.
1966. Cyclomorphosis, turbulence and overwintering in Daphnia.
Verh. Int. Ver. Limnol., 16:1653–1659.

Brooks, J. L.
1968. The effects of prey size selection by lake planktivores.
Syst. Zool., 17:273–291.

Brooks, J. L., and S. I. Dodson.
1965. Predation, body size, and composition of plankton.
Science, 150:28–35.

Brooks, J. L., and G. E. Hutchinson.
1950. On the rate of passive sinking in Daphnia.
Proc. Nat. Acad. Sci., 36:272–277.

Brown, D. H., C. E. Gibby, and M. Hickman.
1972. Photosynthetic rhythms in epipelic algal populations.
Brit. Phycol. Bull., 7:37–44.

Brown, S.-D., and A. P. Austin.
1973a. Diatom succession and interaction in littoral periphyton and plankton.
Hydrobiologia, 43:333–356.

Brown, S.-D., and A. P. Austin.
1973b. Spatial and temporal variation in periphyton and physico-chemical conditions in the littoral of a lake.
Arch. Hydrobiol., 71:183–232.

Brown, S. R.
1968. Bacterial carotenoids from freshwater sediments.
Limnol. Oceanogr., 13:233–241.

Brown, S. R.
1969. Paleolimnological evidence from fossil pigments.
Mitt. Int. Ver. Limnol., 17:95–103.

Brown, S., and B. Colman.
1963. Oscillaxanthin in lake sediments.
Limnol. Oceanogr., 8:352–353.

Brunskill, G. J., and S. D. Ludlam.
1969. Fayetteville Green Lake, New York. I. Physical and chemical limnology.
Limnol. Oceanogr., 14:817–829.

Bryson, R. A., and R. A. Ragotzkie.
1960. On internal waves in lakes.
Limnol. Oceanogr., 5:397–408.

Burgis, M. J.
1970. The effect of temperature on the development time of eggs of Thermocyclops sp., a tropical cyclopoid copepod from Lake George, Uganda.
Limnol. Oceanogr., 15:742–747.

Burkholder, P. R., and G. H. Bornside.
1957. Decomposition of marsh grass by aerobic marine bacteria.
Bull. Torr. Bot. Club, 84:366–383.

Burkholder, P. R., L. M. Burkholder, and J. A. Rivero.
1959. Some chemical constituents of turtlegrass, *Thalassia testudinum*.
Bull. Torr. Bot. Club, *86*:88–93.

Burla, H.
1971. Gerichtete Ortsveränderung bei Muscheln der Gattung *Anodonta* im Zürichsee.
Vierteljahrsschr. Naturf. Gesellschaft Zürich, *116*:181–194.

Burnison, B. K., and R. Y. Morita.
1973. Competitive inhibition for amino acid uptake by the indigenous microflora of Upper Klamath Lake.
Appl. Microbiol., *25*:103–106.

Burns, C. W.
1968a. The relationship between body size of filter-feeding Cladocera and the maximum size of particle ingested.
Limnol. Oceanogr., *13*:675–678.

Burns, C. W.
1968b. Direct observations of mechanisms regulating feeding behavior of *Daphnia* in lakewater.
Int. Rev. ges. Hydrobiol., *53*:83–100.

Burns, C. W.
1969a. Particle size and sedimentation in the feeding behavior of two species of *Daphnia*.
Limnol. Oceanogr., *14*:392–402.

Burns, C. W.
1969b. Relation between filtering rate, temperature, and body size in four species of *Daphnia*.
Limnol. Oceanogr., *14*:696–700.

Burns, C. W., and F. H. Rigler.
1967. Comparison of filtering rates of *Daphnia rosea* in lake water and in suspensions of yeast.
Limnol. Oceanogr., *12*:492–502.

Burns, N. M., and C. Ross.
1971. Nutrient relationships in a stratified eutrophic lake.
Proc. Conf. Great Lakes Res., Int. Assoc. Great Lakes Res., *14*:749–760.

Burris, R. H., F. J. Eppling, H. B. Wahlin, and P. W. Wilson.
1943. Detection of nitrogen-fixation with isotopic nitrogen.
J. Biol. Chem., *148*:349–357.

Burton, J. D., T. M. Leatherland, and P. S. Liss.
1970. The reactivity of dissolved silicon in some natural waters.
Limnol. Oceanogr., *15*:473–476.

Buscemi, P. A.
1958. Littoral oxygen depletion produced by a cover of *Elodea canadensis*.
Oikos, *9*:239–245.

Buscemi, P. A.
1961. Ecology of the bottom fauna of Parvin Lake, Colorado.
Trans. Amer. Microsc. Soc., *80*:266–307.

Bushnell, J. H., Jr.
1966. Environmental relations of Michigan Ectoprocta, and dynamics of natural populations of *Plumatella repens*.
Ecol. Monogr., *36*:95–123.

Bushnell, J. H.
1974. Bryozoans (Ectoprocta). *In* C. W. Hart, Jr., and S. L. H. Fuller, eds. Pollution Ecology of Freshwater Invertebrates.
New York, Academic Press, pp. 157–194.

Butlin, K. R.
1953. The bacterial sulphur cycle.
Research, *6*:184–191.

Buttery, B. R., and J. M. Lambert.
1965. Competition between *Glyceria maxima* and *Phragmites communis* in the region of Surlingham Broad. I. The competition mechanism.
J. Ecol., *53*:163–181.

Buttery, B. R., W. T. Williams, and J. M. Lambert.
1965. Competition between *Glyceria maxima* and *Phragmites communis* in the region of Surlingham Broad. II. The fen gradient.
J. Ecol., *53*:183–195.

Cairns, J., Jr.
1974. Protozoans (Protozoa). *In* C. W. Hart, Jr., and S. L. H. Fuller, eds. Pollution Ecology of Freshwater Invertebrates.
New York, Academic Press, pp. 1–28.

Caldwell, D. E., and J. M. Tiedje
1975. The stricture of anaerobic bacterial communities in the hypolimnia of several Michigan lakes.
Can. J. Microbiol., *21*:377–385.

Cannon, D., J. W. G. Lund, and J. Sieminska.
1961. The growth of *Tabellaria flocculosa* (Roth) Kütz. var. *flocculosa* (Roth) Knuds. under natural conditions of light and temperature.
J. Ecol., *49*:277–287.

Canter, H. M.
1973. A new primitive protozoan devouring centric diatoms in the plankton.
Zool. J. Linn. Soc., *52*:63–83.

Canter, H. M., and J. W. G. Lund.
1948. Studies on plankton parasites. I. Fluctuations in the numbers of *Asterionella formosa* Hass. in relation to fungal epidemics.
New Phytol., 47:238–261.

Canter, H. M., and J. W. G. Lund.
1968. The importance of Protozoa in controlling the abundance of planktonic algae in lakes.
Proc. Linn. Soc. Lond., 179:203–219.

Canter, H. M., and J. W. G. Lund.
1969. The parasitism of planktonic desmids by fungi.
Österr. Bot. Z., 116:351–377.

Capblancq, J.
1973. Phytobenthos et productivité primaire d'un lac de haute montagne dans les Pyrénées centrales.
Ann. Limnol., 9:193–230.

Cappenberg, T. E.
1972. Ecological observations on heterotrophic, methane oxidizing and sulfate reducing bacteria in a pond.
Hydrobiologia, 40:471–485.

Carpenter, G. F., E. L. Mansey, and N. H. F. Watson.
1974. Abundance and life history of *Mysis relicta* in the St. Lawrence Great Lakes.
J. Fish. Res. Bd. Canada, 31:319–325.

Carr, J. F.
1962. Dissolved oxygen in Lake Erie, past and present.
Publ. Great Lakes Res. Div., Univ. Mich., 9:1–14.

Carr, J. L.
1969. The primary productivity and physiology of *Ceratophyllum demersum*. II. Micro primary productivity, pH, and the P/R ratio.
Austral. J. Mar. Freshwat. Res., 20:127–142.

Carroll, D.
1962. Rainwater as a chemical agent of geologic processes—a review.
U.S. Geol. Surv. Water-Supply Pap., 1535-G, 18 pp.

Carter, J. C. H.
1974. Life cycles of three limnetic copepods in a beaver pond.
J. Fish. Res. Bd. Canada, 31:421–434.

Castenholz, R. W.
1960. Seasonal changes in the attached algae of freshwater and saline lakes in the Lower Grand Coulee, Washington.
Limnol. Oceanogr., 5:1–28.

Castenholz, R. W.
1961. An evaluation of a submerged glass method of estimating production of attached algae.
Verh. Int. Ver. Limnol., 14:155–159.

Castenholz, R. W.
1969. Thermophilic blue-green algae and the thermal environment.
Bacteriol. Rev., 33:476–504.

Chamberlain, W. M.
1968. A preliminary investigation of the nature and importance of soluble organic phosphorus in the phosphorus cycle of lakes.
Ph.D. Diss., University of Toronto, Ontario, 232 pp.

Chandler, C. M.
1966. Environmental factors affecting the local distribution and abundance of four speces of stream-dwelling triclads.
Invest. Indiana Lakes Streams, 7:1–56.

Chandler, D. C.
1937. Fate of typical lake plankton in streams.
Ecol. Monogr., 7:445–479.

Chapin, J. D., and P. D. Uttormark.
1973. Atmospheric Contributions of Nitrogen and Phosphorus.
Tech. Rep. Wat. Resources Ctr. Univ. Wis., 73-2, 35 pp.

Chapman, D. W.
1967. Production in fish populations. *In* Gerking, S. D., ed. The Biological Basis of Freshwater Fish Production.
New York, John Wiley & Sons, Inc., pp. 3–29.

Chave, K. E.
1965. Carbonates: Association with organic matter in surface seawater.
Science, 148:1723–1724.

Chave, K. E., and E. Suess.
1970. Calcium carbonate saturation in seawater: Effects of dissolved organic matter.
Limnol. Oceanogr., 15:633–637.

Chawla, V. K., and Y. K. Chau.
1969. Trace elements in Lake Erie.
Proc. Conf. Great Lakes Res., Int. Assoc. Great Lakes Res., 12:760–765.

Chen, K. Y., and J. C. Morris.
1972. Kinetics of oxidation of aqueous sulfide by O_2.
Environ. Sci. Technol., 6:529–537.

Chen, R. L., D. R. Keeney, and J. G. Konrad.
1972. Nitrification in lake sediments.
J. Environ. Quality, 1:151–154.

Chen, R. L., D. R. Keeney, D. A. Graetz, and A. J. Holding.
1972. Denitrification and nitrate reduction in lake sediments.
J. Environ. Quality, 1:158–162.

Chen, R. L., D. R. Keeney, J. G. Konrad, A. J. Holding, and D. A. Graetz.
1972. Gas production in sediments of Lake Mendota, Wisconsin.
J. Environ. Quality, 1:155–158.

Chow, V. T., ed.
1964. Handbook of Applied Hydrology.
New York, McGraw-Hill Book Co.

Chu, S. P.
1942. The influence of the mineral composition of the medium on the growth of planktonic algae. I. Methods and culture media.
J. Ecol., *30*:284–325.

Chu, S. P.
1943. The influence of the mineral composition of the medium on the growth of planktonic algae. II. The influence of the concentration of inorganic nitrogen and phosphate phosphorus.
J. Ecol., *31*:109–148.

Chua, K. E., and R. O. Brinkhurst.
1973. Evidence of interspecific interactions in the respiration of tubificid oligochaetes.
J. Fish. Res. Bd. Canada, *30*:617–622.

Clark, C.
1967. Population Growth and Land Use.
London, Macmillan Press Ltd., 406 pp.

Clarke, F. W.
1924. The Data of Geochemistry. 5th ed.
Bull. U.S. Geol. Surv., *770*, 841 pp.

Clarke, G. L.
1939. The utilization of solar energy by aquatic organisms. *In* Problems in lake biology.
Publ. Amer. Assoc. Adv. Sci., *10*:27–38.

Clémencon, H.
1963. Verbreitung einiger B-Vitamine in Kleingewässern aus der Umgebung von Bern und Untersuchungen über die Abgabe von B-Vitaminen durch Algen.
Schweiz. Z. Hydrol., *25*:157–165.

Clymo, R. S.
1963. Ion exchange in *Sphagnum* and its relation to bog ecology.
Ann. Bot., N.S., *27*:309–324.

Clymo, R. S.
1964. The origin of acidity in *Sphagnum* bogs.
Bryologist, *67*:427–431.

Clymo, R. S.
1965. Experiments on breakdown of *Sphagnum* in two bogs.
J. Ecol., *53*:747–758.

Clymo, R. S.
1967. Control of cation concentrations, and in particular of pH, in *Sphagnum* dominated communities. *In* H. L. Golterman and R. S. Clymo, eds. Chemical Environment in the Aquatic Habitat.
Amsterdam, N.V. Noord-Hollandsche Uitgevers Maatschappij, pp. 273–284.

Clymo, R. S., and E. J. F. Reddaway.
1971. Productivity of *Sphagnum* (bog moss) and peat accumulation.
Hidrobiologia, *12*:181–192.

Clymo, R. S., and E. J. F. Reddaway.
1974. Growth rate of *Sphagnum rubellum* Wils, on pennine blanket bog.
J. Ecol., *62*:191–196.

Coffin, C. C., F. R. Hayes, L. H. Jodrey, and S. G. Whiteway.
1949. Exchange of materials in a lake as studied by the addition of radioactive phosphorus.
Can. J. Res., (Ser. D), *27*:207–222.

Comita, G. W.
1964. The energy budget of *Diaptomus siciloides*, Lilljeborg.
Verh. Int. Ver. Limnol., *15*:646–653.

Comita, G. W.
1972. The seasonal zooplankton cycles, production and transformations of energy in Severson Lake, Minnesota.
Arch. Hydrobiol., *70*:14–66.

Comita, G. W., and G. C. Anderson.
1959. The seasonal development of a population of *Diaptomus ashlandi* Marsh, and related phytoplankton cycles in Lake Washington.
Limnol. Oceanogr., *4*:37–52.

Confer, J. L.
1971. Intrazooplankton predation by *Mesocyclops edax* at natural prey densities.
Limnol. Oceanogr., *16*:663–666.

Confer, J. L.
1972. Interrelations among plankton, attached algae, and the phosphorus cycle in artificial open systems.
Ecol. Monogr., *42*:1–23.

Conrad, H. M., and P. Saltman.
1962. Growth substances. *In* R. A. Lewin, ed. Physiology and Biochemistry of Algae.
New York, Academic Press, pp. 663–671.

Conway, E. J
1942. Mean geochemical data in relation to oceanic evolution.
Proc. Royal Irish Acad. (Ser. B), *48*:119–159.

Cooke, W. B.
1956. Colonization of artificial bare areas by microorganisms.
Bot. Rev., *22*:613–638.

Cooper, C. F.
1969. Nutrient output from managed forests. *In* Eutrophication: Causes, Consequences, Correctives.
Washington, D.C., National Academy of Sciences, pp. 446–463.

Cooper, D. C.
1973. Enhancement of net primary productivity by herbivore grazing in aquatic laboratory microcosms.
Limnol. Oceanogr., *18*:31–37.

Cooper, W. E.
1965. Dynamics and production of a natural population of a fresh-water amphipod, *Hyalella azteca*.
Ecol. Monogr., *35*:377–394.

Costa, R. R., and K. W. Cummins.
1972. The contribution of *Leptodora* and other zooplankton to the diet of various fish.
Amer. Midland Nat., *87*:559–564.

Coughlan, M. P.
1971. The role of iron in microbial metabolism.
Sci. Progress, Oxford, *59*:1–23.

Cowen, W. F., and G. F. Lee.
1973. Leaves as source of phosphorus.
Environ. Sci. Technol., *7*:853–854.

Cowgill, U. M., and G. E. Hutchinson.
1963. El Bajo de Santa Fé.
Trans. Amer. Phil. Soc., N. S., *53*(7), 51 pp.

Cowgill, U. M., and G. E. Hutchinson.
1966. The history of a pond in Guatemala.
Arch. Hydrobiol., *62*:355–372.

Cowgill, U. M., C. E. Goulden, G. E. Hutchinson, R. Patrick, A. A. Raček, and M. Tsukada.
1966. The History of Laguna de Petenxil. A Small Lake in Northern Guatemala.
Mem. Conn. Acad. Arts Sci., *17*, 126 pp.

Cummins, K. W.
1962. An evaluation of some techniques for the collection and analysis of benthic samples with special emphasis on lotic waters.
Amer. Midland Nat., *67*:477–504.

Cummins, K. W.
1964. Factors limiting the microdistribution of larvae of the caddisflies *Pycnopsyche lepida* (Hagen) and *Pycnopsyche guttifer* (Walker) in a Michigan stream (Trichoptera: Limnephilidae).
Ecol. Monogr., *34*:271–295.

Cummins, K. W.
1972. Predicting variations in energy flow through a semi-controlled lotic ecosystem.
Tech. Rept. Inst. Water Res. Mich. State Univ., *19*, 21 pp.

Cummins, K. W.
1973. Trophic relations of aquatic insects.
Ann. Rev. Entomol., *18*:183–206.

Cummins, K. W., and G. H. Lauff.
1969. The influence of substrate particle size on the microdistribution of stream macrobenthos.
Hydrobiologia, *34*:145–181.

Cummins, K. W., R. R. Costa, R. E. Rowe, G. A. Moshiri, R. M. Scanlon, and R. K. Zajdel.
1969. Ecological energetics of a natural population of the predaceous zooplankter *Leptodora kindtii* Focke (Cladocera).
Oikos, *20*:189–223.

Cummins, K. W., and J. C. Wuycheck.
1971. Caloric equivalents for investigations in ecological energetics.
Mitt. Int. Ver. Limnol., *18*, 158 pp.

Cummins, K. W., M. J. Klug, R. G. Wetzel, R. C. Petersen, K. F. Suberkropp, B. A. Manny, J. C. Wuycheck, and F. O. Howard.
1972. Organic enrichment with leaf leachate in experimental lotic ecosystems.
BioScience, *22*:719–722.

Cummins, K. W., R. C. Petersen, F. O. Howard, J. C. Wuycheck, and V. I. Holt.
1973. The utilization of leaf litter by stream detritivores.
Ecology, *54*:336–345.

Cunningham, L.
1972. Vertical migrations of *Daphnia* and copepods under the ice.
Limnol. Oceanogr., *17*:301–303.

Czeczuga, B.
1959. On oxygen minimum and maximum in the metalimnion of Rajgród lakes.
Acta Hydrobiol., *1*:109–122.

Czeczuga, B.
1968a. An attempt to determine the primary production of the green sulphur bacteria, Chlorobium limicola Nads, (Chlorobacteriaceae).
Hydrobiologia, *31*:317–333.

Czeczuga, B.
1968b. Primary production of the green hydrosulphuric bacteria, *Chlorobium limicola* Nads. (Chlorobacteriaceae).
Photosynthetica, *2*:11–15.

Czeczuga, B.
1968c. Primary production of the purple sulphuric bacteria, *Thiopedia rosea* Winogr. (Thiorhodaceae).
Photosynthetica, *2*:161–166.

Czeczuga, B., and E. Bobiatyńska-ksok.
1970. The extent of consumption of the energy contained in the food suspension by *Ceriodaphnia reticulata* (Jurine). *In* Z. Kajak and A. Hillbricht-Ilkowska, eds. Productivity Problems of Freshwaters.
Warsaw, PWN Polish Scientific Publishers, pp. 739–748.

Czeczuga, B., and R. Czerpak.
1967. Obserwacje nad bakterioplanktonem jezior legińskich.
Zesz. Nauk WSR Olszt., 23:35–44.

Czeczuga, B., and R. Czerpak.
1968. Investigations on vegetable pigments in post-glacial bed sediments of lakes.
Schweiz. Z. Hydrol., 30:217–231.

Daan, N., and J. Ringelberg.
1969. Further studies on the positive and negative phototactic reaction of *Daphnia magna* Straus.
Netherlands J. Zool., 19:525–540.

Daley, R. J.
1973. Experimental characterization of lacustrine chlorophyll diagenesis. II. Bacterial, viral and herbivore grazing effects.
Arch. Hydrobiol., 72:409–439.

Daley, R. J., and S. R. Brown.
1973. Experimental characterization of lacustrine chlorophyll diagenesis. I. Physiological and environmental effects.
Arch. Hydrobiol., 72:277–304.

Danforth, W. F.
1962. Substrate assimilation and heterotrophy.
In R. A. Lewin, ed. Physiology and Biochemistry of Algae.
New York, Academic Press, pp. 99–123.

Darnell, R. M.
1964. Organic detritus in relation to secondary production in aquatic communities.
Verh. Int. Ver. Limnol., 15:462–470.

Datsko, V. G.
1959. Organicheskoe Veshchestvo v Vodakh Iuzhnykh Morei SSSR.
Moscow, Izdatel'stvo Akademii Nauk SSSR, 271 pp.

Davies, G. S.
1970. Productivity of macrophytes in Marion Lake, British Columbia.
J. Fish. Res. Bd. Canada, 27:71–81.

Davies, W.
1971. The phytopsammon of a sandy beach transect.
Amer. Midland Nat., 86:292–308.

Davis, C. C.
1963. On questions of production and productivity in ecology.
Arch. Hydrobiol., 59:145–161.

Davis, M. B.
1968. Pollen grains in lake sediments: Redeposition caused by seasonal water circulation.
Science, 162:796–799.

Davis, M. B.
1973. Redeposition of pollen grains in lake sediment.
Limnol. Oceanogr., 18:44–52.

Davis, M. B., and L. B. Brubaker.
1973. Differential sedimentation of pollen grains in lakes.
Limnol. Oceanogr., 18:635–646.

Davis, M. B., and E. S. Deevey, Jr.
1964. Pollen accumulation rates: Estimates from late-glacial sediment of Rogers Lake.
Science, 145:1293–1295.

Davis, R. B.
1974. Stratigraphic effects of tubificids in profundal lake sediments.
Limnol. Oceanogr., 19:466–488.

Davis, S. N.
1964. Silica in streams and ground water.
Amer. J. Sci., 262:870–891.

Davis, S. N., and R. J. M. DeWiest.
1966. Hydrogeology.
New York, John Wiley & Sons, Inc., 463 pp.

Davis, W. M.
1933. The lakes of California.
Calif. J. Mines Geology, 29:175–236.

DeCosta, J.
1968. Species diversity of chydorid fossil communities in the Mississippi Valley.
Hydrobiologia, 32:497–512.

Deevey, E. S., Jr.
1941. Limnological studies in Connecticut. VI. The quantity and composition of the bottom fauna of thirty-six Connecticut and New York lakes.
Ecol. Monogr., 11:414–455.

Delfino, J. J., and G. F. Lee.
1968. Chemistry of manganese in Lake Mendota, Wisconsin.
Environ. Sci. Technol., 2:1094–1100.

Delfino, J. J., and G. F. Lee.
1971. Variation of manganese, dissolved oxygen and related chemical parameters in the bottom waters of Lake Mendota, Wisconsin.
Water Res., 5:1207–1217.

Della Croce, N.
1955. The conditions of sedimentation and their relations with Oligochaeta populations of Lake Maggiore.
Mem. Ist. Ital. Idrobiol., 8(Suppl.):39–62.

DeMarte, J. A., and R. T. Hartman.
1974. Studies on absorption of ^{32}P, ^{59}Fe, and ^{45}Ca by water-milfoil (*Myriophyllum exalbescens* Fernald).
Ecology, 55:188–194.

Dendy, J. S.
1963. Observations on bryozoan ecology in farm ponds.
Limnol. Oceanogr., 8:478–482.

Denny, P.
1972. Sites of nutrient absorption in aquatic macrophytes.
J. Ecol., 60:819–829.

DeWitt, R. M.
1954. Reproduction, embryonic development, and growth in the pond snail, *Physa gyrina* Say.
Trans. Amer. Microsc. Soc., 73:124–137.

Deyl, Z.
1961. Anaerobic fermentations.
Sci. Pap. Inst. Chem. Technol. Prague, Technol. Water, 5(2):131–234.

Dinsdale, M. T., and A. E. Walsby.
1972. The interrelations of cell turgor pressure, gas-vacuolation, and buoyancy in a blue-green alga.
J. Exp. Bot., 23:561–570.

Dodson, S. I.
1970. Complementary feeding niches sustained by size-selective predation.
Limnol. Oceanogr., 15:131–137.

Dodson, S. I.
1972. Mortality in a population of *Daphnia rosea*.
Ecology, 53:1011–1023.

Dodson, S. I.
1974a. Zooplankton competition and predation: An experimental test of the size-efficiency hypothesis.
Ecology, 55:605–613.

Dodson, S. I.
1974b. Adaptive change in plankton morphology in response to size-selective predation: A new hypothesis of cyclomorphosis.
Limnol. Oceanogr., 19:721–729.

Döhler, G., and K.-R. Przybylla.
1973. Einfluss der Temperatur auf die Lichtatmung der Blaualge *Anacystis nidulans*.
Planta, 110:153–158.

Dokulil, M.
1973. Planktonic primary production within the Phragmites community of Lake Neusiedlersee (Austria).
Pol. Arch. Hydrobiol., 20:175–180.

Domogalla, B. P., and E. B. Fred.
1926. Ammonia and nitrate studies of lakes near Madison, Wisconsin.
J. Amer. Soc. Agronomy, 18:897–911.

Domogalla, B. P., E. B. Fred, and W. H. Peterson.
1926. Seasonal variations in the ammonia and nitrate content of lake waters.
J. Amer. Water Works Assoc., 15:369–385.

Dor, I.
1970. Production rate of the periphyton in Lake Tiberias as measured by the glass-slide method.
Israel J. Bot., 19:1–15.

Douglas, B.
1958. The ecology of the attached diatoms and other algae in a small stony stream.
J. Ecol., 46:295–322.

Doyle, R. W.
1968a. The origin of the ferrous ion-ferric oxide Nernst potential in environments containing dissolved ferrous iron.
Amer. J. Sci., 266:840–859.

Doyle, R. W.
1968b. Identification and solubility of iron sulfide in anaerobic lake sediment.
Amer. J. Sci., 266:980–994.

Droop, M. R.
1973. Some thoughts on nutrient limitation in algae.
J. Phycol., 9:264–272.

Droop, M. R.
1974. Heterotrophy of carbon. *In* W. D. P. Stewart, ed. Algal Physiology and Biochemistry.
Berkeley, Univ. of California Press, pp. 530–559.

Drummond, A. J.
1971. Recent measurements of the solar radiation incident on the atmosphere. *In* Space Research XI.
Berlin, Akademie-Verlag, pp. 681–693.

Dubinina, G. A., V. M. Gorlenko, and J. I. Suleimanov.
1973. A study of microorganisms involved in the circulation of manganese, iron, and sulfur in meromictic Lake Gek-Gel'.
Mikrobiologiya, 42:918–924.

Dugdale, R. C., and V. A. Dugdale.
1961. Sources of phosphorus and nitrogen for lakes on Afognak Island.
Limnol. Oceanogr., 6:13–23.

Dugdale, V. A., and R. C. Dugdale.
1962. Nitrogen metabolism in lakes. II. Role of nitrogen fixation in Sanctuary Lake, Pennsylvania.
Limnol. Oceanogr., 7:170–177.

Dumont, H. J.
1972. The biological cycle of molybdenum in relation to primary production and water-bloom formation in a eutrophic pond.
Verh. Int. Ver. Limnol., 18:84–92.

Dunne, T.
1975. Integration of hillslope flow processes. *In* M. J. Kirkby, ed. Hillslope Hydrology. (In press).

Dunne, T., and R. D. Black.
1970a. An experimental investigation of runoff production in permeable soils.
Water Resources Res., 6:478–490.

Dunne, T., and R. D. Black.
1970b. Partial area contributions to storm runoff in a small New England watershed.
Water Resources Res., 6:1296–1311.

Dunne, T., and R. D. Black.
1971. Runoff processes during snowmelt.
Water Resources Res., 7:1160–1172.

Duong, T. P.
1972. Nitrogen fixation and productivity in a eutrophic hard-water lake: *In situ* and laboratory studies.
Ph. D. Disser., Michigan State University, 241 pp.

Durum, W. H., and J. Haffty.
1961. Occurrence of minor elements in water.
Circ. U. S. Geol. Surv., 445, 11 pp.

Dussart, B.
1966. Limnologie. L'Étude des Eaux Continentales.
Paris, Gauthier-Villars, 677 pp.

Dutton, J. A., and R. A. Bryson.
1962. Heat flux in Lake Mendota.
Limnol. Oceanogr., 7:80–97.

Duursma, E. K.
1961. Dissolved organic carbon, nitrogen and phosphorus in the sea.
Netherlands J. Mar. Res., 1:1–148.

Duursma, E. K.
1967. The mobility of compounds in sediments in relation to exchange between bottom and supernatant water. *In* H. L. Golterman and R. S. Clymo, eds. Chemical Environment in the Aquatic Habitat.
Amsterdam, N. V. Noord-Hollandsche Uitgevers Maatsahappij, pp. 288–296.

Dvořák, J.
1970. A quantitative study on the macrofauna of stands of emergent vegetation in a carp pond of south-west Bohemia.
Rozpravy Československ. Akad. Věd, Řada Matem. Přírod. Věd, 80(6):63–110.

Dykyjová, D., and D. Hradecká.
1973. Productivity of reed-bed stands in relation to the ecotype, microclimate and trophic conditions of the habitat.
Pol. Arch. Hydrobiol., 20:111–119.

Eberly, W. R.
1959. The metalimnetic oxygen maximum in Myers Lake.
Invest. Indiana Lakes Streams, 5:1–46.

Eberly, W. R.
1963. Oxygen production in some northern Indiana lakes.
Proc. Indust. Wastes Conf., Purdue Univ., 17:733–747.

Eberly, W. R.
1964. Further studies on the metalimnetic oxygen maximum, with special reference to its occurrence throughout the world.
Invest. Indiana Lakes Streams, 6:103–139.

Edmondson, W. T.
1944. Ecological studies of sessile Rotatoria. Part I. Factors affecting distribution.
Ecol. Monogr., 14:31–66.

Edmondson, W. T.
1945. Ecological studies of sessile Rotatoria. Part II. Dynamics of populations and social structures.
Ecol. Monogr., 15:141–172.

Edmondson, W. T.
1946. Factors in the dynamics of rotifer populations.
Ecol. Monogr., 16:357–372.

Edmondson, W. T.
1948. Ecological applications of Lansing's physiological work on longevity in *Rotatoria*.
Science, 108:123–126.

Edmondson, W. T.
1960. Reproductive rates of rotifers in natural populations.
Mem. Ist. Ital. Idrobiol., 12:21–77.

Edmondson, W. T.
1961. Changes in Lake Washington following an increase in the nutrient income.
Verh. Int. Ver. Limnol., 14:167–175.

Edmondson, W. T.
1965. Reproductive rate of planktonic rotifers as related to food and temperature in nature.
Ecol. Monogr., 35:61–111.

Edmondson, W. T.
1968. A graphical model for evaluating the use of the egg ratio for measuring birth and death rates.
Oecologia, 1:1–37.

Edmondson, W. T.
1969. Eutrophication in North America. *In* Eutrophication: Causes, Consequences, Correctives.
Washington, D.C., National Academy of Sciences, pp. 124–149.

Edmondson, W. T.
1972. Nutrients and phytoplankton in Lake Washington. *In* G. E. Likens, ed. Nutrients and Eutrophication: The Limiting-Nutrient Controversy.
Special Symposium, Amer. Soc. Limnol. Oceanogr., *1*:172–193.

Edmondson, W. T., and G. G. Winberg, eds.
1971. A Manual on Methods for the Assessment of Secondary Productivity in Fresh Waters.
Int. Biol. Program Handbook, *17*. Oxford, Blackwell Scientific Publications, 358 pp.

Edwards, A. M. C., and P. S. Liss.
1973. Evidence for buffering of dissolved silicon in fresh waters.
Nature, *243*:341–342.

Edwards, R. W., and M. Owens.
1960. The effects of plants on river conditions. I. Summer crops and estimates of net productivity of macrophytes in a chalk stream.
J. Ecol., *48*:151–160.

Efford, I. E.
1967. Temporal and spatial differences in phytoplankton productivity in Marion Lake, British Columbia.
J. Fish. Res. Bd., Canada *24*:2283–2307.

Eggleton, F. E.
1956. Limnology of a meromictic, interglacial, plunge-basin lake.
Trans. Amer. Microsc. Soc., *75*:334–378.

Egloff, D. A., and D. S. Palmer.
1971. Size relations of the filtering area of two *Daphnia* species.
Limnol. Oceanogr., *16*:900–905.

Einsele, W.
1936. Über die Beziehungen des Eisenkreislaufs zum Phosphatkreislauf im eutrophen See.
Arch. Hydrobiol., *29*:664–686.

Einsele, W.
1941. Die Umsetzung von zugeführtem, anorganischen Phosphat im eutrophen See und ihre Rüchwirkungen auf seinen Gesamthaushalt.
Zeitsch. f. Fischerei, *39*:407–488.

Eisenberg, D., and W. Kauzmann.
1969. The Structure and Properties of Water.
New York, Oxford University Press, 296 pp.

Eisenberg, R. M.
1966. The regulation of density in a natural population of the pond snail, *Lymnaea elodes*.
Ecology, *47*:889–906.

Eklund, H.
1963. Fresh water: Temperature of maximum density calculated from compressibility.
Science, *142*:1457–1458.

Eklund, H.
1965. Stability of lakes near the temperature of maximum density.
Science, *149*:632–633.

Elgmork, K.
1959. Seasonal occurrence of *Cyclops strenuus strenuus* in relation to environment in small water bodies in southern Norway.
Folia Limnol. Scandinavica, *11*, 196 pp.

Elliott, J. M.
1973a. The diel activity pattern, drifting and food of the leech *Erpobdella octoculata* (L.) (Hirudinea: Erpobdellidae) in a Lake District stream.
J. Anim. Ecol., *42*:449–459.

Elliott, J. M.
1973b. The life cycle and production of the leech *Erpobdella octoculata* (L.) (Hirudinea: Erpobdellidae) in a Lake District stream.
J. Anim. Ecol., *42*:435–448.

Ellis, J., and S. Kanamori.
1973. An evaluation of the Miller method for dissolved oxygen analysis.
Limnol. Oceanogr., *18*:1002–1005.

Ellis, P., and W. D. Williams.
1970. The biology of *Haloniscus searlei* Chilton, an oniscoid isopod living in Australian salt lakes.
Austral. J. Mar. Freshw. Res., *21*:51–69.

Elster, H.-J.
1954b. Über die Populationsdynamik von *Eudiaptomus gracilis* Sars und *Heterocope borealis* Fischer in Bodensee-Obersee.
Arch. Hydrobiol. Suppl., *20*:546–614.

Elster, H.-J.
1954a. Einige Gedanken zur Systematik, Terminologie und Zielsetzung der dynamischen Limnologie.
Arch. Hydrobiol. Suppl., *20*:487–523.

Elster, H.-J., and M. Štěpánek.
1967. Eine neue Modifikation der Secchischeibe.
Arch. Hydrobiol. Suppl., *33*:101–106.

Emerson, S., W. Broecker, and D. W. Schindler.
1973. Gas-exchange rates in a small lake as determined by the radon method.
J. Fish. Res. Bd. Canada, *30*:1475–1484.

von Engeln, C. D.
1961. The Finger Lakes Region: Its Origin and Nature.
Ithaca, N.Y., Cornell University Press, 156 pp.

Epploy, R. W.
1962. Major cations. *In* R. A. Lewin, ed. Physiology and Biochemistry of Algae.
New York, Academic Press, pp. 255–266.

Eppley, R. W., and F. M. MaciasR.
1963. Role of the alga *Chlamydomonas mundana* in anaerobic waste stabilization lagoons.
Limnol. Oceanogr., 8:411–416.

Eppley, R. W., and W. H. Thomas.
1969. Comparison of half-saturation constants for growth and nitrate uptake of marine phytoplankton.
J. Phycol., 5:375–379.

Epstein, E.
1965. Mineral metabolism. *In* J. Bonner and J. E. Varner, eds. Plant Biochemistry.
New York, Academic Press, pp. 438–466.

Eriksen, C. H.
1963a. Respiratory regulation in *Ephemera simulans* Walker and *Hexagenia limbata* (Serville) (Ephemeroptera).
J. Exp. Biol., 40:455–467.

Eriksen, C. H.
1963b. The relation of oxygen consumption to substrate particle size in two burrowing mayflies.
J. Exp. Biol., 40:447–453.

Eriksen, C. H.
1968. Aspects of the limno-ecology of *Corophium spinicorne* Stimpson (Amphipoda) and *Gnorimosphaeroma oregonensis* (Dana) (Isopoda).
Crustaceana, 14:1–12.

Faller, A. J.
1969. The generation of·Langmuir circulations by the eddy pressure of surface waves.
Limnol. Oceanogr., 14:504–513.

Faller, A. J.
1971. Oceanic turbulence and the Langmuir circulations.
Ann. Rev. Ecol. Syst., 2:201–236.

Faust, S. J., and J. V. Hunter, eds.
1971. Organic Compounds in Aquatic Environments.
New York, Marcel Dekker, Inc., 638 pp.

Felföldy, L. J. M.
1960. Experiments on the carbonate assimilation of some unicellular algae by Ruttner's conductiometric method.
Acta Biol. Hung., 11:67–75.

Felföldy, L. J. M.
1961. On the chlorophyll content and biological productivity of periphytic diatom communities on the stony shores of Lake Balaton.
Ann. Biol. Tihany, 28:99–104.

Ferling, E.
1957. Die Wirkungen des erhöhten hydrostatischen Druckes auf Wachstum und Differenzierung submerser Blütenpflanzen.
Planta, 49:235–270.

Fiala, K.
1971. Seasonal changes in the growth of clones of *Typha latifolia* L. in natural conditions.
Folia Geobot. Phytotax. Praha, 6:255–270.

Fiala, K.
1973. Growth and production of underground organs of *Typha angustifolia* L., *Typha latifolia* L. and *Phragmites communis* Trin.
Pol. Arch. Hydrobiol., 20:59–66.

Fiala, K., D. Dykyjová, J. Květ and J. Svoboda.
1968. Methods of assessing rhizome and root production in reed-bed stands. *In* Methods of Productivity Studies in Root Systems and Rhizosphere Organisms.
Leningrad, Publishing House Nauka, pp. 36–47.

Ficke, E. R., and J. F. Ficke.
1974. Ice on Rivers and Lakes: A Bibliographic Essay.
U.S. Geol. Surv. Water-Resources Invest. Report (in press).

Findenegg, I.
1935. Limnologische Untersuchungen im Kärntner Seengebiete. Ein Beitrag zur Kenntnis des Stoffhaushaltes in Alpenseen.
Int. Rev. ges. Hydrobiol., 32:369–423.

Findenegg, I.
1937. Holomiktische und meromiktische Seen.
Int. Rev. ges. Hydrobiol., 35:586–610.

Findenegg, I.
1943. Untersuchungen über die Ökologie und die Produktionsverhältnisse des Planktons im Kärntner Seengebiete.
Int. Rev. ges. Hydrobiol., 43:368–429.

Findenegg, I.
1965. Relationship between standing crop and primary productivity.
Mem. Ist. Ital. Idrobiol., 18(Suppl.):271–289.

Fischer, Z.
1970. Elements of energy balance in grass carp *Ctenopharyngodon idella* Val.
Pol. Arch. Hydrobiol., 17:421–434.

Fish, G. R.
1956. Chemical factors limiting growth of phytoplankton in Lake Victoria.
East African Agricult. J., 21:152–158.

Fisher, D. W., A. W. Gambell, G. E. Likens, and F. H. Bormann.
1968. Atmospheric contributions to water quality of streams in the Hubbard Brook Experimental Forest, New Hampshire.
Water Resources Res., 4:1115–1126.

Fisher, S. G., and G. E. Likens.
1973. Energy flow in Bear Brook, New Hamp-

shire: An integrative approach to stream eco-system metabolism.
Ecol. Monogr., *43*:421–439.

Fitzgerald, G. P.
1972. Bioassay analysis of nutrient availability. *In* Allen, H. E., and J. R. Kramer, eds. Nutrients in Natural Waters.
New York, John Wiley & Sons, Inc., pp. 147–169.

Flaig, W.
1964. Effects of micro-organisms in the transformation of lignin to humic substances.
Geochim. Cosmochim. Acta, *28*:1523–1535.

Flohn, H.
1973. Der Wasserhaushalt der Erde Schwankungen und Eingriffe.
Naturwissenschaften, *60*:340–348.

Fogg, G. E.
1963. The role of algae in organic production in aquatic environments.
Brit. Phycol. Bull., *2*:195–205.

Fogg, G. E.
1965. Algal Cultures and Phytoplankton Ecology.
Madison, University of Wisconsin Press, 126 pp.

Fogg, G. E.
1971a. Nitrogen fixation in lakes.
Plant Soil (Spec. Vol.), *1971*:393–401.

Fogg, G. E.
1971b. Extracellular products of algae in freshwater.
Arch. Hydrobiol. Beih. Ergebn. Limnol., 5, 25 pp.

Fogg, G. E., and J. H. Belcher.
1961. Pigments from the bottom deposits of an English lake.
New Phytol., *60*:129–142.

Fogg, G. E., and A. E. Walsby.
1971. Buoyancy regulation and the growth of planktonic blue-green algae.
Mitt. Int. Ver. Limnol., *19*:182–188.

Fogg, G. E., and D. F. Westlake.
1955. The importance of extracellular products of algae in freshwater.
Verh. Int. Ver. Limnol., *12*:219–232.

Fogg, G. E., W. D. P. Stewart, P. Fay, and A. E. Walsby.
1973. The Blue-Green Algae.
New York, Academic Press, 459 pp.

Forbes, S. A.
1887. The lake as a microcosm.
Bull. Peoria (Ill.) Sci. Assoc. 1887. Reprinted in Bull.
Ill. Nat. Hist. Surv., *15*:537–550 (1925).

Forel, F.-A.
1892. Le Léman: Monographie Limnologique. Tome I. Géographie, Hydrographie, Géologie, Climatologie, Hydrologie.
Lausanne, F. Rouge, 543 pp. (Reprinted Genève, Slatkine Reprints, 1969.)

Forel, F.-A.
1895. Le Léman: Monographie Limnologique. Tome II. Mécanique, Hydraulique, Thermique, Optique, Acoustique, Chemie.
Lausanne, F. Rouge, 651 pp. (Reprinted Genève, Slatkine Reprints, 1969.)

Forel, F.-A.
1904. Le Léman: Monographie Limnologique. Tome III. Biologie, Histoire, Navigation, Pêche.
Lausanne, F. Rouge, 715 pp. (Reprinted Genève, Slatkine Reprints, 1969.)

Francisco, D. E., R. A. Mah, and A. C. Rabin.
1973. Acridine orange-epifluorescence technique for counting bacteria in natural waters.
Trans. Amer. Microsc. Soc., *92*:416–421.

Frank, P. W.
1952. A laboratory study of intraspecific and interspecific competition in *Daphnia pulicaria* (Forbes) and *Simocephalus vetulus* O. F. Müller.
Physiol. Zool., *25*:178–204.

Frank, P. W.
1957. Coactions in laboratory populations of two species of *Daphnia.*
Ecology, *38*:510–519.

Frey, D. G.
1955a. Längsee: A history of meromixis.
Mem. Ist. Ital. Idrobiol., 8(Suppl.):141–164.

Frey, D. G.
1955b. Distributional ecology of the cisco *(Coregonus artedii)* in Indiana.
Invest. Indiana Lakes Streams, *4*:177–228.

Frey, D. G.
1964. Remains of animals in Quaternary lake and bog sediments and their interpretation.
Arch. Hydrobiol Beih. Ergebn. Limnol., 2, 114 pp.

Frey, D. G.
1969a. The rationale of paleolimnology.
Mitt. Int. Ver. Limnol., *17*:7–18.

Frey, D. G.
1969b. Evidence for eutrophication from remains of organisms in sediments. *In* Eutrophication: Causes, Consequences, Correctives.
Washington, D.C., National Academy of Sciences, pp. 594–613.

Frey, D. G.
1974. Paleolimnology.
Mitt. Int. Ver. Limnol., *20*:95–123.

Frink, C. R.
1967. Nutrient budget: Rational analysis of eutrophication in a Connecticut lake.
Environ. Sci. Technol., *1*:425–428.

Fryer, G.
1954. Contributions to our knowledge of the biology and systematics of the freshwater Copepoda.
Schweiz. Z. Hydrol., *16*:64–77.

Fryer, G.
1957a. The food of some freshwater cyclopoid copepods and its ecological significance.
J. Anim. Ecol., *26*:263–286.

Fryer, G.
1957b. The feeding mechanism of some freshwater cyclopoid copepods.
Proc. Zool. Soc. London, *129*:1–25.

Fuhs, G. W., S. D. Demmerle, E. Canelli, and M. Chen.
1972. Characterization of phosphorus-limited plankton algae (with reflections on the limiting-nutrient concept). *In* G. E. Likens, ed. Nutrients and Eutrophication: The Limiting-Nutrient Controversy.
Special Symposium, Amer. Soc. Limnol. Oceanogr., *1*:113–133.

Fuller, S. L. H.
1974. Clams and mussels (Mollusca: Bivalvia). *In* C. W. Hart, Jr., and S. L. H. Fuller, eds. Pollution Ecology of Freshwater Invertebrates.
New York, Academic Press, pp. 215–273.

Gaevskaia, N. S.
1966. Rol' vysshikh vodnykh rastenii v pitanii zhivotnykh presnykh vodoemov.
Moskva, Izdatel'stvo Nauka, 327 pp.
(Translated into English, Nat. Lending Libr. Sci. Technol., Yorkshire, England, 1969, as: The Role of Higher Aquatic Plants in the Nutrition of the Animals of Fresh-Water Basins.)

Gak, D. Z.
1959. Fiziologicheskaia aktivnost' i sistematicheskoe polozhenie mobilizuiushchikh fosfor mikroorganizmov, vydelennykh iz vodoemov pribaltiki.
Mikrobiologiya, *28*:551–556.

Gak, D. Z.
1963. Vertikal'noe raspredelenie mobilizuiushchikh fosfor bakterii v gruntakh Latviiskikh vodoemov.
Mikrobiologiya, *32*:838–842.

Galbraith, M. G., Jr.
1967. Size-selective predation on *Daphnia* by rainbow trout and yellow perch.
Trans. Amer. Fish. Soc., *96*:1–10.

Ganf, G. G.
1974a. Incident solar irradiance and underwater light penetration as factors controlling the chlorophyll *a* content of a shallow equatorial lake (Lake George, Uganda).
J. Ecol., *62*:593–609.

Ganf, G. G.
1974b. Diurnal mixing and the vertical distribution of phytoplankton in a shallow equatorial lake (Lake George, Uganda).
J. Ecol., *62*:611–629.

Garrels, R. M.
1965. Silica: Role in the buffering of natural waters.
Science, *148*:69.

Gates, D. M.
1962. Energy Exchange in the Biosphere.
New York, Harper & Row Publishers, 151 pp.

Gates, F. C.
1942. The bogs of northern Lower Michigan.
Ecol. Monogr., *12*:213–254.

Gaudet, J. J.
1968. The correlation of physiological differences and various leaf forms of an aquatic plant.
Physiol. Plant., *21*:594–601.

Geiling, W. T., and R. S. Campbell.
1972. The effect of temperature on the development rate of the major life stages of *Diaptomus pallidus* Herrick.
Limnol. Oceangr., *17*:304–307.

George, D. G., and R. W. Edwards.
1973. *Daphnia* distribution within Langmuir circulations.
Limnol. Oceanogr., *18*:798–800.

George, D. G., and R. W. Edwards.
1974. Population dynamics and production of *Daphnia hyalina* in a eutrophic reservoir.
Freshwat. Biol., *4*:445–465.

George, M. G., and C. H. Fernando.
1969. Seasonal distribution and vertical migration of planktonic rotifers in two lakes of eastern Canada.
Verh. Int. Ver. Limnol., *17*:817–829.

George, M. G., and C. H. Fernando.
1970. Diurnal migration in three species of rotifers in Sunfish Lake, Ontario.
Limnol. Oceanogr., *15*:218–223.

Gerking, S. D.
1962. Production and food utilization in a population of bluegill sunfish.
Ecol. Monogr., *32*:31–78.

Gerletti, M.
1968. Dark bottle measurements in primary productivity studies.
Mem. Ist. Ital. Idrobiol., *23*:197–208.

Gerloff, G. C.
1969. Evaluating nutrient supplies for the growth of aquatic plants in natural waters. *In* Eutrophication: Causes, Consequences, Correctives.
Washington, D. C., National Academy of Sciences, pp. 537–555.

Gerloff, G. C., G. P. Fitzgerald, and F. Skoog.
1952. The mineral nutrition of *Microcystis aeruginosa*.
Amer. J. Bot., *39*:26–32.

Gerloff, G. C., and F. Skoog.
1957. Availability of iron and manganese in southern Wisconsin lakes for the growth of *Microcystis aeruginosa*.
Ecology, *38*:551–556.

Gessner, F.
1955. Hydrobotanik. Die Physiologischen Grundlagen der Pflanzenverbreitung im Wasser. I. Energiehaushalt.
Berlin, VEB Deutscher Verlag der Wissenschaften, 517 pp.

Gessner, F.
1959. Hydrobotanik. Die Physiologischen Grundlagen der Pflanzenverbreitung im Wasser. II. Stoffhaushalt.
Berlin, VEB Deutscher Verlag der Wissenschaften, 701 pp.

Gessner, F., and A. Diehl.
1951. Die Wirkung natürlicher Ultraviolettstrahlung auf die Chlorophyllzerstörung von Planktonalgen.
Arch. Mikrobiol., *15*:439–453.

Gest, H., A. San Pietro, and L. P. Vernon, eds.
1963. Bacterial Photosynthesis.
Yellow Springs, Ohio, Antioch Press, 523 pp.

Ghiretti, F.
1966. Respiration. *In* Wilbur, K. M., and C. M. Yonge, eds. Physiology of Mollusca. Vol. 2.
New York, Academic Press, pp. 175–208.

Gibbs, R. J.
1970. Mechanisms controlling world water chemistry.
Science, *170*:1088–1090.

Gibbs, R. J.
1973. Mechanisms of trace metal transport in rivers.
Science, *180*:71–73.

Gilbert, J. J.
1967a. Control of sexuality in the rotifer *Asplanchna brightwelli* by dietary lipids of plant origin.
Proc. Nat. Acad. Sci., *57*:1218–1225.

Gilbert, J. J.
1967b. *Asplanchna* and postero-lateral spine production in *Brachionus calyciflorus*.
Arch. Hydrobiol., *64*:1–62.

Gilbert, J. J.
1968. Dietary control of sexuality in the rotifer *Asplanchna brightwelli* Gosse.
Physiol. Zool., *41*:14–43.

Gilbert, J. J.
1972. α-tocopherol in males of the rotifer *Asplanchna sieboldi:* Its metabolism and its distribution in the testis and rudimentary gut.
J. Exp. Zool., *181*:117–128.

Gilbert, J. J.
1973. Induction and ecological significance of gigantism in the rotifer *Asplanchna sieboldi*.
Science, *181*:63–66.

Gilbert, J. J., and H. L. Allen.
1973. Studies on the physiology of the green freshwater sponge *Spongilla lacustris:* Primary productivity, organic matter, and chlorophyll content.
Verh. Int. Ver. Limnol., *18*:1413–1420.

Gilbert, J. J., and C. W. Birky, Jr.
1971. Sensitivity and specificity of the *Asplanchna* response to dietary α-tocopherol.
J. Nutrition, *101*:113–126.

Gilbert, J. J., and G. A. Thompson, Jr.
1968. Alpha tocopherol control of sexuality and polymorphism in the rotifer *Asplanchna*.
Science, *159*:734–736.

Gilbert, J. J., and J. K., Waage.
1967. *Asplanchna, Asplanchna*-substance, and posterolateral spine length variation of the rotifer *Brachionus calyciflorus* in a natural environment.
Ecology, *48*:1027–1031.

Gjessing, E. T.
1964. Ferrous iron in water.
Limnol. Oceanogr., *9*:272–274.

Gjessing, E. T.
1970. Reduction of aquatic humus in streams.
Vatten, *26*:14–23.

Gjessing, E. T., and T. Gjerdahl.
1970. Influence of ultra-violet radiation on aquatic humus.
Vatten, *26*:144–145.

Glob, P. V.
1969. The Bog People. Iron-Age Man Preserved.
Ithaca, N.Y., Cornell University Press, 200 pp.

Glooschenko, W. A., J. E. Moore, and R. A. Vollenweider.
1973. Chlorophyll *a* distribution in Lake Huron and its relationship to primary production.
Proc. Conf. Great Lakes Res., Int. Assoc. Great Lakes Res., *16*:40–49.

Glooschenko, W. A., J. E. Moore, M. Muna-
wata, and R. A. Vollenweider.
1974a. Primary production in lakes Ontario
and Erie: A comparative study.
J. Fish. Res. Bd. Canada, *31*:253–263.

Glooschenko, W. A., J. E. Moore, and R. A.
Vollenweider.
1974b. Spatial and temporal distribution of
chlorophyll *a* and pheopigments in surface
waters of Lake Erie.
J. Fish Res. Bd. Canada, *31*:265–274.

Gocke, K.
1970. Untersuchungen über Abgabe und Aufnahme von Aminosäuren und Polypepti-
den durch Planktonorganismen.
Arch. Hydrobiol., 67:285–367.

Gode, P., and J. Overbeck.
1972. Untersuchungen zur heterotrophen Ni-
trifikation im See.
Z. Allg. Mikrobiol., *12*:567–574.

Godward, M. B. E.
1962. Invisible radiations. *In* R. A. Lewin, ed.
Physiology and Biochemistry of Algae.
New York, Academic Press, pp. 551–566.

Goering, J. J., and V. A. Dugdale.
1966. Estimates of the rates of denitrification
in a subarctic lake.
Limnol. Oceanogr., *11*:113–117.

Goering, J. J., and J. C. Neess.
1964. Nitrogen fixation in two Wisconsin
lakes.
Limnol. Oceanogr., 9:530–539.

Golachowska, J. B.
1971. The pathways of phosphorus in lake
water.
Pol. Arch. Hydrobiol., *18*:325–345.

Goldman, C. R.
1960. Primary productivity and limiting factors in three lakes of the Alaska Peninsula.
Ecol. Monogr., *30*:207–230.

Goldman, C. R.
1960. Molybdenum as a factor limiting primary productivity in Castle Lake, California.
Science, *132*:1016–1017.

Goldman, C. R.
1961. The contribution of alder trees (*Alnus
tenuifolia*) to the primary productivity of Castle Lake, California.
Ecology, *42*:282–288.

Goldman, C. R.
1970. Antarctic freshwater ecosystems. *In* M.
W. Holdgate, ed. Antarctic Ecology.
New York, Academic Press, pp. 609–627.

Goldman, C. R.
1972. The role of minor nutrients in limiting
the productivity of aquatic ecosystems. *In* G.

E. Likens, ed. Nutrients and Eutrophication:
The Limiting-Nutrient Controversy.
Special Symposium, Amer. Soc. Limnol.
Oceanogr., *1*:21–38.

Goldman, C. R., D. T. Mason, and B. J. B.
Wood.
1963. Light injury and inhibition in Antarctic
freshwater phytoplankton.
Limnol. Oceanogr., 8:313–322.

Goldman, C. R., and R. G. Wetzel.
1963. A study of the primary productivity of
Clear Lake, Lake County, California.
Ecology, 44:283–294.

Goldman, J. C., D. B. Porcella, E. J. Middle-
brooks, and D. F. Toerien.
1972. The effect of carbon on algal growth — its
relationship to eutrophication.
Water Res., 6:637–679.

Goldspink, C. R., and D. B. C. Scott.
1971. Vertical migration of *Chaoborus flavicans* in a Scottish loch.
Freshwat. Biol., *1*:411–421.

Goldsworthy, A.
1970. Photorespiration.
Bot. Rev., 36:321–340.

Golterman, H. L.
1960. Studies on the cycle of elements in
fresh water.
Acta Bot. Neerlandica, 9:1–58.

Golterman, H. L., ed.
1969. Methods for Chemical Analysis of Fresh
Waters. Int. Biol. Program Handbook 8.
Oxford, Blackwell Scientific Publications, 172
pp.

Golterman, H. L.
1971. The determination of mineralization
losses in correlation with the estimation of net
primary production with the oxygen method
and chemical inhibitors.
Freshwat. Biol., *1*:249–256.

Golterman, H. L., C. C. Bakels, and J. Jakobs-
Mögelin.
1969. Availability of mud phosphates for the
growth of algae.
Verh. Int. Ver. Limnol., *17*:467–479.

Golubić, S.
1963. Hydrostatischer Druck, Licht und sub-
merse Vegetation im Vrana-See.
Int. Rev. ges. Hydrobiol., *48*:1–7.

Golubić, S.
1973. The relationship between blue-green
algae and carbonate deposits. *In* N. G. Carr
and B. A. Whitton, eds. The Biology of Blue-
Green Algae.
Berkeley, University of California Press, pp.
434–472.

Goodwin, T. W.
1974. Carotenoids and biliproteins. *In* W. D. P. Stewart, ed. Algal Physiology and Biochemistry.
Berkeley, University of California Press, pp. 176–205.

Gophen, M., B. Z. Cavari, and T. Berman.
1974. Zooplankton feeding on differentially labelled algae and bacteria.
Nature, *247*:393–394.

Gorbunov, K. V.
1953. Raspad ostatkov vysshikh vodnykh rastenii i ego ekologicheskaia rol' v vodoemakh nizhnei zony Del 'ty Volgi.
Trudy Vses. Gidrobiol. Obshshestva, 5:158–202.

Gorham, E.
1955. On some factors affecting the chemical composition of Swedish fresh waters.
Geochim. Cosmochim. Acta, 7:129–150.

Gorham, E.
1956. On the chemical composition of some waters from the Moor House Nature Reserve.
J. Ecol., *44*:377–382.

Gorham, E.
1957. The development of peat lands.
Quart. Rev. Biol., *32*:145–166.

Gorham, E.
1958. Observations on the formation and breakdown of the oxidized microzone at the mud surface in lakes.
Limnol. Oceanogr., 3:291–298.

Gorham, E.
1961. Factors influencing supply of major ions to inland waters, with special reference to the atmosphere.
Bull. Geol. Soc. Amer., *72*:795–840.

Gorham, E.
1964. Morphometric control of annual heat budgets in temperate lakes.
Limnol. Oceanogr., 9:525–529.

Gorham, E., and J. Sanger.
1964. Chlorophyll derivatives in woodland, swamp, and pond soils of Cedar Creek Natural History Area, Minnesota, U.S.A. *In* Recent Researches in the Fields of Hydrosphere, Atmosphere and Nuclear Geochemistry.
Tokyo, Mauzen Co., Ltd., pp. 1–12.

Gorham, E., and J. E. Sanger.
1972. Fossil pigments in the surface sediments of a meromictic lake.
Limnol. Oceanogr., *17*:618–622.

Goulden, C. E.
1969a. Interpretative studies of cladoceran microfossils in lake sediments.
Mitt. Int. Ver. Limnol., *17*:43–55.

Goulden, C. E.
1969b. Temporal changes in diversity. *In* G. M. Woodwell and H. H. Smith, eds. Diversity and Stability in Ecological Systems.
Brookhaven Symposia in Biology, *22*:96–102.

Goulder, R.
1969. Interactions between the rates of production of a freshwater macrophyte and phytoplankton in a pond.
Oikos, *20*:300–309.

Goulder, R.
1970. Day-time variations in the rates of production by two natural communities of submerged freshwater macrophytes.
J. Ecol., *58*:521–528.

Goulder, R.
1972. Grazing by the ciliated protozoon *Loxodes magnus* on the alga *Scenedesmus* in a eutrophic pond.
Oikos, *23*:109–115.

Govindjee, and B. Z. Braun.
1974. Light absorption, emission and photosynthesis. *In* W. D. P. Stewart, ed. Algal Physiology and Biochemistry.
Berkeley, University of California Press, pp. 346–390.

Green, E. J., and D. E. Carritt.
1967. New tables for oxygen saturation of seawater.
J. Mar. Res., *25*:140–147.

Green, J., and O. B. Lan.
1974. *Asplanchna* and the spines of *Brachionus calyciflorus* in two Javanese sewage ponds.
Freshwat. Biol., *4*:223–226.

Greenbank, J.
1945. Limnological conditions in ice-covered lakes, especially as related to winter-kill of fish.
Ecol. Monogr., *15*:343–392.

Grenney, W. J., D. A. Bella, and H. C. Curl, Jr.
1973. A theoretical approach to interspecific competition in phytoplankton communities.
Amer. Nat., *107*:405–425.

Grey, D. M., ed.
1970. Handbook on the Principles of Hydrology.
Ottawa, National Research Council of Canada, 676 pp.

Griffiths, M., P. S. Perrott, and W. T. Edmondson.
1969. Oscillaxanthin in the sediment of Lake Washington.
Limnol. Oceanogr., *14*:317–326.

Grim, J.
1952. Vermehrungsleistungen planktischer Algenpopulationen in Gleichgewichtsperioden.
Arch. Hydrobiol. Suppl., 20:238–260.

Groth, P.
1971. Untersuchungen über einige Spurenelemente in Seen.
Arch. Hydrobiol., 68:305–375.

Gruendling, G. K.
1971. Ecology of the epipelic algal communities in Marion Lake, British Columbia.
J. Phycol., 7:239–249.

Gunnison, D., and M. Alexander.
1975. Resistance and susceptibility of algae to decomposition by natural microbial communities.
Limnol. Oceanogr., 20:64–70.

Gusev, M. V., and K. A. Nikitina.
1974. A study of the death of blue-green algae under conditions of darkness. (Trans. Consultants Bureau, Plenum Publ. Corp.)
Mikrobiologiya, 43:333–337.

Guseva, K. A., and S. P. Goncharova.
1965. O vliianii vysshei vodnoi rastitel 'nosti na razvitie planktonnykh sinezelenykh vodoroslei. In Ekologiia i Fiziologiia Sinezelenykh Vodoroslei.
Leningrad, pp. 230–234.

Haan, H. de
1972. Some structural and ecological studies on soluble humic compounds from Tjeukemeer.
Verh. Int. Ver. Limnol., 18:685–695.

Haan, H. de
1974. Effect of a fulvic acid fraction on the growth of a Pseudomonas from Tjeukemeer (The Netherlands).
Freshwat. Biol., 4:301–309.

Hadl, G.
1972. Zur Ökologie und Biologie der Pisidien (Bivalvia: Sphaeriidae) im Lunzer Untersee.
Sitzungsber. Österr. Akad. Wissensch. Math.-nat. Kl., Abt. I, 180:317–338.

Hagedorn, H.
1971. Experimentelle Untersuchungen über den Einfluss des Thiamins auf die natürliche Algenpopulation des Pelagials.
Arch. Hydrobiol., 68:382–399.

Haines, D. A., and R. A. Bryson.
1961. An empirical study of wind factor in Lake Mendota.
Limnol. Oceanogr., 6:356–364.

Halbach, U., and G. Halbach-Keup.
1974. Quantitative Beziehungen zwischen Phytoplankton und der Populationsdynamik des Rotators Brachionus calyciflorus Pallas. Befunde aus Laboratoriumsexperimenten und Freilanduntersuchungen.
Arch. Hydrobiol., 73:273–309.

Hall, D. J.
1964. An experimental approach to the dynamics of a natural population of Daphnia galeata mendotae.
Ecology, 45:94–112.

Hall, D. J., W. E. Cooper, and E. E. Werner.
1970. An experimental approach to the production dynamics and structure of freshwater animal communities.
Limnol. Oceanogr., 15:839–928.

Hall, J. B.
1971. Evolution of the prokaryotes.
J. Theor. Biol., 30:429–454.

Hall, K. J., P. M. Kleiber, and I. Yesaki.
1972. Heterotrophic uptake of organic solutes by microorganisms in the sediment.
Mem. Ist. Ital. Idrobiol., 29(Suppl.):441–471.

Hammer, U. T., and W. W. Sawchyn.
1968. Seasonal succession and congeneric associations of Diaptomus spp. (Copepoda) in some Saskatchewan ponds.
Limnol. Oceanogr., 13:476–484.

Haney, J. F.
1971. An in situ method for the measurement of zooplankton grazing rates.
Limnol. Oceanogr., 16:970–977.

Haney, J. F.
1973. An in situ examination of the grazing activities of natural zooplankton communities.
Arch. Hydrobiol., 72:87–132.

Hansen, K.
1959a. Sediments from Danish lakes.
J. Sediment. Petrol., 29:38–46.

Hansen, K.
1959b. The terms gyttja and dy.
Hydrobiologia, 13:309–315.

Hansen, K.
1961. Lake types and lake sediments.
Verh. Int. Ver. Limnol., 14:285–290.

Hanušová, J.
1962. Ein Beitrag zum Studium des Schwefelkreislaufes während der Sommerstagnation in der Talsperre Sedlice.
Sci. Pap. Inst. Chem. Technol. Prague Technol., Water, 6:177–191.

Hardman, Y.
1941. The surface tension of Wisconsin lake waters.
Trans. Wis. Acad. Sci. Arts. Lett., 33:395–404.

Hardman, Y., and A. T. Henrici.
1939. Studies of freshwater bacteria. V. The distribution of *Siderocapsa treubii* in some lakes and streams.
J. Bact., *37*:97–104.

Hardy, J. T.
1973. Phytoneuston ecology of a temperate marine lagoon.
Limnol. Oceanogr., *18*:525–533.

Hardy, R. W. F., R. C. Burns, and R. D. Holsten.
1973. Applications of the acetylene-ethylene assay for measurement of nitrogen fixation.
Soil Biol. Biochem., *5*:47–81.

Hargrave, B. T.
1969. Epibenthic algal production and community respiration in the sediments of Marion Lake.
J. Fish Res. Bd. Canada, *26*:2003–2026.

Hargrave, B. T.
1970a. The utilization of benthic microflora by *Hyalella azteca* (Amphipoda).
J. Anim. Ecol., *39*:427–437.

Hargrave, B. T.
1970b. Distribution, growth, and seasonal abundance of *Hyalella azteca* (Amphipoda) in relation to sediment microflora.
J. Fish. Res. Bd. Canada, *27*:685–699.

Hargrave, B. T.
1970c. The effect of a deposit-feeding amphipod on the metabolism of benthic microflora.
Limnol. Oceanogr., *15*:21–30.

Hargrave, B. T.
1971. An energy budget for a deposit-feeding amphipod.
Limnol. Oceanogr., *16*:99–103.

Hargrave, B. T.
1972a. A comparison of sediment oxygen uptake, hypolimnetic oxygen deficit and primary production in Lake Esrom, Denmark.
Verh. Int. Ver. Limnol., *18*:134–139.

Hargrave, B. T.
1972b. Oxidation-reduction potentials, oxygen concentration and oxygen uptake of profundal sediments in a eutrophic lake.
Oikos, *23*:167–177.

Hargrave, B. T.
1972c. Aerobic decomposition of sediment and detritus as a function of particle surface area and organic content.
Limnol. Oceanogr., *17*:583–596.

Hargrave, B. T., and G. H. Geen.
1968. Phosphorus excretion by zooplankton.
Limnol. Oceanogr., *13*:332–342.

Harlin, M. M.
1973. Transfer of products between epiphytic marine algae and host plants.
J. Phycol., *9*:243–248.

Harman, W. N.
1974. Snails (Mollusca: Gastropoda). *In* C. W. Hart, Jr., and S. L. H. Fuller, eds. Pollution Ecology of Freshwater Invertebrates.
New York, Academic Press, pp. 275–312.

Harrison, F. W.
1974. Sponges (Porifera: Spongillidae). *In* C. W. Hart, Jr., and S. L. H. Fuller, eds. Pollution Ecology of Freshwater Invertebrates.
New York, Academic Press, pp. 29–66.

Harrison, M. J., R. T. Wright, and R. Y. Morita.
1971. Method for measuring mineralization in lake sediments.
Appl. Microbiol., *21*:698–702.

Harriss, R. C.
1967. Silica and chloride in interstitial waters of river and lake sediments.
Limnol. Oceanogr., *12*:8–12.

Harter, R. D.
1968. Adsorption of phosphorus by lake sediment.
Proc. Soil Sci. Soc. Amer., *32*:514–518.

Hartman, R. T., and D. L. Brown.
1967. Changes in internal atmosphere of submersed vascular hydrophytes in relation to photosynthesis.
Ecology, *48*:252–258.

den Hartog, C., and S. Segal.
1964. A new classification of the water-plant communities.
Acta Bot. Nerl., *13*:367–393.

Haslam, S. M.
1971a. Community regulation in *Phragmites communis* Trin. I. Monodominant stands.
J. Ecol., *59*:65–73.

Haslam, S. M.
1971b. Community regulation in *Phragmites communis* Trin. II. Mixed stands.
J. Ecol., *59*:75–88.

Haslam, S. M.
1973. Some aspects of the life history and autecology of *Phragmites communis* Trin. A review.
Pol. Arch. Hydrobiol., *20*:79–100.

Hasler, A. D.
1947. Eutrophication of lakes by domestic drainage.
Ecology, *28*:383–395.

Hasler, A. D., and W. G. Einsele.
1948. Fertilization for increasing productivity of natural inland waters.
Trans. N. Amer. Wildl. Conf., *13*:527–555.

Hasler, A. D., and E. Jones.
1949. Demonstration of the antagonistic action of large aquatic plants on algae and rotifers.
Ecology, *30*:359–364.

Hatch, M. D., C. B. Osmond, and R. O. Slatyer, eds.
1971. Photosynthesis and Photorespiration.
New York, John Wiley & Sons, Inc., 565 pp.

Haworth, R. D.
1971. The chemical nature of humic acid.
Soil Sci., *111*:71–79.

Hayes, F. R.
1955. The effect of bacteria on the exchange of radiophosphorus at the mud-water interface.
Verh. Int. Ver. Limnol., *12*:111–116.

Hayes, F. R.
1957. On the variation in bottom fauna and fish yield in relation to trophic level and lake dimensions.
J. Fish. Res. Bd. Canada, *14*:1–32.

Hayes, F. R.
1964. The mud-water interface.
Oceanogr. Mar. Biol. Ann. Rev., 2:121–145.

Hayes, F. R., and E. H. Anthony.
1959. Lake water and sediment. VI. The standing crop of bacteria in lake sediments and its place in the classification of lakes.
Limnol. Oceanogr., *4*:299–315.

Hayes. F. R., and A. H. Anthony.
1964. Productive capacity of North American lakes as related to the quantity and the trophic level of fish, the lake dimensions, and the water chemistry.
Trans. Amer. Fish. Soc., *93*:53–57.

Hayes, F. R., and C. C. Coffin.
1951. Radioactive phosphorus and the exchange of lake nutrients.
Endeavor, *10*:78–81.

Hayes, F. R., and M. A. MacAulay.
1959. Lake water and sediment. V. Oxygen consumed in water over sediment cores.
Limnol. Oceanogr., *4*:291–298.

Hayes, F. R., J. A. McCarter, M. L. Cameron, and D. A. Livingstone.
1952. On the kinetics of phosphorus exchange in lakes.
J. Ecol., *40*:202–216.

Hayes, F. R., and J. E. Phillips.
1958. Lake water and sediment. IV. Radiophosphorus equilibrium with mud, plants, and bacteria under oxidized and reduced conditions.
Limnol. Oceanogr., *3*:459–475.

Hayes, F. R., B. L. Reid, and M. L. Cameron.
1958. Lake water and sediment. II. Oxidation-reduction relations at the mud-water interface.
Limnol. Oceanogr., *3*:308–317.

Hayne, D. W., and R. C. Ball.
1956. Benthic productivity as influenced by fish predation.
Limnol. Oceanogr., *1*:162–175.

Hecky, R. E., and P. Kilham.
1973. Diatoms in alkaline, saline lakes: Ecology and geochemical implications.
Limnol. Oceanogr., *18*:53–71.

Hegel, G. W. F.
1807. Phänomenologie des Geistes. System der Wissenschaft.
Bamberg u. Würzburg, J. A. Goebhardt, p. 21.

Hejný, S.
1960. Ökologische Charakteristik der Wasser- und Sumpfpflanzen in den slowakischen Tiefebenen (Donau- und Theissgebiet).
Bratislava, Verlag Slowakischen Akad. Wissenschaften, 487 pp.

Hellebust, J. A.
1974. Extracellular products. *In* W. D. P. Stewart, ed. Algal Physiology and Biochemistry.
Berkeley, University of California Press, pp. 838–863.

Hem, J. D.
1960a. Restraints on dissolved ferrous iron imposed by bicarbonate, redox potential, and pH. *In* Chemistry of Iron in Natural Water.
U. S. Geol. Surv. Water-Supply Pap., *1459-B*:33–55.

Hem, J. D.
1960b. Complexes of ferrous iron with tannic acid. *In* Chemistry of Iron in Natural Water.
U. S. Geol. Surv. Water-Supply Pap., *1459-D*:75–94.

Hem, J. D.
1960c. Some chemical relationships among sulfur species and dissolved ferrous iron. *In* Chemistry of Iron in Natural Water.
U. S. Geol. Surv. Water-Supply Pap., *1459-C*:57–73.

Hem, J. D.
1963. Chemical equilibria and rates of manganese oxidation. *In* Chemistry of Manganese in Natural Water.
U. S. Geol. Surv. Water-Supply Pap., *1667-A*, 64 pp.

Hem, J. D.
1964. Deposition and solution of manganese oxides. *In* Chemistry of Manganese in Natural Water.
U. S. Geol. Surv. Water-Supply Pap., *1667-B*, 42 pp.

Hem, J. D., and W. H. Cropper.
1959. Survey of ferrous-ferric chemical equilibria and redox potentials. *In* Chemistry of Iron in Natural Water.
U. S. Geol. Surv. Water-Supply Pap., *1459-A*:1–30.

Hem, J. D., and M. W. Skougstad.
1960. Coprecipitation effects in solutions con-

taining ferrous, ferric, and cupric ions. *In* Chemistry of Iron in Natural Water.
U. S. Geol. Surv. Water-Supply Pap., *1459-E*:95–110.

Henrici, A. T., and E. McCoy.
1938. The distribution of heterotrophic bacteria in the bottom deposits of some lakes.
Trans. Wis. Acad. Sci. Arts Lett., *31*:323–361.

Hepher, B.
1966. Some aspects of the phosphorus cycle in fish ponds.
Verh. Int. Ver. Limnol., *16*:1293–1297.

Herman, S. S.
1963. Vertical migration of the opossum shrimp, *Neomysis americana* Smith.
Limnol. Oceanogr., *8*:228–238.

Heron, J.
1961. The seasonal variation of phosphate, silicate, and nitrate in waters of the English Lake District.
Limnol. Oceanogr., *6*:338–346.

Hesslein, R., and P. Quay.
1973. Vertical eddy diffusion studies in the thermocline of a small stratified lake.
J. Fish. Res. Bd. Canada, *30*:1491–1500.

Heywood, J., and R. W. Edwards.
1962. Some aspects of the ecology of *Potamopyrgus jenkinsi* Smith.
J. Anim. Ecol., *31*:239–250.

Hickman, M.
1971a. Standing crops and primary productivity of the epipelon of two small ponds in North Somerset, U. K.
Oecologia, *6*:238–253.

Hickman, M.
1971b. The standing crop and primary productivity of the epiphyton attached to *Equisetum fluviatile* L. in Priddy Pool, North Somerset.
Brit. Phycol. J., *6*:51–59.

Hickman, M., and F. E. Round.
1970. Primary production and standing crops of epipsammic and epipelic algae.
Brit. Phycol. J., *5*:247–255.

Hillbricht-Ilkowska, A., I. Spodniewska, T. Weglenska, and A. Karabin.
1970. The seasonal variation of some ecological efficiencies and production rates in the plankton community of several Polish lakes of different trophy. *In* Z. Kajak and A. Hillbricht-Ilkowska, eds. Productivity Problems of Freshwaters.
Warsaw, PWN Polish Scientific Publishers, pp. 111–127.

Hillman, W. S.
1961. The Lemnaceae, or duckweeds.
Bot. Rev., *27*:221–287.

Hobbie, J. E.
1961. Summer temperatures in Lake Schrader, Alaska.
Limnol. Oceanogr., *6*:326–329.

Hobbie, J. E.
1964. Carbon-14 measurements of primary production in two arctic Alaskan lakes.
Verh. Int. Ver. Limnol., *15*:360–364.

Hobbie, J. E.
1967. Glucose and acetate in freshwater: Concentrations and turnover rates. *In* H. L. Golterman and R. S. Clymo, eds. Chemical Environment in the Aquatic Habitat.
Amsterdam, N. V. Noord-Hollandsche Uitgevers Mattschappij, pp. 245–251.

Hobbie, J. E.
1971. Heterotrophic bacteria in aquatic ecosystems; Some results of studies with organic radioisotopes. *In* J. Cairns, Jr., ed. The Structure and Function of Fresh-Water Microbial Communities.
Res. Div. Monogr. 3, Virginia Polytechnic Inst., pp. 181–194.

Hobbie, J. E.
1973. Arctic limnology: A review. *In* M. E. Britton, ed. Alaskan Arctic Tundra.
Tech. Paper No. 25, Arctic Inst. of North America, pp. 127–168.

Hobbie, J. E., and C. C. Crawford.
1969. Bacterial uptake of organic substrate: New methods of study and application to eutrophication.
Verh. Int. Ver. Limnol., *17*:725–730.

Hobbie, J. E., C. C. Crawford, and K. L. Webb.
1968. Amino acid flux in an estuary.
Science, *159*:1463–1464.

Hobbie, J. E., and R. T. Wright.
1965. Competition between planktonic bacteria and algae for organic solutes.
Mem. Ist. Ital. Idrobiol., *18*(Suppl.):175–185.

Hobbie, J. E., R. J. Barsdate, V. Alexander, D. W. Stanley, C. P. McRoy, R. G. Stross, D. A. Bierle, R. D. Dillon, and M. C. Miller.
1972. Carbon Flux Through a Tundra Pond Ecosystem at Barrow, Alaska.
U. S. Tundra Biome Report *72-1*, 28 pp.

Hobbs, H. H., Jr., and E. T. Hall, Jr.
1974. Crayfishes (Decapoda: Astacidae). *In* C. W. Hart, Jr., and S. L. H. Fuller, eds. Pollution Ecology of Freshwater Invertebrates.
New York, Academic Press, pp. 195–214.

Hogetsu, K., Y. Okanishi, and H. Sugawara.
1960. Studies on the antagonistic relationship between phytoplankton and rooted aquatic plants.
Japan. J. Limnol., *21*:124–130.

Holden, A. V.
1961. The removal of dissolved phosphate from lake waters by bottom deposits.
Verh. Int. Ver. Limnol., *14*:247–251.

Höll, K.
1972. Water: Examination, Assessment, Conditioning, Chemistry, Bacteriology, Biology.
Berlin, Walder de Gruyter, 389 pp.

Holme, N. A., and A. D. McIntyre, eds.
1971. Methods for the Study of Marine Benthos. Int. Biol. Program Handbook 16.
Oxford, Blackwell Scientific Publications, 334 pp.

Holm-Hansen, O., and H. W. Paerl.
1972. The applicability of ATP determination for estimation of microbial biomas and metabolic activity.
Mem. Ist. Ital. Idrobiol., *29*(Suppl.):149–168.

Hood, D. W., ed.
1970. Organic matter in natural waters.
Occas. Publ. Inst. Mar. Sci. Univ. Alaska *1*, 625 pp.

Hooper, F. F., and A. M. Elliott.
1953. Release of inorganic phosphorus from extracts of lake mud by protozoa.
Trans. Amer. Microsc. Soc., *72*:276–281.

Horie, S.
1962. Morphometric features and the classification of all the lakes in Japan.
Mem. College Sci. Univ. Kyoto (Ser. B), *29*:191–262.

Horne, A. J., and G. E. Fogg.
1970. Nitrogen fixation in some English lakes.
Proc. Roy. Soc. London (Ser. B), *175*:351–366.

Horne, A. J., and C. R. Goldman.
1972. Nitrogen fixation in Clear Lake, California. I. Seasonal variation and the role of heterocysts.
Limnol. Oceanogr., *17*:678–692.

Horne, A. J., J. E. Dillard, D. K. Fujita, and C. R. Goldman.
1972. Nitrogen fixation in Clear Lake, California. II. Synoptic studies on the autumn *Anabaena* bloom.
Limnol. Oceanogr., *17*:693–703.

Horne, R. A., ed.
1972. Water and Aqueous Solutions. Structure, Thermodynamics, and Transport Processes.
New York, Wiley-Interscience, 837 pp.

Horvath, R. S.
1972. Microbial co-metabolism and the degradation of organic compounds in nature.
Bacteriol. Rev., *36*:146–155.

Hough, J. L.
1958. Geology of the Great Lakes.
Urbana, University of Illinois Press, 313 pp.

Hough, R. A.
1974. Photorespiration and productivity in submersed aquatic vascular plants.
Limnol. Oceanogr., *19*:912–927.

Hough, R. A., and R. G. Wetzel.
1972. A ^{14}C-assay for photorespiration in aquatic plants.
Plant Physiol., *49*:987–990.

Hough, R. A., and R. G. Wetzel.
1975. The release of dissolved organic carbon from submersed aquatic macrophytes: Diel, seasonal, and community relationships.
Verh. Int. Ver. Limnol., *19*: (in press).

Hoyt, P. B.
1966. Chlorophyll-type compounds in soil. II. Their decomposition.
Plant Soil, *25*:313–328.

Hrbáček, J.
1958. Typologie und Produktivität der teichartigen Gewässer.
Verh. Int. Ver. Limnol., *13*:394–399.

Hrbáček, J.
1962. Species composition and the amount of the zooplankton in relation to the fish stock.
Rozpravy Českosl. Akad. Věd, Řada Matem. Prír. Věd, *72*(10):1–114.

Hrbáček, J., M. Dvořákova, V. Kořínek, and L. Procházkóva.
1961. Demonstration of the effect of the fish stock on the species composition of zooplankton and the intensity of metabolism of the whole plankton association.
Verh. Int. Ver. Limnol., *14*:192–195.

Hrbáček, J., and M. Dvořáková-Novotná.
1965. Plankton of four backwaters related to their size and fish stock.
Rozpravy Českosl. Akad. Věd, Řada Matem. Přír. Věd, *75*(13):1–65.

Hrbáček, J., and M. Straškraba.
1966. Horizontal and vertical distribution of temperature, oxygen, pH and water movements in Slapy Reservoir (1958–1960).
Hydrobiol. Stud., *1*:7–40.

Hubley, J. H., J. R. Mitton, and J. F. Wilkinson.
1974. The oxidation of carbon monoxide by methane-oxidizing bacteria.
Arch. Microbiol., *95*:365–368.

Hudec, P. P., and P. Sonnenfeld.
1974. Hot brines on Los Roques, Venezuela.
Science, *185*:440–442.

Hunding, C.
1971. Production of benthic microalgae in the littoral zone of a eutrophic lake.
Oikos, *22*:389–397.

Hunding, C., and B. T. Hargrave.
1973. A comparison of benthic microalgal pro-

duction measured by C^{14} and oxygen methods.
J. Fish. Res. Bd. Canada, 30:309–312.

Hutchinson, G. E.
1937. A contribution to the limnology of arid regions primarily founded on observations made in the Lahontan Basin.
Trans. Conn. Acad. Arts Sci., 33:47–132.

Hutchinson, G. E.
1938. Chemical stratification and lake morphology.
Proc. Nat. Acad. Sci., 24:63–69.

Hutchinson, G. E.
1941. Limnological studies in Connecticut. IV. The mechanisms of intermediary metabolism in stratified lakes.
Ecol. Monogr., 11:21–60.

Hutchinson, G. E.
1944. Nitrogen in the biogeochemistry of the atmosphere.
Amer. Scientist, 32:178–195.

Hutchinson, G. E.
1950. The biogeochemistry of vertebrate excretion.
Bull. Amer. Mus. Nat. Hist., 96, 554 pp.

Hutchinson, G. E.
1957. A Treatise on Limnology. I. Geography, Physics, and Chemistry.
New York, John Wiley & Sons, Inc., 1015 pp.

Hutchinson, G. E.
1961. The paradox of the plankton.
Amer. Nat., 95:137–146.

Hutchinson, G. E.
1964. The lacustrine microcosm reconsidered.
Amer. Sci., 52:334–341.

Hutchinson, G. E.
1967. A Treatise on Limnology. II. Introduction to Lake Biology and the Limnoplankton.
New York, John Wiley & Sons, Inc., 1115 pp.

Hutchinson, G. E.
1973. Eutrophication. The scientific background of a contemporary practical problem.
Amer. Sci., 61:269–279.

Hutchinson, G. E.
1974. De rebus planktonicis.
Limnol. Oceanogr., 19:360–361.

Hutchison, G. E.
1975. A Treatise on Limnology. III. Aquatic Macrophytes and Attached Algae.
New York, John Wiley & Sons, Inc., (in press).

Hutchinson, G. E., and V. T. Bowen.
1947. A direct demonstration of the phosphorus cycle in a small lake.
Proc. Nat. Acad. Sci., 33:148–153.

Hutchinson, G. E., and V. T. Bowen.
1950. Limnological studies in Connecticut.

IX. A quantitative radiochemical study of the phosphorus cycle in Linsley Pond.
Ecology, 31:194–203.

Hutchinson, G. E., and U. M. Cowgill.
1973. The waters of Merom: A study of Lake Huleh. III. The major chemical constituents of a 54 m. core.
Arch. Hydrobiol., 72:145–185.

Hutchinson, G. E., E. S. Deevey, Jr., and A. Wollack.
1939. The oxidation-reduction potentials of lake waters and their ecological significance.
Proc. Nat. Acad. Sci., 25:87–90.

Hutchinson, G. E., and H. Löffler.
1956. The thermal classification of lakes.
Proc. Nat. Acad. Sci., 42:84–86.

Hutchinson, G. E., et al.
1970. Ianula: An account of the history and development of the Lago di Monterosi, Latium, Italy.
Trans. Amer. Phil. Soc., N. S., 60(4), 178 pp.

Hutner, S. H., and L. Provasoli.
1964. Nutrition of algae.
Ann. Rev. Plant Physiol., 15:37–56.

Hyman, L. H.
1951. The Invertebrates: Acanthocephala, Aschelminthes, and Entoprocta. The Pseudocoelomate Bilateria. Vol. III.
New York, McGraw-Hill Book Co., 572 pp.

Hynes, H. B. N.
1960. The Biology of Polluted Waters.
Liverpool, Liverpool University Press, 202 pp.

Hynes, H. B. N.
1963. Imported organic matter and secondary productivity in streams.
Proc. 16th Int. Congr. Zool., 4:324–329.

Hynes, H. B. N.
1970. The Ecology of Running Waters.
Toronto, University of Toronto Press, 555 pp.

Hynes, H. B. N., and B. J. Greib.
1970. Movement of phosphate and other ions from and through lake muds.
J. Fish. Res. Bd. Canada, 27:653–668.

Infante, A.
1973. Untersuchungen über die Ausnutzbarkeit verschiedener Algen durch das Zooplankton.
Arch. Hydrobiol. Suppl., 42:340–405.

Ingram, L. O., J. A. Calder, C. Van Baalen, F. E. Plucker, and P. L. Parker.
1973a. Role of reduced exogenous organic compounds in the physiology of the blue-

green bacteria (algae): Photoheterotrophic growth of a "heterotrophic" blue-green bacterium.
J. Bacteriol., *114*:695–700.

Ingram, L. O., C. Van Baalen, and J. A. Calder.
1973b. Role of reduced exogenous organic compounds in the physiology of the blue-green bacteria (algae): Photoheterotrophic growth of an "autotrophic" blue-green bacterium.
J. Bacteriol., *114*:701–705.

International Association of Limnology.
1959. Symposium on the classification of brackish waters.
Arch. Oceanogr. Limnol., *11*(Suppl.):1–248.

Iversen, H. W.
1952. Laboratory study of breakers. *In* Gravity Waves.
Circ. U. S. Bur. Standards, *521*:9–32.

Iversen, T. M.
1973. Decomposition of autumn-shed beech leaves in a springbrook and its significance for the fauna.
Arch. Hydrobiol., *72*:305–312.

Ivlev, V. S.
1945. The biological productivity of waters.
Uspekhi Sovremennoi Biol., *19*:98–120. (Translat. in J. Fish. Res. Bd. Canada, 23:1727–1759.)

Iyengar, V. K. S., D. M. Davies, and H. Kleerekoper.
1963. Some relationships between Chironomidae and their substrate in nine freshwater lakes of southern Ontario, Canada.
Arch. Hydrobiol., *59*:289–310.

Jackson, W. A., and R. J. Volk.
1970. Photorespiration.
Ann. Rev. Plant Physiol., *21*:385–432.

Jacobs, J.
1962. Light and turbulence as co-determinants of relative growth rates in cyclomorphic *Daphnia*.
Int. Rev. ges. Hydrobiol., *47*:146–156.

Jacobs, J.
1966. Predation and rate of evolution in cyclomorphic *Daphnia*.
Verh. Int. Ver. Limnol., *16*:1645–1652.

James, H. R., with E. A. Birge.
1938. A laboratory study of the absorption of light by lake waters.
Trans. Wis. Acad. Sci. Arts Lett., *31*:1–154.

Jannasch, H. W.
1969. Current concepts in aquatic microbiology.
Verh. Int. Ver. Limnol., *17*:25–39.

Jannasch, H. W.
1974. Steady state and the chemostat in ecology.
Limnol. Oceanogr., *19*:716–720.

Jannasch, H. W., and G. E. Jones.
1959. Bacterial populations in sea water as determined by different methods of enumeration.
Limnol. Oceanogr., *4*:128–139.

Järnefelt, H.
1925. Zur Limnologie einiger Gewässer Finlands.
Ann. Soc. Zool.-Bot. Fennicae Vanamo, 2:185–352.

Jarvis, N. L.
1967. Adsorption of surface-active material at the sea-air interface.
Limnol. Oceanogr., *12*:213–221.

Jarvis, N. L., W. D. Garrett, M. A. Scheiman, and C. O. Timmons.
1967. Surface chemical characterization of surface-active material in seawater.
Limnol. Oceanogr., *12*:88–96.

Jassby, A. D., and C. R. Goldman.
1974. A quantitative measure of succession rate and its application to the phytoplankton of lakes.
Amer. Nat., *108*:688–693.

Jensen, M. L., and N. Nakai.
1961. Sources and isotopic composition of atmospheric sulfur.
Science, *134*:2102–2104.

Jeschke, W. D., and W. Simonis.
1965. Über die Aufnahme von Phosphat- und Sulfationen durch Blätter von *Elodea densa* und ihre Beeinflussung durch Licht, Temperatur und Aussenkonzentration.
Planta, *67*:6–32.

Jewell, M. E.
1935. An ecological study of the fresh-water sponges of northern Wisconsin.
Ecol. Monogr., *5*:461–504.

Jewell, M. E.
1939. An ecological study of the fresh-water sponges of Wisconsin. II. The influence of calcium.
Ecology, *20*:11–28.

Jewell, W. J., and P. L. McCarty.
1971. Aerobic decomposition of algae.
Environ. Sci. Technol., *5*:1023–1031.

Johannes, R. E.
1964a. Uptake and release of phosphorus by a benthic marine amphipod.
Limnol. Oceanogr., *9*:235–242.

Johannes, R. E.
1964b. Uptake and release of dissolved

organic phosphorus by representatives of a coastal marine ecosystem.
Limnol. Oceanogr., 9:224–234.

Johannes, R. E.
1964c. Phosphorus excretion and body size in marine animals: Microzooplankton and nutrient regeneration.
Science, 146:923–924.

Johnson, L.
1964. Temperature regime of deep lakes.
Science, 144:1336–1337.

Johnson, L.
1966. Temperature of maximum density of fresh water and its effect on circulation in Great Bear Lake.
J. Fish Res. Bd. Canada, 23:963–973.

Johnson, M. G., and R. O. Brinkhurst.
1971. Production of benthic macroinvertebrates of Bay of Quinte and Lake Ontario.
J. Fish. Res. Bd. Canada, 28:1699–1714.

Johnson, M. P.
1967. Temperature dependent leaf morphogenesis in *Ranunculus flabellaris*.
Nature, 214:1354–1355.

Johnson, N. M., R. C. Reynolds, and G. E. Likens.
1972. Atmospheric sulfur: Its effect on the chemical weathering of New England.
Science, 177:514–516.

Jónasson, P. M.
1969. Bottom fauna and eutrophication. *In* Eutrophication: Causes, Consequences, Correctives.
Washington, D.C., National Academy of Sciences, pp. 274–305.

Jónasson, P. M.
1972. Ecology and production of the profundal benthos in relation to phytoplankton in Lake Esrom.
Oikos Suppl., 14, 148 pp.

Jónasson, P. M., and H. Mathiesen.
1959. Measurements of primary production in two Danish eutrophic lakes. Esrom Sø and Furesø.
Oikos, 10:137–167.

Jónasson, P. M., and F. Thorhauge.
1972. Life cycle of *Potamothrix hammoniensis* (Tubificidae) in the profundal of a eutrophic lake.
Oikos, 23:151–158.

Jones, S. W., and R. Goulder.
1973. Swimming speeds of some ciliated Protozoa from a eutrophic pond.
Naturalist (Hull, England) (No. 924): 33–35.

Jordan, M., and G. E. Likens.
1975. An organic carbon budget for an oligotrophic lake in New Hampshire, U.S.A.
Verh. Int. Ver. Limnol., 19 (in press).

Jørgensen, E. G.
1957. Diatom periodicity and silicon assimilation.
Dansk Bot. Arkiv, 18(1), 54 pp.

Jørgensen, E. G.
1964. Adaptation to different light intensities in the diatom *Cyclotella Meneghiniana* Kütz.
Physiol. Plant., 17:136–145.

Jørgensen, E. G.
1968. The adaptation of plankton algae. II. Aspects of the temperature adaptation of *Skeletonema costatum*.
Physiol. Plant., 21:423–427.

Jørgensen, E. G.
1969. The adaptation of plankton algae. IV. Light adaptation in different algal species.
Physiol. Plant., 22:1307–1315.

Juday, C.
1921. Quantitative studies of the bottom fauna in the deeper waters of Lake Mendota.
Trans. Wis. Acad. Sci. Arts Lett., 20:461–493.

Juday, C.
1924. The productivity of Green Lake, Wisconsin.
Verh. Int. Ver. Limnol., 2:357–360.

Juday, C.
1940. The annual energy budget of an inland lake.
Ecology, 21:438–450.

Juday, C.
1942. The summer standing crop of plants and animals in four Wisconsin lakes.
Trans. Wis. Acad. Sci. Arts Lett., 34:103–135.

Juday, C., and E. A. Birge.
1931. A second report on the phosphorus content of Wisconsin lake waters.
Trans. Wis. Acad. Sci. Arts Lett., 26:353–382.

Juday, C., and E. A. Birge.
1933. The transparency, the color and the specific conductance of the lake waters of northeastern Wisconsin.
Trans. Wis. Acad. Sci. Arts Lett., 28:205–259.

Juday, C., E. A. Birge, G. I. Kemmerer, and R. J. Robinson.
1927. Phosphorus content of lake waters of northeastern Wisconsin.
Trans. Wis. Acad. Sci. Arts Lett., 23:233–248.

Juday, C., E. A. Birge, and V. W. Meloche.
1938. Mineral content of the lake waters of northeastern Wisconsin.
Trans. Wis. Acad. Sci. Arts Lett., 31:223–276.

Judd, J. H.
1970. Lake stratification caused by runoff from street deicing.
Water Res., *4*:521–532.

Juse, A.
1966. Diatomeen in Seesedimenten.
Arch. Hydrobiol. Beih. Ergebn. Limnol.. *4*, 32 pp.

Kajak, Z.
1970a. Some remarks on the necessities and prospects of the studies on biological production of freshwater ecosystems.
Pol. Arch. Hydrobiol., *17*:43–54.

Kajak, Z.
1970b. Analysis of the influence of fish on benthos by the method of enclosures. *In* Z. Kajak and A. Hillbricht-Ilkowska, eds. Productivity Problems of Freshwaters.
Warsaw, PWN Polish Scientific Publishers, pp. 781–793.

Kajak, Z., and K. Dusoge.
1970. Production efficiency of *Procladius choreus* MG (Chironomidae, Diptera) and its dependence on the trophic conditions.
Pol. Arch. Hydrobiol., *17*:217–224.

Kajak, Z., A. Hillbricht-Ilkowska, and E. Pieczyńska.
1970. The production processes in several Polish lakes. *In* Z. Kajak and A. Hillbricht-Ilkowska, eds. Productivity Problems of Freshwaters.
Warsaw, PWN Polish Scientific Publishers, pp. 129–147.

Kajak, Z., and B. Ranke-Rybicka.
1970. Feeding and production efficiency of *Chaoborus flavicans* Meigen (Diptera, Culicidae) larvae in eutrophic and dystrophic lake.
Pol. Arch. Hydrobiol., *17*:225–232.

Kajosaari, E.
1966. Estimation of the detention period of a lake.
Verh. Int. Ver. Limnol., *16*:139–143.

Kalff, J., and H. E. Welch.
1974. Phytoplankton production in Char Lake, a natural polar lake, and in Meretta Lake, a polluted polar lake, Cornwallis Island, Northwest Territories.
J. Fish. Res. Bd. Canada, *31*:621–636.

Kalinin, G. P., and V. D. Bykov.
1969. The world's water resources, present and future.
Impact of Science on Technology, *19*:135–150.

Karcher, F. H.
1939. Untersuchungen über den Stickstoffhaushalt in ostpreussischen Waldseen.
Arch. Hydrobiol., *35*:177–266.

Kashiwada, K., A. Kanazawa, and S. Tachibanazono.
1963. Studies on organic compounds in natural water. II. On the seasonal variations in the content of nicotinic acid, pantothenic acid, biotin, folic acid and vitamin B_{12} in the water of the Lake Kasumigaura. (In Japanese.)
Mem. Fac. Fish., Kagoshima Univ., *12*:153–157.

Kaushik, N. K., and H. B. N. Hynes.
1971. The fate of the dead leaves that fall into streams.
Arch. Hydrobiol., *68*:465–515.

Keefe, C. W.
1972. Marsh production: A summary of the literature.
Contr. Mar. Sci. Univ. Texas, *16*:163–181.

Keen, R.
1973. A probabilistic approach to the dynamics of natural populations of the Chydoridae (Cladocera, Crustacea).
Ecology, *54*:524–534.

Keeney, D. R.
1972. The fate of nitrogen in aquatic ecosystems.
Literature Rev., 3, Water Resources Center, University of Wisconsin, 59 pp.

Keeney, D. R.
1973. The nitrogen cycle in sediment-water systems.
J. Environ. Quality, *2*:15–29.

Keeney, D. R., R. L. Chen, and D. A. Graetz.
1971. Importance of denitrification and nitrate reduction in sediments to the nitrogen budgets of lakes.
Nature, *233*:66–67.

Keeney, D. R., J. G. Konrad, and G. Chesters.
1970. Nitrogen distribution in some Wisconsin lake sediments.
J. Water Poll. Control Fed., *42*:411–417.

Kellogg, W. W., R. D. Cadle, E. R. Allen, A. L. Lazrus, and E. A. Martell.
1972. The sulfur cycle.
Science, *175*:587–596.

Kern, D. M.
1960. The hydration of carbon dioxide.
J. Chem. Education, *37*:14–23.

Kerr, P. C., D. L. Brockway, D. F. Paris, and J. T. Barnett, Jr.
1972. The interrelation of carbon and phosphorus in regulating heterotrophic and auto-

trophic populations in an aquatic ecosystem, Shriner's Pond. *In* G. E. Likens, ed. Nutrients and Eutrophication: The Limiting-Nutrient Controversy.
Special Symposium, Amer. Soc. Limnol. Oceanogr., *1*:41–62.

Kersting, K., and W. Holterman.
1973. The feeding behaviour of *Daphnia magna*, studied with the Coulter Counter.
Verh. Int. Ver. Limnol., *18*:1434–1440.

Keup, L. E.
1968. Phosphorus in flowing waters.
Water Res., *2*:373–386.

Khailov, K. M.
1971. Ekologicheskii metabolizm v more.
Kiev, Izdatel ' stvo Naukova Dumka, 252 pp.

Kibby, H. V.
1971. Energetics and population dynamics of *Diaptomus gracilis*.
Ecol. Monogr., *41*:311–327.

Kibby, H. V., and F. H. Rigler.
1973. Filtering rates of *Limnocalanus*.
Verh. Int. Ver. Limnol., *18*:1457–1461.

Kilham, P.
1971. A hypothesis concerning silica and the freshwater planktonic diatoms.
Limnol. Oceanogr., *16*:10–18.

Kimball, K. D.
1973. Seasonal fluctuations of ionic copper in Knights Pond, Massachusetts.
Limnol. Oceanogr., *18*:169–172.

King, C. E.
1967. Food, age, and the dynamics of a laboratory population of rotifers.
Ecology, *48*:111–128.

King, D. L., and R. C. Ball.
1966. A qualitative and quantitative measure of *Aufwuchs* production.
Trans. Amer. Microsc. Soc., *85*:232–240.

Kjensmo, J.
1962. Some extreme features of the iron metabolism in lakes.
Schweiz. Z. Hydrol., *24*:244–252.

Kjensmo, J.
1967. The development and some main features of "iron-meromictic" soft water lakes.
Arch. Hydrobiol. Suppl., *32*:137–312.

Kjensmo, J.
1968. Iron as the primary factor rendering lakes meromictic, and related problems.
Mitt. Int. Ver. Limnol., *14*:83–93.

Kjensmo, J.
1970. The redox potentials in small oligo and meromictic lakes.
Nordic Hydrol., *1*:56–65.

Klekowski, R. Z.
1970. Bioenergetic budgets and their application for estimation of production efficiency.
Pol. Arch. Hydrobiol., *17*:55–80.

Klekowski, R. Z., E. Fischer, Z. Fischer, M. B. Ivanova, T. Prus, E. A. Shushkina, T. Stachurska, Z. Stepien, and H. Zyromska-Rudzka.
1970. Energy budgets and energy transformation efficiencies of several animal species of different feeding types. *In* Z. Kajak and A. Hillbricht-Ilkowska, eds. Productivity Problems of Freshwaters.
Warsaw, PWN Polish Scientific Publishers, pp. 749–763.

Klekowski, R. Z., and E. A. Shushkina.
1966. Ernährung, Atmung, Wachstum und Energie-Umformung in *Macrocyclops albidus* (Jurine).
Verh. Int. Ver. Limnol., *16*:399–418.

Klötzli, F.
1971. Biogenous influence on aquatic macrophytes, especially *Phragmites communis*.
Hidrobiologia, *12*:107–111.

Knight, A., R. C. Ball, and F. F. Hooper.
1962. Some estimates of primary production rates in Michigan ponds.
Pap. Mich. Acad. Sci. Arts Lett., *47*:219–233.

Kobayasi, H.
1961. Productivity in sessile algal community of Japanese mountain river.
Bot. Mag. Tokyo, *74*:331–341.

Kogan, Sh. I., and G. A. Chinnova.
1972. Relations between *Ceratophyllum demersum* (L.) and some blue-green algae.
Hydrobiol. J. (USSR; Translation Ser.), *8*:14–25.

Kondrat 'eva, E. N.
1965. Photosynthetic Bacteria.
Moscow, Izdatel 'stvo Akademii Nauk SSSR 1963. (Translated into English, Israel Program for Scientific Translations, Jerusalem).
243 pp.

Kononova, M. M.
1966. Soil Organic Matter. Its Nature, its Role in Soil Formation and in Soil Fertility. 2nd ed.
Oxford, Pergamon Press, 544 pp.

Konrad, J. G., D. R. Keeney, G. Chesters, and K.-L. Chen.
1970. Nitrogen and carbon distribution in sediment cores of selected Wisconsin lakes.
J. Water Poll. Control Fed., *42*:2094–2101.

Korde, N. W.
1966. Algenreste in Seesedimenten. Zur Entwicklungsgeschichte der Seen und umliegenden Landschaften.
Arch. Hydrobiol. Beih. Ergebn. Limnol., 3, 38 pp.

Koreliakova, I. L.
1958. Nekotorye nabluideniia nad raspodom perezimovavshei pribrezhno-vodnoi rastitel'nosti Rybinskogo Vodokhranilishcha.
Bull. Inst. Biol. Vodochranilishch, 1:22–25

Koreliakova, I. L.
1959. O raslade skoshennoi pribrezhno-vodnoi rastitel'nosti.
Bull. Inst. Biol. Vodochranilishch, 3:13–16.

Kořínková, J.
1967. Relations between predation pressure of carp, submerged plant development and littoral bottom-fauna of Pond Smyslov.
Rozpravy Českosl. Akad. Věd, Rada Matem. Přírod. Věd, 77(11):35–62.

Kormondy, E. J.
1968. Weight loss of cellulose and aquatic macrophytes in a Carolina bay.
Limnol. Oceanogr., 13:522–526.

Kowalczewski, A.
1965. Changes in periphyton biomass of Mikolajskie Lake.
Bull. Acad. Polon. Sci. (Cl. II), 13:395–398.

Koyama, T.
1955. Gaseous metabolism in lake muds and paddy soils.
J. Earth Sci. Nagoya Univ., 3:65–76.

Koyama, T.
1964. Gaseous metabolism in lake sediments and paddy soils. In U. Colombo and G. D. Hobson, eds. Advances in Organic Geochemistry.
New York, Macmillan Co., pp. 363–375.

Kozhov, M.
1963. Lake Baikal and Its Life.
The Hague, W. Junk, Publishers, 344 pp.

Kozlovsky, D. G.
1968. A critical evaluation of the trophic level concept. I. Ecological efficiencies.
Ecology, 49:48–60.

Kózmiński, Z., and J. Wisniewski.
1935. Über die Vorfrühlingthermik der Wigry-Seen.
Arch. Hydrobiol., 28:198–235.

Krasheninnikova, S. A.
1958. Mikrobiologicheskie protsessy raslada vodnoi rastitel'nosti v litorali Rybinskogo Vodokhranilishcha.
Bull. Inst. Biol. Vodokhranilishch, 2:3–6.

Kratz, W. A., and J. Myers.
1955. Nutrition and growth of several blue-green algae.
Amer. J. Bot., 42:282–287.

Krause, H. R.
1962. Investigation of the Decomposition of Organic Matter in Natural Waters.
FAO Fish. Biol. Report 34(FB/R34), 19 pp.

Krause, H. R.
1964a. Zur Chemie und Biochemie der Zersetzung von Süsswasserorganismen, unter besonderer Berücksichtigung des Abbaues der organischen Phosphorkomponenten.
Verh. Int. Ver. Limnol., 15:549–561.

Krause, H. R., L. Möchel, and M. Stegmann.
1961. Organische Säuren als gelöste Intermediärprodukte des postmortalen Abbaues von Süsswasser-Zooplankton.
Naturwissenschaften, 48:434–435.

Krauskopf, K. B.
1956. Dissolution and precipitation of silica at low temperatures.
Geochim. Cosmochim. Acta, 10:1–26.

Krauss, R. W.
1958. Physiology of the fresh-water algae.
Ann. Rev. Plant Physiol., 9:207–244.

Krauss, R. W.
1962. Inhibitors. In R. A. Lewin, ed. Physiology and Biochemistry of Algae.
New York, Academic Press, pp. 673–685.

Kriss, A. E., and R. Tomson.
1973. Origin of the warm water near the bottom of Lake Vanda in the Antarctic (25.5-27°): Microbiological data.
Mikrobiologiya, 42:942–943.

Krokhin, E. M.
1960. Vozniknovenie sloia temperaturnogo skachka v ozerakh. (The formation of the thermocline in lakes.)
Izvestiia Akad. Nauk SSSR, Ser. Geograficheskaia, 6:90–97.

Krogh, A.
1939. Osmotic Regulation in Aquatic Animals.
Cambridge, Cambridge University Press, 242 pp.

Krogh, A., and E. Lange.
1932. Quantitative Untersuchungen über Plankton, Kolloide und gelöste organische und anorganische Substanzen in dem Füresee.
Int. Rev. ges. Hydrobiol., 26:20–53.

Kuenzler, E. J.
1970. Dissolved organic phosphorus excretion by marine phytoplankton.
J. Phycol., 6:7–13.

Kuhl, A.
1962. Inorganic phosphorus uptake and metabolism. In R. A. Lewin, ed. Physiology and Biochemistry of Algae.
New York, Academic Press, pp. 211–229.

Kuznetsov, S. I.
1935. Microbiological researches in the study of the oxygenous regimen of lakes.
Verh. Int. Ver. Limnol., 7:562–582.

Kuznetsov, S. I.
1959. Die Rolle der Mikroorganismen im Stoffkreislauf der Seen.
Berlin, VEB Deutsch. Verlag Wissenschaften, 301 pp.

Kuznetsov, S. I.
1964. Biogeochemistry of sulphur. In Lo zolfo in agricoltura.
Simposio Int. Agrochimica (Palermo, Italy), 5:312–330.

Kuznetsov, S. I.
1968. Recent studies on the role of microorganisms in the cycling of substances in lakes.
Limnol. Oceanogr., 13:211–224.

Kuznetsov, S. I.
1970. Mikroflora ozer i ee geokhimicheskaya deyatel'nost'. (Microflora of Lakes and Their Geochemical Activities.) (In Russian.)
Leningrad, Izdatel'stvo Nauka, 440 pp.

Kuznetsov, S. I. and G. S. Karzinkin.
1931. Direct method for the quantitative study of bacteria in water and some considerations on causes which produce a zone of oxygen-minimum in Lake Glubokoje.
Zbl. Bakt., Ser. II, 83:169–174.

Kuznetsov, S. I., and E. M. Khartulari.
1941. Mikrobiologicheskaia kharakteristika protsessov anaerobnogo raspada organicheskogo veshchestva ila Belogo Ozera v Kosine. (Microbiological characteristics of the process of anaerobic decomposition of organic substances of sediments of Beloye Lake in Kosine.)
Mikrobiologiya, 10:834–849.

Kuznetsov, S. I., and V. I. Romanenko.
1963. Mikrobiologicheskoe izuchenie viutrennikh vodoemov. Laboratornoe rukobodstvo.
Moscow, Izdatel'stvo Akademii Nauk SSSR, 129 pp.

Květ, J.
1971. Growth analysis approach to the production ecology of reedswamp plant communities.
Hidrobiologia, 12:15–40.

Květ, J., J. Svoboda, and K. Fiala.
1969. Canopy development in stands of Typha latifolia L. and Phragmites communis Trin. in South Moravia.
Hidrobiologia, 10:63–75.

Landner, L., and T. Larsson.
1973. Indications of disturbances in the nitrification process in a heavily nitrogen-polluted water body.
Ambio, 2:154–157.

Langford, R. R., and E. G. Jermolajev.
1966. Direct effect of wind on plankton distribution.
Verh. Int. Ver. Limnol., 16:188–193.

Langmuir, I.
1938. Surface motion of water induced by wind.
Science, 87:119–123.

Larkin, P. A., and T. G. Northcote.
1969. Fish as indices of eutrophication. In Eutrophication: Causes, Consequences, Correctives.
Washington, D. C., National Academy of Sciences, pp. 256–273.

LaRow, E. J.
1968. A persistent diurnal rhythm in Chaoborus larvae. I. The nature of the rhythmicity.
Limnol. Oceanogr., 13:250–256.

LaRow, E. J.
1969. A persistent diurnal rhythm in Chaoborus larvae. II. Ecological significance.
Limnol. Oceanogr., 14:213–218.

LaRow, E. J.
1970. The effect of oxygen tension on the vertical migration of Chaoborus larvae.
Limnol. Oceanogr., 15:357–362.

Lasenby, D. C., and R. R. Langford.
1972. Growth, life history, and respiration of Mysis relicta in an arctic and temperate lake.
J. Fish. Res. Bd. Canada, 29:1701–1708.

Lasenby, D. C., and R. R. Langford.
1973. Feeding and assimilation of Mysis relicta.
Limnol. Oceanogr., 18:280–285.

Latimer, J. R.
1972. Radiation measurement.
Tech. Manual Ser. Int. Field Year Great Lakes 2, 53 pp.

Laube, H. R., and J. R. Wohler.
1973. Studies on the decomposition of a duckweed (Lemnaceae) community.
Bull. Torr. Bot. Club., 100:238–240.

Laurent, M., and J. Badia.
1973. Étude comparative du cycle biologique de l 'azote dans deux étangs.
Ann. Hydrobiol., 4:77–102.

Lauwers, A. M., and W. Heinen.
1974. Bio-degradation and utilization of silica and quartz.
Arch. Microbiol., 95:67–78.

Lawacz, W.
1969. The characteristics of sinking materials and the formation of bottom deposits in a eutrophic lake.
Mitt. Int. Ver. Limnol., 17:319–331.

Leach, J. H.
1975. Seston composition in the Point Pelee Area of Lake Erie.
Limnol. Oceanogr., *20*: (in press).

Lean, D. R. S.
1973a. Phosphorus dynamics in lake water.
Science, *179*:678–680.

Lean, D. R. S.
1973b. Movements of phosphorus between its biologically important forms in lake water.
J. Fish. Res. Bd. Canada, *30*:1525–1536.

Lean, D. R. S., and F. H. Rigler.
1974. A test of the hypothesis that abiotic phosphate complexing influences phosphorus kinetics in epilimnetic lake water.
Limnol. Oceanogr., *19*:784–788.

Learner, M. A., and D. W. B. Potter.
1974. Life-history and production of the leech *Helobdella stagnalis* (L.) (Hirudinea) in a shallow eutrophic reservoir in South Wales.
J. Anim. Ecol., *43*:199–208.

LeCren, E. D.
1958. Observations on the growth of perch (*Perca fluviatilis* L.) over twenty-two years with special reference to the effects of temperature and changes in population density.
J. Anim. Ecol., *27*:287–334.

Lefèvre, M.
1964. Extracellular products of algae. *In* D. F. Jackson, ed. Algae and Man.
New York, Plenum Press, pp. 337–367.

Lehn, H.
1965. Zur Durchsichtigkeitsmessung im Bodensee.
Schrift. Ver. Geschichte Bodensees Umgebung, *83*:32–44.

Lehn, H.
1968. Litorale Aufwuchsalgen im Pelagial des Bodensee.
Beitr. Naturk. Forsch. Südw.-Dtl., *27*:97–100.

Lellák, J.
1961. Zur Benthosproduktion und ihrer Dynamik in drei böhmischen Teichen.
Verh. Int. Ver. Limnol., *14*:213–219.

Lellák, J.
1965. The food supply as a factor regulating the population dynamics of bottom animals.
Mitt. Int. Ver. Limnol., *13*:128–138.

Lenhard, G., W. R. Ross, and A. du Plooy.
1962. A study of methods for the classification of bottom deposits of natural waters.
Hydrobiologia, *20*:223–240.

Leopold, L. B., M. G. Wolman, and J. P. Miller.
1964. Fluvial Processes in Geomorphology.
San Francisco, W. H. Freeman and Co., 522 pp.

Lerman, A., and M. Stiller.
1969. Vertical eddy diffusion in Lake Tiberias.
Verh. Int. Ver. Limnol., *17*:323–333.

Lewin, J. C.
1962. Silicification. *In* R. A. Lewin, ed. Physiology and Biochemistry of Algae.
New York, Academic Press, pp. 445–455.

Lewis, W. M., Jr.
1973. The thermal regime of Lake Lanao (Philippines) and its theoretical implications for tropical lakes.
Limnol. Oceanogr., *18*:200–217.

Lewis, W. M., Jr.
1974a. Implications of quantum theory for the interpretation of light penetration in aquatic habitats.
Ecology (in press).

Lewis, W. M., Jr.
1974b. Primary production in the plankton community of a tropical lake.
Ecol. Monogr., *44*:377–409.

Li, W. C., D. E. Armstrong, J. D. H. Williams, R. F. Harris, and J. K. Syers.
1972. Rate and extent of inorganic phosphate exchange in lake sediments.
Proc. Soil Sci. Soc. Amer., *36*:279–285.

Likens, G. E., ed.
1972. Nutrients and Eutrophication: The Limiting-Nutrient Controversy.
Special Symposium, Amer. Soc. Limnol. Oceanogr., *1*, 328 pp.

Likens, G. E.
1975. Primary production of inland aquatic ecosystems. *In* H. Lieth and R. H. Whittaker, eds. The Primary Productivity of the Biosphere.
New York, Springer-Verlag, (in press).

Likens, G. E., and F. H. Bormann.
1972. Nutrient cycling in ecosystems. *In* J. A. Weins, ed. Ecosystem Structure and Function.
Corvallis, Oregon State University Press, pp. 25–67.

Likens, G. E., F. H. Bormann, and N. M. Johnson.
1972. Acid rain.
Environment, *14*:33–40.

Likens, G. E., F. H. Bormann, N. M. Johnson, and R. S. Pierce.
1967. The calcium, magnesium, potassium, and sodium budgets for a small forested ecosystem.
Ecology, *48*:772–785.

Likens, G. E., F. H. Bormann, N. M. Johnson, D. W. Fisher, and R. S. Pierce.
1970. Effects of forest cutting and herbicide treatment on nutrient budgets in the Hubbard Brook watershed-ecosystem.
Ecol. Monogr., *40*:23–47.

Likens, G. E., and A. D. Hasler.
1960. Movement of radiosodium in a chemically stratified lake.
Science, *131*:1676–1677.

Likens, G. E., and A. D. Hasler.
1962. Movements of radiosodium (Na²⁴) within an ice-covered lake.
Limnol Oceanogr., *7*:48–56.

Likens, G. E., and N. M. Johnson.
1969. Measurement and analysis of the annual heat budget for the sediments in two Wisconsin lakes.
Limnol. Oceanogr., *14*:115–135.

Likens, G. E., and P. L. Johnson.
1966. A chemically stratified lake in Alaska.
Science, *153*:875–877.

Likens, G. E., and R. A. Ragotzkie.
1965. Vertical water motions in a small ice-covered lake.
J. Geophys. Res., *70*:2333–2344.

Likens, G. E., and R. A. Ragotzkie.
1966. Rotary circulation of water in an ice-covered lake.
Verh. Int. Ver. Limnol., *16*:126–133.

Lindeman, R. L.
1942. The trophic-dynamic aspect of ecology.
Ecology, *23*:399–418.

Littlefield, L., and C. Forsberg.
1965. Absorption and translocation of phosphorus-32 by *Chara globularis* Thuill.
Physiol. Plant., *18*:291–296.

Livingstone, D. A.
1954. On the orientation of lake basins.
Amer. J. Sci., *252*:547–554.

Livingstone, D. A.
1963. Chemical composition of rivers and lakes. Chap. G. Data of Geochemistry. 6th ed.
Prof. Pap. U.S. Geol. Surv., *440-G*, 64 pp.

Livingstone, D. A.
1963. Alaska, Yukon, Northwest Territories, and Greenland. *In* D. G. Frey, ed. Limnology in North America.
Madison, University of Wisconsin Press, pp. 559–574.

Livingstone, D. A., K. Bryan, Jr., and R. G. Leahy.
1958. Effects of an arctic environment on the origin and development of freshwater lakes.
Limnol. Oceanogr., *3*:192–214.

Loden, M. S.
1974. Predation by chironomid (Diptera) larvae on oligochaetes.
Limnol. Oceanogr., *19*:156–159.

Lohuis, D., V. W. Meloche, and C. Juday.
1938. Sodium and potassium content of Wisconsin lake waters and their residues.
Trans. Wis. Acad. Sci. Arts Lett., *31*:285–304.

Lueschow, L. A., J. M. Helm, D. R. Winter, and G. W. Karl.
1970. Trophic nature of selected Wisconsin lakes.
Trans. Wis. Acad. Sci. Arts Lett., *58*:237–264.

Lund, J. W. G.
1949. Studies on *Asterionella*. I. The origin and nature of the cells producing seasonal maxima.
J. Ecol., 37:389–419.

Lund, J. W. G.
1950. Studies on *Asterionella formosa* Hass. II. Nutrient depletion and the spring maximum. (Parts I and II).
J. Ecol., *38*:1–35.

Lund, J. W. G.
1954. The seasonal cycle of the plankton diatom, *Melosira italica* (Ehr.) Kütz. subsp. *subarctica* O. Müll.
J. Ecol., *42*:151–179.

Lund, J. W. G.
1955. Further observations on the seasonal cycle of *Melosira italica* (Ehr.) Kütz. subsp. *subarctica* O. Müll.
J. Ecol., *43*:90–102.

Lund, J. W. G.
1959. Buoyancy in relation to the ecology of the freshwater phytoplankton.
Brit. Phycol. Bull., *1*:1–17.

Lund, J. W. G.
1964. Primary production and periodicity of phytoplankton.
Verh. Int. Ver. Limnol., *15*:37–56.

Lund, J. W. G.
1965. The ecology of the freshwater phytoplankton.
Biol. Rev., *40*:231–293.

Lund, J. W. G., C. Kipling, and E. D. LeCren.
1958. The inverted microscope method of estimating algal numbers and the statistical basis of estimations by counting.
Hydrobiologia, *11*:143–170.

Lund, J. W. G., F. J. H. Mackereth, and C. H. Mortimer.
1963. Changes in depth and time of certain chemical and physical conditions and of the standing crop of *Asterionella formosa* Hass. in the North Basin of Windermere in 1947.
Phil. Trans. Roy. Soc. London (Ser. B), *246*:255–290.

Lund, J. W. G., and J. F. Talling.
1957. Botanical limnological methods with special reference to the algae.
Bot. Rev., 23:489–583.

Lush, D. L., and H. B. N. Hynes.
1973. The formation of particles in freshwater leachates of dead leaves.
Limnol. Oceanogr., 18:968–977.

Lüttge, U.
1964. Mikroautoradiographische Untersuchungen über die Funktion der Hydropoten von Nymphaea.
Protoplasma, 59:157–162.

Lvovitch, M. I.
1973. The global water balance.
Trans. Amer. Geophys. Union, 54:28–42.

Macan, T. T.
1961. Factors that limit the range of freshwater animals.
Biol. Rev., 36:151–198.

Macan, T. T.
1970. Biological Studies of the English Lakes.
New York, American Elsevier Publishing Co., Inc., 260 pp.

MacArthur, J. W., and W. H. T. Baillie.
1929. Metabolic activity and duration of life. I. Influence of temperature on longevity in Daphnia magna.
J. Exp. Zool., 53:221–242.

MacFayden, A.
1948. The meaning of productivity in biological systems.
J. Anim. Ecol., 17:75–80.

MacFayden, A.
1950. Biologische Produktivität.
Arch. Hydrobiol., 43:166–170.

Macgregor, A. N., and D. R. Keeney.
1973. Denitrification in lake sediments.
Environ. Lett., 5:175–181.

Machta, L.
1973. Prediction of CO_2 in the atmosphere. In G. M. Woodwell and E. V. Pecan, eds. Carbon and the Biosphere.
Brookhaven, N.Y., Proc. Brookhaven Symp. in Biol. 24. Tech. Information Center, U. S. Atomic Energy Commission CONF-720510, pp. 21–31.

Maciolek, J. A.
1954. Artificial fertilization of lakes and ponds. A review of the literature.
Spec. Sci. Rep. Fish., 113, 41 pp.

Mackenzie, F. T., and R. M. Garrels.
1965. Silicates: Reactivity with sea water.
Science, 150:57–58.

Mackenzie, F. T., R. M. Garrels, O. P. Bricker, and F. Bickley.
1967. Silica in sea water: Control by silica minerals.
Science, 155:1404–1405.

Mackereth, F. J. H.
1953. Phosphorus utilization by Asterionella formosa Hass.
J. Exp. Bot., 4:296–313.

Mackereth, F. J. H.
1966. Some chemical observations on postglacial lake sediments.
Phil. Trans. Roy. Soc. London (Ser. B), 250:165–213.

Macpherson, L. B., N. R. Sinclair, and F. R. Hayes.
1958. Lake water and sediment. III. The effect of pH on the partition of inorganic phosphate between water and oxidized mud or its ash.
Limnol. Oceanogr., 3:318–326.

Maeda, O., and S. Ichimura.
1973. On the high density of a phytoplankton population found in a lake under ice.
Int. Rev. ges. Hydrobiol., 58:673–685.

Maistrenko, Iu. G.
1965. Organicheskoe Veshchestvo Vody i Donnykh Otlozhenii Rek i Vodoemov Ukrainy (Basseiny Dnepra i Dunaia).
Kiev, Inst. Gidrobiologii, 239 pp.

Malovitskaia, L. M., and Ju. Sorokin.
1961. Eksperimental'noe issledovanie pitaniia Diaptomus (Crustacea, Copepoda) s pomoshch'iu C^{14}.
Trudy Inst. Biol. Vodokhranilishch, 4:262–272.

Malueg, K. W., and A. D. Hasler.
1966. Echo sounder studies on diel vertical movements of Chaoborus larvae in Wisconsin (U. S. A.) lakes.
Verh. Int. Ver. Limnol., 16:1697–1708.

Mann, K. H.
1962. Leeches (Hirudinea). Their Structure, Physiology, Ecology and Embryology.
Oxford, Pergamon Press, 201 pp.

Manny, B. A.
1972a. Seasonal changes in dissolved organic nitrogen in six Michigan lakes.
Verh. Int. Ver. Limnol., 18:147–156.

Manny, B. A.
1972b. Seasonal changes in organic nitrogen

content of net- and nannophytoplankton in two hardwater lakes.
Arch. Hydrobiol., 71:103–123.

Manny, B. A., and R. G. Wetzel.
1973. Diurnal changes in dissolved organic and inorganic carbon and nitrogen in a hard-water stream.
Freshwat. Biol., 3:31–43.

Manny, B. A., and R. G. Wetzel.
1975. Allochthonous dissolved organic and inorganic nitrogen budget of a marl lake.
(In preparation).

Manny, B. A., R. G. Wetzel, and W. C. Johnson.
1975. Annual contribution of carbon, nitrogen, and phosphorus to a hard-water lake by migrating Canada geese (*Branta canadensis interior* L.).
Verh. Int. Ver. Limnol., 19:(in press).

Marzolf, G. R.
1965a. Substrate relations of the burrowing amphipod *Pontoporeia affinis* in Lake Michigan.
Ecology, 46:579–592.

Marzolf, G. R.
1965b. Vertical migration of *Pontoporeia affinis* (Amphipoda) in Lake Michigan.
Publ. Great Lakes Res. Div., Univ. Mich., 13:133–140.

Mason, M. A.
1952. Some observations of breaking waves.
In Gravity Waves.
Circ. U. S. Bur. Standards, 521:215–220.

Mathews, C. P., and D. F. Westlake.
1969. Estimation of production by populations of higher plants subject to high mortality.
Oikos, 20:156–160.

Mattern, H.
1970. Beobachtungen über die Algenflora im Uferbereich des Bodensees (Überlinger See und Gnadensee).
Arch. Hydrobiol. Suppl., 37:1–163.

Matveev, V. P.
1964. O vertikal'nom raspredelenii temperatury v donnykh otlozheniyakh Ozer Dolgogo (Pitkayarvi) i Volochaevskogo (Vuotyarvi). (On the vertical distribution of temperature in the bottom deposits of Lake Dolgom (Pitkayarvi) and Volochaevskom (Vuotyarvi).
In Ozera Karel' skogo Ieresheika.
Moscow, Izdatel' stvo Nauka, pp. 45–50.

Maucha, R.
1932. Hydrochemische Methoden in der Limnologie.
Die Binnengewässer 12, 173 pp.

McCarter, J. A., F. R. Hayes, L. H. Jodrey, and M. L. Cameron.

1952. Movements of materials in the hypolimnion of a lake as studied by the addition of radioactive phosphorus.
Can. J. Zool., 30:128–133.

McCarty, P. L.
1964. The methane fermentation. *In* H. Heukelekian and N. C. Dondero, eds. Principles and Applications in Aquatic Microbiology.
New York, John Wiley & Sons, Inc., pp. 314–343.

McCracken, M. D., T. D. Gustafson, and M. S. Adams.
1974. Productivity of *Oedogonium* in Lake Wingra, Wisconsin.
Amer. Midland Nat., 92:247–254.

McCraw, B. M.
1961. Life history and growth of the snail *Lymnaea humilis* Say.
Trans. Amer. Microsc. Soc., 80:16–27.

McCraw, B. M.
1970. Aspects of the growth of the snail *Lymnaea palustris* (Müller).
Malacologia, 10:399–413.

McDonnell, A. J.
1971. Variations in oxygen consumption by aquatic macrophytes in a changing environment.
Proc. Conf. Great Lakes Res., Int. Assoc. Great Lakes Res., 14:52–58.

McGowan, L. M.
1974. Ecological studies on *Chaoborus* (Diptera, Chaoboridae) in Lake George, Uganda.
Freshwat. Biol., 4:483–505.

McGregor, D. L.
1969. The reproductive potential, life history and parasitism of the freshwater ostracod *Darwinula stevensoni* (Brady and Robertson).
In J. W. Neale, ed. The Taxonomy, Morphology and Ecology of Recent Ostracoda.
Edinburgh, Oliver & Boyd, pp. 194–221.

McIntire, C. D.
1966. Some factors affecting respiration of periphyton communities in lotic environments.
Ecology, 47:918–930.

McKeague, J. A., and M. G. Cline.
1963a. Silica in soil solutions. I. The form and concentration of dissolved silica in aqueous extracts of some soils.
Can. J. Soil Sci., 43:70–82.

McKeague, J. A., and M. G. Cline.
1963b. Silica in soil solutions. II. The adsorption of monosilicic acid by soil and by other substances.
Can. J. Soil Sci., 43:83–96.

McLaren, I. A.
1963. Effects of temperature on growth of

zooplankton, and the adaptive value of vertical migration.
J. Fish. Res. Bd. Canada, 20:685–727.

McMahon, J. W.
1965. Some physical factors influencing the feeding behavior of *Daphnia magna* Straus.
Can. J. Zool., 43:603–612.

McMahon, J. W.
1969. The annual and diurnal variation in the vertical distribution of acid-soluble ferrous and total iron in a small dimictic lake.
Limnol. Oceanogr., 14:357–367.

McMahon, J. W., and F. H. Rigler.
1963. Mechanisms regulating the feeding rate of *Daphnia magna* Straus.
Can. J. Zool., 41:321–332.

McMahon, J. W., and F. H. Rigler.
1965. Feeding rate of *Daphnia magna* Straus in different foods labeled with radioactive phosphorus.
Limnol. Oceanogr., 10:105–113.

McNaught, D. C.
1966. Depth control by planktonic cladocerans in Lake Michigan.
Publ. Great Lakes Res. Div., Univ. Mich., 15:98–108.

McNaught, D. C., and A. D. Hasler.
1961. Surface schooling and feeding behavior in the white bass, *Roccus chrysops* (Rafinesque), in Lake Mendota.
Limnol. Oceanogr., 6:53–60.

McNaught, D. C., and A. D. Hasler.
1964. Rate of movement of populations of *Daphnia* in relation to changes in light intensity.
J. Fish. Res. Bd. Canada, 21:291–318.

McNaught, D. C., and A. D. Hasler.
1966. Photoenvironments of planktonic Crustacea in Lake Michigan.
Verh. Int. Ver. Limnol., 16:194–203.

McNaughton, S. J.
1966a. Light stimulated oxygen uptake and glycolic acid oxidase in *Typha latifolia* L. leaf discs.
Science, 211:1197–1198.

McNaughton, S. J.
1966b. Ecotype function in the *Typha* community-type.
Ecol. Monogr., 36:297–325.

McNaughton, S. J.
1969. Genetic and environmental control of glycolic acid oxidase activity in ecotypic populations of *Typha latifolia*.
Amer. J. Bot., 56:37–41.

McNaughton, S. J.
1970. Fitness sets for *Typha*.
Amer. Nat., 104:337–341.

McNaughton, S. J., and L. W. Fullem.
1969. Photosynthesis and photorespiration in *Typha latifolia*.
Plant Physiol., 45:703–707.

McQueen, D. J.
1969. Reduction of zooplankton standing stocks by predaceous *Cyclops bicuspidatus thomasi* in Marion Lake, British Columbia.
J. Fish. Res. Bd. Canada, 26:1605–1618.

McQueen, D. J.
1970. Grazing rates and food selection in *Diaptomus oregonensis* (Copepoda) from Marion Lake, British Columbia.
J. Fish. Res. Bd. Canada, 27:13–20.

McRoy, C. P., R. J. Barsdate, and M. Nebert.
1972. Phosphorus cycling in an eelgrass (*Zostera marina* L.) ecosystem.
Limnol. Oceanogr., 17:58–67.

Meadows, P. S., and J. G. Anderson.
1966. Micro-organisms attached to marine and freshwater sand grains.
Nature, 212:1059–1060.

Meadows, P. S., and J. G. Anderson.
1968. Micro-organisms attached to marine sand grains.
J. Mar. Biol. Assoc. U. K., 48:161–175.

Meeks, J. C.
1974. Chlorophylls. *In* W. D. P. Stewart, ed. Algal Physiology and Biochemistry.
Berkeley, University of California Press, pp. 161–175.

Megard, R. O.
1972. Phytoplankton, photosynthesis, and phosphorus in Lake Minnetonka, Minnesota.
Limnol. Oceanogr., 17:68–87.

Meinzer, O. E., ed.
1942. Hydrology.
New York, McGraw-Hill Book Co., 712 pp.

Menzel, D. W., and J. P. Spaeth.
1962. Occurrence of ammonia in Sargasso Sea waters and in rain water at Bermuda.
Limnol. Oceanogr., 7:159–162.

Meyers, P. A., and J. G. Quinn.
1971. Interaction between fatty acids and calcite in seawater.
Limnol. Oceanogr., 16:992–997.

Milbrink, G.
1973a. Communities of Oligochaeta as indicators of the water quality in Lake Hjälmaren.
Zoon, 1:77–88.

Milbrink, G.
1973b. On the use of indicator communities of Tubificidae and some Lumbriculidae in the assessment of water pollution in Swedish lakes.
Zoon, 1:125–139.

Milbrink, G.
1973c. On the vertical distribution of oligochaetes in lake sediments.
Rep. Inst. Freshw. Res. Drottningholm,
53:34–50.

Miller, M. C.
1972. The carbon cycle in the epilimnion of
two Michigan lakes.
Ph.D. Diss., Michigan State University, 214
pp.

Mills, E. L., and R. T. Oglesby.
1971. Five trace elements and vitamin B_{12} in
Cayuga Lake, New York.
Proc. Conf. Great Lakes Res., Int. Assoc.
Great Lakes Res., 14:256–267.

Minckley, W. L., and G. A. Cole.
1963. Ecological and morphological studies
on gammarid amphipods (Gammarus spp.) in
spring-fed streams of northern Kentucky.
Occas. Pap. Adams Center Ecol. Stud., 10, 35
pp.

Minder, L.
1922. Über biogene Entkalkung im Zürichsee.
Verh. Int. Ver. Limnol., 1:20–32.

Minder, L.
1923. Studien über den Sauerstoffgehalt des
Zürichsees.
Arch. Hydrobiol. Suppl., 3:107–155.

Mitchell, D. S., and P. A. Thomas.
1972. Ecology of water weeds in the neotro-
pics: An ecological survey of the aquatic
weeds Eichhornia crassipedes and Salvinia
species, and their natural enemies in the neo-
tropics.
Tech. Pap. in Hydrology, UNESCO 12, 50 pp.

Momot, W. T.
1967a. Population dynamics and productivity
of the crayfish, Orconectes virilis, in a marl
lake.
Amer. Midland Nat., 78:55–81.

Momot, W. T.
1967b. Effects of brook trout predation on a
crayfish population.
Trans. Amer. Fish. Soc., 96:202–209.

Monakov, A. B., and Ju. I. Sorokin.
1961. Kolichestvenn'ie dann'ie o pitanii Daf-
nii.
Trudy Inst. Biol. Vodokhranilishch, 4:251–
261.

Mooij-Vogelaar, J. W., J. C. Jager, and W. J.
van der Steen.
1973. Effects of density levels, and changes in
density levels on reproduction, feeding and
growth in the pond snail Lymnaea stagnalis
(L.).
Proc. Nederl. Akad. Wetensc. Amsterdam
(Ser. C), 76:245–256.

Moore, A. W.
1969. Azolla: Biology and agronomic significance.
Bot. Rev., 35:17–34.

Moore, P. D., and D. J. Bellamy.
1974. Peatlands.
London, Elek Science, 221 pp.

Morel, A., and R. C. Smith.
1974. Relation between total quanta and total
energy for aquatic photosynthesis.
Limnol. Oceanogr., 19:591–600.

Morikawa, M., Y. Fukuo, and F. Hirao.
1959. Limnological researches in Lake Biwa
near the mouth of the River Ado. I. Density
distribution of the lake water off the mouth of
the river. (In Japanese.)
Japan. J. Limnol., 20:10–20.

Morris, I.
1967. An Introduction to the Algae.
London, Hutchinson University Library, 189
pp.

Morris, J. C., and W. Stumm.
1967. Redox equilibria and measurements of
potentials in the aquatic environment.
Advances in Chemistry Series, 67:270–285.

Mortimer, C. H.
1941. The exchange of dissolved substances
between mud and water in lakes (Parts I and
II).
J. Ecol., 29:280–329.

Mortimer, C. H.
1942. The exchange of dissolved substances
between mud and water in lakes (Parts III,
IV, summary, and references).
J. Ecol., 30:147–201.

Mortimer, C. H.
1951. The use of models in the study of water
movement in stratified lakes.
Verh. Int. Ver. Limnol., 11:254–260.

Mortimer, C. H.
1952. Water movements in lakes during summer stratification; Evidence from the distribution of temperature in Windermere.
Proc. Roy. Soc. London (Ser. B), 236:355–404.

Mortimer, C. H.
1953. The resonant response of stratified lakes
to wind.
Schweiz. Z. Hydrol., 15:94–151.

Mortimer, C. H.
1954. Models of the flow-pattern in lakes.
Weather, 9:177–184.

Mortimer, C. H.
1955. Some effects of the earth's rotation on
water movements in stratified lakes.
Verh. Int. Ver. Limnol., 12:66–77.

Mortimer, C. H.
1956. The oxygen content of air-saturated fresh waters, and aids in calculating percentage saturation.
Mitt. Int. Ver. Limnol., *6*, 20 pp.

Mortimer, C. H.
1961. Motion in thermoclines.
Verh. Int. Ver. Limnol., *14*:79–83.

Mortimer, C. H.
1963. Frontiers in physical limnology with particular reference to long waves in rotating basins.
Publ. Great Lakes Res. Div., Univ. Mich., *10*:9–42.

Mortimer, C. H.
1965. Spectra of long surface waves and tides in Lake Michigan and at Green Bay, Wisconsin.
Publ. Great Lakes Res. Div., Univ. Mich., *13*:304–325.

Mortimer, C. H.
1971. Chemical exchanges between sediments and water in the Great Lakes—speculations on probable regulatory mechanisms.
Limnol. Oceanogr., *16*:387–404.

Mortimer, C. H.
1971. Large-Scale Oscillatory Motions and Seasonal Temperature Changes in Lake Michigan and Lake Ontario.
Spec. Rept. No. 12, Center for Great Lakes Studies, University of Wisconsin-Milwaukee. Part I, Text, 111 pp. and Part II, Illustrations, 106 pp.

Mortimer, C. H.
1974. Lake hydrodynamics.
Mitt. Int. Ver. Limnol., *20*:124–197.

Mortimer, C. H., and C. F. Hickling.
1954. Fertilizers in fishponds.
Fish. Publ. U. K. Colonial Office, London 5, 155 pp.

Mortimer, C. H., and F. J. H. Mackereth.
1958. Convection and its consequences in ice-covered lakes.
Verh. Int. Ver. Limnol., *13*:923–932.

Mortimer, C. H., D. C. McNaught, and K. M. Stewart.
1968. Short internal waves near their high-frequency limit in central Lake Michigan.
Proc. Conf. Great Lakes Res., Int. Assoc. Great Lakes Res., *11*:454–469.

Morton, S. D., and G. F. Lee.
1968. Calcium carbonate equilibria in lakes.
J. Chem. Educ., *45*:511–513.

Moshiri, G. A., K. W. Cummins, and R. R. Costa.
1969. Respiratory energy expenditure by the predaceous zooplankter *Leptodora kindtii* (Focke) (Crustacea: Cladocera).
Limnol. Oceanogr., *14*:475–484.

Moskalenko, B. K., and K. K. Votinsev.
1970. Biological productivity and balance of organic substance and energy in Lake Baikal.
In Z. Kajak and A. Hillbricht-Ilkowska, eds. Productivity Problems of Freshwaters.
Warsaw, PWN Polish Scientific Publishers, pp. 207–226.

Moss, B.
1968. Studies on the degradation of chlorophyll *a* and carotenoids in freshwaters.
New Phytol., *67*:49–59.

Moss, B.
1969a. Vertical heterogeneity in the water column of Abbot's Pond. II. The influence of physical and chemical conditions on the spatial and temporal distribution of the phytoplankton and of a community of epipelic algae.
J. Ecol., *57*:397–414.

Moss, B.
1969b. Algae of two Somersetshire pools: Standing crops of phytoplankton and epipelic algae as measured by cell numbers and chlorophyll *a*.
J. Phycol., *5*:158–168.

Moss, B.
1972a. Studies on Gull Lake, Michigan. I. Seasonal and depth distribution of phytoplankton.
Freshwat. Biol., *2*:289–307.

Moss, B.
1972b. Studies on Gull Lake, Michigan. II. Eutrophication—evidence and prognosis.
Freshwat. Biol., *2*:309–320.

Moss, B.
1972c. The influence of environmental factors on the distribution of freshwater algae: An experimental study. I. Introduction and the influence of calcium concentration.
J. Ecol., *60*:917–932.

Moss, B.
1973a. The influence of environmental factors on the distribution of freshwater algae: An experimental study. II. The role of pH and the carbon dioxide-bicarbonate system.
J. Ecol., *61*:157–177.

Moss, B.
1973b. The influence of environmental factors on the distribution of freshwater algae: An experimental study. III. Effects of temperature, vitamin requirements and inorganic nitrogen compounds on growth.
J. Ecol., *61*:179–192.

Moss, B.
1973c. The influence of environmental factors

on the distribution of freshwater algae: An experimental study. IV. Growth of test species in natural lake waters, and conclusion.
J. Ecol., *61*:193–211.

Moss, B.
1973d. Diversity in fresh-water phytoplankton.
Amer. Midland Nat., *90*:341–355.

Moss, B., and A. G. Abdel Karim.
1969. Phytoplankton associations in two pools and their relationships with associated benthic flora.
Hydrobiologia, *33*:587–600.

Moss, B., and J. Moss.
1969. Aspects of the limnology of an endorheic African lake (L. Chilwa, Malawi).
Ecology, *50*:109–118.

Moss, B., and F. E. Round.
1967. Observations on standing crops of epipelic and epipsammic algal communities in Shear Water, Wilts.
Brit. Phycol. Bull., *3*:241–248.

Mueller, W. P.
1964. The distribution of cladoceran remains in surficial sediments from three northern Indiana lakes.
Invest. Indiana Lakes Streams, 6:1–63.

Müller, H.
1967. Eine neue qualitative Bestandsaufnahme des Phytoplanktons des Bodensee-Obersees mit besonderer Berücksichtigung der tychoplanktischen Diatomeen.
Arch. Hydrobiol. Suppl., *33*:206–236.

Müller-Haeckel, A.
1965. Tagesperiodik des Siliziumgehaltes in einem Fliessgewässer.
Oikos, *16*:232–233.

Mullin, M. M., P. R. Sloan, and R. W. Eppley.
1966. Relationship between carbon content, cell volume, and area in phytoplankton.
Limnol. Oceanogr., *11*:307–311.

Muscatine, L., and H. M. Lenhoff.
1963. Symbiosis: On the role of algae symbiotic with hydra.
Science, *142*:956–958.

Musgrave, A., M. B. Jackson, and E. Ling.
1972. Callitriche stem elongation is controlled by ethylene and gibberellin.
Nature (New Biol.), *238*:93–96.

Myer, G. E.
1969. A field study of Langmuir circulations.
Proc. Conf. Great Lakes Res., Int. Assoc. Great Lakes Res., *12*:652–663.

Nagasawa, M.
1959. On the dichotomous microstratification of pH in a lake.
Japan. J. Limnol., *20*:75–79.

Nakazawa, S.
1973. Artificial induction of lake balls.
Naturwissenschaften, *60*:481.

Nalewajko, C., and D. R. S. Lean.
1972. Growth and excretion in planktonic algae and bacteria.
J. Phycol., *8*:361–366.

National Academy of Sciences.
1969. Eutrophication: Causes, Consequences, Correctives.
Washington, D.C., National Academy of Sciences, 661 pp.

Naumann, E.
1919. Några synpunkter angående limnoplanktons ökologi med särskild hänsyn till fytoplankton.
Svensk Bot. Tidskr., *13*:129–163. (English translat. by the Freshwater Biological Association, No. 49.)

Naumann, E.
1931. Limnologische Terminologie. Handbuch der biologischen Arbeitsmethoden, Abt. IX, Teil 8.
Berlin, Urban & Schwarzenberg, 776 pp.

Naumann, E.
1932. Grundzüge der regionalen Limnologie.
Die Binnengewässer, *11*, 176 pp.

Nauwerck, A.
1959. Zur Bestimmung der Filterierrate limnischer Planktontiere.
Arch. Hydrobiol. Suppl., *25*:83–101.

Nauwerck, A.
1963. Die Beziehungen zwischen Zooplankton und Phytoplankton im See Erken.
Symbol. Bot. Upsalien., *17*(5), 163 pp.

Neilson, A. H., and R. A. Lewin.
1974. The uptake and utilization of organic carbon by algae: An essay in comparative biochemistry.
Phycologia, *13*:227–264.

Neumann, J.
1959. Maximum depth and average depth of lakes.
J. Fish. Res. Bd. Canada, *16*:923–927.

Newcombe, C. L.
1950. A quantitative study of attachment materials in Sodon Lake, Michigan.
Ecology, *31*:204–215.

Nichols, D. S., and D. R. Keeney.
1973. Nitrogen and phosphorus release from decaying water milfoil.
Hydrobiologia, *42*:509–525.

Niewolak, S.
1970. Seasonal changes of nitrogen-fixing and nitrifying and denitrifying bacteria in the bottom deposits of Ilawa lakes.
Pol. Arch. Hydrobiol., *17*:89–103.

Niewolak, S.
1972. Fixation of atmospheric nitrogen by *Azotobacter* sp., and other heterotrophic oligonitrophilous bacteria in the Ilawa lakes.
Acta Hydrobiol., *14*:287–305.

Nishimura, M., S. Nakaya, and K. Tanaka.
1973. Boron in the atmosphere and precipitation: Is the sea the source of atmospheric boron? *In* Proc. Symposium on Hydrogeochemistry and Biogeochemistry. I. Hydrogeochemistry.
Washington, D. C., Clarke Company, pp. 547–557.

Noble, V. E.
1961. Measurement of horizontal diffusion in the Great Lakes.
Publ. Great Lakes Res. Div., Univ. Mich., 7:85–95.

Noble, V. E.
1967. Evidences of geostrophically defined circulation in Lake Michigan.
Proc. Conf. Great Lakes Res., Int. Assoc. Great Lakes Res., *10*:289–298.

Noland, L. E., and M. Gojdics.
1967. Ecology of Free-Living Protozoa. *In* T. Chen, ed. Research in Protozoology. Vol. 2.
Oxford, Pergamon Press, pp. 215–266.

Northcote, T. G.
1964. Use of a high-frequency echo sounder to record distribution and migration of *Chaoborus* larvae.
Limnol. Oceanogr., 9:87–91.

Novotná, M., and V. Kořínek.
1966. Effect of the fishstock on the quantity and species composition of the plankton of two backwaters.
Hydrobiol. Stud., *1*:297–322.

Nriagu, J. O.
1968. Sulfur metabolism and sedimentary environment: Lake Mendota, Wisconsin.
Limnol. Oceanogr., *13*:430–439.

Nygaard, G.
1938. Hydrobiologische Studien über dänische Teiche und Seen. 1. Teil: Chemisch-Physikalische Untersuchungen und Planktonwägungen.
Arch. Hydrobiol., *32*:523–692.

Nygaard, G.
1949. Hydrobiological studies on some Danish ponds and lakes. Part II. The quotient hypothesis and some new or little known phytoplankton organisms.
Kongel. Danske Vidensk. Selskab Biol. Skrift., 7(1), 293 pp.

Nygaard, G.
1955. On the productivity of five Danish waters.
Verh. Int. Ver. Limnol., *12*:123–133.

Nykvist, N.
1963. Leaching and decomposition of water-soluble organic substances from different types of leaf and needle litter.
Stud. Forest. Suecica *3*, 31 pp.

Oborn, E. T.
1960a. Iron content of selected water and land plants. *In* Chemistry of Iron in Natural Water.
U. S. Geol. Surv. Water-Supply Pap., *1459-G*:191–211.

Oborn, E. T.
1960b. A survey of pertinent biochemical literature. *In* Chemistry of Iron in Natural Water.
U. S. Geol. Surv. Water-Supply Pap., *1459-F*:111–190.

Oborn, E. T.
1964. Intracellular and extracellular concentration of manganese and other elements by aquatic organisms. *In* Chemistry of Manganese in Natural Water.
U. S. Geol. Surv. Water-Supply Pap., *1667-C*, 18 pp.

Oborn, E. T., and J. D. Hem.
1962. Some effects of the larger types of aquatic vegetation on iron content of water. *In* Chemistry of Iron in Natural Water.
U. S. Geol. Surv. Water-Supply Pap., *1459-I*:237–268.

O'Brien, W. J., and F. deNoyelles.
1974. Filtering rate of *Ceriodaphnia reticulata* in pond waters of varying phytoplankton concentrations.
Amer. Midland Nat., *91*:509–512.

Odum, E. P.
1962. Relationships between structure and function in the ecosystem.
Japan. J. Ecol., *12*:108–118.

Odum, E. P.
1963. Primary and secondary energy flow in relation to ecosystem structure.
Proc. 16th Int. Congr. Zool., *4*:336–338.

Odum, E. P., and A. A. de la Cruz.
1963. Detritus as a major component of ecosystems.
Bull. Amer. Inst. Biol. Sci., *13*:39–40.

Odum, H. T.
1957. Trophic structure and productivity of Silver Springs, Florida.
Ecol. Monogr., 27:55–112.

Ogawa, R. E., and J. F. Carr.
1969. The influence of nitrogen on heterocyst production in blue-green algae.
Limnol. Oceanogr., *14*:342–351.

Ohle, W.
1934a. Chemische und physikalische Untersuchungen norddeutscher Seen.
Arch. Hydrobiol., 26:386–464 and 584–658.

Ohle, W.
1934b. Über organische Stoffe in Binnenseen.
Verh. Int. Ver. Limnol., 6(Pt. 2):249–262.

Ohle, W.
1938. Zur Vervollkommnung der hydrochemischen Analyse. III. Die Phosphorbestimmung.
Angew. Chem. 51:906–911.

Ohle, W.
1952. Die hypolimnische Kohlendioxyd-Akkumulation als produktionsbiologischer Indikator.
Arch. Hydrobiol., 46:153–285.

Ohle, W.
1953. Die chemische und elektrochemische Bestimmung des molekular gelösten Sauerstoffs der Binnengewässer.
Mitt. Int. Ver. Limnol., 3, 44 pp.

Ohle, W.
1954. Sulfat als "Katalysator" des limnischen Stoffkreislaufes.
Vom Wasser, 21:13–32.

Ohle, W.
1955a. Ionenaustausch der Gewässersedimente.
Mem. Ist. Ital. Idrobiol., 8(Suppl.):221–245.

Ohle, W.
1955b. Beiträge zur Produktionsbiologie der Gewässer.
Arch. Hydrobiol. Suppl., 22:456–479.

Ohle, W.
1956. Bioactivity, production, and energy utilization of lakes.
Limnol. Oceanogr., 1:139–149.

Ohle, W.
1958a. Die Stoffwechseldynamik der Seen in Abhängigkeit von der Gasausscheidung ihres Schlammes.
Vom Wasser, 25:127–149.

Ohle, W.
1958b. Typologische Kennzeichnung der Gewässer auf Grund ihrer Bioaktivität.
Verh. Int. Ver. Limnol., 13:196–211.

Ohle, W.
1962. Der Stoffhaushalt der Seen als Grundlage einer allgemeinen Stoffwechseldynamik der Gewässer.
Kieler Meeresforschungen, 18:107–120.

Ohle, W.
1964. Kolloidkomplexe als Kationen- und Anionenaustauscher in Binnengewässern.
Vom Wasser, 30(1963):50–64.

Ohle, W.
1965. Nährstoffanreicherung der Gewässer durch Düngemittel und Meliorationen.
Münchner Beiträge, 12:54–83.

Ohle, W.
1972. Die Sedimente des Grossen Plöner Sees als Dokumente der Zivilisation.
Jahrb. Heimatkunde Plön, 2:7–27.

Ohmori, M., and A. Hattori.
1972. Effect of nitrate on nitrogen-fixation by the blue-green alga *Anabaena cylindrica*.
Plant Cell Physiol., 13:589–599.

Ohwada, K., M. Otsuhata, and N. Taga.
1972. Seasonal cycles of vitamin B_{12}, thiamine and biotin in the surface water of Lake Tsukui.
Bull. Japan. Soc. Sci. Fish., 38:817–823.

Ohwada, K., and N. Taga.
1972. Vitamin B_{12}, thiamine, and biotin in Lake Sagami.
Limnol. Oceanogr., 17:315–320.

Ökland, J.
1964. The eutrophic Lake Borrevann (Norway)—an ecological study on shore and bottom fauna with special reference to gastropods, including a hydrographic survey.
Folia Limnol. Scandinavica, 13, 337 pp.

Oláh, J.
1969a. The quantity, vertical and horizontal distribution of the total bacterioplankton of Lake Balaton in 1966/67.
Annal. Biol. Tihany, 36:185–195.

Oláh, J.
1969b. A quantitative study of the saprophytic and total bacterioplankton in the open water and the littoral zone of Lake Balaton in 1968.
Annal. Biol. Tihany, 36:197–212.

Oláh, J.
1972. Leaching, colonization and stabilization during detritus formation.
Mem. Ist. Ital. Idrobiol., 29(Suppl.):105–127.

Olsen, S.
1958a. Phosphate adsorption and isotopic exchange in lake muds. Experiments with P 32; Preliminary report.
Verh. Int. Ver. Limnol., 13:915–922.

Olsen, S.
1958b. Fosfatbalancen mellem bund og vand i Furesø. Forsøg med radioaktivt fosfor.
Folia Limnol. Scandinavica, 10:39–96.

Olsen, S.
1964. Phosphate equilibrium between reduced sediments and water. Laboratory experiments with radioactive phosphorus.
Verh. Int. Ver. Limnol., 15:333–341.

Olson, F. C. W.
1960. A system of morphometry.
Int. Hydrogr. Rev., 37:147–155.

Olson, F. C. W., and T. Ichiye.
1959. Horizontal diffusion.
Science, 130:1255.

Orlov, D. S., I. A. Pivovarova, and N. I. Gorbunov.
1973. Interaction of humic substances with minerals and the nature of their bond—a review.
Agrokhimiya, 1973(9):140–153. (Translat. in Soviet Soil Science, 5:568–581.)

Oswald, G. K. A., and G. de Q. Robin.
1973. Lakes beneath the Antarctic ice sheet.
Nature, 245:251–254.

Otsuki, A., and T. Hanya.
1968. On the production of dissolved nitrogen-rich organic matter.
Limnol. Oceanogr., 13:183–185.

Otsuki, A., and T. Hanya.
1972a. Production of dissolved organic matter from dead green algal cells. I. Aerobic microbial decomposition.
Limnol. Oceanogr., 17:248–257.

Otsuki, A., and T. Hanya.
1972b. Production of dissolved organic matter from dead green algal cells. II. Anaerobic microbial decomposition.
Limnol. Oceanogr., 17:258–264.

Otsuki, A., and R. G. Wetzel.
1972. Coprecipitation of phosphate with carbonates in a marl lake.
Limnol. Oceanogr., 17:763–767.

Otsuki, A., and R. G. Wetzel.
1973. Interaction of yellow organic acids with calcium carbonate in freshwater.
Limnol. Oceanogr., 18:490–493.

Otsuki, A., and R. G. Wetzel.
1974a. Calcium and total alkalinity budgets and calcium carbonate precipitation of a small hard-water lake.
Arch. Hydrobiol., 73:14–30.

Otsuki, A., and R. G. Wetzel.
1974b. Release of dissolved organic matter by autolysis of a submersed macrophyte, Scirpus subterminalis.
Limnol. Oceanogr., 19:842–845.

Overbeck, J.
1962. Untersuchungen zum Phosphathaushalt von Grünalgen. III. Das Verhalten der Zellfraktionen von Scenedesmus quadricauda (Turp.) Bréb. im Tagescyclus unter verschiedenen Belichtungsbedingungen und bei verschiedenen Phosphatverbindungen.
Arch. Mikrobiol., 41:11–26.

Overbeck, J.
1963. Untersuchungen zum Phosphathaushalt von Grünalgen. VI. Ein Beitrag zum Polyphosphatstoffwechsel des Phytoplanktons.
Ber. Deut. Bot. Gesellsch., 76:276–286.

Overbeck, J.
1965. Primärproduktion und Gewässerbakterien.
Naturwissenschaften, 51:145.

Overbeck, J.
1968. Prinzipielles zum Vorkommen der Bakterien im See.
Mitt. Int. Ver. Limnol., 14:134–144.

Overbeck, J., and H.-D. Babenzien.
1964. Bakterien und Phytoplankton eines Kleingewässers im Jahreszyklus.
Z. Allg. Mikrobiol., 4:59–76.

Owens, M., and R. W. Edwards.
1961. The effects of plants on river conditions. II. Further crop studies and estimates of net productivity of macrophytes in a chalk stream.
J. Ecol., 49:119–126.

Owens, M., and R. W. Edwards.
1962. The effects of plants on river conditions. III. Crop studies and estimates of net productivity of macrophytes in four streams in southern England.
J. Ecol., 50:157–162.

Owens, M., and P. J. Maris.
1964. Some factors affecting the respiration of some aquatic plants.
Hydrobiologia, 23:533–543.

Ownbey, C. R., and D. A. Kee.
1967. Chlorides in Lake Erie.
Proc. Conf. Great Lakes Res., Int. Assoc. Great Lakes Res., 10:382–389.

Palmer, J. D., and F. E. Round.
1965. Persistent, vertical-migration rhythms in benthic microflora. I. The effect of light and temperature on the rhythmic behaviour of Euglena obtusa.
J. Mar. Biol. Assoc. U. K., 45:567–582.

Paloheimo, J. E.
1974. Calculation of instantaneous birth rate.
Limnol. Oceanogr., 19:692–694.

Pannier, F.
1957. El consumo de oxigeno de plantas acuaticas en relacion a distintas concentraciones de oxigeno. Parte I.
Acta Cient. Venez., 8:148–161.

Pannier, F.
1958. El consumo de oxigeno de plantas acuaticas en relacion a distintas concentraciones de oxigeno. Parte II.
Acta Cient. Venez., 9:2–13.

Parker, B. C., and M. A. Wachtel.
1971. Seasonal distribution of cobalamins, biotin and niacin in rainwater. *In* Cairns, J., Jr., ed. The structure and function of fresh-water microbial communities.
Res. Div. Monogr., Virginia Polytech. Inst., 3:195-207.

Parker, M., and A. D. Hasler.
1969. Studies on the distribution of cobalt in lakes.
Limnol. Oceanogr., 14:229-241.

Parker, R. A., and D. H. Hazelwood.
1962. Some possible effects of trace elements on fresh-water microcrustacean populations.
Limnol. Oceanogr., 7:344-347.

Parma, S.
1971. *Chaoborus flavicans* (Meigen) (Diptera, Chaoboridae): an autecological study. Ph.D. Diss., University of Groningen.
Rotterdam, Bronder-Offset n.v., 128 pp.

Parsons, T. R., and J. D. H. Strickland.
1962. On the production of particulate organic carbon by heterotrophic processes in sea water.
Deep-Sea Res., 8:211-222.

Patalas, K.
1961. Wind- und morphologiebedingte Wasserbewegungstypen als bestimmender Faktor für die Intensität des Stoffkreislaufes in nordpolnischen Seen.
Verh. Int. Ver. Limnol., 14:59-64.

Patriarche, M. H., and R. C. Ball.
1949. An analysis of the bottom fauna production in fertilized and unfertilized ponds and its utilization by young-of-the-year fish.
Tech. Bull. Mich. State Univ. Agricult. Exp. Stat., Sec. Zool., 207, 35 pp.

Patrick, R., B. Crum, and J. Coles.
1969. Temperature and manganese as determining factors in the presence of diatom or blue-green algal floras in streams.
Proc. Nat. Acad. Sci., 64:472-478.

Patrick, R., M. H. Hohn, and J. H. Wallace.
1954. A new method for determining the pattern of the diatom flora.
Notulae Naturae Acad. Nat. Sci. Philadelphia, 259, 12 pp.

Pearsall, W. H.
1922. A suggestion as to factors influencing the distribution of free-floating vegetation.
J. Ecol., 9:241-253.

Pearsall, W. H.
1932. Phytoplankton in the English lakes. II. The composition of the phytoplankton in relation to dissolved substances.
J. Ecol., 20:241-262.

Pechan'-Finenko, G. A.
1971. Effectiveness of the assimilation of food by plankton crustaceans. (Translat. Consultants Bureau.)
Ekologiia, 2:64-72.

Pechlaner, R.
1970. The phytoplankton spring outburst and its conditions in Lake Erken (Sweden).
Limnol. Oceanogr., 15:113-130.

Pechlaner, R.
1971. Factors that control the production rate and biomass of phytoplankton in high-mountain lakes.
Mitt. Int. Ver. Limnol., 19:125-145.

Pechlaner, R., G. Bretschko, P. Gollmann, H. Pfeifer, M. Tilzer, and H. P. Weissenbach.
1970. The production processes in two high-mountain lakes (Vorderer and Hinterer Finstertaler See, Kühtai, Austria). *In* Z. Kajak and A. Hillbricht-Ilkowska, eds. Productivity Problems of Freshwaters.
Warsaw, PWN Polish Scientific Publishers, pp. 239-269.

Penfound, W. T., and T. T. Earle.
1948. The biology of the water hyacinth.
Ecol. Monogr., 18:447-472.

Pennak, R. W.
1940. Ecology of the microscopic Metazoa inhabiting the sandy beaches of some Wisconsin lakes.
Ecol. Monogr., 10:537-615.

Pennak, R. W.
1944. Diurnal movements of zooplankton organisms in some Colorado mountain lakes.
Ecology, 25:387-403.

Pennak, R. W.
1953. Fresh-Water Invertebrates of the United States.
New York, Ronald Press Co., 769 pp.

Pennak, R. W.
1973. Some evidence for aquatic macrophytes as repellents for a limnetic species of *Daphnia*.
Int. Rev. ges. Hydrobiol., 58:569-576.

Pennington, W.
1974. Seston and sediment formation in five Lake District lakes.
J. Ecol., 62:215-251.

Perfil'ev, B. V., and D. R. Gabe.
1969. Capillary Methods of Investigating Micro-Organisms. (Translat. of 1961 edition by J. M. Shewan.)
Toronto, University of Toronto Press, 627 pp.

Peterson, W. H., E. B. Fred, and B. P. Domogalla.
1925. The occurrence of amino acids and

other organic nitrogen compounds in lake water.
J. Biol. Chem., 23:287–295.

Pfennig, N.
1967. Photosynthetic bacteria.
Ann. Rev. Microbiol., 21:285–324.

Pia, J.
1933. Kohlensäure und Kalk.
Die Binnengewässer 13, 183 pp.

Pieczyńska, E.
1959. Character of the occurrence of free-living Nematoda in various types of periphyton in Lake Tajty.
Ekol. Polska (Ser. A), 7:317-337. (Polish; English summary.)

Pieczyńska, E.
1964. Investigations on colonization of new substrates by nematodes (Nematoda) and some other periphyton organisms.
Ekol. Polska (Ser. A), 12:185–234.

Pieczyńska, E.
1965. Variations in the primary production of plankton and periphyton in the littoral zone of lakes.
Bull. Acad. Polon. Sci., Cl. II, 13:219–225.

Pieczyńska, E.
1968. Dependence of the primary production of periphyton upon the substrate area suitable for colonization.
Bull. Acad. Polon. Sci., Cl. II, 16:165–169.

Pieczyńska, E.
1970. Production and decomposition in the eulittoral zone of lakes. In Z. Kajak and A. Hillbricht-Ilkowska, eds. Productivity Problems of Freshwaters.
Warsaw, PWN Polish Scientific Publishers, pp. 271–285.

Pieczyńska, E.
1972. Rola materii allochtonicznej w jeziorach.
Wiadomości Ekologiczne, 18:131–140.

Pieczyńska, E., E. Pieczyński, T. Prus, and K. Tarwid.
1963. The biomass of the bottom fauna of 42 lakes in the Wegorzewo District.
Ekol. Polska (Ser. A), 11:495–502.

Pieczyńska, E., and I. Spodniewska.
1963. Occurrence and colonization of periphyton organisms in accordance with the type of substrate.
Ekol. Polska (Ser. A), 11:533–545.

Pieczyńska, E., and W. Szczepańska.
1966. Primary production in the littoral of several Masurian lakes.
Verh. Int. Ver. Limnol., 16:372–379.

Pimentel, G. C., and A. L. McClellan.
1960. The Hydrogen Bond.
San Francisco, W. H. Freeman and Co., 475 pp.

Pivovarov, A. A.
1973. Thermal Conditions in Freezing Lakes and Rivers. Israel Program for Scientific Translations of 1972 Russian edition.
New York, John Wiley & Sons, 136 pp.

Poltz, J.
1972. Untersuchungen über das Vorkommen und den Abbau von Fetten und Fettsäuren in Seen.
Arch. Hydrobiol. Suppl., 40:315–399.

Pomeroy, L. R., E. E. Smith, and C. M. Grant.
1965. The exchange of phosphate between estuarine water and sediments.
Limnol. Oceanogr., 10:167–172.

Porter, K. G.
1973. Selective grazing and differential digestion of algae by zooplankton.
Nature, 244:179–180.

Porter, K. G.
1975. Viable gut passage of gelatinous green algae ingested by Daphnia.
Verh. Int. Ver. Limnol., 19 (in press).

Poštolková, M.
1967. Comparison of the zooplankton amount and primary production of the fenced and unfenced littoral regions of Smyslov Pond.
Rozpravy Českosl. Akad. Věd, Řada Matem. Přír. Věd, 77:(11):63–79.

Potts, W. T. W., and G. Parry.
1964. Osmotic and Ionic Regulation in Animals.
Oxford, Pergamon Press, 423 pp.

Potzger, J. E., and W. A. van Engel.
1942. Study of the rooted aquatic vegetation of Weber Lake, Vilas County, Wisconsin.
Trans. Wis. Acad. Sci. Arts Lett., 34:149–166.

Pourriot, R.
1965. Recherches sur l'écologie des rotifères.
Vie et Milieu, Suppl., 21, 224 pp.

Pourriot, R.
1974. Relations prédateur-proie chez les rotifères: influence du prédateur (Asplanchna brightwelli) sur la morphologie de la proie (Brachionus bidentata).
Ann. Hydrobiol., 5:43–55.

Povoledo, D.
1961. Ulteriori studi sulle sostanze organiche disciolte nell'acqua del Lago Maggiore: Frazionamento e separazione delle proteine dai peptidi e dagli aminoacidi liberi.
Mem. Ist. Ital. Idrobiol., 13:203–222.

Prins, H. B. A., and R. W. Wolff.
1974. Photorespiration in leaves of *Vallisneria spiralis*. The effect of oxygen on the carbon dioxide compensation point.
Proc. Akad. van Wetensc. Amsterdam (Ser. C), 77:239–245.

Provasoli, L.
1958. Nutrition and ecology of Protozoa and algae.
Ann. Rev. Microbiol., *12*:279–308.

Provasoli, L.
1963. Organic regulation of phytoplankton fertility. *In* M. N. Hill, ed. The Sea. Vol. 2.
New York, Interscience, pp. 165–219.

Provasoli, L., and A. F. Carlucci.
1974. Vitamins and growth regulators. *In* W. D. P. Stewart, ed. Algal Physiology and Biochemistry.
Berkeley, University of California Press, pp. 741–787.

Provasoli, L., J. J. A. McLaughlin, and I. J. Pintner.
1954. Relative and limiting concentrations of major mineral constituents for the growth of algal flagellates.
Trans. N. Y. Acad. Sci. (Ser. 2), *16*:412-417.

Pugh, G. J. F., and J. L. Mulder.
1971. Mycoflora associated with *Typha latifolia*.
Trans. Br. Mycol. Soc., *57*:273–282.

Putnam, H. D., and T. A. Olson.
1961. Studies on the Productivity and Plankton of Lake Superior.
Rep. School Public Health, University of Minnesota, 24 pp.

Ragotzkie, R. A., and G. E. Likens.
1964. The heat budget of two Antarctic lakes.
Limnol. Oceanogr., *9*:412–425.

Rainwater, F. H., and L. L. Thatcher.
1960. Methods for collection and analysis of water samples.
U. S. Geol. Surv. Water-Supply Pap., *1454*, 301 pp.

Ramsey, W. L.
1962a. Bubble growth from dissolved oxygen near the sea surface.
Limnol. Oceanogr., *7*:1–7.

Ramsey, W. L.
1962b. Dissolved oxygen in shallow nearshore water and its relation to possible bubble formation.
Limnol. Oceanogr., *7*:453–461.

Raspopov, I. M.
1971. Litoralvegetation der Onega- und Ladogaseen.
Hidrobiologia, *12*:241–247.

Rasumov, A. S.
1962. Mikrobial'nyi plankton vody.
Trudy Vsesoyusnogo Gidrobiol. Obshch., *12*:60–190.

Raven, J. A.
1970. Exogenous inorganic carbon sources in plant photosynthesis.
Biol. Rev., *45*:167–221.

Rawson, D. S.
1952. Mean depth and the fish production of large lakes.
Ecology, *33*:515–521.

Rawson, D. S.
1955. Morphometry as a dominant factor in the productivity of large lakes.
Verh. Int. Ver. Limnol., *12*:164–175.

Rawson, D. S.
1956. Algal indicators of trophic lake types.
Limnol. Oceanogr., *1*:18–25.

Rawson, D. S., and J. E. Moore.
1944. The saline lakes of Saskatchewan.
Can. J. Res. (Sec. D), *22*:141–201.

Regier, H. A., and W. L. Hartman.
1973. Lake Erie's fish community:150 years of cultural stresses.
Science, *180*:1248–1255.

Reif, C. B., and D. W. Tappa.
1966. Selective predation: Smelt and cladocerans in Harveys Lake.
Limnol. Oceanogr., *11*:437–438.

Reimers, N., J. A. Maciolek, and E. P. Pister.
1955. Limnological study of the lakes in Convict Creek Basin, Mono County, California.
Fish. Bull. U. S. Fish. Wildl. Serv., *56*:437–503.

Remane, A., and C. Schlieper.
1971. Biology of brackish water. 2nd ed.
Die Binnengewässer, *25*, 372 pp.

Reynoldson, T. B.
1961. Observations on the occurrence of *Asellus* (Isopoda, Crustacea) in some lakes of northern Britain.
Verh. Int. Ver. Limnol., *14*:988–994.

Reynoldson, T. B.
1966. The distribution and abundance of lake-dwelling triclads—towards a hypothesis.
Adv. Ecol. Res., *3*:1–71.

Reynoldson, T. B., and L. S. Bellamy.
1971. Intraspecific competition in lake-dwelling triclads. A laboratory study.
Oikos, *22*:315–328.

Rhee, G.-Y.
1972. Competition between an alga and an aquatic bacterium for phosphate.
Limnol. Oceanogr., 17:505–514.

Rhee, G-Y.
1973. A continuous culture study of phosphate uptake, growth rate and polyphosphate in *Scenedesmus* sp.
J. Phycol., 9:495–506.

Rice, E. L., and S. K. Pancholy.
1972. Inhibition of nitrification by climax ecosystems.
Amer. J. Bot., 59:1033–1040.

Rice, E. L., and S. K. Pancholy.
1973. Inhibition of nitrification by climax ecosystems. II. Additional evidence and possible role of tannins.
Amer. J. Bot., 60:691–702.

Rich, P. H.
1975. Benthic metabolism of a soft-water lake.
Verh. Int. Ver. Limnol., 19 (in press).

Rich, P. H., and R. G. Wetzel.
1969. A simple, sensitive underwater photometer.
Limnol. Oceanogr., 14:611–613.

Rich, P. H., R. G. Wetzel, and N. V. Thuy.
1971. Distribution, production and role of aquatic macrophytes in a southern Michigan marl lake.
Freshwat. Biol., 1:3–21.

Rich, P. H., and R. G. Wetzel.
1975. Detritus in the lake ecosystem. (In preparation.)

Richardson, L. F.
1925. Turbulence and vertical temperature difference near trees.
Phil. Mag., 49:81–90.

Richerson, P., R. Armstrong, and C. R. Goldman.
1970. Contemporaneous disequilibrium, a new hypothesis to explain the "Paradox of the Plankton."
Proc. Nat. Acad. Sci., 67:1710–1714.

Richman, S.
1958. The transformation of energy by *Daphnia pulex*.
Ecol. Monogr., 28:273–291.

Richman, S.
1966. The effect of phytoplankton concentration on the feeding rate of *Diaptomus oregonensis*.
Verh. Int. Ver. Limnol., 16:392–398.

Ricker, W. E.
1934. A critical discussion of various measures of oxygen saturation in lakes.
Ecology, 15:348–363.

Ricker, W. E.
1937. Physical and chemical characteristics of Cultus Lake, British Columbia.
J. Biol. Bd. Canada, 3(4):363-402.

Ricker, W. E.
1952. The benthos of Cultus Lake.
J. Fish. Res. Bd. Canada, 9:204-212.

Rickett, H. W.
1921. A quantitative study of the larger aquatic plants of Lake Mendota, Wisconsin.
Trans. Wis. Acad. Sci. Arts Lett., 20:501–527.

Riggs, L. A., and J. J. Gilbert.
1972. The labile period for α-tocopherol-induced mictic female and body wall outgrowth responses in embryos of the rotifer *Asplanchna sieboldi*.
Int. Rev ges. Hydrobiol., 57:675–683.

Rigler, F. H.
1956. A tracer study of the phosphorus cycle in lake water.
Ecology, 37:550–562.

Rigler, F. H.
1961. The uptake and release of inorganic phosphorus by *Daphnia magna* Straus.
Limnol. Oceanogr., 6:165–174.

Rigler, F. H.
1964a. The phosphorus fractions and the turnover time of inorganic phosphorus in different types of lakes.
Limnol. Oceanogr., 9:511–518.

Rigler, F. H.
1964b. The contribution of zooplankton to the turnover of phosphorus in the epilimnion of lakes.
Can. Fish Culturist, 32:3–9.

Rigler, F. H.
1971. Feeding rates: zooplankton. *In* W. T. Edmondson and G. G. Winberg, eds. A Manual on Methods for the Assessment of Secondary Productivity in Fresh Waters. Int. Biol. Program Handbook 17.
Oxford, Blackwell Scientific Publications, pp. 228–255.

Rigler, F. H.
1973. A dynamic view of the phosphorus cycle in lakes. *In* E. J. Griffith, A. Beeton, J. M. Spencer, and D. T. Mitchell, eds. Environmental Phosphorus Handbook.
New York, John Wiley & Sons, pp. 539–572.

Rigler, F. H., M. E. MacCallum, and J. C. Roff.
1974. Production of zooplankton in Char Lake.
J. Fish. Res. Bd. Canada, 31:637–646.

Riley, G. A.
1939. Limnological studies in Connecticut.

Part I. General limnological survey
Part II. The copper cycle.
Ecol. Monogr., 9:66–94.

Ringelberg, J.
1964. The positively phototactic reaction of
Daphnia magna Straus: A contribution to the
understanding of diurnal vertical migration.
Netherlands J. Sea Res., 2:319–406.

Roback, S. S.
1974. Insects (Arthropoda: Insecta). *In* C. W.
Hart, Jr., and S. L. H. Fuller, eds. Pollution
Ecology of Freshwater Invertebrates.
New York, Academic Press, pp. 313–376.

Robbins, J. A., E. Landstrom, and M. Wahl-
gren.
1972. Tributary inputs of soluble trace metals
to Lake Michigan.
Proc. Conf. Great Lakes Res., Int. Assoc.
Great Lakes Res., 15:270–290.

Robertson, J. D.
1941. The function and metabolism of calcium
in the Invertebrata.
Biol. Rev., 16:106–133.

Robertson, R. N.
1960. Ion transport and respiration.
Biol. Rev., 35:231–264.

Rodewald-Rudescu, L.
1974. Das Schilfrohr *Phragmites communis*
Trinius.
Die Binnengewässer, 27, 302 pp.

Rodgers, G. K.
1966. The thermal bar in Lake Ontario, spring
1965 and winter 1965–66.
Publ. Great Lakes Res. Div., Univ. Mich.,
15:369–374.

Rodhe, W.
1948. Environmental requirements of freshwater plankton algae. Experimental studies in
the ecology of phytoplankton.
Symbol. Bot. Upsalien. 10(1), 149 pp.

Rodhe, W.
1949. The ionic composition of lake waters.
Verh. Int. Ver. Limnol., 10:377–386.

Rodhe, W.
1951. Minor constituents in lake waters.
Verh. Int. Ver. Limnol., 11:317–323.

Rodhe, W.
1955. Can plankton production proceed during winter darkness in subarctic lakes?
Verh. Int. Ver. Limnol., 12:117–122.

Rodhe, W.
1958. Primärproduktion und Seetypen.
Verh. Int. Ver. Limnol., 13:121–141.

Rodhe, W.
1969. Crystallization of eutrophication con-
cepts in Northern Europe. *In* Eutrophication:
Causes, Consequences, Correctives.
Washington, D.C., National Academy of
Sciences, pp. 50–64.

Rodhe, W.
1974. Plankton, planktic, planktonic.
Limnol. Oceanogr., 19:360.

Rodhe, W., R. A. Vollenweider, and A. Nauwerck.
1958. The primary production and standing
crop of phytoplankton. *In* A. A. Buzzati-
Traverso, ed. Perspectives in Marine Biology.
Berkeley, University of California Press, pp.
299–322.

Rodhe, W., J. E. Hobbie, and R. T. Wright.
1966. Phototrophy and heterotrophy in high
mountain lakes.
Verh. Int. Ver. Limnol., 16:302–313.

Rodina, A. G.
1965. Metody Vodnoi Mikrobiologii.
Moscow, Izdatel'stvo Nauka, 363 pp.

Rodina. A. G.
1972. Methods in Aquatic Microbiology.
Translat. and rev. R. R. Colwell and M. S.
Zambruski.
Baltimore, University Park Press, 461 pp.

Rohlich, G. A.
1969. Engineering aspects of nutrient removal. *In* Eutrophication: Causes, Consequences, Correctives.
Washington, D.C., National Academy of
Sciences, pp. 371–382.

Romanenko, V. I.
1966. Microbiological processes in the formation and breakdown of organic matter in the
Rybinsk Reservoir.
Trudy Inst. Biol. Vnutrennikh Vod, 13:137–
158. (Translated into English, Israel Program
for Scientific Translations, Jerusalem, 1969.)

Romanenko, V. I.
1967. Sootnoshenie mezhdu fotosintezom fito-
planktona i destruktsiei organicheskogo veshchestva v vodokhranilishchakh.
Trudy Inst. Biol. Vnutrennikh Vod, 15:61–74.

Romanenko, V. I., and E. G. Dobrynin.
1973. Potreblenie kisloroda, temnovaia
assimiliatsiia CO_2 i intensivnost fotosinteza
v natural'nykh i profil'trovannykh probakh
vody.
Mikrobiologiya, 42:573–575.

Romanenko, V. I., and S. I. Kuznetsov.
1974. Ekologiia Mikroorganizmov Presnykh
Vodoemov. Laboratornoe Rukovodstvo.
Leningrad, Izdatel'stvo Nauka, 194 pp.

Ross, H. H.
1963. Stream communities and terrestrial biomes.
Arch. Hydrobiol., 59:235–242.

Rossolimo, L.
1935. Die Boden-Gasausscheidung und das Sauerstoffregime der Seen.
Verh. Int. Ver. Limnol., 7:539–561.

Rosswall, T., ed.
1973. Modern Methods in the Study of Microbial Ecology.
Bull. Ecological Research Committee, Swedish Natural Science Research Council 17, 511 pp.

Roth, J. C.
1968. Benthic and limnetic distribution of three Chaoborus species in a southern Michigan lake (Diptera, Chaoboridae).
Limnol. Oceanogr., 13:242–249.

Round, F. E.
1957a. Studies on bottom-living algae in some lakes of the English Lake District. Part I. Some chemical features of the sediments related to algal productivities.
J. Ecol., 45:133–148.

Round, F. E.
1957b. Studies on bottom-living algae in some lakes of the English Lake District. Part II. The distribution of Bacillariophyceae on the sediments.
J. Ecol., 45:343–360.

Round, F. E.
1957c. Studies on bottom-living algae in some lakes of the English Lake District. Part III. The distribution on the sediments of algal groups other than the Bacillariophyceae.
J. Ecol., 45:649–664.

Round, F. E.
1960. Studies on bottom-living algae in some lakes of the English Lake District. Part IV. The seasonal cycles of the Bacillariophyceae.
J. Ecol., 48:529–547.

Round, F. E.
1961a. Studies on bottom-living algae in some lakes of the English Lake District. Part V. The seasonal cycles of the Cyanophyceae.
J. Ecol., 49:31–38.

Round, F. E.
1961b. Studies on bottom-living algae in some lakes of the English Lake District. Part VI. The effect of depth on the epipelic algal community.
J. Ecol., 49:245-254.

Round, F. E.
1964a. The ecology of benthic algae. In D. F. Jackson, ed. Algae and Man.
New York, Plenum Press, pp. 138–184.

Round, F. E.
1964b. The diatom sequence in lake deposits: Some problems of interpretation.
Verh. Int. Ver. Limnol., 15:1012–1020.

Round, F. E.
1965. The epipsammon; A relatively unknown freshwater algal association.
Brit. Phycol. Bull., 2:456–462.

Round, F. E.
1971. The growth and succession of algal populations in freshwaters.
Mitt. Int. Ver. Limnol., 19:70–99.

Round, F. E.
1972. Patterns of seasonal succession of freshwater epipelic algae.
Brit. Phycol. J., 7:213-220.

Round, F. E., and J. W. Eaton.
1966. Persistent, vertical-migration rhythms in benthic microflora. III. The rhythm of epipelic algae in a freshwater pond.
J. Ecol., 54:609–615.

Round, F. E., and C. M. Happey.
1965. Persistent, vertical-migration rhythms in benthic microflora. IV. A diurnal rhythm of the epipelic diatom association in non-tidal flowing water.
Brit. Phycol. Bull., 2:463–471.

Ruck, P.
1965. The components of the visual system of a dragonfly.
J. Gen. Physiol., 49:289–307.

Rudescu, L., C. Niculescu, and I. P. Chivu.
1965. Monografia Stufului din Delta Dunării.
Bucharest, Editura Acad. Rep. Soc. Romania, 542 pp.

Ruschke, R.
1968. Die Bedeutung von Wassermyxobakterien für den Abbau organischen Materials.
Mitt. Int. Ver. Limnol., 14:164–167.

Rusness, D., and R. H. Burris.
1970. Acetylene reduction (nitrogen fixation) in Wisconsin lakes.
Limnol. Oceanogr., 15:808–813.

Ruttner, F.
1931. Hydrographische und hydrochemische Beobachtungen auf Java, Sumatra und Bali.
Arch. Hydrobiol. Suppl., 8:197–454.

Ruttner, F.
1933. Über metalimnische Sauerstoffminimum.
Die Naturwissenschaften, 1933(21–23):401–404.

Ruttner, F.
1947. Zur Frage der Karbonatassimilation der Wasserpflanzen. I. Die beiden Haupttypen der Kohlenstoffaufnahme.
Öst. Bot. Z., 94:265–294.

Ruttner, F.
1948. Zur Frage der Karbonatassimilation der Wasserpflanzen. II. Das Verhalten von *Elodea canadensis* und *Fontinalis antipyretica* in Lösungen von Natrium- bzw. Kaliumbikarbonat.
Öst. Bot. Z., 95:208–238.

Ruttner, F.
1960. Über die Kohlenstoffaufnahme bei Algen aus der Rhodophyceen-Gattung *Batrachospermum*.
Schweiz. Z. Hydrol., 22:280–291.

Ruttner, F.
1963. Fundamentals of Limnology. (Translat. D. G. Frey and F. E. J. Fry.)
Toronto, University of Toronto Press, 295 pp.

Ruttner-Kolisko, A.
1964. Über die labile Periode im Fortpflanzungszyklus der Rädertiere.
Int. Rev. ges. Hydrobiol., 49:473-482.

Ruttner-Kolisko, A.
1972. Rotatoria. *In* Das Zooplankton der Binnengewässer. 1. Teil.
Die Binnengewässer, 26(Pt. 1):99–234.

Ryhänen, R.
1968. Die Bedeutung der Humussubstanzen im Stoffhaushalt der Gewässer Finnlands.
Mitt. Int. Ver. Limnol., 14:168–178.

Safferman, R. S., and M.-E. Morris.
1963. Algal virus: Isolation.
Science, 140:679–680.

Sager, P. E., and A. D. Hasler.
1969. Species diversity in lacustrine phytoplankton. I. The components of the index of diversity from Shannon's formula.
Amer. Nat., 103:51–59.

Sandner, H., and J. Wilkialis.
1972. Leech communities (Hirudinea) in the Mazurian and Bialystok regions and the Pomeranian Lake District.
Ekol. Polska, 20:345–365.

Sandercock, G. A.
1967. A study of selected mechanisms for the coexistence of *Diaptomus* spp. in Clarke Lake, Ontario.
Limnol. Oceanogr., 12:97–112.

Sanger, J. E.
1971a. Identification and quantitative measurement of plant pigments in soil humus layers.
Ecology, 52:959–963.

Sanger, J. E.
1971b. Quantitative investigations of leaf pigments from their inception in buds through autumn coloration to decomposition in falling leaves.
Ecology, 52:1075–1089.

Sanger, J. E., and E. Gorham.
1970. The diversity of pigments in lake sediments and its ecological significance.
Limnol. Oceanogr., 15:59–69.

Sanger, J. E., and E. Gorham.
1973. A comparison of the abundance and diversity of fossil pigments in wetland peats and woodland humus layers.
Ecology, 54:605–611.

Sauberer, F.
1939. Beiträge zur Kenntnis des Lichtklimas einiger Alpenseen.
Int. Rev. ges. Hydrobiol., 39:20–55.

Sauberer, F.
1950. Die spektrale Strahlungsdurchlässigkeit des Eises.
Wetter u. Leben, 2:193–197.

Sauberer, F.
1962. Empfehlungen für die Durchführung von Strahlungsmessungen an und in Gewässern.
Mitt. Int. Ver. Limnol., 11, 77 pp.

Saunders, G. W.
1957. Interrelations of dissolved organic matter and phytoplankton.
Bot. Rev., 23:389–410.

Saunders, G. W.
1963. The biological characteristics of fresh water.
Publ. Great Lakes Res. Div., Univ. Mich., 10:245–257.

Saunders, G. W.
1969. Some aspects of feeding in zooplankton. *In* Eutrophication: Causes, Consequences, Correctives.
Washington, D.C., National Academy of Sciences, pp. 556–573.

Saunders, G. W.
1971. Carbon flow in the aquatic system. *In* J. Cairns, Jr., ed. The Structure and Function of Fresh-Water Microbial Communities. Res. Div. Monogr. 3, Virginia Polytechnic Inst., pp. 31–45.

Saunders, G. W.
1972a. The transformation of artificial detritus in lake water.
Mem. Ist. Ital. Idrobiol., 29 (Suppl.):261–288.

Saunders, G. W.,
1972b. Summary of the general conclusions of the symposium.
Mem. Ist. Ital. Idrobiol., 29(Suppl.):533–540.

Saunders, G. W.
1972c. Potential heterotrophy in a natural population of *Oscillatoria agardhii* var. *isothrix* Skuja.
Limnol. Oceanogr., *17*:704–711.

Saunders, G. W., and T. A. Storch.
1971. Coupled oscillatory control mechanism in a planktonic system.
Nature (New Biol.), *230*:58–60.

Savostin, P.
1972. Microbial transformation of silicates.
Z. Pflanzenernährung Bodenkunde, *132*:37–45.

Sawyer, R. T.
1974. Leeches (Annelida: Hirudinea). *In* C. W. Hart, Jr., and S. L. H. Fuller, eds. Pollution Ecology of Freshwater Invertebrates.
New York, Academic Press, pp. 81–142.

Schelske, C. L.
1962. Iron, organic matter, and other factors limiting primary productivity in a marl lake.
Science, *136*:45–46.

Schelske, C. L., L. E. Feldt, M. A. Santiago, and E. F. Stoermer.
1972. Nutrient enrichment and its effects on phytoplankton production and species composition in Lake Superior.
Proc. Conf. Great Lakes Res., Int. Assoc. Great Lakes Res., *15*:149–165.

Schelske, C. L., F. F. Hooper, and E. J. Haertl.
1962. Responses of a marl lake to chelated iron and fertilizer.
Ecology, *43*:646–653.

Schelske, C. L., and E. F. Stoermer.
1971. Eutrophication, silica depletion, and predicted changes in algal quality in Lake Michigan.
Science, *173*:423–424.

Schelske, C. L., and E. F. Stoermer.
1972. Phosphorus, silica, and eutrophication of Lake Michigan. *In* G. E. Likens, ed. Nutrients and Eutrophication.
Special Symposium, Amer. Soc. Limnol. Oceanogr., *1*:157–171.

Schiegl, W. E.
1972. Deuterium content of peat as a paleoclimatic recorder.
Science, *175*:512–513.

Schindler, D. W.
1968. Feeding, assimilation and respiration rates of *Daphnia magna* under various environmental conditions and their relation to production estimates.
J. Anim. Ecol., *37*:369–385.

Schindler, D. W.
1970. Production of phytoplankton and zoo-
plankton in Canadian Shield lakes. *In* Z. Kajak and A. Hillbricht-Ilkowska, eds. Productivity Problems of Freshwaters.
Warsaw, PWN Polish Scientific Publishers, pp. 311–331.

Schindler, D. W.
1974. Eutrophication and recovery in experimental lakes: Implications for lake management.
Science, *184*:897–899.

Schindler, D. W., G. J. Brunskill, S. Emerson, W. S. Broecker, and T.-H. Peng.
1972. Atmospheric carbon dioxide: Its role in maintaining phytoplankton standing crops.
Science, *177*:1192–1194.

Schindler, D. W., V. E. Frost, and R. V. Schmidt.
1973. Production of epilithiphyton in two lakes of the Experimental Lakes Area, northwestern Ontario.
J. Fish. Res. Bd. Canada, *30*:1511–1524.

Schmidt, W.
1915. Über den Energie-gehalt der Seen. Mit Beispielen vom Lunzer Untersee nach Messungen mit einem enfachen Temperaturlot.
Int. Rev. Hydrobiol. Suppl., *6*. (Not seen in original.)

Schmidt, W.
1928. Über Temperatur und Stabilitätsverhältnisse von Seen.
Geographiska Annaler, *10*:145–177.

Schnitzer, M.
1971. Metal-organic matter interactions in soils and waters. *In* S. J. Faust and J. V. Hunter, eds. Organic Compounds in Aquatic Environments.
New York, Marcel Dekker, Inc., pp. 297–315.

Schnitzer, M., and S. U. Khan.
1972. Humic Substances in the Environment.
New York, Marcel Dekker, Inc., 327 pp.

Schönborn, W.
1962. Über Planktismus und Zyklomorphose bei *Difflugia limnetica* (Levander) Penard.
Limnologica, *1*:21–34.

Schreiter, T.
1928. Untersuchungen über den Einfluss einer Helodeawucherung auf das Netzplankton des Hirschberger Grossteiches in Böhmen in den Jahren 1921 bis 1925 incl.
Sborník Výzk. Úst. Zeměd. RČS *61*, 98 pp.

Schröder, R.
1973. Die Freisetzung von Pflanzennährstoffen im Schilfgebiet und ihr Transport in das Freiwasser am Beispiel des Bodensee-Untersees.
Arch. Hydrobiol., *71*:145–158.

Schultz, D. M., and J. G. Quinn
1973. Fatty acid composition of organic detritus from *Spartina alterniflora*.
Estuarine and Coastal Mar. Sci., *1*:177–190.

Schwoerbel, J., and G. C. Tillmanns.
1964a. Konzentrationsabhängige Aufnahme von wasserlöslichem PO₄-P bei submersen Wasserpflanzen.
Naturwissenschaften, *51*:319–320.

Schwoerbel, J., and G. C. Tillmanns.
1964b. Untersuchungen über die Stoffwechseldynamik in Fliessgewässern. I. Die Rolle höherer Wasserpflanzen: *Callitriche hamulata* Kütz.
Arch. Hydrobiol. Suppl., *28*:245–258.

Schwoerbel, J., and G. C. Tillmanns.
1964c. Untersuchungen über die Stoffwechseldynamik in Fliessgewässern. II. Experimentelle Untersuchungen über die Ammoniumaufnahme und pH-Änderung im Wasser durch *Callitriche hamulta* Kütz. und *Fontinalis antipyretica* L.
Arch. Hydrobiol. Suppl., *28*:259–267.

Schwoerbel, J., and G. C. Tillmanns.
1972. Ammonium-Adaptation bei submersen Phanerogamen in situ.
Arch. Hydrobiol. Suppl., *42*:139–141.

Scott, J. T., G. E. Myer, R. Stewart, and E. G. Walther.
1969. On the mechanism of Langmuir circulations and their role in epilimnion mixing.
Limnol. Oceanogr., *14*:493–503.

Scott, W.
1924. The diurnal oxygen pulse in Eagle (Winona) Lake.
Proc. Indiana Acad. Sci., *33*(1923):311–314.

Sculthorpe, C. D.
1967. The Biology of Aquatic Vascular Plants.
New York, St. Martin's Press, 610 pp.

Sebestyén, O.
1949. Studies of detritus drifts in Lake Balaton.
Verh. Int. Ver. Limnol., *10*:414–419.

Sederholm, H., A. Mauranen, and L. Montonen.
1973. Some observations on the microbial degradation of humous substances in water.
Verh. Int. Ver. Limnol., *18*:1301–1305.

Šesták, Z., J. Čatský, and P. G. Jarvis, eds.
1971. Plant Photosynthetic Production Manual of Methods.
The Hague, W. Junk N. V. Publishers, 818 pp.

Shannon, E. L.
1953. The production of root hairs by aquatic plants.
Amer. Midland. Nat., *50*:474–479.

Shapiro, J.
1957. Chemical and biological studies on the yellow organic acids of lake water.
Limnol. Oceanogr., 2:161–179.

Shapiro, J.
1960. The cause of a metalimnetic minimum of dissolved oxygen.
Limnol. Oceanogr., 5:216–227.

Shapiro, J.
1964. Effect of yellow organic acids on iron and other metals in water.
J. Amer. Water Wks. Assoc., 56:1062–1082.

Shapiro, J.
1966. The relation of humic color to iron in natural waters.
Verh. Int. Ver. Limnol., *16*:477–484.

Shapiro, J.
1969. Iron in natural waters—its characteristics and biological availability as determined with the ferrigram.
Verh. Int. Ver. Limnol., *17*:456–466.

Shelford, V. E.
1918. Conditions of existence. In Freshwater Biology. H. B. Ward and G. C. Whipple, eds. New York, John Wiley & Sons, Inc., pp. 21–60.

Shilo, M.
1971. Biological agents which cause lysis of blue-green algae.
Mitt. Int. Ver. Limnol., *19*:206–213.

Shtegman, B. K., ed.
1966. Produtsirovanie i Krugovorot Organicheskogo Veshchestva vo Vnutrennikh Vodoemakh. (Production and circulation of organic matter in inland waters.)
(Translated into English, Israel Program for Scientific Translations, Jerusalem, 1969.)
Trudy Inst. Biol. Vnutrennikh Vod *13*, 287 pp.

Shukla, S. S., J. K. Syers, J. D. H. Williams, D. E. Armstrong, and R. F. Harris.
1971. Sorption of inorganic phosphate by lake sediments.
Proc. Soil Sci. Soc. Amer., 35:244–249.

Shulman, M. D., and R. A. Bryson.
1961. The vertical variation of wind-driven currents in Lake Mendota.
Limnol. Oceanogr., 6:347–355.

Siebeck, O.
1960. Untersuchungen über die Vertikalwanderung planktischer Crustaceen unter Berücksichtigung der Strahlungsverhältnisse.
Int. Rev. ges. Hydrobiol, *45*:381–454.

Siebeck, O.
1968. "Uferflucht" und optische Orientierung pelagischer Crustaceen.
Arch. Hydrobiol. Suppl., 35:1–118.

Siebeck, O., and J. Ringelberg.
1969. Spatial orientation of planktonic crustaceans. I. The swimming behaviour in a horizontal plane. 2. The swimming behaviour in a vertical plane.
Verh. Int. Ver. Limnol., *17*:831–847.

Simpson, J. H. and J. D. Woods.
1970. Temperature microstructure in a freshwater thermocline.
Nature, *226*:832–834.

Sjörs, H.
1961. Surface patterns in Boreal peatland.
Endeavour, *20*:217–224.

Sládeček, V.
1973. System of water quality from the biological point of view.
Arch. Hydrobiol. Beih. Ergebn. Limnol., 7, 218 pp.

Sládeček, V., and A. Sládečková.
1963. Limnological study of the Reservoir Sealice near Želiv. XXIII. Periphyton production.
Sborník Vysoké Školy Chem.-Technol. Praze, Technol. Vody, 7:77–133.

Sládeček, V., and A. Sládečková.
1964. Determination of the periphyton production by means of the glass slide method.
Hydrobiologia, 23:125–158.

Sládečková, A.
1960. Limnological study of the Reservoir Sedlice near Želiv. XI. Periphyton stratification during the first year-long period (June 1957-July 1958).
Sci. Pap. Inst. Chem. Technol. Prague, Faculty of Technol. Fuel Water, 4:143–261.

Sládečková, A.
1962. Limnological investigation methods for the periphyton ("Aufwuchs") community.
Bot. Rev., 28:286–350.

Slobodkin, L. B.
1954. Population dynamics in *Daphnia obtusa* Kurz.
Ecol. Monogr., 24:69–88.

Slobodkin, L. B.
1960. Ecological energy relationships at the population level.
Amer. Nat., 94:213–236.

Slobodkin, L. B.
1962. Energy in animal ecology.
Adv. Ecol. Res., 1:69–101.

Smayda, T. J., and B. J. Boleyn.
1965. Experimental observations on the flotation of marine diatoms. I. *Thalassiosira* cf. *nana, Thalassiosira rotula* and *Nitzschia seriata.*
Limnol. Oceanogr., 10:499–509.

Smayda, T. J., and B. J. Boleyn.
1966a. Experimental observations on the flotation of marine diatoms. II. *Skeletonema costatum* and *Rhizosolenia setigera.*
Limnol. Oceanogr., *11*:18–34.

Smayda, T. J., and B. J. Boleyn.
1966b. Experimental observations on the flotation of marine diatoms. III. *Bacteriastrum hyalinum* and *Chaetoceros lauderi.*
Limnol. Oceanogr., *11*:35–43.

Smith, I. R., and I. J. Sinclair.
1972. Deep water waves in lakes.
Freshwat. Biol., 2:387–399.

Smith, M. W.
1969. Changes in environment and biota of a natural lake after fertilization.
J. Fish. Res. Bd. Canada, 26:3101–3132.

Smith, R. C., and W. H. Wilson, Jr.
1972. Photon scalar irradiance.
Appl. Optics, *11*:934–938.

Smith, R. C., J. E. Tyler, and C. R. Goldman.
1973. Optical properties and color of Lake Tahoe and Crater Lake.
Limnol. Oceanogr., *18*:189–199.

Smith, S. H.
1970. Species interactions of the alewife in the Great Lakes.
Trans. Amer. Fish. Soc., 99:754–765.

Smith, S. H.
1972a. Factors of ecologic succession in oligotrophic fish communities of the Laurentian Great Lakes.
J. Fish. Res. Bd. Canada, 29:717–730.

Smith, S. H.
1972b. The future of salmonid communities in the Laurentian Great Lakes.
J. Fish, Res. Bd. Canada, 29:951–957.

Smyly, W. J. P.
1973. Bionomics of *Cyclops strenuus abyssorum* Sars (Copepoda: Cyclopoida).
Oecologia, *11*:163–186.

Soeder, C. J.
1965. Some aspects of phytoplankton growth and activity.
Mem. Ist. Ital. Idrobiol., *18*(Suppl.):47–59.

Sokolova, G. A.
1961. Rol' zhalezobakterii v dinamike zheleza v Glubokom Ozere.
Trudy Vses. Gidrobiol. Obshch., *11*:5–11.

Solski, A.
1962. Mineralizacja roślin wodnych. I. Uwalnianie fosforu i potasu przez wymywanie.
Pol. Arch. Hydrobiol., *10*:167–196.

Sommers, L. E., R. F. Harris, J. D. H. Williams, D. E. Armstrong, and J. K. Syers.

1970. Determination of total organic phosphorus in lake sediments.
Limnol. Oceanogr., 15:301–304.

Sorokin, Ju. I.
1964a. On the primary production and bacterial activities in the Black Sea.
J. Cons. Int. Explor. Mer, 29:41–60.

Sorokin, Ju. I.
1964b. On the trophic role of chemosynthesis in water bodies.
Int. Rev. ges. Hydrobiol., 49:307–324.

Sorokin, Ju. I.
1965. On the trophic role of chemosynthesis and bacterial biosynthesis in water bodies.
Mem. Ist. Ital. Idrobiol., 18(Suppl.):187–205.

Sorokin, Ju. I.
1966. Vzaimosviaz mikrobiologicheskikh protsessov krugovorota sepy i ugleroda v meromikticheskom Ozere Belovod. In Plankton i Bentos.
Trudy Inst. Biol. Vnutrennikh Vod, 12:332–355.

Sorokin, Ju. I.
1968. The use of ^{14}C in the study of nutrition of aquatic animals.
Mitt. Int. Ver. Limnol., 16, 41 pp.

Sorokin, Ju. I.
1970. Interrelations between sulphur and carbon turnover in meromictic lakes.
Arch. Hydrobiol., 66:391–446.

Sorokin, Y. I., and H. Kadota, eds.
1972. Techniques for the Assessment of Microbial Production and Decomposition in Fresh Waters. Int. Biol. Program Handbook 23.
Oxford, Blackwell Scientific Publications, 112 pp.

Sorokin, Ju, I. and E. B. Paveljeva.
1972. On the quantitative characteristics of the pelagic ecosystem of Dalnee Lake (Kamchatka).
Hydrobiologia, 40:519–552.

Sparrow, F. K., Jr.
1968. Ecology of freshwater fungi. In G. C. Ainsworth and A. S. Sussman, eds. The Fungi. Vol. III. The Fungal Population.
New York, Academic Press, pp. 41–93.

Spence, D. H. N., and J. Chrystal.
1970a. Photosynthesis and zonation of freshwater macrophytes. I. Depth distribution and shade tolerance.
New Phytol., 69:205–215.

Spence, D. H. N., and J. Chrystal.
1970b. Photosynthesis and zonation of fresh water macrophytes. II. Adaptability of species of deep and shallow water.
New Phytol., 69:217–227.

Spence, D. H. N., R. M. Campbell, and J. Chrystal.
1971. Productivity of submerged freshwater macrophytes.
Hidrobiologia, 12:169–176.

Sprules, W. G.
1972. Effects of size-selective predation and food competition on high altitude zooplankton communities.
Ecology, 53:375–386.

Stadelmann, P.
1971. Stickstoffkreislauf und Primärproduktion im mesotrophen Vierwaldstättersee (Horwer Bucht) und im eutrophen Rotsee, mit besonderer Berücksichtigung des Nitrats als limitierenden Faktors.
Schweiz. Z. Hydrol., 33:1–65.

Stahl, J. B.
1959. The developmental history of the chironomid and Chaoborus faunas of Myers Lake.
Invest. Indiana Lakes Streams, 5:47–102.

Stahl, J. B.
1966a. The ecology of Chaoborus in Myers Lake, Indiana.
Limnol. Oceanogr., 11:177–183.

Stahl, J. B.
1966b. Coexistence in Chaoborus and its ecological significance.
Invest. Indiana Lakes Streams, 7:99–113.

Stangenberg, M.
1968. Bacteriostatic effects of some algae- and Lemna minor extracts.
Hydrobiologia, 32:88–96.

Stangenberg-Oporowska, K.
1967. Sodium contents in carp-pond waters in Poland.
Pol. Arch. Hydrobiol., 14:11–17.

Stanier, R. Y.
1973. Autotrophy and heterotrophy in unicellular blue-green algae. In N. G. Carr and B. A. Whitton, eds. The Biology of Blue-Green Algae.
Berkeley, University of California Press, pp. 501–518.

Stanley, R. A.
1972. Photosynthesis in Eurasian watermilfoil (Myriophyllum spicatum L.)
Plant Physiol., 50:149–151.

Steele, J. H., and I. E. Baird.
1968. Production ecology of a sandy beach.
Limnol. Oceanogr., 13:14–25.

Steemann Nielsen, E.
1944. Dependence of freshwater plants on quantity of carbon dioxide and hydrogen ion concentration.
Dansk Bot. Ark., 11:1–25.

Steemann Nielsen, E.
1947. Photosynthesis of aquatic plants with special reference to the carbon sources.
Dansk Bot. Ark., 12:5–71.

Steemann Nielsen, E.
1962a. Inactivation of the photochemical mechanism in photosynthesis as a means to protect the cells against too high light intensities.
Physiol. Plant., 15:161–171.

Steemann Nielsen, E.
1962b. On the maximum quantity of plankton chlorophyll per surface unit of a lake or the sea.
Int. Rev. ges. Hydrobiol., 47:333–338.

Steemann Nielsen, E., and E. G., Jørgensen.
1962. The physiological background for using chlorophyll measurements in hydrobiology and a theory explaining daily variations in chlorophyll concentration.
Arch. Hydrobiol., 58:349–357.

Steemann Nielsen, E., and E. G., Jørgensen.
1968a. The adaptation of plankton algae. I. General part.
Physiol. Plant., 21:401–413.

Steemann Nielsen, E., and E. G., Jørgensen.
1968b. The adaptation of plankton algae. III. With special consideration of the importance in nature.
Physiol. Plant., 21:647–654.

Steemann Nielsen, E., and M. Willemoës.
1971. How to measure the illumination rate when investigating the rate of photosynthesis of unicellular algae under various light conditions.
Int. Rev. ges. Hydrobiol., 56:541–556.

Stein, J. R., ed.
1973. Handbook of Phycological Methods. Culture Methods and Growth Measurements. Cambridge, Cambridge University Press, 448 pp.

Stemler, A., and Govindjee.
1973. Bicarbonate ion as a critical factor in photosynthetic oxygen evolution.
Plant Physiol., 52:119–123.

Štěpánek, M.
1959. Limnological study of the reservoir Sedlice near Želiv. IX. Transmission and transparency of water.
Sci. Pap. Inst. Chem. Technol., Prague, Fac. Technol.
Fuel Water, 3(Pt. 2):363–430.

Stewart, K. M.
1973. Detailed time variations in mean temperature and heat content of some Madison lakes.
Limnol. Oceanogr., 18:218–226.

Stewart, R., and R. K. Schmitt.
1968. Wave interaction and Langmuir circulations.
Proc. Conf. Great Lakes Res., Int. Assoc. Great Lakes Res., 11:496–499.

Stewart, W. D. P.
1968. Nitrogen input into aquatic ecosystems.
In D. F. Jackson, ed. Algae, Man, and the Environment.
Syracuse, N.Y., Syracuse University Press, pp. 53–72.

Stewart, W. D. P.
1969. Biological and ecological aspects of nitrogen fixation by free-living micro-organisms.
Proc. Roy. Soc. London (Ser. B), 172:367–388.

Stewart, W. D. P.
1973. Nitrogen fixation. In N. G. Carr and B. A. Whitton, eds. The Biology of the Blue-Green Algae.
Berkeley, University of California Press, pp. 260–278.

Stöber, W.
1967. Formation of silicic acid in aqueous suspensions of different silica modifications. In Equilibrium concepts in natural water systems.
Adv. Chem. Ser., 67:161–182.

Stocking, C. R.
1956. Vascular conduction in submerged plants. In W. Ruhland, ed. Handbuch der Pflanzenphysiologie.
Band 3. Pflanze und Wasser. Berlin, Springer-Verlag, pp. 587–595.

Stockner, J. G., and W. W. Benson.
1967. The succession of diatom assemblages in the recent sediments of Lake Washington.
Limnol. Oceanogr., 12:513–532.

Stommel, H.
1949. Horizontal diffusion due to oceanic turbulence.
J. Mar. Res., 8:199–225.

Stout, G. E., ed.
1967. Isotope Techniques in the Hydrologic Cycle.
Geophys. Monogr. Ser. 11. Washington, D.C., American Geophysical Union, 199 pp.

Straškraba, M.
1963. Share of the littoral region in the productivity of two fishponds in southern Bohemia.
Rozpravy Českosl. Akad. Věd, Řada Matem. Přír. Věd, 73(13), 64 pp.

Straškraba, M.
1965. The effect of fish on the number of invertebrates in ponds and streams.
Mitt. Int. Ver. Limnol., 13:106–127.

Straškraba, M.
1967. Quantitative study on the littoral zoo-plankton of the Poltruba Backwater with an attempt to disclose the effect on fish.
Rozpravy Ćeskosl. Akad. Vĕd, Řada Matem.
Přír. Vĕd, 77(11):7–34.

Straškraba, M.
1968. Der Anteil der höheren Pflanzen an der Produktion der stehenden Gewässer.
Mitt. Int. Ver. Limnol., 14:212–230.

Straškraba, M., J. Hrbáček, and P. Javornický.
1973. Effect of an upstream reservoir on the stratification conditions in Slapy Reservoir.
Hydrobiol. Studies, 2:7–82.

Straškraba, M., and E. Piezyńska.
1970. Field experiments on shading effect by emergents on littoral phytoplankton and peri-phyton production.
Rozpravy Ceskosl. Akad. Ved, Rada Matem.
Přír. Vĕd, 80(6):7–32.

Strayer, R. F.
1973. The in situ rate of methane production in a small eutrophic, hard-water lake.
M. Sc. Thesis, Michigan State University, East Lansing, 40 pp.

Strekal, T. A., and W. F. McDiffett.
1974. Factors affecting germination, growth, and distribution of the freshwater sponge, Spongilla fragilis Leidy (Porifera).
Biol. Bull., 146:267–278.

Strickland, J. D. H.
1958. Solar radiation penetrating the ocean. A review of requirements, data and methods of measurement, with particular reference to photosynthetic productivity.
J. Fish. Res. Bd. Canada, 15:453–493.

Strickland, J. D. H.
1960. Measuring the production of marine phytoplankton.
Bull. Fish. Res. Bd. Canada, 122, 172 pp.

Strickland, J. D. H., and T. R. Parsons.
1972. A practical handbook of seawater analysis. 2nd ed.
Bull. Fish. Res. Bd. Canada, 167, 310 pp.

Strøm, K. M.
1933. Nutrition of algae. Experiments upon: The feasibility of the Schreiber Method in fresh waters; the relative importance of iron and manganese in the nutritive medium; the nutritive substance given off by lake bottom muds.
Arch. Hydrobiol., 25:38–47.

Strøm, K. M.
1945. The temperature of maximum density in fresh waters. An attempt to determine its low-ering with increased pressure from observations in deep lakes.
Geofysiske Publikasjoner, 16(8), 14 pp.

Strøm, K. M.
1947. Correlation between χ^{18} and pH in lakes.
Nature, 159:782–783.

Strøm, K. M.
1955. Waters and sediments in the deep of lakes.
Mem. Ist. Ital. Idrobiol., 8(Suppl.):345–356.

Strong, D. R., Jr.
1972. Life history variation among popula-tions of an amphipod (Hyalella azteca).
Ecology, 53:1103–1111.

Stross, R. G.
1966. Light and temperature requirements for diapause development and release in Daph-nia.
Ecology, 47:368–374.

Stross, R. G., and J. C. Hill.
1965. Diapause induction in Daphnia requires two stimuli.
Science, 150:1462–1464.

Stuiver, M.
1967. The sulfur cycle in lake waters during thermal stratification.
Geochim. Cosmochim. Acta, 31:2151–2167.

Stuiver, M.
1968. Oxygen-18 content of atmospheric pre-cipitation during last 11,000 years in the Great Lakes region.
Science, 162:994–997.

Stull, E. A., E. de Amezaga, and C. R. Goldman.
1973. The contribution of individual species of algae to primary productivity of Castle Lake, California.
Verh. Int. Ver. Limnol., 18:1776–1783.

Stumm, W.
1966. Redox potential as an environmental parameter; conceptual significance and opera-tional limitation.
3rd Int. Conf. on Water Poll. Res., Water Poll. Control Federation. Sec. 1, Paper 13, 16 pp.

Stumm, W., and J. O. Leckie.
1971. Phosphate exchange with sediments; Its role in the productivity of surface waters.
Proc. Water Poll. Res. Conf. III, Art. 26, 16 pp.

Stumm, W., and G. F. Lee.
1960. The chemistry of aqueous iron.
Schweiz. Z. Hydrol., 22:295–319.

Stumm, W., and J. J. Morgan.
1970. Aquatic Chemistry. An Introduction Emphasizing Chemical Equilibria in Natural Waters.
New York, Wiley-Interscience, 583 pp.

Suess, E.
1968. Calcium carbonate interaction with organic compounds.
Ph.D. Diss., Lehigh University, Bethlehem, Pa.

Suess, E.
1970. Interaction of organic compounds with calcium carbonate. I. Association phenomena and geochemical implications.
Geochim. Cosmochim. Acta, 34:157–168.

Sugawara, K.
1961. Na, Cl and Na/Cl in inland waters.
Japan. J. Limnol., 22:49–65.

Swain, F. M.
1963. Geochemistry of humus. In I. A. Breger, ed. Organic Geochemistry.
New York, Macmillan Company, pp. 87–147.

Swain, F. M.
1965. Geochemistry of some Quaternary lake sediments of North America. In H. E. Wright, Jr., and D. G. Frey, eds. The Quaternary of the United States.
Princeton, N. J., Princeton University Press, pp. 765–781.

Swüste, H. F. J., R. Cremer, and S. Parma.
1973. Selective predation by larvae of Chaoborus flavicans (Diptera, Chaoboridae).
Verh. Int. Ver. Limnol., 18:1559–1563.

Symons, F.
1972. On the changes in the structure of two algal populations: Species diversity and stability.
Hydrobiologia, 40:499–502.

Symposium on the Classification of Brackish Waters.
1959. Venice, 1958. Societas Internationalis Limnologiae.
Arch. Oceanogr. Limnol. Suppl., 11, 248 pp.

Szczepańska, W.
1973. Production of helophytes in different types of lakes.
Pol. Arch. Hydrobiol., 20:51–57.

Szczepański, A.
1965. Deciduous leaves as a source of organic matter in lakes.
Bull. Açad. Polon. Sci., Cl., II, 13:215–217.

Szczepański, A.
1968. Scattering of light and visibility in water of different types of lakes.
Pol. Arch. Hydrobiol., 15:51–77.

Szczepański, A.
1969. Biomass of underground parts of the reed Phragmites communis Trin.
Bull. Acad. Polon. Sci., Cl., II, 17:245–246.

Szczepański, A., and W. Szczepańska.
1966. Primary production and its dependence on the quantity of periphyton.
Bull. Acad. Polon. Sci., Cl., II, 14:45–50.

Szilágyi, M.
1973. The redox properties and the determination of the normal potential of the peat-water system.
Soil Sci., 115:434–437.

Szumiec, M.
1961. Eingangsmessungen der Strahlungsintensität in Teichen.
Acta Hydrobiol., 3:133–142.

Takahashi, M., and S. Ichimura.
1968. Vertical distribution and organic matter production of photosynthetic sulfur bacteria in Japanese lakes.
Limnol. Oceanogr., 13:644–655.

Talling, J. F.
1960. Self-shading effects in natural populations of a planctonic diatom.
Wetter u. Leben, 12:235–242.

Talling, J. F.
1965. The photosynthetic activity of phytoplankton in East African lakes.
Int. Rev. ges. Hydrobiol., 50:1–32.

Talling, J. F.
1969. The incidence of vertical mixing, and some biological and chemical consequences, in tropical African lakes.
Verh. Int. Ver. Limnol., 17:998–1012.

Talling, J. F.
1971. The underwater light climate as a controlling factor in the production ecology of freshwater phytoplankton.
Mitt. Int. Ver. Limnol., 19:214–243.

Talling, J. F.
1973. The application of some electrochemical methods to the measurement of photosynthesis and respiration in fresh waters.
Freshwat. Biol., 3:335–362.

Talling, J. F., and I. B. Talling.
1965. The chemical composition of African lake waters.
Int. Rev. ges. Hydrobiol., 50:421–463.

Talling, J. F., R. B. Wood, M. V. Prosser, and R. M. Baxter.
1973. The upper limit of photosynthetic productivity by phytoplankton: Evidence from Ethiopian soda lakes.
Freshwat. Biol., 3:53–76.

Tan, T. L.
1973. Physiologie der Nitratreduktion bei Pseudomonas aeruginosa.
Z. Allg. Mikrobiol., 13:83–94.

Tan, T. L., and J. Overbeck.
1973. Ökologische Untersuchungen über nitratreduzierende Bakterien im Wasser des Pluss-sees (Schleswig-Holstein).
Z. Allg. Mikrobiol., 13:71–82.

Tappa, D. W.
1965. The dynamics of the association of six limnetic species of Daphnia in Aziscoos Lake, Maine.
Ecol. Monogr., 35:395–423.

Taub, F. B., and A. M. Dollar.
1968. The nutritional inadequacy of *Chlorella* and *Chlamydomonas* as food for *Daphnia pulex*.
Limnol. Oceanogr., *13*:607–617.

Teal, J. M.
1957. Community metabolism in a temperate cold spring.
Ecol. Monogr., *27*:283–302.

Tessenow, U.
1966. Untersuchungen über den Kieselsäurehaushalt der Binnengewässer.
Arch. Hydrobiol. Suppl., *32*:1–136.

Thienemann, A.
1925. Die Binnengewässer Mitteleuropas. Eine limnologische Einführung.
Die Binnengewässer, *1*, 255 pp.

Thienemann, A.
1926. Der Nahrungskreislauf im Wasser.
Verh. Deutsch. Zool. Gesell., *31*:29–79.

Thienemann, A.
1927. Der Bau des Seebeckens in seiner Bedeutung für den Ablauf des Lebens im See.
Verh. Zool.-Bot. Ges. Wien, 77:87–91.

Thienemann, A.
1928. Der Sauerstoff im eutrophen und oligotrophen See. Ein Beitrag zur Seetypenlehre.
Die Binnengewässer, *4*, 175 pp.

Thienemann, A.
1931. Der Productionsbegriff in der Biologie.
Arch. Hydrobiol., *22*:616–622.

Thienemann, A.
1954. *Chironomus*. Leben, Verbreitung und wirtschaftliche Bedeutung der Chironomiden.
Die Binnengewässer, *20*, 834 pp.

Thomas, E. A.
1951. Sturmeinfluss auf das Tiefenwasser des Zürichsees im Winter.
Schweiz. Z. Hydrol., *13*:5–23.

Thomas, E. A.
1960. Sauerstoffminima und Stoffkreisläufe im ufernahen Oberflächenwasser des Zürichsees (Cladophora- und Phragmites-Gürtel).
Monatsbull. Schweiz. Ver. Gas- Wasserfachmännern, *1960*(6): 1–8.

Thunmark, S.
1945. Zur Soziologie des Süsswasserplanktons. Eine methodologisch-ökologische Studie.
Folia Limnol. Scandin., *3*, 66 pp.

Thut, R. N.
1969. A study of the profundal bottom fauna of Lake Washington.
Ecol. Monogr., *39*:79–100.

Tillman, D. L., and J. R. Barnes.
1973. The reproductive biology of the leech *Helobdella stagnalis* (L.) in Utah Lake, Utah.
Freshwat. Biol., 3:137–145.

Tilzer, M.
1972. Dynamik und Produktivität von Phytoplankton und pelagischen Bakterien in einem Hochgebirgssee (Vorderer Finstertaler See, Österreich).
Arch. Hydrobiol. Supp., *42*:201–273.

Tilzer, M. M.
1973. Diurnal periodicity in the phytoplankton assemblage of a high mountain lake.
Limnol. Oceanogr., *18*:15–30.

Tilzer, M. M., C. R. Goldman, and E. de Amezaga.
1975. The efficiency of photosynthetic light energy utilization by lake phytoplankton.
Verh. Int. Ver. Limnol., *19* (in press)

Timm, T.
1962a. Maloshchetnikovie chervi Chudsko Pskovskogo Ozera.
Gidrobiol. Issledovaniia, 3:106–108.

Timm, T.
1962b. O rasprostranenii maloshchetnikovikh chervei (Oligochaeta) v ozerakh Estonii.
Gidrobiol. Issledovaniia, 3:162–168.

Tippett, R.
1970. Artificial surfaces as a method of studying populations of benthic micro-algae in fresh water.
Brit. Phycol. J., 5:187–199.

Toetz, D. W.
1973a. The limnology of nitrogen in an Oklahoma reservoir: nitrogenase activity and related limnological factors.
Amer. Midland Nat., 89:369–380.

Toetz, D. W.
1973b. The kinetics of NH_4 uptake by *Ceratophyllum*.
Hydrobiologia, *41*:275–290.

Toetz, D. W.
1974. Uptake and translocation of ammonia by freshwater hydrophytes.
Ecology, 55:199–201.

Tressler, W. L.
1957. The Ostracoda of Great Slave Lake.
J. Wash. Acad. Sci., 47:415–423.

Trifonova, N. A., et al.
1969. Materialy k Soveshchaniiu po Prognozirovaniiu Soderzhaniia Biolennykh Elementov i Organicheskogo Veshchestva v Vodokhranilishchakh.
Rybinsk., Inst. Biol. Vnutrennikh Vod, 175 pp.

Trojanowski, J.
1969. Biological degradation of lignin.
Int. Biodetn. Bull., 5:119–124.

Truesdale, G. A., A. L. Downing, and G. F. Lowden.
1955. The solubility of oxygen in pure water and sea-water.
J. Appl. Chem., 5:53–62.

Trussell, R. P.
1972. The percent un-ionized ammonia in aqueous ammonia solutions at different pH levels and temperatures.
J. Fish. Res. Bd. Canada, 29:1505–1507.

Tucker, A.
1957. The relation of phytoplankton periodicity to the nature of the physico-chemical environment with special reference to phosphorus. I. Morphometrical, physical and chemical conditions. II. Seasonal and vertical distribution of the phytoplankton in relation to the environment.
Amer. Midland Nat., 57:300–370.

Tutin, W.
1969. The usefulness of pollen analysis in interpretation of stratigraphic horizons, both Late-glacial and Post-glacial.
Mitt. Int. Ver. Limnol., 17:154–164.

Tyler, J. E.
1961a. Sun-altitude effect on the distribution of underwater light.
Limnol. Oceanogr., 6:24–25.

Tyler, J. E.
1961b. Scattering properties of distilled and natural waters.
Limnol. Oceanogr., 6:451–456.

Tyler, J. E.
1968. The Secchi disc.
Limnol. Oceanogr.,13:1–6.

Tyler, J. E., and R. C. Smith.
1970. Measurements of Spectral Irradiance Underwater.
New York, Gordan and Breach Science Publ., 103 pp.

Úlehlová, B.
1970. An ecological study of aquatic habitats in northwest Overijssel, the Netherlands.
Acta Bot. Neerl.,19:830–858.

Úlehlová, B.
1971. Decomposition and humification of plant material in the vegetation of Stratiotes aloides in NW Overijssel, Holland.
Hidrobiologia, 12:279–285.

Vallentyne, J. R.
1956. Epiphasic carotenoids in post-glacial lake sediments.
Limnol. Oceanogr., 1:252–262.

Vallentyne, J. R.
1957a. Principles of modern limnology.
Amer. Sci., 45:218–244.

Vallentyne, J. R.
1957b. The molecular nature of organic matter in lakes and oceans, with lesser reference to sewage and terrestrial soils.
J. Fish. Res. Bd. Canada, 14:33–82.

Vallentyne, J. R.
1960. Fossil pigments. In M. B. Allen, ed. Comparative Biochemistry of Photoreactive Systems.
New York, Academic Press, pp. 83–105.

Vallentyne, J. R.
1962. Solubility and the decomposition of organic matter in nature.
Arch. Hydrobiol., 58:423–434.

Vallentyne, J. R.
1967. A simplified model of a lake for instructional use.
J. Fish. Res. Bd. Canada, 24:2473–2479.

Vallentyne, J. R.
1969. Sedimentary organic matter and paleolimnology.
Mitt. Int. Ver. Limnol., 17:104–110.

Vallentyne, J. R.
1970. Phosphorus and the control of eutrophication.
Can. Res. Development (May-June, 1970): 36–43, 49.

Vallentyne, J. R.
1972. Freshwater supplies and pollution: Effects of the demophoric explosion on water and man. In N. Polunin, ed. The Environmental Future.
London, Macmillan Press Ltd., pp. 181–211.

Vallentyne, J. R.
1974. The algal bowl – lakes and man.
Ottawa, Misc. Special Publ. 22, Dept. of the Environment, 185 pp.

Van Baalen, C.
1968. The effects of ultraviolet irradiation on a coccoid blue-green alga: Survival, photosynthesis, and photoreactivation.
Plant Physiol., 43:1689–1695.

Verber, J. L.
1964. Initial current studies in Lake Michigan.
Limnol. Oceanogr., 9:426–430.

Verber, J. L.
1966. Inertial currents in the Great Lakes.
Publ. Great Lakes Res. Div., Univ. Mich., 15:375–379.

Verduin, J.
1959. Photosynthesis by aquatic communities in northwestern Ohio.
Ecology, *40*:377–383.

Verduin, J.
1961. Separation rate and neighbor diffusivity.
Science, *134*:837–838.

Vernon, L. P.
1964. Bacterial photosynthesis.
Ann. Rev. Plant Physiol., *15*:73–100.

Vinberg, G. G.
1934. K voprosu o metalimnianom minimume kisloroda.
Trudy Limnol. Sta. Kosine, *18*:137–142.

Vinberg, G. G.
1960. Pervichnaya Produktsiia Voedoemov. Minsk, Izdatel'stvo Akademii Nauk, 329 pp. (English translat. 1963. The Primary Production of Bodies of Water.
U. S. Atomic Energy Comm., Div. Tech. Info. AEC-tr-5692, 601 pp.)

Vinberg, G. G., and V. P. Liakhnovich.
1965. Udobrenie prudov. (Fertilization of fish ponds.)
Moscow, Izdatel'stvo "Pishchevaia Promyshlennost,'" 271 pp. (English translat. 1969. Fish. Res. Bd. Canada Translation Ser. No. 1339, 482 pp.)

Vinberg, G. G., et al.
1972. Biological productivity of different types of lakes. *In* Z. Kajak and A. Hillbricht-Ilkowska, eds. Productivity Problems of Freshwaters.
Warsaw, PWN Polish Scientific Publishers, pp. 382–404.

Visser, S. A.
1964a. Origin of nitrates in tropical rainwater.
Nature, *201*:35–36.

Visser, S. A.
1964b. Oxidation-reduction potentials and capillary activities of humic acids.
Nature, *204*:581.

Vivekanandan, E., M. A. Haniffa, T. J. Pandian, and R. Raghuraman.
1974. Studies on energy transformation in the freshwater snail *Pila globosa*. 1. Influence of feeding rate.
Freshwat. Biol., *4*:275–280.

Vlymen, W. J.
1970. Energy expenditure of swimming copepods.
Limnol. Oceanogr., *15*:348–356.

Vollenweider, R. A.
1950. Ökologische Untersuchungen von planktischen Algen auf experimenteller Grundlage.
Schweiz. Z. Hydrol., *12*:193–262.

Vollenweider, R. A.
1955. Ein Nomogramm zur Bestimmung des Transmissionskoeffizienten sowie einige Bemerkungen zur Methode seiner Berechnung in der Limnologie.
Schweiz. Z. Hydrol., *17*:205–216.

Vollenweider, R. A.
1964. Über oligomiktische Verhältnisse des Lago Maggiore und einiger anderer insubrischer Seen.
Mem. Ist. Ital. Idrobiol., *17*:191–206.

Vollenweider, R. A.
1965. Calculation models of photosynthesis-depth curves and some implications regarding day rate estimates in primary production measurements.
Mem. Ist. Ital. Idrobiol., *18*(Suppl.):425–457.

Vollenweider, R. A.
1968. Scientific Fundamentals of the Eutrophication of Lakes and Flowing Waters, with Particular Reference to Nitrogen and Phosphorus as Factors in Eutrophication.
Paris, Rep. Organisation for Economic Cooperation and Development, DAS/CSI/68.27, 192 pp.; Annex, 21 pp.; Bibliography, 61 pp.

Vollenweider, R. A.
1969a. Möglichkeiten und Grenzen elementarer Modelle der Stoffbilanz von Seen.
Arch. Hydrobiol, *66*:1–36.

Vollenweider, R. A., ed.
1969b. A Manual on Methods for Measuring Primary Production in Aquatic Environments. Int. Biol. Program Handbook 12.
Oxford, Blackwell Scientific Publications, 213 pp.

Vollenweider, R. A.
1972. Input-Output Models.
Mimeographed report, Burlington, Ontario, Can. Cent. Inland Waters, 40 pp.

Vollenweider, R. A., M. Munawar, and P. Stadelmann.
1974. A comparative review of phytoplankton and primary production in the Laurentian Great Lakes.
J. Fish. Res. Bd. Canada, *31*:739–762.

Wachs, B.
1967. Die Oligochaeten-Fauna der Fliessgewässer unter besonderer Berücksichtigung der Beziehungen zwischen der Tubificiden-Besiedlung und dem Substrat.
Arch. Hydrobiol., *63*:310–386.

Walker, K. F., W. D. Williams, and U. T. Hammer.
1970. The Miller method for oxygen determination applied to saline waters.
Limnol. Oceanogr., *15*:814–815.

Walsby, A. E.
1972. Structure and function of gas vacuoles.
Bacteriol. Rev., *36*:1–32.

Walsh, G. E.
1965a. Studies on dissolved carbohydrate in
Cape Cod waters. I. General survey.
Limnol. Oceanogr., *10*:570-576.

Walsh, G. E.
1965b. Studies on dissolved carbohydrate in
Cape Cod waters. II. Diurnal fluctuation in
Oyster Pond.
Limnol. Oceanogr., *10*:577-582.

Walsh, G. E.
1966. Studies on dissolved carbohydrate in
Cape Cod waters. III. Seasonal variation in
Oyster Pond and Wequaquet Lake, Massachu-
setts.
Limnol. Oceanogr., *11*:249–256.

Wangersky, P. J.
1963. Manganese in ecology. *In* V. Schultz and
A. W. Klement, Jr., eds. Radioecology. First
National Symposium on Radioecology.
New York. Reinhold Publishing Corp., pp.
499–508.

Ward, A. K., and R. G. Wetzel.
1975. Sodium: A factor in growth of blue-green
algae.
J. Phycol. (in press).

Warren, C. E., and G. E. Davis.
1971. Laboratory stream research: Objectives,
possibilites, and constraints.
Ann. Rev. Ecol. Systematics, *2*:111-114.

Watt, W. D., and F. R. Hayes.
1963. Tracer study of the phosphorus cycle in
sea water.
Limnol. Oceanogr., *8*:276-285.

Wavre, M., and R. O. Brinkhurst.
1971. Interactions between some tubificid
oligochaetes and bacteria found in the sediments of Toronto Harbour, Ontario.
J. Fish. Res. Bd. Canada, *28*:335–341.

Weast, R. C., ed.
1970. Handbook of Chemistry and Physics.
51st Ed.
Cleveland, Ohio, Chemical Rubber Co., 2367
pp.

Weibel, S. R.
1969. Urban drainage as a factor in eutrophication. *In* Eutrophication: Causes, Consequences, Correctives.
Washington, D.C., National Academy of
Sciences, pp. 383–403.

Weinmann, G.
1970. Gelöste Kohlenhydrate und andere
organische Stoffe in natürlichen Gewässern
und in Kulturen von *Scenedesmus quadricauda*.
Arch. Hydrobiol. Suppl., *37*:164-242.

Welch, H. E.
1968. Relationships between assimilation efficiencies and growth efficiencies for aquatic
consumers.
Ecology, *49*:755–759.

Welch, P. S.
1948. Limnological Methods.
Philadelphia, Blakiston Co., 381 pp.

Welch, P. S., and F. E. Eggleton.
1932. Limnological investigations on northern
Michigan lakes. II. A further study of depression individuality in Douglas Lake.
Pap. Mich. Acad. Sci. Arts Lett., *15*(1931):491-
508.

Wells, L.
1960. Seasonal abundance and vertical movements of planktonic Crustacea in Lake Michigan.
Fish. Bull. U. S. Fish. Wildl. Serv., *60*(172):
343-369.

Wells, L.
1968. Daytime distribution of *Pontoporeia affinis* off bottom in Lake Michigan.
Limnol. Oceanogr., *13*:703-705.

Wentz, D. A., and G. F. Lee.
1969. Sedimentary phosphorus in lake
cores—analytical procedure.
Environ. Sci. Technol., *3*:750-754.

Werner, E. E.
1974. The fish size, prey size, handling time
relation in several sunfishes and some implications.
J. Fish. Res. Bd. Canada, *31*:1531-1536.

Werner, E. E., and D. J. Hall.
1974. Optimal foraging and the size selection
of prey by the bluegill sunfish (*Lepomis
macrochirus*).
Ecology, *55*:1042-1052.

Westlake, D. F.
1963. Comparisons of plant productivity.
Biol. Rev., *38*:385-425.

Westlake, D. F.
1965a. Theoretical aspects of comparability of
productivity data.
Mem. Ist. Ital. Idrobiol., *18*(Suppl.):313-322.

Westlake, D. F.
1965b. Some basic data for investigations of
the productivity of aquatic macrophytes.
Mem. Ist. Ital. Idrobiol., *18*(Suppl.):229-248.

Westlake, D. F.
1965c. Some problems in the measurement of
radiation under water: A review.
Photochem. Photobiol., *4*:849–868.

Westlake, D. F.
1966. The biomass and productivity of
Glyceria maxima. I. Seasonal changes in
biomass.
J. Ecol., *54*:745-753.

Westlake, D. F.
1967. Some effects of low-velocity currents on the metabolism of aquatic macrophytes.
J. Exp. Bot., *18*:187-205.

Westlake, D. F.
1968. Methods used to determine the annual production of reedswamp plants with extensive rhizomes. *In* Methods of Productivity Studies in Root Systems and Rhizosphere Organisms.
Leningrad, Publishing House Nauka, pp. 226-234.

Wetzel, R. G.
1960. Marl encrustation on hydrophytes in several Michigan lakes.
Oikos, *11*:223-236.

Wetzel, R. G.
1964. A comparative study of the primary productivity of higher aquatic plants, periphyton, and phytoplankton in a large, shallow lake.
Int. Rev. ges. Hydrobiol., *49*:1-64.

Wetzel, R. G.
1965a. Nutritional aspects of algal productivity in marl lakes with particular reference to enrichment bioassays and their interpretation.
Mem. Ist. Ital. Idrobiol., *18*(Suppl.):137-157.

Wetzel, R. G.
1965b. Techniques and problems of primary productivity measurements in higher aquatic plants and periphyton.
Mem. Ist. Ital. Idrobiol., *18*(Suppl.):249-267.

Wetzel, R. G.
1966a. Variations in productivity of Goose and hypereutrophic Sylvan lakes, Indiana.
Invest. Indiana Lakes Streams, 7:147–184.

Wetzel, R. G.
1966b. Productivity and nutrient relationships in marl lakes of northern Indiana.
Verh. Int. Ver. Limnol., *16*:321-332.

Wetzel, R. G.
1967. Dissolved organic compounds and their utilization in two marl lakes. *In* Problems of organic matter determination in freshwater.
Hidrológiai Közlöny, 47:298–303.

Wetzel, R. G.
1968. Dissolved organic matter and phytoplanktonic productivity in marl lakes.
Mitt. Int. Ver. Limnol., *14*:261-270.

Wetzel, R. G.
1969. Factors influencing photosynthesis and excretion of dissolved organic matter by aquatic macrophytes in hard-water lakes.
Verh. Int. Ver. Limnol., *17*:72-85.

Wetzel, R. G.
1970. Recent and postglacial production rates of a marl lake.
Limnol. Oceanogr., *15*:491-503.

Wetzel, R. G.
1972. The role of carbon in hard-water marl lakes. *In* G. E. Likens, ed. Nutrients and Eutrophication: The Limiting-Nutrient Controversy.
Special Symposium, Amer. Soc. Limnol. Oceanogr., *1*:84-91.

Wetzel, R. G.
1973. Productivity investigations of interconnected lakes. I. The eight lakes of the Oliver and Walters chains, northeastern Indiana.
Hydrobiol. Stud., *3*:91-143.

Wetzel, R. G.
1975a. General Secretary's Report—19th Congress of the Societas Internationalis Limnologiae.
Verh. Int. Ver. Limnol., *19*:(in press).

Wetzel, R. G.
1975b. Primary production. *In* B. A. Whitton, ed. River Ecology.
Cambridge, Cambridge University Press, pp. 230-247.

Wetzel, R. G., and H. L. Allen.
1970. Functions and interactions of dissolved organic matter and the littoral zone in lake metabolism and eutrophication. *In* Z. Kajak and A. Hillbricht-Ilkowska, eds. Productivity Problems of Freshwaters.
Warsaw, PWN Polish Scientific Publishers, pp. 333-347.

Wetzel, R. G., and R. A. Hough.
1973. Productivity and role of aquatic macrophytes in lakes: An assessment.
Pol. Arch. Hydrobiol., *20*:9-19.

Wetzel, R. G., and B. A. Manny.
1972a. Secretion of dissolved organic carbon and nitrogen by aquatic macrophytes.
Verh. Int. Ver. Limnol., *18*:162-170.

Wetzel, R. G., and B. A. Manny.
1972b. Decomposition of dissolved organic carbon and nitrogen compounds from leaves in an experimental hard-water stream.
Limnol. Oceanogr., *17*:927-931.

Wetzel, R. G., and D. G. McGregor.
1968. Axenic culture and nutritional studies of aquatic macrophytes.
Amer. Midland Nat., *80*:52-64.

Wetzel, R. G., and A. Otsuki.
1974. Allochthonous organic carbon of a marl lake.
Arch. Hydrobiol., *73*:31-56.

Wetzel, R. G., and P. H. Rich.
1973. Carbon in freshwater systems. *In* G. M. Woodwell and E. V. Pecan, eds. Carbon and the Biosphere. Proc. Brookhaven Symp. in Biol. 24.
Brookhaven, N.Y., Tech. Information Center,

U. S. Atomic Energy Commission CONF-720510, pp. 241-263.

Wetzel, R. G., and D. F. Westlake.
1969. Periphyton. *In* R. A. Vollenweider, ed. A Manual on Methods for Measuring Primary Production in Aquatic Environments. Int. Biol. Program Handbook 12.
Oxford, Blackwell Scientific Publications, pp. 33-40.

Wetzel, R. G., P. H. Rich, M. C. Miller, and H. L. Allen.
1972. Metabolism of dissolved and particulate detrital carbon in a temperate hard-water lake. Mem. Ist. Ital. Idrobiol., 29(Suppl.):185-243.

Wetzel, R. G., B. A. Manny, W. S. White, R. A. Hough, and K. R. McKinley.
1975. Wintergreen Lake: A study in hypereutrophy.
(In preparation.)

Whipple, G. C.
1898. Classifications of lakes according to temperature.
Amer. Nat., 32:25-33.

Whipple, G. C.
1927. The Microscopy of Drinking Water.
4th ed. New York, John Wiley and Sons, Inc., 586 pp.

White, W. S.
1974. Role of calcium carbonate precipitation in lake metabolism.
Ph.D. Diss., Michigan State University. East Lansing, 141 pp.

White, W. S., and R. G. Wetzel.
1975. Nitrogen, phosphorus, particulate and colloidal carbon content of sedimenting seston of a hard-water lake.
Verh. Int. Ver. Limnol., *19*:(in press).

Whitehead, H. C., and J. H. Feth.
1961. Recent chemical analyses of waters from several closed-basin lakes and their tributaries in the western United States.
Bull. Geol. Soc. Amer., 72:1421-1426.

Whiteside, M. C.
1969. Chydorid (Cladocera) remains in surficial sediments of Danish lakes and their significance to paleolimnological interpretations. Mitt. Int. Ver. Limnol., *17*:193-201.

Whitney, L. V.
1937. Microstratification in the waters of inland lakes in summer.
Science, 85:224-225.

Wierzbicka, M.
1966. Les résultats des recherches concernant l'état de repos (resting stage) des Cyclopoida. Verh. Int. Ver. Limnol., *16*:592-599.

Wiessner, W.
1962. Inorganic micronutrients. *In* R. A. Lewin, ed. Physiology and Biochemistry of Algae. New York, Academic Press, pp. 267-286.

Williams, J. D. H., and T. Mayer.
1972. Effects of sediment diagenesis and regeneration of phosphorus with special reference to lakes Erie and Ontario. *In* H. E. Allen and J. R. Kramer, eds. Nutrients in Natural Waters.
New York, John Wiley and Sons, Inc., pp. 281-315.

Williams, J. D. H., J. K. Syers, R. F. Harris, and D. E. Armstrong.
1970. Adsorption and desorption of inorganic phosphorus by lake sediments in a 0.1M NaCl system.
Environ. Sci. Technol., 4:517–519.

Williams, J. D. H., J. K. Syers, S. S. Shukla, R. F. Harris, and D. E. Armstrong.
1971a. Levels of inorganic and total phosphorus in lake sediments as related to other sediment parameters.
Environ. Sci. Technol., 5:1113–1120.

Williams, J. D. H., J. K. Syers, R. F. Harris, and D. E. Armstrong.
1971b. Fractionation of inorganic phosphate in calcareous lake sediments.
Proc. Soil Sci. Soc. Amer., 35:250-255.

Williams, J. D. H., J. K. Syers, D. E. Armstrong, and R. F. Harris.
1971c. Characterization of inorganic phosphate in noncalcareous lake sediments.
Proc. Soil Sci. Soc. Amer., 35:556-561.

Williams, W. D., and H. F. Wan.
1972. Some distinctive features of Australian inland waters.
Water Res., 6:829-836.

Willoughby, L. G.
1974. Decomposition of litter in fresh water. *In* C. H. Dickinson and G. J. F. Pugh, eds. Biology of Plant Litter Decomposition.
Vol. 2. New York, Academic Press, pp. 659-681.

Wilson, M. S., and H. C. Yeatman.
1959. Free-living Copepoda. *In* W. T. Edmondson, ed. Fresh-Water Biology.
2nd ed. New York, John Wiley and Sons, Inc., pp. 735-861.

Wilson, L. R.
1937. A quantitative and ecological study of the larger aquatic plants of Sweeney Lake, Oneida County, Wisconsin.
Bull. Torr. Bot. Club, 64:199-208.

Wilson, L. R.
1941. The larger aquatic vegetation of Trout Lake, Vilas County, Wisconsin.
Trans. Wis. Acad. Sci. Arts Lett., 33:135-146.

Winberg, G. G.
1970. Some interim results of Soviet IBP investigations on lakes. *In* Z. Kajak and A. Hillbricht-Ilkowska, eds. Productivity Problems of Freshwaters.
Warsaw, PWN Polish Scientific Publishers, pp. 363-381.

Winberg, G. G., ed.
1971. Methods for the Estimation of Production of Aquatic Animals.
New York, Academic Press, 175 pp.

Winberg, G. G., et al.
1970. Biological productivity of different types of lakes. *In* Z. Kajak and A. Hillbricht-Ilkowska, eds. Productivity Problems of Freshwaters.
Warsaw, PWN Polish Scientific Publishers, pp. 383-404.

Winberg, G. G., et al.
1971. Symbols, Units and Conversion Factors in Studies of Freshwater Productivity.
Int. Biol. Program, Sec. PF—Productivity of Freshwaters, 23 pp.

Winberg, G. G., et al.
1973. The progress and state of research on the metabolism, growth, nutrition, and production of fresh-water invertebrate animals.
Hydrobiol. J., 9:77-84.

Wohler, J. R.
1966. Productivity of the duckweeds.
M.Sc. Thesis, University of Pittsburgh, 69 pp.

Wohlschlag, D. E.
1950. Vegetation and invertebrate life in a marl lake.
Invest. Indiana Lakes Streams, 3:321-372.

Wolk, C. P.
1968. Movement of carbon from vegetative cells to heterocysts in *Anabaena cylindrica*.
J. Bacteriol., 96:2138-2143.

Wolk, C. P.
1973. Physiology and cytological chemistry of blue-green algae.
Bacteriol. Rev., 37:32-101.

Woodwell, G. M., and R. H. Whittaker.
1968. Primary production in terrestrial ecosystems.
Amer. Zool., 8:19-30.

Woodwell, G. M., P. H. Rich, and C. A. S. Hall.
1973. The carbon cycle of estuaries. *In* G. M. Woodwell and E. V. Pecan, eds. Carbon and the Biosphere. Proc. 24th Brookhaven Symposium in Biology.
Brookhaven, N.Y., U. S. Atomic Energy Commission. Symp. Ser. CONF-720510, pp. 221-240.

Wong, B., and F. J. Ward.
1972. Size selection of *Daphnia pulicaria* by yellow perch (*Perca flavescens*) fry in West Blue Lake, Manitoba.
J. Fish. Res. Bd. Canada, 29:1761-1764.

Worthington, E. B.
1931. Vertical movements of fresh-water macroplankton.
Int. Rev. ges. Hydrobiol., 25:394-436.

Wright, J. C.
1965. The population dynamics and production of *Daphnia* in Canyon Ferry Reservoir, Montana.
Limnol. Oceanogr., 10:583-590.

Wright, R. T.
1964. Dynamics of a phytoplankton community in an ice-covered lake.
Limnol. Oceanogr., 9:163-178.

Wright, R. T., and J. E. Hobbie.
1966. Use of glucose and acetate by bacteria and algae in aquatic ecosystems.
Ecology, 47:447-464.

Wright, S.
1955. Limnological Survey of Western Lake Erie.
Spec. Sci. Rept. Fish. U. S. Fish Wildl. Serv., 139, 341 pp.

Wuim-Andersen, S.
1971. Photosynthetic uptake of free CO_2 by the roots of *Lobelia dortmanna*.
Physiol. Plant, 25:245-248.

Wunderlich, W. O.
1971. The dynamics of density-stratified reservoirs. *In* G. E. Hall, ed. Reservoir Fisheries and Limnology.
Washington, D.C., Spec. Publ. 8, Amer. Fish. Soc., pp. 219-231.

Yentsch, C. S.
1960. The influence of phytoplankton pigments on the color of sea water.
Deep-Sea Res., 7:1-9.

Yoshimura, S.
1935. Relation between depth for maximum amount of excess oxygen during the summer stagnation period and the transparency of freshwater lakes of Japan.
Proc. Imperial Acad. Japan, 11:356-358.

Yoshimura, S.,
1936a. A contribution to the knowledge of deep water temperatures of Japanese lakes. I. Summer temperatures.
Jap. J. Astronomy Geophys., 13:61-120.

Yoshimura, S.
1936b. Contributions to the knowledge of iron dissolved in the lake waters of Japan. Second report.
Japan. J. Geol. Geogr., 13:39-56.

Young, J. O.
1973. The prey and predators of *Phaenocora typhlops* (Vejdovsky) (Turbellaria: Neorhabdocoela) living in a small pond.
J. Anim. Ecol., 42:637-643.

Zafar, A. R.
1959. Taxonomy of lakes.
Hydrobiologia, 13:287-299.

Zaitsev, Yu. P.
1970. Marine neustonology.
Kiev. Akad. Nauk Ukrainskoi SSR, Naukova Dumka, 207 pp.
(Translat. into English, Programs for Scientific Translation, Jerusalem, 1971.)

Zaret, T. M.
1972a. Predator-prey interaction in a tropical lacustrine ecosystem.
Ecology, 53:248-257.

Zaret, T. M.
1972b. Predators, invisible prey, and the nature of polymorphism in the Cladocera (Class Crustacea).
Limnol. Oceanogr., 17:171-184.

Zicker, E. L., K. C. Berger, and A. D. Hasler.
1956. Phosphorus release from bog lake muds.
Limnol. Oceanogr., 1:296-303.

ZoBell, C. E.
1964. Geochemical aspects of the microbial modification of carbon compounds. *In* U. Colombo and G. D. Hobson, eds. Advances in Organic Geochemistry.
New York, Macmillan Co., pp. 339-356.

ZoBell, C. E.
1973. Microbial biogeochemistry of oxygen. *In* A. A. Imshenetskii, ed. Geokhimicheskaia Deiatel'nost' Mikroorganizmov v Vodoemakh i Mestorozhdeniiach Poleznykh Iskopaemykh.
Moscow, Tipografiia Izdatel'stva Sovetskoe Radio, pp. 3-76.

Züllig, H.
1961. Die Bestimmung von Myxoxanthophyll in Bohrprofilen zum Nachweis vergangener Blaualgenentfaltungen.
Verh. Int. Ver. Limnol., 14:263-270.

Zumberge, J. H.
1952. The Lakes of Minnesota. Their Origin and Classification.
Minneapolis, University of Minnesota Press, 99 pp.

Zygmuntowa, J.
1972. Occurrence of free amino acids in pond water.
Acta Hydrobiol., 14:317-325.

index

Acidity, 172–173
Aestival lakes, 652
Air-water interface, 13, 300–301
Algae. *See also* Phytoplankton; Littoral algae.
  bacterial competition for phosphorus, 237–239
  bacterial interactions, 398–402
  carbonate precipitation, 171
  elemental demands, 641
  food for zooplankton, 430–431, 447–450
  heterotrophy, 182–184, 313–314
  littoral, 388–402
    as a source of phytoplankton, 404–405
    bacterial interactions, 400–402
    distribution, 391–392
    growth, 394–395
    light availability, 404
    photosynthesis, 392–393
    productivity, 405–418
      measurement, 393–394, 405–407, 412–413
      phytoplankton, 408
      rates of macrophytes, 411–416
      spatial-temporal variations, 407–411
    seasonal population dynamics, 393–397
    succession, 395–397
    terminology, 389–391
    zonation, 389–391
  macrophyte complex, 371–372, 398–402
  osmotic adaptations, 163
  paleolimnological record, 633–635
  phosphorus requirements, 235–237
  pigments, 289–290, 301, 332–337
  secretion of organic matter, 344–346, 555
  silicon utilization, 283–286
  streams, 396
  systematics, 288–296
  use of carbon, 182–185
  vitamin requirements, 311
Alkalinity, 172–173
  above sediments, 223–225
  in littoral zone, 403
Allochthonous,
  defined, 42
  organic matter. *See* Carbon, organic.
Amictic lakes, 77
Amino acids, and carbon flux, 564–565
Amphipoda, 506–509

Animals. *See* specific groups.
  food utilization, 450–452, 484
  osmotic adaptations, 163–165
  paleolimnological record, 635–636
  phosphorus cycling, 227
Annelida, 495–501
  Hirudinea (leeches), 499–501
  Oligochaeta, 495–499
Antibiotics,
  in natural waters, 314–315
Aquaculture, 660
Aquatic plants. *See* Macrophytes.
Area, 30
Arheic lake regions, 39
Assimilation efficiency, 450, 518ff
Autochthonous,
  defined, 42
Autolysis of organisms, 552–554

Bacteria,
  algal competition for phosphorus, 237–239
  carbon monoxide metabolizing, 599–600
  chemosynthetic, 277, 279
  decomposition, 574–589
    littoral, 583–589
    pelagial, 579–583
  denitrification, 200–202, 209
  distribution, 571–574
  gaseous metabolism, 597–602
  hydrogen metabolizing, 599
  in lakes of differing productivity, 572
  iron metabolizing, 259–261
  manganese metabolizing, 259–261
  metabolism of phosphorus, 226
  methane metabolizing, 597–599
  nitrate reduction, 200–202, 209
  nitrification, 198–200, 209
  nitrogen fixation, 192–194
  photosynthetic, 274–275, 277–279
  plates, 276
  respiration, 127
  sediments, 592–602
  sulfur metabolizing, 259–260, 271–279
Benthic fauna, 488–532
  composition in lakes of differing productivity, 521

733